The Botany and Natural History of Panama:

La Botánica e Historia Natural de Panamá

Editors **William G. D'Arcy** **Mireya D. Correa A.**
Editores Missouri Botanical Garden Universidad de Panamá

República de Panamá

ELEVATIONS

0–100 m
0–333 ft

100–800 m
333–2,666 ft

800–3,475 m
2,666–11,580 ft

Republic of Panama

The Botany and Natural History of Panama:

La Botánica e Historia Natural de Panamá

COVER:
Top Left: Tree lizard *(Anolis frenatus),* Barro Colorado Island. photo N. Smythe
Top Center: Kuna women, Ustuppo Island. photo W.G. D'Arcy
Top Left: Blue-gray Tanager *(Thraupis episcopus),* central Panama. photo N. Smythe
Middle: Cattle grazing, Azuero Peninsula. photo Carol Todzia
Bottom Left: Hanging fruits *(Solanum siparunoides),* El Llano-Carti road. photo W.G. D'Arcy
Bottom Left-Center: Plant collecting near top of El Baru. photo W.G. D'Arcy
Bottom Right-Center: Panama Hat Plant *(Carludovica palmata),* beside Gatun Lake, Canal Area. photo W.G. D'Arcy
Bottom Right: Sawyers cleaning up after forest removal, La Fortuna watershed, Chiriqui Province. photo Hugh Churchill

The Botany and Natural History of Panama:

La Botánica e Historia Natural de Panamá

Editors
Editores

William G. D'Arcy
Missouri Botanical Garden

Mireya D. Correa A.
Universidad de Panamá

Missouri Botanical Garden
Saint Louis, Missouri
The United States of America
1985

MONOGRAPHS IN SYSTEMATIC BOTANY
from the Missouri Botanical Garden
Vol. 10. 1985

The Botany and Natural History of Panama: La Botánica e Historia Natural de Panamá.

Edited by W.G. D'Arcy, Missouri Botanical Garden, P.O. Box 299, St. Louis, MO 63166, U.S.A.
and
Mireya D. Correa A., Herbario, Escuela de Biología, Facultad de Ciencias Naturales y Farmacia, Estafeta Universitaria, Universidad de Panamá, Panamá, República de Panamá.

Published simultaneously in Canada.
Printed in the United States of America by
Braun-Brumfield, Inc., Ann Arbor, Michigan

Library of Congress Cataloging in Publication Data

The Botany and Natural History of Panama.

(Monographs in systematic botany, ISSN 6161-1542; vol. 10)
English and Spanish
1. Botany—Panama. 2. Natural history—Panama.
I. D'Arcy, William G., 1931- . II. Correa A., Mireya D.
III. Title: Botańica e historia natural de Panamá. IV. Series:
Monographs in systematic botany; v. 10.
QK221.B68 1985 574.97287 85-4906
ISBN 0-915-279-03-7 (pkb.)

Publication of this book was assisted by a grant from the Dirección Nacional de Recursos Renovables (RENARE) del Ministerio de Agricultura (MIDA), Panamá.

El apoyo financiero de la Dirección Nacional de Recursos Naturales Renovables (RENARE) del Ministerio de Desarollo Agropecuario (MIDA), como una de las Instituciónes patrocinadoras del Simposio de Botánica e Historia Natural celebrado en abril de 1980 en la Universidad de Panamá, fue específicamente destinada, a solicitud de esa Institución, para la publicación de las actas resultantes del evento.

Copies of *The Botany and Natural History of Panama: La Botánica e Historia Natural de Panamá* can be ordered from: Department 11, Missouri Botanical Garden, Box 299, St. Louis, MO 63166. U.S.A. (314) 577-5170

ISBN: 0-915279-03-7 (Paperback)

Contents

PART I: THE ORGANISMS—LOS ORGANISMOS

PLANT GROUPS—GRUPOS DE PLANTAS, 1

ANIMAL GROUPS—GRUPOS DE ANIMALES, 67

PLANTS AND GEOGRAPHY—PLANTAS Y GEOGRAFIA, 115

PART III: THE HUMAN ASPECT—
EL ASPECTO HUMANO

PLANTS THAT AFFECT MAN—
PLANTAS QUE AFECTAN AL HOMBRE, 289

INDIGENOUS PEOPLES—POBLACION INDIGENA, 315

MODERN MAN AND THE LANDSCAPE—
HOMBRE MODERNO Y AREAS NATURALES, 367

Authors' Addresses

Ackerman, James D. 225
Department of Biology
University of Puerto Rico
Rio Piedras, PR 00931
U.S.A.

Aiello, Annette 263
Smithsonian Tropical Research Institute
Apartado Box 2072
Balboa
REPUBLICA DE PANAMA

Araúz, Reina Torres de 291
Dirección de Patrimonio Histórico
Panamá,
REPUBLICA DE PANAMA

Augspurger, Carol K. 235
Department of Botany
University of Illinois
Urbana, IL 61801
U.S.A.

Bennett, Charles F. 117
Department of Geography
University of California at Los Angeles
Los Angeles, CA 90024
U.S.A.

Burger, William C. 125
Field Museum of Natural History
Roosevelt Road and Lake Shore Drive
Chicago, IL 60605
U.S.A.

Cintron, Gilberto 53
Department of Natural Resources
P.O. Box 5887
San Juan, PR 00906
U.S.A.

Coley, Phyllis D. 263
Smithsonian Tropical Research Institute
Apartado 2072
Balboa
REPUBLICA DE PANAMA

Croat, Thomas B. 5
Missouri Botanical Garden
Box 299
St. Louis, MO 63166
U.S.A.

Crosby, Marshall R. 49
Missouri Botanical Garden
Box 299
St. Louis, MO 63166
U.S.A.

Davidse, Gerrit 13
Missouri Botanical Garden
Box 299
St. Louis, MO 63166
U.S.A.

Denslow, Julie Sloan 245
Department of Botany
University of Wisconsin at Madison
Madison, WI 53706
U.S.A.

Díaz H., Irving R. 441
Director, Dirección Nacional
de Recursos Naturales Renovables (RENARE)
Panamá
REPUBLICA DE PANAMA

Dressler, Robert L. 25
Smithsonian Tropical Research Institute
Apartado 2072
Balboa
REPUBLICA DE PANAMA

Duke, James. A. 299
Economic Botany Laboratory
(EBL) Building 256, BARC-East
Beltsville, MD 20705
U.S.A.

Dwyer, John D. 179
Missouri Botanical Garden
Box 299
St. Louis, MO 63166
and
Department of Biology
St. Louis University
St. Louis, MO 63103
U.S.A.

Elvin-Lewis, Memory P.F. 309
School of Dental Medicine
Washington University
St. Louis, MO 63130
U.S.A.

Escobar, Novencido 305
Apartado 9258
Panamá 6
REPUBLICA DE PANAMA

Forero, Enrique 137, 185
Sección de Botánica
Instituto de Ciencias Naturales
de la Universidad Nacional
Apartado 7495
Bogotá
COLOMBIA

Foster, Robin 255
Missouri Botanical Garden
Box 299
St. Louis, MO 63166
and
Field Museum of Natural History
Roosevelt Road and Lake Shore Drive
Chicago, IL 60605
U.S.A. (mailing address)
and
Smithsonian Tropical Research Institute
Apartado 2072
Balboa
REPUBLICA DE PANAMA

Garber, Paul A. 69
Department of Anthropology
University of Illinois
Urbana, IL 61801
U.S.A.

Gentry, Alwyn H. 29, 147
Missouri Botanical Garden
Box 299
St. Louis, MO 63166
U.S.A.

Graham, Alan 161
Department of Biological Sciences
Kent State University
Kent, OH 44242
U.S.A.

Heckadon Moreno, Standley 397
Apartado 1462
Panamá 1
REPUBLICA DE PANAMA

Howe, James 317
Anthropology-Humanities Program
Massachusetts Institute of Technology
Cambridge, MA 02139
U.S.A.

Jaén Suarez, Omar 379
Dirección Ejecutiva para
Asuntos del Tratado
Apartado 3403
Panamá 4
REPUBLICA DE PANAMA

Joly, Luz Graciela 369
Apartado 4726
David, Chiriquí
REPUBLICA DE PANAMA

Karr, James. R. 77
Smithsonian Tropical Research Institute
Aprtado 2072
Balboa
REPUBLICA DE PANAMA

Keating, Richard C. 193
School of Science
Southern Illinois University
Edwardsville, IL 62026
U.S.A.

Kress, W. John 267
Marie Selby Botanical Gardens
800 S. Palm Avenue
Sarasota FL 33577
U.S.A.

Lellinger, David B. 43
United States National Herbarium
Smithsonian Institution
Washington, DC 20560
U.S.A.

Lewis, Walter H. 309
Department of Biology
Washington University
St. Louis, MO 63130
U.S.A.

Lu, Karen 263
Department of Biology
Humboldt State University
Arcata, CA 95521
U.S.A.

Lugo, Ariel E. 53
Institute of Tropical Forestry
Southern Forest Experiment Station
P.O. Box AQ
Rio Piedras, PR 00928
U.S.A.

MacFarland, Craig 423
Centro Agronómico Tropical
de Investigación y Enseñanza
(CATIE)
Turrialba
COSTA RICA

Martinez, Ramon 423
Department of Natural Resources
P.O. Box 5887
San Juan, PR 00906
U.S.A.

McKay, Alberto 393
Escuela de Geografía
Universidad de Panamá
Estafeta Universitaria
Panamá
REPUBLICA DE PANAMA

Méndez, Eustorgio 415
Laboratorio Conmemorativo Gorgas
Apartado 6991
Panamá
REPUBLICA DE PANAMA

Moermond, Timothy C. 245
Department of Zoology
University of Wisconsin at Madison
Madison, WS 53706
U.S.A.

Montalvo, Arlee 263
Smithsonian Tropical Research Institute
Apartado 2072
Balboa
REPUBLICA DE PANAMA

Moore, Alexander 333
Department of Anthropology
University of Southern California
Los Angeles, CA 90007
U.S.A.

Morales, Roger 423
Centro Agronómico Tropical
de Investigación y Enseñanza
(CATIE)
Turrialba
COSTA RICA

Putz, Francis E. 263
Department of Botany
University of Florida
Gainesville, FL 32611
U.S.A.

Raven, Peter H. 433
Missouri Botanical Garden
Box 299
St. Louis, MO 63166
U.S.A.

Read, Robert G. 411
Department of Meteorology
San Jose State University
San Jose, CA 95192
U.S.A.

Reyna R., Narciso 419
Casa 2J
Urbanización La Colonia
Santo Domingo, Heredia
COSTA RICA

Rockwood, Larry L. 197
Department of Biology
George Mason University
4400 University Drive
Fairfax, VA 22030
U.S.A.

Salazar, Noris A. 49
Departmento de Botánica
Universidad de Panamá
Estafeta Universitaria
Panamá
REPUBLICA DE PANAMA

Sexton, Owen J. 95, 345
Department of Biology
Washington University
St. Louis, MO 63130
U.S.A.

Sherzer, Dina 347
Department of Languages
University of Texas at Austin
Austin, TX 78712
U.S.A.

Sherzer, Joel 351
Department of Anthropology
University of Texas at Austin
Austin, TX 78712
U.S.A.

Stiles,, F. Gary 273
Escuela de Biología
Universidad de Costa Rica
Ciudad univeritaria
COSTA RICA

Sugden, Andrew M. 207
Botany School
South Parks Road
Oxford OXI 3RA
UNITED KINGDOM

Thorne, Barbara L. 111
Museum of Comparative Zoology
Harvard University
Cambridge, MA 02138
U.S.A.

Young, Philip D. 357
Department of Anthropology
University of Oregon
Eugene, OR 97403
U.S.A.

Preface

In April l980 a symposium was held at the University of Panama to celebrate completion of the Flora of Panama and to form the basis of a natural history volume. The long time between the symposium and the appearance of this volume was partly taken in actual completion of the Flora (at the time of the symposium it was 'nearly' complete) and with editing of the manuscripts. To those who have waited overly long to see their efforts in print we apologize. In the meantime, a few of the papers presented at the symposium were published elsewhere: their inclusion here makes them available to a wider audience. Almost none of the papers suffer from loss of timeliness: almost all are still the current word on their subjects.

The Flora of Panama was begun as a formal scientific series in the Annals of the Missouri Botanical Garden in l943, following more than two decades of work by Garden botanists in Panama. In the years until its completion (l981), 102 scientists from many different countries contributed treatments to the Flora. The Flora of Panama was the first attempt to catalogue all the plants occurring in Panama, and it included descriptions, keys, illustrations, and notes on localities, ranges, habitats, and sometimes the uses of over 5000 species. It is an invaluable tool for all those who work with Panamanian plants and is also an important resource for neighboring countries that do not yet have published floras. The Flora was prepared by the Missouri Botanical Garden, and since l957 it received financial support from the National Science Foundation. In Panama, it received assistance from the University of Panama, the Smithsonian Tropical Research Institute, the United States Army, and Gorgas Memorial Laboratory, as well as many other agencies and many private citizens and residents.

In l970 the Missouri Botanical Garden established a field station in Panama with a resident botanist who collected plants throughout Panama and curated Summit Herbarium. Summit Herbarium was later moved to facilities provided by the Smithsonian Tropical Research Institute, and the station continues as a facility operated jointly by the Missouri Botanical Garden and the Smithsonian Tropical Research Institute.

In 1967 the University of Panama established the Department of Botany and this marked the beginning of participation by the university in botanical investigations in the country. In 1969 the herbarium was begun as an instrument for teaching, research and extention service. Over the years the Department has trained over 70 botanists who now work in various institutions in the country. They have participated in floristic and ecological evaluations of state environmental and hydroelectric projects and in many other activities related to the study of plants.

Although publication of the Flora of Panama was completed in l981, the Flora of Panama project continues. The richness of plant life on the Isthmus was repeatedly underestimated, and each part of the published Flora of Panama omitted many species that were found subsequently. New species are still being found at a rate undiminished from previous years, and it is essential that collecting activity continue before the remaining forests are gone. The Missouri Botanical Garden and the University of Panama continue to study and collect the Panamanian Flora. A computerized updated checklist of the flora is to be published by the Missouri Botanical Garden, and it will be reissued from time to time.

The symposium signalling completion of the Flora took place 14-17 April l980, and it was a good success to judge from remarks of the participants. It was attended by about 250 people. The symposium was organized by the two editors of this volume; William G. D'Arcy, Missouri Botanical Garden, and Mireya D. Correa A., University of Panama, who were admirably supported by the several committees and sponsors listed on page xvii. Although the theme of the symposium was the Botany and Natural History of Panama, the actual content was shaped partly by the kinds of studies which were coming to the fore at the time. This modified Botany and Natural History is mirrored in the present volume. A theme that ran through nearly all papers presented was the urgency for studies at this time. This theme stemmed from fear that the natural conditions necessary for such studies will not last much longer under the requirements of man to support increasing populations, better standards of living and to realize the benefits of modern technologies.

Since the symposium, studies are continuing and natural areas continue to fall before development needs; no early change in this situation is predicted, although curtailment of natural landscapes must eventually restrict or greatly alter natural history studies. The beginning of this can already be seen.

Prefacio

En abril de 1980 se realizó, en la Universidad de Panamá, un simposio para celebrar la finalización de la Flora de Panamá y para fundamentar un libro de historia natural. El tiempo que ha pasado desde la realización del simposio hasta la publicación de este volumen se ha debido principalmente a que, la Flora fue realmente completada (1981) después de la celebración del simposio. Cabe enfatizar que la labor de editar los manuscritos también consumió mucho tiempo. Nos disculpamos ante aquellas personas que han esperado tanto para ver plasmados sus esfuerzos en esta publicación. Algunas ponencias presentadas en el simposio han sido ya publicadas en otros lugares, sin embargo su inclusión en este volumen las hace accesibles a una mayor audiencia. A pesar de la demora, la mayoría de los manuscritos no han perdido validez y aún son la palabra corriente en su materia.

En 1943, el Jardín Botánico de Missouri, en sus Anales, inició la publicación de una serie científica formal, la Flora de Panamá. Dicha serie fue la continuación del trabajo botánico que por más de dos décadas llevaron a cabo los científicos de esa Institución en Panamá. Durante los años de su publicación hasta su finalización, 102 científicos de diversos países contribuyeron con monografías a la Flora. La Flora de Panamá es el primer intento de catalogar todas las plantas que ocurren en Panamá. Esta incluye descripciones, claves, ilustraciones y notas sobre localidades, distribución, hábitat y algunos usos de más de 5000 especies. Es un instrumento de incalculable valor para aquellos que trabajan con plantas panameñas, al igual que es un importante recurso para los países adjacentes que aún no poseen una Flora publicada. Cabe resaltar que en el Nuevo Mundo, sólo Cuba, Jamaica y Guatemala poseen floras nacionales, modernas y completas. La Flora fue preparada por el Jardín Botánico de Missouri y desde 1957 recibió el apoyo financiero de la Fundación para la Ciencia de los Estados Unidos (NSF). En Panamá cooperaron instituciones como la Universidad de Panamá, el Instituto Smithsoniano de Investigación Tropical, la Fuerza Armada de los Estados Unidos y el Laboratorio Conmemorativo Gorgas, al igual que otras agencias y muchos ciudadanos y residentes.

En 1970, el Jardín Botánico de Missouri estableció una estación experimental en Panamá con un botánico en residencia. Este colectó plantas a través del país y adicionalmente administró el Herbario del Summit. Posteriormente, este Herbario fue mudado al edificio del Instituto Smithsoniano de Investigación Tropical que proveyó las facilidades necesarias.

En 1967, la Universidad de Panamá fundó el Departamento de Botánica y este hecho marcó el comienzo de la participación de la Universidad en las investigaciones botánicas del país. En 1969, esta Institución brindó todo el apoyo para iniciar un herbario como instrumento de apoyo de las actividades de docencia, investigación y extensión. Actualmente, tanto el Departamento de Botánica como el Herbario aún continuan funcionando. En el transcurso de los años se han preparado más de 70 botánicos, quienes al momento trabajan en diversas instituciones del país; se ha participado en evaluaciones florísticas y ecológicas de proyectos ambientales e hidroeléctricos del estado, en asesorías y en muchas otras actividades que parcial o totalmente están relacionadas con el estudio de las plantas.

A pesar de que la Flora de Panamá fue completada en 1981, el proyecto aún continúa. La riqueza de la vida vegetal en el Istmo ha sido repetidamente subestimada, y cada parte de la Flora de Panamá publicada omitió muchas especies que han sido encontradas subsecuentemente. Hoy día, constantemente se descubren especies nuevas a la misma tasa que durante años previos y es esencial que la actividad de colecta continue antes de que los bosques, que aún existen, desaparezcan. El Jardín Botánico de Missouri, al igual que la Universidad de Panamá prosiguen con actividades de colecta y de estudio de la Flora de Panamá. Una lista actualizada de la Flora será publicada, de vez en cuando, por el Jardín Botánico de Missouri. Un simposio que señaló la finalización de la Flora se realizó del 14–17 de Abril de 1980 y a juzgar por los comentarios de los participantes fue un éxito. Asistieron aproximadamente 250 personas. El simposio fue organizado por los editores de este volumen: William G. D'Arcy del Jardín Botánico de Missouri y Mireya D. Correa A. de la Universidad de Panamá, quienes fueron admirablemente apoyados por varios comités y por las instituciones patrocinadoras listadas en las primeras páginas. A pesar de que el tema del evento fue la Botánica e Historia Natural de Panamá, el contenido real consistió, al menos parcialmente, en los estudios que en ese momento estaban en sus etapas finales. Consecuentamente, esta visión modificada de la Botánica e Historia Natural es reflejada en este volumen. Una preocupación que estuvo latente en casi todas las ponencias fue la urgencia de esos estudios en esa ocasión, especialmente debido al miedo de que las condiciones naturales necesarias para éstos cambiaran como resultado de los requerimentos del hombre para sostener los incrementos de población, los mejores estándares de vida y la noción de las beneficios de la tecnología moderna. Después del simposio, las investigaciones continuan, así mismo las áreas naturales prosiguen disminuyendo ante las necesidades del desarrollo; no se visualiza ningún cambio inmediato de esa situación, no obstante, la reducción de los paisajes naturales debe, al menos eventualmente, restringir o alterar excesivamente los estudios de historia natural y el comienzo de ésto ya se puede observar.

Symposium on The Botany and Natural History of Panama

Simposio Sobre la Botánica e Historia Natural de Panamá

14-17 April 1980 — 14-17 Abril 1980

ORGANIZING COMMITTEE — COMITE ORGANIZADOR

Dra. Reina Torres de Araúz
Directora de Patrimonio Histórico
Dra. Deyanira Barnett H.
Universidad de Panamá (Zoología)
Prof. Luis Carrasquilla
Universidad de Panamá (Botánica)
Ing. Irving R. Díaz H.
Director de RENARE (MIDA)
Dr. Robert L. Dressler
Smithsonian Tropical Research Institute
Prof. Novencido Escobar
Universidad de Panamá (Botánica)
Dr. Félix Nuñez
Universidad de Panamá (Zoología)
Prof. Claudia de Peralta
Universidad de Panamá (Botánica)
Dr. Ceferino Sánchez
Vice Rector Académico, Universidad de Panamá
Dr. Nicholas Smythe
Smithsonian Tropical Research Institute
Dr. Alfredo Soler B.
Decano, Facultad de Ciencias Naturales y Farmacia, Universidad de Panamá
Prof. Francisca de Sousa
Universidad de Panamá (Zoología)
Dr. Alberto S. Taylor B.
Universidad de Panamá (Coordinador de Botánica)

SCIENTIFIC COMMITTEE — COMITE CIENTIFICO

Dra. Reina Torres de Araúz
Directora de Patrimonio Histórico
Prof. Mireya D. Correa A.
Universidad de Panamá (Botánica)
Dra. Deyanira Barnett H.
Universidad de Panamá (Zoología)
Dr. William G. D'Arcy
Missouri Botanical Garden
Dr. Robert L. Dressler
Smithsonian Tropical Research Institute
Dr. Peter H. Raven
Director, Missouri Botanical Garden
Dr. Ira Rubinoff
Director, Smithsonian Tropical Research Institute
Dr. Nicholas Smythe
Smithsonian Tropical Research Institute

SPONSORING ORGANIZATIONS — INSTITUCIONES PATROCINADORAS

Dirección de Patrimonio Histórico del INAC
Dirección Nacional de Recursos Naturales Renovables (RENARE)
Instituto de Recursos Hidraúlicos y Eléctricos (IRHE)
Missouri Botanical Garden
National Science Foundation
Smithsonian Tropical Research Institute
Universidad de Panamá

HONORARY PRESIDENT — PRESIDENTE HONORARIO

Dr. Diógenes Cedeño Cenci
Rector, Universidad de Panamá

JOINT CHAIRMEN — CO-PRESIDENTES

Dr. William G. D'Arcy
Missouri Botanical Garden

Prof. Mireya D. Correa A.
Universidad de Panamá

VICE PRESIDENT — VICE PRESIDENTE

Lic. Pedro Salazar
Vice Rector Administrativo, Universidad de Panamá

TREASURER — TESORERA

Prof. Noris Salazar A.
Universidad de Panamá

Panama Information

GEOGRAPHY:

Area: 77,082 km^2 (29,202 mi^2).

Population: 1,830,175 (1980).

Capital: Panama City (pop. 498,624, 1980).

Other Cities: Colon (117,227, 1980), David (80,371, 1980).

TOPOGRAPHY:

The long axis of the country runs east and west. Mostly mountainous and hilly inland. Highest mountains (3,475 m, 11,583 ft) in the west. Rainfall mostly heavy and aseasonal along the Atlantic, less (but still heavy, 125 inches) and with a pronounced dry season along the Pacific. Soils tend to be of fertile volcanic materials, but are mostly short-lived after clearing because of rapid erosion under torrent conditions and high temperatures under direct sunlight.

VEGETATION:

Most of Panama was at one time covered with dense forests. Lowland forests along the Caribbean (northern) side tend to have high rainfall and be non seasonal. Along the Pacific (southern) side forests are deciduous during the dry season from mid December to May. Much of the forest has been cut for cattle raising, farming, and urbanization, and widespread use of fire prevents reestablishment of forests. A large part of the forest has already been removed. A considerable part of eastern Panama (Darien Province) is occupied by swamp forest. Upland areas are forested except at the top of the highest peaks where shrubs and herbs are present.

PEOPLE:

Mostly of Spanish, Indian and Negro ancestry. Separate enclaves of English speaking Negro (Colon, Bocas del Toro) and three main groups of Indians, Choco (Darien), Kuna (San Blas and Darien), and Guaymi (Veraguas and Chiriqui) persist. Spanish is the dominant language, and English a widely spoken second language.

HISTORY:

1000 A.D. Maize culture prevails and population increases

1501 Rodrigo de Bastidas discovers Panama.

1505 Spanish exploration and colonization begins.

1513 Balboa sights and visits the Pacific Ocean.

1525 Central Panama firmly under European rule, indigenous population reduced to outlying remnants. Formerly cultivated areas begin to grow back into forest.

1532 Peru comes under Spanish influence from Panama. Gold movement across Panama to Spain begins in earnest.

1671 Panama Vieja destroyed by Henry Morgan and soon relocated 20 km west.

1821 Panama achieves independence from Spain, soon joins Gran Colombia.

1850's Immigration of Choco Indians from Colombia and population of offshore islands by Kuna.

1855 Panama Railway built

1882-1900 French attempt canal construction

1899-1902 War of a Thousand Days reduces rural population and cattle numbers

1903 Panama proclaims independence from Colombia.

1903-1914 Building of Panama Canal

1966-67 All weather road completed between Chiriqui and Panama City.

1977 Treaty ratified returning the Canal Zone to Panama.

ECONOMY:

Major agricultural crops/exports are bananas (western lowlands), sugar (central lowlands), coffee (western hills and mountains). Cattle is a dominant industry in many parts, and dairying and poultry are important in the uplands. The timber industry is based on deforestation to support abortive colonization of remaining forested lands. Fishing, especially shrimp, is important for both internal and export demand. Much of rural Panama continues near the subsistence level. There is a striking and sometimes discordant contrast with these mainly rural pursuits and activity geared to modern international trade.

The Panama Canal services international shipping, and international banking, petroleum handling, and tourism have grown rapidly in recent years. Panama City is largely geared to international activities with cosmopolitan wage levels and consumer demands.

Información sobre Panamá

GEOGRAFIA:

Area: 77.082 km^2 (29,202 mi^2).

Población: 1.830.175 (1980).

Capital: Ciudad de Panamá, pob. 498.664 (1980).

Otras ciudades: Colón, pob. 117.227 (1980); David, pob. 80.371 (1980).

TOPOGRAFIA

El eje longitudinal del país está dispuesto de este a oeste. El país es primordialmente montañoso en la parte central y plano hacia las costas. La montaña más alta, 3.475 m, está localizada en el oeste, en la provincia de Chiriquí. A lo largo del Atlántico, la precipitación es copiosa, de distribución uniforme durante el año y sin periodicidad notable, mientras que en el Pacífico es menos intensa, pero aún abundante (125 pulg.), con una estación seca pronunciada. Los suelos generalmente son fértiles de origen volcánico, pero de vida corta después de la deforestación, debido a la erosión rápida bajo condiciones torrenciales y a la temperatura alta por exposición directa a la luz solar.

VEGETACION

En el pasado, Panamá estuvo cubierta, casi enteramente, de bosques densos. Los bosques de las tierras bajas, a lo largo del Caribe, usualmente presentan lluvias intensas y carecen de estación seca. En el Pacífico (Sur), los bosques son deciduos durante la estación seca o sea desde la mitad de diciembre hasta mayo. Muchos bosques se han cortado para utilizar la tierra para la cría de ganado, la agricultura y para urbanizar. Una costumbre, ampliamente establecida, es el uso del fuego para prevenir la regeneración del bosque. Una gran parte del bosque ya ha sido removido. Una parte considerable del este de Panamá (provincia del Darién) está ocupada por bosques pantanosos. Las tierras altas aún poseen bosques, excepto en la cima de los picos más altos, en donde sólo existen hierbas y arbustos.

POBLACION

Los ancestros de la población son principalmente españoles, indios y negros. Hay enclaves separados de negros que hablan inglés (Colón, Bocas del Toro) y aún persisten tres grupos principales de indios, los chocoes (Darién), los cunas (San Blas y Darién) y los guaymíes (Veraguas y Chiriquí). El español es el idioma dominante y el inglés se habla ampliamente como segundo idioma.

HISTORIA

1000 A.C. La cultura del maíz es importante y la población aumenta.

1501 Rodrigo de Bastidas descubre Panamá.

1505 Exploración y colonización por los españoles comienza.

1513 Balboa descubre el océano Pacífico.

1525 El centro de Panamá regido firmemente por los europeos y la población indígena notablemente diezmada. Areas previamente cultivadas empezaron a regenerarse y convertirse en bosques.

1532 Desde Panamá, el Perú recibe la influencia española. El transporte de oro hacia España, a través del Istmo comienza.

1671-1673 Panamá Vieja es destruida por Henry Morgan y reconstruida a 20 km oeste.

1821 Panamá se independiza de España y casi inmediatamente se une a Colombia.

1850 Inmigración de chocoes desde Colombia mientras que los cunas habitan las islas cercanas a la costa de San Blas.

1855 Inauguración del Ferrocarril transístmico.

1882-1890 Intento de construcción de un Canal por los franceses.

1899-1902 La guerra de los Mil Días reduce la población y el ganado, en el Istmo.

1903 Panamá proclama su Independencia de Colombia.

1903-1914 Construcción del Canal de Panamá.

1966-67 Finalización de la carretera Interamericana que unió la ciudad de Panamá con la de David y con la frontera con Costa Rica.

1977 Firma del nuevo Tratado del Canal de Panamá entre los Estados Unidos y Panamá.

ECONOMIA

Los productos agrícolas mayormente exportados son los bananos (tierras bajas del oeste), azúcar (tierras bajas centrales) y café (tierras de altitud mediana y alta, del oeste). El ganado es la industria dominante en muchas áreas, al igual que lecherías y crías avícolas. La industria maderera está basada en la deforestación para apoyar la colonización abortiva de las tierras que aún poseen bosques. La pesca, especialmente de camarones, es importante al nivel nacional e internacional. La economía rural de Panamá está fundamentada en actividades de subsistencia. Estas actividades rurales contrastan notablemente con las generadas por el comercio internacional. En años recientes, los servicios internacionales de transporte a través del Canal de Panamá, la banca internacional, el manejo del petróleo y del turismo han crecido rápidamente. Panamá encaja en las actividades internacionales con niveles salariales y demandas del consumidor de tipo cosmopolita.

PART I

THE ORGANISMS—LOS ORGANISMOS

Plant Groups

Living organisms are one of the foundations of Natural History studies, and Panama is unusually rich in plants and animals. Biological diversity tends to increase as one moves from the poles to the equator, and this is true for plant life. The total number of vascular species in Panama is believed to be between 8,000 and 10,000 species. This number of species in Panama compares with the approximately 2440 species in the state of Missouri which is about 1.6 times the size of Panama or with 3360 species in North and South Carolina, an area almost three times that of Panama. There are also large numbers of non-vascular species in Panama which have had scant investigation. Tropical forests differ from temperate forests not only in numbers of species but also in appearance and structure. Noteworthy are the conspicuous epiphytes and lianas supported by the structure of trees, and these elements add greatly to the size of the total flora. For example, the flora of the Carolinas includes only 89 species of ferns and 53 orchids, nearly all of these terrestrial. In Panama there are over 687 species of ferns and about 1000 species of orchids, a great proportion of them aerial. Another feature that determines the appearance of tropical forests is the abundance of conspicuous large monocotyledonous plants which are mostly lacking in the temperate zone. The following papers discuss diversity in some of Panama's major plant groups and formations.

The usual abundance of large monocotyledons is a characteristic difference between the humid tropics and other parts of the world. They are frequent in open areas as well as in forests. The area near the ground, which in the temperate forest is occupied by seasonal herbs and small shrubs, is in the tropical forest often an array of conspicuous, broad-leaved monocots. The monocots were the first families to be treated in the Flora of Panama when it began to appear some four decades ago. Croat discusses the large monocots by family and compares each family with its state of knowledge when it was treated in the Flora. One group of large monocots, the Heliconias, are discussed further later in papers by Kress and Stiles, and that most important group, the grasses, includes many species of large size.

Grasses, so conspicuous in temperate climates, are also a mainstay of agriculture in the tropics, where they are a large and diverse group of plants. Davidse discusses the wealth of Panamanian species and how they relate to the grasses of other areas, speculating on where most of Panama's grasses might have originated. Gentry, elsewhere in this volume presents similar reflections on floras of upland and lowland Panama.

Orchids are especially diverse in Panama, and they occur in Panamanian forests at all levels. It is appropriate that Panama has chosen one of them, *Peristeria elata*, the Holy Ghost Orchid, as a national emblem. Although they too are monocots, few of them are so large as those considered by Croat. Dressler discusses exploration for orchids in Panama and notes many of the interesting species.

Another group of plants borne aerially are the lianas and vines. These are of diverse families and usually abundant in neotropical forests. They are often less studied than other plant groups, and Gentry provides a valuable analysis of their distribution, ranges and 'life styles'.

Ferns, both epiphytic and terrestrial, are especially numerous and conspicuous in the humid tropics. Lellinger indicates that many species are restricted to a single life zone. A large portion of Panama's ferns also occur in Costa Rica.

Mosses abound in moist cloud forests as well as other parts of Panama. Salazar & Crosby report on the many species that have been recorded from Panama, including some new records, and they make clear that the serious study of Panama's bryophytes has just begun.

Mangroves are characteristic plant formations of tropical seacoasts. Cintron & Lugo point out that unlike inland tropical forests, mangrove formations are poor in species, and organisms in them are influenced to a greater degree by physical conditions than by biological interactions. They outline the effects of differing environmental conditions on the formation of mangrove vegetation.

Grupos de Plantas

Los seres vivientes son una de las bases para los estudios de historia natural y Panamá es sumamente rica en plantas, animales y otros organismos. La diversidad biológica tiende a crecer a medida que se avanza de los polos hacia el ecuador. El total de especies de plantas vasculares de Panamá se estima entre 8.000 a 10.000. El número de especies de Panamá comparado con las 2.440 del estado de Missouri cuya área es 1.7 veces mayor y con las 3.360 de Carolina del Norte y del Sur, con un área 3 veces mayor, es significativamente más rico. Asimismo existen innumerables especies de plantas no vasculares, las cuales han sido escasamente investigadas. Los bosques tropicales difieren de los de la zona templada no sólo en el número de especies sino también en apariencia y estructura. En los bosques de los trópicos húmedos son notables las epífitas y lianas sostenidas por la estructura de los árboles y estos elementos constituyen una porción extraordinaria de la flora total. Así, por ejemplo, la flora de las Carolinas incluye sólo 89 especies de helechos y 53 de orquídeas, casi todos son terrestres. En Panamá están representadas un poco más de 687 especies de helechos y aproximadamente 1000 especies de orquídeas, la mayoría aéreas. Otro rasgo que determina la apariencia de los bosques tropicales es la abundancia de monocotidedóneas grandes y conspicuas; estas plantas están mayormente ausentes en la zona templada. Los trabajos que se presentan a continuación discuten la diversidad de algunos de los principales grupos de plantas de Panamá.

La abundancia usual de monocotiledóneas grandes en los bosques de los trópicos húmedos es una característica fundamental que los diferencia de los bosques de otras regiones del mundo. Estas son frecuentes en áreas abiertas, al igual que en los bosques. El área próxima al suelo, en la zona templada, está ocupada por hierbas de vida corta y por arbustos pequeños; en el bosque tropical este área está cubierta por monocotildóneas conspicuas de hojas anchas. Las familias de monocotiledóneas fueron de las primeras en publicarse en la Flora de Panamá, al inicio de ésta, hace aproximadamente cuatro décadas. Croat discute las monocotiledóneas grandes y conspicuas por familia, compara el conocimiento actual sobre cada una de éstas con aquel publicado en la Flora. Más adelante, Kress y Stiles revisan un grupo notable de monocotiledóneas, las *Heliconia*, mientras que Davidse discute uno muy importante, las gramíneas.

Las gramíneas que son tan notables en los climas templados, también son la base de la agricultura en los trópicos, en donde es un grupo extenso y muy diverso. Davidse discute la riqueza de las especies panameñas y las relaciona con gramíneas de otras regiones, especulando sobre el origen de las especies de Panamá. Gentry en su trabajo de fitogeografía, en otro de los capítulos de este volumen, presenta reflecciones similares entre las floras de las tierras altas y las bajas de Panamá.

En Panamá, las orquídeas son especialmente diversas y éstas ocurren a todos los niveles de los bosques panameños. A pesar de que son monocotiledóneas, si hay alguna grande, ninguna lo es tanto como las plantas anteriores. Es muy apropiado que Panamá haya elegido una orquídea, *Peristeria elata*, la flor del Espíritu Santo como la flor nacional. Dressler discute las exploraciones para obtener orquídeas en Panamá y señala muchas de las especies interesantes.

Otro grupo de plantas aéreas lo son las lianas y bejucos. Este representa diversas familias y usualmente son abundantes en los bosques neotropicales. Frecuentemente son menos estudiadas que cualquier otro grupo de plantas y Gentry proporciona un valioso análisis de su distribución y de su forma de vida.

En los trópicos húmedos, los helechos, tanto los epífitos como los terrestres, son especialmente abundantes. Lellinger indica que muchas especies están restringidas a una sola zona de vida. Numerosos helechos panameños también ocurren en Costa Rica.

Los musgos son exuberantes en los bosques nebulosos y en otras áreas de Panamá. Salazar y Crosby reportan muchas especies que se han registrado por primera vez en Panamá, al igual que enfatizan que las investigaciones serias de las briófitas de Panamá sólo han comenzado.

Los manglares son formaciones vegetales características de las costas tropicales. Cintrón y Lugo señalan que a diferencia de los bosques del interior de un país, los manglares son pobres en especies y están más influenciados por las condiciones físicas que por las interacciones biológicas. Cintrón y Lugo esbozan los efectos que las diferentes condiciones ambientales pueden ejercer sobre los manglares.

The Large Monocots of Panama

Thomas B. Croat

Las monocotiledóneas grandes caracterizan la vegetación tropical. Su gran número, su tamaño y arquitectura exclusiva hacen de ellas una característica sorprendente de la vegetación. Las monocotiledóneas son mayormente abundantes por debajo de los 1000 a 2000 m y en zonas de vida muy húmedas. Muchas poseen flores muy vistosas. En Panamá, se conocen 8 familias y muchas especies nuevas se han descrito después de publicada la Flora de Panamá.

One of the principal features that characterizes tropical vegetation is habit or growth form diversity. It is the large monocots comprising fewer than one dozen families, more than any other group, that contribute to growth form diversity in tropical forests.

The large monocots, which are lacking in temperate forests, best typify tropical vegetation, and along with woody lianas and tree buttressing they form the immediate "tropical impression" upon one's entrance into the tropics. Because of their unique architecture, their frequently large size, and occasional dominance, no group of plants is more conspicuous than the family Palmae. Almost as conspicuous and frequently more abundant in some life zones are the families Musaceae, Marantaceae, and Araceae. Araceae are particularly abundant in some areas, where in terms of individuals they may be more numerous than any other family. At higher elevations Bromeliaceae are frequently abundant and sometimes even the dominant plant group. The families Cyclanthaceae and Zingiberaceae, though rarely ever rich in number of species in a local area, are nevertheless among the most conspicuous because of their beautiful inflorescences, or because of their unusual leaf shapes and unusual growth patterns. Two other relatively uncommon but conspicuous plant families are the Cannaceae and Agavaceae.

Las monocotiledóneas grandes caracterizan la vegetación tropical. Palmae, Musaceae, Marantaceae y Araceae son conspicuas. Las bromeliáceas son abundantes en lugares elevados. Las especies de Cyclanthaceae y Zingiberaceae si no son numerosas, al menos son conspicuas. Igualmente, las especies de Agavaceae y Cannaceae son poco comunes, pero notables.

Most of these families were treated in early issues of the Flora of Panama and as a result they are currently among the most poorly known members of the flora of Panama. Most are difficult to collect and preserve, and thus they are poorly represented even among modern collections. This paper introduces these families and presents an analysis of their occurrence in Panama.

Agavaceae

The Agavaceae is the rarest family of large monocots in Panama, represented there by three genera with one species each. Generally when we mention the Agavaceae we think of the very large and conspicuous genera *Agave* or *Yucca* with a basal rosette of leathery sword-shaped leaves and often towering inflorescence with large white flowers. *Furcraea*, with a rotate rather than funnel-form corolla like *Agave*, also has its leaves in a basal rosette. The genus *Dracaena* has strap-shaped cauline leaves which are closely spiralled on the more or less woody stems. The inflorescences are large terminal panicles and, unlike those of *Agave*, they have small but otherwise typical lily-like flowers. *Dracaena* was not discovered in Panama until 1972 when I found it during a National Geographic Society expedition to the Burica Peninsula, where in an isolated region of tropical wet forest near the Costa Rican border it was quite abundant.

La familia Agavaceae está representada, en Panamá, por tres especies. Agave (o Yucca) y Furcraea son grandes y presentan sus hojas organizadas en una roseta basal. Dracaena presenta hojas ensiformes espiralmente organizadas sobre tallos leñosos.

Araceae

The Araceae is one of the most conspicuous tropical monocotyledonous groups. In most moist or wet life zones where forest still exists they are everywhere abundant. In wetter forests where little or no dry season occurs and especially in regions of cloud forest, the family becomes almost dominant. In any case, the family may be easily the largest family in the local flora in terms of numbers of individuals, with virtually every tree housing one to scores of individual aroids. In some cases, the species involved are large and usually quite showy so that they frequently dominate the view of the forest. Individuals of any species may be ubiquitous or exceedingly rare, but species diversity at any given wet forest or moist forest site is usually quite high. The family ranges from sea level to almost 3000 m in Panama with the greatest concentrations of species at middle elevations, especially in the 500 to 1500 m range.

The family is distinguished by having its flowers congested into spikes and at least initially surrounded by a spathe which subtends the spadix. Leaves are diverse, and at first sight many are not typically monocot-like. Most are relatively large and net-veined and, while some genera have closely parallel veins, few have grass-like or lily-like leaves. Aroids have a diverse array of habits including free floating aquatics like *Pistia*, rooted aquatics like *Montrichardia* and *Urospatha*, and there are mostly terrestrial genera like *Caladium*, *Dieffenbachia*, *Dracontium*, *Homalomena*, *Spathiphyllum*, *Urospatha*, and *Xanthosoma*. Most of the larger genera are hemiepiphytic climbers or vines like *Heteropsis*, *Monstera*, *Rhodospatha*, *Syngonium*, and most *Philodendron*. *Stenospermation*, most *Anthurium*, and many *Philodendron* are true epiphytes.

Las familia Araceae es abundante en individuos y en especies, especialmente en las zonas húmedas. Estas plantas son especialmente más abundantes entre los 500 a los 1500 m de altitud. Muchas son epífitas o hemiepífitas, sin embargo Pistia es acuática y flota libremente sobre la superficie del agua, mientras que Montrichardia y Urospatha que son igualmente acuáticas están arraigadas al sustrato.

Flowers of many genera, including *Anthurium*, *Heteropsis*, *Monstera*, *Stenospermation* and *Rhodospatha*, are bisexual and have spadices uniform throughout. The remaining genera in Panama are monoecious with the unisexual flowers separated on the spadix. The staminate flowers are always situated toward the apex and the pistillate flowers at the base. In the monoecious genera neuter flowers sometimes occupy a central position.

La mayoría de los géneros son monoicos con las flores estaminadas en la parte superior del espádice y las pistiladas en la inferior, sin embargo muchos Anthurium, Heteropsis, Monstera, Stenospermation y Rhodospatha poseen flores bisexuales.

The spathe plays an important role in the taxonomic separation of the genera. In *Anthurium* it usually envelopes the spadix only in bud and folds away even before anthesis of the flowers. In other genera, such as *Monstera*, *Rhodospatha*, *Stenospermation* and *Heteropsis*, the spathe usually falls free from the base after anthesis, whereas in *Caladium*, *Syngonium*, and *Xanthosoma* only the blade (upper part over the staminate part of the spadix) falls off after flowering. In several genera, including *Philodendron*, *Homalomena*, and *Dieffenbachia*, the spathe recloses around the entire spadix after flowering, and the staminate apical part rots away inside. The latter three genera are pollinated at night by ruteline scarab beetles. *Anthurium* is pollinated by euglossine bees, flies or sphecid wasps. *Spathiphyllum* is pollinated, at least in part, by euglossine bees. Fruit dispersal is by animals, *Anthurium* fruits being adapted to bird dispersal, *Philodendron* to birds or mammals and perhaps even leaf cutter ants. *Montrichardia* and some *Spathiphyllum* fruits are probably water dispersed.

En su tratado de las especies de Araceae de la Flora de Panamá, Standley presentó 77 especies distribuidas en 17 géneros. Actualmente se conocen por lo menos 302 especies. Muchos de los nombres que Standley empleó han cambiado.

The Araceae is proving to be one of the most poorly known of all the families in the Flora of Panama. Standley (1944) treated 17 genera with a total of 77 species. The actual number was only 68 species because a number of synonyms were also included. Today there is only one additional genus, *Chlorospatha*, (though *Anepsias* has been dropped and *Heteropsis* added), but the number of species has already increased to at least 302 and others are expected. The overall increase is thus 344% in the almost 40 years since the Flora of Panama treatment was published. Owing to the difficulty of collecting and preserving specimens, the taxonomic difficulty of the group, and the results of recent intensive investigations, the treatment for the Araceae is seen to be one of the most inadequate of all families done for the Flora. For example, the genus *Anthurium* with 37 names reported by Standley (actually representing only 29 species) now has approximately 150 species, an increase of 417%. The quality of Standley's nomenclatural judgements is particularly bad. For example, only 17 of his 37 names in *Anthurium* are still in use today in Panama. The nomenclatural situation for *Philodendron* is better, with only 4 of the 12 species names having to be changed, but the degree of representation is even worse; at least 75 species of *Philodendron* are now known from Panama, an increase of 525%.

Bromeliaceae

The Bromeliaceae is the most ecologically diverse family among the large monocots, ranging from the driest life zones to the wettest, from sea level to the highest mountain tops in Panama. The ecological range of a single species may in fact be broader than most tropical families. There are several species such as *Tillandsia usneoides* (L.) L. and *T. recurvata* (L.) L. which range from sea level to 3000–4000 m above sea level. Many species range from sea level to above 1000 m. In fact, the average elevational range for all species in the family is approximately 1100 m, greater surely than the average for any other family.

Las especies de Bromeliaceae se extienden desde las zonas de vida muy secas hasta las muy húmedas y algunas especies se encuentran desde el nivel del mar hasta los 3000 a 4000 m de altitud. Muchas se encuentran por encima de 1000 m.

Species of the family may be relatively common at lower elevations and species diversity quite high, although bromels are usually not an overly conspicuous element of the flora except when in flower. When in flower they are generally conspicuous and frequently beautiful as well. Both species diversity and usually numbers of individuals increase to some extent with elevation. For example, less than 1% of Panamanian bromels are restricted to elevations below 1000 m, whereas 22 species (26%) are always found above 1000 m, and 37 species (44%) are always found above 500 m. At higher elevations in Panama, the forest can even be dominated by Bromeliaceae, with every tree bearing many tank epiphytes in the same way trees are dominated by Araceae at middle elevations. Eventually, going upward, the epiphyte flora of mountain forests may become almost exclusively bromeliads.

The Bromeliaceae is a very natural family rarely confused with any other. Plants are characterized by their rosulate habit and have spirally arranged, frequently strap-shaped or slender, pointed leaves that are generally closely imbricate at the base and often form a water-tight basin. These tank epiphytes commonly hold not only water but considerable insect life as well, and the liquid into which the roots may grow gives important nutrients to the plant. The water reserve is helpful in withstanding drought. Another adaptation of many bromels to prevent desiccation are the stellate trichomes or specialized scales which are capable of absorbing water. Bromels also have principally non-functional stomates on the lower blade surfaces. The plants can store and exchange the gases produced by photosynthesis and respiration through intercellular aerating canals which prevents loss of the gases to the atmosphere. Because of these unique adaptations bromels can withstand prolonged drought better than epiphytes in other families such as Araceae, and they are consequently found in even the driest types of forest. For example, on parts of the Azuero Peninsula where the dry season is particularly severe despite relatively high overall rainfall, epiphytic aroids are almost non-existant but bromeliads are relatively common.

Las bromeliáceas presentan hojas alargadas, puntiagudas, organizadas espiralmente y en rosetas que forman a manera de una vasija central que contiene agua. Muchas poseen tricomas esteliformes los cuales pueden absorber agua. Las bromeliáceas resisten largos períodos de sequía.

The bromeliad inflorescence—usually a spike or raceme or a panicle of racemes—arises from the middle of the rosette and is often densely bracteate. The flowers are also subtended by closely fitting bracts and emerge from beneath the bracts at anthesis. They are usually more or less tubular and colorful, but sometimes zygomorphic. Pollinators are bees, bats, and hummingbirds. Red flowers are relatively rare, but often the attractant for the birds is the bright red bracts associated with frequently yellow flowers.

Most bromel fruits have comose seeds and are dispersed by the wind, but fruits in some genera are berries and are animal dispersed, usually by birds.

Smith, en la Flora de Panamá, incluyó 94 es-

pecies en 10 géneros. Actualmente, se conocen más de 160 especies.

The treatment for the Bromeliaceae for the *Flora of Panama* completed in 1944 by L.B. Smith included 10 genera and 94 species. Although the Flora Neotropica revision of the Bromeliaceae by Smith & Downs (1974–79) has been published and some new material from Panama was used, relatively little of the material collected during the past 15 years was cited in these works. This is unfortunate since the collections made during this period outnumber all material collected in Panama during all preceding years of botanical activity. After studying both the species cited in the Flora Neotropica and the annotated material in the herbarium at MO, I estimate that there are now approximately 160 species in Panama, an increase of 60% over the number in the Flora. This is a smaller increase than most of the large monocot families. The initial taxonomic treatment by Smith was good, and few species have changed names since the Flora of Panama treatment was published.

Cannaceae

With the exception of the Agavaceae, the Cannaceae, consisting of a single genus in Panama, is the smallest family of large monocots in Panama. Cultivated *Canna* are common ornamentals. The Cannaceae is a family of large monocots most commonly found in disturbed areas and usually found associated with man. I have never seen *Canna* in undisturbed forest areas in Panama. Although species of *Canna* range from sea level to about 2000 m, they are relatively rare where they occur and are seldom found in large stands except where cultivated. Sterile plants of *Canna* can easily be confused with *Heliconia* as they have similar, broad, banana-like leaves with stout midribs and numerous close parallel veins. The family is readily distinguished in flower from the Musaceae, however, by the structure of the inflorescence and by the flowers. The inflorescence consists of a bracteate raceme or panicle with the flowers borne in pairs. The inflorescences are never subtended by conspicuous boat-shaped bracts as in *Heliconia*. Flowers consist of 3 inconspicuous sepals and 3 small petals. The showy part of the flower consists of a petaloid stamen and 2 or 3 petaloid staminodia which are markedly more conspicuous than the petals. Fruits are conspicuous, large, warty or spiny capsules which ultimately weather to permit release of the several round hard seeds.

Nueve especies de Canna se conocen de Panamá. Algunas especies son cultivadas como ornamentales.

The Flora of Panama treatment by Woodson & Schery (1945c) dealt with 6 species of *Canna*. In an as yet unpublished revision of the Cannaceae for the flora Neotropica, P. Maas has concluded that there are only 9 species in *Canna*. Only two of these occur as natives in the Panamanian flora, namely *C. glauca* L., a yellow flowered species of low elevations frequenting swamps, and *C. tuerckheimii* Kranzl. a species ranging from 500 to 2000 m. *Canna indica* L. is often escaped from cultivation and is perhaps the species most frequently encountered. *Canna flaccida* Salisb., a U.S. gulf coast species, has also been cultivated in Panama.

Cyclanthaceae

As a whole the family Cyclanthaceae plays a lesser role than other large monocots such as Araceae, Musaceae, Marantaceae, Zingiberaceae and Palmae. However at low elevations the family is generally more abundant than Bromeliaceae and in some places, especially along forest margins in areas exposed to the sun, *Carludovica palmata* R. & P. can even be the dominant species. In some wet forest sites *Asplundia* can become the dominant understory plant. The majority of the species are relatively small and epiphytic, but owing to their unusual leaf shape they are always immediately obvious.

Species diversity drops off rapidly with increase in elevation or decrease in precipitation, and the epiphytic members of the family are uncommon in areas where the dry season is severe or the total annual rainfall is low.

The family consists of perennial, usually epiphytic vines or short-stemmed epiphytic creepers or terrestrial plants with short stems. In appearance plants are palm-like, having simple, entire or cleft pleated leaves with palmate venation, but only the terrestrial *Carludovica* strongly resembles a palm. Cyclanths are distinguished from palms in having conspicuous, elongate staminodia, several, usually deciduous spathes subtending the simple spadix, and a fleshy syncarpous fruit with numerous small seeds.

The unisexual flowers (monoecious) are interspersed on the spadix. The long, usually white staminodia act as an attractant both by their appearance and the fragrance they produce on the first day of anthesis. Pollinators are insects, most probably weevils *(Coleoptera)* according to Harling (1958).

Leaf shape plays an important role taxonomically—whether the blade is entire (*Ludovia*), flabelliform-parted (*Carludovica*), or simple and bifid (all other genera). The disposition of the leaves is also important. In *Sphaeradenia* and *Ludovia* they are distichous whereas in other genera they are dispersed. The pistillate flowers are connate in *Asplundia* and *Thoracocarpus* and free in *Evodianthus* and *Dicranopygium*. The number of spathes present is also important taxonomically, with *Asplundia* and *Thoracocarpus* having 3–11 spathes and *Evodianthus*, *Dicranopygium*, and *Sphaeradenia* having usually 2–3 spathes.

Las plantas que se incluyen en la familia Cyclanthaceae son muy similares a las palmas y poseen flores unisexuales. En Panamá, esta familia está representada por aproximadamente 30 especies.

Perhaps no other group of the large monocots has been so drastically revised following completion for the Flora of Panama than the *Cyclanthaceae*. Woodson & Schery (1943) included 2 genera: *Cyclanthus* with a single species, *C. bipartitus* Poir., and *Carludovica* with 12 species. But five of the 12 species of *Carludovica* were actually misidentified. Harling (1958) reported 8 genera and 22 species for Panama. Most species of *Carludovica* were segregated into *Asplundia*, *Ludovia*, *Dicranopygium*, *Evodianthus*, *Sphaeradenia*, and *Thoracocarpus*, and only 3 species of the 12 species of *Carludovica* in Panama still have the same name. Additional species of Cyclanthaceae have been discovered since Harling's revision, and the current number of species is estimated at about 30 for an increase of 131%. Despite the work by Harling the family remains poorly known in Panama and more new species are to be expected.

Marantaceae

The family Marantaceae is a characteristic and often conspicuous element in wet and moist forests of Panama from sea level to 1500 (rarely to 2000 m). Species of Marantaceae occur in a wide range of habitats from semideciduous forest to cloud forest and rain forest. Both numbers of species and numbers of individuals are greater at lower elevations. Some species, such as *Calathea marantifolia* Standley and *C. indecorea* Woodson, form almost solid stands in open areas and along forest margins, but most species occur as scattered individuals or small clumps within the forest. A few species prefer wet situations. *Calathea lutea* (Aubl.) Schult. may occasionally be found growing in standing water. *Thalia geniculata* L. is strictly a rooted aquatic.

Las marantáceas son abundantes en los bosques húmedos y muy húmedos. Sus hojas anchas y delgadas poseen venas transversales paralelas que se conectan con las venas longitudinales sigmoídeas. El ápice del pecíolo frecuentemente presenta un ensanchamiento.

The family consists of rhizomatous perennials that occasionally have highly reduced (internodes) stems. Species of the family are easily distinguished even when immature by their broad thin leaves with parallel transverse veins connecting the sigmoid longitudinal veins and by the discolored often swollen area of specialized cells at the apex of the terete petiole (the pulvinus). The flowers are subtended by conspicuous bracts and are usually aggregated into spikes. The inflorescences of some genera, such as *Calathea*, are conelike, whereas in other genera, such as *Thalia*, *Maranta* and *Stromanthe*, the inflorescences can be relatively diffuse to branched.

Flowers of the Marantaceae are unique and highly evolved for the order Scitamineae (Kennedy, 1978). They are irregular and consist of unequal petals, a single fertile stamen and petaloid staminodia which usually form the attractive part of the flower. The most unusual feature of the flower is its pollinating mechanism which has explosive, secondary pollen presentation. Prior to pollination the flowers are "spring-loaded." The pollinators (usually bees) trigger the flower, causing the style to spring forward. The pollen, which is deposited in a depression in the style before anthesis, is carried with the style as it springs forward and strikes the pollinator.

Hoy día se conocen cincuenta y ocho especies de Marantaceae de Panamá.

Woodson & Schery (1945d), in the treatment of the Marantaceae for the Flora of Panama, included 6 genera and 23 species. The genus *Calathea*, though treated relatively well taxonomically—only 5 of the 14 species have had name changes—has had many new species added. Now approximately 37 species are known for Panama and other new species are expected. Nine were recognized as distinct from *Myrosma* (no longer in the flora). *Ischnosiphon* was split into two genera to include *Pleiostachya*. Relatively few new species have been added to other genera, but the total number of species for the family is now approximately 58. The overall increase of 152% is not so large in comparison to many of the other large monocot families.

Musaceae

The Musaceae, represented in Panama only by the cultivated genus *Musa* and by *Heliconia*, is a conspicuous member of the tropical flora. The family is particularly abundant at lower elevations, but *Heliconia* ranges to about 2000 m. Though relatively poor in species at higher elevations, *Heliconia* is always a conspicuous and generally colorful member of the flora. At lower elevations some species, such as *H. mariae* Hook. f. and *H. latispatha* Benth., may form large, nearly closed stands in disturbed areas, but most species occur as individuals or more frequently large to medium-sized clumps in open areas of the forest, along streams or at the edges of clearings. In growth form, the family is closest to the Marantaceae among the families of large monocots. Ecologically, the Musaceae are close to the Zingiberaceae which are also strictly terrestrial plants occurring in similar habitats.

La familia Musaceae incluye el banano y el plátano, ambos introducidos y cultivados, e igualmente incluye el género Heliconia, que es bastante grande, conspicuo y abundante en altitudes bajas. Heliconia tiene flores tubulares y alargadas las cuales son polinizadas por colibríes.

The Musaceae are rhizomatous herbs with a stemlike structure formed by closely imbricated sheathing petioles. The broad, mostly oblong leaves with stout midribs and close, fine, straight lateral veins characterize the family. In this respect the family is similar to the Marantaceae, but sterile plants of the Musaceae can be distinguished by having the petioles sulcate adaxially rather than terete. The long tubular flowers are bisexual in *Heliconia* and unisexual (monoecious) in *Musa*. In *Heliconia* the flowers are in dense aggregations and subtended or enveloped by conspicuous, colorful, distichous bracts. The color, shape, and disposition of these bracts, as well as flower color, provide the principal taxonomic characteristics for separation of species.

Pollination of *Heliconia* is by hummingbirds. Aggregated species such as *H. latispatha* Benth. are pollinated by territorial hummingbirds, and isolated forest species are pollinated by nonterritorial species (Linhart, 1973; Stiles, 1975). Fruits are bird dispersed, chiefly by manakins (Stiles, 1975).

Cerca de 30 especies de Heliconia se conocen de Panamá.

The Flora of Panama treatment by Woodson (1945a) included only 11 species of *Heliconia*, and the majority of these still have the same names. Smith (1968) included 20 species for Panama. Current estimates indicate 27 species, and more are likely to be discovered when an expert in the family concentrates in Panama. Although the genus has recently been revised for Costa Rica (Daniels & Stiles, 1979), no major effort has been made to revise the Panamanian species.

Palmae

The family Palmae forms a conspicuous and often dominant element of all but the driest life zones throughout Panama. Because of their unique growth form, the larger palms always represent the most conspicuous element in tropical forests. Palms range from sea level to about 2000 m but are most abundant at lower elevations. However, individuals of a single or a few species may be even more prevalent than at lower elevations. Palms often form the dominant part of the vegetation in special situations, such as *Colpothrinax cookii* Read on Cerro Jefe and *Dictyocaryum* aff. *platysepalum* Burret along the cloud forest ridge between Cerro Pirre and Altos de Nique near the Columbian border. At lower elevations also some species are abundant, and species such as *Scheelea zonensis* L. H. Bailey may be particularly conspicuous in some parts of lowland tropical moist forest. Though most palms occur as individuals, some genera such as *Bactris* may form large stands. Though most species occur on well drained soils, a few, such as *Elaeis oleifera* (H.B.K.) Cort. *Raphia taedigera* Mart., *Manicaria saccifera* Gaertn., and *Phytelephas seemannii* Cook occur principally in marshes.

Las palmas son conspicuas en los bosques panameños y primordialmente se encuentran en altitudes inferiores a los 2000 m. La mayoría de las palmas panameñas poseen hojas divididas, pero Colpothrinax, Crysophila y Sabal las tienen palmatilobadas.

Most palm genera in Panama have few species

(mostly 1 or 2) but some genera, such as *Bactris*, *Geonoma*, and *Chamaedorea*, have many species.

The Palmae is a natural family and is characterized by leaf architecture with plicate blades and always woody, unbranched stems, some without emergent stems, elsewhere a few are branched, with often conspicuous leaf scars or with the old petiole bases persisting. The family can be confused only with the Cyclanthaceae. Leaves of both families are pleated or the individual pinnae V-shaped. The Palmae can be distinguished from the Cyclanthaceae by the closed leaf sheath in bud, and by having usually 6 perianth segments, 1 ovule per locule, and frequently compound inflorescences with a prophyll and 1 or more, sometimes persistent peduncular bracts, included staminodes, and usually 1–3-seeded fruits.

While most palms in Panama have pinnately divided leaves, three genera, *Colpothrinax*, *Cryosophila*, and *Sabal*, have palmately lobed leaves. Although entire leaves are common for juveniles of many species, few species have entire leaves as adults, the exceptions in Panama being *Asterogyne martiana* Wendl. ex Burret and a few species of *Chamaedorea*, *Bactris*, and *Geonoma*. Palm flowers are sometimes inconspicuous but are always more or less aggregated onto simple or branched inflorescences that are occasionally massive. Flowers may be unisexual or bisexual, and plants may be monoecious or dioecious.

The inflorescences bear a prophyll and one or more peduncular bracts which, in some cocosoid genera, may be conspicuous and boat-shaped, frequently persisting with the spadix. Pollination is principally by insects, beetles, and syrphid flies (Essig, 1971, 1973; Uhl & Moore, 1977), but some species elsewhere are also believed to be pollinated by bees, drosophilid flies, wasps, and wind (Uhl and Moore, 1977). The fruits vary from minute to very large as in the coconut, *Cocos nucifera* L. They are primarily bird or mammal dispersed.

The Flora of Panama treatment by L. H. Bailey (1943) included 27 genera with 70 species. Twenty of the genera had a single species, and three additional genera had only 2 species. This is remarkably few species per genus (2.6) compared to other monocot families in the flora. The average number of species drops to 2.3 if we consider the 7 new genera since added to the flora, 6 of which have only 1 species and the other 2 species.

No systematic revision of any major genus in the flora has been completed subsequent to 1943 nor has any palm expert spent a significant amount of time collecting palms in Panama since the work by Bailey. As a result general ignorance of this group is greater than for any other monocot family.

Many species of palms new to the flora have been collected since Bailey's work but they remain largely unidentified or at least with questionable identifications. However, the flora now contains 34 genera with approximately 100 species, an increase of 43% from the number in the Flora of Panama treatment. Some genera with which I am more familiar, for example *Bactris*, have been expanded by larger percentages. For *Bactris* Bailey reported 13 names in the Flora of Panama treatment which actually represented 10 species. I concentrated on collecting *Bactris* for several years, and the number of *Bactris* species is now closer to 20, a 100% increase. The genus *Chamaedorea*, already the largest genus of palms in Panama, will probably show the largest increase following field work by a palm expert.

La flora de palmas de Panamá es poco conocida y hasta el momento se han registrado cerca de 100 especies distribuidas en 34 géneros.

The difficulty of collecting the taller palms explains why collections of them are generally poorer than those of the average member of the flora. The family is badly in need of more active field work. Because the family has been less well studied than most of the other large monocot families, the percentage of increase suggested here for the palms will probably be an underestimate.

Zingiberaceae

Despite its relatively small size (ca 35 species), the Panamanian Zingiberaceae often forms a conspicuous element of tropical forests from sea level to about 2000 m. Because of their striking growth form and usually beautiful inflorescence, species of Zingiberaceae are usually showy. They occur along trails, in tree fall openings and at forest edges, but because of light requirements relatively few species are found in dense understory areas. While the majority are of mediocre size, some species attain truly immense size for monocotyledonous plants without any secondary thickening. *Costus laevis* R. & P., for example, may attain a height of 6 m. With the exception of the weedier species (*Dimerocostus strobilaceous* Kuntze, *Costus villosissimus* Jacq., *Costus guanaiensis* Rusby and a few others), they are usually not common except on a local basis. On the other hand, *Costus woodsonii* Maas, hardly a weedy species, may form large stands in open areas in the strand vegetation along the Caribbean coast. Species of *Renealmia* are mostly rare and of sporadic occurrence. In lowland forest they are most frequently found in the first stages of succession in tree fall areas. In cloud forest areas they form a more regular part of the vegetation where light is adequate, but even here they are of infrequent occurrence. But some species, such as *Renealmia cernua* (R. & S.) Macbride, are common even in more mature forest.

Las zingiberáceas son conspicuas y viven por debajo de los 2000 m. Sus flores son vistosas y su forma de crecimiento es sorprendente.

The family is readily distinguished by the densely bracteate inflorescences which suggest those of the Marantaceae. Inflorescences may be borne at the apex of the unbranched stems, or more frequently they are borne on separate stalks arising directly from the rhizome. Flowers commonly appear one or a few at a time on the species with large showy flowers, but *Renealmia*, usually with more numerous small flowers, may have several flowers open at a time. Here individual flowers are often not individually showy, and attraction is more because of the aggregate inflorescence and the showy bracts. Pollination is mostly by bees, but some species with tubular red flowers are pollinated by hummingbirds. In most cases the showy part of the flower is not the petals but rather the labellum, a modified stamen.

Dimerocostus, with a single species *D. strobilaceus* O. Ktze. in Panama, differs from other genera in having a bicarpellate ovary which is exserted above the bracts. *Costus* and *Renealmia* have tricarpellate ovaries that are hidden by the bracts.

Members of the Zingiberaceae are perennial herbs with horizontal tuberous rhizomes. Stems are unbranched, and leaves are either distichously (*Renealmia*) or spirally arranged (*Dimerocostus* and *Costus*). The unique spiral arrangement of the leaves of the latter genera, often appearing like a spiral stairs along the stem, makes them appealing and readily recognizable.

Las zingiberáceas están representadas, en Panamá, por aproximadamente 35 especies.

The family was written for the Flora of Panama by Woodson (1945b). Like most large monocotyledonous families that were treated in earlier portions of the Flora of Panama, the Zingiberaceae treatment is inadequate. Many species have been added by the revisions of Maas (1972; 1977). Woodson's treatment dealt with only 19 species whereas the revision by Maas in the Flora Neotropica reports 35 species, an increase of 84%. These include a single species of *Dimerocostus*, 19 species of *Costus* and 15 species of *Renealmia*. In Woodson's treatment at least 5 species were reported under incorrect names (not merely names now considered synonymous).

Bailey, L.H. 1943. Palmae *In* R.E. Woodson, Jr. et al. Flora of Panama. Ann. Missouri Bot. Gard. 30: 327–96.

Daniels, G.S. & F.G. *Stiles*. 1979. The *Heliconia* taxa of Costa Rica, keys and descriptions. Brenesia 15, Suppl: 1–150.

Essig, F.B. 1971. Observations of pollination of *Bactris*. Principes 15: 20–24.

———. 1973. Pollination of some New Guinea palms. Principes 17: 75–83.

Harling, G. 1958. Monograph of Cyclanthaceae. Hakan Ohlssouns Bok Tryckeri, Lund, Sweden.

Kennedy, H. 1978. Systematics and pollination of the "closed-flowered" species of *Calathea* (Marantaceae). Univ. Calif. Publ. Bot. 71: 1–90.

Linhart, Y.B. 1973. Ecological and behavioral determinants of pollen dispersal in hummingbird pollinated *Heliconia*. Amer. Nat. 107: 511–23.

Maas, P.J.M. 1972. Costoideae (Zingiberaceae). Flora Neotropica Monogr. 8. Hafner, New York.

———. 1977. Zingiberoideae (Zingiberaceae). Flora Neotropica Monogr. no. 18. Hafner, New York.

Smith, L.B. 1944. Bromeliaceae *In* R.E. Woodson, Jr. et al. Flora of Panama. Ann. Missouri Bot. Gard. 31: 73–137.

———. & R. J. *Downs*. 1974–1979. Tillandsioideae (Bromeliaceae). Flora Neotropica Monogr. 14(1–3): 1–2142. Hafner.

Smith, R.R. 1968. A taxonomic revision of the genus *Heliconia*. Ph.D. diss. Univ. Florida, Gainesville.

Standley, P.C. 1944. Araceae *In* Woodson et al. Flora of Panama, Ann. Missouri Bot. Gard. 31: 1–60.

Stiles, G. 1975. Ecology, flowering phenology and hummingbird pollination of some Costa Rican *Heliconia* species. *Ecology* 56: 285–301.

Uhl, N.W. & H.E. *Moore*. 1977. Correlations of inflorescence, flower structure and floral anatomy with pollination in some palms. Biotrop. 9: 170–190.

Woodson, R.E. & R.W. *Schery*. 1943. Cyclanthaceae *In* Flora of Panama. Ann. Missouri Bot. Gard. 30: 396–403.

———. 1945a. Musaceae *In* Flora of Panama. Ann. Missouri Bot. Gard. 32: 48–57.

———. 1945b. Zingiberaceae *In* Flora of Panama. Ann. Missouri Bot. Gard. 32: 57–73.

———. 1945c. Cannaceae *In* Flora of Panama. Ann. Missouri Bot. Gard. 32: 74–80.

———. 1945d. Marantaceae *In* Flora of Panama. Ann. Missouri Bot. Gard. 32: 81–105.

The Phytogeographic Relationships of the Panamanian Grasses

Gerrit Davidse

En Panamá, la Gramineae, representada por más de 348 especies y 107 géneros, se considera entre las familias más grandes de plantas con flores, ya que ocupa el quarto lugar. Aproximadamente 76 especies, cerca de 24%, se han adicionado desde la publicación de la familia en la Flora de Panamá en 1943. Los principales elementos en la flora de gramíneas panameñas son las especies introducidas (12%), las especies cosmopolitas (8%), las especies distribuidas en América Central y del Sur (57%), aquellas afines a especies originadas al Norte de Panamá (11%), aquellas afines a las del Sur (5%) y las endémicas (6%). La mayor parte de la flora de gramíneas panameñas es parte de la flora de América tropical ampliamente distribuida.

The grasses, or Gramineae, ranks fourth among the flowering plant families of Panama, being exceeded only by the Rubiaceae, Leguminosae, and Orchidaceae. There are now known to be at least 348 species in 107 genera. In 1943 when the family treatment was published for the *Flora of Panama*, Swallen (1943) recognized 279 species in 87 genera. Seventy-six additional species, an increase of 24%, are now known, fifty-six of these native species and the remainder cultivated or alien weedy species (Table 1). Seven species recognized by Swallen (1943) are now reduced to synonymy. Of the additional 20 genera now known, 14 have been newly discovered (Table 1), and 6 are simply due to different generic concepts (Table 2).

En Panamá, la Gramineae representada actualmente por más de 348 especies y 107 géneros, se considera entre las familias más grandes de las plantas con flores, ya que ocupa el quarto lugar. A partir de 1943 cuando se publicó el reporte en la Flora de Panamá, ha habido un aumento aproximado de 24% en el número de especies.

It is likely, based on known distributions in neighboring countries, that an additional 30 species representing 20 genera (8 genera of which are not now known in Panama) will be found (Table 3). Furthermore, the discovery of entirely unexpected or undescribed species, plus the likelihood of additional introductions of alien species, seems likely to push the total number near 400 by the end of the century. Several new, undescribed species are already known and several others are represented by collections inadequate for determination. These have not been included in the following account.

Panamanian Grassland

Since grasslands are best developed on plains in the interior of large continents in climates with distinctly seasonal rainfall (Clayton 1975), it is readily understood that natural grasslands occur on only a relatively small portion of Panama. Panama is, or was before man's destructive influence, truly a country of forests. The important kinds of grassland communities in Panama are summarized here:

(1). Most important are the savannas which occur in the tropical dry forest life-zone of the Holdridge Life-Zone System, primarily in the Canal Area, and the provinces of Panama, Cocle, Herrera, Los Santos, Veraguas, and Chiriqui south of the central cordillera (Holdridge & Budowski 1959). When developed to its full potential, the vegetation in this life zone forms a deciduous forest with a closed canopy in which grasses play an ecologically minor role. However, significant portions of this life zone, with lower soil nutrients and drier soils (Sarmiento & Monasterio

TABLE 1
GRASSES RECORDED SINCE THE FLORA OF PANAMA[1]

Taxa	Status
AGROSTIDEAE	
Agrostis toluensis H.B.K.	Native
ANDROPOGONEAE	
*Bothriochloa bladhii (Retz.) S.T. Blake	Introduced weed, naturalized
B. pertusa (L.) Camus	Introduced weed, naturalized
*Chrysopogon aciculatus (Retz.) Trin.	Introduced weed
Imperata brasiliensis Trin.	Native
Ischaemum latifolium (Spr.) Kunth	Native
*Rottboellia exaltata L.f.	Introduced weed, naturalized
Saccharum spontaneum L.	Introduced ornamental?, naturalized
BAMBUSEAE s.l.	
Bambusa amplexifolia (Presl) Schult. f.	Native
B. arundinacea Willd.	Cultivated
B. balcooa Roxb.	Cultivated
B. glaucescens (Willd.) Sieb. ex Munro	Cultivated
B. guadua H. & B.	Native
B. latifolia H. & B.	Native
B. multiplex (Lour.) Raeusch.	Cultivated
B. nana Roxb.	Cultivated
B. paniculata (Munro) Hack.	Native
Chusquea pittieri Hack.	Native
C. virgata Hack.	Native
*Dendrocalamus latiflorus Munro	Cultivated
D. strictus (Roxb.) Nees	Cultivated
*Neurolepis pittieri McClure	Native
Rhipidocladum parviflorum (Trin.) McClure	Native
R. pittieri (Hack.) McClure	Native
CENTOSTECEAE	
*Calderonella sylvatica Soders. & Decker	Native
Zeugites pittieri Hack.	Native
CHLORIDEAE	
Aegopogon tenellus (DC.) Trin.	Native
*Pentarraphis annua Swallen	Native
Spartina spartinae (Trin.) Hitchc.	Native
ERAGROSTEAE	
Eragrostis mexicana (Hornem.) Link	Native
E. unioloides (Retz.) Nees ex Steud.	Introduced weed
Leptochloa longa Griseb.	Native
L. scabra Nees	Native
OLYREAE	
*Arberella costaricensis (Hitchc.) Soders. & Calderón	Native
*A. dressleri Soders. & Calderón	Native
Cryptochloa concinna (Hook. f.) Swallen	Native
*Maclurolyra tecta Calderón & Soders.	Native
Olyra caudata Trin.	Native
O. ecaudata Doell	Native
O. lateralis (Presl) Chase	Native
O. standleyi Hitchc.	Native
ORYZEAE	
Oryza alta Swallen	Native
PANICEAE	
Axonopus affinis Chase	Native
A. poiophyllous Chase	Native
*Chaetium bromoides(Presl) Benth. ex Hemsl.	Native
Digitaria bicornis (Lam.) R. & S.	Native?
D. panicea Willd. ex Steud.	Native
D. sanguinalis (L.) Scop.	Introduced weed
D. violascens Link	Introduced weed
Ichnanthus nemoralis (Schrad.) Hitchc. & Chase	Native

TABLE 1 Continued

Taxa	Status
Homolepis isocalycia (Meyer) Chase	Native
Lasiacis linearis Swallen	Native
L. nigra Davidse	Native
L. rhizophora (Fourn.) Hitchc.	Native
L. rugelii var. pohlii Davidse	Native
*Melinis minutiflora Beauv.	Introduced as forage or weed, naturalized
Oplismenus setarius (Lam.) R. & S.	Native
Panicum elephantipes Nees	Native
P. glutinosum Sw.	Native
Paspalum blodgetii Chapm.	Native
P. brunneum Mez	Native
P. millegrana Schrad.	Native
P. pulchellum Kunth	Native
P. serpentinum Hochst.	Native
P. squamulatum Fourn.	Native
Pennisetum purpureum Schumach.	Introduced as forage, naturalized
*Rhynchelytrum repens (Willd.) C.E. Hubb.	Introduced weed, naturalized
PARIANEAE	
Pariana aurita Swallen	Native
P. campestris Aubl.	Native
POEAE	
*Holcus lanatus L.	Introduced weed, naturalized
Trisetum tonduzii Hitchc.	Native
SPOROBOLEAE	
Muhlenbergia implicata (H.B.K.) Kunth	Native
Sporobolus purpurascens (Sw.) Hamilt.	Native
S. pyramidatus (Lam.) Hitchc.	Native
S. tenuissima (Schrank) Kuntze	Native? or introduced weed

*New generic record.
[1]Swallen 1943.

TABLE 2
CHANGES IN GENERIC CONCEPTS SINCE THE FLORA OF PANAMA[1]

Flora of Panama	Current	Net change in no. of genera
Guadua	Bambusa	-1
Arthrostylidium	Rhipidocladum	0
Chusquea	Chusquea & Swallenochloa	1
Trichachne	Digitaria	-1
Panicum	Panicum & Dichanthelium, Paspalidium, Acroceras, Brachiaria (in part)	3
Raddia	Raddiella	0
Andropogon	Andropogon, Schizachyrium, Bothriochloa, Hypogynium, Euclasta	4
Manisuris	Coelorachis	0
	Total increase in genera	6

[1]Swallen 1943.

TABLE 3
GRASSES EXPECTED IN PANAMA

Grasses not yet reported from Panama but suspected of occurring there based on nearby distributions.
Andropogon hypogynous Hack.
Aristida adscensionis L.
Brachiaria reptans (L.) Gard. & C.E. Hubb.
Briza minor L.
Calamagrostis spp.
Cenchrus pilosus H.B.K.
Chloris ciliata Sw.
Elytrostachys clavigera McClure
Eragrostis tenuifolia (A. Rich.) Hochst. ex Steud.
Gouinia sp.
Heteropogon contortus (L.) Beauv. ex R. & S.
Hierochloe sp.
Merostachys sp.
Muhlenbergia ciliata (H.B.K.) Kunth
Muhlenbergia diversiglumis Trin.
Panicum cyanescens Nees ex Trin.
Panicum hirtum Lam.
Panicum parvifolium Lam.
Panicum venezuelae Hack.
Paspalum botteri (Fourn.) Chase
Paspalum dilatatum Poir.
Paspalum langei (Fourn.) Nash
Paspalum paspaloides (Michx.) Scribn.
Paspalum plenum Chase
Paspalum trachycoleon Steud.
Paspalum urvillei Steud.
Pennisetum bambusiforme (Fourn.) Hemsl. ex Jacks.
Pereilema beyrichianum (Kunth) Hitchc.
Vulpia sp.

1975), develop a typical savanna vegetation. Such savannas characteristically have a continuous herb layer dominated by grasses and scattered individual or small clumps of trees, especially *Byrsonima crassifolia* (L.) H.B.K. and *Curatella americana L.* The dominant grass genus is *Trachypogon*, and species of *Andropogon, Schizachyrium, Axonopus, Paspalum*, and *Leptocoryphium lanatum* (H.B.K.) Nees are common.

Las comunidades de gramíneas más importantes son las de sabanas naturales de tierra baja.

(2). On the upper slopes of the Chiriqui highlands, grasslands dominated by poöid bunch grasses have developed. Common species are *Festuca dolichophylla* Presl, *F. amplissima* Rupr., and species of *Agrostis* and *Trisetum*. Surprisingly, *Calamagrostis*, so common and dominant in similar montane grasslands in other parts of Central and South America, has not yet been collected. Grasslands of similar physiognomy are common at upper elevations of mountains in both the Old and New Worlds. The Panamanian grasslands are closely related to the alpine grasslands in Costa Rica, Guatemala, and Mexico, as well as the paramos of the South American Andes. At least part of the Chiriqui highland grasslands, especially those without woody vegetation, occupy areas affected by relatively recent volcanic activity.

Las comunidades de gramíneas de tierras altas, parcialmente afectadas por reciente vulcanismo, también son importantes.

A few other areas of even more limited extent have grassland communities.

(3). At intermediate or lower elevations, cinder slopes, lava flows, or exposed rocky slopes formed through volcanic activity may have communities where grasses are important. One of the most spectacular examples of this kind of grassland in Central America is found on the southwestern slopes of Volcan Rincon de la Vieja in Costa Rica. Here occurs an almost pure grassland of *Axonopus volcanicus* Pohl, an endemic species only recently described (Pohl 1980) and apparently very recently evolved after volcanic eruptions had cleared these slopes of forests. A somewhat similar area is the famous collecting locality near the town of El Hato del Volcan in Chiriqui where grasses form an important part of plant cover on the old lava flow. Here the bunch grasses *Muhlenbergia lehmanniana* Henr. and *Trachypogon plumosus* (Willd.) Nees are the dominant grasses, and the Panamanian endemic, *Digitaria curtigluma* Hitchc. was discovered there recently. A similar grassland appears on nearby Cerro Vaca, also in Chiriqui province. Three endemic Panamanian grasses, *Ctenium concissum* Swallen, *Thrasya gracilis* Swallen, and *T. villosa* Hitchc. are known from here as well as *Aristida planifolia* Swallen, known only from Panama and Colombia. Such areas appear to be good examples of newly opened habitats that provide opportunity for rapid speciation, and they certainly merit detailed study.

Comunidades de gramíneas sobre material volcánico reciente sugieren especiación rápida.

(4). On the sandy, saline soils of the littoral zone, there is often a narrow strip of grassy vegetation. In Panama the common species are *Sporobolus virginicus* (L.) Kunth, *Jouvea straminea* Fourn., *Uniola pittieri* Hack., and *Paspalum vaginatum* Sw.

(5). Grasses may form an important or dominant

part of the aquatic vegetation along river banks, lakes, and marshes. Most prominent in this respect are *Paspalum repens* Berg., *P. fasciculatum* Willd., *Echinochloa polystachya* (H.B.K.) Hitchc., *Panicum grande* Hitchc. & Chase, *P. stagnatile* Hitchc. & Chase, *Leersia hexandra* Sw., *Hymenachne* species, and *Gynerium sagittatum* (Aubl.) Beauv.

Las gramíneas ocurren en fajas estrechas, cerca del mar al igual que a lo largo de ríos, cerca de lagos y de pántanos.

In the primarily forested areas of Panama grasses clearly play a secondary role. Typically they are forest-edge plants, often occurring in tree-fall clearings, landslide areas, and similarly naturally exposed sites. Bamboos may sometimes form an important understory layer in the forest, and the herbaceous bambusoid grasses such as *Arberella*, *Crytochloa*, *Pariana*, and *Pharus* are adapted to the shady habitat of undisturbed rainforests. Margins of streams and lakes and steep slopes in mountainous terrain are other natural habitats for grasses. The characteristics which enable grasses to occupy such naturally open and secondary communities also preadapted them for the enormous open and disturbed areas created by man.

Phytogeographic Analysis

The following analysis is primarily based on herbarium records, especially those at MO, and published floristic works, especially the *Flora of Panama* (Swallen 1943), the *Flora of Guatemala* (Swallen 1955), and *Flora Costaricensis* (Pohl 1980), plus recent revisions and monographs. The agricultural literature has not been consulted extensively, so the number of purely cultivated grasses may be underestimated, especially since these are rarely collected by botanical collectors. For each Panamanian species the entire distributional range was determined. For the main part of the analysis distributions were generalized into four categories, and each species was scored for its presence in: (1) Central America (Costa Rica to tropical Mexico), (2) North America (United States and subtropical Mexico), (3) South America (including Trinidad), and (4) the West Indies. Eleven of the 15 possible combinations of these four categories were found, and these are listed in Table 4 as nos. 4 through 14, along with the other following categories: cosmopolitan, cultivated, alien weeds, endemics (restricted to Panama, Panama and Costa Rica, or Panama and Colombia), plus miscellaneous other distributions. To simplify the discussion the 18 categories were reduced to 7 combined categories (Table 4). The distinction between Central America and North America is imprecise but in fact caused few problems in the analysis. Throughout this paper the use of Central America and North America are as defined above.

Introduced Species

Forty species owe their presence in Panama to the direct or indirect influence of man (Table 4). They are the non-native cultivated species (cereals, sugar cane, forage grasses, ornamental grasses, turf grasses) and the weedy species. Although less interesting from the phytogeographic point of view since the agency and circumstance of their introduction is known in a general way (although the precise historical details of most species are not) they, in fact, are the most important grasses to man's economy since they are a very important part of his diet, food for his animals, sources of industrial products, erosion control, ornamental and turf plants, and weeds of his fields.

Maize, rice, and sugar cane are clearly the most important grasses. Many of the large Asian bamboos have been included in this category, but it is unclear how many are widely distributed and used. Many may actually be only experimental plants and garden introductions for testing and evaluations. Many forage grasses have become widely naturalized and act like natives of secondary vegetation, most important being *Panicum maximum* Jacq., *Pennisetum purpureum* Schumach., *Hyparrhenia rufa* (Nees) Stapf, and *Melinis minutiflora* Beauv. All are African in origin, and it is entirely possible that they first reached Panama as weeds rather than as deliberate forage introductions. Among the ornamental grasses, *Coix lacryma-jobi* L., whose highly modified leaf sheaths are beads used in making necklaces and rosaries, is also included here. Although no doubt originally introduced into the Western Hemisphere as a source of beads, the species is now widely naturalized in marshy areas and is rarely deliberately cultivated. *Axonopus scoparius* (Fluegge) Kuhlm. *(zacate imperial)*, a large succulent grass used as cut forage in middle to upper elevations, is sometimes thought to be native to Central America, but this has not been clearly established. It seems to escape cultivation rarely in Central America and is probably native only in the Andes.

Maíz, arroz y caña de azúcar, constituyen los cultivos más importantes de gramíneas.

Among the alien weeds are some that still have

TABLE 4
DISTRIBUTION PATTERNS OF PANAMANIAN GRASSES

Category[a]	No. of Species	Combined Categories	No. of Species	Percentage
1. Cultivated	22	Introduced	40	12%
2. Alien Weeds	18			
3. Cosmopolitan[b]	27	Cosmopolitan	27	8%
4. CA & WI & NA	5			
5. CA & WI	2			
6. CA & NA	2	Northern	39	11%
7. WI & NA	1			
8. CA	29			
9. SA & WI	3	Southern	19	5%
10. SA	16			
11. CA & SA & WI & NA	30			
12. CA & SA & WI	88	Southern-		
13. CA & SA & NA	4	Northern		
14. CA & SA	76	Combination	198	57%
15. Panama	11			
16. Panama & Costa Rica	8	Endemic Element	22	6%
17. Panama & Colombia	3			
18. Other	3	Other	3	1%
		TOTAL	348	100%

[a]Abbreviations: CA = Central America (Costa Rica—Tropical Mexico); NA = North America (United States to subtropical Mexico); SA = South America (including Trinidad); WI = West Indies.
[b]Includes all native American species distributed on at least one Old World continent.

very local distributions. For example, *Eragrostis unioloides* (Retz.) Trin. is probably a recent introduction into America as the only known localities in tropical America are in Panama and in Venezuela in the Gran Sabana area. Other alien weeds are so ubiquitous and thoroughly naturalized that it is difficult to know for certain whether they are alien or native. For example, *Eleusine indica* (L.) Gaertn., one of the most common pantropical weeds that also extends into temperate areas, is judged to be an introduction primarily because of its weedy nature and because the entire genus of nine species is native to Africa (Phillips 1972) except for one dubious species of southern South America (*E. tristachya* (Lam.) Lam.

Saccharum spontaneum L., a wild sugar cane, is now a common weed in the Canal Zone. It seems likely from the lack of earlier herbarium specimens that it has been introduced within the last 20 years, perhaps as an ornamental valued for its handsome, fluffy inflorescences. It is likely to become widely distributed in tropical America as it seems to be aggressive.

A recently introduced alien weed may be *Chrysopogon aciculatus* Retz.(Trin.) which has become troublesome in Canal Area lawns. It could have been introduced as a contaminant of grass seed of *Polytrias amaura* (Buese ex Miq.) Kuntze, an Asian turf species used in the Canal Area.

Muchas gramíneas son malezas en Panamá y están ampliamente distribuídas. Algunas son locales.

The role that shipping played in transporting grasses is unknown. Although most shipping uses the Panama Canal only as a passage and not as a final destination, the Canal is so heavily used that occasional chance introductions may occur.

Cosmopolitan Species

Included in this category are all native American grasses that are distributed on at least one Old World continent. Also included are species whose distribution is so general that it is impossible to say with certainty where they originated; for example, through natural means, *Phragmites australis* (Cav.) Steud., *Leersia hexandra* Sw., *Paspalidium geminatum* (Forssk.) Stapf, or possibly through man's agency, *Echinochloa colonum* (L.) Link.

One of the interesting subgroups among these cosmopolitan grasses consists of six species with an African-American distribution. All but one are andropogonoid savanna grasses with prominently awned, hairy disseminules that presumably are readily dispersed. The four annuals, *Diectomis fastigiata* (Sw.) Beauv., *Euclasta condylotricha* (Hochst.) Stapf, *Schizachyrium brevifolium* (Sw.) Buese, *Sorghastrum incompletum* (Presl) Nash, may owe their wide distribution to man's activities in historical times, especially since they show some weedy tendencies. The perennial *Hyparrhenia bracteata* (Willd.) Stapf, on the other hand, may well represent an instance of long-distance dispersal to America from Africa, the center of diversity of this genus which includes 52 species (Clayton 1969). The sixth species, *Olyra latifolia L.*, is a herbaceous bambusoid grass whose hard, bony florets are well adapted to internal bird dispersal. It is the only species with such a distribution in the otherwise strictly American genus of 22 species (Calderon & Soderstrom 1980), and it probably also represents a example of natural long-distance dispersal in the opposite direction from others in the subgroup. However, Soderstrom (1981), based on its weedy tendencies, believes that the species was spread by man.

27 especies son cosmopolitas y entre éstas está un grupo interesante de especies africanas.

Of the remaining species almost all are aquatic or sea-shore grasses (8 species) or weedy plants of open, secondary communities (10 species). Most of the latter have spikelets with obvious structures for efficient dispersal. The aquatic and sea-shore species often have wide distributions. Two of the weedy species are basically temperate-zone grasses (*Poa annua L.*, *Digitaria sanguinalis (L.)* Scop.) that occur only from middle to upper elevations in Panama. Both of these and the tropical weeds such as *Digitaria ciliaris* (Retz.) Koel. and *Paspalum conjugatum* Berg. owe most of their wide distributions to man's inadvertant help. *Pseudechinolaena polystachya* (H.B.K.) Stapf, a forest or forest-edge species has large reflexed multicellular hairs on the spikelets which readily attach to animals or clothing, and it is also probably introduced.

Northern Species

Thirty-nine Panamanian grasses whose distribution is strictly north of Panama in Central or North America or the West Indies are included in this category (Table 4). This group includes such species as *Eustachys petraea* (Sw.) Desv., whose distribution includes the West Indies, Central America, and North America. Other species have distributions that are primarily Central American but may also extend into Mexico and the United States, for example, *Aegopogon tenellus* (DC.) Trin., which ranges from Arizona south through Central America, reaching its southern limit in Panama, or they may extend into the West Indies as in *Digitaria argillacea* (Hitchc. & Chase) Fernald. One species, *Sacciolepis striata* (L.) Nash, is known from North America and the West Indies but only from Panama in Central America. However, the majority of species in this northern category have a strictly Central American distribution, as exemplified by *Lasiacis linearis* Swallen. The species of this northern category belong to nine genera (*Aegopogon*, *Chaetium*, *Dichanthelium*, *Jouvea*, *Lasiacis*, *Muhlenbergia*, *Pentarrhaphis*, *Tripsacum*, *Zeugites*) that clearly have their center of diversity and presumably their center of origin in Panama-Central America-North America. However, all nine genera except *Jouvea*, which occurs from Central America to Baja California, do have one or more species that have reached the South American continent, and in *Muhlenbergia* India, and in *Dichanthelium* the Hawaiian Islands. Of the remaining 14 genera all are worldwide in distribution or clearly centered in South America, sometimes with a distant species complex (such as the *Panicum arundinarae* group) in Central America. These are represented in the northern region by a minority of their species and usually by only a small part of the total diversity in the genus. However, of the worldwide genera in this category the poöid genera *Trisetum*, *Festuca*, and *Bromus* must ultimately be regarded as having a north-temperate origin.

39 especies están distribuidas sólo hacia el norte de Panamá y sólo algunas de ellas llegan a las Antillas. Muchas de ellas pertenecen a géneros muy distribuídos.

Southern Species

Nineteen species are included in this category, three of which also range into the West Indies (Table 4). Of the genera composing this category, only

Neurolepis and *Raddiella*, both bambusoid grasses, are entirely limited to South America and Panama. Of the remaining 10 genera, six have their centers of diversity in South America (*Paspalum*, *Olyra*, *Pariana*, *Luziola*, *Ichnanthus*, *Thrasya*), while one is primarily Central American (*Tripsacum*), and three (*Arundinella*, *Leptochloa*, and *Panicum*) are worldwide.

Northern-Southern Species

The large bulk of the Panamanian grasses, a group five times larger than any other grouping in this analysis, have distributions that include large parts of Central and South America (Table 4). Many species extend to various degrees into North America and the West Indies as well. Such widespread species reaching all the four combined regions used in the analysis are termed American, while those found in the tropical areas, that is, all regions except North America, are designated tropical American. Predictably, the widespread American and tropical American species tend to belong to widely distributed genera, all 33 of which also occur on at least one of the Old World continents. The only exceptions are 6 strictly American genera: *Dichanthelium*, also known from the Hawaiian Islands, *Lasiacis*, *Bouteloua*, *Gynerium*, *Lithachne*, *Leptocoryphium*, and *Pharus*. However, *Axonopus*, *Ichnanthus*, and *Paspalum* are primarily American, each having one or a few species outside the American continents. A subgroup of this category that is found only in Central and South America, and not in North America or the West Indies, contains a higher number (14) of strictly American genera as exemplified by *Thrasya* and *Zeugites*.

El mayor grupo de gramíneas panameñas, 51%, está distribuido en América Central y del Sur. Este grupo se subdivide en 4 subgrupos.

A more detailed examination was made of this category in order to determine more precisely whether any directional distributional tendencies are evident within the large Central and South American area that these species occupy.

Four subgroups could be distinguished. The largest was composed of species (59%) that do in fact occupy most of the suitable habitats from subtropical Argentina, Brazil, and Peru to southern Mexico. *Eragrostis hypnoides* (Lam.) B.S.P., an annual river-bank species, *Andropogon leucostachyus* H.B.K., a savanna grass, and *Lithachne pauciflora* (Sw.) Poir., a bambusoid forest grass, are examples. Those species occurring in Central America and only northern tropical America composed 33% of the species. Some examples are: *Andropogon angustatus* (Presl) Steud., an annual savanna grass known from Cuba and southern Mexico to northern Brazil; *Paspalum serpentinum* Hochst., a perennial savanna species with a similar distribution. Much rarer (4%) is the distribution in which the species is known from all or most of South America but has penetrated Central America only to Costa Rica. An example is the aquatic *Hymenachne donacifolia* (Raddi) Chase.

Muchas de éstas (59%) ocurren en hábitat óptimos desde América del Sur subtropical hasta el sur de México. Otro subgrupo (33%) está limitado a América Central y norte de América del Sur. Pocas gramíneas (4%) de Sur América llegan tan sólo hasta Costa Rica.

The differences between these first three subgroups are only trends and not absolute distinctions since the variation between the species distributions is probably more nearly continuous than restricted to the discrete areas within tropical America employed in this analysis.

The fourth subgroup (4%) is more definite because it is formed almost entirely by montane species. Almost all follow the main Central American cordillera and extend south to various degrees in the main Andean ranges. Several have managed to reach outlying mountains such as the coastal cordillera of Venezuela (*Zeugites mexicana* (Kunth) Steud.) or the Guiana Highlands (*Olyra lateralis* (Presl) Chase), but such links, especially the latter, are not numerous. At higher elevations, all species are poöid grasses.

Endemic Species

El cuarto subgrupo (4%) está constituido por especies de las montañas.

The 22 species endemic to the Panamanian grass flora comprise 6% of the total (Table 5). Of these, 11 species (3%) are known only from Panama. As many of these Panamanian endemics may also be found in neighboring countries a broad concept of endemism is used in this analysis.

The most notable subgroup of the endemics is that of the forest grasses, including both herbaceous bambusoids and centosthecoids. The forest grasses (excluding forest panicoids) constitute only 8% of all the Panamanian grasses, but endemic forest grasses constitute 45.5% of all the endemic grasses (Table 6).

TABLE 5
SUMMARY OF PANAMANIAN ENDEMIC SPECIES
Including Panamanian-Costa Rican and Panamanian-Colombian

Taxon	Panamanian Prov./Country	Habitat	Elevation (m)
AGROSTIDEAE			
Agrostis hoffmanii Mez	Chiriqui, Costa Rica	paramo high, montane grassland	above 2,800
ARISTIDEAE			
Aristida planifolia Swallen	Chiriqui, Costa Rica	savanna	900–1,200
BAMBUSEAE			
Chusquea virgata Hack.	Chiriqui, Costa Rica	montane forests	1,800
Swallenochloa vulcanalis Soderst. & Calderón	Chiriqui, Costa Rica	paramo	3,000
CENTOSTECEAE			
Calderonella sylvatica Soders. & Decker	Colon, San Blas	lowland rainforest	200
Zeugites panamensis Swallen	Panama	cloud forest margins	600-800
CHLORIDEAE			
Ctenium concissum Swallen	Cocle, Chiriqui	savanna	900–1,200
OLYREAE			
Arberrela costaricensis (Hitchc.) Soders. & Calderon	Colon, Costa Rica	lowland rainforest	above 100
A. dressleri Soders. & Calderon	Colon, Costa Rica	cloud forests	500
Cryptochloa concinna (Hook. f.) Swallen	Veraguas, Costa Rica		900–1,300
C. variana Swallen	Canal Area, Cocle, Colon, Panama, San Blas, Veraguas, Colombia	lower-middle montane forest	200–1,800
Maclurolyra tecta Calderon & Soders.	Cocle, Colon, Panama	lower montane forest	200–800
Olyra standleyi Hitchc.	Panama, Costa Rica	lower-middle montane forest	1,000
PANICEAE			
Digitaria hirsuta Swallen	Canal Area, Panama	savanna	1,500–1,800
Panicum sublaeve Swallen	Panama	savanna	0–1,200
Paspalum jimenezii Chase	Panama, Costa Rica	river banks	above 200
Thrasya ciliatifolia Swallen	Canal Area, Chiriqui	savanna	0–1,200
T. gracilis Swallen	Chiriqui, Panama	savanna	700–1,200
T. villosa Hitchc.	Chiriqui	savanna	700–1,200
PARIANEAE			
Pariana strigosa Swallen	Panama	lowland rainforest	1–100
PHAREAE			
Pharus longifolius Swallen	Panama, Colombia	lowland rainforest	0–200
POEAE			
Festuca chiriquiensis Swallen	Chiriqui, Costa Rica	middle montane forest openings	1,300–2,600

Two of the forest grass genera, *Calderonella* (Soderstrom & Decker 1973), and *Maclurolyra* (Calderon & Soderstrom 1973) were discovered and described within the last 10 years and are still known only from Panama. Both are likely to be found in the Choco of Colombia, however. A third genus, *Cryptochloa*, ranges from Mexico to Peru and Brazil. Of the remaining four genera, only *Zeugites* is primarily Central American, while *Pariana*, *Pharus*, *Olyra*, and perhaps *Arberella* are most highly developed in South America.

22 especies son endémicas al área de Panamá y 11 (3%) sólo a Panamá. La mayoría son especies de los bosques, pero algunas ocurren en las sabanas.

Most of the remaining Panamanian endemics are savanna species. All are members of either cosmopolitan genera or genera that reach their greatest development in South America. A single lowland species, *Paspalum jimenezii* Chase, occurs along open alluvial river banks and stream beds. The remainder

TABLE 6
PROPORTION OF ENDEMIC GRASSES IN PANAMA'S ENTIRE GRASS FLORA
Total endemics 22; total flora 348

Taxa	Endemic Species No.	Endemic Species Percent	Entire Flora No.	Entire Flora Percent
Forest Grasses	10	45.5%	30	8%
Herbaceous Bambusoids	8	36.4%	22	7%
Centostecoids	2	9.1%	5	1%
Bambuseae	2	9.1%	23	7%
Panicoids	6	27.2%	216	62%
Chloridoids	2	9.1%	53	15%
Pooid	2	9.1%	16	5%

inhabit upper elevations in the mountains and are either poöid grasses with northern affinities or bamboos with southern affinities.

Other Affinities

The three species included here are in reality only special cases of categories previously discussed. *Spartina spartinae* (Trin.) Hitchc. occurs basically around the Caribbean coast from Florida to Panama and in South America only in Paraguay and Argentina. This general type of distribution is also known for six genera of grasses—*Monanthochloe*, *Blepharidachne*, *Munroa*, *Tridens*, *Erioneuron*, *Scleropogon* (Burkhart 1975).

Digitaria singularis Mez (Panama and Paraguay) and *Paspalum subciliatum* Chase (Panama and Brazil) also have decidedly disjunct ranges. Although not unprecedented in grasses, for example, *Luziola fragilis* Swallen is disjunct between Mato Grosso and Costa Rica (Pohl & Davidse 1971), the ranges probably reflect more a collecting than a distributional gap.

Discussion and Conclusions

From this brief review of the distribution of Panamanian grasses one overriding conclusion emerges, namely, that the major portion of the Panamanian grass flora forms part of a widely distributed tropical American grass flora, especially one that is centered in northern South America and southern Central America. This conclusion is inescapable since 70% of the native, noncosmopolitan Panamanian species have distributions extending through major portions of tropical Central and South America. In view of the central position of Panama and since most of the major tropical life zones occur in the country (although in a long and narrow form compared to their South American counterparts), such a pattern is not unexpected. In this large tropical area it is difficult to delineate any one area as the most important in the evolution of the tropical American grass flora except to generalize that ultimately a net migration and radiation must have occurred from a tropical core area in South America northward toward the climatic limits formed by the subtropical and temperate zones. For example, for savanna species the central Brazilian plateau is undoubtedly the core area, with many species radiating to the northwest—either unchanged or newly evolved, but clearly related to Brazilian savanna relatives.

La mayoría de las gramíneas de Panamá es parte de la flora tropical americana, ampliamente distribuida, que está centrada en el norte de América del Sur y en el sur de América Central.

This general pattern of lowland, tropical Central America being dominated by plants of South American origin is one that is well known for many tropical plant families (Raven & Axelrod 1974, 1975) but has not always been fully appreciated for grasses. That this pattern involves the tropical portion of the family is evident from the fact that of the native, noncosmopolitan species, 96% belong to genera that are fundamentally tropical, although some of these species, especially the annual ones, extend into the temperate zone as well, for example, *Eragrostis hypnoides* (Lam.) B.S.P., *Leptochloa filiformis* (Lam.) Beauv., and *Panicum dichotomiflorum* Michx. The generic proportion is 82% tropical, although some of these basically tropical genera may have secondarily derived temperate species, for example, *Andropogon gerardii* Vitm. and *Sorghastrum nutans* (L.) Nash.

Despite this overall pattern of migration, speciation in place throughout the tropical American region has been important in enriching overall diversity. The existence of endemic species throughout the entire region supports this.

Migration is, however, not always unidirectional, and it is sometimes difficult to generalize about migration patterns. For example, in *Lasiacis* the primitive species are characterized by the creeping habitat common to most panicoid forest grasses. These primitive species are centered primarily in Central America. The advanced, erect woody species occur widely in both Central and South America. It seems simplest to hypothesize that *Lasiacis* evolved in Central America from a generalized, creeping, forest panicoid grass that had reached Central America from its South American center of origin. Erect woody species subsequently evolved in Central America, exploiting the forest-edge habitat and ecological secondary seral stages to a much greater degree than their prostrate relatives, and then migrated back into South America. After migration and further speciation, five woody species: *L. sorghoidea* (Desv.) Hitchc. & Chase, *L. ligulata* Hitchc. & Chase, *L. divaricata* (L.) Hitchc., and *L. ruscifolia* (H.B.K.) Hitchc., are now common and widespread in South America (Davidse 1979).

A fundamental differentiation in the Gramineae between cold-temperate species and tropical-subtropical species has long been recognized. Although the Gramineae probably originated in Gondwanaland, there likely was an early adaptation and radiation into cold-temperate climates by progenitors of the subfamily Pooideae. The poöid grasses have continued to exploit these cool climatic zones and their counterparts in the high tropical mountains almost exclusively ever since their origin. In Panama this distribution is also evident, the poöid genera (*Agrostis*, *Polypogon*, *Bromus*, *Brachypodium*, *Festuca*, and *Trisetum*) inhabiting only upper elevations in the mountains, rarely descending below 2,000 m and reaching optimal development above 3,000 m. Although the tendency for a northern distribution among the Panamanian poöid species is twice as great as for the flora in general, on the basis of modern distributions *per se*, this northern-temperate relationship would not be evident from a purely statistical analysis since these species share the same basic Central and South American distributions of many of the lowland tropical species. Their ultimate northern relationship must be deduced primarily from their adaptation to a temperate habitat and their taxonomic relationships.

La gramíneas "poöid" de las zonas templadas frescas sólo ocurren en las tierras altas de Panamá.

Of the relatively small part (11%) of the Panamanian grasses with a distribution north of Panama and which can be considered as a distinctive Central American element, most are related to fundamentally tropical species that have differentiated in the northern limit of their range, rather than temperate species adapting to the tropics. At the generic level, only 3 of the 26 genera (*Trisetum*, *Festuca*, *Bromus*) have a north-temperate origin; all the others are tropical. Of the fundamentally tropical genera, nine (*Pentarraphis*, *Muhlenbergia*, *Aegopogon*, *Jouvea*, *Tripsacum*, *Zeugites*, *Chaetium*, *Lasiacis*) have their greatest diversity and, presumably, origin in Central America. Especially notable in this respect are the chloridoid genera (*Pentarraphis*, *Muhlenbergia*, *Aegopogon*, and *Jouvea*). Although the subfamily Chloridoideae is cosmopolitan, especially in tropical and subtropical hot, dry areas, there is a secondary development of genera centered in Mexico, where they are often associated with saline and alkaline soils (Hartley & Slater 1960; Clayton 1975; Rzedowski 1975). This seems to be a case where primitive, tropical, chloridoid stock reached North America relatively early compared with the bulk of panicoid taxa, which allowed the development of highly endemic taxa, some of these subsequently migrating south into Central America and beyond when migration became easier and more direct, and differentiating new taxa along the way.

La mayoría de las especies de gramíneas endémicas de América Central son de origen tropical.

The endemic Panamanian grass flora is relatively small, but this is not surprising since the country is small and grasses tend to be easily dispersed. Many grasses occupy secondary ecological stages and tend to be especially widespread. Of the three main endemic groups, rain forest, savanna, and mountain, the forest endemics are especially notable. Most have been closely associated in their evolutionary history with the well-known Panama-Choco refugium (Soderstrom & Calderon 1974; Calderon & Soderstrom 1973). The mountain endemics are mostly local species that have evolved in isolation as a result of the patchiness and limited extent of their montane habitats. Some of the savanna endemics may also have evolved in response to limited patches of savanna, and others may have recently evolved in response to new habitats created by volcanic activity. These patterns suggest that forest endemics are more likely to be relictual and savanna and mountain endemics recently evolved endemics, but such speculations are highly tentative.

La mayoría de las especies endémicas se localizan en los bosques y están asociadas con el refugio del Chocó y probablemente son remanentes. Las especies endémicas de sabanas probablemente han evolucionado más recientemente.

Burkart, A. 1975. Evolution of grasses and grasslands in South America. Taxon 24: 53–66.

Calderón, C.E. & T.R. *Soderstrom*. 1973. Morphological and anatomical considerations of the grass subfamily Bambusoideae based on the new genus *Maclurolyra*. Smithsonian Contr. Bot. 11: 1–55.

——— & ———. 1980. The genera of Bambusoideae (Poaceae) of the American continent: Keys and comments. Smithsonian Contr. Bot. 44:1–27.

Clayton, W.D. 1969. A revision of the genus *Hyparrhenia*. Kew Bull. Add. Ser. 2: 1–196.

———. 1975. Chorology of the genera of Gramineae. Kew Bull. 30: 111–132.

Davidse, G. 1979. A systematic study of the genus *Lasiacis* (Gramineae: Paniceae). Ann. Missouri Bot. Gard. 65: 1133–1254.

Hartley, W. & C. *Slater*. 1960. Studies on the origin, evolution, and distribution of the Gramineae. III. The tribes of the subfamily Eragrostoideae. Austral. J. Bot. 8: 256–276.

Holdridge, L.R. & G. *Budowski*. 1959. Mapa ecológico de Panamá. Instituto Interamericano de Ciencias Agrícolas, Turrialba, Costa Rica.

Phillips, S.M. 1972. A survey of the genus *Eleusine* Gaertn. (Gramineae) in Africa. Kew Bull. 27: 251–270.

Pohl, R.W. 1980. Family 15, Gramineae. *In* Flora Costaricencis. Fieldiana, Bot., n.s., 4: 1–608.

——— & G. *Davidse*. 1971. Chromosome numbers of Costa Rican grasses. Brittonia 23: 293–324.

Raven, P.H. & D.L. *Axelrod*. 1974. Angiosperm biogeography and past continental movements. Ann. Missouri Bot. Gard. 61: 539–673.

——— & D.L. *Axelrod*. 1975. History of the flora and fauna of Latin America. Amer. Sci. 63: 420–429.

Rzedowski, J. 1975. An ecological and phytogeographic analysis of the grasslands of Mexico. Taxon 24: 67–80.

Sarmiento, G. & M. *Monasterio*. 1975. A critical consideration of the environmental conditions with the occurrence of savanna ecosystems in tropical America. Chap. 16, *In* F.B. Golley & E. Medinas, ed., Tropical Ecological Systems. Springer-Verlag, New York.

Shantz, H.L. 1954. The place of grasslands in the earth's cover of vegetation. Ecology 35: 143–145.

Soderstrom, T.R. 1981. Some evolutionary trends in the Bambusoideae (Poaceae). Ann. Missouri Bot. Gard. 68: 15–47.

——— & C.E. *Calderón*. 1974. Primitive forest grasses and evolution of the Bambusoideae. Biotrop. 6: 141–153.

——— & H.F. *Decker*. 1973. *Calderonella*, a new genus of grasses and its relationships to the centostecoid genera. Ann. Missouri Bot. Gard. 60: 427–441.

Swallen, J.R. 1943. Gramineae. *In* Flora of Panama. Ann. Missouri Bot. Gard. 30: 104–280.

———. 1955. Grasses of Guatemala. *In* Flora of Guatemala. Part 2. Fieldiana, Bot. 24(2): 1–390.

The Orchid Flora of Panama

Robert L. Dressler

Panamá es sumamente rica en orquídeas, con muchas especies y una magnífica representación de la diversidad de éstas. A pesar de que muchas son endémicas, otras son de distribución amplia. Tanto las especies saprófitas, como las asociadas a hormigas, al igual que los sistemas de polinización especializados, son causa del interés en su biología y estudio en sus localidades naturales.

The orchid family is the largest family of flowering plants, and it is very well represented in Panama. Indeed, as tiny as Panama appears on a map of the world, it has about a 25th of the world's orchid flora, which I find very impressive, even though—in numerical terms—Panama trails far behind Colombia, Ecuador or Papua-New Guinea. When the orchid volume of the Flora of Panama was published in 1949 (Williams & Allen 1946–1949) there were 470 species of orchid recorded for Panama, but the sample was still quite inadequate, especially for the mountains and the wetter caribbean slope, which are the richest areas. Last year Bill D'Arcy asked me for a list of additions, and I tabulated 310 species. Since sending this list to D'Arcy, 17 further additions have appeared, giving us nearly 800 species in about 156 genera. This clearly is not the final word; novelties are still easy to find in many areas, vast areas are still virtually untouched, and I have at least a quarter of a herbarium cabinet full of material that I cannot identify with the resources available on the Isthmus. Therefore, an estimate of 1000 orchid species for Panama is really conservative.

Panamá es sumamente rica en orquídeas con aproximadamente 1000 especies. Desde la publicación del tratamiento de Orchidaceae en 1949, en la Flora de Panamá, se han encontrado más de 325 especies adicionales.

Within Panama, the greatest concentration of species is in moderately wet forests and especially above about 750 m. We may mention a few areas of special interest. The hills north of El Valle de Anton are, of course, the classical orchid locality in Panama, and we are still finding surprises in this area, though human activity is seriously damaging the habitat. I took part in the environmental survey of the rich, wet Fortuna valley in Chiriqui, which is to be the site of a hydroelectric project. In about ten days of field work, I found 136 species of orchid in the valley, and further field work will undoubtedly reveal others. In the area between Cerro Tute and the continental divide, north of Santa Fe de Veraguas, I found sixty orchid species *in flower* in three days of collecting. The area surely has well over a hundred orchid species. But a visit to the same area now is disappointing. The trees that were cut to make the road have all rotted away, and there are few easily accessible orchids. They have not disappeared from the area, they are merely inaccessible. Finally, about four days of field work near Corriente Grande, on the Changuinola River, in Bocas del Toro, yielded sixty-five orchid species, a rich flora for an area of only 50–100 m in elevation, but Corriente Grande is an area without any dry season.

El Valle de Antón es el lugar clásico para las orquídeas, sin embargo hay muchas otras áreas de interés tales como el valle de Fortuna en Chiriquí, Cerro Tute cerca de Santa Fe en Veraguas y el río Changuinola en Bocas del Toro.

A quick analysis of the species listed in the Flora of Panama shows that about one half of the species are known only from Panama and Costa Rica. Of the remainder, about one half (one quarter of the total) range up through Central America or into Mexico, while the other quarter ranges into South America. There is an interesting pattern here, with the species of drier habitats tending to range east into Venezuela, while the species of wetter habitats tend to range southward into the Choco or western Ecuador. But one should not go too far with such a crude analysis.

Aproximadamente la mitad de las orquídeas panameñas son endémicas de Panamá y Costa Rica, un cuarto se extiende hacia el norte y el resto hacia América del Sur.

Popular belief considers the orchids as rare and endemic to very restricted habitats. In fact, such rarity seems almost always to be an illusion. A few years ago, I would have cited *Houlletia tigrina* as one of the rarest orchid species in Panama. There are, or used to be, a few plants to be found on the Mesa north of El Valle de Anton, and I had never seen a Panamanian plant in flower. Now, we find the species to be frequent in some areas of Darien. It is common to find a new species in Panama and have the same new species appear more or less simultaneously in Colombia or Nicaragua. We do have a few Panamanian orchid species that—on present evidence, at least—really seem to be scarce. *Epidendrum pendens*, *Sievekingia butcheri* and *Trichopilia leucoxantha* are all known only from the area north of El Valle de Anton, and all three seem to be scarce. *Epidendrum caligarium* has been collected only twice, and only once with locality data. It may still occur somewhere in Veraguas, or it may be quite extinct. Several orchid species now scarce in the known localities in Panama which have a wider distribution in Central or South America, such as *Chysis aurea*, *Epidendrum pseudepidendrum*, *Koellensteinia kellneriana*, *Neomoorea irrorata and Oncidium kramerianum*. The "Holy Ghost" or "flor de espiritu santo," *Peristeria elata*, is by no means in danger of extinction at present. Indeed, it finds a congenial habitat on road cuts and other areas of human disturbance, and it is probably more abundant now than before human alteration of the Panamanian landscape when it must have been limited to steep slopes and open, rocky areas. Still, as the national flower, some degree of protection seems appropriate. Many plants are gathered for sale to tourists from Panama City, and nearly all of these are doomed to a lingering death, as it is not an easy plant for the uninitiated to cultivate. In the long run, the species in greatest danger are probably those of the Pacific slope forests, such as *Oncidium ampliatum*. Much of the Pacific slope forests have been cut, and the lowland forests of Darien are now being cut rapidly. Classifying these orchids as "endangered species" will not help them in the slightest unless an adequate sample of their habitats can be protected from destruction. Indeed, if the forests are to be cut and burned, it would be far better to export the orchids; some economic benefit would be derived, and some of the plants might survive in cultivation.

Las orquídeas no son particularmente raras, ya que pueden ocurrir en otros lugares y presentar distribución amplia. Pocas de las orquídeas panameñas son escasas. La deforestación representa el mayor peligro en la supervivencia de las especies de orquídeas.

We have in Panama representatives of all the subfamilies except the southeast Asian Apostasioideae. Further, we have several primitive genera such as *Selenipedium (Cypripedioideae)*, *Corymborkis* (Spiranthoideae), and *Palmorchis* and *Vanilla* (Epidendroideae). There are only a few saprophytes, which are troublesome taxonomically, but *Uleiorchis*, and both species of *Wullschlaegelia* have appeared within the last couple of years, and others may also appear. I consider the saprophytes troublesome because their vegetative morphology is quite unlike that of their relatives—stomatal development is an important taxonomic feature here, but what can we do with plants that lack leaves?—and most saprophytes are self-pollinating and bloom for a short time, so we must get the flower at just the right moment to make out the floral details that are important in classification.

Las pocas especies de saprófitas que se hallan en Panamá son, desde el punto de vista taxonómico, un problema.

Several other aspects of ecology warrant comment. We find two sorts of ant plants, "myrmecophytes," among the orchids. In *Caularthron* and in *Schomburgkia* section *Chaunoschomburgkia* (or the genus *Myrmecophila*; I am told that plants of this taxon occur in Bocas del Toro, but I have not seen them) we find hollow pseudobulbs with an aperture at the base. These pseudobulbs are usually inhabited by large and unfriendly ants that probably give the plant some protection from herbivores, as well as contributing something to the plant's mineral nutrition. However, these hollow-bulbed myrmecophytes survive very well in cultivation without any ants. *Coryanthes* and *Epidendrum imatophyllum*, on the other hand, always grow on "ant gardens" with species of *Aechmea* , *Anthurium*, *Epiphyllum* and *Peperomia*. Again, the ants probably give some protection as well as improved mineral nutrition, while the plants' roots serve to hold the ant nest together. These plants are very difficult to grow without ants, but an acid medium, fertilizer and the judicious use of insecticides may solve the problem. It is interesting, too, that these "ant gardens" often include two different ant genera in the same nest. There are other orchid species, such as *Epidendrum baumannianum* and *Sievekingia fimbriata*, that often occur in ant gardens and seem to grow most vigorously there, but they are not limited to this habitat.

Orquídeas que mantienen nidos de hormigas en sus tallos pueden cultivarse sin hormigas, mientras que aquellas que crecen sobre nidos de hormigas son difíciles de mantener en cultivo.

Another interesting phenomenon is that of gregarious flowering, in which many plants of the same species flower on the same day, the short-lived flowers withering within a few hours. In the well-studied "pigeon orchid," *Dendrobium crumenatum*, of Malaysia (Coster 1926), the flower buds stop development at a certain size, and then all buds resume development when a rainstorm causes a sudden drop in temperature. Thus, all these buds open on the same morning, nine days after the temperature drop. In Panama we find similar behavior in *Sobralia*, *Triphora* and *Psilochilus*, but none of these has been studied.

En algunas orquídeas panameñas como Sobralia, Triphora y Psilochilus, la floración es gregaria.

Pollination mechanisms make up a major theme of orchid evolution, and some aspects of pollination have been relatively well studied. We know a great deal about pollination by euglossine bees which pollinate the Catasetinae, Stanhopeinae, Zygopetalinae and some members of other groups. In this relationship, the male bees are strongly attracted by floral perfumes which they gather and apparently use as pheromones or scent signals. In some cases, the orchid perfume attracts only a single species of pollinator; in others, several bee species serve as pollinators for a single orchid species. In these euglossine-pollinated orchids, we find such bizarre systems as that of *Coryanthes*, in which the bees fall into a water-filled cup with smooth sides, where they must climb up a convenient "stairway" leading them under the stigma and the anther. The exit is a tight squeeze, and the bees must struggle to escape, yet so great is the attraction of these perfumes that the bees often return to deposit pollinia in a second flower. While these complex mechanisms are impressive, we should remember that there are also simpler mechanisms involving euglossine bees, and there are also many other orchid pollinators besides the euglossine bees. We find hummingbird pollination in a number of orchids, and these are usually characterized by bright colors, red, pink or yellow. Hummingbird pollination seems to be especially important at higher elevations where insects become inactive under unfavorable weather conditions, while the endothermic or warm blooded hummingbirds continue to forage and pollinate every day. Characteristic hummingbird flowers are *Rodriguezia lanceolata*, *Comparettia falcata*, *Elleanthus*, *Hexisea* and *Stenorrhynchos*. Dark, bluish or purplish pollinia are of frequent occurrence in hummingbird pollinated orchids (Dressler 1971). Presumably, when a bird sees a bright yellow object on its beak, it often brushes the object away, thus providing the selective pressure for dark, slaty pollinia instead of yellow.

Las orquídeas presentan sistemas de polinización muy diversos e interesantes. Los perfumes de algunas orquídeas atraen a las abejas Euglosinas, mientras que los colores brillantes a los colibríes.

The Diptera, or flies, make up a neglected group of pollinators but an important one (van der Pijl & Dodson 1966). The Asiatic *Bulbophyllum* species whose flowers have both the color and the perfume of rotting flesh are clearly fly flowers, and, indeed, are usually covered with flies when in flower, but there are many other kinds of flies. So far, it appears that the orchids of the subtribe Pleurothallidinae, of which there are at least 250 species in Panama, are almost all pollinated by small, *Drosophila*-like flies, but we still know very little about this system of pollination. The flowers of the recently named genus, *Dracula*, offer a striking case of mimicry. The lip of this genus looks like a small fungus, the flower smells like a fungus, and Vogel (1978) suggests that these flowers are pollinated by fungus gnats that are deceived by the mimicry. Flies are also involved in the one clear case of pseudocopulation in tropical America. The flowers of the Andean *Trichoceros* look like bristly flies, so much so that the flowers are pollinated by male tachinid flies that mistake them for female flies. Andres Maduro reports that in Chiriqui flowers of the related genus *Telipogon* are also visited by these bristly tachinids.

En Panamá, moscas pequeñas parecidas a las Drosophila polinizan cerca de 250 especies de Pleurothallidinae, mientras que Dracula que

presenta flores similares a un hongo atrae a otras moscas.

The Lepidoptera are important pollinators, and the vast majority of *Epidendrum* species seem to be pollinated by moths or butterflies. The species of the *Epidendrum paniculatum* complex form a striking parallel with the flowers pollinated by male euglossine bees. These flowers are visited by male ithomiid butterflies, the butterflies apparently collecting pyrrolizidine alkaloids that they later modify and use as pheromones.

A theme that appears frequently in orchid pollination is what I call "empty promises." Flowers that "promise" rotting meat or fungus fruiting bodies ripe for fly eggs are examples of this. In less obvious cases a flower looks like another flower but lacks nectar. These may be flowers which mimic another specific kind of flower with nectar or edible pollen, or they may not. In the case of the orchid, *Cochleanthes lipscombiae*, my colleague James Ackerman finds that both males and females of *Eulaema meriana* and *E. bombiformis* occasionally visit the flowers, and they seem to expect nectar, but the flower has none. Jim finds that only young and inexperienced bees visit these flowers. Older bees know better.

Muchas orquídeas aparentan ofrecer recompensas al polinizador, cuando en realidad no poseen nada que brindar.

Finally, we may note that the best orchid habitats are not necessarily the best pollinator habitats. The bees, as a group, are best developed in dry, sunny habitats. The orchids reach their greatest diversity in wet, drizzly cloud forests. This is surely one reason that hummingbirds (Cruden 1972) and flies become more important at higher elevations. The euglossine bees are still frequent at about 1000 m, but they are active only in good weather, which is, at best, intermittent in cloud forests. Some euglossine-pollinated orchid flowers last longer in cloud forests than at lower elevations. Others put out a strong and non-specific perfume. When I first analyzed pollinator specificity, I concluded that selection favored more specific pollination relationships when two or more species of the same genus were sympatric. While this fits beautifully with evolutionary theory, I now believe that I had it backwards. The isolated species are often those of cloud forests where selection favors a very strong and less selective perfume.

Los bosques nublados son ricos en orquídeas, sin embargo no son adecuados para las abejas que son las polinizadoras potenciales.

At present, we are only beginning to learn something of the complexities in the evolution, ecology and pollination of this large and varied family. This is not something that can be done in herbaria or in temperate zone greenhouses. We must study the plants in their natural habitat, the tropical forests.

Coster, C. 1926. Periodische Blütenerscheinungen in den Tropen. Ann. Jard. Bot. Buitenzorg 35: 125–162.

Cruden, R.W. 1972. Pollinators in high-elevation ecosystems: relative effectiveness of birds and bees. Science 176: 1439–1440.

Dressler, R.L. 1968. Pollination by euglossine bees. Evolution 22: 202–210.

———. 1971. Dark pollinia in hummingbird-pollinated orchids, or do hummingbirds suffer from strabismus? Amer. Nat. 105: 80–83.

van der Pijl, L. & C.H. *Dodson*. 1966. Orchid Flowers: their Pollination and Evolution. Coral Gables. Univ. of Miami Press.

Vogel, S. 1978. Pilzmückenblumen als Pilzmimeten. Flora 167: 329–398.

Williams, L.O. & P.H. *Allen*. 1946–1949. Orchidaceae *In* Flora of Panama. Ann. Missouri Bot. Gard. 33(1,4), 36(1,2); reprinted with amplification by R.L. Dressler, 1980.

An Ecotaxonomic Survey of Panamanian Lianas

Alwyn H. Gentry

Las plantas trepadoras son de los componentes más importantes de los bosques tropicales. Estructuralmente constituyen gran parte de la vegetación y, florísticamente, igualmente forman un alto porcentaje del número total de especies. Frecuentemente ellas compiten con los árboles por luz y nutrimentos y estos últimos están en desventaja. La mayoría de las lianas de Panamá se incluyen en las familias Bignoniaceae, Sapindaceae, Leguminosae, Malpighiaceae y Apocynaceae. Las lianas presentan muchas adaptaciones especializadas para trepar, tales como zarcillos, raíces adventicias y hojas opuestas y ásperas. Además, muchas poseen tallos con estructura altamente especializada. Las lianas son más numerosas en los trópicos y son muy variables y más abundantes en los bosques secundarios de edad avanzada. Los bosques secos tienden a presentar pocas hemiepífitas. Algunas lianas, primordialmente las que están en el dosel del bosque, florecen masivamente, mientras que otras, especialmente aquellas que están en el sotobosque, poseen estrategias de floración más estables. La dispersión mediante el viento es predominante, a pesar de que en algunos hábitat es común la dispersión por medio del agua. Las relaciones fauna-flora que incluyen mimetismo y sistemas químicos son comunes entre las trepadoras.

Climbing plants are an important component of both Panamanian and other tropical forests. Woody lianas are such a conspicuous and characteristic life form in tropical plant communities that their presence has been called the single most important physiognomic feature differentiating tropical from temperate forests (Croat 1978). Forty three to 50% of the trees over 15 (-20) cm dbh on Barro Colorado Island (BCI) have lianas in their crowns (Montgomery & Sunquist 1978; Putz 1982), and 42% of the trees over 10 cm dbh at San Carlos de Rio Negro, Venezuela have lianas (Putz 1982).

Yet the importance of lianas in tropical forest ecosystems is only beginning to be recognized. For example, in Gabon Charles-Dominique (1977) showed that the prosimian *Euoticus elegantulus* feeds on the gum of the liana *Entada gigas* to the extent that its distribution is exactly determined by that of its host liana. In Amazonian Peru the gymnospermous liana *Gnetum* is one of the most important food plants of the pygmy marmoset (Soini, pers. comm.). Lianas make up from 13 to 35% of the plant species utilized for food by various tropical primates (average 21%) and almost half (42.8%) of the food species of African elephants (Emmons & Gentry 1983). Even where lianas are not directly important as food sources, they may be very significant to forest animals. For example, Montgomery and Sunquist (1978) found that sloths on Barro Colorado Island selected trees largely on the basis of presence or absence of lianas in their crown. They concluded that the single most important factor affecting tree selection by two-toed sloths was the presence and density of lianas in their crowns; three-toed sloths also preferentially selected trees with lianas. We (Emmons & Gentry 1983) have suggested that intercontinental differences in liana densities have been the major selective pressure leading to evolution of prehensile tails in neotropical vertebrates versus gliding adaptations in Asia and to neither of these in Africa.

Las lianas son un componente importante de los bosques tropicales y juegan un papel importante en la vida de los mamíferos del bosque. Los primates y monos perezozos, frecuentemente dependen de las lianas.

Lianas are also important structural components of tropical forests. In an extensive series of neotropical sites (Gentry 1981, 1982b) they represent 20% of the plants over 1″ diameter; on these sites there are more lianas than there are large (10 cm dbh.) trees, which are usually exclusively considered in studies of tropical forest structure and composition. By comparison lianas in three temperate zone forests averaged less than 4% of the plants 1″ or more in diameter (Gentry 1982). Many tropical forests are literally tied together with lianas (Fox 1968; Jacobs 1976; Gentry 1983). Another measure of the ecological importance of lianas in tropical forest communities is Hladik's (1974, 1978) study of leaf production and biomass in tropical Africa, which reveals that over a third (36%) of all leaf litter is due to lianas. In a Thailand forest lianas accounted for 32% of leaf biomass (Ogawa et al. 1965). In a Mexican dry forest several of the most important species contributing to leaf biomass are lianas (Sarukhan, pers. comm.). Although no comparable figures are available from Panama, it seems likely that lianas account for an equal portion of the leaf biomass and photosynthetic capacity of lowland Panamanian forests.

Las lianas son elementos estructurales que entrelazan los árboles y que constituyen parte importante de la biomasa y de la hojarasca.

Lianas also make a major contribution to the taxonomic diversity of Panamanian forests. For example, on BCI, 13% (171) of the vascular plant species are lianas and 20% (265) are climbers (Croat 1978). In the forest canopy lianas are an even more important floristic component. In their analysis of sloth modal trees on BCI, Montgomery and Sunquist (1978) encountered 55 different liana species, and an average of more than four liana species per tree species.

Florísticamente, las plantas trepadoras probablemente constituyen el 20% de la flora vascular de la IBC, mientras que las lianas sólo forman un 13%.

Lianas compete with trees for light and nutrients. Putz (1980) has noted that competition with lianas exerts a strong selective pressure on tropical trees. Apparently many trees have evolved special growth strategies to avoid or minimize "the ravages of lianas". Ogawa et al. (1965) suggest that competition with lianas is an important cause of death of large trees. They find that liana-covered trees have much less foliage than liana free ones of equivalent size, the additional leaves in effect replacing those of the tree so total leaf biomass of the tree-plus-liana unit remains constant. Even the taxonomic and architectural diversity of tropical forest trees may be due in part to avoidance of invasion of lianas; by ensuring that adjacent trees are of different species, there may be greater likelihood of out of phase movement so that lianas cannot so easily invade one tree from the next (Putz 1980).

Las lianas compiten con los árboles por luz y nutrimentos.

Vines do not form any kind of natural taxonomic unit. Rather, climbing and liana habits have evolved repeatedly and independently in numerous evolutionary lineages, presumably as a response to similar selective pressures. Competition for light is certainly the main selective force leading to evolution of a climbing habit. Development of a thin, flexible stem capable of rapid elongation without the necessity of massive support tissue gives the juvenile liana a decided advantage in its race to the canopy. Even a scrambling herbaceous climber that clambers over the tops of shrubby vegetation often completely covers the supporting plants, presumably as a result of its capacity for rapid growth. A canopy liana can enjoy the benefits of full-sunshine photosynthesis without the necessity of diverting much of its photosynthate to a supporting trunk.

El hábito trepador y el de liana han evolucionado repetida e independientemente a través de muchas líneas evolutivas.

Evolution for flexibility—anatomically, spatially, and temporally has been a major theme in liana evolution, and the advantages of such flexibility are attested by the prevalence of the lianas in so many evolutionary lineages and so many tropical forests. In general lianas are more flexible than trees in both space and time; in space they are capable of moving from tree to tree through the canopy, covering much larger areas than even the most wide-spreading tree; in time they can regenerate vegetatively from branches that frequently trail across the forest floor, periodically sending up shoots that may become independent individuals outliving both the parent liana and its arborescent support. Sometimes lianas of a given species are in nearly linear arrangements in the forest representing the sprouts from a fallen ancestral vine, which suggests that lianas may outlive the trees which support them. Putz (pers. comm.) suspects from such patterns that all the *Uncaria* lianas on BCI derive from vegetative regeneration of only two or three original individuals.

Las lianas son más flexibles en el espacio y el tiempo que sus árboles hospederos y pueden sobrevivirles.

What is a Vine?

How are vines and lianas to be defined? A convenient but sometimes subjective morphological-ecological division of climbing Panamanian plants recognizes four types of climbers: (a) Lianas are woody, relatively thick-stemmed climbers that begin life as terrestrial seedlings and are capable of growth in mature forests. (b) Vines are herbaceous and subwoody thin-stemmed climbers or clamberers that begin life as terrestrial seedlings and generally grow in disturbed habitats or at the forest edge; prostrate terrestrial herbs that do not climb over other vegetation are omitted, although many of the included "vines" are alternatively capable of climbing or growing prostrate over the ground. Basically epiphytic plants make up two other kinds of climbers. One group (c) is the woody hemiepiphytes, like most of the Marcgraviaceae, which frequently begin life as epiphytic seedlings and which develop woody climbing stems growing appressed to supporting trees in mature forests. Although these plants have roots that reach the ground, at least at maturity, the branching of their climbing main stems as they descend and ascend the supporting tree suggests their hemiepiphytic nature. Large epiphytes that send down free-hanging roots which reach the ground but do not develop climbing main stems are excluded from this group as are stranglers (only *Ficus* and *Coussapoa* in Panama). The final category of Panamanian climbers is that of (d) herbaceous epiphytes, as represented by many species of *Peperomia*. All of those herbaceous plants that typically grow appressed to tree trunks and limbs belong to this category. These plants invariably have adventitious roots for attachment but rarely have roots reaching the ground. All those species that typically crawl over tree trunks are included, even though some of these may occur also as terrestrial creepers in cloud forest areas.

Las lianas son leñosas, de tallos gruesos y trepadoras, cuyas semillas germinan en el suelo y crecen en el bosque maduro. Las enredaderas son trepadoras de tallos delgados, cuyas semillas germinan en el suelo y que, primordialmente, crecen en los bordes de bosques ya trabajados.

These categories are not sharply delimited, and there are many intermediates between, for example, climbing and erect epiphytes, and between herbaceous vines and woody lianas. Nevertheless, most Panamanian climbers fall into one of these four habit classes. I know in the field most Panamanian climbers and virtually all climbing genera and when in doubt have relied on general impressions gained from this first hand experience to assign species to categories. Whether a species typically occurs in mature forest or not has been an especially important criterion in separating lianas from vines and in assigning parasitic herbaceous vines to the vine category rather than the herbaceous epiphyte one. When in doubt about a particular species, I have tended to assume that the ambiguous species shares the habit of the majority of its congeners unless there is clear evidence to the contrary. The main exception to this is in the case of lianas where any woody species known to me to be capable of developing a definitely climbing growth form is included as a liana, even some that are usually shrubs. The exception to this exception is *Malvaviscus arboreus*, a shrub or small tree that very rarely is distinctly scandent and whose inclusion would add a whole family to the roster of Panamanian lianas. In the case of strangler figs, that may be more or less climbing when young, I have included only *Ficus schippii* which is the only Panamanian *Ficus* species that appears to remain a woody vine at maturity. Climbing parasites are included in the habit class in which they otherwise best fit—Loranthaceae as woody hemiepiphytes, *Cassytha* and *Cuscuta* as herbaceous vines. In some of the analyses below, various of these classes are combined as indicated (woody climbers & lianas plus woody hemiepiphytes). In general lianas, the ultimate in development of a climbing habit, are emphasized.

Taxonomic Survey

Families

Climbing plants are distributed among 71 families in Panama (Table 1). Twenty seven Panamanian families are entirely or predominantly climbers based on per cent of climbing versus non-climbing *genera*, and 21 Panamanian families are predominantly climbers based on per cent of climbing versus non-climbing *species*.

The biggest Panamanian vine families, based on

TABLE 1
MAIN CLIMBING FAMILIES IN PANAMA

NUMBERS OF PREDOMINANTLY CLIMBING GENERA

Bignoniaceae	26(-27)	Araceae	5
Leguminosae	24(-29)	Dilleniaceae	4
Cucurbitaceae	19	Euphorbiaceae	4
Apocynaceae	12	Marcgraviaceae	4
Compositae	11(-16)	Solanaceae	3(-5)
Hippocrateaceae	11	Amaranthaceae	3(-4)
Convolvulaceae	10	Connaraceae	3
Rubiaceae	8(-13)	Cyclanthaceae	3
Asclepiadaceae	8	Gesneriaceae	3
Malpighiaceae	8	Passifloraceae	3
Menispermaceae	7(-8)	Polygonaceae	3
Sapindaceae	5	Gramineae	3
Cactaceae	5		

NUMBERS OF PREDOMINANTLY CLIMBING SPECIES

Leguminosae	105	Solanaceae	27	Ericaceae	9
Araceae	102	Menispermaceae	17	Cactaceae	9
Bignoniaceae	58	Marcgraviaceae	16	Connaraceae	8
Sapindaceae	56	Rubiaceae	16	Euphorbiaceae	8
Convolvulaceae	53	Hippocrateaceae	15	Loranthaceae	8
Piperaceae	42	Aristolochiaceae	14	Smilacaceae	8
Asclepiadaceae	38	Gramineae	13	Verbenaceae	8
Malpighiaceae	38	Acanthaceae	12	Loganiaceae	6
Cucurbitaceae	37	Dilleniaceae	11	Amaryllidaceae	5
Passifloraceae	33	Dioscoreaceae	11	Begoniaceae	5
Apocynaceae	31	Melastomataceae	11	Boraginaceae	5
Compositae	31	Vitaceae	11	Combretaceae	5
Gesneriaceae	27	Amaranthaceae	9	Polygonaceae	5

Numbers in parentheses include additional generally non-climbing genera with one or more scandent Panamanian species.

TABLE 2
PANAMANIAN FAMILIES WITH *REAL* WOODY LIANAS
Based on number of species
(excluding slender vines and hemiepiphytes)

Bignoniaceae	53	Aristolochiaceae	8	Euphorbiaceae	2
Sapindaceae	45	Loganiaceae	6	Sterculiaceae	2
Leguminosae	44	Vitaceae	6	Phytolaccaceae	2
Malpighiaceae	35	Combretaceae	5	Trigoniaceae	2
Apocynaceae	23	Passifloraceae	5	Palmae	2
Compositae	17	Verbenaceae	5	Ranunculaceae	2
Asclepiadaceae	16	Boraginaceae	4	Cyclanthaceae	1
Hippocrateaceae	15	Piperaceae	4	Celastraceae	1
Solanaceae	12	Convolvulaceae	3	Icacinaceae	1
Dilleniaceae	11	Polygalaceae	3	Melastomataceae	1
Cucurbitaceae	11	Polygonaceae	3	Nyctaginaceae	1
Menispermaceae	9	Rhamnaceae	3	Ulmaceae	1
Acanthaceae	9	Smilacaceae	3	Urticaceae	1
Rubiaceae	9	Amaranthaceae	2	Gnetaceae	1
Connaraceae	8	Dichapetalaceae	2	Violaceae	1

45 families 400 species 1/2 of species belong to 5 families

TABLE 3
PANAMANIAN FAMILIES WITH WOODY HEMIEPIPHYTIC CLIMBERS
Number of Species

Marcgraviaceae	16	Bignoniaceae	5
Melastomataceae	10	Guttiferae	2
Ericaceae	9	Saxifragaceae	2
Loranthaceae	8	Rubiaceae	1
Solanaceae	6	Moraceae	1

60 species

number of species, are Leguminosae (105 species), Araceae (102), Bignoniaceae (58), Sapindaceae (56), Convolvulaceae (53), Piperaceae (42), Asclepiadaceae (38), Malpighiaceae (38), Cucurbitaceae (37), Passifloraceae (33), Apocynaceae (31), Compositae (31), Gesneriaceae (27), Solanaceae (27), Menispermaceae (17), Marcgraviaceae (16), Hippocrateaceae (15), and Aristolochiaceae (14). Nineteen other families have between five and 12 species of climbers in Panama, and the other 35 families with Panamanian vines have only one or two species each.

En Panamá, 71 familias incluyen especies de trepadoras. La Leguminosae (105 especies) y la Araceae (102 especies), poseen el mayor número de especies, y en el país hay 279 géneros con especies trepadoras.

If only woody lianas are considered (Table 2, 3), five families clearly predominate in Panama, accounting for over half of the species. Bignoniaceae is by far the most important with 53 species—58 species if woody but hemiepiphytic *Schlegelia* is included. The other four families are Sapindaceae (45 species), Leguminosae (40), Malpighiaceae (35), and Apocynaceae (23). Six more families have between ten and 17 species; Compositae, Asclepiadaceae, Hippocrateaceae, Solanaceae, Dilleniaceae, and Cucurbitaceae. Thirty-four additional families contribute to the liana flora of Panama but have fewer than ten species. Thus only 45 (60%) of the families with climbing species in Panama are represented by woody lianas. Araceae constitute the great majority of herbaceous and epiphytic climbers (Table 4).

Las familias Bignoniaceae, Sapindaceae, Leguminosae, Malpighiaceae y Apocynaceae son las que incluyen la mayoría de las lianas leñosas.

TABLE 4
PANAMANIAN FAMILIES WITH HERBACEOUS EPIPHYTIC CLIMBERS AND HEMIEPIPHYTIC CLIMBERS
Number of Species

Araceae	101	Orchidaceae	4
Piperaceae	36	Cyclanthaceae	3
Cactaceae	7	Rubiaceae	2
Begoniaceae	5		
	158 species		

Genera

Two hundred and seventy nine Panamanian genera have scandent species. Of these 245 are predominantly climbing. The largest vine genera in Panama are *Philodendron* (73 climbing species), *Paullinia* (37), *Peperomia* (36), *Passiflora* (31) and *Ipomoea* (29). Twenty-two other genera representing fifteen different families have between 8 and 14 climbing species (Table 5).

Although 204 genera are wholly represented by climbing species in Panama, most (106) of these are monotypic. Only one large genus in Panama is entirely scandent—*Paullinia* with 37 species. All of the other uniformly scandent genera have 14 or fewer species. Clearly the adaptive shift between climbing and erect habit is not a major one.

Species

Even at the species level climbing habit is not always constant. Some shrubs have weak, reclining to scrambling stems or subscandent branches, and many liana species grow facultatively as shrubs in open areas. Even inside the mature forest many canopy liana species start out as shrubs and become scandent only after they are well established.

TABLE 5
LARGEST PANAMANIAN VINE GENERA
Number of Species

Philodendron	73	Solanum	12	Vigna	10
Paullinia	37	Dioscorea	11	Matelea	9
Peperomia	36	Drymonia	11	Canavalia	9
Passiflora	31	Prestonia	11	Mendoncia	9
Ipomoea	29	Machaerium	10	Dioclea	8
Syngonium	14	Monstera	10	Smilax	8
Mikania	14	Arrabidaea	10	Lycianthes	8
Aristolochia	14	Cissus	10	Marcgravia	8
Serjania	14	Columnea	10		

Trends

A major point is that liana habits have independently evolved repeatedly in many different evolutionary lineages. Not only do most vine families and even genera have non-scandent members, but 37% of all Panamanian families include at least one scandent member in Panama. On a world basis Schenck (1892) and Darwin (1867) estimated that about half of all flowering plant families contain climbers. The evolution of liana habit in so many unrelated groups suggests that there are strong evolutionary pressures favoring a climbing habit in tropical plant communities, as noted long ago by Darwin (1867).

El hábito de liana ha evolucionado independientemente a partir de muchas y diferentes líneas evolutivas.

Special Adaptations of Lianas

Climbing Mechanisms

Among the most striking adaptations to a scandent habit are the development of specialized climbing mechanisms. Often different taxonomic groups have evolved very different climbing mechanisms, but a given climbing mechanism tends to be typical of all scandent members of a given taxonomic group.

The most striking and probably the most effective climbing mechanisms of scandent plants are specialized tendrils (Darwin 1867); although only 10 Panamanian plant families have well developed tendrils, these families account for 44% of the Panamanian liana species. The tendrils of different scandent families have very different ontogenies, but all may be broadly classified as evolved from modifications of individual leaves or from specialized branches.

Las trepadoras con zarcillos componen el 44% de las lianas de Panamá. Algunos zarcillos han evolucionado a partir de hojas y otros provienen de ramas. Algunos poseen discos adherentes en el ápice y otros pueden enrollarse alrededor del hospedero.

True leaf-derived tendrils are found in only four Panamanian families; an additional three families have more or less twining petioles that function as tendrils. The climbing family par excellence of the neotropics, Bignoniaceae, is uniformly characterized by tendrils evolved from the terminal leaflet(s) of a compound leaf. All scandent neotropical Bignoniaceae except the hemiepiphytic *Schlegelia*, which may belong in Scrophulariaceae, have tendrils of this type. These tendrils may be variously divided apically or their tip modified into disks which adhere to the bark of the supporting tree; they provide important taxonomic characters for generic delimitation. Similar tendrils are found in *Cobaea* of the Polemoniaceae and *Entada* of the Leguminosae. The fourth Panamanian family with leaf-derived tendrils is Smilacaceae. The tendrils of *Smilax* arise in pairs from near the base of the petiole and are actually modified stipules. Climbing members of Tropaeolaceae, Ranunculaceae (*Clematis*), and *Hidalgoa* (Compositae) have sensitive petioles which may function as tendrils.

Eight Panamanian families have scandent genera with tendrils derived from modified branches; Cucurbitaceae, Passifloraceae, Sapindaceae, Vitaceae, Rhamnaceae (*Gouania*), Polygonaceae (*Antigonum*), Loganiaceae (*Strychnos*), and Leguminosae (*Bauhinia*, *Dalbergia*, *Dioclea*). In several of these families, especially Sapindaceae, Polygonaceae, and Rhamnaceae, the tendrils are often associated with inflorescences and here tendrils, even on vegetative branches, may represent modified inflorescences.

Three families with branch-derived tendrils—Passifloraceae, Cucurbitaceae, and Vitaceae tend to have long helically flexible coiling tendrils capable of catching a support and then contracting to draw a branch of the climber toward that support. The other five families have more rigid usually shorter axillary tendrils with spiral tips. Passifloraceae and Cucurbitaceae are now considered to be closely related evolutionarily and their tendrils probably share a common origin. The branch tendrils of the other families, despite their generally remarkably similar forms, are independently evolved structures. Even within the single family Leguminosae tendrils have been independently evolved in several unrelated genera.

There are many other modifications to facilitate a scandent habit. Many vines and lianas have spines that aid in climbing by catching against the supporting tree. Probably the most spectacular development of spines occurs in *Desmoncus*, the only neotropical climbing palm or rattan, that has both a densely spiny stem and paired terminal leaflets of a compound leaf developed into spines which act as grappling hooks. Other Panamanian families in which spines or thorns are prominent in at least some scandent genera include Rosaceae (*Rubus*), Solanaceae (*Solanum*), Smilacaceae (*Smilax*), Sterculiaceae (*Byttneria*), and Leguminosae (*Acacia*, *Mimosa*, *Caesalpinia*). These have small reflexed spines scattered along stems and leaves. Another group of spiny Panamanian climbers appears to have stipules modified into larger spines. These include Ulmaceae (*Celtis*, usually only a single spine developed per node), Rubiaceae (*Uncaria*, *Randia*, the latter usually with only one spine developed per node), and Leguminosae (*Machaerium*, *Pithecellobium*). Another kind of spine, apparently representing reduced branches, is found in some climbing species of Combretaceae (*Combretum*) and Nyctaginaceae (*Pisonia*).

Algunas lianas poseen espinas para trepar, mientras que otras presentan hojas ásperas para prevenir el deslizamiento. Hojas opuestas, ramas fuertemente dirigidas hacia atrás y curvadas, corteza externa suave, gruesa, áspera o con corcho pueden también ser adaptaciones para trepar.

One whole family of Panamanian lianas, Dilleniaceae, is characterized by rough leaves and juvenile branches which prevent slippage in a manner similar to spines. Putz (pers. comm.) finds that *Tetracera volubilis* of this family is able to climb at a steeper angle than would be expected from the rigidity of its stem, presumably because of its rough surface. Three other genera—*Petrea* (Verbenaceae), *Prionostemma* (Hippocrateaceae) and *Scleria* (Cyperaceae) are also characterized by rough leaves and stems which aid in climbing. Climbing members of *Scleria* ("cortadera") are notorious for their sharp and asperous cutting leaf edges and stems.

Another modification that may facilitate climbing and prevent slipping out of the canopy is development of a soft, thick, rough, corky, bark, especially noteworthy in climbing members of at least six Panamanian families—Aristolochiaceae, Apocynaceae, Cucurbitaceae, Menispermaceae, Passifloraceae, and Vitaceae. Many of these groups climb predominantly by spiralling around a supporting tree, and the thick bark may be advantageous to keep the coils of such a species from slipping gradually down a tree trunk. All of these plants are relatively soft-wooded although the adaptive significance of this, if any, is unclear, and it may reflect development from herbaceous ancestors rather than an actual adaptation for liana habit. This group of climbers tends not to grow over the canopy but rather to remain in close proximity to the larger branches of supporting trees.

Strongly reflexed tightly looped branches are the primary climbing mechanism of several groups of Panamanian lianas. Three important families of lianas—Hippocrateaceae, Malpighiaceae, and Connaraceae have looping reflexed branches as their major adaptation for climbing. The only two liana genera of Polygalaceae which reach Panama also climb by reflexed branches, as do some Leguminosae.

Opposite branching, often associated with opposite leaves, may be a preadaptation to climbing habit. There is a preponderance of opposite leaves in lianas as compared to the flora as a whole. Seventeen of the 46 families with woody lianas in Panama have opposite leaves, at least in their liana members. Six of the eight most important liana families in number of Panamanian species have opposite leaves. Probably opposite leaves and stiff strongly divaricating opposite branches help maintain a liana's position in the canopy. Croat (1978) has suggested that conspicuously swollen nodes in such opposite-leaved families as Gnetaceae and Bignoniaceae may be a similar adaptation to climbing.

A final climbing mechanism used by Panamanian vines is adventitious roots that fasten the climber to the supporting tree. These roots are especially important in epiphytic and hemiepiphytic climbers and are the major adaptation for climbing in these groups, including many species of Araceae, Begoniaceae, Cactaceae, Cyclanthaceae, Guttifereae, Marcgraviaceae, Melastomataceae, Moraceae, Piperaceae, Rubiaceae, Saxifragaceae, and miscellaneous genera of other families. Adventitious roots may be developed in climbing genera of many families, often in conjunction with another climbing mechanism. For example the two species of *Macfadyena* (Bignoniaceae) have both uncate "cat's claw" tendrils and adventitious roots.

Las raíces adventicias contituyen un medio frecuente de trepar.

Many Panamanian climbers lack specialized climbing mechanisms other than twining stems (the spiral twiners of Darwin 1867). Especially among the thin-stemmed herbaceous vines, specialized climbing organs are typically absent. Nearly all of these groups are twiners that climb by winding around a support, although some are merely scramblers that grow haphazardly over the rest of the vegetation. Panamanian liana families with no other climbing mechanism than a twining growth form (sometimes accompanied by corky bark) include Acanthaceae, Amaranthaceae, Apocynaceae, Aristolochiaceae, Asclepiadaceae, Boraginaceae, Celastraceae, Compositae, (except *Hidalgoa*), Convolvulaceae, Dichapetalaceae, Euphorbiaceae (in *Omphalea* the stem may actually form a kind of tendril), Icacinaceae, Menispermaceae, Phytolaccaceae, Solanaceae (a few climbing species of *Solanum* have spines and *Juanulloa* has adventitious roots), Urticaceae, Verbenaceae, and Violaceae. In addition most of the herbaceous vines lack specialized climbing mechanisms, as do miscellaneous genera even in the more specialized liana families.

Muchas trepadoras carecen de mecanismos especializados y sólo trepan mediante los tallos, los cuales se enroscan alrededor del hospedero.

Wood Anatomy

One striking way in which lianas have adapted to a way of life facilitated by stem flexibility is through development of anomalous stem anatomy. Many lianas are characterized by the unusual patterns of their stem cross sections and many of these patterns are family-specific (Schenck 1892).

Bignoniaceae lianas are characterized by failure of the secondary xylem to develop in a continuous cylinder which results in a radially symmetric cross-shaped pattern formed by the secondary phloem in cross section. In some genera there are four phloem wedges, in others these proliferate to 8, 16, or even more. The number of phloem wedges and their shape, especially whether they are of a uniform width or become broader peripherally, are important generic characters in the family. In general the more phloem wedges, the greater the stem flexibility.

A more common peculiarity of liana stems is development of very thick and well-defined concentric rings of secondary xylem, prominent in families like Dilleniaceae, Gnetaceae, and Menispermaceae. These may be useful in family recognition. In some families, notably Menispermaceae, such rings may be markedly asymmetrical and the liana stem may be almost flat. Asymmetric one-sided stem growth is also typical of lianas in several families that lack unusually conspicuous growth rings; some members of Icacinaceae and Convolvulaceae provide good examples. A more extreme form of stem flattening characterizes the common liana genus *Bauhinia*. In *Bauhinia* the stem is not only conspicuously flattened but may have the center of the strap-shaped stem thinner than the margins, sometimes resulting in a peculiar ladder-like segmental effect commonly termed "escalera del mono"; *Bauhinia* stems are also characterized by lobed and fragmented xylem.

Another widespread peculiarity of liana stems is a tendency to become fasciculate that results in a twisted multistranded cable-like stem or "bush rope". Species forming multi-stranded bushropes at maturity are usually characterized by lobed or otherwise interupted and anomalous xylem in younger stems. Many species of Sapindaceae may have the stem in distinct fascicles even when very young. Compound-wooded species of *Paullinia* usually have a characteristic 3–angled form, at least when young, with the central stem paralleled on three sides by subsidiary vascular cylinders. Other families which may have older stems compounded of numerous more or less separate twisted strands include Malpighiaceae, Leguminosae, and Bignoniaceae, all of which have lobed xylem when younger.

La anatomía del leño de muchas trepadoras es especializada; muchas presentan proyecciones adicionales a partir de los anillos concéntricos de floema-xilema. Muchas presentan el leño en fascículos similares a sogas y otras poseen el tallo con estructura extremadamente compleja.

Some lianas, especially those belonging to predominently herbaceous groups, can have incredibly complex stem structure. *Mendoncia* (Acanthaceae), Compositae, *Securidaca* (Polygalaceae), Aristolochiaceae, and Cucurbitaceae are especially notable among Panamanian lianas. These are among the softest-wooded and most flexible of all Panamanian lianas, further supporting the idea that anomalous secondary growth and stem structure are correlated with flexibility.

Physiology

The driving force behind the liana strategy—maximization of photosynthesis—is a physiological one. Many liana groups have also developed specific kinds of growth peculiarities which presumably reflect individual physiological peculiarities.

Heteromorphy, with distinctive juvenile forms, is characteristic of a number of climbers. The juvenile forms, which grow in the dark forest understory, usually have leaves which grow appressed against tree trunks and are arranged so that adjacent leaves have almost no overlap and entirely avoid shading each other. These juvenile leaves may also be distinctively colored or darker green than the mature leaves and are presumably adapted to maximize photosynthesis at low light levels. The juvenile leaves are always smaller, presumably because the stems (and thus the bark) are small and relatively cheap energetically (Putz, pers. comm.); all available energy can thus be concentrated on reaching the canopy quickly. Among families that include such juvenile forms are Araceae, Bignoniaceae, Marcgraviaceae, Saxifragaceae, Solanaceae. Other genera, notably *Peperomia*, *Pilea*, and several members of Rubiaceae, have groups of species remarkably similar at maturity to these heteromorphic juvenile forms and may represent a kind of neotenous condition.

Las formas juveniles pueden diferir muchísimo de las maduras; las plantas jóvenes pueden poseer sus hojas adheridas al tronco de un árbol y muchas veces éstas son coloreadas. Algunas Aráceas crecen en la sombra y no en el sol, como es lo usual en la mayoría de las plantas.

One unusual trait of many of these forms is a striking skototropic response as discussed by Strong & Ray (1975) for a member of the Araceae. The noteworthy phenomenon here is that the plants grow toward shade, the exact reverse of the phototropic response of most lianas and even mature plants of some of the same species. Thus the vine seedlings can crawl along the ground directly toward the nearest tree trunk which casts a darker shadow than do the surroundings. Darwin (1867) observed that individual tendrils of some climbing plants show a similar skototropic response, presumably aiding them to reach for and grasp appropriate supports.

Many lianas are characterized by having latex, although many tropical trees also have latex. A circumstantial correlation of latex and the liana habit comes from families like Sapindaceae and Leguminosae in which most lianas have some kind of latex whereas latex is absent or occurs less frequently in arboreal members of the same families. Liana families characterized by milky latex include Apocynaceae, Asclepiadaceae, and Sapindaceae. Most Panamanian legume lianas, especially members of the Papilionatae, have red sap, although this is not always conspicuous; *Machaerium* and many *Diocleas* have especially noticeable red sap. One Hippocrateaceae genus, *Hylenaea*, also has reddish sap. Most Euphorbiaceae lianas in Panama have a cloudy (but not milky) latex which may turn strikingly purplish or bluish on exposure to air *(Omphalea)*. *Gnetum* has a viscous syrupy latex.

Muchas lianas presentan látex y flores características.

Another characteristic of lianas is their odors. While I have not systematically investigated liana odors, they are apparently just as diversified and taxonomically useful as are the odors of trees which are so widely used as identification aids by foresters. In Bignoniaceae, for example, many liana genera can be recognized easily by their characteristic vegetative odors. *Tynanthus* has a clove odor, *Tanaecium* an almond or cyanide odor, *Mansoa* (including *Pachyptera*) an onion or garlic odor, and *Paragonia* a distinctive sweetish odor. It is likely that similar biochemical differentiation has also taken place in other groups of lianas (see below).

Ecology

Diversity Patterns

Although lianas and other climbers are an important component of virtually all neotropical forests, they also show major distributional and ecological differences between different forest types and plant communities. The following analysis is based on data from 1000 m^2 samples of 16 different neotropical plant communities including three Panamanian ones (Curundu Forest, Madden Forest, Pipeline Road wet forest area) (Gentry 1982b).

The number of liana individuals per unit area is remarkably constant in lowland dry, moist, and wet forest communities. This differs markedly from the common impression that dry forests are "vinier" than are mature moist and wet forests. Number of liana individuals per 1000 m^2 in dry forest sites averaged 61, in moist forest sites 55, in wet forest sites 65. In contrast, an average of only six lianas per 1000 m^2 was found in similar samples of temperate zone forest. My subjective impression is that montane forests have fewer lianas than lowland neotropical ones.

Las lianas son más numerosas en los bosques tropicales que en los de las zonas templadas. También son más numerosas en los bosques maduros que en los bosques jóvenes.

Liana diversity, unlike density, shows a clear correlation with precipitation. Wetter lowland tropical forests have many more species of lianas than do drier ones. The rate of increase of liana diversity with precipitation is linear on a log-log plot (Figure 1). The dry forest average is 12 liana species per 1000 m^2. The comparable moist forest and wet forest figures are 25 species per 1000 m^2 and 31 species per 1000 m^2, respectively. The temperate zone communities sampled averaged two species of liana per 1000 m^2. This pattern parallels that of tree species diversity.

Both density and diversity of lianas are apparently lower in old second growth forest than in mature forest. The lianas of a mature tropical forest may be older than the trees and thus it takes longer to accumulate a full liana complement as a forest regenerates. An interesting paradox is that lianas, often considered typical of second growth vegetation, would seem to be a better indicator of climax forest than the trees.

While overall liana density is more or less constant in dry, moist, and wet forest vegetations, the physiognomic, as well as taxonomic, composition of the vine community does change. While virtually all climbers of a mature tropical dry forest are free-climbing lianas, hemiepiphytic and epiphytic climbers become progressively more important in wetter conditions. In dry forest Santa Rosa National Park in Costa Rica 8% of the species are lianas and an additional 10% are small vines (data mostly extracted from Janzen & Liesner 1980). In moist forest Barro Colorado Island 11% of the species are lianas and an additional 9% of the species are small vines (data from Croat 1978, including only woody climbers with 1″ stem diameter). In wet forest Rio Palenque, Ecuador, 8% of the species are lianas and another 8% are small vines (Dodson & Gentry 1978); at a moist forest site in western Ecuador (Jauneche), 10% of the species are lianas and 10% vines (Dodson & Gentry in prep.). While these percentages are as similar as predicted above (absolute species numbers cannot be compared because of the very different sizes of the three areas), breakdown of the "lianas" shows that hemiepiphytic lianas are a much more important component of the wet forest community. None of the Santa Rosa lianas is hemiepiphytic, but three (2%) of the Barro Colorado species and seven (10%) of the Rio Palenque species are woody hemiepiphytes. A similar trend prevails among the thin vines, here also including woody vines that are always less than 1″ diameter, with 9% of the Rio Palenque vines but none of the Santa Rosa ones being hemiepiphytic. Similarly herbaceous epiphytic climbers are much more prevalent in wetter communities. Similar trends take place in upland cloud forest communities.

La composición de la comunidad de trepadoras cambia a medida que el bosque madura. Las hemiepífitas leñosas forman una mayor proporción en los bosques húmedos que en los secos.

Pollination Ecology

Lianas generally reflect the general diversity of pollination strategies and pollen vectors that characterize other elements of the neotropical flora. Various Panamanian lianas are pollinated by bees, butterflies, flies, hawkmoths, bats, hummingbirds, and wasps. Phenologically they may be grossly divided into mass flowerers—most canopy lianas—and

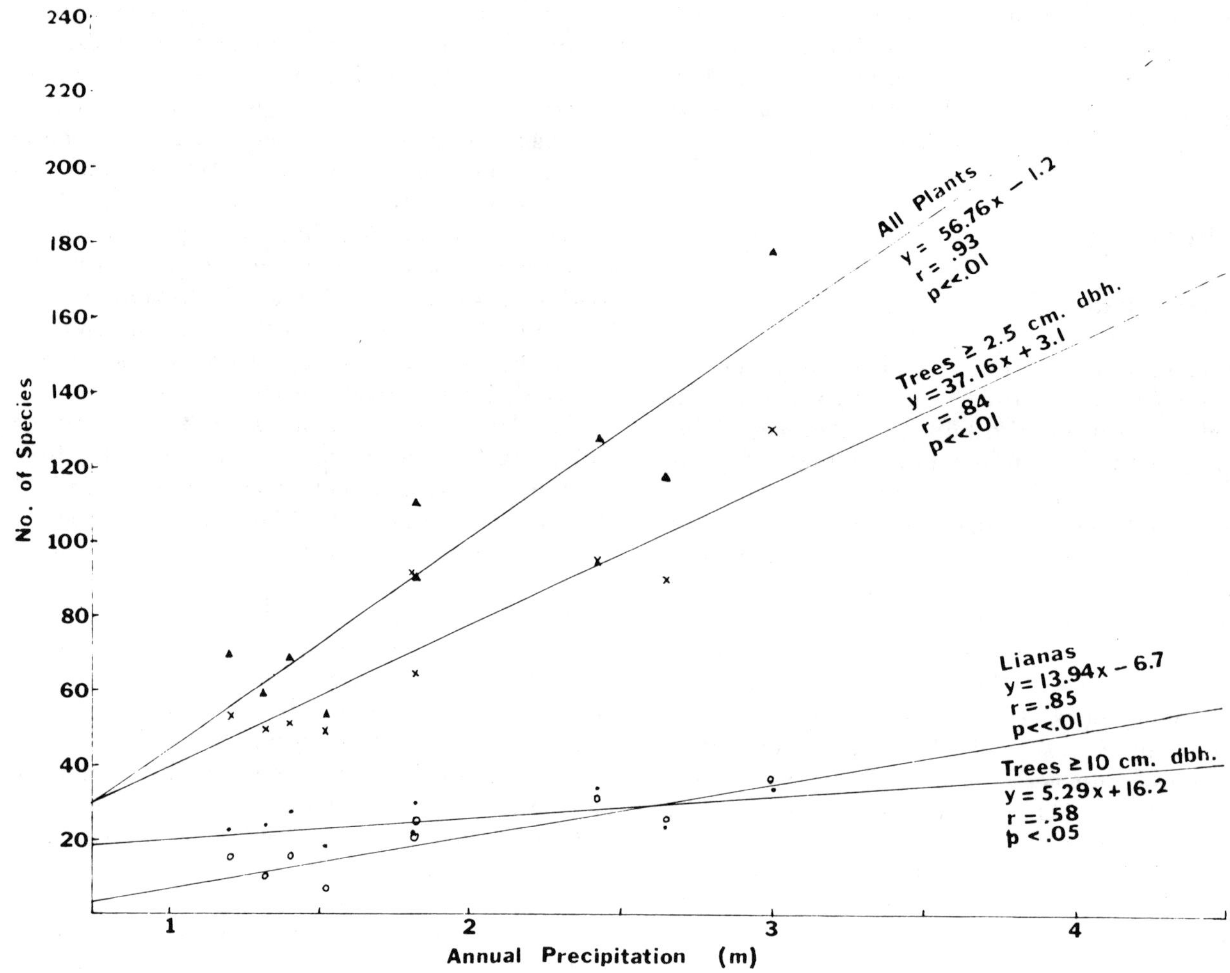

Fig. 1. Increase in species richness with precipitation for lianas and other habit groups (plants 2.5 cm dbh only) in 100 m² samples. From Gentry 1982b.

steady state flowerers—most subcanopy cauliflorous species.

Las lianas pueden ser polinizadas por abejas, mariposas, moscas, murciélagos, colibríes y avispas. Florecen masivamente, especialmente en el dosel, y menos debajo del dosel.

The pollination syndromes of Bignoniaceae are probably the best studied among Panamanian vines (Gentry 1974, 1976) and may be taken as a paradigm of the diversification of liana pollination strategies. Especially striking in the Bignoniaceae system is the apparent niche specificity of sympatric Bignoniaceae with about 20 species of Bignoniaceae (including trees), each occupying a different pollination niche, occurring in each Panamanian forest community. A few Bignoniaceae lianas are pollinated by hawkmoths, by hummingbirds, and by a diverse group of butterflies and small bees. Usually one species using each of these vectors (plus a bat-pollinated tree species) occurs in each plant community along with about fifteen species that are pollinated by a generalized group of large and medium-sized bees, mostly female euglossines and anthophorids. In each community this latter group of species includes one or two species that have "steady state" phenologies, one or two with "big bang" phenologies", one or two non-nectar-producing mimics with "multiple bang" phenologies, and about ten species with "cornucopia" phenologies. Each sympatric "cornucopia" species tends to have a different month of peak flowering. A few other liana families have conspicuous tubular flowers like those of Bignoniaceae-Apocynaceae, Convolvulaceae, *Mendoncia*, *Petrea*, and *Corynostylis*, but their pollination strategies and how they fit into the plant-pollinator community are not known.

Aristolochiaceae have fly-trap flowers which mimic

rotting meat. Malpighiaceae are mostly pollinated by bees collecting oils and waxes from their calyx (Anderson 1979). Legume lianas are diversified in pollination system as in everything else and include species that are probably bat-pollinated (some *Mucuna* species), steady state species with single large bee-pollinated flowers (species of *Clitoria*, *Centrosema*), mass-flowering bee-pollinated species (*Dioclea*, *Lonchocarpus*, *Machaerium*) and small-flowered species presumably pollinated mostly by small bees. Polygalaceae lianas look strikingly like mass-flowering papilionate legumes and are probably also pollinated by bees. Compositae vines are likely bee-pollinated also.

Passifloraceae, of special interest for an apparent niche specificity similar to that of Bignoniaceae (Gilbert 1979; Benson 1979) are mostly bee-pollinated, but one Panamanian species is hummingbird-pollinated (Janzen 1968; Gentry 1980). In most wet and moist forest communities one hummingbird-pollinated species, one large-flowered large-bee-pollinated species, and several small-flowered, small-bee-pollinated species co-occur.

Cucurbitaceae include two mostly orange-flowered genera—*Gurania* and *Psiguria* which are pollinated mostly by *Heliconius* butterflies (Ehrlich & Gilbert 1973; Gilbert 1979). The plants are steady state flowerers and have unusually large thin-walled pollen which is collected by the butterflies. Pollinators of most other cucurbits are unknown although some are probably pollinated by oligolectic bees.

Most of the rest of the Panamanian liana species have the small inconspicuous whitish or greenish flowers that are so prevalent in wet tropical forest tree species and whose presumably generalized pollinators mostly remain to be discovered.

Dispersal Ecology

Although vegetative reproduction plays a major role in the reproductive biology of many lianas, seed dispersal is also an important part of liana life histories. Woody lianas are predominantly wind-dispersed. Figure 2 compares the percentage of wind-dispersed species of trees and lianas for a series of neotropical plant communities, including three Panamanian ones. 100% of the lianas in one dry forest sample are wind-dispersed, and all the dry forest sites have over 70% wind-dispersed species. When mode of dispersal is evaluated on an individual basis, 82% to 100% of all dry forest liana samples are found to be wind-dispersed. Even in moist and wet forest sites wind dispersal is generally predominant among lianas, and the absolute number of species of wind-dispersed lianas is greater in wetter communities, despite the

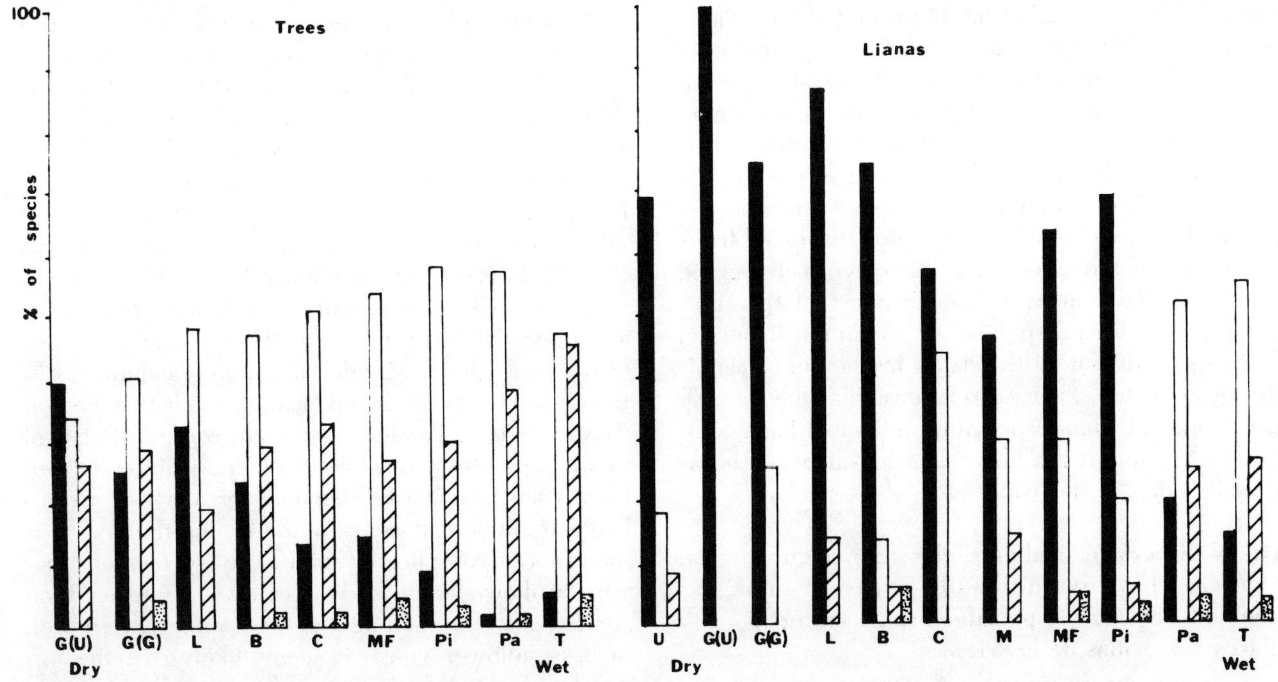

Fig. 2. Dispersal modes for trees and lianas in 1000 m^2 samples. Black bar = wind-dispersed. Open bar = bird-dispersed. Cross-hatched bar = mammal-dispersed. Dotted bar = autochorous and water-dispersed. Dry forest on left, pluvial forest on right. U, Boca de Uchire, Venezuela; Gu, Guanacaste upland, Costa Rica; Gg, Guanacaste gallery forest, Costa Rica; L, Estacion Biologico de los Llanos, Venezuela; Bl, Blohm Ranch, Venezuela; C. Curundu, Panama; M, Manaus, Brazil; MF, Madden Forrest, Panama; Pi, Pipeline Road, Panama; RP, Rio Palenque, Ecuador; T, Tutuendo, Colombia. From Gentry, 1982b.

decrease in percent of wind dispersal. The exception to these generalizations, the wet forest Rio Palenque site in Ecuador, is probably related to marked cloud forest effects on the physiognomy and composition of the forest with an accompanying shift of the "liana" component, as measured in these samples, to include a large proportion of hemiepiphytic climbers rather than entirely free-climbing lianas.

Las semillas de lianas son primordialmente dispersadas por el viento, aun en los bosques húmedos y los muy húmedos.

Wind dispersal correlates with wide geographic ranges, suggesting the effectiveness of the mechanism, when compared to the more localized distributions of many mammal-dispersed species (Gentry 1979). The diaspores of wind-dispersed lianas tend to be larger than those of wind-dispersed vines, and they also differ in usually having distinct wings to facilitate dispersal. A larger seed body is probably related to the need for the seedling to begin life in the relatively limited light gaps of the mature forest understory.

A subsidiary evolutionary pattern is for wind-dispersed lianas to give rise to water-dispersed derivatives that tend to occupy such specialized habitats as mangroves and seasonally inundated river margins and swamps. Good examples are found in Bignoniaceae (*Phryganocydia*) and Hippocrateaceae. There are also parallel evolutionary switches in a number of unrelated families from animal-dispersal to wind-dispersal (or vice versa) correlated with development of the liana habit. In Malpighiaceae, for example, the trees have bird-dispersed berry-like fruits while the lianas are wind-dispersed. In Hippocrateaceae the relatively few trees all have mammal-dispersed fruits whereas most of the lianas have wind-dispersed seeds. Similarly in Apocynaceae many of the trees have fleshy bird-dispersed or mammal-dispersed fruits or seeds but all the lianas have plumed wind-dispersed seeds. Such correlations, like the general prevalence of wind-dispersal in individual liana communities, support the idea that wind-dispersal is especially adaptive for lianas.

De la dispersión mediante el viento, algunas especies han evolucionado a dispersión mediante al agua, especialmente en los manglares y a orillas de los ríos.

Vines, for which I have not compiled actual statistics, also have a diversified array of dispersal types. Autochory, which is rare among lianas, is common, especially among legume vines. Wind-dispersed vine seeds tend to be smaller-bodied and to have their seeds plumed (most Asclepiadaceae and Compositae) rather than winged.

Among epiphytic and hemiepiphytic climbers bird dispersal predominates. Berries with small seeds that pass through a bird's digestive tract or small sticky diaspores which may stick to a bird's feet or feathers (*Peperomia*) are especially well adapted for deposit on the tree branches and upper trunks where such plants must germinate. These groups have a much higher rate of local endemism than do wind-dispersed groups (Gentry 1981).

La dispersión de semillas mediante aves es predominante en las trepadoras epífitas y hemiepífitas. En algunos casos, las aves comen las semillas y, en otros, éstas se adhieren a sus cuerpos.

Coevolution with Herbivores and Predators

Several of the most striking co-evolutionary stories that have been worked out feature climbing plants. Gilbert's (1972) Cucurbitaceae-*Heliconius* pollination story has already been noted. An even more intricately coevolved interaction has taken place between Heliconiine butterflies, probably the largest and most complex Muellerian mimicry complex in the tropics, and their obligate foodplants of the Passifloraceae. Gilbert (1975) has suggested that the incredible diversity of leaf forms shown by Passifloraceae results from selective pressures exerted by egg-laying female *Heliconius* butterflies that use visual cues like leaf shape, as well as chemical ones, in locating oviposition sites. Benson (1979) has shown how intricately interrelated these butterflies and vines are in most neotropical communities, with many *Heliconius* species specific to a single *Passiflora* species. Another of the best known Muellerian mimicry complexes in the neotropics is the group of swallowtails that feed on *Aristolochia* as larvae. The single representative of this complex that reaches the United States is well known as the distasteful model for a whole assemblage of mimicry systems; in the tropics the system is much more complicated with many species of pipevine swallowtails all feeding as larvae exclusively on *Aristolochia* vines. In light of the similar pattern seen in passionflower vines, it seems likely that the unusual diversity of leaf form in *Aristolochia* is more than coincidence.

Many other prominent members of the neotropical insect fauna are intimately interrelated with other vines and lianas. The ithomiine butterflies, yet another large and important Muellerian mimicry complex, are specific to Solanaceae, a group with a large representation of lianas. The leaf-mining buprestid genus *Hyleogea* has 15 species in Panama, all but one of which are specific to Bignoniaceae lianas (Hespenheide, pers. comm.). The chief food plants of the *Morpho* butterflies, so conspicuous as to be considered emblematic of the neotropics, are lianas of the genera *Machaerium* and *Paragonia*. The conspicuous day-flying *Urania* moth feeds exclusively on the liana *Omphalea* (N.Smith, pers. comm.).

La fauna de insectos tropicales está estrechamente relacionada con las enredaderas y lianas. El mimetismo y los atributos químicos están frecuentemente asociados.

It is tempting to conclude from the wealth of such coevolutionary stories that involve lianas that the liana habit, in general, is somehow uniquely related to the intricate biotic interactions that are so much a characteristic of the tropics. Perhaps lianas are preadapted for the development of coevolutionary radiations due to their need to emphasize specific kinds of biochemical defenses rather than such generalized defenses as the tannins and sclerified tissues that typify many tropical trees. This would seem a logical correlation of a life history featuring rapid growth and thus the general prevalence of soft young edible vegetative parts. In any case, lianas play a major role in tropical ecosystems.

Summary

The main generalization is that the climbing habit is a successful one that has evolved independently many times in many different tropical plant groups. Different vine groups have evolved different specialized adaptations for climbing with the most successful liana families characterized by various kinds of tendril and anomalous xylem formation in their stems. Wind dispersal characterizes a majority of free-climbing lianas, while bird dispersal characterizes hemiepiphytic climbes. Vines and lianas are an integral part of neotropical and Panamanian forest communities, and their role in the complex, co-evolved biotic interactions that characterize the neotropics seems especially important.

The study of lianas has been neglected. An example of the misconception that can result from such neglect is the recent suggestion of Jordan & Murphy (1978) that temperate trees are more efficient than tropical ones because the ratio of leaf biomass to wood biomass is greater in tropical than in temperate forests. This discrepancy is surely due largely to the different structures of the two kinds of forest, with lianas and their intrinsically high ratios of leaf to wood biomass accounting for most of the cited differences (Gentry 1983c). Failure to appreciate the importance of lianas led Jordan and Murphy to misinterpret their data.

Like other components of tropical forests, lianas are threatened by widespread habitat destruction. However, Jacobs (1974) has pointed out that lianas often are subjected to additional deliberate persecution by foresters who tend to regard then as weeds because they compete with trees and cause damage during logging (Fox 1968). Accordingly Jacobs (1974) considers lianas to be "the most threatened life form amongst plants".

High diversity and intrinsic evolutionary interest, lack of knowledge, complex and fascinating biotic interactions, and urgent conservational imperatives are among the most cogent arguments advanced to support additional study of tropical forests. From any of these perspectives lianas are especially worthy of additional study. In this context it is gratifying to see the recent interest among tropical biologists in lianas as important components of tropical forests.

Acknowledgments

I thank Francis Putz for reviewing the manuscript and the National Geographic Society and the National Science Foundation for supporting the research which made it possible.

Anderson, W. 1979. Floral conservation in neotropical Malpighiaceae. Biotrop. 11: 219–223.

Benson, W.W. 1979. Resource partitioning in passion vine butterflies. Evolution 32: 493–518.

Charles-Dominique, P. 1977. Ecology and Behavior of Nocturnal Primates. Duckworth's, London.

Croat, T.B. 1978. Flora of Barro Colorado Island. Stanford Univ. Press, Stanford, CA.

Darwin, C. 1867. On the movements and habits of climbing plants. J. Linn. Soc. (Bot.) 9: 1–118.

Dodson, C. & A. *Gentry*. 1978. Flora of the Río Palenque Science Center. Selbyana 4: 1–628.

Ehrlich, P.R. & L. *Gilbert*. 1973. Population structure and dynamics of the tropical butterfly *Heliconius ethilla*. Biotrop. 5: 69–82.

Emmons, L. H. & A. H. *Gentry*. 1983. Tropical forest structure and the distribution of gliding and prehensile tailed vertebrates. Am. Nat. 121: 513–524.

Fox, J.E.D. 1968. Logging damage and the influence of climber cutting prior to logging in the lowland dipterocarp forest of Sabah. Mal. Forest. 31: 326–347.

Gentry, A.H. 1974. Coevolutionary patterns in Central American Bignoniaceae. Ann. Missouri Bot. Gard. 61: 533–537.

———. 1976. Bignoniaceae of southern Central America: distribution and ecological specificity. Biotrop. 8: 117–131.

———. 1979. Distribution patterns of Neotropical Bignoniaceae. Pp. 339–354, *In* K. Larsen & L. Holm-Nielsen, eds., Tropical Botany. Academic Press, London.

———. 1981. Distributional patterns and an additional species of the *Passiflora vitifolia* complex: Amazonian species diversity due to edaphically differentiated communities. Plant Syst. Evol. 137: 95–105.

———. 1982a. Phytogeographic patterns in northwest South America and southern Central America as evidence for a Choco refugium. Pp. 112–136, *In* G. Prance, ed., The Biological Model of Diversification in the Tropics. Columbia Univ. Press, New York.

———. 1982b. Patterns of neotropical plant species diversity. Evol. Biol. 15: 1–84.

———. 1983c. Lianas and the "paradox" of contrasting latitudinal gradients in wood and litter production. Trop. Ecology. 24: 63–67.

Gilbert, L. 1972. Pollen feeding and the reproductive biology of *Heliconius* butterflies. Proc. Nat. Acad. Sci. 69: 1403–1407.

———. 1975. Ecological consequences of a coevolved mutualism between butterflies and plants. Pp. 210–240, *In* L. Gilbert & P. Raven, eds., Coevolution of Animals and Plants. Univ. Texas Press, Austin.

———. 1979. Food web organization and the conservation of neotropical diversity. Pp. 11–33, *In* O. Soule & B. Wilcox, eds., Conservation Biology: an Evolutionary Ecological Perspective. Sinauer, Sunderland, ME.

Hlakik, A. 1974. Importance des lianes dans la production foliaire de la foret équatoriale du Nord-Est du Gabon. Compt. Rend. Acad. Sc. Paris 278; sér. D, 2527–2530.

———. 1978. Phenology of leaf production in rain forest of Gabon: distribution and composition of food for folivores. Pp. 51–71, *In* G. Montgomery, ed., The Ecology of Arboreal Folivores. Smithsonian Instit. Press, Washington, DC.

Jacobs, M. 1974. The study of lianas. Flora Males. Bull. 29: 2610–2618.

Janzen, D. 1968. Reproductive behavior in the Passifloraceae and some of its pollinators in Central America. Behavior 32: 33–48.

——— & R. *Liesner*. 1980. Annotated checklist of plants of lowland Guanacaste Province, Costa Rica, exclusive of grasses and non-vascular cryptogams. Brenesia 18: 15–90.

Jordan, C.F. & P.G. *Murphy*. 1978. A latitudinal gradient of wood and litter production and its implication regarding competition and species diversity in trees. Am. Midl. Nat. 99: 415–434.

Montgomery, G.G. & M.E. *Sunquist*. 1978. Habitat selection and use by two-toed and three-toed sloths. Pp. 329–359, *In* G. Montgomery, ed., The Ecology of Arboreal Folivores. Smithsonian Instit. Press.

Ogawa, H., K. *Yoda*, K. *Ogino* & T. *Kira*. 1965. Comparative ecological studies on three main types of forest vegetation in Thailand. II. Plant biomass. Nature Life SE Asia 4: 49–80.

Putz, F.E. 1980. Lianas vs. trees. Biotrop. 12: 224–225.

———. 1982. Natural History of Lianas and their Influences on Tropical Forest Dynamics. Ph.D. thesis, Cornell University.

Schenck, H. 1892. Beitrage zur Biologie und Anatomie der Lianen im Besonderen der in Brasilien einheimischen Arten. Bot. Mitt. Tropen 4: 1–253.

Strong, D.R. & T. *Ray*, Jr. 1975. Host tree location behavior of a tropical vine *(Monstera gigantea)* by skototropism. Science 190: 804–806.

The Distribution of Panama's Pteridophytes

David B. Lellinger

Actualmente, la rica flora de helechos de Panamá incluye 687 especies o variedades, 51% de ellas restringidas a una sola zona. Los helechos existen en todas las regiones del país, pero están mayormente representados en los bosques premontanos. El porcentaje de especies comunes con Costa Rica es de 88%, mientras que con Colombia es de 63%, especialmente en las áreas fronterizas en donde Panamá comparte hábitat con esos países.

Los hábitat acuáticos presentan algunos helechos: Acrosticum que crece detrás de los manglares, Lygodium y algunas especies de Thelypteris subg. Meniscum que habitan pantanos asociados a los ríos en las tierras bajas. Ocasionalmente Azolla, Ceratopteris, Marsilea, y Salvinia se encuentran en charcos, lagos y quebradas que, característicamente, carecen de corrientes fuertes.

At latest count, the pteridophyte flora of Panama contains 687 species and infraspecific taxa. This number is correlated with the diversity of altitudes and physiography, which is less than that of Costa Rica (estimated to have about 800 species of pteridophytes) and more than that of Colombia's Departmento del Choco (393 species recorded in 1975).

Information on the physiography, climate, and soils of Panama as related to its vegetation was aptly summarized by Porter (1973). To place the various pteridophyte species into a classification of vegetation, I have found it convenient in part to utilize Holdridge's life zone system (Holdridge et al. 1971, pp. 6–17)—despite Myers' (1969, pp. 9–12) critique—plus portions of Page's classification (1979). Holdridge and Budowski (1959) have published a map summarizing this classification for Panama.

Pteridophytes occur in almost all habitats found in Panama. Of course, they are least frequent in aquatic and xeric habitats and most frequent in mesic forests, especially pre-montane cloud forests.

Aquatic habitats: Very few ferns are adapted to aquatic habitats, which are characterized by strong sunlight. At sea level behind mangrove swamps grow *Acrostichum aureum* and *A. danaeifolium*. The latter species also grows in open, brackish and fresh water swamps above sea level. Lowland swamps associated with rivers often occur inland from the mangrove swamps and are home to *Lygodium* and species of *Thelypteris* subg. *Meniscium*. Ponds and slow-moving streams, especially below 200 m, have *Azolla*, *Ceratopteris*, *Marsilea*, and *Salvinia*, although none of these is common.

Xeric habitats: Because of the pronounced dry season on the Pacific side of Panama, large areas are now in grassland or savanna. Man's activities have extended and maintained such areas, which are generally inhospitable to ferns. Some species persist in seasonally riparian forests, where *Adiantum* is especially common. The savannas of Panama contain several species of *Anemia*, which often occur together and so yield taxonomically confusing hybrid plants. In areas of bare ground, whether natural or man-made, *Gleichenia* and *Dicranopteris* occur as pioneers or near-pioneers. *Hypolepis* and *Pteridium* also occur. The scrambling habit has developed in both the Gleicheniaceae and the Dennstaedtiaceae, as have long-creeping rhizomes, both characters which contribute to success in colonizing disturbed ground. A special fern flora develops on rocky exposures in dry country savannas and light forests. Most of the species are cheilanthoid ferns, especially *Cheilanthes* and *Notholaena*. Cheilanthoid ferns are best represented in the dry regions of Mexico, and a few species apparently are moving southward through Central America, presumably as aridity there increases due to climatic change or deforestation. Luis Diego Gomez and Bruce McAlpin both have found cheilanthoid species new to Costa Rica and Panama in the Nicoya and Chiriqui regions in recent years.

En las áreas secas de Panamá algunas especies de Adiantum existen en bosques "ribereños" y varias especies de Anemia cohabitan las sabanas del Pacífico. Gleichenia, Dicranopteris, Hypolepis y Pteridium son especies pioneras. Rocas expuestas y secas albergan Cheilanthes y Notholaena.

Mesic habitats: Because of the lack of a pronounced dry season, mesic habitats have a great number of pteridophyte species. Holdridge's life zones are a convenient way to divide the mesic habitats in Panama. In general, the life zones correlate well with elevation: lowland tropical forests (0–500 m), premontane forests (500–1500 m), lower montane forests (1500–2500 m), montane forests (2500–3250 m), and subalpine areas (3250 m, according to Myers, 1969, p. 3).

Lowland tropical forests occur from sea level to about 500 m elevation. In Panama they are moist or very moist (especially on the Caribbean side) with a small portion that is dry (Pacific side), or they are part of a transition between the two types (Darien, Pacific side). As far as the pteridophytes are concerned, lowland tropical moist forests are divided into two distinct habitats, terrestrial and epiphytic. The forest floor, at least in undisturbed places, is likely to be very dark. Some dark green or even bluish-green species of *Danaea* and *Selaginella* occur. The ferns here and in less dense forests often are pinnate or pinnate-pinnatifid and have broad pinnae, presumably an adaptation for intercepting the maximum amount of light. On steep slopes or where trees have fallen, considerable clay soil is exposed. The rhizomes of terrestrial species, even small ones like *Trichomanes*, often are erect, enabling the plants to survive floods of mud around their roots. In addition, many species, especially of *Diplazium* and *Tectaria*, reproduce by plantlets budded from the rachis; reproduction by spores is not likely to be successful, for gametophytes are readily buried by mud before sporophytes can be produced. Various genera of tree ferns, especially *Cyathea* and *Trichopteris*, grow at forest margins.

Bosques húmedos de tierras bajas proveen hábitat terrestres y epífitos. Danaea, Selaginella y Trichomanes crecen sobre el suelo del bosque, especialmente en donde hay poca luz. Las especies de estos géneros presumiblemente están adaptadas para interceptar la máxima luz posible y algunas especies de Trichomanes pueden sobrevivir inundaciones de lodo alrededor de sus raíces, debido a que sus rizomas son frecuentemente erectos. Algunas veces se encuentran epífitas sobre troncos de árboles y el género Campyloneurum se establece sobre las ramas de éstos.

Trunk epiphytes occur, but not in abundance. *Campyloneurum* tends to perch in crotches where major tree branches join the trunk. Various species of *Trichomanes*, especially those in subg. *Didymoglossum*, grow in fissures of the bark. Those found in the deepest shade often are dark green like their terrestrial counterparts. A few *Trichomanes* species are hemiepiphytes, remaining rooted in the ground but with rhizomes long-creeping up tree trunks; their laminae, which are one cell thick, are adherent to the smooth tree trunks on which they grow. Perhaps they catch water flowing down the tree trunk, or perhaps they lose less water by having one surface of the lamina adherent.

Branch epiphytes can withstand more sunlight and drying and grow in more exposed places than do trunk epiphytes. The diversity of branch epiphytes in lowland tropical forests is less than at higher elevations. *Dicranoglossum* and *Microgramma* are typical.

Pre-montane forests occur above the lowland tropical forests, from about 500 m to 1500 m elevation. In Panama they range from very wet (mostly Caribbean slopes plus the highlands adjacent to Colombia) and very moist (mainly Pacific slopes plus the highlands adjacent to Colombia) to moist and dry (Pacific slopes of the Provinces Cocle, Herrera, and adjacent Los Santos). Among the pteridophytes, terrestrial species and both trunk and branch epiphytes are most diverse in pre-montane forests, especially in regularly cloudy areas (Sugden & Robins 1979). *Blechnum occidentale* and *Adiantum concinnum* are common along roadsides and in disturbed areas in well drained soil. *Pteridium aquilinum* invades after deforestation, especially after burning, and can be a serious weed.

Los bosques premontanos, desde 500 a 1500 m, albergan la mayor diversidad de helechos, especialmente en las áreas nubladas.

Large species in *Blechnum* subg. *Lomaridium*, *Elaphoglossum*, and *Niphidium* are especially common trunk epiphytes. However, genera with small fronds, like *Oleandra*, *Polypodium* subg. *Goniophlebium*, and *Trichomanes* also occur. *Peltapteris* and various genera of the Polypodiaceae are especially common branch epiphytes.

Lower montane forests occur at 1500–2500 m. In Panama, these forests are moist, wet, or very wet and occur at or near the top of the cordillera extending east from Costa Rica and in the highlands adjacent to Colombia. On windward slopes, clouds drift in and are a more or less continual source of water. The very wet parts of Panama need more botanical exploration, but living and collecting there are difficult. Fewer species are known to occur than in pre-montane forests. Terrestrial ferns include various species of *Blechnum* subg. *Lomaria* and *Gleichenia*. Tree trunks support many species of *Asplenium* and *Polypodium*. *Hymenophyllum* replaces *Trichomanes* at these and higher elevations. Branch epiphytes are abundant and include *Elaphoglossum*, *Peltapteris*, and *Polypodium* subg. *Marginaria*.

Los bosques montanos bajos, 1500–2500 m, son áreas húmedas y se extienden hacia Costa Rica y Colombia. Éstos tienen muchos helechos, pero indudablemente, menos que el bosque premontano.

Montane forests have far fewer species of pteridophytes because of their harsher, colder environment at 2500–3250 m. They occur at the top of the higher portions of the cordillera. In Panama, these forests are always wet. At the edges of montane forests, several genera of the Cyatheaceae and *Pteris* species that are large and tree-like occur, along with *Histiopteris* and *Lophosoria*, both of which also are found in the Andes and in Costa Rica. There is less distinction between trunk and branch epiphytes than at lower elevations. Both provide a similar, moss and liverwort-festooned habitat. Fewer pteridophyte species occur than at lower elevations. The division between terrestrial and epiphytic habitats also tends to break down at this elevation. When a branch falls to the ground, the plants often continue to grow; this is seen especially in *Elaphoglossum*. Probably the mossy branch or trunk epiphytic habit is necessary for gametophyte establishment, but the established sporophytes can continue to grow terrestrially.

El bosque montano, por encima de 2500 m, presenta menos especies, pero algunas habitan por debajo de 3200 m en donde las condiciones subalpinas excluyen casi todos los helechos, exceptuando a Plecosorus speciosissimus, a Lycopodium contiguum, a Pityrogramma chrysoconia y a Grammitis moniliformis.

Subalpine areas occur in Panama only at the very top of Volcan Chiriqui, or Baru. Such areas in Costa Rica are mostly above 3500 m, although the top of Baru is about 250 m below this. The vegetation is shrubby with patches of grasses, and ferns are very few. *Plecosorus speciosissimus*, *Lycopodium contiguum*, and *Pityrogramma chrysoconia* occur terrestrially here and in Costa Rica. *Grammitis moniliformis* has been collected as a rock grower, or epipetric plant, rather than as an epiphyte, as at lower elevations.

The numerical distribution of pteridophytes in Panama is summarized in Figure 1. The following conclusions can be drawn from the figure and its supporting data: (1) Of 687 species or infraspecific taxa, 353 (51%) occur in a single zone and 334 (49%) occur in two or more zones. About half the pteridophytes, therefore, have a fairly wide ecological amplitude. (2) The greatest number of pteridophytes occurs in the pre-montane zone, more than in the tropical zone and many more than in the lower montane zone. (3) Only 36 (5%) of the species or infraspecific taxa occur in three zones or in two non-adjacent zones. Few pteridophytes have a wide ecological amplitude, and of these some are aggressive and weedy.

The percentages of pteridophyte species and infraspecific taxa of Panama also found in parts of the New World are shown in Figure 2. The percentage in common with Costa Rica—88%—is by far the highest; the two countries share continuous habitats across their common border. Colombia is second, with 63%. For the most part, the percentage of species in common remains high as far north as Mexico and as far south and east as Peru and Venezuela. The percentages are similar to those for Choco pteridophytes (Lellinger 1975, p. 113), although Panama has more taxa in common with Mexico and Guatemala than has Choco, while it has fewer in common with Colombia, Venezuela, Colombia, Ecuador, Peru, and the Greater and Lesser Antilles. The affinities of the Panamanian fern flora are more decidedly with Mexico and Central America, whereas the affinities of the Choco flora are less so.

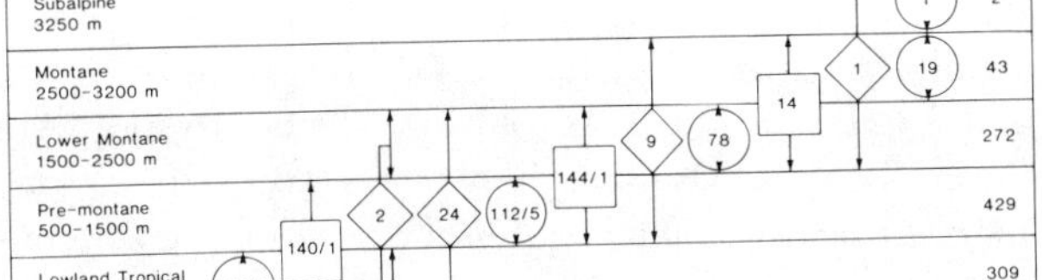

Fig. 1. Distribution of Panamanian pteridophytes into tropical, pre-montane, lower montane, montane and subalpine life zones.

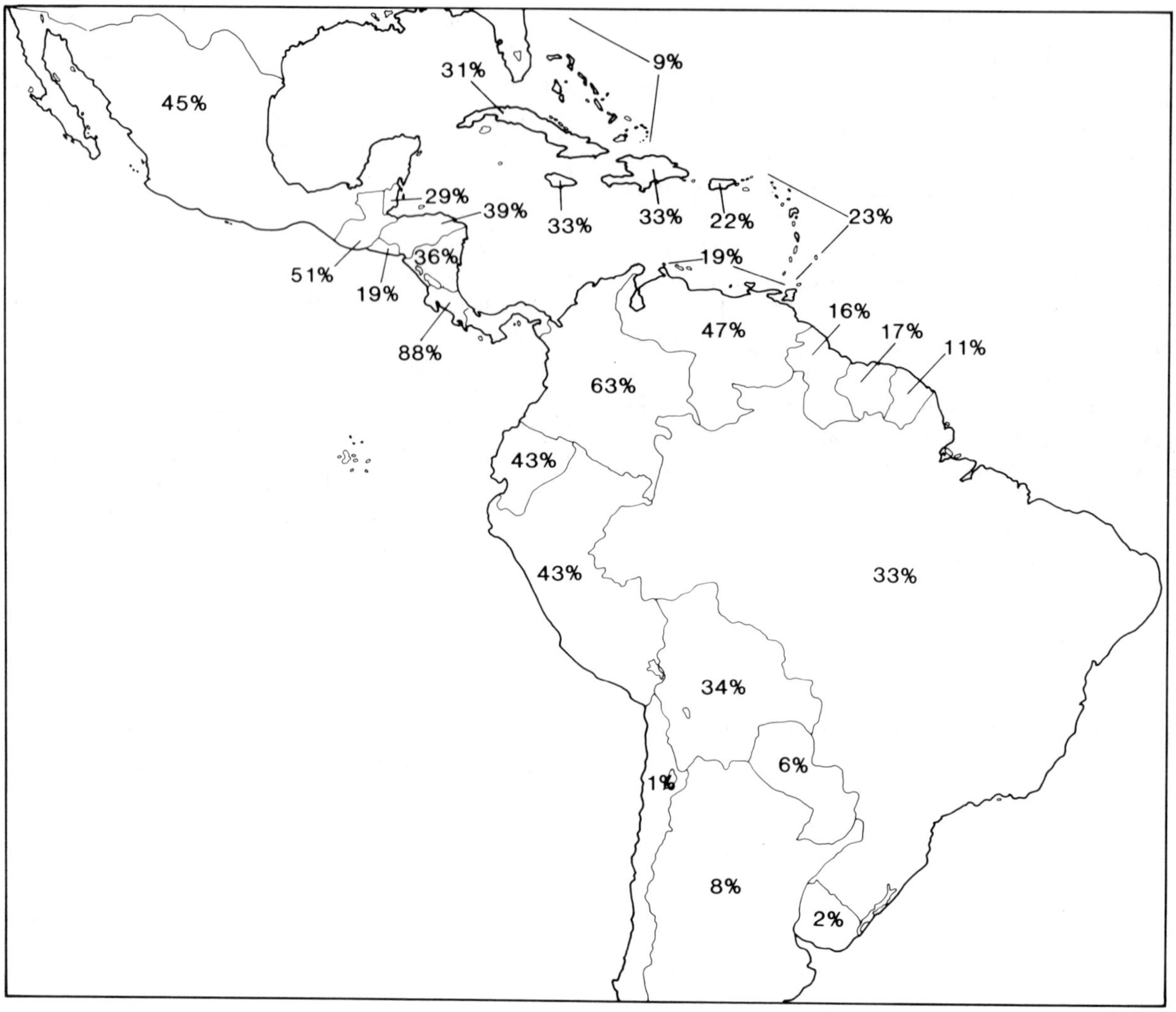

Fig. 2. Map showing percentage of Panamanian pteridophytes also known from other countries or Antillean island groups of the New World.

Panamá posee muchos helechos en común con Costa Rica (88%) y con Colombia (63%), sin embargo Panamá posee más especies en común con México y Guatemala que con el Chocó de Colombia.

Panama differs from Costa Rica and the Choco in having few endemic species, only 12 out of 687 taxa. This is only 2%, a figure comparable to a continental flora, for example, Texas (Smith 1972, p. 6), and much smaller than the 11 or 12% to 71% recorded by Smith and by Tryon (1979, p. 58) for island pteridophyte floras. Panama also falls between Tryon's (1972) primary Andean regional center of pteridophyte concentration, which includes the Choco, and his secondary Central American center, which includes Costa Rica, but not Panama. Five of the endemics are found in the tropical lowlands (*Adiantum obliquum* x *petiolatum*, *Cyathea conformis*, *Thelypteris clypeatum*, and *Pleopeltis panamensis*), one is found in lowland and pre-montane forests (*Thelypteris cumingiana*), five are exclusively pre-montane (*Asplenium psilacrum*, *Cnemidaria cocleana*, *Cyathea impar*, *Grammitis podocarpa*, and *Oleandra panamensis*), and one is both pre-montane and lower montane (*Cnemidaria mutica* var. *chiricana*). Four of the twelve are members of the Cyatheaceae. Tryon and Gastony (1975) suggest that local endemics, which comprise 44% of the species in the Cyatheaceae, are ecologically specialized and have evolved as a result of local speciation in a stable mosaic of slightly different environments. Six of the endemics are terrestrial, five epiphytic, and one aquatic. Perhaps the preponderance of endemics at low elevations, which is the reverse of the situation prevailing in Costa Rica, is a result of new habitats in the recently emerged lowlands.

Panamá sólo tiene 12 especies endémicas, mucho menos que Costa Rica o que el Chocó y algo menos que la mayoría de las islas. Cinco de las endémicas son de tierras bajas y cuatro son árboles de la familia Cyatheaceae.

Holdridge, L.R., et al. 1971. Forest Environments in Tropical Life Zones, A Pilot Study. Pergamon Press, Oxford & New York.

——— & G. *Budowski*. 1959. Mapa ecológico de Panamá (1:1,000,000). Instituto Interamericano de Ciencias Agrícolas, Turrialba, Costa Rica. (Reprinted as plate 27 in the Atlas Nacional de Panamá.)

Lellinger, D.B. 1975. A phytogeographic analysis of Chocó pteridophytes. Britt. Fern Gaz. 11: 105–114.

——— & *Myers*, C.W. 1969. The ecological geography of cloud forest in Panama. Amer. Mus. Novitates 2396: 1–52.

Page, C.N. 1979. The diversity of ferns, an ecological perspective. Chapter 2 *In* A.F. Dyer, ed. The Experimental Biology of Ferns. Academic Press, London & New York.

Porter, D.M. 1973. The vegetation of Panama: a review. *In* A. Graham, ed. Vegetation and Vegetational History of Northern Latin America. Elsevier, Amsterdam, London & New York.

Smith, A.R. 1972. Comparison of fern and flowering plant distributions with some evolutionary interpretations for ferns. Biotropica 4: 4–9.

Sugden, A.M. & R.J. *Robins*. 1979. Aspects of the ecology of vascular epiphytes in Colombian cloud forests, I. The distribution of the epiphytic flora. Biotropica 11: 173–188.

Tryon, R.M., Jr. 1972. Endemic areas and geographic speciation in tropical American ferns. Biotropica 4: 121–131.

———. 1979. Biogeography of the Antillean fern flora. Chapter 4 *In* D. Bramwell, ed. Plants and Islands. Academic Press, London & New York.

—— & J.G. *Gastony*. 1975. The biogeography of endemism in the Cyatheaceae. Britt. Fern Gaz. 11: 73–79.

Distribución y Diversidad de la Flora de Musgos de Panamá

Noris Salazar A. and Marshall R. Crosby

A survey of the literature and holdings at the bryophyte herbarium of the University of Panama yielded 304 species of mosses occurring in Panama, 32 new reports. Most reports are from Chiriqui, followed by the province of Panama, which reflects collecting interest. Five species are restricted to Panama, many fewer than were reported earlier.

Este trabajo es un estudio general, preliminar, de la flora de musgos de Panamá. Los datos presentados se basan, en gran medida, en informaciones obtenidas de la revisión de la literatura y de los datos anotados en las etiquetas de los especímenes del Herbario de Briófitas de la Universidad de Panamá. La revisión bibliográfica se basa principalmente en los trabajos de Welch (1974), Crosby (1969, 1971) y Breen & Reese (1971).

This study is based mainly upon the available literature and data from specimens at the moss herbarium at the University of Panama (PMA).

Indudablemente, una revisión cuidadosa de los especímenes que reposan en otros herbarios del mundo permitirá una apreciación más exacta de la flora briológica del Istmo y, quizás, brinde nuevas adiciones a la misma.

El número total de especies, que es de 304, se basa en el recuento de todas aquellas reportadas hasta el presente, aún cuando se desconozca la localidad de recolecta. Treinta y dos de las trescientas cuatro especies no han sido reportadas oficialmente y son parte de un listado elaborado por los autores en el transcurso de sus estudios taxonómicos (Cuadro 1).

The total yield was 304 species, 32 of these new reports.

A fin de obtener una idea general de la distribución de especies de acuerdo a la altitud, se ideó una escala arbitraria que comprende desde el nivel del mar hasta alturas de más de 2600 m. La escala fue diseñada de la siguiente manera: para baja altitud se consideraron alturas entre el nivel del mar y los 700 m, para altitud media de 800 a 2600 m y para la alta, 2600 m o más (Fig. 1).

Distribution of species is correlated with 3 arbitrarily established elevational ranges: low (0–700 m), medium (800–2600 m), and high (above 2600 m).

El mayor número de especies, más de la mitad, se encuentra en Chiriquí. La provincia de Panamá también está representada por un número considerable de especies con relación al resto de las otras provincias (Fig. 2).

La proporción relativamente alta de especies reportadas de estas provincias se debe al hecho de que las mismas han sido y son sitios escogidos de recolecta por la riqueza y accesibilidad de las áreas de interés. La provincia de Chiriquí, por ejemplo, posee una variada topografía con una diversidad de hábitat, desde tierras a nivel del mar hasta los 3475 m de altura en la cima del Volcán Barú y es, quizás, la provincia, aparte de Panamá y Colón, de mayor afluencia turística, particularmente en las áreas de clima fresco, como lo son el Volcán, Boquete, Cerro Punta y regiones adyacentes.

The relatively high proportion of reported species in some provinces (Chiriqui and Panama) may be due to their richness and accessibility.

Herrera y Los Santos son provincias con un alto grado de deforestación y utilización de la tierra en actividades de pastoreo. Estas condiciones son poco

CUADRO 1
NUEVAS ADICIONES A LA FLORA DE MUSGOS DE PANAMA

Barbella cubensis (Mitt.) Broth.
Darién: Ladera oeste del macizo del Tacarcuna. *Mori & Gentry 4419* (MO, PMA).

Brachymenium wrightii (Sull.) Broth. var. *mnioides* (Besch.) Florsch.
Darién: Cerro Pirre. *Bristan 1241* (NY).

Breutelia tomentosa (Brid.) Jaeg.
Chiriquí: Cerro Punta, a 600 m de la escuela, después del puente. *Salazar et al. 1562* (PMA).

Callicostella oerstediana (C. Muell.) Jaeg.
Panamá: Isla de Barro Colorado. *Montalvo 380* (PMA).—det. Dana Griffin.

Cryphaea patens Hornsch.
Chiriquí: Cerro Punta. *Salazar 485* (PMA).—det. Dana Griffin.

Cyclodictyon varians (Sull.) O. Kuntze
Chiriquí: NE del campamento de Fortuna. *Correa et al. 2456* (MO, PMA).
Panamá: Bayano. *Salazar 990* (PMA).

Dicranoloma brittonae Bartr.
Panamá: Cima de Cerro Jefe. *Crosby 10903* (MO, PMA).

Dicranium frigidum C. Muell.
Chiriquí: Cima del Volcán Barú. *Miranda 2c* (MO, PMA); *Salazar & Béliz 980, 985* (PMA); *Maxon 5309* (MO, PMA).
Vecinidades de Las Nubes al oeste de Cerro Punta. *Croat 22428A* (MO).
Darién: Localidad de Caná. *Williams 5309* (MO).

Fissidens weirii Mitt.
Chiriquí: a 5 m del río Caldera. *Salazar et al. 843* (PMA).

Holomitrium falcatum Bartr.
Panamá: Cerro Campana. *Crosby 4365* (MO, PMA).

Holomitrium longifolium Hampe
Chiriquí: Trocha 3 de noviembre. *Salazar 639a* (PMA). Cima del Volcán Barú. *Salazar & Béliz 966* (PMA). Las Cumbres near Cerro Punta. *Croat 16092* (MO, PMA).
Panamá: Cerca de la cima de Cerro Jefe. *Stimson 5413* (MO).

Holomitrium pulchellum Mitt.
Chiriquí: Palo Alto, Boquete. *Salazar & Wilson 972* (PMA).

Hookeriopsis cruegeriana (C. Muell.) Jaeg.
Chiriquí: NE del sitio del campamento en Fortuna. *Salazar et al. 579* (PMA).

Hookeriopsis subfalcata (Hampe) Jaeg.
Chiriquí: Trocha 3 de noviembre entre Alto Guayabo y la División Continental. *Salazar 930* (PMA). A lo largo de la trocha Holcomb hacia Cerro Horqueta, N de Boquete. *Crosby 4017, 4024A* (MO).
Panamá: Cerro Jefe, *Croat 13100A* (MO).

Hookeriopsis variabilis (Mitt.) Jaeg.
Panamá: Isla Majé, Bayano. *Salazar 739A* (PMA).

Isopterygium diminutivum Bartr.
Chiriquí: Santa Marta, Bugaba. *Salazar & Vinda 1521* (PMA).

Lepidopilum stolonaceum C. Muell.
Colón: Sta. Rita Ridge. *Crosby 10370* (MO).

Lepyrodontopsis trichophylla (Hedw.) Broth.
Veraguas: Valle de riachuelo, 10.4 km a lo largo de la carretera a la Escuela Agrícola Alto de Piedra. *Crosby 10131* (MO, PMA).
Coclé: N of El Valle de Antón. *Crosby 4433* (MO).

Leucobryum polakowskyi (Besch.) Card.
Chiriquí: NO del campamento de Fortuna. *Salazar et al. 527* (PMA).

Leucomium lignicola Spruce
Bocas del Toro: NE del campamento Changuinola 1. Cerro Brachá. *Salazar et al. 1654* (PMA).
Chiriquí: NO del campamento de Fortuna. *Salazar et al. 533* (PMA).

Leucophanes calymperatum C. Muell.
Colón: A lo largo del camino al final de la ruta Santa Rita Ridge Road. *Crosby 10322* (MO).
Veraguas: Valle de riachuelo, 10.4 km hacia el NO desde la Escuela Agrícola alto de Piedra. *Crosby 10160* (MO).

Macromitrium stratosum Mitt.
Chiriquí: Palo Alto, Boquete. *Salazar 842* (PMA).

Neckera ehrenbergii C. Muell.
Chiriquí: El Aguacatal, Volcán. *Salazar 1142* (FLAS, NY, PMA).

Orthodontium pellucens (Hook.) B.S.G.
Panamá: Al NE de la cima de Cerro Jefe. *Mori & Kallunki 3621* (MO).

Papillaria penicillata (Dozy & Molk.) Broth.
Chiriquí: Cerro Punta. *Salazar 485 (PMA)*. Las Cumbres near Cerro Punta. *Croat 13645a* (MO).

Papillaria wagneri Lor.
Localidad desconocida. *Wagner 1858* (NY).

Plagiothecium scalpellifolium (C. Muell.) Bartr.
Colón: Ladera izquierda del río Guanche, SO de Portobelo. *Crosby 10778* (MO, PMA).

Pogonatum robustum Mitt.
Chiriquí: Las Cumbres. *Croat 16153, 16081A, 16160B* (MO, PMA). Cerro Pando. *Mori & Bolten 7283* (MO, PMA). Boquete, sendero a Cerro Punta. *Tyson 6610* (MO, PMA).

Porotrichodendron superbum (Tayl.) Broth.
Chiriquí: Palo Alto, Boquete. *Salazar 831* (PMA).

Prionodon dichotomus Hampe
Chiriquí: Las Cumbres near Cerro Punta. *Croat 16091B* (MO).

Sphagnum magellanicum Brid.
Darién: Cerro Malí. *Mori & Gentry 4353* (MO).
Panamá: Cerro Jefe en el camino que lleva al norte desde la cima. *Witherspoon 8506; Mori et al. 3762* (MO, PMA).
Veraguas: Cerro Tute. *Mori et al. 7604* (MO, PMA).

Thamniopsis pendula (Hook.) Fleisch.
Darién: Al Oeste del macizo del Tacarcuna. *Crosby 10434* (MO).

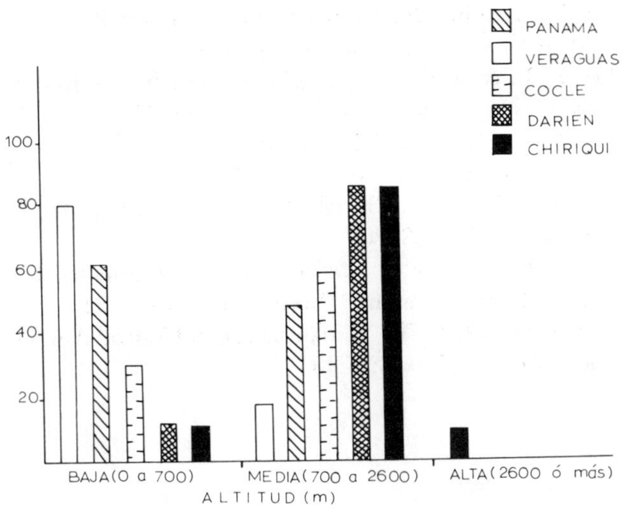

Fig. 1. Distribución de las especies por provincia segun la altitud. Alta 2600 m o más, Media 800 a 2600 m y Baja desde el nivel del mar (cero metros) hasta los 700 m. Los números indican el total de géneros especies para cada provincia.

favorables para el crecimiento de briófitas. La recolecta en las áreas montañosas de estas provincias podría dar mayor información sobre la vegetación de musgos que, probablemente, se extendía, hace algunos años, mucho más allá de su perímetro actual.

In Herrera and Los Santos, partly due to deforestation, mosses are poorly represented. In San Blas the only known specimen was collected at the coast.

En cuanto a San Blas, el único especimen que se tiene es del área de la costa. Se desconoce de recolectas en las áreas montañosas que colindan con las provincias de Panamá y Darién.

La relativa inaccesibilidad a ciertas áreas tales como las selvas de la provincia de Bocas del Toro, del norte de Veraguas y del Darién, se reflejan en el número de géneros y especies recolectados hasta el presente. En la última década, con el desarrollo de proyectos hidroeléctricos y viales a nivel nacional, se han abierto nuevas rutas de acceso a sitios botánicamente poco visitados. Esto ha beneficiado a instituciones nacionales asociadas a la investigación científica, así como a grupos de científicos de varias partes del mundo que frecuentemente visitan nuestro país en proyectos relacionados con la flora neotropical.

The inaccessibility of areas such as the forest of Bocas del Toro, the northern part of Veraguas, and the Darien is reflected in the small number of genera and species.

En las áreas del Istmo en donde predominan las tierras bajas con áreas de sabana y vegetación pionera, así como también en regiones de bosque húmedo tropical (Atlas Nacional de Panamá, 1975), se observan con frecuencia géneros tales como *Octoblepharum*, *Syrrhopodon*, *Orthostichopsis* y *Neckeropsis*, entre otros.

La mayor cantidad de las especies reportadas en Chiriquí y Darién se encuentran distribuídas entre los 800 y los 2600 m de altitud (Fig. 1). Esto corresponde, en la escala propuesta, a las regiones de altitud media.

In Chiriqui and Darien most of the species reported are distributed between 800–2600 m elevation. Some species apparently have restricted distribution.

Algunas especies parecen tener una distribución restringida a estas áreas de altitud media, como es el caso de *Prionodon densus* (Hedw.) C. Muell., *Schliephackea meteorioides* (Williams) Broth. y *Zygodon reinwardtii* (Hornsch.) A. Braun, entre otras. Ninguna de estas especies, hasta el presente, ha sido reportada de las zonas de baja altitud.

El número de especies que se conoce sólo de Panamá es de 5, lo cual es mucho más bajo que el de 11 reportado por Crum y Steere (1950). Estas especies son: *Callicostella grossiretis* Bartr. in Willis, *Cyclodictyon maxonii* Williams, *Crossomitrium acuminatum* Bartr., *Hookeriopsis callicostelloide* Herz. et Thér. y *Squamidium chiriquense* Crosby. Algunas de éstas podrían ser endémicas.

More studies and collecting are necessary for exact and detailed information about the richness and distribution of the Panamanian moss flora.

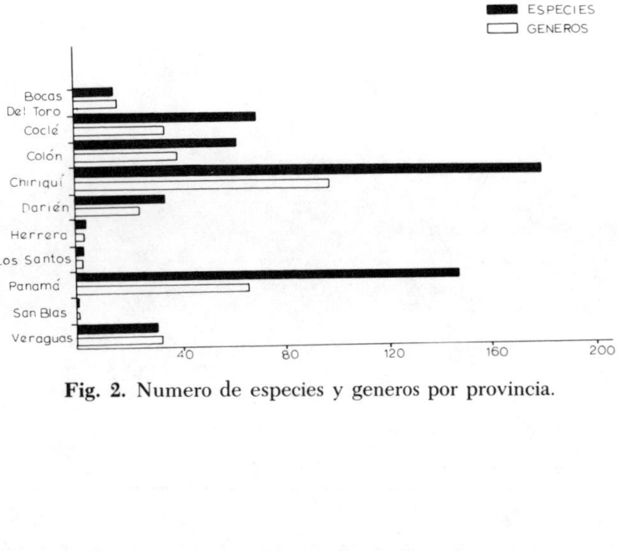

Fig. 2. Numero de especies y generos por provincia.

Es indudable que se necesitan más estudios sobre la flora de musgos de Panamá a fin de obtener una visión más exacta y detallada de su riqueza y distribución. Se espera contar con la colaboración y buena voluntad de las autoridades e instituciones interesadas en este tipo de trabajo de manera que se puedan llevar a cabo recolectas e investigaciones que coadyuven al conocimiento, preservación y conservación de nuestros valiosos recursos naturales.

Atlas Nacional de Panamá. 1975. Mapa de Vegetación. Lámina 28. Comisión del Atlas de Panamá. Ministerio de Educación. Panamá.

Breen, R. & W.D. *Reese*. 1971. A contribution to the muscology of Panama. Bryol. 74: 33–36.

Crosby, M.R. 1969. The mosses reported from Panama. Bryol. 72(4): 513–521.

———. 1971. Additional Panamanian bryophytes. Bryol. 58(2): 258–260.

Crum, H.A. & W.C. *Steere*. 1950. Additions to the moss flora of Panama. Bryol. 53(2): 139–152.

Welch, W. 1974. The Hookeriaceae of Central America. Bryol. 77(3): 328–404.

Structural and Functional Properties of Mangrove Forests

Gilberto **Cintron**, Ariel E. **Lugo**, and Ramon **Martinez**

Los manglares no contienen una flora muy diversa, pero aún así son muy útiles al hombre. Estos pueden clasificarse en tres o seis tipos generales, sin embargo los más importantes son los localizados a orillas de los ríos, los que bordean las costas marinas o los que se desarrollan en cuerpos de agua influenciados por agua de mar. Cada tipo presenta diferentes características de tamaño, de riqueza de especies y de tasa de procesos ecológicos que pueden modificarse bajo condiciones óptimas o extremas. Los manglares del Darién son especialmente grandes y ricos.

All mangrove ecosystems are estuarine and open systems subjected to periodic flooding by sea water and by freshwater. In general, they have a net import of nutrients, freshwater, and sediments from terrestrial environments and a net outflow of organic matter and water to marine or estuarine waters. Mangroves also exchange salts with the sea and organisms with estuarine and marine waters and with adjacent terrestrial ecosystems. These common properties of mangroves can be found in any mangrove stand, but stands differ in the stature of the vegetation, types of animal populations, in complexity and diversity, and in the rate at which processes occur. Exchanges of matter and energy with adjacent ecosystems also vary depending upon the conditions at the site.

Being swampy and salty, mangrove ecosystems do not support a diverse flora. In Darien Province, for example, in an area of 4,164 km^2 of which 284 km^2 were mangroves, a quarter hectare of mangroves had only 19 plant species, but in the same area of Tropical Moist Forest there were 399 species (Golley et al. 1975).

The low floristic richness of mangrove forests does not imply a low value (ecological or otherwise). In fact, mangrove ecosystems are being studied with more interest worldwide because of their economic importance in support of commercial fisheries alone. Because of other intensive uses in Asia and other parts of the world (including production of fuel wood), mangroves are ecosystems that in their natural state have considerable economic value to humans. Ecologically, mangroves contribute to estuarine and terrestrial food chains, to water quality, and to the maintenance of coastal geomorphology.

Many forces act at different rates and intensities on the mangrove environment. The sum of all the forces that act on any ecosystem may be termed its "energy signature" and this "energy signature" dictates the types of organisms that will survive and the speed of ecological processes. It is now difficult to quantify the energy signature of mangrove ecosystems, but Cintrón et al. (1978a) and Lugo et al. (1981) have identified the most important forces altering mangrove structure and developed data to show mangrove response to these external factors. Since the structure and function of mangroves reflects the nature and intensity of these forces, it should be possible to classify mangroves into a few discreet types that correspond to natural forcing functions. This would allow the conceptualization of ecosystems in relation to the natural forces that regulate their behavior. In addition, the approach avoids the tendency to treat all mangroves as if they were uniform. We know that mangroves vary from the giant Panamanian Pacific Coast forests dominated by *Rhizophora*, reaching 40 m in height and containing 571 tons of organic matter per ha (t/ha), to forests of the same species that are only 1 m tall and contain 1/50th the biomass.

Mangroves can be classified by a number of classification systems. Snedaker's (Lugo & Snedaker 1974) system of mangrove classification uses the geomorphologic configuration of the site to separate mangroves into six general types. In this paper we review the structural and functional characteristics of mangrove types and recommend a simplification of the classification system. Mangrove ecologists should use this simple classification to organize structural and functional information on mangroves and as a means of developing insight into forests for which no quantitative data are available but whose general category can be inferred from simple field observation.

Types of Mangrove By Geomorphology

The six types of mangroves suggested by Lugo and Snedaker (1974) were riverine, basin, fringing, overwash, hammock, and dwarf. After comparing structural and functional data for these mangrove types, we delete the overwash mangrove type and incorporate stands previously classified in this category with the fringe mangrove category. An overwash mangrove forest is in fact a fringe mangrove located in an island. Due to a lack of information, this review deals mostly with fringe, basin, and riverine mangroves. Dwarf and hammock mangroves are discussed briefly but not in depth. The types of mangrove forests are compared in Table 1.

Riverine Forests

These mangroves develop along the edges of rivers, often as far as the inland toe of the saline intrusion. The best growth occurs in the lower course of the estuary. Here, water flows are moderate to high and high nutrient inputs prevail. The combination of adequate fresh water and nutrient inputs develops luxuriant stands. The delivery of natural energies to this forest type is termed a power line (Brown et al. 1979) because water flow is confined to a channel or floodway and thus concentrated along a corridor.

On the periphery of the forest the dominant species is *Rhizophora* sp. This is an area of high kinetic energy due to the combination of river flow and tidal motion. The complex root systems of *Rhizophora* trees with their extraordinarily developed adventitious "stilt" roots, support large trees on unstable, soft soils. In the inner riverine forest, where natural energies are diluted, mixed stands of *Laguncularia* sp. and *Avicennia* sp. occur.

The interstitial salinities in riverine forests are generally lower than those of other forest types, usually on the order of 10,000–20,000 parts per million (ppm) or less. In the dry season, when rivers run low, higher salinities are the rule, due to greater influence of the salt intrusions. As rivers swell during the wet season, salt wedges are driven seaward and salts are leached from the sediments.

Fringe Forests

Los manglares que ocurren a orillas de los ríos generalmente se localizan tierra adentro. Combinación de agua dulce y de nutrimentos que varían asociados a las estaciones apoyan este crecimiento.

Fringing mangroves occur along the seaward edge

TABLE 1
CHARACTERISTICS OF MANGROVE TYPES

Type of Mangrove	Energy Delivery	Nutrient Concentrations	Relative Water and Nutrient Turnover	Relative New Primary Productivity	Relative Structural Development
Riverine	power line[1]	Eutrophic	high	highest	highest
Fringes	power front[2] or power point[3]	Oligotrophic	high	moderately high	moderately high
Basins	power sheet[4]	Eutrophic	moderate to low	high to low	high to low
Dwarf	power sheet[4] power front[2] or power point[3]	Oligotrophic	low	lowest	lowest

[1]The main component delivered over a small area as a force moving parallel to the surface.
[2]The main component delivered over a broad area as a force parallel with the surface.
[3]The main component delivered in a concentrated form over a small area perpendicular to the surface.
[4]Main component delivered over a broad area perpendicular to the surface.

of protected shorelines or around overwashed islands. Fringes usually have a strong horizontal gradient in topography as well as a gradient in turbulence and tidal amplitude. Tidal and wave energy dissipates rapidly with distance from shore. Energy is thus delivered to the forest along a frontal area, termed a power front (Brown et al. 1979). In karst regions, fringe mangroves may be found inland where tidal waters emerge through ocean holes. In these cases energy delivery to the forest resembles a power point because it is concentrated on a small area (Brown et al. 1979).

The inland reduction of flushing results in a salinity gradient with higher salinity prevailing inland. Toward the outer edge of the fringe, where greater energy prevails, the large, heavy seedlings of *Rhizophora* become established. In the higher, less flushed parts of the fringe forest, *Avicennia* becomes dominant. The interstitial salinity in *Avicennia* zones in Puerto Rico, Florida, and other Caribbean sites averages 59,000 ppm whereas in the red (*Rhizophora*) mangrove belt, the mean of the interstitial salinities was only 39,000 ppm (Cintrón et al. 1978b).

Basin Forests

Los manglares que se localizan hacia el mar, protegidos por la línea de la costa, o alrededor de islas, presentan en la parte externa, donde la energía es mayor, especies de Rhizophora y en las áreas internas especies de Avicennia.

These mangrove forests develop in inland basins influenced by salt water. Basin mangroves are characterized by sheet flows over wide areas of small topographic relief. The energy delivery to the forest is thus diluted and termed a power sheet (Brown et al. 1979). The forests generally receive substantial freshwater and nutrients during the wet season, and in the dry season salinity increases as the salt water intrudes again. Due to the homogeneous sheet flows, there may not be strong gradients in salinity within the basin. Species of *Laguncularia* and *Avicennia* are usually dominant in the basin mangroves, although where the sheet flows converge into channels (power lines), the banks of these become lined with *Rhizophora*. *Avicennia* dominate basin forests where high salinities prevail (over 50,000 ppm), whereas *Laguncularia* dominate the low salinity basins. Mixed forests occur at intermediate salinities (30,000–40,000 ppm).

Structural development of basin mangroves depends on hydroperiod. Where weak but constant flows occur, there is oxygenation, nutrient transport, and remineralization but no accumulation of toxic substances like H_2S and salts. Forest growth is significant under these conditions. In stagnant basins there is oxygen depletion, salinization, and mineralization, and nutrient recycling is slowed down. In these basins forest growth is reduced. Basin forests are also characterized by large quantities of pneumatophores and other adventitious roots.

Los bosques de mangles que ocurren asociados a cuerpos de agua que se localizan tierra adentro y que son influidos por el agua salada, generalmente reciben agua dulce durante la estación lluviosa. Sin embargo, los gradientes no son muy grandes.

Other Forest Types

Under extreme conditions dwarfed mangrove forests develop. Their structural development is limited by edaphic factors, probably the unavailability of nutrients. Energy delivery to these forests may be concentrated at power points, diluted in power fronts, or power sheets. Mature trees in these forests are usually less than 2 m tall. We have not seen them along watercourses. Salinity does not appear to be the cause of dwarfness because leaf size is not reduced as usually happens when mangroves are under the influence of a chronic stress such as high salinity or low temperature (Lugo et al. 1981).

This type of forest is thus different from scrub forests which develop under the influence of a chronic stress (Lugo & Patterson-Zucca 1977; Cintrón et al. 1978a).

The dominant species in dwarf forests in Florida and Puerto Rico is *Rhizophora mangle* but *Laguncularia racemosa* dominated dwarf forests are found in southern Brazil (Schaeffer-Novelli et al. 1980).

Manglares enanos de Rhizophora mangle (Florida y Puerto Rico) o de Laguncularia racemosa (Brasil) existen bajo condiciones extremas.

Structural Characteristics of Mangrove Forests

A summary of structural parameters for mangrove forests located throughout the world is given by geomorphologic type in Tables 2–4. This information includes location and latitude, number of tree species, tree density and basal area for trees with diameters at breast height (dbh) equal to or greater than 2.5

TABLE 2
STRUCTURAL CHARACTERISTICS OF RIVERINE MANGROVE FORESTS
Units on a per hectare basis. Trees measured at dbh.

Stand location	Ref.	Latitude (N)	No. of Tree Species	Mean dbh cm	Number of trees 2.5 cm	Number of trees 10 cm	Basal Area (m^2) 2.5 cm	Basal Area (m^2) 10 cm	Stand Height m	Complexity Index 2.5 cm	Complexity Index 10 cm
Florida											
Ten Thousand Islands	30	25°50′	2	11.1	4000	600	38.5	21.7	9.0	27.7	2.3
Mexico											
Isla La Palma	30	22°35′	3	18.1	2360	1450	60.8	55.9	17.0	73.2	41.3
Roblitos	30	22°31′	2	13.0	2240	910	29.6	24.1	8.0	10.6	3.5
Rio de las Canas	30	″	3	20.3	1790	1030	57.8	56.1	16.0	49.7	27.7
Laguna de Terminos	(a)	18°40′	3	11.4	3360	—	34.2	—	—	—	—
Puerto Rico											
Rio Cocal	26	18°28′	1	19.1	400	400	11.5	11.5	7.5	0.4	0.3
	26	″	2	7.4	4670	670	20.0	14.0	13.0	24.3	2.4
Vacia Talega	30	18°27′	3	11.9	1890	980	20.9	17.1	13.0	15.4	6.5
Rio Espiritu Santo	26	″	2	18.0	1460	1460	37.0	37.0	16.0	17.3	17.3
	26	″	2	18.1	680	680	17.5	17.5	12.3	2.9	2.9
	26	″	2	11.3	3230	1090	32.5	20.0	16.5	34.6	7.2
	26	″	3	16.2	700	510	14.5	13.5	14.3	4.4	3.0
	28	″	2	12.3	2340	—	28.0	—	—	—	—
	28	″	2	16.6	1610	—	34.7	—	—	—	—
	28	″	2	17.3	1910	—	45.0	—	—	—	—
	28	″	2	19.8	1660	—	51.3	—	—	—	—
Rio Anton Ruiz	26	18°11′	3	10.4	3000	580	25.5	17.5	19.8	45.4	6.0
	26	″	3	15.5	1140	820	21.5	20.0	21.8	16.0	10.7
	26	″	3	6.9	4070	290	15.0	8.0	18.0	33.0	1.3
Cano Corazones	26	18°11′	2	12.1	2270	1100	26.0	20.0	14.3	16.9	6.3
	26	″	1	11.1	1660	650	16.0	13.0	11.8	3.1	1.0
	26	″	1	12.6	1560	1280	19.5	18.0	11.3	3.4	2.6
	26	″	1	9.3	2040	570	14.0	6.5	10.6	3.0	0.4
	26	″	1	13.4	1520	880	21.5	19.0	20.0	6.5	3.3
	26	″	2	14.6	1010	1010	17.0	17.0	12.8	4.4	4.4
Costa Rica											
Moin	30	10°00′	4	29.9	1370	1180	96.4	95.3	16.0	84.5	72.0
Boca Barranca	30	″	3	19.5	1100	660	32.9	31.8	9.5	10.3	6.0
Osa site 8C	16		1	—	360	—	12.5	—	34.0	1.5	—
Panama											
	15	—	—	—	—	410	0	14.3	—	—	—
Darien	11	8°30′	1	—	—	—	0	13.6	41	—	—
	27	″	—	—	—	313	0	25.1	24	—	—
Ecuador											
Isla de Ancon	1	1°	1	—	—	52	34.0	34.0	34	—	—
Estero de Molina	1	″	1	—	—	110	78.6	78.6		—	—
Canal de Changuaral	1	″	1	—	—	210	100.8	100.8		—	—
Isla de Tatabrero	1	″	1	—	—	220	24.0	24.0		—	—
La Tolita	1	″	1	—	—	120	46.3	46.3	—	—	—
Vuelta del Pailon	1	″	1	—	—	400	90.5	90.5	—	—	—
Naranjal	1	″	1	—	—	170	60.2	60.2	—	—	—

(a) Day, J.H. pers. comm.

TABLE 3
STRUCTURAL CHARACTERISTICS OF FRINGE MANGROVE FORESTS
Units on a per hectare basis. Trees measured at dbh.

Stand location	Ref.	Latitude (N)	No. of Tree Species	Mean dbh cm	Number of trees 2.5 cm	Number of trees 10 cm	Basal Area (m^2) 2.5 cm	Basal Area (m^2) 10 cm	Stand Height m	Complexity Index 2.5 cm	Complexity Index 10 cm
Florida											
Rookery Bay	19	25°55′	3	7.7	—	3400	—	15.9	6.5	10.5	—
Ten Thousand Islands	30	25°50′	2	10.3	1800	600	15.0	11.0	6.3	3.4	0.8
	30	"	2	10.3	2800	800	23.5	12.4	7.3	9.6	1.4
Mexico											
Isla La Palma	30	22°35′	3	15.5	1480	910	28.5	26.2	8.0	10.1	5.7
Laguna Terminos	(a)	18°40′	3	6.3	7510	—	23.3	—	—	—	—
Puerto Rico											
Ceiba	30	18°15′	2	6.1	5690	260	16.7	3.4	8.5	16.2	0.2
Bahia Medio Mundo	26	"	1	5.7	16760	550	43.0	5.0	9.8	70.6	0.3
	26	"	1	12.1	280	1890	33.0	28.0	16.0	14.8	8.5
Ensenada Honda	26	18°14′	1	3.2	7490	0	6.0	0	5.7	2.6	0
	26	"	3	4.7	11830	0	20.5	0	7.2	52.4	0
Laguna Joyuda	26	18°08′	1	24.9	440	440	21.5	21.5	22.0	2.1	2.5
	26	"	1	11.2	1780	1150	17.5	15.5	13.8	4.3	2.5
	26	"	2	18.9	1180	1180	33.0	33.0	20.0	15.6	15.6
	26	"	1	8.4	1820	110	10.0	3.0	8.3	1.5	0.1
Laguna Las Salinas	26	17°58′	2	9.3	3860	1600	26.0	16.5	9.0	18.0	4.8
Bahia de Guanica	26	"	2	10.8	2510	1130	23.0	16.5	10.8	12.5	4.0
Caballo Blanco	26	"	1	14.2	630	330	10.0	8.0	14.8	0.9	0.4
	26	"	1	13.1	1400	1040	19.0	17.5	14.3	3.8	2.6
Cayo Enrique	26	17°58′	1	7.2	6590	640	27.0	6.0	9.2	16.4	0.4
Bahia Sucia	26	"	1	14.1	1930	1500	30.0	28.0	15.0	8.7	6.3
	26	"	1	7.3	5200	330	21.5	3.0	7.0	7.8	0.1
Punta Pitahaya	26	17°57′	1	8.2	2830	500	15.0	8.0	8.8	3.7	0.4
Bahia Salinas	26	"	1	14.5	1750	1750	29.0	29.0	12.0	6.1	6.1
	26	"	2	11.3	2080	1390	21.0	18.0	10.8	9.4	5.4
Punta Gorda	30	"	1	7.0	1780	260	6.9	3.1	7.0	0.9	0.0
Aguirre	30	"	3	8.9	3670	110	22.6	13.9	12.0	29.9	0.6
Punta Pitahaya	26	"	1	11.1	2770	1700	27.0	23.0	12.2	9.1	4.8
	26	"	2	5.4	7980	100	18.5	1.0	6.0	17.7	0.1
	26	"	1	12.8	2940	2140	38.0	35.0	13.8	15.4	10.3
Bahia de Jobos	30	"	3	4.1	11650	260	15.5	3.2	4.8	26.0	0
Cayos Caribe	26	17°55′	2	13.1	2210	1910	30.0	28.0	9.0	11.9	9.6
	26	"	2	10.2	2550	1020	21.0	14.0	9.0	9.6	2.6
Costa Rica											
Santa Rosa	30	10°45′	2	16.8	1050	800	23.2	22.2	10.0	4.9	3.6

(a) Day, J.H. pers. comm.

and 10 cm, canopy height, average tree dbh, and Complexity Index for trees with dbh equal to or greater than 2.5 and 10 cm. The Complexity Index was defined by Holdridge (1967) and applied to mangroves by Pool et al. (1977). Average dbh of trees is given because numerous correlations can be established between this parameter and other structural and functional characteristics of stands (Fig. 1). Quantitative information for comparing mangrove ecosystems is still sparse and inconsistent.

Tables 2–4 reflect information for 38 riverine sites (Table 2), 33 fringe sites (Table 3), and 31 basin sites (Table 4). Some of the problems encountered during the analysis of the data were inconsistent methodology; not enough information to locate study sites and describe their environments and the type of mangrove ecosystem being studied; lack of information about stand age or history; incomplete data sets; and failure to report observations quantitatively or in units that can be easily transformed to other units.

TABLE 4
STRUCTURAL CHARACTERISTICS OF BASIN MANGROVE FORESTS
Units on a per hectare basis. Trees measured at dbh.

Stand location	Ref.	Latitude (North)	No. of Tree Species	Mean dbh cm	Number of trees 2.5 cm	Number of trees 10 cm	Basal Area (m²) 2.5 cm	Basal Area (m²) 10 cm	Stand Height m	Complexity Index 2.5 cm	Complexity Index 10 cm
Florida											
Rookery Bay	30	25°55′	3	6.6	5900	660	20.3	14.4	6.5	23.4	1.9
	19	"	2	8.0	3080	—	15.6	—	6.5	6.2	—
	19	"	2	7.0	5320	—	20.3	—	6.5	14.0	—
	19	"	2	7.9	2600	—	12.9	—	6.5	4.4	—
	19	"	2	7.9	2600	—	12.9	—	6.5	4.4	—
	19	"	3	8.1	4480	—	23.1	—	6.5	20.2	—
	19	"	3	11.5	3400	—	35.5	—	6.5	23.5	—
	19	"	2	6.8	6500	—	23.4	—	6.5	19.8	—
	19	"	3	6.7	5500	—	19.3	—	6.5	20.7	—
	19	"	3	5.4	9400	—	21.3	—	6.5	39.0	—
Brasil											
Cananeia, Sao Paulo	32	25°S	3	10.1	3500	—	27.9	—	4.0	11.7	—
	32	"	2	8.1	3600	—	18.4	—	5.0	6.6	—
Mexico											
El Calon	30	22°44′	2	7.9	3120	610	15.2	8.3	9.0	8.5	0.9
Puerto Rico											
Pinones	30	18°27′	3	10.1	1970	570	15.8	3.7	14.0	13.1	0.9
	30	"	3	8.2	3210	460	16.9	5.6	11.5	18.7	0.9
	30	"	3	12.5	1380	680	16.9	15.5	16.0	11.2	5.1
	30	"	3	9.8	2850	810	21.6	9.6	13.0	24.0	3.0
Vista Mar	26	18°27′	3	7.9	6710	1070	32.5	15.0	14.1	92.2	6.8
	26	"		17.3	1230	1230	29.0	29.0	14.8	15.8	15.8
	26	"		7.6	2230	370	10.0	4.0	10.5	7.0	0.5
Ensenada Honda	26	18°14′	1	3.3	9130	0	8.0	0	4.8	3.5	0
	26	"	1	11.5	1680	910	17.5	13.0	13.6	4.0	01.6
Punta Pitahaya	26	17°57′	3	4.7	3180	170	5.5	2.0	4.5	2.4	0
Bahia Sucia	26	17°58′	1	5.0	3330	0	6.5	0	4.1	0.9	0
	26	"	1	4.0	3750	0	4.0	0	3.0	0.4	0
Laguna Joyuda	26	18°08′	3	6.7	5350	210	19.0	3.0	11.8	35.9	0.2
	26	18°08′	1	12.9	1530	1090	20.0	18.5	11.8	3.6	2.4
	26	"	1	6.1	2060	0	6.0	0	7.0	0.9	0
	26	"	1	9.5	850	210	6.0	3.0	6.3	0.3	0.1
Penon Brusi	26	18°29′	3	13.7	1220	600	18.0	16.5	13.8	9.1	4.1
	26	"	2	25.9	590	590	31.0	31.0	11.8	4.3	4.3
Mona Island	30	18°06′	2	12.3	2930	1790	35.0	29.7	15.0	30.8	15.9

The information presented is from published and unpublished results of studies on structure and function of mangroves all over the world. However, the data strongly reflects the mangroves of Puerto Rico and Florida where most of the quantitative work has been done.

Detailed examination of this data will be done later but it is important to make the information available to stimulate analysis and comparisons. Table 5 is a summary of the structural information of Tables 2–4, and the reader is directed to that table for the statistical means of structural characteristics of mangrove forest types.

Number of Species

Basin mangrove forests support more species than riverine mangroves which, in turn, support more species than fringe mangrove forests. Fringe forests develop in more constant conditions of salinity and

TABLE 5
STATISTICAL VALUES FOR THREE TYPES OF MANGROVE FORESTS

Mean and (one standard error) n: number of observations values on a one hectare basis

Type of Mangrove Forest	No. of Tree Species	Number of Trees		Basal Area (m^2)		Stand Height	Complexity Index[1]	
		2.5 cm dbh	10 cm dbh	2.5 cm dbh	10 cm dbh	m	2.5 cm dbh	10 cm dbh
Riverine	1.9(0.1)	1979(209)	661(71)	30.4(3.5)	32.6(4.7)	17.0(1.6)	21.2(4.8)	10.4(3.6)
n:	36	28	32	28	32	26	23	22
Fringe	1.7(0.1)	4005(642)	852(115)	22.2(1.5)	14.6(1.9)	10.5(0.7)	13.3(2.6)	3.2(0.7)
n:	33	33	31	33	31	32	32	31
Basin	2.3(0.1)	3599(400)	573(102)	18.5(1.6)	10.6(2.2)	9.0(0.7)	15.4(3.2)	3.1(1.0)
n:	31	31	21	31	21	31	31	21

[1]Based on 0.1 ha.
Individual stand observations are given in Tables 2–4.

nutrients and are exposed to higher kinetic energy regimes. These conditions select for dominance of one or few species. In the Caribbean this is usually *Rhizophora mangle*. In basin and riverine forests, nutrients and salinity may oscillate widely, and the kinetic energy of water flows is much lower. The hydroperiod of basin forests is relatively short and perhaps this accounts for their higher species richness.

During periods of flood, considerable tidal and water flow energies are dissipated in riverine sites, but these periods are short in comparison with the constant tidal motion in the fringe forests. As a result of the oscillating conditions and more favorable nutrient balances, riverine and basin sites are able to support more species. These two mangrove types in fact usually grade into comparatively species-rich freshwater forested wetlands, and species from these forests enrich the mangrove stands.

Los manglares de tierra adentro presentan más especies que los que se localizan a orillas de los ríos y éstos a su vez más especies que los de las costas. Las condiciones oscilantes y balance favorable de nutrimentos conllevan a la riqueza de especies.

Tree Densities

For trees larger than 10 cm dbh, densities tend to be higher in fringe forests followed by riverine, and lastly, basin forests. A different order results if smaller diameter trees are considered, but the age of stands may obscure real trends when smaller diameters are used. Mangrove basins are characterized by low tree density. The higher tree density in fringe forests may be a response to intensive tidal and wave action.

Los árboles grandes son más densos en los manglares de las costas, luego en los que están a orillas de los ríos y los menos densos son los de tierra adentro.

Basal Area

Riverine forests develop the largest basal area by far. Basal area is low in basin forests as a result of the lower density of trees over 10 cm dbh. The basal area in fringe forests is intermediate.

Tree Height

Riverine mangroves contain the tallest trees followed by fringe and basin forests. There does not appear to be a significant difference in the height of the last two mangrove types.

Los árboles más altos se encuentran en los manglares localizados a orillas de los ríos, luego en los que bordean las costas y los de menor tamaño están representados en los manglares de tierra adentro.

Average Tree Diameter

The average dbh of the stand was highest in the riverine mangrove forests, 14.7 cm. The other two forest types exhibited a similar average stand dbh but basin mangroves were at the lowest end of the range—9.1 cm vs. 10.4 cm for fringe mangroves.

Complexity Index

The Complexity Index—which integrates structural information—shows that riverine mangroves develop the most structural complexity followed by fringe and basin forests (Table 5).

Hay un gradiente de mayor a menor complejidad estructural de los manglares a orillas de los ríos, a los de las costas, a los de tierra adentro.

Biomass

Biomass data cannot be arranged easily in a pattern by geomorphologic type because of the difficulties with the data base mentioned earlier. Yet, from the summary in Table 6 one can see that the highest aboveground biomass value is for riverine forests (166 t/ha vs. 100 t/ha for fringe forests), while a dwarf forest contained the least aboveground biomass (7.9 t/ha). The data from Thailand is for plantations with abnormally high tree densities and was not used in this comparison.

Por encima del suelo, la biomasa parece ser

TABLE 6
BIOMASS OF MANGROVE FORESTS

				Biomass(t/ha)				
Location	Ref.	Latitude (N)	Age	Above ground	Sub surface roots	Total plants	Litter	Total for stand
RIVERINE								
Australia	3	33°-5′		128.4	153.8	282.2		
Florida	21	25°-26°		98.2				42.9
Florida	21	25°-26°		173.9			33.9	
Florida	21	25°-26°		86.2			22.7	
Malaysia	17	12°-13°	5	16.0				
Malaysia	17	12°-13°	15	257.4				
Malaysia	17	12°-13°	28	286.8				
Panama	11	8°-30′		279.2	189.8	469.0	102.1	571.0
FRINGE								
Florida	21	25°-26°		129.6			17.3	
Florida	21	25°-26°		119.6			14.0	
Florida	21	25°-26°		117.5			60.2	
Florida	21	25°-26°		152.9			98.4	
Florida	21	25°-26°	40–50	49.0				
Florida	21	25°-26°	5	8.1	14.1	22.2	0.3	22.5
Puerto Rico	10	18°		62.8	50.0	112.8		
Thailand	5	8°	15	159.1			0.3	
DWARF								
Florida	21	25°		7.9	8.0	15.9	1.1	17.0
HAMMOCK								
Florida	34			98.4				
PLANTATION[1]								
Thailand	2	12°-13°	3	20.8			0.3	
Thailand	2	12°-13°	6	50.0				
Thailand	2	12°-13°	9	93.1				
Thailand	2	12°-13°	11	116.3				
Thailand	2	12°-13°	12	149.3				
Thailand	2	12°-13°	13	167.4				
Thailand	2	12°-13°	14	187.9				
Philippines	9	12°-13°		45.9				

[1]Plantation trees stocked at 10,000 stems per hectare, all stands unthinned.

mayor en los manglares a orillas de los ríos y menor en los bosques enanos.

Leaf Area Index

There are few reports of Leaf Area Indices for mangrove forests (Pool et al. 1975; Lugo & Patterson-Zucca 1977; Lugo & Snedaker 1975). Values range from 5.1 for a riverine forest to 0.2 for a dwarf forest. The mean for 8 stands is 3.8. These are low values compared to Indices of over 6 for complex tropical forest ecosystems. Statistical differences cannot be shown, but trends suggest that the higher values are for fringe and riverine forests which are the types that receive light on two sides of the forests (top and the fringe side). Basin forests tend to exhibit lower values, but the lowest are found in stressed forests such as dwarf mangroves and scrubby temperature-stressed forests.

In conclusion, riverine mangrove forests are more developed in structure than any other mangrove type. High structural development is reflected in stand basal area, height, mean dbh of trees in the stand, biomass, and in Complexity Index. This is in spite of the lower tree densities which is a reflection of the dominance of fewer but much larger trees.

Mangrove fringes and mangrove basins cannot be precisely separated with available data and appear to share similar levels of structural development. However, there is a tendency for basin forests to be at an extreme from riverine forests and for fringe forests to be intermediate (e.g. basal area, mean tree diameter, height, and Complexity Index).

In a given geographic area the energy signature may contribute to the greater development of structure in one of these types and not in the others. Thus, basins (which are runoff and rainfall subsidized) may be well developed structurally in high precipitation areas and highly stressed and poorly developed in arid areas. Fringe forests develop in conditions which are stable over wide geographic areas and are similar to riverine forests in being less dependent on local rainfall and runoff. Riverine systems which derive a large measure of their energy signature from high and constant fresh water flows (often derived from inland regions in different Life Zones) are more independent of local conditions and more constant in structural characters.

For example, in such dry areas as the Ecuadorian coast, there are extensive and highly developed riverine forests in spite of the aridity, whereas basins give way to extensive "salitrales" or salt flats. The riverine forests develop in the estuaries of streams that descend from the wetter Andean highlands to the east. When all stands are averaged together relationships such as the ones described are lost. For this reason, data for individual stands are given in Tables 2–4. Average values should be interpreted with caution. In a similar analysis of freshwater forested wetlands, Brown et al. (1979) could not separate flowing and still water wetlands using structural parameters. Only tree density (higher in still water wetlands) exhibited a significant difference between types.

Comparison With Other Forested Areas

Brown et al. (1979) summarized structural data for freshwater forested wetlands. A comparison of their results with the data presented here for mangroves reveals that mangroves are structurally less complex than freshwater forested wetlands. Only riverine mangrove forests approach the least complex freshwater forested wetlands. The difference is due to the lower floristic richness of mangroves which in turn is probably due to the high ambient salinity. However, in terms of aboveground biomass, individual mangrove forests accumulate as much biomass as freshwater forested wetlands. The average for 17 stands in Table 6 (excluding plantations and the dwarf forests) is 131 t/ha compared to 191 to 260 t/ha for still water and flowing water forested freshwater wetlands.

Los manglares son estructuralmente menos complejos que los bosques de agua dulce de las tierras húmedas.

Compared to mangrove forests in Florida and the Caribbean, mangroves in Darien Province are taller, contain more biomass, and are floristically richer (Table 7). Floristic richness occurs because both Pacific and Atlantic mangrove species are represented on the respective coastlines of the country (West 1956). Lugo and Snedaker (1974) attributed the greater stature of Panamanian mangroves to the lack of hurricane disturbance and to the favorable conditions for mangrove growth in this region.

Los manglares del Darién son más altos, contienen más biomasa y son florísticamente mucho más ricos que los de la Florida y del Caribe. Tanto las especies del Atlántico como las del Pacífico están representadas.

Functional Characteristics of Mangrove Forests

By functional characteristics we mean the rates of primary production, mineral cycling, and litterfall; re-

TABLE 7
STRUCTURAL DATA FOR RIVERINE MANGROVE FORESTS
For trees over 10 cm dbh

Parameter	Panamanian Mangroves	All Other Riverine Mangroves
Tree density (Number/ha)	356	623
	408[a]	
	407[a]	
	313[b]	
Basal area (m^2/ha)	13.6	32.4
	26.9[a]	
	14.6[a]	
Average tree dbh (cm)	19.4	14.6
Maximum height (m)	41	17
	25[a]	
	35[a]	
Prop root biomass (t/ha)	116	13
Total above ground biomass (t/ha)	279	155
Root biomass (t/ha)	190	154
Total plant biomass (t/ha)	469	322
Litter biomass (t/ha)	102	33
Total stand biomass (t/ha)	571	355

[a]Holdridge et al. 1964.
[b]Mayo 1965.
Panama data from Golley et al. 1975

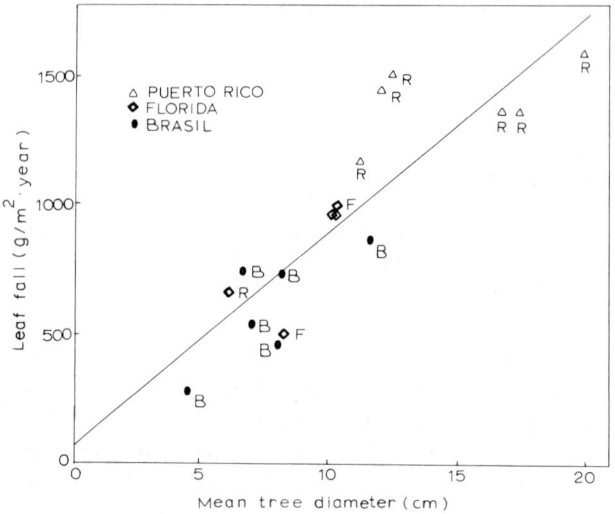

Fig. 1. Relationship between litterfall and mean tree diameter in mangrove forest ecosystems. R is riverine, B is basin, and F is fringe mangrove forests. Lines along the lower portion of the diameter axis show one standard error for the range of mean stem diameters by forest type.

sponses to stressors; export of organic matter; succession; and so on. The amount of information on the functional characteristics of mangrove forests is sparse and biased towards new world mangroves—particularly Florida and the Caribbean.

Litterfall

Pool et al. (1975) suggested that the rate of litterfall in mangroves is related to increasing turnover of freshwater. This leads to highest rates of litterfall in riverine forests and lowest in dwarf forests. In Fig. 1 we exclude dwarf mangrove forest data to show that basin mangrove forests fall at the lower end of the range of litterfall while fringing forests exhibit intermediate rates between basin and riverine forests which produce the most.

La producción primaria de los manglares está asociada al vuelco de agua dulce.

Primary Productivity

Net primary productivity for riverine mangroves in Florida was 9.6, 15.0, and 13.2 g OM/m^2 per day; a Florida fringe forest exhibited a rate of 8.8 g OM/m^2 per day, and for a basin forest the corresponding value was 5.6 g OM/m^2 per day (Lugo & Brinson 1979). However, the basin forest had higher respiration rates, and thus its gross primary productivity was higher than the gross primary productivity of the fringe forest (18 vs. 12.6 g OM/m^2 per day).

Decomposition and Export of Organic Matter

The few data available (Table 8) show that *Avicennia* leaves decompose 3 times faster than *Rhizophora* leaves (Lugo et al. 1981a). *Avicennia* is the predominant species in basin forests, while *Rhizophora* predominates in riverine and fringe forests. They found a faster rate of leaf decomposition in red mangrove litter in a high flushing environment compared to a low flushing environment. These species differences lead to faster turnover of litter (due to decomposition alone) in black *(Avicennia)* mangrove-dominated basin mangroves in comparison to riverine or fringe forests (Table 8). The difference in decomposition rates due to species also leads to a greater export of dissolved substances from basin forests relative to riverine and fringe forests. In these two types, export of organic matter is mostly in the particulate form. The timing and amount of organic export also differs between systems. In basin forests rain pulses are an important forcing function for organic

TABLE 8
LITTER DEGRADATION RATES IN MANGROVE FORESTS

Forest/litter type	Location	Decay rate k(yr[-1])	Half life (days)	Source
Fringe				
Rhizophora mangle	Florida	8.39	30	12
	Florida	2.55	99	12
	Florida	1.46	173	12
	Florida	1.10	231	12
	Florida	0.85	346	21
	Puerto Rico	2.55	99	—
	Mexico	1.42	178	a
	Brazil	1.23	206	31
		MEAN HALF LIFE: 170 + 34		
Fringe				
Laguncularia racemosa	Brazil	3.08	82	31
Basin				
Rhizophora mangle	Florida	0.30	—	19
	Florida	1.22	231	
	Florida	2.01	139	19
	Florida	1.28	231	8
	Florida	1.63	173	8
(Fast flushing)	Florida	1.90	133	19
(Slow flushing)	Florida	0.97	260	19
		MEAN HALF LIFE: 194 + 22		
Basin				
Avicennia germinans				
(Fast flushing)	Florida	6.04	60	19
(Slow flushing)	Florida	3.89	65	19
		MEAN HALF LIFE: 63 + 3		
Riverine	Mexico	2.55	99	a
	Colombia	2.81	87	13
		MEAN HALF LIFE: 93 + 6		
Other Types				
Dwarf	Florida	1.46	173	34
	Florida	2.29	115	34
Hammock	Florida	2.29	115	34
		MEAN HALF LIFE: 134 + 19		

[a]J.H. Day, personal communication

export. On the other hand, daily tidal cycles dominate export in the fringe forest. Riverine forests export during floods by spring tides or during storm events.

Las hojas de Avicennia se descomponen más rápido que las de Rhizophora y éstas a su vez más rápido en medios ambientes que intercambian agua.

Tree Growth and Succession

Tree growth and succession are relatively fast processes in most mangrove types with the exception of those under severe salinity, cold, or low nutrient stress. Lugo (1981b) concluded that these processes are faster in riverine conditions but vary in the other mangrove ecosystem types depending upon the components of the energy signature. Rainfall, freshwater runoff, and high nutrient concentrations accelerate tree growth and succession.

Ratios of gross primary productivity to biomass (yielding turnovers/year) were calculated for riverine (0.36–0.73), fringe (0.67), and dwarfed (1.3) mangroves (Lugo 1980). These, and high P/R ratios, suggest that mangroves are steady state rather than successional ecosystems.

Responses to Stressors

Lugo et al. (1980) classified the responses of mangrove ecosystems to human induced and natural stressors into five categories based on their point of attack in the ecosystem. In humid environments equal intensities and duration of such stressors as diking, damming, channelization, and levee and road construction have a more severe impact on mangroves than fires, harvesting, or moderate grazing. The response of the mangroves of the former group of stressors is exponential or logarithmic in nature, linear to the latter group. The impact of other stressors is intermediate in severity.

Los medios ambientes húmedos, extremos debidos a construcción de diques, presas, canales y caminos, ejercen mayor y más severa influencia que los fuegos, las cosechas o el pacer del ganado.

The susceptibility of mangroves to certain stressors also varies with the type of forest. Mangroves growing in arid climates, mostly fringe forests, are not so susceptible to alterations of terrigenous factors as those in humid environments where the ecosystem is more dependent on terrestrial runoff (riverine and basin forests). In arid environments hypersalinity is a severe natural stressor and its effects are aggravated by human actions that reduce water flushing. For example, with chronic exposure to oil in a fringe mangrove forest located in an arid climate, leaves are the ecosystem component where stress first becomes visible. Oil appears to interfere with gaseous exchange of roots and sediments which then affect leaves. Mangroves without frequent tidal flushing were more severely affected by oil than those with frequent tidal flushing (Lugo et al. 1981b).

La exposición crónica a aceites constituye un severo agente limitante para los manglares.

Mineral Cycling

In a comparative mineral cycling study of forests in Darien Province, Golley et al. (1975) described the riverine mangroves as having the following salient features (in comparison to terrestrial forests):

—composed of few large trees;

—a great portion of their high biomass in roots and litter;

—low biomass in fruits and understory and epiphytic plants;

—lower concentration of N, P, K, and Ca in tissues, but higher concentrations of Mg, Na, B, Fe, and Sr;

—muds with highest concentrations of P, Co, Cs, Fe, and Sr;

—roots and leaves contained highest concentrations of most elements tested; and

—highest percentage of most elements tested was sequestered in organic matter.

This and the work of Lugo and Snedaker (1975) in Florida are the only ecosystem level work on mangrove mineral cycling. Lugo and Snedaker (1975) showed that basin mangrove forests improved water quality through rapid nutrient and elemental sequestering in soil and biomass. The results of Golley et al. (1975) document this role for riverine forests as did Sell (1977) in terms of sewage uptake by trees and mud.

Mangrove Under Extreme Conditions

Extreme humidity, aridity, low temperature, oligotrophy, or eutrophy result in unusually large or unusually short mangroves. The giant mangroves of Panama, Colombia and Ecuador are exposed to extreme tides, fresh water runoff, and high nutrient concentrations. The dwarfs of Florida grow on limestone while those in Puerto Rico grow on oligotrophic peats. Scrubbiness occurs with low and high temperatures or high salinity. In general, mangroves respond quickly to environmental change. For example, Figs. 2 and 3 illustrate the responses in mangrove height and leaf sizes to latitudinal and stress gradients. In Fig. 2 the size of red mangrove leaves is shown to

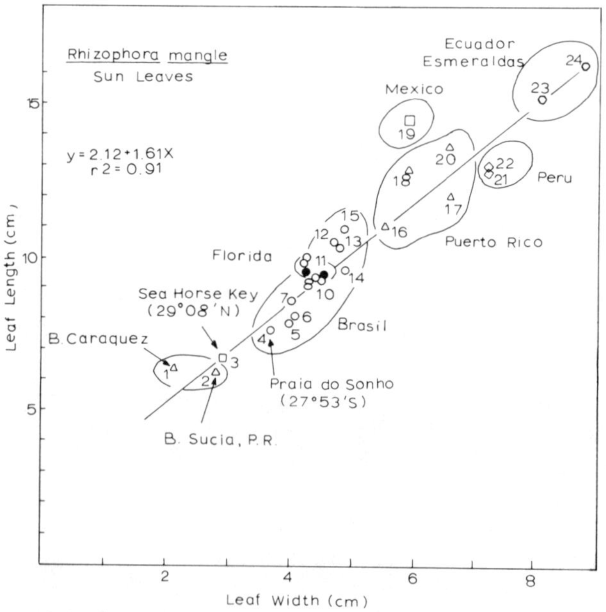

Fig. 2. Relationship between length and width of red mangrove leaves in New World mangrove forests. Forests with smaller leaves reflect conditions of stress, while those with larger leaves grow in more favorable conditions.

vary threefold between the arid or cold environments of Bahia Sucia (arid) in Puerto Rico and Sea Horse Key (cold) in Florida to the near optimal conditions in Equador. In Fig. 3 maximum tree height changes ten-fold along a latitudinal gradient. Similar regressions for number of species and structural characteristics have been described for the Gulf of Mexico (Lugo & Patterson-Zucca 1977), demonstrating the plasticity of mangroves and their ability to colonize extreme habitats.

Las mareas extremas, la cantidad de agua dulce y elevada cantidad de nutrimentos conducen a la existencia de los manglares gigantes de Panamá, Colombia y Ecuador.

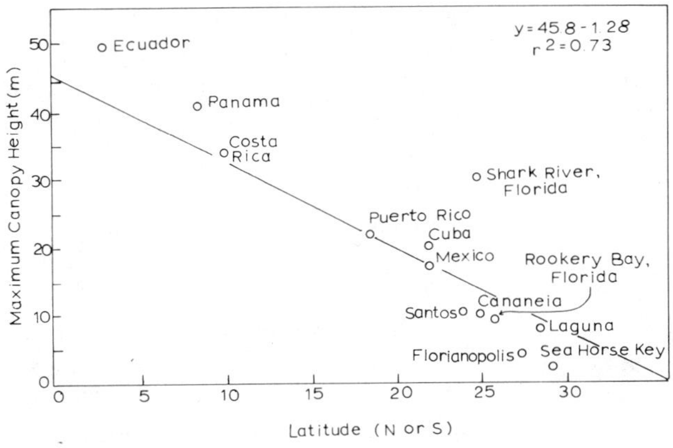

Fig. 3. Relationship between maximum canopy height and latitude in mangrove forests of the New World. The two regressions may separate mangroves growing under servere stressors (for example hurricanes, frost, or hypersalinity) from those growing under optimal conditions.

Conclusion

At the outset we suggested that the "energy signature" of mangroves could be approximated by observing the structural and functional characteristics of the various types of mangrove forests. The idea was to group mangroves into a few general types that could serve as points of reference for developing hypotheses, management strategies, or generalizations, even in the absence of quantitative information. The data base is still too sparse to be conclusive but several trends are clear. For example, riverine mangroves are always the most productive, most complex, and the systems with the fastest rates of ecological processes. Mangroves growing under extreme conditions exhibit the slowest rates of ecological processes and least structural complexity. Basin mangrove forests exhibit more structural complexity and higher rates of ecological processes than dwarf and scrubby (stressed) mangroves but are much lower and slower in these features than riverine forests. Fringe forests tend to be intermediate in structural and functional characteristics between riverine and basin mangrove forests.

Agrupando los manglares en los tipos generales mencionados previamente demuestra que éstos difieren en productividad, complejidad e incrementos de sus procesos y que éstos son modificados por condiciones extremas.

We recommend a greater effort in research on mangroves and orientation of research towards a more general understanding of the structural and functional responses of these ecosystems to external forces.

Acosta-Solis, M. 1959. Los manglares del Ecuador. Contr. Inst. Ecuator. Cienc. Nat. 29.

Aksornkoae, S. 1975. (unpubl.). Structure, regeneration, and productivity of mangroves in Thailand. Ph.D. disser. Michigan State University.

Briggs, S.V. 1977. Estimates of biomass in a temperate mangrove community. Austral. J. Ecol. 2: 369–373.

Brown, S., M.M. *Brinson* & A.E. *Lugo*. 1979. Structure and function of riparian wetlands. Pp. 17–31. *In* R.R. Johnson & J.F. McCormick, eds. Symp. Strategies Produc. Managem. Flood Plain. U.S.D.A. Forest Service, Washington, D.C.

Christensen, B. 1978. Biomass and primary production of *Rhizophora apiculata* Bl. in a mangrove in southern Thailand. Aquat. Bot. 4: 43–52.

Cintrón, G., A.E. *Lugo*, D.J. *Pool* & G. *Morris*. 1978a. Mangroves of arid environments in Puerto Rico and adjacent islands. Biotrop. 10: 110–121.

Cintrón, G., C. *Goenaga* & J. *González-Liboy*. 1978b. Ecología del manglar en una costa árida: exposición al oleaje y estructura del manglar. Memor. 5 Simpos. Latinoamer. Oceanogr. Biológ. Sáo Paulo, Brasil.

Courtney, C.M., J.C. *Kinch* & J.M. *Hatcher*. 1976. The production of organic detritus in an impounded black mangrove forest. Marco Applied Marine Ecology Station. Mimeo.

De la Cruz, A.A. & J.F. *Banaag*. 1967. The ecology of a small mangrove patch in Matabungkay Beach, Batangas Province. Univ. Philipp. Nat. Appl. Sci. Bull.

Golley, F.B., H.T. *Odum* & R.F. *Wilson*. 1962. The

structure and metabolism of a Puerto Rican red mangrove forest in May. Ecol. 43: 9–19.

Golley, F.B., J.T. *McGinnis*, R.G. *Clements*, G.I. *Child* & M.J. *Duever*. 1975. Mineral Cycling in a Tropical Moist Forest Ecosystem. Univ. Georgia Press, Athens, Georgia, U.S.A.

Heald, E. 1971. The production of organic detritus in a south Florida estuary. Univ. Miami. Sea Grant Tech. Bull. 6.

Hernández, A.R. & K.P. *Mullen*. 1975. Observaciones preliminares sobre la productividad primaria neta en un ecosistema de manglar-estuario (Guapi-Colombia). Mem. 2 Simp. Latinoamer. Oceanogr. Biol. Univ. Oriente, Venezuela.

Holdridge, L.R. 1967. Life zone ecology. Tropical Science Center, San José, Costa Rica.

Holdridge, L.R., J.A. *Duke*, W.A. *Finch* & W.C. *Grenke*. 1964. El Real Environmental Survey, Darien Province, Republic of Panama, 1962, TRECOM Tech. Rep. 63–72, U.S. Army Transport. Res. Command, Ft. Eustis, Virginia.

Holdridge, L.R., W.C. *Grenke*, W.H. *Hatheway*, T. *Liang*, and J.A. *Tosi*, Jr. 1971. Forest Environments in Tropical Life Zones. Pergamon Press, New York.

Jing-Eong, O., G. *Wooi-Khoon*, W. *Chee-Hoong* & G. *Dhanarajan*. 1979. Productivity of a mangrove forest in West Malaysia. Internat. Conf. "Trends in Applied Biology in S.E. Asia", U.S.N. Penang. Mimeo.

Lugo, A.E. 1980. Mangrove ecosystems: successional or steady state? Biotrop. 12 suppl. 2: 65–72.

Lugo, A.E., R. *Twilley* & C. *Patterson-Zucca*. 1981a. The role of black mangrove forests in the productivity of coastal ecosystems in south Florida. Final rep. U.S. Environm. Protect. Agency, Corvallis, Environm. Res. Lab., Oregon.

Lugo, A.E. & S.C. *Snedaker*. 1974. The ecology of mangroves. Annual Reviews Ecology and Systematics. 5: 39–64.

Lugo, A.E. & S.C. *Snedaker*. 1975. Properties of a mangrove forest in southern Florida. Pp. 170–212 *In* G.E. Walsh, S.C. Snedaker & H.J. Teas, eds. Proc. Internat. Symp. Biol. Managem. Mangroves. IFAS, Univ. Florida, Gainesville.

Lugo, A.E. & C. *Patterson-Zucca*. 1977. The impact of low temperature stress on mangrove structure and growth. Trop. Ecol. 18(2): 149–161.

Lugo, A.E., G. *Evink*, M. *Brinson*, A. *Broce* & S.C. *Snedaker*. 1975. Diurnal rates of photosynthesis, respiration, and transpiration in mangrove forests of Florida. Ch. 22 *In* F.B. Golley & E. Medina, eds. Trop. Ecol. Syst. Springer-Verlag, New York.

Lugo, A.E. & M.M. *Brinson*. 1979. Calculation of the value of salt water wetlands. Pp. 120–130 *In* P.E. Greeson, J.R. Clarke & J.E. Clark, eds. Wetland Functions and Values: the State of our Understanding. Amer. Water Res. Assoc. Minneapolis, Minnesota.

Lugo, A.E., G. *Cintrón* & C. *Goenaga*. 1981b. Mangrove ecosystems under stress. Pp. 129–153 *In* G.W. Barrett & R. Rosenberg, eds. Stress Effects on Natural Ecosystems, John Wiley & Sons, Sussex, England.

Martínez, R., G. *Cintrón*, & L.A. *Encarnación*. 1979. Mangroves in Puerto Rico: A structural inventory. Final rep. Off. Coastal Zone Managem. NOAA. Dep. Nat. Res., Area Sci. Res. Govt. Puerto Rico.

Mayo, M.E. 1965. Algunas caracteristicas ecológicas de los bosques inundables de Darién, Panamá, con miras a ser posible su utilización. Turrialba 15: 336–347.

Mosquera, R.A. 1979. Química del agua intersticial y varios parámetros estructurales cuantificados en un manglar riverino, en el Río Espíritu Santo, Río Grande, Puerto Rico. Center Energy & Environm. Res. Univ. Puerto Rico. & U.S. D. Energy. Pub. No. 18.

Pool, D.J., A.E. *Lugo* & S.C. *Snedaker*. 1975. Litter production in mangrove forests of southern Florida and Puerto Rico. *In* G.E. Walsh, S.C. Snedaker & H. J. Teas, eds. Proc. Internat. Symp. Biol. Managem. Mangroves. IFAS, Univ. Florida, Gainesville.

Pool, D.J., S.C. *Snedaker*, A.E. *Lugo*. 1977. Structure of mangrove forests in Florida, Puerto Rico, Mexico, and Costa Rica. Biotrop. 9: 195–212.

Schaeffer-Novelli, Y. & G. *Cintrón*. 1980. Estimados de producción de hojas, hojarasca y tasas de descomposición foliar en un manglar de Cananeia, Sáo Paulo, Brasil. Mimeo.

Schaeffer-Novelli, Y., G. *Cintrón* & R.R. *Adaime*. 1980. Algumas características dos manguezais de regiao de Cananeia, S. Paulo, Brasil. Semi. Biol. Marinha. Acad. Brasil. Cienc. Inst. Biol. Marinha U.S.P. Abstr.

Sell, M.G. 1977. Modeling the response of mangrove ecosystems to herbicide spraying, hurricanes, nutrient enrichment and economic development. Ph.D. dissert. Univ. Florida, Gainesville.

Snedaker, S.C. 1975. Reports on Grant R-803340–02. Univ Florida, Instit. Food & Agric. Sci.

West, R.C. 1956. Mangrove swamps of the Pacific coast of Colombia. Ann. Assoc. Amer. Geogr. 46: 98–121.

Animal Groups

Animal life in Panama is diverse and interesting. One recent report records trapping 717 different species of beetle from a single locality on Barro Colorado Island. Over 900 different species of birds make up Panama's bird fauna. And there is a wealth of mammalian life, including coatis, monkeys, agoutis, opossums, mice, deer, raccoons, bats. pumas, sloths, and others. These many different organisms all have different adaptations to habitat and feeding possibilities, and defense against predators and environmental hazards also requires adaptation. Many of these evolutionary responses to the environment are highly specialized, and are often not evident without meticulous scientific study. The following three papers provide insight into the kinds of diversity present in Panama and to the kinds of factors to which organisms must respond.

Karr's paper is a condensed catalogue of all the birds occurring in Panama, with an analysis of migrations and environmental determinants of bird movement.

Sexton examines the factors determining the distribution of groups of lizards, and he uses comparative methods to arrive at determinations in a situation where the more usual scientific method of experimentation with controls would be impossible.

Garber examines the feeding behavior of groups of small monkeys, finding that they are obligate on a single food plant to provide satisfactory dietary levels of calcium, and that their distribution in Panama is limited by the distribution of this lowland tree species.

Thus Karr, Sexton and Garber examine the same question, the distribution requirements and limitations of specific animals, and because of the complexity of plant and animal life in the area, careful field observations and analysis of results obtained elsewhere, some on quite different organisms, are necessary to provide a plausible hypothesis for the patterns observed. In each case, part of the distribution story lies in the response to other organisms.

Grupos de Animales

La vida animal en Panamá es diversa e interesante. En una sola localidad de la Isla de Barro Colorado, un reporte reciente registra 717 especies diferentes de escarabajos. Un poco más de 900 especies constituyen la fauna de aves de Panamá. La riqueza de los mamíferos incluye coatíes, monos, agoutis, zorras, ratones, mapaches, murciélagos, pumas, monos perezozos y otros. Todos estos animales presentan diferentes adaptaciones al ambiente y distintas posibilidades de alimentación. La defensa contra predadores y eventualidades ambientales también requieren adaptación. Numerosas de estas respuestas evolutivas al ambiente son extraordinariamente especializadas y frecuentemente no son evidentes a menos que se realice un meticuloso estudio científico. En el presente capítulo tres trabajos proveen conocimiento profundo de distintas clases de diversidad faúnica en Panamá.

El estudio de Karr es un catálogo condensado de todas las aves que ocurren en Panamá, con un análisis de las migraciones y los factores ambientales que determinan sus movimientos.

Sexton examina las causas que determinan la distribución de las lagartijas usando métodos comparativos y alcanza determinaciones en una situación en donde los métodos usuales científicos experimentales con controles hubiesen sido imposibles.

Garber examina el comportamiento de alimentación en grupos de monos pequeños, encontrando que éstos se alimentan de exudados vegetales y obtienen así niveles de calcio satisfactorios. En Panamá, la distribución de esos monos está restringida por la distribución de la especies de árboles que producen el exudado.

Los trabajos de Karr, Sexton y Garber examinan los requerimentos de distribución y las limitaciones de grupos específicos de animales. Como quiera que la vida animal y vegetal del área es compleja, observaciones cuidadosas en el campo y análisis de resultados obtenidos en otros lugares, algunas veces en grupos de organismos muy diferentes, son imprescindibles para enunciar las hipótesis plausibles que expliquen los patrones observados. En cada caso, la historia de la distribución descansa en gran medida en relación a la respuesta de otros organismos.

Plant Exudates in the Diet of the Panamanian Tamarin *(Saguinus oedipus geofroyi)*

Paul A. **Garber**

Estudios de campo de los monos tití de Panamá, que habitan las tierras bajas del Pacífico, han demostrado que éstos se alimentan principalmente de insectos y frutos, al igual que de exudados vegetales de dos especies de Anacardiaceae. Los exudados les proveen calcio y algo de proteína y son alimentos obligatorios para estos monos pequeños.

The Callitrichidae are a family of New World primates that have experienced a successful and diversified radiation in the tropical forests of Central and South America. This radiation, which includes the marmoset genera *Cebuella* and *Callithrix*, and the tamarin genera *Leontopithecus* and *Saguinus*, is characterized by small-bodied insectivorous monkeys inhabiting southeastern Costa Rica and Panama, east through the Amazon Basin, and continuing as far as southeastern Brazil (Hershkovitz 1977).

Los monos tití y las marmotas son monos pequeños e insectívoros del Nuevo Mundo.

Field data on the ecology and behavior of wild tamarins and marmosets are extremely limited. Many of our current ideas concerning tamarin behavior are derived from information collected on captive specimens, occasional sightings, or short-term field observation. This lack of detailed information has failed to provide a clear framework within which to understand callitrichid adaptations.

The evolution of claw-like nails or *tegulae* (Weber 1928) in the Callitrichidae is related to a specialized feeding adaptation (Garber 1980b). Specifically, claw-like tegulae enable the Panamanian tamarin to cling to large vertical and sharply inclined tree trunks while feeding on plant exudates such as gums and resins. Exudates are an important part of the diet of these monkeys (Garber 1980a,b).

Los monos tití panameños se cuelgan de los troncos verticales de árboles grandes mediante uñas similares a garras, mientras se alimentan de los exudados de esos árboles.

Why are plant exudates such an attractive and important resource to these callitrichid primates? So important in fact, that this dietary habit appears to be directly associated with the evolution of such characters as: (1) procumbent or specialized incisors, (2) claw-like tegulae or laterally compressed and pointed nails, (3) small body size, (4) ability to cling to large vertical trunks, and (5) insect protein a significant part of the diet.

The nutritional composition of plant exudates and its importance in the tamarin diet are studied here. Rather than serving simply as a substitute for fruits, exudates provide these animals with a high energy resource rich in calcium and protein; the high calcium content in certain exudates may explain their nutritional value for these primates.

Methods

Behavioral Data

From January through August 1978, 850 hours was spent studying the ecology and feeding behavior of free-ranging populations of the Panamanian tamarin *(Saguinus oedipus geoffroyi)* in a dry tropical forest on the Pacific coast of Panama (Garber 1980a,b). The field site, the Rodman Naval Fuel Farm, was a 600 acre forest under the supervision of the United States Southern Military Command. Rodman supports a large and fairly dense tamarin population, and access to this region was restricted solely to authorized personnel. The undisturbed nature of the Rodman forest provided an excellent opportunity to observe the natural behavior of tamarin monkeys.

Estos monos también se estudiaron en la parte central de Panamá, principalmente para conocer su dieta y comportamiento alimenticio.

Observations used an instantaneous time sampling technique. Individual activity records (IAR'S) were collected on focal animals and recorded at 2 1/2 minute intervals throughout the day. The following information was collected:

1) Nature and structure of the substrate exploited.
2) Activities on the supports and types of resources procured.
3) Body posture involved in the use of these supports.
4) Diet.

Dietary Data

Information was collected on general ecology and diet as well as data on activity patterns and behavior. Plant specimens were collected, labeled, dried, and sent to the Missouri Botanical Garden for taxonomic identification. Samples of plant exudates consumed by *Saguinus oedipus* were sent to the Langmuir Laboratories, Cornell University and the A and L Great Lakes Agricultural Laboratories, Fort Wayne, Indiana, for nutritional analysis.

Exudate Feeding in *Saguinus oedipus*

The diet of the Panamanian tamarin is composed of three distinct types of food resources: insects (39.4%), fruits (38.4%), and plant exudates (14.4%)(Table 1). When feeding on insects and fruits, tamarins restrict much of their locomotor activity to dense vine tangles and fragile branches in the forest canopy. About 66% of feeding and 85% of foraging occurred on branches less than 10 cm around (Table 2). Exudates, however, are taken from a very different set of arboreal supports. About 96% of all cases recorded as large-branch feeding—over 50 cm around—involve clinging to vertical trunks while feeding on plant exudates. Claw-like tegulae enable this primate to grip vertical and sharply inclined supports that are too large to be spanned by its tiny appendages. Use of large vertical trunks by this monkey is associated principally with its exudativorous feeding habit. Particularly during the months of June and July the Panamanian tamarin spent numerous hours clinging to the trunks of "espave," *Anacardium excelsum*, feeding on a dark reddish-brown exudate oozing from the bark. The exudate from this single species of tree accounted for over 23 per cent of the non-insect portion of the tamarian diet.

TABLE 1
DIET OF THE PANAMANIAN TAMARIN
Percentage of time spent feeding and foraging on various types of food items

Type of food	Number of cases	Percent composition of the diet
Insects	597	39.4%
Fruit	582	38.4%
Exudates	218	14.4%
Leafy material	35	2.3%
Buds	26	1.7%
Seeds	22	1.5%
Flowers	02	0.1%
Unknown	32	2.0%
TOTAL	1514	99.8%

TABLE 2
SUPPORT PREFERENCE DURING FEEDING
IN THE PANAMANIAN TAMARIN

Support Circumference	Number of Observations	Frequency
Under 2.5 cm	158	20.7%
2.5–5.0 cm	139	18.2%
5.0–10.0 cm	208	27.3%
10–25 cm	58	7.6%
25–50 cm	27	3.5%
over 50 cm	172	22.5%
TOTAL	762	99.8%

SUPPORT PREFERENCE DURING FORAGING
IN THE PANAMANIAN TAMARIN

Support Circumference	Number of Observations	Frequency
Under 2.5 cm	248	36.9%
2.5–5.0 cm	175	26.0%
5.0–10.0 cm	148	22.0%
10–25 cm	46	6.8%
25–50 cm	14	2.0%
over 50 cm	40	5.9%
TOTAL	671	99.6%

A typical gum feeding session involves 2–4 group members clinging to moistened areas of the trunk licking, scraping, and pulling at accumulations of exudate.

La dieta de los monos tití comprende 39% de insectos, 38% de frutos y 14% de exudados vegetales. Estos animales se alimentaron del exudado oscuro del espavé, Anacardium excelsum, durante bastante tiempo.

After feeding for a few minutes, individuals either adjust their position on the trunk and continue to feed, or retire to some other part of the canopy. Those who leave are immediately replaced by other group members. Feeding bouts such as these may last for up to 45 minutes and occur principally in the early morning (36.2%) and late afternoon (43.6%; Fig. 1).

Activities surrounding exudate feeding are highly coordinated and stereotypic. That is, individuals enter the "gum" tree as a group and remain near the tree until all have finished. There were 7 large espave trees in the home range of the main study group. Three of these were in the group's core area, and only these were used and reused extensively. For example, during June, the group visited the same two espave trees twice a day for 10 consecutive days. Each time gum was taken from the same region of the trunk, an area where there had been severe damage to the bark. Thus the availability and distribution of suitable trees appears to have had a direct effect of group movement.

Unlike marmosets, tamarins do not gouge holes in the trunk to stimulate the flow of exudates. Rather, gum flow is the result of external factors such as natural fissures in the bark, disease, and especially the parasitic activity of certain wood-boring aphids, beetles, and moths. Charles-Dominique (1977: 41) states, "Homopteran bugs are responsible for a great deal of gum-exudation, which represents the plant's natural defensive response against the bite-wounds inflicted by the bugs' mouth parts." It is likely that by disrupting the wound and removing the semi-hardened exudate seal, tamarins stimulate the renewed flow of gum. In this regard, the Panamanian tamarin is active in an ecological interaction with wood-boring insect parasites and exudate-producing trees. This co-evolutionary relationship has not been studied.

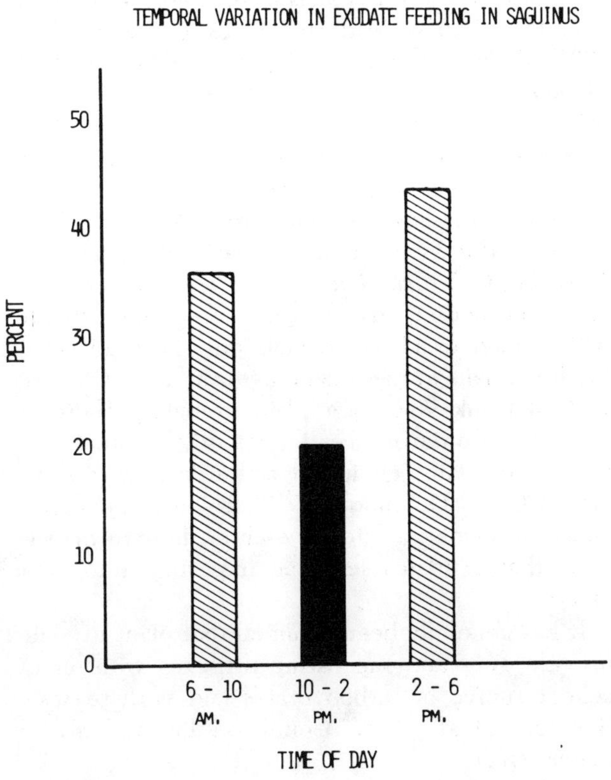

Fig. 1. Temporal variation in exudate feeding in *Saguinus*.

Nutritional Composition of *Anacardium* Exudate

In this study the Panamanian tamarin was observed to eat exudates from the trunks of two kinds of tree, hog plum, *Spondias mombin*, and espave, *Anacardium excelsum*. Both species are members of the Anacardiaceae, an important family of gum producing trees (Howes 1949). The exudate of *S. mombin* was eaten sparingly and accounted for only 2.7% of the exudate portion of the tamarin diet. The remaining 97.3% consisted solely of espave gum. This extreme dependence on the exudate of a single species of tree is noteworthy and suggests that this resource satisfies certain essential nutritional requirements of this primate. A chemical analysis of espave gum samples collected on 2 December 1979 and on 14 May 1980 showed the presence of several minerals including calcium, phosphorus, sodium, magnesium, and potassium. Most of these minerals are present in small amounts. However, espave gum has an outstandingly high calcium content (Table 3). All samples tested had calcium values of between 310mg/100g and 710mg/100g. High calcium content appears to characterize other plant exudates as well. Bearder and Martin (1980) report that the exudate of *Acacia karroo* is calcium-rich and consumed by lesser bushbabies, *Galago senegalensis*, in the Northern Transvaal of South Africa. Although no data are available on cal-

TABLE 3
NUTRITIONAL COMPOSITION OF ANACARDIUM EXCELSUM EXUDATE

	Percent Water Content	Percent Dry Matter	Percent Crude Protein*
Sample 1	55.5%	44.5%	1.8 %
Sample 2	58.5%	41.5%	1.6%
Sample 3	45.2%	54.8%	9.35%
Sample 4	40.3%	59.7%	9.94%

	Mineral Content* (Mg/100G)					
	CA	P	MG	NA	K	CA/P Ratio
Sample 1	710	05	176	0.1	46	142:1
Sample 3	620	10	160	0.2	400	62:1
Sample 4	310	10	130	0.2	480	31:1

Samples 1 and 2 were collected on December 02, 1979 in Panama.
Samples 3 and 4 were collected on May 14, 1980 in Panama.
*Values expressed in terms of dry weight and based on a protein conversion factor of 6.25.

TABLE 4
PRIMATE SPECIES OBSERVED TO FEED ON PLANT EXUDATES

Aotus trivirgatus	Macaca radiata
Aloutatta palliata	Microcebus coquereli
Callithrix jacchus	Microcebus murinus
Cebuella pygmaea	Pan troglodytes
Cebus capucinus	Papio cynocephalus
Cercopithecus aethiops	Perodicticus potto
Cercopithecus nictitans	Phaner furcifer
Evoticus elegantulus	Presbytis entellus
Galago crassicaudatus	Propithecus verreauxi
Galago demidovii	Saguinus fuscicollis
Galago senegalensis	Saguinus mystax
Lemur catta	Saguinus oedipus geoffroyi
Lemur fulvus	Saguinus oedipus oedipus

cium requirements of wild tamarins, values found in espave gum appear to exceed daily needs. Robinson (1980) recommends a daily allowance of calcium for adult *Homo sapiens* of approximately 746 mg.

Los monos tití también se alimentaron de exudados de jobo, Spondias mombin, aunque durante menos tiempo. El jobo, al igual que el espavé, es una Anacardiaceae.

Exudates are generally thought to be protein deficient (Coimbra-Filho 1972; Coimbra-Filho and Mittermeier 1977; Bearder and Martin 1980). However, samples collected in May (Table 3; samples 3 and 4) had a relatively high protein content—approximately 10% dry weight. In contrast, samples collected in December (samples 1 and 2) were nitrogen poor and characterized by protein levels of less than 2%. This seasonal variation in the protein content of espave gum coincides with seasonal patterns of gum feeding in the monkey.

Exudate Feeding in Primates

A number of species of primates feed on mucilaginous secretions that collect on trees. These exudates attract interest because it now appears that they are important to the diet of many primates. Coimbra-Filho and Mittermeier (1977) review exudate feeding in primates. In all, 18 genera and 24 species of primates have been observed to feed on plant exudates (Table 4). For the vast majority of these species exudate feeding is infrequent and apparently of little nutritional importance. However, for a small number of species exudates are a major component of the diet.

Extensive observations of exudate feeding have been reported for *Cebuella pygmaea* (Kinzey et al., 1975; Moynihan 1976; Castro and Soini 1977; Ramirez et al., 1977), *Callithrix jacchus* (Coimbra-Filho 1972; Coimbra-Filho & Mittermeier 1977; Lacher et al. 1981), *Galago senegalensis* (Bearder & Martin 1980), *Phaner furcifer* (Petter 1978; Petter et al., 1975), *Euoticus elegantulus* (Charles-Dominique 1974, 1977), *Perodicticus potto* (Charles-Dominique 1974, 1977), and *Microcebus coquereli* (Hladik 1979). Many of these animals show characters remarkably similar to those in the Panamanian Tamarin. In particular, *Cebuella*, *Callithrix*, *Phaner*, *Euoticus*, and *Saguinus* are small-bodied insect-exudate feeding primates with keeled or claw-like nails on nearly all digits. Modified nails in these genera permit access to large vertical trunks and their plant exudates. These resources are often inaccessible to other primates exploiting the same environment but lacking modified nails (Charles-Dominique 1977). The extreme dependence on plant exudates in each of these forms suggests that exudates serve an important nutritional function.

It has generally been assumed that plant exudates provide gum and sap eating primates with an excellent source of carbohydrates and perhaps water (Ramirez et al., 1977; Coimbra-Filho and Mittermeier 1977).

In the case of the Panamanian tamarin and perhaps

in other gum feeders too, there is no direct evidence that exudates serve merely as a replacement for fruits. Exudates may be a high-energy carbohydrate resource, but, in addition, they provide essential minerals or other nutrients (amino acids) that are generally unavailable in fruits. Thus exudate feeding occurs in addition to fruit feeding and not only as a substitute.

Most observations of gum eating by the Panamanian tamarin occurred during a time of the year when fruits were abundant. Although exudates were a preferred food during this period, fruit and insect consumption continued as well (Fig. 2). Therefore gum feeding is not simply a response to competition from other fruit eating inhabitants of the forest. About 38% of the tamarin diet is composed of fruit. These primates are adept at obtaining fruit, and there was no evidence whatsoever that frugivory was constrained by competition. Instead, exudate feeding is *obligatory* for these small-bodied callitrichids and it provides them with a high-energy resource rich in calcium and protein.

Discussion

Los exudados proveen nutrimentos que no se encuentran en frutos, por consiguiente éstos son alimento obligatorio para estos animales.

The diet of the Panamanian tamarin is heavily dependent on both insects and exudates. Insects provide the Panamanian tamarin with an excellent source of proteins and lipids throughout most of the year. These nutrients are vital to the diet of this highly active small monkey. However in general, insects are a poor source of dietary calcium as much of the calcium is bound up in the chitinous exoskeleton and is nutritionally unavailable to *Saguinus* (see Garber 1984). Thus insectivorous feeding alone may not satisfy the mineral requirements of this animal. This may of particular importance during certain 'critical' periods when calcium requirements are significantly elevated; during these "critical" times exudate feeding becomes a necessary dietary adaptation.

The exudates consumed by the Panamanian tamar-

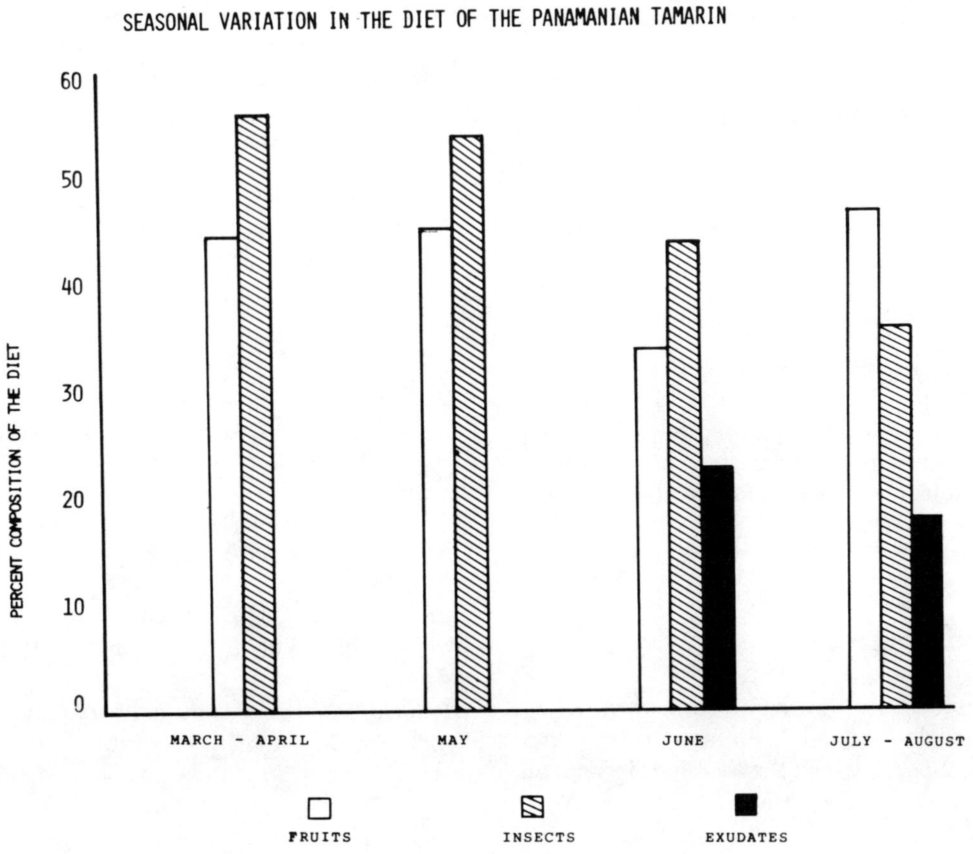

Fig. 2. Seasonal variation in the diet of the Panamanian tamarin.

in (principally *Anacardium excelsum*) were found to be rich in both protein and mineral fractions. Seasonal variation in the protein content of espave gum coincides with seasonal gum feeding patterns in the tamarin. The Panamanian tamarin consumed exudates at a time of the year when protein levels were approximately 10% dry weight. The protein and amino acid content of this food appears to be important to the tamarin diet.

Espave gum has calcium, phosphorus, potassium, magnesium, sodium and other minerals, in concentrations which vary dramatically. It is calcium-rich and provides a large and available supply of this important mineral. Dietary calcium may be of particular concern to the tamarin during periods of pregnancy and lactation. Daly (1979: 338) reports that in mammals, "depletion of maternal calcium stores for reproductive purposes is probably typical." In these small-bodied monkeys, problems of skeletal calcium depletion may be acute owing to the callitrichid reproductive strategy of multiple births coupled with extreme birth weight. In tamarins and marmosets, newborn animals average between 14 and 23% of the mothers body weight (Leutenegger 1979). These values are higher than for any other group of anthropoid primates. During these "critical" periods of the reproductive cycle, increased calcium demands require calcium-rich food such as gums.

Los monos tití recién nacidos son grandes en proporción al peso de la madre y ésto, además de los nacimientos múltiples normales, constituye un problema en relación a la acumulación de calcio en estos animales. Los exudados vegetales ayudan a satisfacer, al menos en parte, esta necesidad.

Data on the seasonality of births, calcium requirements, and serum calcium levels in wild callitrichids are unavailable. However the proposed relationship between tamarin reproductive strategies, exudate feeding, and nutritional composition of plant exudates appears to be important.

Acknowledgments

I thank Robert W. Sussman for his support, intellectual stimulation and friendship during my graduate education, of which this research formed an important part, also Robert Sussman, Michael Watson and Lynette Norr for useful comments on the paper. Finally, without the unselfish devotion of my father during his lifetime, all this would have been meaningless.

This research was supported by grants from the National Science Foundation (Dissertation Improvement Award BNS77–24043), the National Institute of Health (56219B), and from Sigma Xi.

Bearder, S.K. & R.D. *Martin*. 1980. *Acacia* gum and its use by lesser bushbabies, *Galago senegalensis* (Primates: Lorisidae). Int. J. of Primat. 1: 103–128.

Castro, R. & P. *Soini*. 1977. Field studies on *Saguinus mystax* and other callitrichids in Amazonian Peru. pp. 73–78. *In*: D.G. Kleiman, ed. The Biology and Conservation of the Callitrichidae. Smithsonian Instit. Press, Washington, D.C.

Charles-Dominique, C.P. 1974. Ecology and feeding behavior of 5 sympatric lorisids in Gabon. Pp. 135–150. *In*: R.D. Martin, A. Walker, & G. Doyle, eds. Prosim. Biol. Duckworth, London.

———. 1977. Ecology and behavior of nocturnal primates. Columbia Univ. Press, New York.

Coimbra-Filho, A.F. 1972. Aspectos ineditos do comportamento de saguis do genero *Callithrix* (Callitrichidae, Primates). Rev. Brazil Biol. 32: 505–512.

——— & R.A. *Mittermeier*. 1977. Tree-gouging, exudate-eating and the 'short-tusked' condition in *Callithrix* and *Cebuella*. pp. 105–115. *In*: D.G. Kleiman, ed. The Biology and Conservation of the Callitrichidae. Smithsonian Inst. Press, Washington, D.C.

Daly, M. 1979. Why don't male mammals lactate? J. Theor. Biol. 78: 325–345.

Garber, P.A. 1980a. Locomotor behavior and feeding ecology of the Panamanian tamarin (*Saguinus oedipus geoffroyi*, Callitrichidae, Primates). Ph.D. thesis, Washington University, St. Louis, Missouri.

———. 1980b. Locomotor behavior and feeding ecology of the Panamanian tamarin (*Saguinus oedipus geoffroyi*, Callitrichidea, Primates). Int. J. of Primat. 1: 185–201.

———. 1984. Proposed nutritional importance of plant exudates in the diet of the Panamanian tamarin, *Saguinus oedipus geofroyi*. Int. J. Primatol. 5:1–16.

Hershkovitz, P. 1977. Living new world monkeys (Platyrrhini) Vol. 1. Univ. Chicago Press, Chicago, Illinois.

Hladik, C.M. 1979. Diet and ecology of prosimians. pp. 307–357. *In*: R.D. Martin & G. Doyle, eds. The Study of Prosimian Behavior. Academic Press, New York.

Howes, F.N. 1949. Vegetable Gums and Resins. Chronica Botanica Co., Boston, MA.

Kinzey, W.G., A.L. *Rosenberger* & M. *Ramirez*. 1975. Vertical clinging and leaping in a Neotropical anthropoid. Nature 255: 327–328. *Lacher*, T.E., & G.A.B. *Da Fonseca*, & C. *Alves*. 1981. Exudate-eating, scent-marking, and territoriality in wild populations of marmosets. Anim. Behav. 29: 306–307. *Leutenegger*, W. 1979. Evolution of litter size in primates. Am. Nat. 114: 525–531.

Moynihan, M. 1976. The new world primates. Princeton Univ. Press, Princeton, NJ.

Petter, J.J. 1978. Ecological and physiological adaptations of 5 sympatric nocturnal lemurs to seasonal variations in food production. Pp. 211–223. *In*: D.J. Chivers & J. Herbret, eds. Recent Advances in Primatology Vol. 1: Behavior. Academic Press. New York.

———, A. *Schilling* & G. *Pariente*. 1975. Observations on the behavior and ecology of *Phaner furcifer*. Pp. 209–218. *In*: I. Tattersal & R.W. Sussman, eds. Lemur Biology. Plenum Press, New York.

Ramirez, M.F., C.H. *Freese* & J. *Revilla*. 1977. Feeding ecology of the pygmy marmoset, *Cebuella pygmaea*, in Northern Peru. Pp. 91–104. *In*: D.G. Kleiman, ed. The Biology and Conservation of the Callitrichidae. Smithsonian Instit. Press, Washington, D.C.

Robinson, C.H. 1980. Basic Nutrition and Diet Therapy. 4th ed., MacMillan Publishing Co., New York.

Weber, M.W.C. 1928. Die Saugethiere: Einfuhrung in die Anatomie und Systematik der recenten und fossilen Mammalia. Vol. 1. Gustaf Fischer, Jena.

Birds of Panama: Biogeography and Ecological Dynamics

James R. Karr

La rica avifauna de Panamá comprende 905 especies distribuidas en 21 órdenes y 84 familias. La mayoría de estas aves provienen de América del Sur. Los paserinos constituyen el grupo más grande, e immediatamente les siguen las aves de agua dulce. Las áreas que se encuentran en etapas sucesionales complejas y avanzadas son de las más ricas en aves y la zona que está a pocos metros del suelo es la más rica de todas. Muchos pájaros emigran en relación a las estaciones. En Barro Colorado ha disminuido la riqueza de algunos grupos de aves, debido al hecho de que es una isla.

The Republic of Panama, a country of only 75,600 km^2, has an unusually rich bird fauna for an area its size. Three major factors are especially important in producing that rich avifauna. First and foremost is the geographic location between 7° and 10° N in the biotically rich tropical belt. A second major factor is its location as a linking and mixing zone between North and South America. Finally, ecological diversity is exceptional for such a small geographic area—the country contains twelve life zones as compared to only one in the Netherlands and ten for the eastern U.S. (Table 1).

Panama is topographically diverse, with a central mountainous area, highest in the west and decreasing in elevation eastward to the lowlands of central Panama (the Canal area). Eastern Panama also has several high mountains, but the highlands are more fragmented than in the west. Flat coastal plains of varying widths are located to the north and south of the centrally located mountains. The Pacific coastal plain is broader and drier than that on the Atlantic slope, especially in the west . This complex of biogeographic position and ecological diversity has created a biota of considerable interest to the amateur as well as the professional biologist.

General Survey of the Avifauna

Over the last century over 900 species of birds have been reported from Panama and its adjacent seas. Ridgley (1976) listed 883, and 21 more have been added. These plus the Rock Dove bring the total to 905 species of 21 orders and 84 families; that is about 75% of the extant orders and 48% of the extant families of the birds of the world. (scientific and common names, following Ridgely 1976, of all species mentioned in the text are included in Appendix I.)

Se han registrado 883 especies de aves en Panamá y desde el año de 1976 se han adicionado 22. Estas especies representan 21 órdenes y 84 familias, ésto es, 75% y 48% de las órdenes y familias existentes a nivel mundial, respectivamente. Se piensa que aproximadamente 693 especies, o sea 77%, se reproducen en el país.

TABLE 1
NUMBER OF LIFE ZONES REPRESENTED IN SELECTED GEOGRAPHIC AREAS

Area	Number of Life Zones
World(total)	123
Tropical	39
Subtropical	31
Warm Temperate	23
Cool Temperate	16
Panama	12
The Netherlands	1
USA(East of 102°W)	10

Of these, 693 species, or 77%, are residents thought to breed in the country (Appendix II). The remaining 208 species are divided as follows:

	No. of species	Percent of fauna
Migrants	129	14
Occasional or rare visitors	34	4
Pelagic visitors	13	1
Introduced species	4	<1
Hypothetical	32	4

Most of the migrants are birds which breed to the north of Panama, generally in temperate North America, and spend the north temperate winter in the neotropics. Most breed in eastern North America since migrant western North American species winter, for the most part, further to the north and west in Central America (Keast & Morton 1980). At least one of the migrants (Brown-chested Martin) breeds far to the south in South America and thus is a migrant to Panama in the Austral winter. Other probably regular South American migrants to Panama include the Patagonian Swallow, usually considered a subspecies (*patagonica*) of the Blue-and-White Swallow (*Pygochelidon cyanoleuca*) and the southern race of the Southern Martin (*Progne modesta elegans*) at least occasionally reaches Panama. The South American Swift (*Chaetura andrei meridionalis*) may also be a regular migrant to Panama. Finally, some of the species listed by Ridgely as pelagic visitants may be regular migrants from the southern hemisphere. These migrants can be expected more or less regularly each year. Some spend extended periods throughout the north temperate winter in Panama, while others occur as transients enroute to South America in August to November and on their return to the north in January to May. The precise status of many of these species is not well documented and, of course, varies from region to region within the country.

The occasional or rare visitor group includes migratory species which rarely occur in Panama (Cinnamon Teal) and vagrants (Jabiru). These as well as the pelagic visitors are more sporadic in their presence in the Republic. Most breed in other regions and have been recorded as stragglers from time to time. The pelagic visitor group is probably the least known because of the inadequate surveys of marine environments adjacent to Panama.

The four species introduced to Panama are Rock Dove, Saffron Finch, House Sparrow, and Tropical Mockingbird. Rock Doves are semi-domestic and well established as commensals of man in urban areas. Saffron Finch is known from the Caribbean coast of the Canal area, having first been sighted in 1951 (Ridgely 1976). Tropical Mockingbird was introduced earlier this century (first record 1932) and continues to spread east and west from the Canal area of central Panama. The House Sparrow is reported only in the last few years and the "success" of its spread from introductions seems clear with apparent breeding in David, Panama City, and in Bocas del Toro (Ridgely, pers. comm.). Records of other exotics include a Common Starling at Albrook Air Force Base in 1979 and a Raven (species?) at Gatun Lake for several years (N.G. Smith, pers. comm.). The latter two species have not established breeding populations in Panama.

Las 208 especies que no se reproducen en Panamá son en su mayoría aves migratorias que provienen del Norte, primordialmente del este de América del Norte y es allí en donde se reproducen. Algunas son aves extraviadas provenientes de otras regiones o del mar abierto. Cuatro especies se han registrado recientemente y no se conoce que se reproduzcan en Panamá. Se requiere confirmar los registros inciertos que existen de 32 especies.

The 32 hypothetical species are included although there is no incontrovertable proof of occurrence in Panama. These are mostly—17 of 32 species—marine or freshwater species. A few (Galapagos Penguin) may have reached Panama with the help of man, perhaps on ships. Ridgely listed species in this group when no confirmed specimens, photographs, or independent sight reports were available. Many of these species are especially difficult to identify.

Finally, of the 693 presumed breeding species, at least five are known to breed in Panama and then depart for South America. Resident breeding races of these species seem to be missing from the country in late wet season and for the first few weeks of the dry season. Several other species (Lesser Nighthawk, Fork-tailed Flycatcher and Streaked Flycatcher) seem to migrate but may not leave the country entirely.

Cinco especies se reproducen en Panamá y en seguida parten hacia América del Sur.

The fact that nearly two dozen new species have been added to the Panama list since *A Guide to the Birds of Panama* was published in 1976 demonstrates that much remains to be learned. It seems unlikely that any species new to science will be discovered within the Republic. However, as our knowledge of the avifauna of remote areas (western Bocas del Toro, the San Blas Coast, Darien, and the Azuero Peninsula) improves, additions can be expected. Further, natural range expansions are always to be expected. Recent examples of this include the White-tailed Kite (Eisenmann 1971) and White-collared Manakin. A Costa Rican mountain endemic—Peg-billed Finch—expanded its range in 1979 into the Cerro Punta area in response to flowering and seed production of bamboos. The species was not seen in 1980 when bamboos did not flower.

Aunque se registrarán especies nuevas a medida que éstas se descubran o que se trasladen a Panamá, probablemente la lista de aves de Panamá continuará declinando debido a la reducción de los hábitat.

Finally, massive modification of habitats by man will add new species as dispersal corridors are opened. The reverse of that may also be expected. Range restrictions or even extinctions within the Republic are to be expected as areas of natural habitat, especially forest, are reduced.

Biogeographic Origin

As noted earlier, the richness of the Panama avifauna is enhanced by its location as a mixing zone for faunas from the north and south. At the family level, the geographic origins of most birds are reasonably well-known (Mayr 1946), although some are disputable. Following Mayr, I have summarized the patterns of origin for the Panama avifauna (Fig. 1). Clearly, the major origin of the avifauna is South America, with 41% of 905 species. Smaller contributions derive from North America (11% of species) and the Old World (12%). In most cases the latter group reached Panama by a dispersal route taking them through North America. A few species (5%) are of pantropical origin, especially the parrots and trogons. A final and substantial portion of the avifauna—34%—is of uncertain origin. The latter are mostly associated with freshwater or marine environments which are hard to classify using a terrestrial, continental perspective.

La mayoría de las aves panameñas provienen de América del Sur (41%), algunas de América del Norte (11%), otras del viejo mundo (12%) y unas cuantas, como los loros y otros, son pantropicales.

Of the 53 land bird families known from Panama, 23 (50% of species) have South America as the region of presumed evolutionary origin (Fig. 2). Only the North American and Old World elements comprise significant portions of the remaining avifauna. The majority of species of North American origin are present in Panama only as migrants.

Finally, the only species in the monotypic family Zeledonidae is endemic to Costa Rica and western Panama. For simplicity I included it as of South American origin. A few species or genera in other families appear to have developed and diversified after reaching middle America.

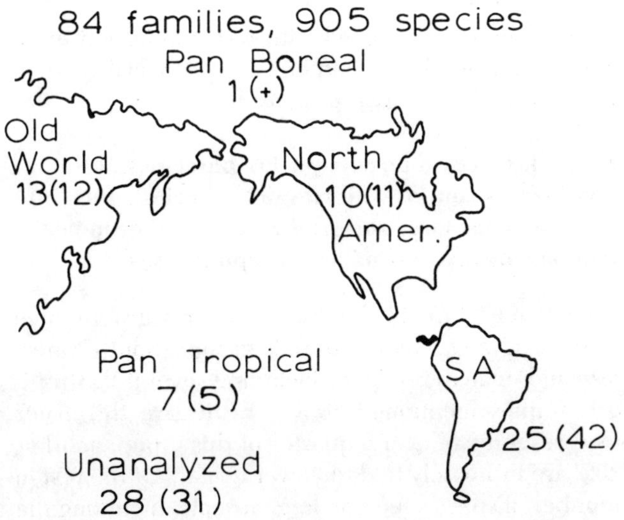

Fig. 1. Geographic origins of all families and species of birds known from the Republic of Panama. After Mayr 1946. Numbers indicate number of families (% of species) with origin as indicated.

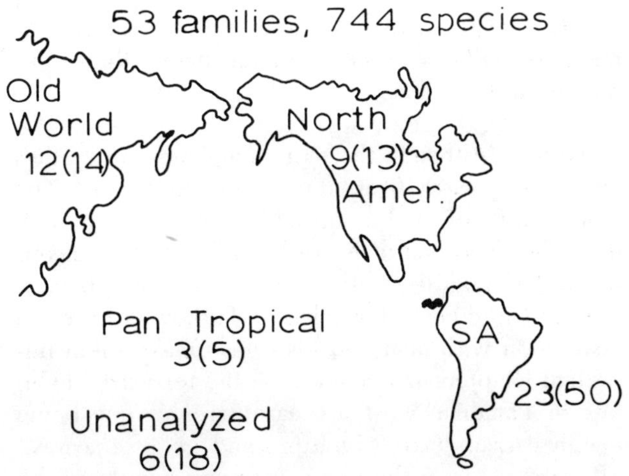

Fig. 2. Geographic origin of the land bird avifauna of the Republic of Panama. After Mayr 1946. Numbers indicate number of families (% of species) with origin as indicated.

Broad Eco-Taxonomic Groups

En América Central, pocos géneros y especies son endémicos y ésto incluye, entre otros, la única especie de la Familia Zeledonidae.

In his monumental work on African birds, Moreau (1966) partitioned that bird fauna into five major groups. In this discussion I follow Moreau except that his aquatic bird group is here divided into two classes: freshwater and marine. With the exception of the Cracidae, all families fall into single eco-taxonomic groups.

Freshwater Birds. This second largest group of families (fourth by number of species) includes species associated with freshwater wetlands and seasonally flooded areas, as well as the shores of lakes and rivers. Several diverse families (shorebirds, gulls and terns, herons) dominate the group, many families containing only a few species. Many of these species, especially the gulls, terns, and shorebirds, are migrants.

Las aves de agua dulce constituyen el segundo grupo en número de familias y el tercero en cantidad de especies.

Marine Birds. As mentioned before, a disproportionate share of these species are on the hypothetical list (10 of 32) because of insufficient surveys. The four dominant families (Procellaridae, Hydrobatidae, Stercocariidae, and Diomedeidae) contain 18 species, 8 of these hypothetical. This group, and to some extent the freshwater birds, are underemphasized in the rest of this presentation.

Las 32 especies de aves marinas, 8 de éstas aún no confirmadas, están distribuidas primordialmente en 4 familias.

Raptors. Although the raptors include few families (5), many species (70) are represented in Panama. The dominant family is the diurnal hawks, eagles, and allies. The nocturnal owls are second with the falcons third. Although the vultures have only four species, the high numbers of Black and Turkey Vultures in association with man and his refuse make them important. Raptors are found in all the terrestrial habitats in Panama. Most are carnivores, a few being mainly carrion feeders (vultures and 2 of 3 caracaras). The size range of the group is impressive: from the smallest, the 14 cm Least Pygmy Owl (estimated at 50–60 g) to the magnificent Harpy Eagle at over 1 m (4.25 kg).

Las 5 familias de aves rapaces incluyen 70 especies. Los halcones constituyen el grupo dominante, luego siguen las lechuzas, buhos y otros. Los buitres (4 especies) son muy conspicuos.

Ground Birds. This group of three families (Cracidae—1 species, Phasianidae, and Tinamidae) is taxonomically diverse with convergence in body form and ecology. The tinamous and cracids are neotropical-derived groups, while the Phasianidae have the Old World as their origin and major center of evolution. After reaching the New World tropics, an offshoot of this group *(Odontophorus)* diversified rapidly in forests of both lowland and highland areas.

Las aves que viven cerca del suelo son taxonómicamente diversas, pero similares en apariencia y ecología.

Other Nonpasserines. This catch-all group of families contains the second largest number of species (193), behind only the passerines which comprise over twice as many species (465). By far the dominant family is the Trochilidae (hummingbirds) with 52 species. Their specialization on flowers has resulted in a rapid diversification into all habitats from sea level to the high mountains—wherever suitable flowering plants are available. Insects also seem to be essential for most hummingbirds. The next two families are largely fruit specialists. Most of the Columbidae (pigeons, doves) are frugivores except the more open country species which feed primarily on grass and other seeds. The Psittacidae (parrots) also feed on seeds and fruits. Woodpeckers are a close fourth in this group of families, with 20 species. Most feed on insects associated with tree trunks and branches, but several also feed on fruits, especially woodpeckers of the genus *Centurus*.

Entre las aves diferentes a los paserinos, los dominantes son los colibríes (52 especies). Las palomas y los loros son frugívoros. Este grupo también incluye los pájaros carpinteros.

Over one-third of the species in this group come from each of two biogeographic groups: South American and unanalyzed land, including several Pantropical families mentioned above. Fruits are the major food resource of over a quarter of this group; another 25% are intimately tied to flower nectar. Of the rest, a number of species feed on large insects, including the puffbirds and cuckoos.

Passerines. This largest of the avian orders includes over half (51%) of the birds of Panama (465 species);

62% of these are in families of South American origin and 20% are of North American origin. Thus, it is in the passerines that we can best see the merging of two major New World avifaunas. In this group too the importance of ecological diversity is clearly displayed.

Los paserinos y aves de percha constituyen el orden más amplio en especies y éste abarca el 51% de las especies panameñas. Las aves insectívoras y las tángaras (gorriones y cerrojillos madereros) son numerosas.

The largest family of Panamanian birds, Tyrannidae (Flycatchers), contains 84 species. This family includes species specialized to environments from wetlands (Pied Water-Tyrannt) to mountain streams (Torrent Tyrannulet), from sea level (Great Kiskadee) to mountain top (Black Phoebe), and from grassland (Fork-tailed Flycatcher) to forest (Sirystes). Indeed, within many of these habitats upwards of 15–20 tyrannids may coexist. Most are insectivores, but a few are specialized frugivores (Ochre-bellied Flycatcher, Olive-striped Flycatcher); many switch from fruit to insects or back with shifts in food abundance.

The second largest passerine family, the Thraupidae (Tanagers), feed primarily on fruits, and a few feed on insects. The Parulidae (Wood Warblers) and the old world Fringillidae (Sparrows) are the next largest families. Most (67%) of the Parulidae are migrants or vagrants from the north; their primary food is insects on their breeding areas, but some eat fruits and even nectar during their period in Panama. The fringillids are a complex and unsettled taxonomic entity, many of which feed on seeds. Some authorities subdivide this group into several families or subfamilies. Among those in Panama, only the Carduelinae, represented by *Spinus*, are basically Old World in origin. The fifth largest family is the neotropical Formicariidae (Antbirds) with 38 species. All are insectivores, and most are found in second-growth and forest environments in the lowlands.

Among these five largest passerine families, two are mostly north temperate in origin, while three are clearly of South American origin.

The Static Avifauna

The four primary gradients in Panama's bird distributions are: elevational, successional, moisture, and within-habitat. These, of course, are far from distinct gradients, especially the successional and moisture gradients.

Altitudinal Gradients

Each species is classified as follows:

Lowland—A species with its main distribution below 1000 m. Many species in this group range up to 1200–1500 m but they tend to be more abundant at lower elevations or, for many, their abundance at higher elevations depends on the activities of man, especially forest clearing and conversion to second growth or "garden" conditions.

Middle elevations—This includes species distributed from 1000–1700 m.

Highland—Species whose distribution is above 1700 m.

Many species do not obey these distributional classes; some cross over two, or even three, of the elevation classes, but if a more refined system of classification were used—including splitting of species among classes—the main conclusions would not be altered.

All four land bird groups have over half their species in the lowlands (Table 2). Both the raptors and the ground feeders decline in species with increase in elevation. However, the nonpasserine and passerine groups both decline from the lowlands to middle elevations (1000 m–1700 m) but increase into the highlands. Overall, two-thirds of land bird species are lowland birds and one-fifth occur most commonly at high elevations. Thirteen percent occur at middle elevations.

Among the suboscine (and more broadly neotropical) passerines, most tend to be concentrated in the lowlands, and fewer species occur in higher regions (Table 3A). One general exception to this is the furnarriids (Ovenbirds), which have the greatest number of species in highland areas.

In contrast, among groups originating in North America (considering residents only for the groups with migrants), the thrushes and warblers are centered in the highlands, while the wrens and vireos are more common in the lowlands (Table 3B).

TABLE 2
ALTITUDINAL DISTRIBUTION OF PANAMA LAND BIRDS (EXCLUDING MIGRANTS) BY ECO-TAXONOMIC GROUPS

Group	Number of Species	Percent of Species		
		Lowland	Middle	Highlands
Raptors	62	92	5	3
Ground	12	58	25	17
Other Nonpasserines	185	70	11	18
Passerines	377	62	14	24
TOTAL	636	67	13	20

TABLE 3
ALTITUDINAL DISTRIBUTION OF PANAMA LAND BIRDS FOR SELECTED FAMILIES OF NEOTROPICAL (A) AND NEARCTIC (B) ORIGIN (RESIDENT SPECIES ONLY)

			Percent of Species		
	Family	Species	Lowland	Middle	Highland
A	Trogonidae	11	64	9	27
	Trochilidae	52	50	15	35
	Dendrocolaptidae	17	65	24	12
	Furnariidae	23	35	13	52
	Formicariidae	38	71	21	8
	Pipridae	11	82	18	—
	Cotingidae	20	85	—	15
B	Troglodytidae	21	76	10	14
	Turdidae	12	8	33	58
	Vireonidae	7	71	—	29
	Parulidae	14	29	7	64

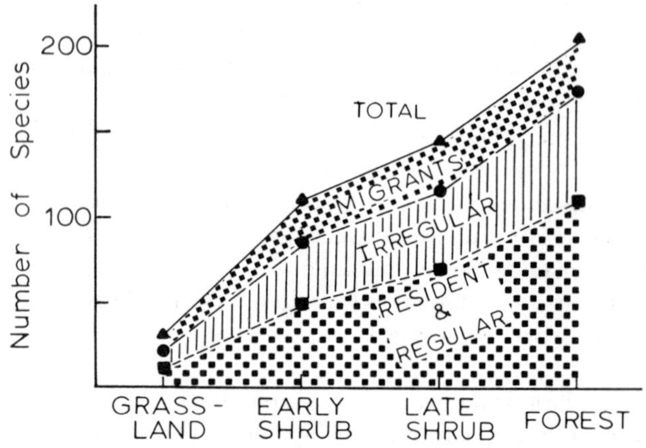

Fig. 3. Changes in species richness (idealized) along a successional gradient in central Panama. Note that the number of resident, and regularly and irregularly occurring species increases along the successional gradient while the number of migrants is similar along the gradient.

Successional Gradients

Ideally the study of a successional gradient involves a study of the same location over a long period of time. However, the more general approach is to select a series of study areas with different times since disturbance and thus different vegetation structures.

Several classic studies of avian succession have been published for North America (Johnston & Odum 1956), but only one has been completed in Panama (Karr 1971, 1976a). Karr censused a series of study plots ranging from grasslands in the area of Pacora to forest in the Canal area. Variation in rainfall in these areas no doubt affects avifauna structure. Rainfall near the grassland sites averages 1900 mm, while that at the forest is probably closer to 2600 mm. However, the differences due to rainfall are likely small compared to those due to vegetation structure.

In one year of research on each of these areas (using four classes), the greatest number of species was found in complex late successional habitats (Fig. 3). Besides changes in species number, there were shifts in food habits along the successional gradient. Grass seeds and small insects dominate in grassland. As soon as a few fruit bearing shrubs appear, one or two fruit eating species is added. Often these fruit-eaters are present during short periods when the pioneer shrubs bear fruit. At about the same time a few nectar-and-insect-feeding hummingbirds begin to use the area, again seasonally.

Las áreas que se encuentran en etapas sucesionales avanzadas y complejas son las más ricas en aves. Las especies y los hábitos de alimentación cambian a medida que la sucesión evoluciona. Las especies de aves que se alimentan de semillas son comunes durante las etapas herbáceas de la sucesión, mientras que las insectívoras, las que se alimentan de néctar y las frugívoras son más importantes durante las etapas más avanzadas.

As one moves into more complex habitat, the seed-eating share of the avifauna declines, while small insectivores, nectarivores, and frugivores increase in importance. Somewhat later, species eating large insects and other arthropods increase in importance.

Successional gradients may also be considered for smaller taxonomic units. A total of eleven wrens are known from the Canal area. These segregate in a successional sequence from the House Wren, which can be found in open garden areas around human habitation, to the Nightingale Wren found in deep forest undergrowth (Fig. 4). Intermediates have an array of habitat (vegetation) requirements, including the White-breasted Woodwren and the Song Wren which seem to be partial to treefall areas in mature forest.

Similarly, among the antshrikes a habitat series exists from the Barred Antshrike, found in relatively open areas, to the Russet Antshrike (a different genus), which frequents middle levels of mature forest (Fig. 5). The width of habitats varies among species, with the Fasciated Antshrike having a very narrow range—habitats associated with vine tangles and thickets, especially along areas such as road cuts—and the Slaty Antshrike common to a wider array of

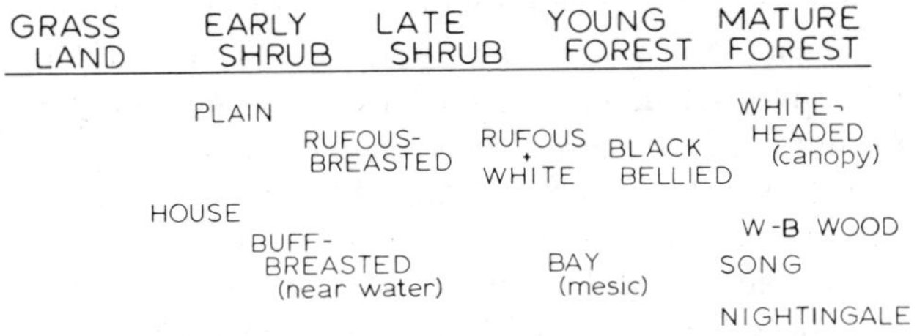

Fig. 4. Distribution of wrens along a successional gradient in central Panama.

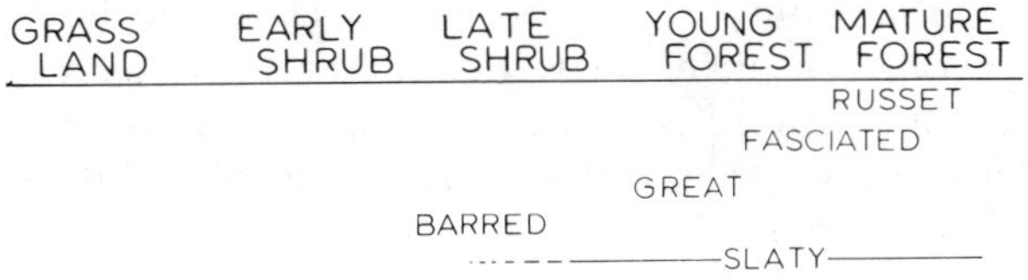

Fig. 5. Distribution of antshrikes along a successional gradient in central panama.

vegetation types. Great Antshrikes are typically at the border of damp woodland.

This pattern of segregation along successional gradients is repeated for many bird families, although not always with the greatest number of species in the more structurally complex habitats. The fringillids (sparrows), for example, dominate the faunas of grasslands and other more open areas and have fewer species in forest habitat.

Moisture Gradients

In the wrens (Fig. 4), several species are distributed along a moisture gradient within habitats of similar structural complexity. The Plain Wren typically occurs at waters edge while the Rufous-breasted Wren is found in drier thicket areas. Among the flycatchers of open country, the Pied Water-Tyrant is associated with wet areas, but the Fork-tailed Flycatcher is more characteristic of dry grassland.

Perhaps the most complex distribution of species along a moisture gradient is along a gradient from dry to wet forest. Even casual observations demonstrate the trend for fewer species to occupy dry than wet lowland forest. The greater richness of the wet forest avifauna in the regions of Achiote Road and Rio Piedras in comparison to Fort Clayton or the Farfan Beach area is an obvious example from central Panama.

Within-habitat Gradients

Gradients in species richness within complex habitats are also striking. My own experience with mist nets suggests that bird richness and activity is greatest in the few meters immediately above the ground. In a survey of the birds of a 2 ha forest study plot along the Pipeline Road near Gamboa I noted 203 species. Of these, 76 species (37%) are characteristically found in the lower levels of the forest, 31 (15%) are more typical of forest canopy areas, and the remainder, 96 (47%), are concentrated in intermediate levels.

Los lugares secos están ocupados por un número menor de especies. La zona más rica está a pocos metros del suelo. Cada familia tiene su distribución propia en relación a la altitud.

Each family has its own species distribution among these levels. Of the five woodpeckers of the study site, four are generally found in the canopy (Cinnamon, Lineated, Crimson-crested, and Black-checked), while one (Crimson-bellied) feeds at low and intermediate levels. Among the antbirds, a few are associated with the ground (antpittas, antthrushes), some are common in the undergrowth (Spotted Antbird), while others are characteristic of intermediate levels (Russet Antshrike). Other families are even more wide-ranging (especially flycatchers), while some are more narrowly restricted (tinamous—ground; manakins—undergrowth and to some extent middle levels; psittacids—canopy).

These spatial and microhabitat gradients, along with gradients in food resource selection within and between families, result in a complex array of bird distributions and, in the aggregate, a forest avifauna of exceptional diversity.

A Dynamic Avifauna

The diversity of dynamics in ecological time is central to an understanding of the Panama avifauna today and to the prognosis for its future. The seasonal influx of migrants from the north temperate zone and bird movements within Panama as a result of the alternation of wet and dry periods are two examples. Although these are a part of a continuum of seasonal range occupation (Karr 1980a), they are separated here for convenience of discussion.

Migrants from the Temperate Zone

As noted earlier, well over 100 species migrate to and from Panama each year. The earliest arrivals begin in late July, with the peak arrival of land birds generally occurring in October. Many species establish "winter territories" on arrival (Kentucky Warbler), while others are more nomadic in their search for food. The thrushes seem to fall into the latter category. In some cases, territories established upon arrival of migrants in the late rainy season are abandoned later as the dry season commences and food becomes more difficult to find. These birds then join mixed species flocks. Some (Tennessee Warbler) even feed on nectar.

En agosto comienza la migración de aves hacia Panamá y el clímax de este proceso ocurre en octubre. Algunas aves presentan patrones establecidos y otras son nómades.

If there is one major lesson from studies of migrants to Panama and other neotropical regions, it is that no single "migrant strategy" is exhibited by all migratory species. The approaches to nonbreeding season survival are as varied as the birds which migrate. For comprehensive, recent review of migrant distribution, behavior, and ecology see Keast and Morton (1980).

Movements within Panama

Many permanent residents in Panama also shift their geographic distribution on a more or less regular basis. Elevational shifts are well known for a number of groups. Hummingbirds (DesGranges 1979), cotingids (Slud 1964), and a wide variety of fruit-eaters seasonally follow their major food sources. In at least one case (some cotingids) populations shift from highland to mangrove forest for the seasonal availability of fruits. However, although these movements have been noted by many ornithologists and birdwatchers, little is known about them: distances involved, variation among years and what determines it, as well as what proportion of the population of a species actually moves and for how long.

Muchas aves residentes se mudan periódicamente a diferentes hábitat o altitudes.

Another group of species with well-known movement patterns is the species of grassland and open savanna. In some cases there are local movements from riparian environments to grassland with scattered shrubs; Lesser Elaenia colonized grasslands for several months in the late dry season to exploit fruit production by *Curatella* (Karr 1976a). In other cases, Ruddy-breasted Seedeater, for example, the species seems to all but disappear from central Panama during most of the dry season. As one moves into habitats with more structural complexity of vegetation, the magnitude of these movements decreases.

But even the forest habitats of the moist lowlands experience shifts in bird abundances over time. One of the earliest and most detailed studies of spatial distribution in tropical forest birds is the work of Willis (1966). In the past twenty years Willis has clarified the behavior and ecology of bird exploitation of army ant swarms (Willis & Oniki 1978 for a detailed review). More recently, Buskirk 1976, Munn & Terborgh 1979, and Gradwohl & Greenberg 1980 have contributed to our understanding of the mixed insectivorous flocks of tropical forest.

Intensive studies by Karr et al. (Karr 1980b; Karr et al. 1982) along the Pipeline Road near Gamboa show the complexity of patterns involved. Detailed samples collected with mist nets in the forest undergrowth are now available for a number of years. Typically, 30 to 38 species are represented in the first 100 captures from this forest. When several dry season samples are compared, the number of species is inversely correlated with the severity of the dry season; more species are captured in dry years.

At least five factors may be responsible for these and other shifts in the use of space by resident birds of central Panama.

1. Shifts between lowland forest and nearby foothill areas are quite regular.

2. More species are captured in mist nets in the forest undergrowth in dry than in wetter years. One explanation for this may be downward shifts in foraging heights for a number of species making them more susceptible to capture in ground-level mist nets (Karr 1980b).

3. Other shifts occur along the moisture gradient across the isthmus. Dry forest species such as Blue Ground Dove and Clay-colored Robin vary in abundance from year to year, apparently correlated with the severity of the dry season. During the driest years they are more common at the moist forest site, presumably due to the more rigorous environment in the drier Pacific coastal forest.

4. A fourth factor in shifting species composition and abundance in this area is treefall dynamics. The importance of treefalls and allied disturbance in the evolution of tropical forest complexity has been the subject of considerable work in recent years (Connell 1978, Foster 1980). At Limbo Hunt Club several changes in species and habitat use are correlated with treefalls (Karr 1980b; Karr et al. 1982; Schemske and Brokaw 1981).

5. Recently, a study began of bird use of four study

plots in an area of about 1 km^2 near Limbo Hunt Club. The four habitats range from an elevated, exposed flat area with forest to forest in a deep, sheltered valley. Results suggest that major habitat selection is going on. For example, some species occur in all habitats throughout the year (Red-capped Manakin) while others shift. During wet periods the latter (Olivaceous Flatbill, Olive-striped Flycatcher) are found over a range of habitats, but in the dry season they are present only in the more sheltered, moist sites.

Ocurren 5 patrones diferentes: 1) Traslados regulares de los bosques de tierras bajas hacia las colinas al pie de las montañas 2) Durante los años más secos se pueden capturar más especies en el sotobosque 3) Algunas especies se trasladan de acuerdo a la precipitación pluvial anual. 4) Algunos cambios de especies están relacionados con las caídas de los árboles. 5) Algunos traslados ocurren como respuesta a la periodicidad de la lluvia.

Similar patterns are clear when undergrowth species are considered in the aggregate. One study area (Limbo Hunt Club) consists of a flat area along a stream, a small rise of about 8 m and a flat upland. Capture rates of some species are higher in one area than in the other (Table 4). Also, significantly, the distribution of birds as reflected by capture rates of all species varies (Table 5). During an especially dry dry season capture rates were significantly higher in the low area than in the upper area. Only three months later, during the wet season, capture rates were identical. During the following relatively wet dry season, there were no differences in the capture rates in the "uplands" and "lowlands". Again recall that these capture rates are for areas only 8 m apart in elevation.

In addition to these small-scale spatial dynamics, avian use of habitats shifts within the day. When I divide the day into three equal 4-hour periods, the ridgetop site has high early morning capture rates but low midday and afternoon capture rates (Fig. 6). In contrast, the mixed site has high capture rates in the morning and late afternoon but lower rates at midday. The deep valley is not an especially good foraging area, so morning and midday capture rates are low. Late afternoon capture rates are quite high due to the use of the area by many species for obtaining water. At this site I have captured an unusual number of forest canopy species (Keel-billed Toucan, Chestnut-headed Oropendula, Scarlet-rumped Cacique, Green Honeycreeper) which seemed to be descending to ground level in search of water.

In another study the species composition of the undergrowth avifauna was sampled for 10–12 month periods nearly a decade apart (Karr, Schemske, and Brokaw 1982). A few of the most abundant species changed in abundance between years. In some cases, shifts seem due to habitat changes (increased area of treefall gaps) while others shift as a response to year-to-year variation in severity of the dry season. In others, seasonal movement of temperate-tropical migrants are responsible. Other seasonal changes included higher capture rates of nectarivores (hummingbirds) in the dry season, a period of peak flowering activity (Croat 1978). Species that follow army ants were captured more rapidly in the wet season.

These examples of dynamics in the Panama avifauna barely scratch the surface in the diverse array of temporal and spatial dynamics occurring daily, seasonally, and from year-to-year at a single forest site. Many other complexities in the Panama avifauna await discovery. Clearly, the popular notion of population and community stability in tropical ecosystems should not be extrapolated to all species, groups, and communities.

TABLE 4
CAPTURE RATES OF SELECTED SPECIES
Species caught significantly more often in upland or lowland portions of Limbo Hunt Club study plot, Parque Nacional la Soberania, Panama.

Species	Lowland	Upland
Chestnut-backed Antbird	6	0
Song Wren	22	4
Red-capped Manakin	63	162
Number of net hours	1805	2277

TABLE 5
CAPTURE RATES FOR BIRDS IN UNDERGROWTH MIST NETS IN UPLAND AND LOWLAND LOCATIONS AT LIMBO HUNT CLUB STUDY PLOT

Year	Season	Region	Number of Net Hours	Number of Captures	Captures Per 100 Net Hours
1979	Dry	Lowland	149	59*	39.6
		Upland	209	46*	22.0
	Wet	Lowland	247	50	20.2
		Upland	256	50	19.5
1980	Dry	Lowland	161	59	36.6
		Upland	187	47	25.1

*significance by X_2, p less than 0.05

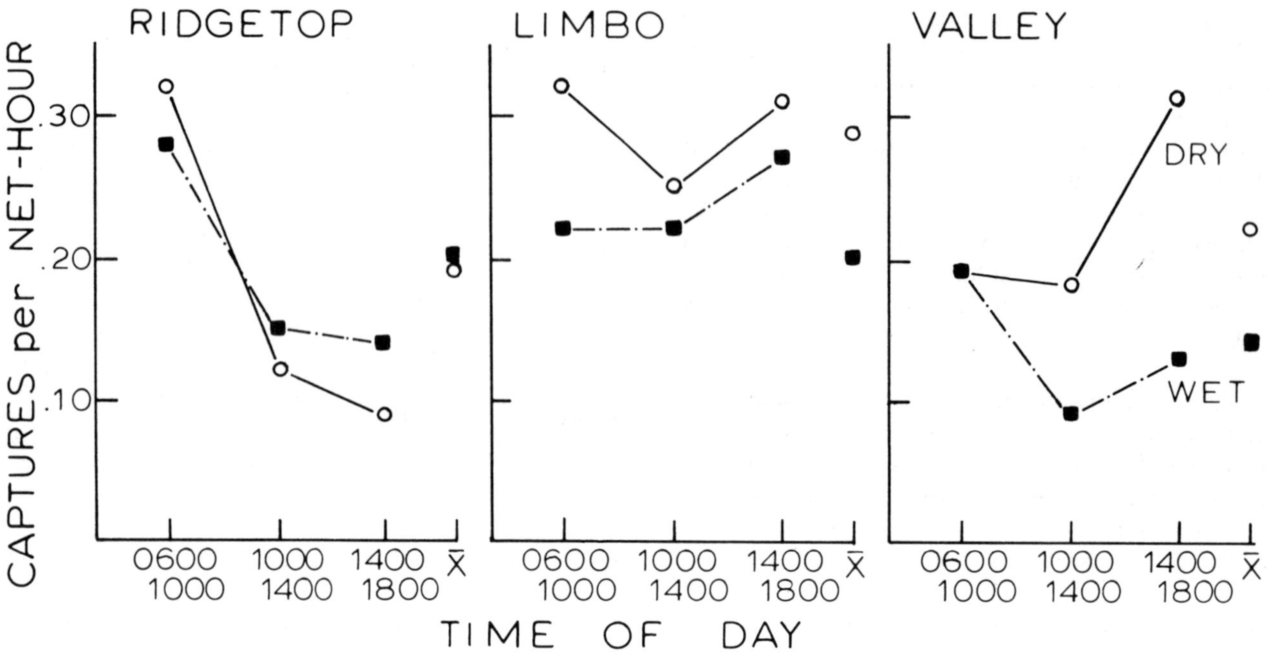

Fig. 6. Mist-net capture rates as a function of time of day in 3 forest study plots near Limbo Hunt Club, parque Nacional "La Soberania," Republic of Panama. Data from dry (March) and wet (July) seasons, 1979.

Ecological Dynamics and the Need for Conservation

The lessons of island biogeography for the design of nature preserves are a major advance in the ecological sciences in the past two decades. While not all controversy has been resolved (Simberloff & Able 1976), it seems clear that with a little foresight we can avert many extinctions.

Early statements on the role of island size as a determinant of species richness included Darlington (1957) and Preston (1962). Shortly thereafter, MacArthur and Wilson (1963, 1967) pioneered examination of the central factors determining the number of species that can survive on an oceanic island. The number of species is determined primarily by an equilibrium established between colonization and extinction rates.

El área disponible determina la riqueza de especies.

Within a few years biologists were applying that theory to the dynamics of species in mainland areas as appropriate habitat is decreased by human or other influences. The development of such information as well as increased recognition of the importance of mosaic phenomena in ecological systems (Pickett & Thompson 1978, Connell 1978, 1980) has provided more insights into the extinction to be expected as habitat area declines.

Barro Colorado Island (BCI) in central Panama has played a central role in understanding the importance of area of preserve in determining the equilibrium number of species (Willis 1974, Willis 1979, Willis & Eisenmann 1979, Wilson & Willis 1975). The most recent published analysis of the BCI avifauna (Willis & Eisenmann 1979) suggests that at least 51 species have been extirpated from BCI since the island was formed early this century. Thirty-four of those species seem to have disappeared because forest growth has replaced their preferred second growth habitat. Sixteen additional species have disappeared despite the continued presence of their required forest habitat. Species in the latter group are indicators of extinction patterns produced by forest fragmentation. A general pattern seems to be what is termed "ecological truncation." Within guilds of similarly foraging birds the largest and most specialized are the first to disappear. In addition, the probability of extinction varies among guilds. In Brazil, for example, large frugivores, large insectivores, large raptors, and some specialized small insectivores are most likely to disappear from small woodlots (Willis 1979).

Desde que Barro Colorado fue convertida en

una isla, ha perdido 51 especies de aves, algunas debido al crecimiento del bosque y otras a causa de su condición insular.

Recently, birds of the undergrowth on BCI were sampled with the same procedures used in studies along the Pipeline Road. A comparison of BCI and Pipelines Road mist-net results yields the following (Table 6, 7):

1. When similar sample efforts are compared, capture rates are about 50% higher on the mainland than on BCI.

2. The number of species captured in similar sample periods is twice as high on the mainland as on BCI.

3. Of nine regularly captured forest undergrowth guilds, six are captured at essentially the same rates (both number of species and individuals) on BCI and the mainland.

4. Capture rates for three other guilds are reduced. Those guilds are the insectivore-nectarivores, foliage insectivores, and ant followers. Major depressions in both species richness (40 to 100%) and individuals (26 to 100%) occur in these groups.

TABLE 6
MIST NET CAPTURE DATA FOR DRY SEASON, 1980, SAMPLES AT BARRO COLORADO ISLAND (BCI) AND LIMBO HUNT CLUB, PIPELINE ROAD, PANAMA

	BCI	Limbo Hunt Club
Number of Net Hours	361	348
Number of Captures	72	106
Number of Species	15	30
Captures per 100 Net Hours	19.9	30.5

TABLE 7
NUMBER OF SPECIES (AND INDIVIDUALS) BY GUILD FOR MIST-NET SAMPLES DURING 1980 DRY SEASON AT BARRO COLORADO ISLAND (BCI), AND LIMBO HUNT CLUB, PIPELINE ROAD, PANAMA

Guild	BCI	Limbo Hunt Club
Terrestrial Frugivores	1 (2)	1 (1)
Arboreal Frugivores	2 (32)	3 (33)
Arboreal Insectivore-Frugivores	2 (3)	2 (3)
Insectivore-Nectarivores	— (—)	4 (15)
Bark Insectivores	2 (4)	1 (1)
Terrestrial Insectivores	1 (2)	1 (1)
Foliage Insectivores	2 (3)	10 (17)
Sallying Insectivores	2 (6)	3 (8)
Ant—followers	3 (20)	5 (27)

Al comparar los datos obtenidos se pueden señalar los siguientes conceptos: 1) Al tomar muestras similares en tierra firme y en la isla de Barro Colorado, un 50% más de aves se capturaron en tierra firme que en ésta última. 2) 100% más especies se obtuvieron en tierra firme que en la isla de Barro Colorado. 3) Seis de las nueve especies de aves capturadas regularmente en tierra firme son igualmente frecuentes en la isla Barro Colorado. 4) Barro Colorado presenta una gran declinación en la riqueza de especies de aves insectívoro- nectarívoras, insectívoras de follaje y seguidoras de hormigas.

Willis has already documented the decline in ant-followers in the forests of BCI. The decline in foliage insectivores probably reflect declining availability and reliability of insect resources on BCI. The same may be true for the nectarivore-insectivores. Local nectar availability is reduced during the dry season because the mountaintop nature of the BCI environment is generally poor for the major food plant of the hermits and other undergrowth nectarivorous hummingbirds. An absence of wetter areas that serve as refuges in dry periods may produce local extinctions which cannot be replaced by recolonization from the nearby mainland (Willis 1974, Karr 1982a,b).

This situation is similar to that on my dry site at the Pipeline Road except that the difficulty of recolonization is minimized on the mainland. Recall the lowered species richness and capture rates on the drier mainland area when residents make seasonal and even daily shifts in foraging areas to survive dry season extremes.

During the dry season virtually all BCI streams are dry. Thus the only moist sites are those in close proximity to the lake margin, areas more likely to support second growth (disturbed) vegetation. This is tragic for populations close to threshold densities (Willis 1979), a common occurrence in tropical forest avifaunas (Karr 1977).

I conclude on a somewhat contradictory note. Year-to-year variation in weather conditions probably plays a major role in allowing coexistence of the many species in Panamanian forest. Inevitable local extinctions are replaced quickly by recolonization from nearby source areas.

Año tras año las fluctuaciones del clima son importantes como causantes de la riqueza de especies en los bosques panameños. Sin embargo, estas fluctuaciones también causan la extinción de especies en áreas pequeñas. Las

modificaciones que el hombre ha hecho al hábitat comprimen el tiempo de evolución a décadas, lo cual puede resultar en la extinción masiva de la avifauna.

However, as man's activities increase habitat modification and fragmentation, the essential mosaic nature of these forests is reduced. As a result the important habitat requirements are no longer available for many species; local extinctions result. In the absence of nearby sources of colonists the long term prognosis is increasing rates of species extinction.

Thus a dynamic avifauna is especially susceptible to the effects of man's modification of environments. By destruction and fragmentation of habitat, man compresses evolutionary time events into periods of a decade or two. The long term prognosis, without careful stewardship, is increasing rates of extinction in what has been an extroardinarily rich avifauna.

Extinctions due to habitat fragmentation and destruction are also important to the segment of the Panama avifauna which spends part of the year in other areas. Recall that nearly 130 species of birds occur as regular visitors to Panama, mostly from North America. The importance of the isthmus as a wintering area is well-documented (Willis 1966, Karr 1976b, Keast & Morton 1980). In addition, the narrow neck of the isthmus serves as a concentraion area of transients migrating to and from South America. Among the raptors, for example, nearly the entire population of Swainson's Hawk and most of the migratory population of the Broad-Winged Hawk pass within a 5 km corridor near Panama City (Smith 1980). Nearly a million individuals of these two species and the Turkey Vulture pass over Panama City each year. Many millions of individuals of other species cannot be counted, even estimated, with any reliability, but their numbers emphasize the importance of the Isthmus of Panama to the avifauna of the New World. Destruction of important stopover areas in Panama would be catastrophic for the avifauna of North America.

The fragile status of several other segments of the avifauna of Panama also deserves brief mention here. These segments include 1) the colonial seabirds which nest on a number of offshore islands; 2) wintering shorebirds which depend on upwelling areas in the Bay of Panama; and 3) the freshwater birds of Gatun Lake. All have been affected by human activities to varying degrees and deserve special protective measures.

Panama's nesting seabirds are concentrated on coastal islands mostly along the Pacific Coast, especially in the Pearl Archipelago. The most abundant large species include Brown Pelican, Blue-footed and Brown Boobies, Neotropic Cormorant, Magnificent Frigatebird, plus several herons and egrets. Historically, these have been little disturbed by human activities but in recent years two problems have developed. The nesting colonies sometimes occur on islands being developed for recreation and tourism. Vacation cottages, golf courses, and other construction are reducing some colonies. Even relatively isolated small islands are influenced by man as boaters land and walk through nesting colonies. Many eggs and nestlings are lost each year to these wasteful and unnecessary disturbances.

The nutrient rich waters of the Bay of Panama provide ideal conditions for a variety of birds. The seabirds mentioned above and a variety of gulls, terns, and shorebirds find ample nourishment as a result of an upwelling of cool nutrient waters during the dry season (December to April). This upwelling results from the offshore displacement of surface water by northerly winds. Increased nutrient availability stimulates increased phytoplankton production which moves through higher trophic levels. Large fish-eating birds such as cormorants, pelicans, frigatebirds, and herons benefit as do the gulls and terns. In addition, coastal areas in the Bay of Panama provide ideal feeding areas for a wealth of transient and nonbreeding visitors of many species, especially shorebirds such as sandpipers, plovers, and others. These food chains also play a role in nourishing the economic system of the Republic of Panama; shrimp, fish and allied products rank as the third largest export product for the Republic. Degradation of those marine systems will result in catastrophic declines in both harvestable marine fish and shellfish and a variety of marine and shore birds.

Finally, I would like to call attention to the catastrophic consequences of apparently minor manipulations of biological systems. Late in the 1960's the Peacock Bass *(Cichla ocellaris)* was introduced into Gatun Lake in central Panama to provide a more viable sport fishery (Zaret & Paine 1973). The initial effect was a sharp reduction in a variety of secondary consumers. Zooplankton populations increased as small fish were consumed by the new predatory fish. The decline in small fish species was immediately followed by decreases in a number of fish-eating birds. Black Terns, for example, used to be common along the boat route from Frijoles to Barro Colorado Island, but now they are rarely seen. The spread of Peacock Bass in Gatun Lake eliminated many herons, kingfishers, terns and gulls from the vicinity of Barro

Colorado Island (Willis & Eisenmann 1979). Many other such examples of negative effects of man's activities on the avifauna could be listed.

It is also worthwhile to mention that man has been instrumental in increasing populations of many species. Black Vultures have benefitted from increasing human populations as have the few introduced species noted above. Among the resident species a number that thrive in modified or urban areas have no doubt benefitted from man's activities. A number of opportunistic omnivores such as the Clay-colored Robin and several flycatchers might be mentioned. Several doves and seed-eating finches which thrive in open, grassy areas have also increased. Many of these and others are more tolerant of disturbance by man to food resources, nesting sites, and other habitat requirements. Finally, examples of range extention like that of the White-tailed Kite (Eisenmann 1971) may, in part, be due to man's influence. For every such species there are many others that decline markedly in the presence of man. Societal concern and management efforts should focus on preservation of the widest array of species, and not be complacent about the success of groups of species with peculiar characteristics which preadapt them to living in association with man.

A great deal is already known about the avifauna of Panama. The people of Panama—and the world—are fortunate to have a field guide of exceptional quality (Ridgely 1976) and a multivolume monographic treatment of birds of their country (Wetmore 1965, 1968, 1973). But much remains to be learned. I hope that the fauna will be preserved by a policy of enlightened societal development so that more can be learned, and future generations of Panamanians can bask in their beautiful natural heritage.

Acknowledgments

My thanks to the people of Panama for aiding me in many ways, including providing a friendly atmosphere in which to work. My research in Panama has been supported by the Smithsonian Tropical Research Institute, the National Geographic Society, the University of Illinois Research Board, and Earthwatch. Helpful comments on this paper were made by E.E. Eisenmann, N. Gale, R.S. Ridgley, and N.G. Smith.

Buskirk, W.H. 1976. Social systems in a tropical forest avifauna. American Naturalist. 110: 293–310.

Connell, J.H. 1978. Diversity in tropical rain forests and coral reefs. Science. 199: 1302–1310.

———. 1980. Tropical rain forests and coral reefs as open non-equilibrium systems. Pp. 141–163. *In* R.M. Anderson, B.D. Turner & L. R. Taylor, eds. Population Dynamics. 20th Sympos. British Ecolog. Soc.

Croat, T. 1978. Flora of Barro Colorado Island. Stanford Univ. Press, Stanford, CA.

Darlington, P.J. 1957. Zoogeography: The Geographical Distribution of Animals. J. Wiley, New York.

DesGranges, J.L. 1979. Organization of a tropical nectar feeding bird guild in a variable environment. The Living Bird. 17: 199–236.

Eisenmann, E. 1971. Range expansion and population increase in North and Middle America of the White-tailed Kite (*Elanus leucurus*). Amer. Birds. 25: 529–536.

Faaborg, J. 1979. Qualitative patterns of avian extinction on neotropical land-bridge islands: Lessons for conservation. J. Appl. Ecol. 16: 99–107.

Foster, R.B. 1980. Heterogeneity and disturbance in tropical vegetation. Pp. 75–92. *In* M.E. Soule & B. Wilcox, eds. Conservation Biology: An Evolutionary-Ecological Perspective. Sinauer Assoc., Sunderland, MA.

Gradwohl, J. & R. *Greenburg*. 1980. The formation of antwren flocks on Barro Colorado Island, Panama. Auk. 97: 385–395.

Johnston, D.W. & E.P. *Odum*. 1956. Breeding bird populations in relation to plant succession on the Piedmont of Georgia. Ecology. 37: 50–62.

Karr, J.R. 1971. Structure of avian communities in selected Panama and Illinois habitats. Ecol. Monogr. 41: 207–233.

———. 1976a. Seasonality, resource availability, and community diversity of neotropical birds. Amer. Nat. 110: 973–994.

———. 1976b. On the relative abundance of migrants from the north temperate zone in tropical habitats. Wilson Bull. 88: 433–453.

———. 1977. Ecological correlates of rarity in a tropical forest avifauna. Auk. 95: 240–247.

———. 1980a. Patterns in the migration systems between the north temperate zone and the tropics. Pp. 529–543. *In* A. Keast and E. S. Morton, eds. Migrant Birds in the American Tropics: Distribution, Ecology, Behavior and Conservation. Smithsonian Instit. Press, Washington, DC.

———. 1980b. Geographical variation in the avifaunas of tropical forest undergrowth. Auk. 97: 283–289.

———. 1982a. Avian extinctions on Barro Colorado

Island, Panama: a reassessment. Amer. Natur. 119: 220–239.

———. 1982b. Population variability and extinction in the avifauna of a tropical land bridge island. Ecology 63: 1957– 1978.

———, D.W. *Schemske* & N. *Brokaw*. 1982. Temporal variation in the undergrowth bird community of a tropical forest. Pp. 441–453 *In* E.G. Leigh, Jr., A.S. Rand, and D.M. Windsor, eds. The Ecology of a Tropical Forest: Seasonal Rhythms and Long-term Changes. Smithsonian Inst. Press, Washington, DC. ed. Seasonal Rhythms in a Tropical Forest. Smithsonian Instit. Press, Washington, DC.

Keast, A & E.S. *Morton*, eds. 1980. Migrant Birds in the American Tropics: Distribution, Ecology, Behavior, and Conservation. Smithsonian Instit. Press, Washington, DC.

MacArthur, R.H. and E.O. *Wilson*. 1963. An equilibrium theory of insular zoogeography. Evolution. 17: 373–387.

———. & E.O. *Wilson*. 1967. The Theory of Island Biogeography. Monographs in Population Biology. No. 1. Princeton Univ. Press, Princeton, NJ.

Mayr, E. 1946. History of the North American bird fauna. Wilson Bull. 58: 3–41.

Moreau, R.E. 1966. The Bird Faunas of Africa and Its Islands. Academic Press, London.

Munn, C.A. & J.W. *Terborgh*. 1979. Multi-species territoriality in neotropical foraging flocks. Condor. 81: 338– 347.

Pickett, S.T.A. & J.N. *Thompson*. 1978. Patch dynamics and the design of nature preserves. Biol. Conserv. 13: 27–37.

Preston, F.W. 1962. The canonical distribution of commonness and rarity, Part II. Ecology. 43: 410–432.

Ridgely, R.S. 1976. A Guide to the Birds of Panama. Princeton Univ. Press, Princeton, NJ.

Schemske, D.W. & N. *Brokaw*. 1981. Treefalls and the distribution of understory birds in a tropical forest. Ecology. 62: 938–945.

Simerloff, D.S. & L.G. *Abele*. 1976. Island biogeography theory and conservation practice. Science 191: 285–286.

Slud, P. 1964. The birds of Costa Rica: distribution and ecology. Bull. Amer. Mus. Nat. Hist. 128: 1–430.

Smith, N.G. 1980. Hawk and vulture migrations in the neotropics. Pp. 51–65. *In* A. Keast & E.S. Morton, eds, Migrant Birds in the Neotropics: Ecology, Behavior, Distribution, and Conservation. Smithsonian Instit. Press, Washington. DC.

Wetmore, A. 1965–1973. The Birds of the Republic of Panama. 1. Tinamidae (Tinamous) to Rhynchopidae (Skimmers). 2. Columbidae (Pidgeons) to Picidae (Woodpeckers). 3. Passeriformes: Dendrocolaptidae (Woodcreepers) to Oxyrundidae (Sharpbills). Smithsonian Misc. Coll. 150(1,2,3).

Willis, E.O. 1966. The behavior of bicolored antbirds. Univ. Calif. Publ. in Zool. 79: 1–127.

———. 1974. Populations and local extinctions of birds on Barro Colorado Island, Panama. Ecol. Monogr. 44: 153–169.

———. 1979. The composition of avian communities in remanescent woodlots in southern Brazil. Papeis Avulsos de Zool. 33: 1–25.

——— & E. *Eisenmann*. 1979. A revised list of birds of Barro Colorado Island, Panama. Smithsonian Misc. Contr. No. 291.

——— & Y. *Oniki*. 1978. Birds and army ants. Ann. Rev. Ecol. Syst. 9: 243–263.

Wilson, E.O. & E.O. *Willis*. 1975. Applied biogeography. Pp. 522–534. *In* M.L. Cody and J.M. Diamond, eds. Ecology and Evolution of Communities. Harvard, Cambridge, MA.

Zaret, T.M. & R.T. *Paine*. 1973. Species introductions in a tropical lake. Science 182: 449–455.

APPENDIX I
COMMON AND SCIENTIFIC NAMES OF ALL BIRD SPECIES MENTIONED IN TEXT
Nomenclature and order follows Ridgely 1976

Galapagos Penguin	*Spheniscus mendiculus*
Brown Pelican	*Pelecanus occidentalis*
Blue-footed Booby	*Sula nebouxii*
Brown Booby	*S. leucogaster*
Neotropic Cormorant	*Phalacrocorax olivaceus*
Magnificent Frigatebird	*Fregata magnificens*
Jabiru	*Jabiru mycteria*
Cinnamon Teal	*Anas cyanoptera*
Black Vulture	*Coragyps atratus*
Turkey Vulture	*Cathartes aura*
White-tailed Kite	*Elanus leucurus*
Swallow-tailed Kite	*Elanoides forficatus*
Plumbeous Kite	*Ictinia plumbea*
Swainsons Hawk	*Buteo sainsoni*
Broad-winged Hawk	*B. platypterus*
Harpy Eagle	*Harpia harpyja*
Blue Ground-Dove	*Claravis pretiosa*
Rock Dove	*Columba livia*
Least Pygmy Owl	*Glaucidium minutissimum*
Lesser Nighthawk	*Chordeiles acutipenis*
Common Nighthawk	*C. minor*
Ashy-tailed Swift	*Chaetura andrei meridionalis*
Cinnamon Woodpecker	*Celeus loricatus*
Lineated Woodpecker	*Dryocopus lineatus*
Black-cheeked Woodpecker	*Melanerpes pucherani*
Crimson-crested Woodpecker	*Campephilus melanoleucos*
Crimson-bellied Woodpecker	*C. haematogaster*
Fasciated Antshrike	*Cymbilaimus lineatus*
Great Antshrike	*Taraba major*
Barred Antshrike	*Thamnophilus doliatus*
Slaty Antshrike	*T. punctatus*
Russet Antshrike	*Thamnistes anabatinus*
Chestnut-backed Antbird	*Myrmeciza exsul*
Spotted Antbird	*Hylophylax naevioides*
Red-capped Manakin	*Pipra mentalis*
White-collared Manakin	*Manacus candei*
Black Phoebe	*Sayornis nigricans*
Sirystes	*Sirystes sibilator*
Pied Water-Tyrant	*Fluvicola pica*
Fork-tailed Flycatcher	*Muscivora tyrannus*
Piratic Flycatcher	*Legatus leucophaius*
Streaked Flycatcher	*Myiodynastes maculatus*
Great Kiskadee	*Pitangus sulphuratus*
Olivaceous Flatbill	*Rhynchocyclus olivaceous*
Torrent Tyrannulet	*Serpophaga cinerea*
Lesser Elaenia	*Elaenia chiriquensis*
Olive-striped Flycatcher	*Mionectes olivaceous*
Southern Martin	*Progne modesta elegans*
Brown-chested Martin	*Phaeoprogne tapera*
Blue-and-white (Patagonian) Swallow	*Pygochelidon cyanoleuca patagonica*
Raven	*Corvus* sp?
White-headed Wren	*Campylorhynchus albobrunneus*
Buff-breasted Wren	*Thryothorus leucotis*
Plain Wren	*T. modestus*
Rufous-and-white Wren	*T. rufalbus*
Bay Wren	*T. nigricapillus*
Black-bellied Wren	*T. fasciatoventris*
Rufous-breasted Wren	*T. rutilus*
Southern House Wren	*Troglodytes musculus*
White-breated Wren	*Henicorhina leucosticta*
Song Wren	*Cyporhinus phaeocephalus*
Nightingale Wren	*Microcerculus marginatus*
Tropical Mockingbird	*Mimus gilvus*
Clay-colored Robin	*Turdus grayi*
Yellow-green Vireo	*Vireo flavoviridis*
Tennessee Warbler	*Vermirora peregrina*
Kentucky Warbler	*Oporornis formosus*
Common Starling	*Sturnus vulgaris*
House Sparrow	*Passer domesticus*
Ruddy-breated Seedeater	*Sporophila minuta*
Saffron Finch	*Sicalis flaveola*
Peg-billed Finch	*Acanthidops bairdi*

APPENDIX II
SUMMARY OF PANAMANIAN BIRDS

Group-Family	No. of Species	Resident	Migrant	Rare visitor	Pelagic visitor	Introd.	Hypoth.	Geographic Origin
MARINE								
Diomedeidae	3						3	U-O
Fregatidae	1	1						U-O
Haematopidae	1	1						U-S
Hydrobatidae	5				4		1	U-O
Pelicanidae	1	1						U-O
Phaethontidae	1	1						U-O
Phalacrocoracidae	2	1					1	U-O
Phalaropodidae	2		1		1			PB
Procellaridae	6	1			2		3	U-O
Rhynchopidae	1		1					PT
Spheniscidae	1						1	U-O
Stercorariidae	4				3		1	U-O
Sulidae	4	2			2			U-O
Subtotal	32	8	2		12		10	

APPENDIX II Continued

Group-Family	No. of Species	Resident	Migrant	Rare visitor	Pelagic visitor	Introd.	Hypoth.	Geographic Origin
FRESHWATER								
Alcedinidae	6	5	1					OW
Anatidae	15	5	6	2			2	U-FW
Anhingidae	1	1						PT
Aramidae	1	1						NA
Ardeidae	19	16	1	1			1	U-FW
Charadriidae	8	3	4	1				U-S
Ciconiidae	2	1		1				U-FW
Cochlearidae	1	1						SA
Eurypygidae	1	1						SA
Heliornithidae	1	1						PT
Jacanidae	2	2						PT
Laridae	23	3	10	6	1		3	U
Pandionidae	1		1					U-FW
Podicipedidae	2	2						U-FW
Rallidae	12	10	2					U-FW
Recurvirostridae	1		1					U-S
Scolopacidae	27		23	4				U-S
Threskiornithidae	6	3	1	1			1	U-FW
Subtotal	129	55	50	16	1		7	
CARNIVORES								
Accipitridae	38	32	5	1				U-L
Cathartidae	4	4						NA
Falconidae	13	10	3					U-L
Strigidae	14	13		1				OW
Tytonidae	1	1						OW
Subtotal	70	60	8	2				
GROUND BIRDS								
Cracidae (in part)	1	1						SA
Phasianidae	7	7						OW
Tinamidae	4	4						SA
Subtotal	12	12						
OTHER NONPASSERINES								
Apodidae	11	8	1	2				U-L
Bucconidae	8	8						SA
Capitonidae	3	3						PT
Caprimulgidae	10	7	1	1			1	U-L
Columbidae	25	23				1	1	OW
Cracidae (in part)	3	3						SA
Cuculidae	13	9	2				2	OW
Galbulidae	4	4						SA
Momotidae	4	4						SA
Nyctibiidae	2	2						SA
Picidae	20	19	1					U-L
Psittacidae	22	22						PT
Rhamphastidae	7	7						SA
Steatornithidae	1	1						SA
Trochilidae	52	51		1				SA
Trogonidae	11	11						PT
Subtotal	196	182	5	4		1	4	
PASSERINES								
Soboscine								
Cotingidae	21	20					1	SA
Dendrocalaptidae	17	17						SA
Formicariidae	38	38						SA
Furnariidae	23	23						SA
Oxyruncidae	1	1						SA
Pipridae	12	12						SA
Rhinocryptidae	3	3						SA
Tyrannidae	84	69	12	2			1	SA
Subtotal	199	183	12	2			2	

APPENDIX II Continued

Group-Family	No. of Species	Resident	Migrant	Rare visitor	Pelagic visitor	Introd.	Hypoth.	Geographic Origin
Oscine								
Bombycillidae	1		1					NA
Cinclidae	1		1					NA
Coerebidae	10	10						SA
Corvidae	4	4						OW
Cyclaridae	1	1						SA
Fringillidae	44	36	5	1		1	1	U-L
Hirundinidae	14	5	5	2			2	OW
Icteridae	21	18	3					SA
Mimidae	3	1	1			1		NA
Motacillidae	1	1						OW
Parulidae	51	16	27	4			4	NA
Ploceidae	1							OW
Ptiligonidae	2	2						NA
Sturnidae	1					1		OW
Sylviidae	4	4						OW
Tersinidae	1	1						SA
Thraupidae	54	51	2	1				SA
Troglodytidae	21	21						NA
Turdidae	16	12	4					OW
Vireolaniidae	2	2						SA
Vireonidae	13	7	3	2			1	NA
Zeledonidae	1	1						SA
Subtotal	267	193	52	10	—	3	9	
TOTAL	905	693	129	34	13	4	32	

FW—Freshwater; *L*—Land; *NA*—North American; *O*—Oceanic; *OW*—Old World; *PB*—Panboreal; *PT*—Pantropical; *S*—Shorebird; *SA*—South American; *U*—Unanalyzed.

Reproductive Cycles of Panamanian Reptiles

Owen J. Sexton

La reproducción de las lagartijas provee una oportunidad de análisis comparativo, el cual puede suplir criterios para la manipulación experimental en sistemas tropicales holísticos. Los sistemas comparten algunas características y difieren en otras y pueden analizarse mediante comparación. Numerosos científicos han conducido censos de lagartijas de diversos tipos, utilizando diferentes características relacionadas con la reproducción y efectuados a distinto tiempo y en diversos lugares. Los nacimientos de las lagartijas usualmente están ligados a las estaciones. En Panamá existen distintos regímenes de lluvias. En algunas especies la producción de huevos es mayor durante condiciones de mayor humedad. Diferentes especies de Anolis tienen hábitat variados y presentan distintos ciclos reproductivos. La humedad parece ser el factor más importante y opera asociada a la necesidad que posee el huevo en absorber humedad de su medio ambiente. Alimento y predación también pueden influir en el sistema y variar el tamaño de la población.

Investigative techniques of holistic biology are often perceived to be inelegant in comparison to those employed in reductionistic biology. This perception is based upon the conceit that intact, complex, interactive systems are not amenable to experimental manipulation. Yet two major approaches have been used to obtain understanding of holistic systems. The first is the use of mathematical models, and the second is the use of comparative studies. The comparative approach is employed in the present treatment.

Comparative studies stress the use of systems, be they taxa, physiological responses, seasons, habitats, areas, successional stages or ecosystems, which share certain characteristics but which differ in others. The tropics offer a unique opportunity to exploit the comparative approach. Some tropical systems are among the most complex biological communities on earth, and as such, they present one end of a continuum of biological complexity, the other end historically being the supposed simplicity of arctic communities. Just as important has been the realization that tropical systems themselves consist of a mosaic of community types. This diversity provides the framework not only for comparative studies within the tropics proper but between tropical and higher latitude systems.

An influential paper by Dobzhansky (1950) suggested that tropical systems are organized primarily by biotic agencies reflecting the interactions of the diverse species present. Much of the recent work in tropical biology has stressed two subsets of biotic interactions: competition and predation. Dobzhansky further suggested that temperate and arctic systems tend to be regulated by abiotic factors. But physical factors are significant in organizing at least components of tropical systems.

Temporal Variation of Reproductive Cycles

A number of papers (Table 1) have presented information about the occurrence of reptile reproduction which was accumulated largely in the course of other studies. Since such efforts produce point data, their usefulness in setting seasonal limits to the reproductive season is limited.

Non-comparative Studies

In the next level of study some aspect of reproduction is followed within one population for at least a

TABLE 1
CASUAL OBSERVATIONS ON REPRODUCTION OF PANAMANIAN REPTILES

Order or Suborder	Family	Species	Location	Activity	Time	Authority
Testudines	Emydidae	*Chrysemys scripta*	Rios Tuquesa and Chucunaque (Darien)	Nesting	4–24 Mar. '24	Breder '46
Crocodilia	Crocodilidae	*Crocodylus acutus*	Rios Tuquesa and Chucunaqne	Eggs and hatching	Mar-Apr '24	Breder '46
		Crocodylus acutus	Barro Colorado Island (=BCI)	Nesting	Feb, '66	Rand '68
Sauria	Sphaero-dactylidae	*Gonatodes albogularis*	BCI	Hatching	29 Nov, '4	Swanson '45
		Gonatodes albogularis	Colon	Eggs	late Jan '24	Breder '46
		Gonatodes albogularis	Rio Canclon (Darien)	Eggs	late May-mid July '61	Sexton et al. '64
	Iguanidae	*Anolis fraseri*	Rio Canclon	Eggs	mid June '61	Sexton et al. '64
		Basiliscus basiliscus	BCI	Hatched eggs	7–8 Mar, '34	Netting '36
		Basiliscus basiliscus	BCI	Nesting	24 June '6	Ortleb '65
		Corytophanes cristatus	BCI	Eggs laid	5 Feb '62	Ream '65
		Iguana iguana		Eggs	24 Feb	Hallinan '20
		Iguana iguana		Nesting	early Feb-March	Swanson '50
Serpentes	Colubridae	*Leimadophis epinephelus*	BCI	Gravid females	late June '62	Sexton and Heatwole '65
		Leptophis ahaetula	BCI	Gravid females	27 July '62	Sexton and Heatwole '65
		Leptophis ahaetula	BCI	Eggs	25 June '68	Rand '69
		Oxybelis aeneus	Rio Canclon (Darien)	Eggs	13 July '61	Sexton and Heatwole '65
		Oxybelis brevirostris	Cerro Campana	Gravid female	18 Mar '6	Sexton and Heatwole '65
	Viperidae	*Bothrops atrox*	BCI	Birth	15 April '44	Swanson '45
		Bothrops atrox	Rio Canclon	Enlarged ova	23 Dec '61	Sexton and Heatwole '65

year. Most such studies have concentrated upon monitoring the female cycle since gross examination of their reproductive organs by autopsy is relatively easy. Studies of the male cycle must make recourse to microscopic techniques. Additional aspects which have been monitored are signs of nesting activities and the presence of eggs or of recently hatched or born young.

The general procedure followed can be illustrated by using the female cycle of an egg laying species. Two criteria must first be established. The first is the sequence of reproductive events, from the enlargement of the ovarian follicles (including deposition of yolk), through ovulation, to migration of the ova down the oviduct with the concomitant production of the investing eggshell. The numbers of reproductive units at each stage, as well as the number of clutches per annum, are obviously important. Characteristics of other reproductive organs (diameter of the oviduct) can be followed. The second criterion is the determination of body size at sexual maturity. A series of females of differing body lengths—usually snout-vent length, (SVL)—or weights is examined by autopsy. On the basis of the minimum size at which females exhibit some positive sign of gonadal or oviducal activity, a minimum value for adult size is established. Criteria 1 and 2 can be used for m_x values in a life table.

Licht (1967) devised a hierarchial scheme for the state of sexual readiness of male lizards based upon maturation and location of the male gametes (Table 2). Ayala & Spain (1975) developed a blood smear technique which can partially replace the autopsy procedure so that animals need not be killed, and the reproductive status of individuals can be followed. Blood taken from lizards is smeared on a slide and stained with a Giemsa stain. The blood of males and of non-ovigerous females stains a bluish-grey color while that of females undergoing oogenesis is stained a rose to brick-red color. This differential staining is associated with the high levels of lipoprotein in the blood plasma during yolk production. The technique does not provide information about the number of eggs nor of their stage of development.

El estadio sexual de las hembras de las lagartijas puede establecerse mediante medidas de longitud y de peso, al igual que con observaciones de frotis de sangre.

TABLE 2
CLASSIFICATION OF SPERMATOGENIC STAGES IN *ANOLIS CAROLINENSIS*[a]

	Histological condition	
Stage	Seminiferous tubules	Epididymis
1	Involuted, with only spermatogonia	atrophic, empty
2	Primary spermatocytes appearing	atrophic, empty
3	Secondary spermatocytes and early spermatids abundant	atrophic, empty
4	Transforming spermatids with a few spermatozoa	atrophic, empty
5	Spermatids and spermatozoa abundant	hypertrophied, empty
6	Spermatozoa abundant (maximal level of spermiogenesis	hypertrophied, with many sperm
7	Spermatozoa abundant but spermatids and spermatocytes greatly reduced	hypertrophied, with many sperm

[a]Modified from Licht (1967) and used with the kind permission of the Wistar Press.

After the two criteria have been established, the next step is to tabulate the reproductive condition of animals collected at regular intervals. The basic data, the number of adults within a reproductive category, are converted to percentages or frequencies and are usually plotted against some aspect of time.

La condición reproductiva de los animales puede tabularse a intervalos y correlacionarse con las estaciones y con otros factores.

Sexton and Turner (1971) studied the seasonal variation in reproduction of the sphaerodactylid lizard *Gonatodes albogularis* at Ancon over a thirteen month period. Both the reproductive condition of males and females and the presence of oviposited eggs were monitored. Males were always sexually active. Ovigerous females were most common in the May-August interval and least common or absent in February-April. Oviposition sites (tree crevices, flower pots, etc.) were censused periodically for eggs (Fig. 1), and egg counts were minimal in the January-March period. Thereafter, they increased until maximal levels were attained in September-October and then began to decline. Telford (1971) corroborated these data.

Telford and Campbell (1970) examined reproduction in the viviparous xantusiid lizard *Lepidophyma flavimaculatum*, primarily near Achiote, Colon Province. All 50 individuals collected were females, suggesting that this population is parthogenetic. The young are born in late April-early June.

Rand (1968) reported on the nesting aggregations of the iguana *(Iguana iguana)* over a three year period on the islet of Slothia off Barro Colorado Island (BCI). The gravid females swam from BCI to Slothia to oviposit eggs in nests dug in a small clearing on the islet. Nesting began at the end of January or early in February and terminated in early March. Hatchlings first appeared about 1 May.

En las lagartijas Gonatodes albogularis y Lepidophyma flamiaculatum, al igual que en la iguana (Iguana iguana) la producción de huevos y los nacimientos están asociados a las estaciones. Qué regula los ciclos ligados a las estaciones?

These temporal studies indicate that the production of eggs and the birth of young are seasonal for these species. What factor(s) then might regulate the seasonal nature of the cycles, and how is such regulation mediated?

The abiotic factors potentially involved include var-

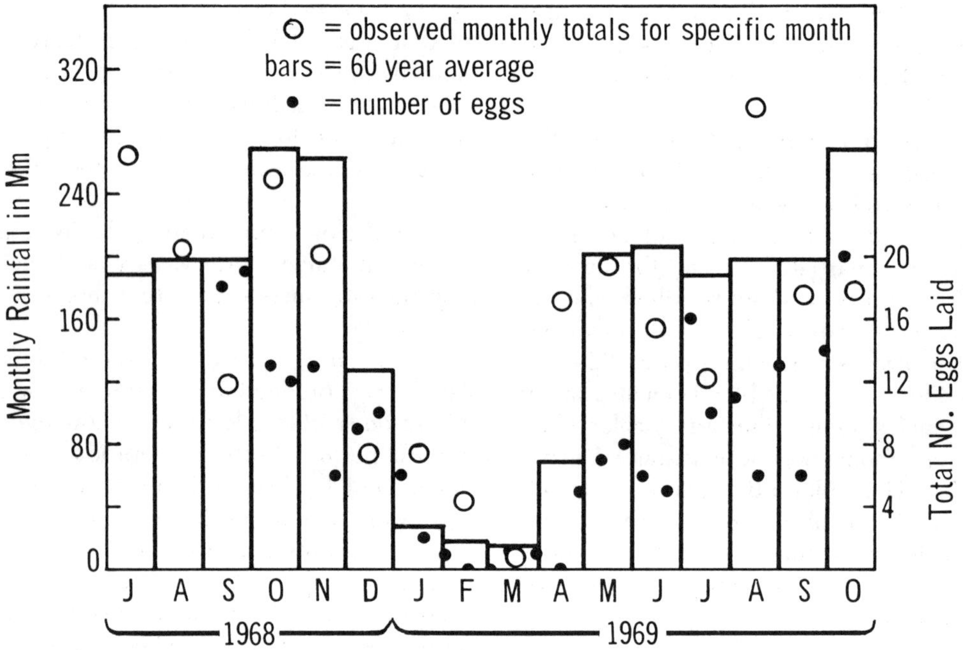

Fig. 1. Summary of annual precipitation and the reproductive cycle of *Gonatodes albogularis* as measured by the number of eggs oviposited in seven selected sites at Balboa Heights, Canal Area. Solid dots = number of eggs found; open circles = monthly rainfall observed at Balboa Heights between 1898 and 1964 (data from Meteorological and Hydrographic Branch of the Panama Canal Company).

ious features of climate. The biotic equivalents include the availability of food and the intensity of predation. Long-term records of the Meteorological and Hydrographic Branch of the Panama Canal Co. provide data to assess potential links between meteorological changes and reproductive events. Sexton et al. (1971) demonstrated the annual variation in rainfall at three locales extending across the Isthmus of Panama (Table 3). Balboa Heights, on the Pacific side, has the lowest mean annual rainfall (1779 mm). Barro Colorado Island, in the Caribbean drainage but near the center of the Isthmus, averages 2729 mm per year. The Gatun records, from a Caribbean coastal locale, average 3140 mm. Thus, a rainfall gradient extends south to north across the Isthmus. Equally important is the monthly variation in rainfall at each locale (Table 3). January-April tend to be dry months at each locale; March is generally a very dry month at all three sites. Beginning in May rainfall increases to high levels through November. October is wettest on the Pacific side but November is wettest at the other two. Both Caribbean locales may have a "little dry season" or "veranillo" during September. December is transitional between wet and dry seasons. Thus, there is a two dimensional precipitation gradient, in locale and in time, across the Isthmus.

La información obtenida en tres sitios de colecta demuestra que el gradiente de precipitación se extiende del sur hacia el norte a través del área central de Panamá. La mayor precipitación se presenta en el norte. Enero a marzo tiende a ser seco y el mes de marzo muy seco. Las lluvias comienzan en mayo y aumentan hasta noviembre. En el Pacífico, octubre es el mes más húmedo, pero noviembre es el más húmedo en dos de los sitios de colecta del norte. En Panamá la precipitación varía asociada al tiempo y al área.

A frequency or probability distribution of total rainfall per month is another estimate of the predictability of precipitation (Fig. 2). Data for these distributions were also collected by the Panama Canal Co. over the 40 year span, 1925–1965. There was a total of 40 samples for each month. Each set of 40 samples was apportioned among rainfall classes of one inch interval (0–0.99, 1.0–1.99, 2.0–2.99 inches of total rain per month, etc.). As an example (Fig. 2), at Balboa Heights in February the frequency of occurrence of total rainfall between 0 and 0.99 inches is approximately 0.85 (34 months out of 40), and more than 1.0 inches is 0.15. These observed frequencies can then

TABLE 3
SUMMARY OF ANNUAL PRECIPITATION UNTIL 1964 ALONG A TRANSECT ACROSS THE ISTHMUS OF PANAMA IN THE CENTRAL ZONE AREA[a]

	Balboa Heights		Barro Colorado		Gatun	
	Mean monthly prec. (mm)	Per cent of annual total	Mean monthly prec. (mm)	Per cent of annual total	Mean monthly prec. (mm)	Per cent of annual total
Jan	28	1.6	56	2.1	84	2.7
Feb	18	1.0	33	1.2	48	1.5
March	15	0.8	31	1.1	46	1.5
April	69	3.9	86	3.2	119	3.8
May	201	11.3	279	10.2	329	10.5
June	206	11.6	288	10.6	295	9.4
July	188	10.6	297	10.9	325	10.4
Aug	198	11.1	318	11.7	343	10.9
Sept	198	11.1	262	9.6	285	9.1
Oct	269	15.1	356	13.0	412	13.1
Nov	262	14.7	454	16.6	539	17.2
Dec	127	7.1	269	9.9	315	10.0
Annual total	1779	99.9	2729	100.1	3140	100.1

[a]Data were modified from records of the Meterological and Hydrographic Branch of the Panama Canal Company. Length of records are: Balboa Heights, 66 years; Barro Colorado Island, 39 years; Gatun, 60 years. Data are to the nearest millimeter.

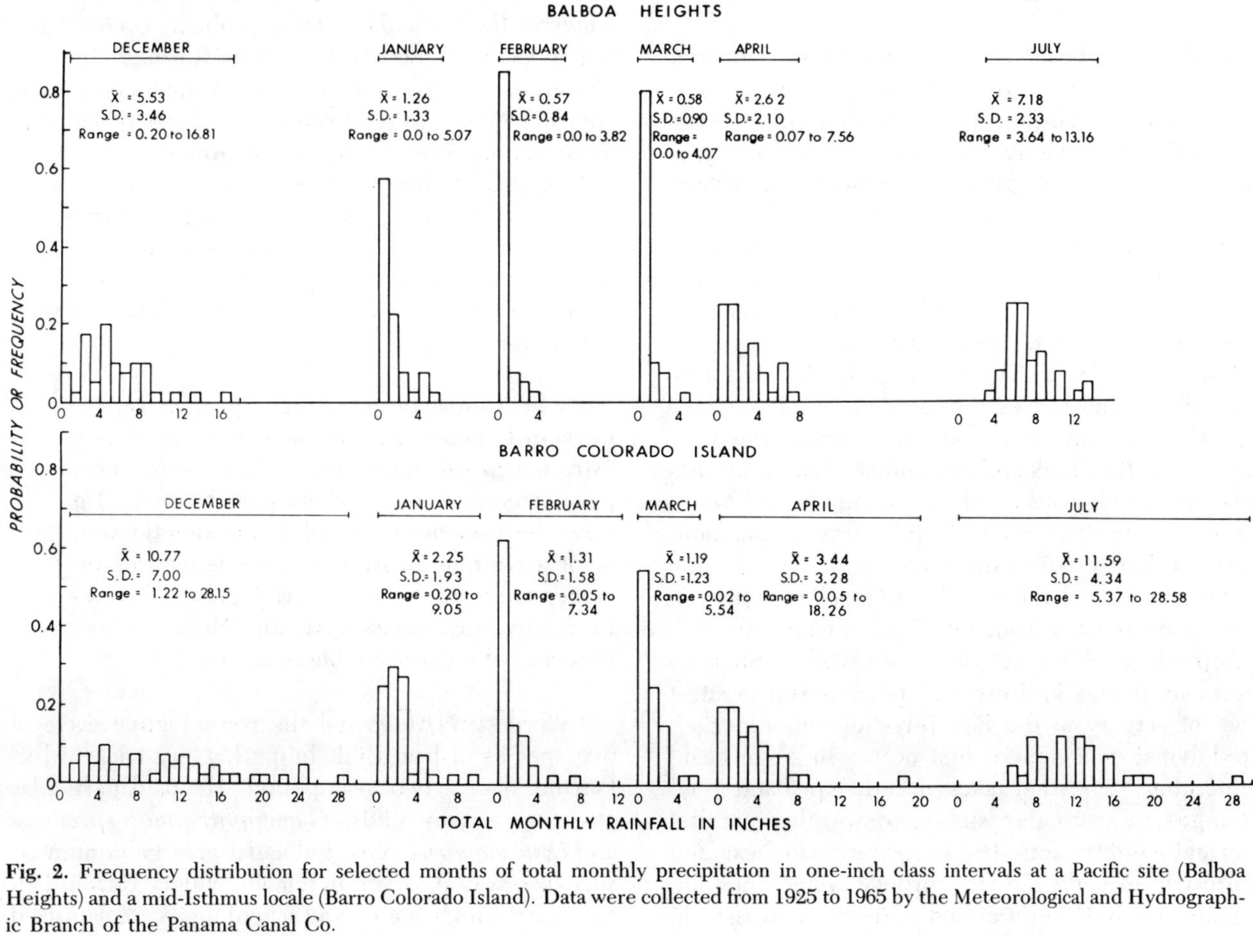

Fig. 2. Frequency distribution for selected months of total monthly precipitation in one-inch class intervals at a Pacific site (Balboa Heights) and a mid-Isthmus locale (Barro Colorado Island). Data were collected from 1925 to 1965 by the Meteorological and Hydrographic Branch of the Panama Canal Co.

be used to estimate rainfall expected in any month. It is easy to see that there is a higher probability of more rainfall at BCI than at Balboa Heights in any of the months shown.

The reproductive cycles of the three species previously discussed can be re-examined in terms of wet (May–December) and dry (January–April) seasons. Egg production in *Gonatodes albogularis* is high in wet season and low or non-existant in dry season. The male cycle seems not to be so adversely affected. The young of *Lepidophyma flavimaculatum* are born during the late dry season and early wet season. The eggs of *Iguana iguana* are laid in mid-dry season and hatch in early wet season.

Temporal Studies

En Gonatodes, la producción de huevos es mayor durante la estación húmeda, sin embargo el ciclo del macho no parece estar afectado. Los Lepidophyma nacen durante el final de la estación seca y comienzo de la lluviosa, mientras que la iguana deposita sus huevos a mediados de la estación seca y éstos nacen a comienzo de la lluviosa.

Studies have been carried out on the reproductive biology of two or more species inhabiting the same area. These concentrate, to varying degrees, on comparing the reproductive responses of different populations to the same gross environment. The groups selected may be at various taxonomic levels.

One of the first and best of such studies was on the life history of three species of Panamanian turtles (Moll & Legler 1971). The greatest effort was spent on the emydid neotropical slider, *Chrysemys (Pseudemys) scripta*. The population inhabited a lagoon of the Rio Chagres at Juan Mina. Analysis of the reproductive status was made from two nearby populations, at Red Tank and at Gamboa. Data from sliders collected elsewhere in Panama and Central America were sometimes used. Samples from populations of two other species, the kinosternid *Kinosternon leucostomum*, and the emydid *Rhinoclemys funera*, were also collected from Red Tank for analysis.

Reproduction is seasonal in *Chrysemys*. Spermatogenesis begins in June and reaches the greatest level of activity in the late July-November period. Epididymal enlargement first occurs in August and September; regression takes place in April and May. Changes in testicular sizes and weights parallel spermatogenetic activity. Testes are smallest and lightest from early January through May. They are biggest in July-November and undergo regression in December and January.

Moll and Legler divided the ovarian cycle into four phases: 1) follicular enlargement; 2) ovulation and intrauterine period; 3) oviposition; 4) a period of quiescence or of ovarian inactivity. Follicular enlargement begins in late August and September with the first set of developing follicles attaining ovulatory size in late November and December. There is continuous follicular development from then until late May although the greatest amount of follicular activity is in the early January-March period. No large follicles were observed after the second half of April. Through the microscopic examination of ovaries, the authors concluded that females laid more than one clutch per nesting season, perhaps as many as five. The number of clutches is positively related to the size of the female.

Observations of nesting were made in 1966. The peak month was February when 45 percent of the 232 observations of nesting were made. Eighty nine percent of the ovipositions occurred in January through March. Overall, the season lasted from mid-December until the end of May. Moll and Legler also studied the other two species using autopsies and examinations of reproductive tracts of specimens from Panama and Central America. In *Rhinoclemys funerea*, the testicular cycle is probably cyclic. Nesting most likely occurs from April through July. In *Kinosternon leucostomum* active spermatogenesis and spermiogenesis were observed in all except January. The ovarian cycle is probably continuous.

Egg production in these three species can be summarized: *Chrysemys scripta*, eggs laid from late wet season through dry season, into early wet season; *Rhinoclemys funerea*, at least from late dry season almost to mid-wet season; *Kinosternon leucostomum*, continuous.

En Chrysemys (antes Psuedemys) la reproducción también está ligada a las estaciones y éstos hacen sus nidos durante la estación seca, pero mayormente en el mes de febrero. En otras dos especies estudiadas al mismo tiempo, se observó que en Kunostermon leucostomum la reproducción estaba casi ligada a las estaciones, mientras que en Rhinoclemys funerea, el ciclo probablemente es cíclico.

Telford (1971) followed the reproductive cycle of five species of lizard inhabiting his city backyard in Panama over a two year period. He paid particular attention to two teiids, *Gymnophthalmus speciosus* and *Leposoma rugiceps*. Follicular enlargement in *G. speciosus* was observed in females collected from December through March and hatchlings were captured from February through May (mid-dry season to early wet season). Females of *L. rugiceps* had active ovarian follicle growth and yolk deposition in samples collected in May through October. Nests and hatchlings were found in August–November. Recruitment in this species occurs more or less in the later half of the wet season. Telford briefly discussed the relationship between the temporal displacement of the reproductive cycles of these two species and potential competition for food between the hatchlings.

Campbell (1973) reported on reproduction in two semi-aquatic species of the iguanid genus *Anolis*. *Anolis poecilopus* lives along clear, rapidly flowing, permanent streams under a complete forest canopy. *Anolis lionotus* inhabits banks along smaller seasonally intermittent streams flowing through forested areas. He carried out his studies in the central area of the Isthmus 6 to 8 km north of Gamboa for somewhat less than one year, December through August, 1968 and April through May, 1969.

Campbell found that *A. poecilopus* hatchlings entered the population in every month of his study. He monitored the presence of oviducal eggs in adult females by palping living individuals and he believed that females with shelled eggs were present throughout the year; the population showed no seasonal change in the proportion of gravid females. Contrarily, *A. lionotus* does seem to undergo some seasonal diminution of reproductive activity in April and May. Females of two populations lacked oviducal eggs in April, and those of one of them also lacked them in May.

En Anolis poecilopus, habitante permanente de las riberas de riachuelos, los nacimientos ocurrieron mensualmente, sin embargo en A. lionotus que habita en riachuelos ligados a las estaciones, la reproducción disminuyó al final de la estación seca, es decir, en abril y mayo.

Campbell associated the difference in reproductive activity in the two species during the end of the dry season to a difference in the availability of water. *A. poecilopus*, the inhabitant of banksides along the permanent streams, possibly suffered less water stress than its congener, a resident of streams which were seasonally reduced to isolated pools.

Comparisons of Taxa

One of the advantages in using the comparative approach is that the investigator can estimate the effect of a particular parameter by examining the response of a biological system along a gradient of that parameter. The preceeding studies have shown that reproductive cycles of a number of species are not continuous, each species exhibiting a characteristic pattern. The patterns can be analyzed in terms of the days, weeks and months of a year or in terms of the annual variation in meteorological patterns. The basic hypothesis of the next series of papers is that reproductive activity in some species of the iguanid lizard *Anolis* is associated with rainfall. The results suggest reasons for the coupling between reproductive activity and moisture.

The initial research (Sexton et al. 1963) was carried out on a forest species, *A. limifrons*, at several sites along the Rio Canclon (Canglon), a tributary of the Rio Chucunaque in the Pacific drainage of Darien. The climate is very seasonal. The lizards were sampled three times during a year: (1) late May until mid-July, 1961 (early wet season); (2) late December, 1961 until mid-January, 1962; (3) late March and early April, 1962 (mid-dry season). Estimates of reproductive activity were obtained incidental to data collected for a structural analysis of the vegetation (Sexton et al. 1964). The number of eggs and the number of hatched individuals associated with randomly selected leaf litter quadrats, shrubs and trees were counted. In addition, the SVL and sex of the captured animals were recorded and the reproductive condition noted by autopsy and microscopic examination of gonadal tissue.

Leaf litter samples were made in wet seasons and mid-dry seasons in two different vegetational stands. In the former period eleven eggs were found in the 75 quadrats of stand 1 and ten in the 120 quadrats of stand 2. No eggs were discovered in mid-dry season. The same trend was observed for eggs associated with the sapling and tree samples. Seven eggs were associated with the 40 saplings and 40 trees of stand 1 and 9 eggs with the 80 saplings and 80 trees of stand 2 during the wet season.

Censos de Anolis limifrons se efectuaron en un sitio del Darién muy influido por las estaciones. Esta especie es anual y al principio de la estación lluviosa sólo se encontraron adultos grandes y jóvenes, pero al final de esta estación los adultos grandes desaparecieron y animales de tamaño mediano eran comunes. Se encontraron muchos huevos durante la estación lluviosa, pero ninguno al final de la estación seca.

The demographic structure of the population (Fig. 3) indicates that recruitment occurs primarily during the wet season and that *A. limifrons* is functionally an annual species. Early in the wet season the population consists of large adults and newly recruited young. In the wet-to-dry season transition, most of the large adults have disappeared and intermediate-

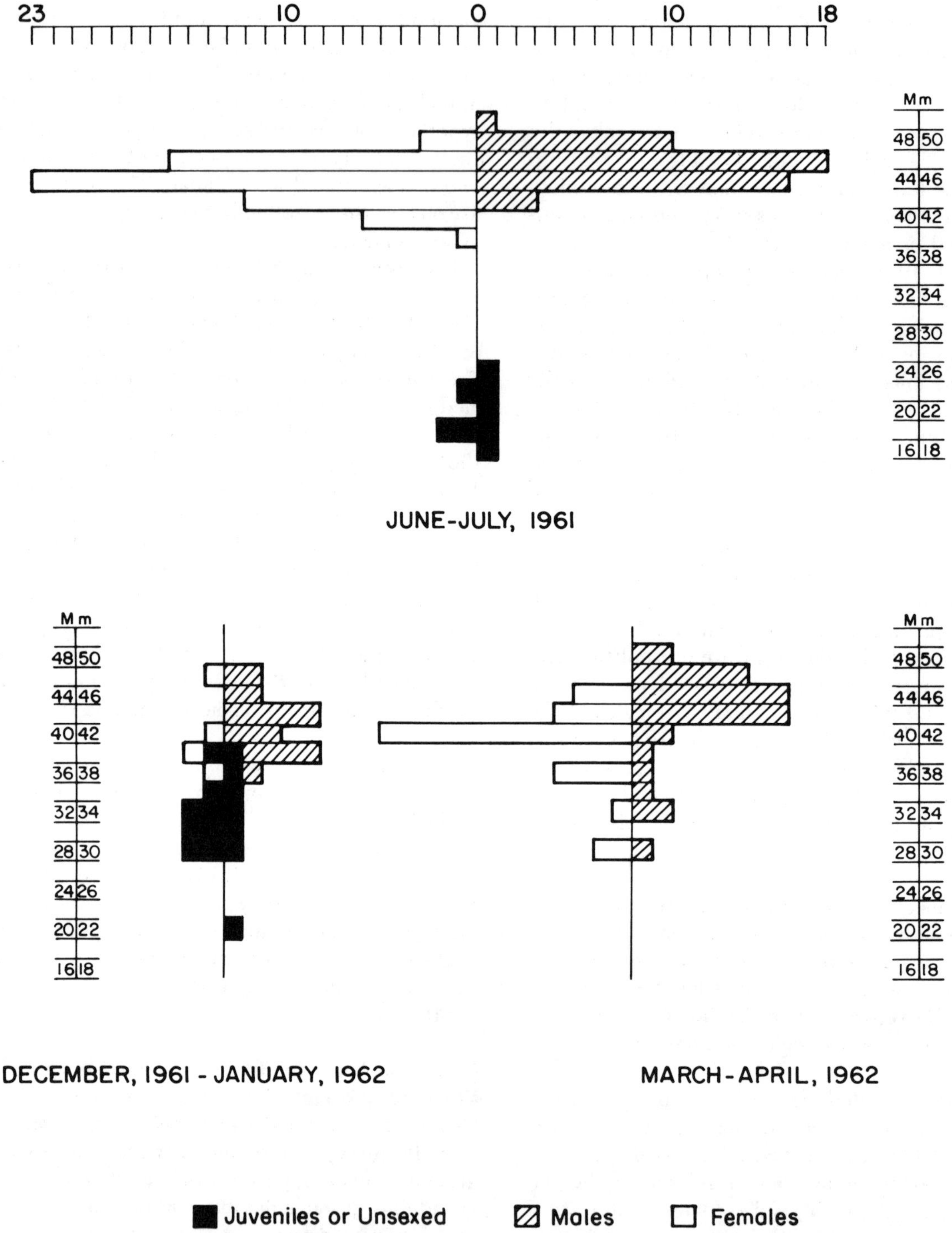

Fig. 3. A comparison of the demographic organization of the *Anolis limifrons* population at the Canclon site during the 3 periods of study. Values along the abscissa are numbers of individuals; those along the ordinate are snout-vent lengths in millimeters (mm).

size animals are common; recruitment is relatively low. In mid-dry season there is no recruitment and the mean population SVL increases. Examination by autopsy indicated that some adult females had enlarged ovarian follicles but that none contained oviductal eggs.

This paper established that the Canclon population of *A. limifrons* was a seasonal breeder. The next paper (Heatwole & Sexton 1966) demonstrated that reproduction in the same species could proceed throughout the year on BCI. The demographic structure showed the same general trends as observed previously (Fig. 4), but hatchlings and eggs are found in all seasons. The overall conclusion was that reproduction was still seasonal, but the decline during the dry season was not so pronounced on BCI as in Darien.

En IBC la reproducción de Anolis limifrons está asociada a las estaciones, pero no tan estrechamente como en el Darién.

A third study (Sexton 1967) estimated the numbers of hatched *A. limifrons* on an equal number of 8 × 8 m quadrats on BCI in the dry seasons of 1963 and 1965. The mean number of lizards per quadrat differed significantly between the two years. In 1963 density averaged 7.5 anoles per quadrat while in 1965 it was 2.5. This contrast was associated with two dry seasons differing in amount of rainfall; that of 1963 was relatively wet while that of 1965 was quite dry. Thus, population densities seemed to have some relationship to the precipitation regime.

These three papers all suggest that the reproductive cycle of *Anolis limifrons* was opportunistic in that reproductive activity could vary as a function of precipitation, both within and between locales. This simplistic hypothesis was tested three dimensionally. The three independent variables were based on areal and temporal differences in rainfall and on species differences in desiccation rates of adult lizards (Sexton et al. 1971). The dependent variable in each case was the reproductive condition of adults of three species

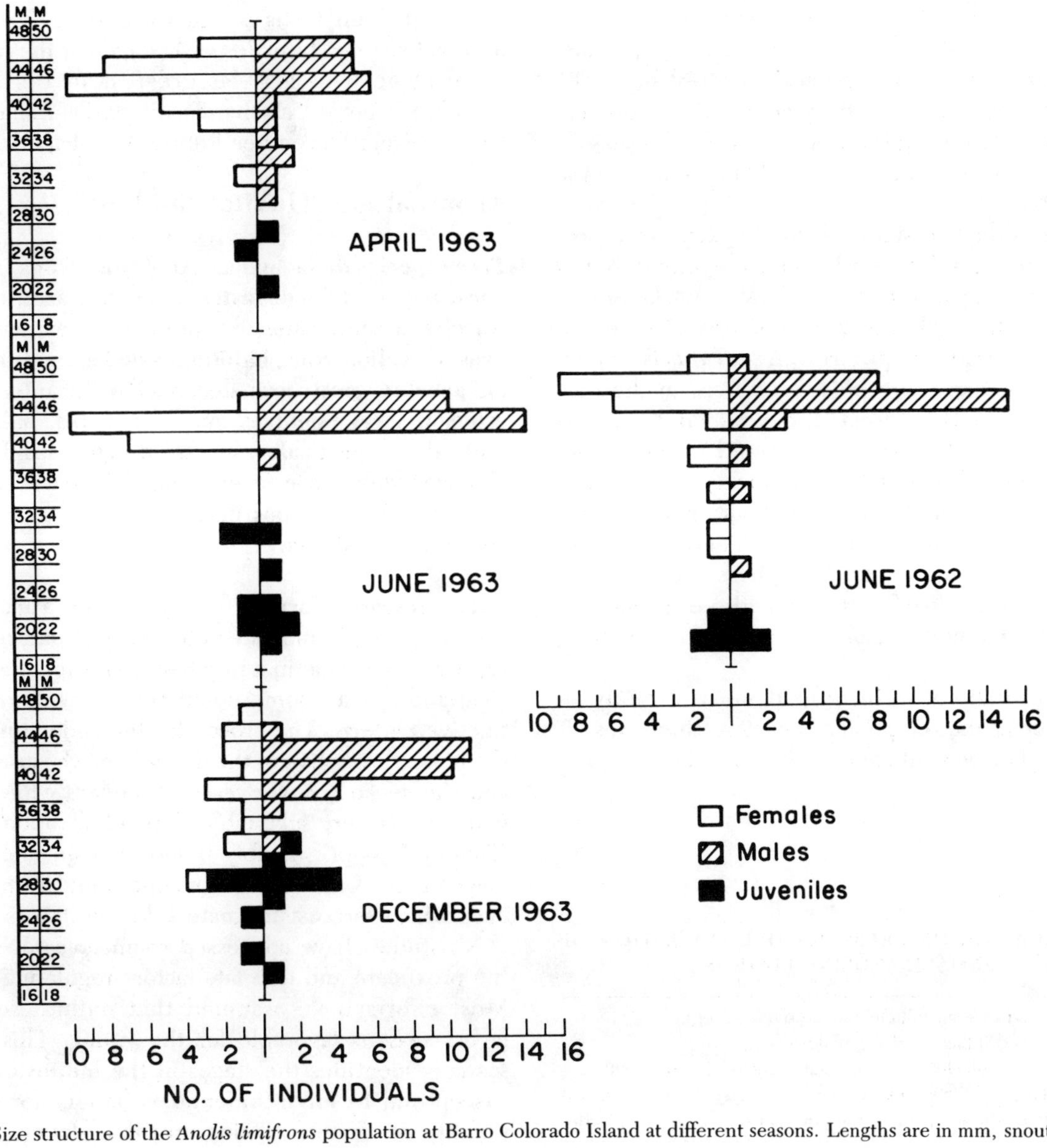

Fig. 4. Size structure of the *Anolis limifrons* population at Barro Colorado Island at different seasons. Lengths are in mm, snout to vent.

of *Anolis*. The male condition was judged by the criteria established by Licht (1967). That of the females was based on criteria established by Licht and Gorman (1970) and Sexton et al. (1971). The clutch of anoline lizards consists of but a single egg although two clutches, differing in maturity, are sometimes carried simultaneously. The four hierarchically arranged categories A through D are presented in Table 4. It is assumed that reproductive activity increases from A through D. The status of a population was estimated by determining the percentage of animals in each category.

The hypothesis was tested geographically by examining the populations at three points along the precipitation gradient extending across the Isthmus. Cocoli near the Pacific coast was the driest site, Barro Colorado Island at mid-Isthmus was intermediate in rainfall, and the Gatun-Pina area near the Caribbean coast was the wettest. It was assumed that reproductive activity would be greater at the wettest sites. The temporal aspect was tested by contrasting a population at one site during the wet versus the dry season. Thus, at any one site, activity would be greater in the wet season.

The reproductive status of three species of lizard, each occupying a different habitat, was monitored as an indirect assay of the status of water availability in their respective habitats. *Anolis limifrons* is an inhabitant of closed-canopy forest, *A. auratus* is a grassland species, and *A. tropidogaster* lives in the ecotone between the forest and grassland. It was assumed that water availability would be most constant in the forest and least in the grassland. The prediction was that at any one time and place, reproductive activity would be most prominent in *A. limifrons* and least in *A. auratus*. These studies were carried out in the Pacific area for all three species and in the mid-Isthmus region for *A. limifrons* and *A. auratus*.

The predictions were upheld most strongly in females along the temporal dimension; there was always a higher percentage of ovigerous (classes C and D) females of all three species in the wet season than in the dry season at all locales tested. Among the two species studied at two *(A. auratus)* or three *(A. limifrons)* locales, there was support for the hypothesis that reproductive activity should reflect differences in rainfall along the precipitation gradient. While both species reproduce maximally during the wet season at all locales, they differed in their response during the dry season. Both species at the Pacific locale declined to very low levels during the dry season but only *A. auratus* repeated the same pattern at the wetter mid-Isthmus area. *Anolis limifrons* exhibited much less decline.

Less information was available from males. Most males of *A. limifrons* from BCI were in class 6 regardless of season, as were dry season males from Cocoli. Changes in testis weight were recorded from both sites with high testis weights occurring in early to mid-wet season. In males of *A. auratus*, the frequency of stage 6 males was greatest in the March-November period at the Pacific. Fewer males collected during December-February had spermatozoa.

TABLE 4
CLASSIFICATION OF REPRODUCTIVE CONDITION OF FEMALE ANOLINE LIZARDS

A,	no enlarged ovarian follicle and no oviducal egg.
B,	one enlarged follicle and no oviducal egg.
C,	one enlarged follicle and a contralaterally located oviducal egg.
D,	one enlarged follicle and two oviducal eggs, the older on the same side as the follicle and the younger on the opposite side.

Arboreal and Habitat Studies

Tres especies de lagartijas, Anolis limifrons, A. auratus y A. tripodogastes se compararon en relación a condiciones de humedad. Las hembras de A. limifrons, habitantes de las copas de los árboles, mostraron declinación durante la estación seca, mientras que A. auratus, habitante de los pastizales, mostró mucho más declinación durante esta estación, aún en áreas húmedas. Los machos presentaron un patrón reproductivo similar.

The research carried out to date on Panamanian reptiles strongly indicates the seasonal nature of the reproductive cycle in a number of species. Each species, and at least some populations, has its own distinctive pattern. The crocodile, the slider turtle and the iguana are active in the dry season, the grass anole and the gecko are wet season breeders while *Anolis limifrons* is very adaptable depending upon locale. Each cycle is presumably timed by a specific assemblage of environmental factors operating as proximate cues which forecast ultimate selective forces.

No studies have addressed themselves directly to the proximate and ultimate factors regulating cycles. Most efforts have assumed that natural selection favors seasons favorable for the young. This truism scarcely identifies the stages in the life history most susceptible to environmental variation nor does it identify the factor(s) acting on those stages. The remainder of this paper deals with the effects of seasonality of productive success of *A. limifrons* because it is one of the best known species of tropical lizard. In addition to our knowledge of its reproductive cycle, we have information about its growth (Andrews 1976) and reproductive effort (Andrews 1979; Andrews & Rand, 1974); egg morphology (Sexton et al. 1979), habitat (Sexton et al. 1964, Scott et al. 1976), its thermal (Ballinger et al. 1970) and water balance requirements (Sexton & Heatwole 1968), diet (Sexton et al. 1972) and energy utilization (Andrews & Asato 1977), and signalling behavior (Jenssen & Hover 1976).

Selective Forces Moulding Reproductive Patterns

Anolis limifrons es una especie de lagartija muy conocida y existen estudios sobre su crecimiento, sus esfuerzos reproductivos, la morfología de sus huevos, su hábitat, sus necesidades en el balance de agua y temperatura, su dieta, la utilización de energía y su comportamiento.

Given that the reproductive cycle of *Anolis limifrons* is positively associated with local precipitation, what are links coupling the cycle to rainfall? A logical one is that increased rainfall predicts an abundance of insects suitable as prey for both hatchlings and reproductively active females. Mayhew (1966) considered that this link explained the relationship between rainfall and egg production of desert lizards. The inference is that increased rainfall induces renewed plant growth which, in turn, promotes the increase of primary consumers. Secondary or higher order consumers then have an abundance of prey.

But three separate sets of data argue against this reasonable hypothesis. The first (Sexton et al. 1972) comes from a study where stomach contents of intermediate-size and of adult *A. limifrons* were collected at three intervals: 1) an early wet season (4–17 June, 1963) sample; 2) an early dry season (11–25 January, 1970, an unusually wet period early in the normal dry period); and 3) a later dry season (30 March–5 April, 1970) group. Much to our surprise, we could find no support for the hypothesis. The frequency of food items per stomach was similar for all three samples. Indeed, the mean size of dry season prey was larger than that of wet season prey while the mean lizard size in the population showed the reverse trend.

En Anolis limifrons la relación entre la humedad y la reproducción no está ligada a la disponibilidad de alimento.

The second item was obtained during the study of the reproductive activity of *A. limifrons* on BCI (Sexton et al. 1971). During that investigation we measured the total lipid content as a percent of body weight of adult lizards at four periods during a year; the highest lipid level was achieved in late dry season (April) and the lowest value in early wet season (June). Licht and Gorman (1970) showed much the same in their study of Caribbean anolines. If food were scarce in the dry season when the animals are active, not dormant, we should expect to find low lipid levels, not high ones.

Andrews (1976) provided the third and most direct refutation of the food-linkage hypothesis. She compared the growth rates of wild and captive hatchling and juvenile *A. limifrons*. Wild individuals on BCI ate whatever food was available in the field, and the captive group was hatched from eggs of captive females originating from BCI and fed ad libitum. There was no difference in growth rate between the two groups. These three sets of data not only refute the rainfall-insect prey linkage hypothesis but indicate that *A. limifrons* is not regularly food limited. Andrews (1979) went on to compare certain life history parameters of a mainland species—*A. limifrons* in lowland, coastal Costa Rica to comparable ones of an insular species—*A. oculatus* on Dominica. Both populations inhabited cacao plantations. She found that the insular population was more food-limited than the mainland one.

A recent study, however, (Andrews & Rand 1982) forces a modification in the view that the BCI populations of *A. limifrons* are not food-limited. They argued that intraspecific competition for food occurs infrequently because: (1) the diets of males and females overlap; (2) the species are relatively r-selected and (3) population density was correlated with rainfall (to be discussed later). They then reasoned that intraspecific competition would be expected at times of high population densities. They proceeded to test this hypothesis by examining food intake of individuals from two subpopulations on BCI at a period of extremely high lizard density.

Sin embargo, en algunos casos la competencia por comida puede influir.

One of their study areas (Lutz) was located in an area of young forest and the second was in an area of old forest (Armour). Population densities were estimated through the use of periodic capture-release techniques over a year period. Densities at Lutz were as high as 1150 individuals per hectare in the late wet season of 1976 and declined to lows of 260 individuals per hectare in the early wet season of 1977. Corresponding values for the Armour area were 1860 and 660 individuals respectively. Food intake was estimated on the basis of scat production (Andrews 1979). Andrews and Rand (1982) found that intake was inversely related to population density. The Lutz site lizards grew faster than those of the Armour site. They concluded that intraspecific competition for food was intermittent.

Temporal Aspects

An alternative to the food hypothesis is based on the critical role of water in the development of embryos of certain oviparous squamate reptiles. The eggshells of oviparous species belong to one of two categories (Bustard 1968). The first type is the parchment-shelled egg whose shell is composed of concentric zones of fibrils and which is not highly calcified. Anoline eggs belong to this category. The second type is the highly calcareous eggshell. Eggs of certain groups of geckos and their relatives are of this kind (*Gonatodes albogularis* of the Sphaerodactylidae). It is well known that eggs of the parchment type must obtain water from their surroundings for development (see review by Packard et al. (1977) for details). This alternative hypothesis is that the seasonal nature of the reproductive cycle of such species as *A. limifrons* is dependent on the availability of soil moisture for the developing embryo. Its egg must be particularly sensitive because of the type of sites where it is laid. Sexton et al. (1964) found eggs in the interface between soil and leaf litter, at the base of tree trunks, in crevices within tree trunks, in dead leaves in the crotches of trees, on or under logs, etc.

Los huevos de algunas lagartijas están pobremente calcificados y sólo presentan áreas calcificadas en la parte externa, mientras que otros están altamente calcificados.

Andrews and Sexton (1981) have compared the water uptake of eggs of *A. limifrons* to that of the grassland species, *A. auratus*. They presumed that the egg of the latter species is laid in a more protected spot than that of the former species. Eggs of both species hatched successfully under experimental conditions when subjected to soils having a water potential of −1.48 MPa or more. Eggs exposed to soils with a water potential between −2.13 and −2.74 MPa initially increased in weight but eventually lost water. The eggs of *A. limifrons* both gained and lost water at rates faster than eggs of *A. auratus*, so that eggs of the latter species are better buffered against moisture changes. Hence, eggs of the forest species were more susceptible to moisture changes than eggs of the grassland species.

Las lagartijas Anolis cuyos huevos presentan áreas calcificadas obtienen el agua para el embrión de su medio ambiente.

Since the only physical link between the embryo and the water source (the soil) and between the embryo and the sink (the atmosphere) is the inert eggshell, an examination of its structure might throw light on the importance of moisture. Sexton et al. (1979) investigated the eggshell morphology of *A. limifrons* using transmission and scanning electron microscopy. The eggshell of *A. auratus* (Andrews & Sexton 1982) is similar in gross morphology to that of *A. limifrons* except that the perifibrillar layer is denser and the eggshell somewhat thicker (Fig. 5). These morphological differences reinforce the difference noted in water uptake.

Los huevos de Anolis limifrons ganan y pierden agua más rápido que los de A. auratus y son más susceptibles a los cambios de humedad. La cubierta de los huevos de A. auratus es más densa, observada con el microscopio electrónico.

It is my current view that the dynamic relationship between the egg and its water source and sink is responsible for the seasonal nature of the reproductive cycle of *A. limifrons*. The control(s) (unknown at present) of egg production is sensitive and flexible enough to track local moisture conditions. Thus, at both Cocoli and BCI, reproduction is heavy during the wet season, but it ceases almost entirely at the former site during the dry season while at the latter site it declines. Andrews and Rand (1974) have confirmed this observation by measuring the inter-clutch interval of captive BCI females of *A. limifrons*. In the wet season the interval was 7.7 days but in the dry season it was 19.8 days. This ability to track moisture conditions gives rise to the seasonal demographic structure (Sexton et al. 1963) and population levels (Sexton 1967) discussed earlier and to the dynamics of a population under long term study by Andrews and Rand (1982) to be discussed later.

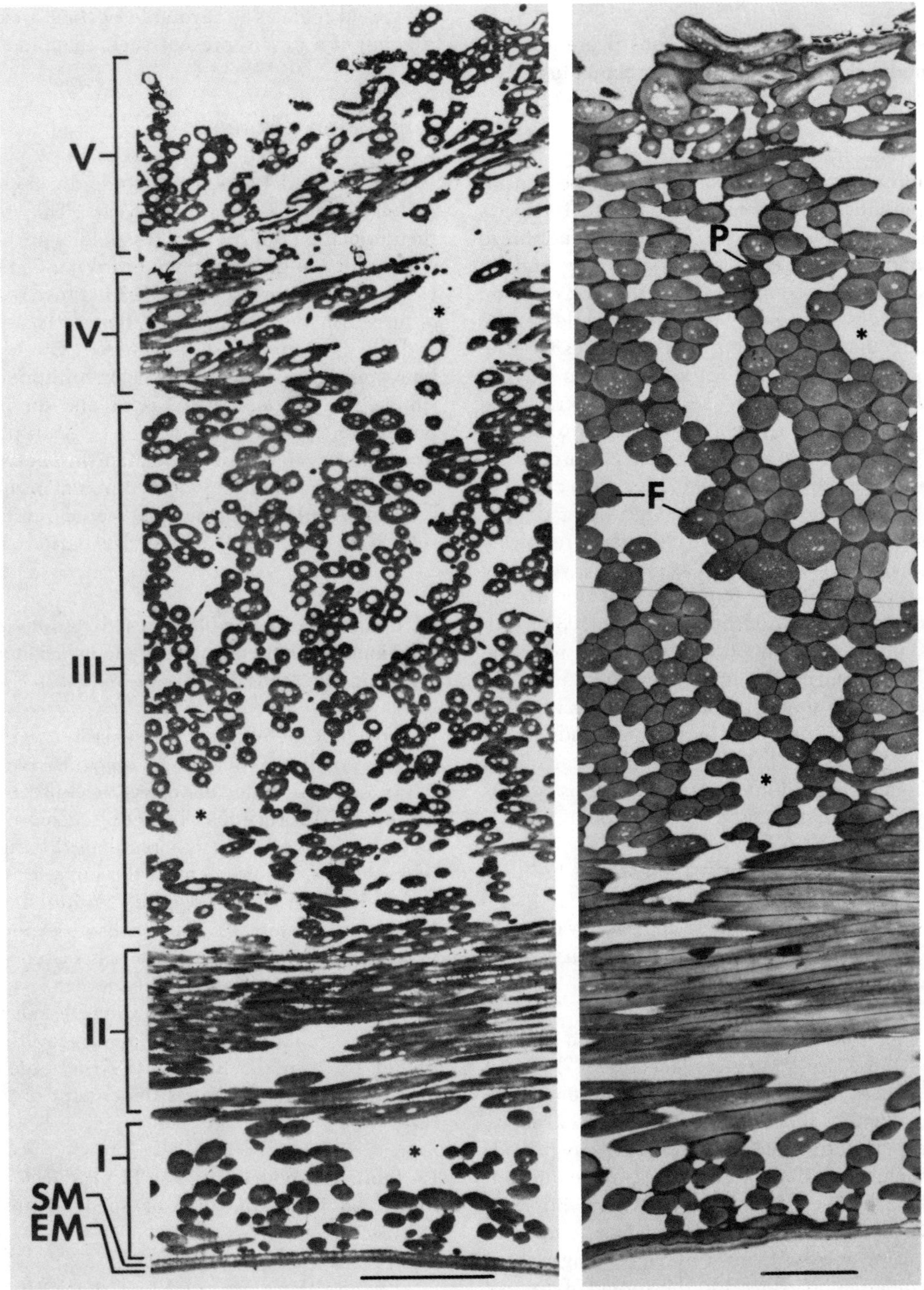

Fig. 5. Transmission electron micrograph section through the eggshells of *Anolis limifrons* on the left (x 4,050) and *A. auratus* on the right (x 2,775). The five concentric zones of fibrils are labeled from I at the inner surface of the eggshell to V at the external surface of the eggshell. The thickness of the eggshells shown is 48 μm for *A. limifrons* and 62 μm for *A. auratus*. The scale at the bottom of each figure is 5 μm in length. Symbols: SM = shell membrane; EM = egg membrane; F = fibril, P = perifibrillar layer, * = interfibrillar space.

Other Aspects of Reproduction

En la estación seca, Anolis limifrons presentó un promedio de 19.8 días entre nacimientos, mientras que sólo de 7.7 días durante la estación lluviosa.

The clutch of any species of lizard can be considered by potential predators as a food cache. Two ways for a female lizard to sequester her eggs to minimize predation on them are by scatter-hoarding and by larder-hoarding (Bossema 1979). In the first case the individual eggs are widely scattered while in the second one they are concentrated in one site. The genus *Anolis* (and perhaps sphaerodactylids such as *Gonatodes albogularis)* are scatter-hoarders while species such as *Iguana iguana*, *Crocodylus acutus* and *Chrysemys scripta* represent larder-hoarders. *A. limifrons* is an extreme scatter-hoarder. The clutch of one egg is laid beneath leaf litter and other debris. The interval between successive clutches is relatively short so that a female that lives as long as a year may deposit 40 eggs (Andrews & Rand 1974). Incubation approximates 44 days (Andrews & Sexton 1982). Presumably, the clutch is deposited within or near the female's home range. The iguana represents the opposite extreme. In some populations females travel as far as 3,000 m to communal nesting grounds (Rand 1968; Montgomery et al. 1973). The multi-egg clutch is buried mainly in mid-dry season in a hole as deep as 65 cm below the earth surface (Rand 1968).

Algunas lagartijas depositan sus huevos en lugares separados, mientras que otras los depositan en un solo sitio debajo de hojarasca y de otros debris. En el último caso, la estrategia es la de esconder o enterrar los huevos.

Some experimental data may indicate the relative success of these two strategies. Janzen (1978) placed clutches consisting either of a single egg or of five eggs of a quail on the rain forest floor at La Selva, C.R. The eggs were monitored over an 18 day period. At the conclusion half of the single eggs were uneaten but none of the five-egg clutches survived. Bossema (1979) did an equivalent experiment in a temperate situation using acorns. Here scatter-hoarding was no better than larder hoarding, but the spread of the risk was better for scatter-hoarding.

These studies indicate that larder-hoarded clutches deposited on the ground in the same fashion as scatter-hoarded ones would be selected against. For larder-hoarding to succeed would call for special adaptations to hide the clutch from predators. Travel by gravid females to favorable nesting areas and the digging of a nest represent such adaptations.

Population Changes

Andrews and Rand (1982) studied a population of *A. limifrons* on BCI for eight years. They found that the population density varied sixfold over this period, with adults ranging from one lizard per 58 m^2 in 1976 to one lizard per 10 m^2 in 1979. Most recruitment occurred in the beginning of the wet season. Adult and juvenile survivorship averaged 75% to 73% per four weeks, respectively. Survivorship appeared to be similar for wet and dry seasons and for periods of population increase and decrease. Multiple regression analyses uncovered a relationship between rainfall and population size for a combination of December and April rainfall, and indicated recruitment is related to the length of the dry period, not to its severity.

El tamaño de las poblaciones de Anolis está relacionado con la duración de la estación seca y no con su intensidad.

I conclude by offering a speculative, verbal model of how population size of *A. limifrons* is regulated. Recruitment is under the direct or indirect control of moisture. Moisture may be acting at more than one level. On one level it may regulate the initiation or regression of enlargement of the oocycte, ovulation, and deposition of the eggshell within the oviduct. Brown and Sexton (1973) have shown experimentally that egg production is positively associated with moisture in *A. sagrei* of Belize. On another level the eggs may be retained by females during periods when soil moisture is suboptimal, and laying may then be induced by rainfall. Stamps (1976) suggested that oviposition in *A. aeneus* is dependent on soil moisture.

La humedad podría regular la producción de huevos al igual que el proceso de depositar huevos.

Although there are predators specialized for prey such as *A. limifrons*—the vine snake, *Oxybelis aeneus*,—this lizard suffers predation from a wide variety of generalized predators. This diffuse predation depresses population growth sufficiently that intraspecific competition is only occasionally important in regulation. All stages in the life history are susceptible to predation. Eggs are subject to fungal attack or to predation by foraging animals such as the teid lizard, *Ameiva* sp., or the coati. Hatchlings and juveniles are eaten by a wide variety of invertebrates such as arachnids, predacious katydids, scorpions, army ants, etc. This same class of prey plus adults are eaten by other lizards, especially *Anolis frenatus*, many snakes (*Bothrops atrox, Oxybelis aeneus, Leptophis ahaetula*, etc.), various birds (puffbirds, toucans, small forest hawks, ant birds, etc.) and a variety of mammals (white-faced monkeys, coatis, armadillos, etc.). Guerrero et al. (1977) discussed the incidence and severity of malarial infection in *A. limifrons* and suggested that this may have an impact on population dynamics.

La población disminuye debido a la presencia de diferentes tipos de predadores. Así mismo influye la interacción entre la variación en la producción de huevos, el proceso de depositarlos y la presencia de predadores.

Population size at any time is the sum of variable egg production and of a relatively constant loss due to predation. This interaction accounts for the great fluctuations noted by Andrews and Rand (1982).

Anolis limifrons has been treated in detail merely as a model. Each species or population must be treated in its own right, and evidence from all levels must be brought to bear in identifying the selective forces acting to regulate the reproductive cycles of tropical species.

Acknowledgments

This paper is dedicated to Robert Woodson of the Botany Department of Washington University and of the Missouri Botanical Garden who first aroused my interest in Panama. The National Science Foundation, The National Institutes of Health, and Washington University contributed support. At a crucial early period, Sr. Tomas Guardia, Jr. of the Darien Subcommittee provided support. The Smithsonian Tropical Research Institute was always helpful, and approximately 50 students, colleagues and friends helped with the field work. Drs. M. Robin Andrews and A. Standley Rand, two friendly and cooperative rivals, provided manuscripts to fill out the picture presented here. And finally, I thank my wife, Pidge, and our four children for the many sacrifices they made so I could conduct this work.

Andrews, R.M. 1976. Growth rate in island and mainland anoline lizards. Copeia 1976: 477–482.

———. 1979. Reproductive effort of female *Anolis limifrons* (Sauria: Iguanidae). Copeia 1979: 620–626.

——— & T. *Asato*. 1977. Energy utilization of a tropical lizard. Comp. Biochem. & Physiol. 58A: 57–62.

——— & A.S. *Rand*. 1974. Reproductive effort in anoline lizards. Ecology 55: 1317–1327.

———. & A.S. *Rand*. 1982. Long term changes in population density of the lizard *Anolis limifrons*. *In* Seasonal Rhythms and the Ecology of A Tropical Forest, E.G. Leigh, Jr., Ed. Smithsonian Instit. Press, Washington, DC.

——— & O.J. *Sexton*. 1981. Water relations of the eggs of *Anolis auratus* and *Anolis limifrons*. Ecology 62: 556–562.

Ayala, S.C. & J.L. *Spain*. 1975. Annual oogenesis in the lizard *Anolis auratus* determined by a blood smear technique. Copeia 1975: 138–141.

Ballinger, R.E., K.R. *Marion* & O.J. *Sexton*. 1970. Thermal ecology of the lizard, *Anolis limifrons* with comparative notes on three additional Panamanian anoles. Ecology 51: 246–254.

Bossema, I. 1979. Jays and oaks: an eco-ethological study of a symbiosis. Behaviour 70: 1–117.

Breder, C.M., Jr. 1946. Amphibians and reptiles of the Rio Chucunaque drainage, Darien, Panama, with notes on their life histories and habits. Bull. Amer. Mus. of Nat. Hist. 86.

Brown, K.M. & O.J. *Sexton*. 1973. Stimulation of reproductive activity of female *Anolis sagrei* by moisture. Physiol. Zool. 46: 168–172.

Bustard, H.R. 1968. The egg-shell of gekkonid lizards: A taxonomic adjunct. Copeia 1968: 162–164.

Campbell, H.W. 1973. Ecological observations on *Anolis lionotus* and *Anolis poecilopus* (Reptilia, Sauria) in Panama. Amer. Mus. Novit. 2516: 1–29.

Dobzhansky, T. 1950. Evolution in the tropics. Amer. Sci. 38: 209–222.

Guerrero, S., C. *Rodríguez* & S.C. *Ayala*. 1977. Prevalencia de hemopar̃sitos en largartijas de la Isla Barro Colorado, Panamá. Biotrop. 9: 118–123.

Hallinan, T. 1920. Notes on lizards of the Canal Zone, Isthmus of Panama. Copeia 1920: 45–49.

Heatwole, H. & O.J. *Sexton*. 1966. Herpetofaunal comparisons between two climatic zones in Panama. Amer. Midl. Nat. 75: 45–60.

Janzen, D.H. 1978. Predation intensity on eggs on the ground in two Costa Rican forests. Amer. Midl. Nat. 100:467–470.

Jenssen, T.A. & E.L. *Hover*. 1976. Display analysis of the signature display of *Anolis limifrons* (Sauria: Iguanidae). Behaviour 57: 227–240.

Licht, P. 1967. Environmental control of annual testicular cycles in the lizard *Anolis carolinensis*. I. Interaction of light and temperature in the initiation of testicular recrudescence. J. Exp. Zool. 165: 505–516.

Licht, P. & G.C. *Gorman*. 1970. Reproductive and fat cycles in Caribbean *Anolis* lizards. Univ. Calif. Publ. Zool. 95: 1–52.

Mayhew, W.S. 1966. Reproduction in the arenicolous lizard *Uma notata*. Ecology 47: 9–18.

Moll, E.O. & J.M. *Legler*. 1971. The life history of a neotropical slider turtle *Pseudemys scripta* (Schoepff) in Panama. Bull. Los Angeles County Mus. Nat. Hist. Sci. 11.

Montgomery, G.G., A.S. *Rand* & M.E. *Sunquist*. 1973. Post-nesting movements of iguanas from a nesting aggregation. Copeia 1973: 620–622.

Netting, M.G. 1936. Notes on a collection of reptiles from Barro Colorado Island, Panama Canal Zone. Ann. Carnegie Mus. 25: 113–120.

Ortleb, E.P. 1964. Hatching of basilisk eggs. Herpetol. 20: 277–279.

Packard, G.C., C.R. *Tracy* & J.T. *Roth*. 1977. The physiological ecology of reptilian eggs and embryos, and the evolution of viviparity within the class Reptilia. Biol. Rev. 52: 71–105.

Rand, A.S. 1968. A nesting aggregation of iguanas. Copeia 1968: 552–561.

———. 1969. *Leptophis ahaetula* eggs. Copeia 1969: 402.

———. 1983. Ecological effects of malaria on the populations of the lizard *Anolis limifrons* on Barro Colorado Island, Panama. *In* A.G.J. Rhodin & K. Migata, eds. Advances in Herpetology and Evolutionary Biology. Museum of Comparative Zoology, Harvard Univ., Cambridge, MA.

Ream, C.H. 1965. Notes on the behavior and egg laying of *Corythophanes cristatus*. Herpetol. 20: 239–242.

Scott, N.J., Jr., D.E. *Wilson*, C. *Jones* & R.M. *Andrews*. 1976. The choice of perch dimensions by lizards of the genus *Anolis* (Reptilia, Lacertilia, Iguanidae). J. Herpetol. 10: 75–84.

Sexton, O.J. 1967. Population changes in a tropical lizard *Anolis limifrons* on Barro Colorado Island, Panama Canal Zone. Copeia 1967: 219–222.

———, H.F. *Heatwole* & E.H. *Meseth*. 1963. Seasonal population changes in the lizard, *Anolis limifrons*, in Panama. Amer. Midl. Nat. 69: 482–491.

———, H.F. *Heatwole* & D. *Knight*. 1964. Correlation of the micro-distribution of some Panamanian amphibians and reptiles with the structural organization of the habitat. Carib. J. Sci. 4: 261–295.

——— & H.F. *Heatwole*. 1965. Life history notes on some Panamanian snakes. Carib. J. Sci. 5: 39–43.

——— & H.F. *Heatwole*. 1968. An experimental investigation of habitat selection and water loss in some anoline lizards. Ecol. 49: 762–767.

——— & O. *Turner*. 1971. The reproductive cycle of a neotropical lizard. Ecol. 52: 159–164.

———, E.P. *Ortleb*, L.M. *Hathaway*, R.E. *Ballinger* & P. *Licht*. 1971. Reproductive cycles of three species of anoline lizards from the Isthmus of Panama. Ecol. 52: 201–215.

———, J. *Bauman* & E.P. *Ortleb*. 1972. Seasonal food habits of *Anolis limifrons*. Ecol. 53: 183–186.

———, G.M. *Veith* & D.M. *Phillips*. 1979. Ultrastructure of the eggshell of two species of anoline lizards. J. Exp. Zool. 207: 227–236.

Stamps, J.A. 1976. Egg retention, rainfall and egg laying in a tropical lizard, *Anolis aeneus*. Copeia 1976: 759–764.

Swanson, P.L. 1945. Herpetological notes from Panama. Copeia 1945: 210–216.

———. 1950. The iguana *Iguana iguana iguana* (L.). Herpetol. 6: 187–193.

Telford, S.R., Jr. 1971. Reproductive patterns and relative abundance of two microteiids species in Panama. Copeia 1971: 670–675.

——— & H.W. *Campbell*. 1970. Ecological observations on an all female population of the lizard *Lepidophyma flavimaculatum* (Xantusiidae) in Panama. Copeia 1970: 379–381.

Las Termitas Neotropicales como Colonizadoras de Montículos Creados por Arboles Caídos: Informe Preliminar

Barbara L. Thorne

Nests of six species of termites were found in uproot mounds of fallen trees in the forest on Barro Colorado Island. This uproot soil is an attractive habitat for soil dwelling termites because the mound has an abundance of unoccupied soil, it connects to the heartwood of the fallen tree, the soil is well drained and available for construction of nests and galeries, and after initial erosion of the mound, the rootlet system aids the termites in the task of anchoring the soil. Possible effects of the termites on such mounds include alteration of the soil profile and changes in the amount and distribution of organic material. Because they feed on plant tissue and modify the substrate, the termites may also influence plants colonizing the mounds.

Las caídas de árboles son un componente notable de la dinámica de las selvas neotropicales (Hartshorn 1978, 1980). Al caerse, el árbol puede romperse a cierta distancia de la base, o bien el árbol puede quedar con las raíces completamente descubiertas (reseña en Putz et al. este volumen). Al desarraigarse, el árbol suele dejar sólo un hoyo que descubre la posición orginal porque hay un montículo de suelo que se pega a las raíces del árbol al derrumbarse. El presente informe examina el uso y la conveniencia de tales montículos como sitio para los nidos de termitas del suelo en la isla de Barro Colorado (IBC) de la República de Panamá.

Estudio del Sitio e Informe Sobre los Antecedentes

La isla de Barro Colorado (9°09' N, 79°51' W) se formó entre 1911 y 1914 cuando se creó el lago Gatún, parte del canal de Panamá, después de la construcción de la presa en el río Chagres. IBC (15.6 km^2) se clasifica como selva tropical húmeda con el sistema de zonas vitales según Holdridge (Holdridge et al. 1971). El suelo de la IBC se describe como un barro de tipo "Frijoles" (Bennett 1929) con zonas dispersas de "gley" (Knight 1975).

La fauna de termitas de la IBC abarca tres familias de Isoptera (Kalotermitidae, Rhinotermitidae, Termitidae) e incluye aproximadamente 30 especies (Snyder 1925; Jaén, en prep.). Los sitios para los nidos de estas termitas varían. Las Kalotermitidas—termitas de madera seca—habitan la madera que está entera, seca y dura y no utilizan el suelo para llenar las galerías vacías del tronco. El Rhinotermitida *Coptotermes niger* vive en tocones muertos, en el duramen de troncos de árboles o en troncos grandes (Dietz & Snyder 1923; pers. obs.). Las entradas a los nidos de las *C. niger* suelen cubrirse de capas gruesas de tierra, manifiestamente construidas por las termitas. Las Termitidae exhiben una serie de costumbres aún más diversas dentro del hábitat. Las termitas del género *Nasutitermes* construyen nidos grandes de cartón oscuro en los troncos y en las ramas de árboles en todas partes del bosque IBC. La *Microcerotermes arboreus* también construye nidos inconfundiblemente arbóreos (normalmente más cerca del suelo que los nidos de la *Nasutitermes*), aunque en los comienzos de un segundo crecimiento, la *M. arboreus* construye nidos epígeos que no lindan con árboles (Thorne & Ventocilla, en prep.). Las otras termitidas que se encuentran en IBC (la mayoría de las especies de termitas en la isla) hacen sus nidos en montículos, troncos, tocones o debajo del suelo. El componente que tienen en común para crear los nidos es el suelo:

los nidos epígeos se construyen ante todo de suelo; las cavidades o galerías vacías de troncos descompuestos o tocones se llenan de tierra o están forradas de esta tierra. Tales nidos fácilmente pueden ser reparados, aumentados o cambiados porque el suelo es fácil de manipular para las termitas.

The termite fauna of BCI spans three families of Isoptera, Kalotermididae, Rhinotermitidae, Termitidae and includes about 30 species. Kalotermitiae—dry wood termites—inhabit dry, hard wood and do not fill empty galleries with soil. Coptotermes niger (Rhinotermidida) lives in dead stumps, in the heartwood of dead trunks, or in large logs. Entrances to C. niger nests are usually covered with coarse soil plate, apparently constructed by the termites. The Termididae show more diverse habitat patterns. Nasutitermes build large carton nests on trunks and branches throughout the BCI forest. Microcerotermes arboreus also builds arboreal nests. All species make use of soil in their nests.

Observaciones

El suelo que se pega al pie de los árboles caídos a menudo contiene por lo menos una colonia de termitas. En los montículos recientes que tienen menos de uno o dos años raramente se encuentran termitas y si están, probablemente son forrajeadoras de un nido establecido en otra parte. La tierra debe estar descubierta durante por lo menos dos estaciones húmedas (que coinciden con los vuelos de las aladas) para entonces ser colonizada por un joven rey y una reina. Si la tierra descubierta es poco profunda o mal pegada, puede ser arrastrada por el agua o erosionarse de tal manera que no conviene para las termitas.

Las siguientes termitas se encontraron haciendo nidos en la tierra pegada a la parte inferior de árboles caídos del IBC: *Heterotermes* sp. (Rhinotermitidae) y *Amitermes beaumonti*, *Anoplotermes* sp., *Armitermes chagresi*, *Orthognathotermes wheeleri*, y *Subulitermes kirbyi* (Termitidae). A veces más de una especie se encuentra en un montículo, aunque la coexistencia puede ser transitoria porque la agresión explícita entre las especies y dentro de una misma especie puede ser aguda (Dropkin 1946; Stuart 1970; Thorne, datos no publicados).

The following termites were found nesting in soil attached to uprooted tree bases: Heterotermes sp. (Rhinotermitidae), Amitermes beaumonti, A. chagresi, Anoplotermes sp., Orthognathotermes wheeleri and Subulitermes kirbyi (Termididae).

Es poco probable que los montículos de raíces de árboles caídos sean el único sitio disponible para cualquiera de las especies, pero para las que se sabe que tienen una alternativa para situar sus nidos, los montículos sirven como hábitat consistente. La *Armitermes chagresi*, por ejemplo, construye montículos firmes de tierra en los lados de tocones pequeños o encima de troncos pequeños y parcialmente descompuestos en el suelo del bosque. Las galerías del nido son enteramente de suelo: el soporte de madera se usa sólo como armazón arquitectónico. En los montículos, los obreros de la *A. chagresi* simplemente excavan las galerías y pegan el suelo que las rodea para hacer las paredes del nido.

La *Anoplotermes* también construye montículos independientes normalmente en el lado de un árbol (al nivel del suelo) o en un tronco pequeño. Las galerías de la *Anoplotermes* son mucho más pequeñas que las de la *Armitermes*, pero, de nuevo, es el suelo y no la madera que se utiliza para formar el centro efectivo del nido. Al construir sus nidos en los montículos, la *Anoplotermes* no cementa la tierra tan rígidamente como la *Armitermes*, pero hay evidentes galerías y manipulación de la tierra.

Discusión

Las condiciones que permiten que las termitas hagan nidos en montículos esperan futuros estudios, pero se deberían considerar las siguientes ventajas del hábitat:

1. Un montículo es una nueva "isla" que probablemente no será colonizada por otras termitas antes de que sea descubierta por la perturbación de la caída del árbol.
2. El montículo ofrece un acceso fácil a una tierra descubierta, suelta e aireada.
3. La base de suelo en un extremo del árbol provee una entrada directa al duramen muerto de un árbol caído.
4. Un montículo es indicio definitivo de que existe alimento futuro (al morir y descomponerse un árbol).
5. La presencia de un nido en un montículo es garantía de que hay buen desagüe porque está elevado.
6. Hay una abundancia de tierra para la construcción de galerías de nidos y para tapar cavernas des-

pués de usarlas para forrajear. La *Armitermes*, por ejemplo, llena la madera que come de tierra (con la excepción de los pasillos), quizá para reducir el acceso de animales de rapiña u otros competidores al vaciarse el tronco.

7. Las pequeñas raíces del árbol caído proveen una estructura que ayuda en pegar la tierra al pie del árbol.

The uproot mounds of fallen trees offer several advantages to termites: 1) The mound is a new, uncolonized site. 2) The soil is accessible, loose, and aerated. 3) It provides direct access to the dead heartwood of the fallen tree. 4) The tree is a source of future food as it dies and rots. 5) The mound is elevated and assured of good drainage. Abundant soil is available for construction purposes. 7) The rootlets aid in holding the soil to the tree base.

Aunque no se sabe la escala de la modificación, las termitas en IBC probablemente influyen en el estado del suelo, en particular en las regiones donde tienen nido o termitaria. Lee & Wood (1971) revisan los efectos de las termitas en el suelo y hacen el siguiente resumen de los posibles cambios mayores: perturbación de los perfiles del suelo, depósitos de "laterita" en galerías desocupadas (un procedimiento conocido y documentado en el trópico del Mundo Antiguo), alteraciones en la cantidad y distribución de materia orgánica en el suelo, el cementar la superficie del suelo, y cambios en la topografía a causa de los montículos. La presencia de las termitas en el suelo puede también influenciar la vegetación porque por un lado consumen cierta materia viva y muerta de las plantas, y por otro lado, las manipulaciones alteran el subsuelo para influenciar el crecimiento de plantas futuras (Lee & Wood 1971).

Termites in the soil can influence the vegetation by altering the subsoil.

Agradecimientos

Esta investigación fue posibilitada por una beca predoctoral del Instituto de Investigaciones Tropicales Smithsonian. Julio Jaén y Jorge Ventocilla fueron muy amables en permitir que yo citara datos todavia no publicados.

Bennett, H.H. 1929. Soil reconnaisance of the Panama Canal Zone and contiguous territory. U.S.D.A. Tech. Bull. 94: 1–46.

Dietz, H.F. & T.E. *Snyder*. 1923. Biological notes on the termites of the Canal Zone and adjoining parts of the Republic of Panama. J. Agric. Res. 26: 279–302.

Dropkin, V.H. 1946. The use of mixed colonies of termites in the study of host-symbiont relations. J. Parasitol. 32: 247–251.

Hartshorn, G.S. 1978. Treefalls and tropical forest dynamics. Pp. 617–630. *In*: Tomlinson, P.B. & M.H. Zimmermann, eds. Tropical Trees as Living Systems. Cambridge, New York.

Hartshorn, G.S. 1980. Neotropical forest dynamics. Biotropica 12 (Suppl.): 23–30.

Holdridge, L.R., W.G. *Grenke*, W.H. *Hatheway*, T. *Liang* & J.A. *Tosi*, Jr. 1971. Forest Environments in Tropical Life Zones: A Pilot Study. Pergamon Press, San Francisco.

Knight, D.H. 1975. A phytosociological analysis of species-rich tropical forest on Barro Colorado Island, Panama. Ecol. Monogr. 45: 259–284.

Lee, K.E. & T.G. *Wood*. 1971. Termites and Soils. Academic Press, New York.

Snyder, T.E. 1925. New termites and hitherto unknown castes from the Canal Zone, Panama. J. Agric. Res. 24(4): 179–193.

Stuart, A.M. 1970. The role of chemicals in termite communication. Pp. 79–106. *In*: Johnson, J.W. Jr., D.G. Moulton & A. Turk, eds. Advances in Chemoreception Vol. I: Communication by Chemical Signals. Appleton-Century-Crofts, New York.

Plants and Geography

In the distant past, North America and South America were widely separated, and the history of their coming together and forming the Central American land mass is complex. Today there is a mixing of the biotas of North and South America along the isthmian corredor. Although Panama provides a continuous land bridge between North and South America, this is essentially a lowland bridge. From the elevations of 3000 m (10,000 ft) in the uplands of Costa Rica and Chiriqui to similar elevations in South America, there is a gap of over 1300 km (800 mi). And at the 1000 m level, there is also a wide gap, but this is interrupted by the small discontinuities of Cerro Tacarcuna, and Cerro Pirre in eastern Panama. These gaps pose a barrier to movement of plants or animals which require cooler temperature and other features of upland climates. The following papers discuss some of the features of past and present biological movements in this context.

Graham's paper gives data on Cenozoic fossils from Panama that show strong affinity with North America and very little with South America. And he notes that an important concept, that of refugia, is as important for Panama as for other regions.

Bennett notes that the Central American isthmus is the only important biological bridge in existence today. But for understanding the role of climate and vegetation types in geographic analysis, he clarifies some confusing terminology.

One result of the mixing of biotas over the meoamerican land bridge is an unusual diversty of plant and faunal life. Burger examines the diversity of plant life: while the mixing of floras may explain the origin of this richness, it does not explain how it can be maintained. He proposes several other factors in the environment which permit so many species to coexist.

The isthmian land bridge is far from homogeneous. Gentry notes how plant groups of the uplands are different from those of the lowlands, and he traces how the groups of the uplands and lowlands appear to have differing origins and relationships with the floras of other regions. Elsewhere in this volume Davidse presents a similar analysis for the grasses.

The survey of Colombian geographic and floristic regions by Forero is useful for extrapolating from conditions in Panama. Colombia is a much larger country, and not surprisingly it presents a greater variety of terrain.

Plantas y Geografía

En el pasado, América del Norte y del Sur estaban ampliamente separadas y la formación de América Central, la masa de tierra que las unió es realmente compleja. Actualmente, a lo largo de ese corredor ístmico hay una amalgama de las biotas de América del Norte y del Sur. A pesar de que Panamá provee un puente terrestre continuo entre América del Norte y del Sur, éste es fundamentalmente de baja altitud. De elevaciones de 3000 m (10.000 pies) en las tierras altas de Costa Rica y de Chiriquí en Panamá a similares altitudes en América del Sur hay una brecha de más de 1300 km (800 millas). Asimismo, a 1000 m sobre el nivel del mar también existe esa laguna, pero ésta es interrumpida por Cerro Tacarcuna y Cerro Pirre en el este de Panamá. Esta discontinuidad montañosa constituye una barrera a la migración de plantas o animales, los cuales requieren temperaturas frescas y otras características de los climas de tierras altas. Las investigaciones que se presentan a continuación, discuten algunos rasgos de los movimientos biológicos del pasado y del presente.

El estudio de Graham aporta información sobre los fósiles cenozoicos de Panamá que demuestran fuerte afinidad con América del Norte y muy poca con América del Sur. Graham nota que el concepto de refugio es tan válido para Panamá, como para otras regiones.

Bennett señala que el Istmo centroamericano es el único puente biológico de importancia que existe en el presente. El aclara alguna de la terminología confusa, para comprender el papel que juegan los tipos de vegetación y de clima en el análisis geográfico.

Una consecuencia de la mezcla de biotas sobre el puente terrestre mesoamericano, es la exuberante diversidad vegetal y animal. Burger analiza la diversidad vegetal e indica que a pesar de que la mezcla de las floras puede explicar su riqueza, definitivamente no aclara cómo se mantiene esa exuberancia. El describe algunos factores ambientales que inciden y permiten la coexistencia de tantas especies.

La vida sobre el puente ístmico está lejos de ser homogénea. Gentry indica cómo los grupos de plantas de las tierras altas son diferentes de aquellos de las tierras bajas. El examina cómo esos dos grupos parecen tener diferentes orígenes y relaciones con las floras de otras regiones. Davidse realiza análisis similar de la flora de gramíneas en donde los patrones son algo diferentes.

La revisión de las regiones geográficas y florísticas de Colombia es provechosa, porque permite extrapolar información a Panamá. Colombia es un país mucho más grande por lo tanto no es sorprendente que presente mayor variedad de formaciones telúricas.

The Panama Land Bridge: A Brief Review

Charles F. Bennett

El Istmo de Mesoamérica y Panamá es el único puente terrestre de importancia biológica que existe en el presente. En épocas pasadas, los climas eran diferentes, sin embargo la terminología comúnmente empleada generalmente causa confusión.

The Panama land bridge—here considered to be the Isthmus of Panama and all the land northward to the Isthmus of Tehuantepec in the Republic of Mexico—is one of the most interesting and challenging parts of this planet to study with regard to questions of past and present biogeography. This is the only extant land bridge of biogeographical significance, and in past geological times its importance as a passageway for flora and fauna between continents may have been equalled only by the Bering land bridge.

El Istmo de Mesoamérica y Panamá es el único puente terrestre de importancia biológica que existe en el presente. Terminología poco precisa ha causado confusión en relación a los diferentes climas del pasado. La distribución presente de muchos organismos es enigmática. Cualquier discusión de corredores de dispersión debe considerar al hombre como un elemento importante en el escenario del Istmo.

It would be impossible to review all the major aspects of this land bridge in anything less than a large monograph and even that length would allow for little more than cursory treatment of many topics. Therefore, this paper is confined principally to a narrow spectrum of topics and concerns: a critique of some of the terminology, methodology and conceptualizations that have wide employment in the ecological and biogeographical literature refering to Pleistocene-Holocene times on the land bridge; the still largely neglected subject of past and present anthropogenic influences on the ecosystems of the land bridge, and finally, lines of field investigation which might illuminate some of the aspects of contemporary biogeography as well as paleoecology and paleobiogeography which are still largely in the realm of conjecture.

Studies of fossil pollen in the northern Andes and in a few locales on the land bridge indicate that during the Pleistocene, climates were different than they are now at least at the elevations and sites where the fossil pollen has been obtained. The most notable published palynological analysis for Panama relates to cores from the Gatun Basin (Bartlett & Barghoorn 1973). This valuable analysis indicates that there were vertical-altitudinal displacements of tree species during the past 12,000 years and it was suggested that from 11,300-9,600 years ago the displacement of the mean air temperature for a given altitude was 2.5° C. This last seems to have been arrived at by estimating the mean air temperature now occurring at the lower altitudinal limits of certain floral elements no longer present in the Gatun area but occurring elsewhere and at higher elevations in Panama. The 2.5° C is based upon an assumed regular atmospheric temperature lapse rate of approximately 0.5° C per 100 m change in altitude. Sea level fluctuated during the past approximately 35,000 years and was lowest at approximately 49 m below today's sea level at around 35,500 years ago.

El estudio del polen ha indicado cambios climáticos, pero la evidencia obtenida no es suficiente como para permitir un panorama total de la situación climática del pasado.

The estimates of temperature change just cited agree fairly well with estimates given by Martin for a site in Costa Rica (Martin 1964).

The palynological investigations of van der Hammen and Urrego (1978) and others in the northern Andes, while not directly related to Panama and the land bridge, have provided valuable insights to Pleistocene macroclimates and indicate, as also do the available land bridge palynological studies, that at the given sites there were reductions of air temperatures at times during the Pleistocene (van der Hammen & Urrego 1978).

The scarcity of published studies of fossil pollen on the land bridge tends to stimulate much conjecture regarding paleoclimates and other paleoenvironmental conditions including postulating a continuous Pleistocene corridor of something called savanna through the length of the land bridge (Webb 1978).

The changing occurrence over time of plant species represented by pollen in a core sample *may* point toward climatic change but perhaps one ought not to adopt the so-called standard temperature lapse rate in the atmosphere as the sole measure of such temperature fluctuations or, indeed to assume that altitudinal displacement of floral elements over time *always* is indicative of temperature change. The "standard" atmospheric temperature lapse rate of about 0.5° C/100m is a very crude approximation and while suited to general discussion ought to be used with caution when attempting to account for past or present floral (or faunal) distributions. There is seldom if ever a regular decrease or increase in air temperature between elevations and relatively large temperature changes may occur over comparatively minor elevational changes. Moreover, one can obtain significantly different air temperature readings for a given altitude depending on whether the slope is oriented to windward or to leeward.

It must also be noted that *mean* air temperature is not seen by many ecologists as being the principal factor controlling the distribution of living plant or animal species, and it follows that this was true for the past as well.

Perhaps before other than the most cautious attempts are made to reconstruct the Pleistocene ecology of the land bridge far more research ought to be undertaken with respect to present ecological and distributional phenomena on the land bridge and elsewhere in the low latitudes of the neotropics. But before any such effort can be undertaken a concerted attack must be mounted upon the often confusing and vague climatological and vegetational terminology that is contained in a large share of the extant literature of the past and present biogeography of the neotropics.

Examples of this vagueness and confusion abound. For climate one encounters such terms as: "arid tropics"; "tropical climate"; "mesic conditions"; "humid conditions"; "savanna climate"; "deterioration of climate"; "amelioration of climate"; "temperate climate"; "rain forest climate"; "wet season"; "dry season". Seldom are these macroclimatic terms accompanied by quantitative information or if so accompanied the data are presented in terms of monthly or annual averages.

El uso corriente de términos tales como "tropical", "trópicos", "sabanas" y "vegetación de sabana", en sentido amplio, abarca diferentes conceptos. Los datos climáticos, frecuentemente, son promedios y no los valores extremos que son los que mayormente determinan el lugar en donde ocurren los organismos.

Examples of vague or imprecise vegetation terminology include: "savanna"; "park savanna"; "woodland savanna"; "rich woodland savanna"; "open country vegetation"; "scrub"; "dense virgin forest"; "rain forest"; "lush vegetation". These terms may or may not be accompanied by the botanical or vernacular names of the more common plant species but almost never by data indicating degree of dominance by individual species or other ecologically significant details.

Several of the climatic and vegetational terms mentioned above are so frequently but varyingly employed that it may be useful to comment upon them here. The terms are: "tropical" used in an adjectival sense; "tropics" used as a synonym for a specific kind of environment as well as a location; "savanna"; "savanna climate"; "rain forest"; "rain forest climate"; "wet season"; "dry season".

There is no available straightforward definition of "tropics" nor its adjectival form "tropical" that will satisfy both astronomy and biogeography. Astronomically, the tropics refer to the Tropics of Cancer and Capricorn located approximately 23.5 degrees north and 23.5 degrees south, respectively, of the equator. The two Tropics *are not* biogeographic boundaries (Bennett 1966). Nevertheless, there seems to be a widely held belief that the region between the two Tropics possesses a marked environmental uniformness. Of course, no such uniformity exists. Macro-and microclimates vary enormously as does also the phytophysiognomy, not to mention elevation and other macro features. Perhaps the only partially unifying feature is photoperiod, but even this is constant only at the equator.

In spite of its inappropriateness as an ecological term the word "tropical" lives on in the literature with a peculiar perseverence. As long as the term is merely employed to *locate* a map or region its continued use is justified, but it is my impression that its use is frequently not so limited. Much confusion could be avoided if the term "low latitudes" were used instead of "tropical region(s)". As a corollary to that suggestion I would urge that the adjective "tropical" no longer be accepted as adequate to describe the physical *ecology* of any part of the world.

Few vegetational terms are more frequently employed in the literature of the past and present ecology of the low latitudes of the neotropics than "savanna" and its companion, "savanna climate". This duo marches across the pages of the literature with such frequency that one might conclude that they have been joined in ecological matrimony. Alas, however, their's is an illicit relationship and the scientific literature will be the better once they have been exposed for the ecological sinners they are.

The word savanna—in Spanish, sabana seems to be of Caribbean Indian derivation—possibly Taino and was probably first heard by Columbus and his crew on Hispaniola in the late 15th century (Sauer 1966). The Indian word was applied to almost any phytophysiognomic situation not dominated by trees. Thus, any area with a considerable grass cover could be so-designated as could mixes of grass, palms, dispersed trees, shrubs and so on. This included a broad range of possibilities, and present-day usage is no different in that regard. In fact, so broad is the range of use that unless there is an accompanying physical and floristic description the term has but limited value for relating vegetation to faunal distributions and dispersals.

Confusing though the use of the vegetational term savanna is, the confusion pales when compared with it's mate, "savanna climate". First, it must be understood that *ecologically speaking*, there is no such thing as *a* savanna climate. There may be dozens of such climates if by savanna climate one is speaking of the supposed relationships between precipitation, air temperatures and low latitude vegetation associations having grassy components. Climatologists use the term "tropical savanna climate" to refer, within broad and highly qualified limits, to low latitude climate(s) which have average annual and monthly precipitation and air temperature values that separate them from other macro-climatic types such as rainforest climate and monsoon climate. What serves the broad descriptive needs of the climatologists often does not serve the much more narrowly defined needs of the ecologist. Macroclimatological categories, relying as they do on (usually) biogeographically meaningless monthly and annual averaging of two climatic parameters—precipitation and air temperature do not equate with the bioclimatic conditions in which plants and animals live and of which their distributions are, in varying degrees, reflective. In the widely known Koeppen and Geiger climate classification only very general assumptions are made respecting the relationships assumed to exist between the occurrence of savanna vegetation (sensu lato) and average rainfall and air temperature. Thus, in this climatic classification, a site can be correctly classified as *Aw*—tropical savanna climate—even though air temperatures of 0° C and lower have been known to occur.

Not a few of the many types of vegetation sometimes described as savanna owe their occurrence far less to seasonal precipitation and air temperature than to edaphic conditions, anthropogenic factors, fire, wind-flow patterns, availability of tree seed sources and possibly the lack of mycorrhizal spores required to allow the establishment of certain tree species.

A perusal of a fairly large portion of the *corpus* of biogeographic literature that pertains to the present and past of lowland neotropica might lead the unaware to conclude that a single vegetational dichotomy exists, viz., the just discussed savanna and *rainforest.* Ever since Schimper (1903) introduced the term *regenwald*, "rainforest" has been very broadly applied. Part of the resulting semantic confusion relates to the fact that different scientific disciplines have different needs for application of the term. In the Koeppen-Geiger system a rainforest climate, *Af*, is one in which the mean temperature of the coolest month is not less than 18° C and the mean precipitation of the dryest month is not less than 60 mm. One may encounter difficulty, however, when attempting to link those two values directly to many floral and faunal distributions. Is there empirical evidence that those two arithmetic mean values actually relate directly to biological distributions? The answer is, of course, negative. It *may* be that a mean monthly temperature of 18° C is biologically meaningful, but until there are supporting data one is permitted to reserve judgment. One *should* ask what are the *maximum and minimum temperatures of record*, what is the length of the record, what is and was the nature of the instrumentation employed and what was the *precise location* of the data collection site (Schulz 1960)?

What is the basis of the almost universal willingness to assume and assert that from month to month and from year to year there are "no significant" variations in the range of air temperature and precipitation in rain forests? How does one relate data collected in clearings, or on roof tops or whatever, to the microclimatic phenomena that directly influence life in the forest (or other) ecosystems? By what extrapolative flight can one leap from a clearing-sited weather instrument to a forest canopy?

There may well be forests in the low latitudes where water required for physiological processes is never so scarce as to be a limiting environmental factor, but the location of such awaits the collection of more sophisticated bioclimatic data than we now possess. The widespread notion that a rainforest extends unbroken all along the Caribbean slope from eastern Panama north to Mexico—as is shown on many maps does not cast light upon many questions respecting forest vegetation and animal distributions either today or in a Pleistocene past. Present-day ecology (and paleoecology) would be better served were a rigorous bioclimatic definition of rainforest developed and adhered to. Such a definition is not likely to be derived from the macroclimatic data now available. What is needed are studies that will make it possible to relate relevant climatic parameters to known distributions of animals and plants.

What about the forested areas which do experience at least occasional episodes of water limitation as well as air temperature variations? While some such areas might be acceptably lumped by climatologists into a tropical monsoon climate or the like, the biologist needs a set of climatic units based upon microclimates and the geographical distribution of individual plant and animal taxa. Again, the averaging of monthly and annual weather data will not suffice. Annual and monthly variations of rainfall *may* be the rule rather than the exception in almost all, if indeed not all, low latitude ecosystems, and these variations should be looked for and when discovered be examined with care. An example of this is provided by Barro Colorado Island, Panama. Many authors have described the macroclimate of the island and its vegetation as "rainforest". However, the more than four-decade-long rainfall record for the island clearly establishes that during most years there is a marked reduction of rainfall for one or two months but the magnitude and duration vary from year to year and during a lesser number of years such a reduction does not occur (Bennett 1964). Biologists familiar with the island are aware that, to a marked degree, the forest is facultatively deciduous. That is, the degree to which there is a seasonal leaf fall is related to the amount and periodicity of rainfall that occurs during the so-called dry season. In some years the forest looks as though it were about to dry up and blow away on the trade winds but such a year may be followed by one in which water is apparently never a limiting factor in the environment and there is almost no seasonal leaf fall. This variation of so-called dry season rainfall is well known to Canal officials who have to modify canal operations in years when much lower than average rainfall occurs on the canal's watershed. As long ago as 1931, Dunn (1931) suggested that the lack of certain amphibian species on the island might be due to the occurrence of an annual dry season (Dunn 1931). Whether he was correct or incorrect, anyone who has witnessed the occasional "unusual" extreme dry season at the island as well as elsewhere in Central America must be aware that such events have potential biogeographical significance.

It has long seemed to me that one of the great errors made by many biologists trained in the middle latitudes is that of assuming that certain gross climatic concepts like the winter and summer of the middle latitudes can be applied to the ecosystems of the low latitudes. This is not to suggest that there is a lack of awareness that in these latitudes "winter never comes," (to use the phrasing of Marston Bates 1952), but that some biologists appear to believe that only the more obvious seasonal changes are ecologically important and that these can be viewed as a simplistic wet season-dry season dichotomy.

The terms wet season and dry season are useful general descriptive words but their continued unqualified use in the ecological literature of this region should not be encouraged. There is such a wide annual range of variation of rainfall and other climatic conditions during the periods of reduced rainfall that to lump all this together is to possibly miss what might be some of the most important environmental cues governing phenological behavior, not to mention questions regarding the local distribution and abundance of plant and animal taxa. In this context it should be added that in view of our often crude knowledge regarding the micro and macroclimates of the present day a major degree of caution ought to be exercised when discussing similar environmental parameters thousands or millions of years in the past.

To conclude my remarks about confusion associated with the use of certain climatological and vegetational terms it should be noted that the beginnings of rational vegetational classifications are already available. One need only cite the work of Beard, Holdridge, Porter, and Gomez-Pompa, to suggest that the real problem may be the greater desire to invent one's own idiosyncratic classification rather than adhere to an already established system (Beard 1955; Holdridge 1957; Porter 1973; Gomez-Pompa 1973). Perhaps a congress should be assembled for the specific purpose of developing an agreed-upon vegetation classification that can be used only by biologists working the low latitudes of the neotropics.

The problem of establishing a usable microclimatological system is greater because the available data base is still meager. However, a start can be made with regard to the use of macroclimatic data so long as they are not misapplied to present or past biogeographical phenomena. One of the outstanding treatments for Central American is by Portig (1976). Published bioclimatic investigations in the Central American region can only be described as scarce. It is a measure of retardation in this important area that Allee's publications dating back more than half a century still represent a significant percentage of the bioclimatic studies thus far accomplished on the land bridge (Allee 1926).

Turning to present-day plant and animal distributions on the land bridge it can be stated that the *precise* distributions of even the more common plant and animal taxa are all but unknown. I do not refer to range maps—which have their value—but to distributional data that can be plotted on maps of topographic scale of 1:50,000 or larger. Such plottings, accompanied by as much detail respecting other organisms, microclimates, soil, light conditions, and structural characteristics at the *edge* of distributions should at the very least get us well along the road to identifying and evaluating those factors which currently prevent or stimulate further dispersal of the observed organisms.

Se carece de datos de distribución precisa, aun de las plantas y animales más comunes.

Before we can speak with even a bare modicum of confidence about the nature of dispersal corridors in the past and the ecological phenomena that allowed or prevented the passage of plant and animal taxa over the land bridge we must gain a far more sophisticated understanding of these elements as they exist at present. Such investigation might conveniently begin with easily observed animal taxa and in this regard the primates now living on the bridge appear to be good candidates. The location of primates on the land bridge today certainly raises a number of questions regarding what influences their lack of uniform distribution in this region. There are six genera present, *Ateles*, *Alouatta*, *Cebus*, *Saimiri*, *Aotus* and *Saguinus*. *Ateles* and *Alouatta* occur as far north as southern Mexico, *Cebus* occurs as far north as the Nicaraguan-Honduran border region, *Saimiri* is restricted to southwestern Costa Rica and southwestern Panama, *Aotus* seems not to occur north of Panama as is also the case with *Saguinus*. Of course, many other possibilities for study, including plant species, are available. The point being made here is that it is at the *extreme geographical limits* of an organism's distribution that one is most likely to discover the factors which limit the organism's further spread.

Actualmente, el estudio de los primates aún presenta muchas incógnitas en relación a las influencias que determinaron su distribución en el Istmo.

Some of the animals now having disjunct mountain or upland distributions through Central America (and sometimes extending into South America) and which are thought to have dispersed over the land bridge during the Pleistocene might now occupy physical habitats and particularly microclimatic situations generally similar to those prevailing at or near sea level during parts of the Pleistocene. Thus, detailed studies of the habitats and the ecologies of some of these animals might shed light upon sea-level environmental conditions which prevailed during parts of the Pleistocene. Among such animals are shrews of the genus *Cryptotis*. Their present distribution in Central and South America is very disjunct and their tendency to occur at ever higher altitudes with increasing proximity to the equator suggest these animals might be worthy of such investigation (Cabrera 1957–1961; Choate 1970).

There is reluctance on the part of some investigators to include an evaluation or even a recognition of anthropogenic factors that may have influenced certain vegetation changes and also to have influenced certain animal dispersals and distributions on the land bridge. Any consideration of late Pleistocene as well as Holocene plant and animal geography on the land bridge *must* include some consideration of the possible influences of *Homo sapiens* (Bennett 1968). Although there is still a lack of agreement as to when our species first arrived in South America an estimation of between 15,000 and 20,000 years ago is now generally considered to be fairly conservative. For around 12,000 years ago there seems to be relatively abundant and well-documented evidence of the presence of humans to the south of the land bridge. There are no comparable dates available for Central America but the dates for South America permit one to assert a similar if not an even greater time depth for the earliest human presence on the land bridge.

Cualquier consideración de la biogeografía del puente terrestre en el pasado debe incluir al hombre, quien probablemente estaba presente, junto con el fuego, desde hace más de 20.000 años.

There is general agreement that even the first humans to have passed over the bridge possessed fire, since human control of fire is known to have occurred in Asia some 350,000 or more years earlier. Fire is thought to have been widely and perhaps universally employed by hunters, and there is no reason to suggest this was not the case in this region. However, no tangible evidence for this has yet been discovered in Central America and thus one cannot unequivocally assert that these people exerted any influence upon isthmian ecosystems through the use of fire.

"Incipient" agriculture, may have been present in parts of Panama by 6,500 years ago (Willey 1960). Materials contained in some archaeological sites suggest that agriculture may have become widely distributed on the land bridge by 3,500 years ago. The ecological significance of pre-Columbian agricultural activities on the Panama Land Bridge can scarcely be over-stated. Currently and in the recent past investigators have been accumulating data which merit the attention of most biologists concerned with the past and present biogeography of this region (Denevan 1961; Johannessen 1963; Linares & Ranere 1980; Turner 1960). At the time of the first European contacts in the beginning of the 16th century, Panama may have possessed an Amerind population of as much as two million. However, the great Spanish chronicler, Gonzalo Fernandez de Oviedo y Valdez who made that numerical estimate hedged it by saying "or they were innumberable" (Oviedo 1851–1853). These were farming folk who also engaged, when time permitted, in hunting and fishing—the former including well-documented fire drives for deer (Andagoya 1892).

Cuando los europeos vinieron, existía una gran población Amerindia, la cual había removido árboles de áreas de considerable tamaño.

Spanish accounts of the early 16th century contain ecological descriptions of parts of Panama which indicate that some areas now under forest cover were then sufficiently free of trees to allow mounted horsemen to pass without hinderance (Espinosa 1892). Much of Darien was then a farming region dominated by cultivated fields and plots of secondary vegetation in brush-fallow (Altolaguirre & Duvale 1914; Espinosa 1892). Descriptions of one of the deer species then hunted in eastern Panama tally with the white tail, *Odocoileus virginianus*, an animal not now found in mature forest in eastern Panama.

Within a century, the once numerous Amerinds seldom survived as more than remnant populations and some groups seem to have disappeared entirely in parts of the land bridge (Denevan 1976). This great mortality was caused partly by diseases introduced accidentally by the invaders and for which the Amerinds had little resistance. Thus, large areas were freed from the millenia-long ecological pressures that had been exerted by humans and began to undergo ecological succession, in many instances, back to forest. It was these secondary forests that were later to lead some biologists to speak of the "virgin" or "primeval" forest—a concept that has persisted in some quarters to the present day in spite of evidence to the contrary.

Inmediatamente después de la llegada de los europeos, la población humana fue diezmada y, consecuentemente, en las áreas deforestadas se inició la sucesión para constituir el bosque nuevamente.

The implications for faunal dispersals are considerable. For example in the pre-Columbian period it may be that farmers inadvertently opened up corridors along which a few vertebrate taxa including white-tail deer and cottontail rabbits *(Sylvilagus floridanus)* were able to reach South America (Bennett 1968). There can be little doubt that certain animal distributions, and numbers *on* the land bridge itself have been modified by human acts (Bennett 1968; Cooke 1977; Daugherty 1972; Heatwole 1966). This latter topic requires more attention than it has thus far received.

Al abrir corredores, posiblemente se ha provisto vías para la migración de venados y conejos. El ganado, el fuego y el aumento de la población han reabierto corredores y sus efectos presentes, además de su influencia sobre el bosque, necesitan investigarse.

The arrival of Europeans and Africans ushered in another ecological chapter on the land bridge and one that continues to unfold to the present day. Perhaps the single most important impact was the introduction of livestock—principally horses and cattle. Areas having the least annual rainfall were first favored for stock raising and this meant that this activity was concentrated on the Pacific side of the land bridge. Fire was used to manage and extend the area of pasture lands. Thus, the Amerind practice of burning was continued on parts of the bridge although often for different reasons. Livestock numbers increased over the centuries chiefly in response to intra-regional market demand that grew more or less with the growth of human population. Beef export outside the region was not important until this century and then only assumed major importance during the past couple of decades. Today, cattle raising oriented toward export is having a profound and disastrous effect upon the remaining forest ecosystems in many parts of the land bridge. Not only have the forest ecosystems been reduced in area but the pastures created are often dominated by exotic grasses of African origin (Parsons 1976).

These activities are in response to rapid human population growth which averages approximately 3 percent per year for Central America and Panama, increased per-capita demand for imported goods, rapid acceleration of petroleum prices, and, above all, the need to obtain hard currencies to pay for the imports. While one may say that these nations are destroying their forested ecosystems to pay for their imports one must also be aware that no politically acceptable alternative appears to be available to them.

The biogeographic implications for these massive ecological changes tax one's imagination. It is probable that a continuous forested corridor no longer exists north-south from Mexico to Colombia and thus the movements of non-volant arboreal animals in this region have been sharply circumscribed. However, major opportunities for the dispersal of non-forest taxa have been created and it is likely that not a few vertebrate and invertebrate taxa have already taken advantage of these opportunities.

Although these non-forested lands are anthropogenic "savannas" they deserve study to see what kinds of faunal movements do or do not take place within them. Similarly, the forest remnants, if given sufficient protection, might provide useful insights into certain aspects of island biogeography. How limited an area, for example, retains a full compliment of the taxa that once ranged over much larger geographical areas? What are the relationships between a given reduction of forest-area and the remaining level of species diversity? Are pollination systems affected by forest area reductions and loss of species diversity? How rapidly and by what species does a *forest ecosystem* reinvade, that is disperse, into adjacent areas if allowed to do so? The answers to these and many other questions might also shed some useful light upon aspects of Pleistocene biogeography.

Estudios del medio ambiente en los Neotrópicos requieren percepción de su fragilidad.

To summarize: a more sophisticated approach must be made with regard to the investigation of contemporary aspects of microclimates and other environmental factors in the low latitudes of the neotropics. There must be detailed investigations of the ecological conditions occurring at the limits of the distributions of living plant and animal taxa if we are to ever approach a degree of accurate understanding of biogeographic processes in the Pleistocene. We must strive to erradicate the use of vague and confusing terminologies. To achieve these and related aims it is necessary that we perceive the neotropical ecosystems not only as being fragile but, more importantly, as being *subtle*.

Allee, W.C. 1926. Measurement of environmental factors in the tropical rain forest of Panama. Ecology 7: 273–302.

Altolaguirre y *Duvale*, Ángel de. 1914. Vasco Nuñez de Balboa. Impr. Patronato Huérfanos Intendencia, Madrid.

Andagoya, Pascual de. 1892. Relación muy circunstanciada de los sucesos de Pedrarias Dávila, etc., vol. 2, *In* A. Cuervo, Documentos Inéditos sobre la Geografía y la Historia de Colombia.

Bartlett, A.S. & E.S. *Barghoorn*. 1973. Phytogeographic history of the Isthmus of Panama during the past 12,000 years: A history of vegetation, climate, and sea-level change. Pp. 203–299, *In* A. Graham, ed., Vegetation and Vegetational History of Northern Latin America. Elsevier, New York.

Bates, Marston. 1952. Where Winter Never Comes. Chas. Scribner's, New York.

Beard, J.S. 1955. The classification of tropical American vegetation types. Ecology 36: 89–100.

Bennett, C.F. 1964. Rainfall variability and the macroclimate at Barro Colorado Island, Canal Zone, Panama. Bull. Ecol. Soc. Amer. 45: 93.

———. 1966. On the location of the nearctic-neotropical boundary in Mexico. Rev. Geograf. (Brasil) 64: 7–16.

———. 1968. Human influences on the zoogeography of Panama. Ibero-Americana 51: 1–112. (también transl. en Esp., Editorial Universitaria, Panamá, 1976.)

Cabrera, Angel. 1957–1961. Catálogo de los Mamíferos de América del Sur. Impr. Casa Editor. Coni, Buenos Aires.

Choate, J.R. 1970. Systematics and zoogeography of Middle American shrews of the genus *Cryptotis*. Univ. Kansas Pub. Mus. Nat. Hist. 193: 195–317.

Cooke, Richard. 1977. Los impactos de las comunidades agrícolas precolombianas sobre los ambientes del trópico estacional: datos del Panamá prehistórico. Act. IV Sympos. Ecol. Trop., Panamá, 3: 919–973. Univ. Panama/INAC, Panama.

Daugherty, H.E. 1972. The impact of man on the zoogeography of El Salvador. Biol. Cons. 4: 273–278.

Denevan, W.M. 1961. The upland pine forests of Nicaragua. Univ. Calif. Publ. Geog. 12: 251–320.

———. (ed.) 1976. The Native Population of the Americas in 1492. Univ. Wisconsin Press.

Dunn, E.R. 1931. The amphibians of Barro Colorado Island. Boston Soc. Nat. Hist., Occas, Papers 5: 403–421.

Espinosa, Gaspar de. 1892. Relación hecha por Gaspar de Espinosa, etc., vol. 2, *In* A. Cuervo, Documentos Inéditos sobre la geografía y la historia de Colombia.

Gómez-Pompa, A. 1973. Ecology of the vegetation of Veracruz. Pp. 73–148, *In* A. Graham, ed., Vegetation and Vegetational History of Northern Latin America. Elsevier.

Heatwole, H. 1966. The effect of man on distribution of some reptiles and amphibians in eastern Panama. Herpetol. 22: 55– 59.

Holdridge, L.R. 1957. The vegetation of mainland Middle America. 8th Pacif. Sci. Congr., Proc., 4: 148–161.

Johannessen, C.L. 1963. Savannas of interior Honduras. Ibero-Americ. 46.

Linares, O.F. & A.J. *Ranere*, eds. 1980. Adaptive radiations in prehistoric Panama. Peabody Museum (Harvard Univ.) Monogr. 5.

Martin, P.S. 1964. Paleoclimatology and a tropical pollen profile. Intern. Quatern. Congr. rep. 4(2): 319–323.

Oviedo Fernández de, Gonzalo. 1851–1853. Historia general y natural de las Indias, Islas, y Tierra Firme del Mar Océano. Impr. Real Acad. Hist., Madrid.

Parsons, J. 1976. Forest to pasture: development or destruction? Rev. Biol. Trop. 24: suppl. 1: 121–138.

Porter, D.M. 1973. The vegetation of Panama: a review. (see reference to Bartlett and Barghoorn: 167–201.)

Portig, W.R. 1976. The climate of Central America. Pp. 405–478, *In* W. Schwerdtfeger, ed., Climates of Central and South America. Elsevier.

Sauer, C.O. 1966. The Early Spanish Main. Univ. Calif. Press.

Schimper, A.F.W. 1898. Pflanzengeographie auf physiologischer Grundlage. Jena.

Schulz, J.P. 1960. Ecological studies on rain forest in northern Suriname. Noord-Hollandsche Uitgevers, Amsterdam.

Turner, B.L. 1974. Prehistoric intensive agriculture in the Mayan lowlands. Science 185: 118–124.

Van der Hammen, T. & G.C. *Urrego*. 1978. Prehistoric man on the sabana de Bogotá; data for an ecological prehistory. Paleogeog., Paleoclim., Paleoecol. 25: 179–190.

Webb, D.S. 1978. A history of savanna vertebrates in the New World. Part II: South America and the great interchange. Ann. Rev. of Ecol. Syst. 9: 393–426.

Willey, G.R. 1960. New World prehistory. Science 131: 73–86.

Why Are There So Many Kinds of Flowering Plants in Costa Rica?

William C. Burger

Costa Rica, un país pequeño, tiene mayor variedad de plantas con flores que otras áreas y las razones de ésto, son muchas y complejas. Diferencias periódicas en los patrones de lluvia y las cadenas montañosas discontinuas proveen condiciones de temperatura y humedad variables, además de que las erupciones volcánicas recientes, han provisto suelos ricos. Fenómenos naturales como lluvia de cenizas volcánicas, deslizamiento de tierras e inundaciones causan continua reiniciación de las etapas sucesionales. El sistema es dinámico y conlleva a una gran riqueza en un área de alta precipitación pluvial. La exuberancia de la flora de epífitas sugiere que estas condiciones de humedad se han dado durante largo tiempo.

La precisa división de nichos en un clima generalmente estable, permite que muchas especies compartan los mismos hábitat y adicionalmente, disturbios menores e impredecibles añaden un componente de azar, a las posibilidades de reemplazo de los elementos vegetales después de accidentes, enfermedades y acción de animales de presa. Estos efectos reducen la tendencia de exclusión de especies mediante competencia y permiten que haya mayor posibilidad de inmigración exitosa. Además de esta situación local y dinámica, la cercanía de áreas continentales ricas en especies y la historia de Costa Rica como archipiélago en el pasado, han contribuido a su diversidad florística.

Costa Rica, the size of Denmark or West Virginia, has half as many flowering plants as are found in all of North America north of Mexico. The reasons why so many species are found in this small area are many and complex. In summarizing the sources of Costa Rica's floristic diversity I have used many recent papers, the more important of which have been: Baker (1970), Buzas (1972), Connell (1978), Hubbell (1979), Huston (1979), Johnson and Raven (1970), Pianka (1966), Pielou (1975), and Whittaker (1972, 1977). Most of these discussions of species richness have been written by ecologists. Perhaps someone working on the systematic aspects of a rich tropical flora can provide a slightly different viewpoint. The words diversity and richness are used here interchangeably for the number of taxa living in an area. (Some ecologists use diversity differently to include evenness or the relative abundances of the taxa in an area).

Species richness generates two paramount general problems, namely: what environmental factors support so many species, and how do so many species manage to live together in the same area? This second question is concerned with alpha diversity: how a single habitat supports a large number of species needing what appears to be a similar set of environmental conditions. Whittaker (1977 and earlier) has distinguished between the diversity found within a particular habitat (alpha diversity), the added diversity found in a different habitat or on an environmental gradient (beta diversity), and the combined diversity of several habitats in a larger community (gamma diversity). First the factors that contribute to beta and gamma diversity.

What Environmental Factors Support So Many Species?

The primary factor responsible for Costa Rica's extraordinary floristic richness is its tropical setting between 8° and 11° north latitude, where freezing temperatures occur only above about 3000 m elevation. Also, with daylength varying less than an hour over the year and because of the moderating effects of two adjacent oceans, temperatures rarely exceed 38° C. Thus, we find severely cold conditions only on the highest mountains at night, and the highest temperatures of the dry season in the lowlands are not especially severe.

En Costa Rica, la variación de la longitud del día es insignificante y las áreas frías son pocas. La precipitación varía enormemente, algunas regiones tienen una estación seca pronunciada y otras presentan condiciones perennes, pero las condiciones en cada uno de estos lugares son constantes.

The driest areas of the country regularly receive more than 1200 mm of rain during the year. No areas are treeless due to insufficient rainfall. While Caribbean slopes and lowlands have rainfall throughout the year, the Pacific slope and lowlands have a definite dry season. In the northeastern Pacific lowlands of Guanacaste Province the dry season usually begins in November and ends in early May, supporting deciduous forest formations. During these five months some areas often receive less than 20 mm of rainfall. The southern half of the Pacific lowlands, with a shorter dry season, supports partly deciduous and evergreen formations. A lack of well defined dry periods allows evergreen formations to dominate the Caribbean side of the country where annual rainfall ranges from about 2000 mm near Panama to over 5000 mm at the Nicaraguan border. The prevailing easterly winds, forced up against the Central Highlands, cause frequent cloudiness and misting that support a vegetation especially rich in epiphytes all along the continental divide and eastern slopes of the country. These varied rainfall patterns superimposed on an elevation range of 3800 m produce a great variety of temperature and rainfall conditions, with the result that Costa Rica supports 12 of the 37 possible kinds of life zones in the Holdridge system. Robert Lawton (pers. comm.), having worked in the Monteverde Cloud Forest Reserve, suggests that the very consistent patterns of wind direction, wind speeds, and precipitation allow vegetational gradients to remain stable even over short distances. This produces fine grained vegetational heterogeneity, a major factor permitting such a small region to support so great a number of species (Baker 1970, Holdridge 1976, Sawyer & Lindsey 1971, Slack 1977, Whittaker 1977, Whitmore 1975).

The mountains of Costa Rica run the length of the country and separate the Pacific and Caribbean lowlands by a barrier ranging from 500–3000 m high. Only near the border with Nicaragua does this barrier drop below 400 m. In addition, the central mountain ranges are somewhat discontinuous with three distinct highland areas: the northwestern volcanic chain including the Sierra de Guanacaste and Sierra de Tilaran, the central volcanic chain and associated Meseta Central, and the Cordillera de Talamanca. A large percentage of Costa Rica's endemic plants are found only on these highlands. Rivers and streams, descending to the lowlands, have cut deep valleys impeding migration and helping to isolate populations.

Las zonas montañosas costarricenses impiden la migración de plantas y causan el aislamiento de especies. Las erupciones volcánicas recientes proveen suelos con alto contenido de nutrimentos, pero la erosión, los temblores y las laderas de las montañas poco estables destruyen la vegetación madura y abren nuevas áreas de colonización.

Much of Costa Rica's topography is geologically very young and is made up of unconsolidated volcanic materials. The recent and continuing vulcanism, mostly in the form of ash falls, has maintained high levels of mineral nutrients in Costa Rican soils. Since many areas have high rainfall, ash falls are probably important in counteracting affects of almost continuous leaching. This may be one reason why Costa Rica has a more litter-dwelling amphibians and reptiles than comparable habitats in Borneo and the Philippines (Scott 1976). But the vulcanism and geologically young highlands have an additional effect on species richness. Slopes are often steep and unstable; erosion is severe; earthquakes and landslides are frequent. These processes, together with occasional flooding destroy mature vegetation and open new areas to colonization. Thus, there are endemic species, such as the large-leaved gunneras, specifically adapted as pioneers on recent landslides. The continual reinitiation of successional processes aids in maintaining a high species richness (Louck 1970).

The Diversity of Epiphytes: a Key to the Past

If the warm moist climate superimposed on varied topography is a primary factor supporting the rich flora found in Costa Rica, we can hypothesize that Costa Rica or its source areas has had similarly benign conditions for a long time. There is only a small portion of this flora that can be expected to survive severe drought. In order to support the idea that either southern Central America or adjacent source areas and refugia in northern South America have had continuously moist conditions over a long period of time, I have compared the numbers of dicotyledonous epiphytes of Costa Rica with those of Java and the Flora of Tropical West Africa area (Table 1).

Costa Rica, comparada con otras áreas tropicales, tiene una flora rica en epífitas, lo que sugiere que el área ha presentado un alto nivel de humedad durante largo tiempo.

The reason for comparing epiphytes is that these plants usually live in situations of extreme drainage and are especially sensitive to drought conditions. In addition, the probability of evolving a plant body and dispersal system adapted to the epiphytic life style should be proportional to the period of time over which favorable conditions have prevailed. This need not hold true for plants with a "pre-adapted" habit such as the water-holding tank bromeliads or the succulent orchids with their minute and easily dispersed seeds. The comparison is restricted to dicotyledons, in part to avoid the effect of the bromeliads and orchids and in part because information on monocotyledons is not so complete. Not many dicotyledonous families have succeeded in entering the epiphytic niche space. Getting to the epiphytic site via an appropriate dispersal system, staying there with an appropriate root system, and surviving water and nutrient deficiences are problems relatively few lineages have been able to solve (Madison 1977). I believe that the number of dicotyledonous epiphytes in a flora can tell us something significant about the past; if not for the area under discussion, then certainly for the near-by source areas.

Estas epífitas podrían haber llegado a Costa Rica procedentes de regiones cercanas con suficiente humedad.

While Costa Rica has about 25% more species of dicotyledons than Java and 25% fewer than the Flora of Tropical West Africa (Table 1), it has over twice as many epiphytic dicot species as Java and over four times as many as in the Flora of West Tropical Africa. What seems significant is that, while all three areas have about the same number of dicot families, Costa Rica has a much larger number of families represented by epiphytes. The fact that a greater number of families have been able to evolve epiphytes in the areas of Costa Rica and adjacent South America, I believe, supports the contention that this region or its refugia has not suffered from severe widespread drought in the recent geological past (in the last 10 to 20 million years).

TABLE 1
INDIGENOUS AND NATURALIZED DICOTYLEDONS AND OF DICOTYLEDONOUS EPIPHYTES IN THREE FLORAS

Dicotyledons	Families	Genera	Species
Costa Rica	150	1008	3920
Java	149	1002	3130
Fl. Trop. W. Africa	157	1298	4969
Dicotyledonous Epiphytes			
Costa Rica	20	51	247
Java	12	22	107
Fl. Trop. W. Africa	9	16	58

	Area in sq. km.	*Altitude Range*
Costa Rica	50,700	0–3820 m.
Java	126,500	0–3680 m.
Fl. Trop. W. Africa	4,500,000	0–4070 m.

Sources: Costa Rica-Standley 1938, Java-Backer & Van den Brink 1963, 1965, Tropical West Africa-Keay 1958, and Hepper 1963, 1964.

The low number of epiphytic species, genera, and families in the Flora of West Tropical Africa probably reflects historical conditions as well as current low rainfall over a large proportion of the area and a general lack of high mountains. Severe past dry periods have been postulated to explain the poverty of Africa's flora at the species level (Richards 1973). The argument here is essentially the converse of the African explanation: that the richness of Costa Rica's flora or that of nearby South America is due to equable moist conditions over a long period.

The Latitudinal Gradient in Species Diversity

The disparity between the species richness of the wet tropics and the temperature zone has been the subject of much discussion. It has been argued that tropical floras have not been subjected to catastrophies such as glaciations in the recent geological past (Baker 1970, Pianka 1966). Moreover, many lineages of plants cannot survive severe frost, and this must also be a factor. More important may be the reduced energy of severely seasonal environments because of the reduced time during which photosynthates are formed (White 1978). In addition, with conditions favorable for growth and reproduction throughout the year, the wet tropics permit the exploitation of resources and a reduction in niche breadth in ways that seasonally frigid areas cannot (Baker 1970, Fischer 1960, Janzen 1967, MacArthur 1969, Pianka 1966, Whittaker 1972, 1977, Wilson 1973). In mammals, for example, the major addition to species richness as one proceeds from northern North America to southern Central America is the increased number of bat species (Wilson 1974). The tropical bats include fruit and nectar feeding bats; life styles that are not possible in more northerly areas. Likewise, the dark floor of the tropical rain forest supports a great many more species of green plants than a similarly dark understory in a northern forest. Low light levels during most of the short growing season may restrict both the numbers and kinds of plants that can exploit this kind of habitat in a northern forest. An insufficiency of pollinators may also limit angiosperm diversity in temperate areas (del Moral & Standley 1979, Feinsinger 1978).

Biological heterogeneity is expressed by a number of life styles—epiphytes, lianas, stranglers, etc.—that are rare or absent in more severe climates. Physical heterogeneity is exemplified by the range of microclimates in Costa Rica, from continuously warm lowlands to cool mountaintops and from seasonally dry to continuously moist areas. The wet tropics may be more "permissive", with a consequent reduction in the likelihood of extinction; the differences in extinction rates resulting in a "pumping" of species from severe climates into the wet tropics (Baker 1970, Valentine 1969).

Osman and Whitlatch (1978) argue that the greater number of large isolated regions in the tropics, as opposed to the temperate and polar parts of the earth, produces a latitudinal gradient in species richness regardless of other physical or biological factors. While this point has some validity, I do not see how the authors can extend this to the conclusion that "species diversity does not necessarily imply anything about the biology of the species or of the environment in which they live". Surely the fact that all of North America north of Mexico supports only twice as many flowering plant species as Costa Rica,—350 times the area—implies that there are profound differences in the biological organization of these communities. If we look at some other large taxa of terrestrial plants the picture remains the same; Costa Rica supports almost twice as many species of ferns as are found in all of North America north of Mexico. This brings us to our second major question: how do some small areas in Costa Rica support so many coexisting species?

How Can So Many Species Live Together?

We have just discussed how a great many different environments and their successional stages allow a great many species to live in Costa Rica. However, many habitats support many coexisting species, and we are faced with the question of why competitive factors have not been more effective in reducing this diversity. The hypothesized factors allowing coexistence tend to fall into two general categories: (1) finer niche-partitioning as a means of avoiding competition, and (2) processes of rarefaction that suppress dominance and reduce the effects of competitive exclusion. Among the first of these, the stability-time hypothesis (Ashton 1969, Sanders 1968, Valentine 1969, Whittaker 1972, 1977) argues that long term environmental equability (stability) has permitted highly refined competitive interactions which have led to more precise niche diversification in the tropics than is possible in more severe (unstable) environments. This hypothesis accounts for much of the species richness encountered in the wet tropics.

En Costa Rica muchas plantas están sincronizadas para cambiar en el medio ambiente y algunas sólo tienen éxito en estados específicos de la sucesión vegetal, mientras que otras requieren claros del bosque de tamaño específico en donde haya suficiente luz. Muchas especies están restringidas, mientras que otras, especialmente los árboles, poseen ámbitos amplios sobre variados terrenos.

Many species are found only in specific kinds of sites or only in certain successional stages within particular elevation and geographical limits. For example, Gomez-Pompa (1971) has shown how species of the genus *Piper* occur in different successional stages in Veracruz, Mexico, each species apparently best adapted to compete in only certain seral stages of the vegetation. Looking at a different environmental parameter, Gentry (1976) has shown how Bignoniaceae are represented in southern Central America by usually twenty species in any particular habitat; the twenty sympatric species apparently utilize different pollination strategies and different flowering phenologies in order to reduce competition for pollinators. Similarly, a study of three palm species that grow in rain forest light gaps has shown how specialized their requirements are: if the light gap is too large they are overgrown, if too small they die (Vandermeer et al. 1974).

However, while there are many ecologically restricted and specialized species in the tropics, there are also a many species with broad geographic ranges that extend over a variety of soils and micro-climates. This is especially true for many of the canopy trees of lowland rain forest formations. Not only do these species have broad ranges over varied terrain, but they share these ranges and ecological parameters with many other canopy tree species, often inhabiting similar slopes with similar soils and exposure in the same forest formations. It seems reasonable to suggest that these trees are not all specialists that have effectively divided the habitat resource space. Rather, two general explanations can account for the diversity and apparent interchangeability of large tree species in the rain forest. The first of these is seed predation and herbivore pressure on early stages of replacement; the second, is that of stochastic (random) effects due to unpredictable aspects of replacement in a mild but subtly unpredictable environment.

For plants, the most significant cause of rarefaction or numbers reduction in the tropics (apart from the activities of technological man) is undoubtedly herbivore pressure on the reproductive and early regeneration phases of the life cycle. Janzen (1970) points out that the moist tropics enable insect herbivores to maintain year-round activities and thus play a more important role than in strongly seasonal environments, while Connell (1971) maintains that predators are more vulnerable to environmental fluctuations than their prey and that prey-specialization is less likely to be a successful strategy in severely unpredictable climates.

Muchos factores reducen el número de plantas. Las enfermedades, los herbívoros, los parásitos, los animales que se alimentan de semillas y los accidentes que tienden a mantener bajas las poblaciones de muchas especies.

For plants, herbivore pressure, seed predation, parasites, and pathogens are thought to play a key role in maintaining dispersed populations in species-rich tropical communities (Bullock 1967, Connell 1971, Gillett 1962, Harper 1969, Janzen 1970, 1973, Jones 1976, Whittaker 1977). The disastrous effects of insect pests on large scale tropical tree plantations is especially well documented in the mahogany family (Grijpma 1976). The role of predation and disruption is, likewise, an important factor in maintaining species-rich animal communities (Caswell 1978, Dayton & Hessler 1972, Huffaker 1971, Levin & Paine 1974, Lloyd et al. 1968, Paine 1966, 1974). The greater pressure of parasite, predator, or disease in the tropics has generally been considered a major factor reducing dominance and keeping populations low; *Homo sapiens* was similarly affected before the advent of modern medicine and sanitation.

Though niche division, disease and predation are necessary factors for maintaining species richness in the tropics; the role of random chance should also be considered. The process of replacement is the primary arena in which community diversity is determined (Grubb 1977). Replacement of a recently deceased plant, whether forest giant or streamside herb, is in large part a matter of chance; that is, the probability of having disseminule, seedling, or sapling in the right place at the right time. Predation and competition are important factors determining success but chance is, in a sense, the first determinant.

Sale (1977) has recently hypothesized that unpredictability allows high species richness among some guilds of fish in the tropical reef environment. He argues that space is the most critical limiting factor and that storms and other disturbances which open new spaces to immigration are unpredictable, thus making interspecific competition primarily a lottery which no single species can consistently win. Sale contrasts this view with the more general consensus that tropical communities are equilibria in which species coexist by having reduced niche-breadth. A review of the distribution and habits of these reef fish leads Sale to reject the reduced-niche equilibrium model, and he emphasizes instead the importance of minor unpredictability in an environment lacking severe fluctuations. He hypothesizes that the species of a guild are "competing in a lottery for living space in which larvae are tickets and the first arrival at a vacant site wins that site." Since vacant sites are generated by unpredictable events, the situation on the reef is comparable to a tropical rain forest with its tree-falls. When chance, rather than competitive superiority, is a major determinant of success there can be many more winners. The popularity of the roulette wheels in Las Vegas and Monte

Carlo is based on the same general premise; one needn't be an expert to win.

Para algunos individuos, el espacio es el factor limitante más importante y la imprevisible aparición de éstos permite que esas especies reemplacen los ocupantes anteriores.

The importance of random tree falls in determining replacement success is implied by the life history of a neotropical species that has apparently devised a strategy to circumvent the unpredictability of tree falls. This species, *Tachigalia versicolor* (Leguminosae), produces only a single seed crop; the tree then dies and in subsequent months comes crashing to the ground, clearing an opening for its seedlings (Foster 1977). This unusual strategy is testimony to the importance of the light-gap in canopy tree replacement and gives no support to Ricklefs' suggestion (1977) that tropical light-gaps have special characteristics that allow more precise niche-partitioning than is possible in temperate forest clearings. While the *Tachigalia* "suicide stategy" can be viewed as an unusual competitive ploy, I believe that it supports the impression that unpredictability of tree falls is important in maintaining species richness among the larger tropical wet forest trees. Evidence for the importance of random events in rainforest replacement has been given by Knight (1975), Poore (1968), Webb et al. (1972) and Whitmore (1975). Historical chance and other stochastic events are important in other natural communities as well (Hubbell 1979, Levin & Paine 1974, Slack 1977, Sutherland 1974). On the other hand, if stochastic factors are so important in replacement, trees producing large numbers of seed should have been selected for and should predominate. However, such an argument fails to recognize the great variety of strategies involved in tree replacement; larger—and therefore fewer—seeds usually have more nutritive or protective components than smaller seeds, likewise, immature trees stand in the understory for many years "waiting" for a nearby canopy tree to fall (Ashton 1969). In the latter case replacement success may be determined decades after the seed was dispersed and, here too, success may be finally determined by chance.

Further support for the contention that random effects play an important role in maintaining high levels of species richness in the tropical lowland rainforest may be gained by considering diversity at generic and family rank among the tall trees.

One of the most impressive aspects of the lowland rainforest is the great diversity of species of large trees. This is especially noteworthy when one considers that large organisms, plants or animals, are usually much less diverse than their small counterparts, both as regards kind and number (VanValen 1973). But the diversity of these tall wet forest trees extends to generic and family rank. In Costa Rica's lowland Caribbean rainforest at La Selva there are 304 species of dicotyledonous trees belonging to 60 families (Table 2, based on Hartshorn 1976), while on Barro Colorado Island in Panama a sample of 192 trees species represented 48 families (Knight 1975), and in a one hectare stand near Manaus in the Amazon basin 212 trees species represented 40 families (Prance et al. 1976). The three most prolific families of the trees sampled at Barro Colorado island contained 38 genera and 54 species (including the legumes as one family). A similar proportion of genera and species is found among the three most prolific families at La Selva (Table 2).

En Costa Rica, la diversidad de la vida vegetal se extiende al nivel de género al igual que en las familias y es más marcado entre los árboles, que entre los arbustos del sotobosque. Los géneros de arbustos tienden a ser menos, sin embargo el número de sus especies es mayor. La riqueza de árboles en las categorías taxonómicas superiores es más bien un resultado del azar.

In contrast, the three most prolific families of dicotyledonous shrubs at La Selva contain only 20 genera but a total of 108 species (Table 2). The trees are much more diverse at family and generic rank than are the understory shrubs of the same forest. The larger number of species in relatively fewer genera of understory shrubs probably implies competitive superiority. In contrast, the diversity of the trees at the family level and the fact that no tree fami-

TABLE 2
DICOTYLEDONOUS TREES AND SHRUBS AT LA SELVA, COSTA RICA

	Families	*Genera*	*Species*	*Spp./Genus*
Trees, total	60	171	304	1.78
Trees-commonest families	3	42	85	2.02
Shrubs-commonest families	3	20	108	5.40
Shrubs, total	28	57	175	3.07

Source: Hartshorn (1976) for trees and from the Organization for Tropical Studies.

Note: The Leguminosae are treated as a single family.

y or genus includes a large portion of the species is, I believe, a reflection of the lottery effect.

Some extreme situations such as swamps, poor soils, or seasonally cold areas are often poor in family diversity, implying that few families have been able to evolve solutions to the problems of these particular habitats. That a great many families have successfully adapted as rainforest trees, I believe, is due to the importance of chance in canopy tree replacement. If competitive ability or special adaptive traits were the overriding factors, diversity at higher taxonomic rank should be much reduced. On the other hand, the diversity of trees at the family level may reflect major chemical differences in predator defense: the lowland forest thus presenting a chemical mosaic to specialized herbivores. But if this were a primary factor, why are some families and genera so successful under the canopy of these same forests; and why are some unusual soils or areas of poor drainage dominated by a few genera and species? In this regard it is interesting to note that in a Malaysian dipterocarp forest area with 375 tree species (representing 52 families and 139 genera) the Dipterocarpaceae family itself represented only 4% of the genera and 8% of the species but an impressive 27% of the larger individual trees (based on Poore 1968). A comparison of species-dominance curves for trees in Central Amazonia and Borneo suggest that the Amazonian species-distributions differ in being "largely governed by chance" according to Brunig and Kluge (1977), though they caution that the Amazonian plots had soils superior to those of Borneo.

Strong (1977) has recently suggested that tropical wet forest diversity is in large part due to the increased number of tree falls caused by heavy vine and epiphyte loads. Species richness, he argues, is a consequence of the persistent pre-climax state of these forests. This follows from the finding that most plant communities have the greatest species richness before the climax formation is fully established, while late secondary species are still present (Goff & Zedler 1968, Louck 1970, Peet 1978, Shafi & Yarranton 1973, Whittaker 1969). Strong's hypothesis (1977) contrasts with the argument that tree falls are unpredictable and hence few species can out-compete others in an arena of chance replacement. Perhaps the two views are complementary and not mutually exclusive, but I believe that the generic and family diversity of the canopy strata favors the importance of chance replacement. Successional habitats in Central America are often dominated by effective colonizers and competitors such as members of the Melastomataceae, Rubiaceae, and Solanaceae families as well as individual genera such as *Heliconia* and *Piper*. The pre-climax hypothesis does not account for the high diversity at family and generic rank among wet lowland canopy tree species. In addition, montane cloud forests with high epiphyte loads and frequent tree falls do not have as great a diversity of tree species (Sawyer & Lindsey 1971) and these cool wet highland forests are often dominated by a single family or genus.

Viewing species-rich wet tropical forests as dynamic systems with continuous small-scale gap replacement explains why we see so few large trees in these forests and why so few of the current canopy members appear to be replacing themselves (Hartshorn 1978). The gap-replacement (micro-successional) mosaic based on relatively small wind-falls differs from areas with severe wide spread cyclone damage where reduced structural and species diversity are found (Whitmore 1978). In general, habitats with little perturbation will suffer species loss through competitive exclusion while habitats with severe disturbance will suffer high extinction rates (Connell 1978, Huston 1979). Osman (1977) has shown in an epifaunal marine community that the degree of disturbance has an optimum where either higher or lower levels of disturbance would result in a reduction of species richness. The same is probably true for most plant communities with the corollary that tropical wet forests can support a greater number of species at their disturbance optimum than seasonally drier or colder formations. It is interesting that Hartshorn (1978) estimates a turnover rate of 118 years at La Selva, a rate much higher than for most north temperate forests. Hartshorn also estimates that 75% of the canopy trees in this lowland rainforest require light gaps formed by tree falls to achieve reproductive maturity. It is my impression that there are similarly high turnover rates in the herbaceous and shrubby strata of many tropical habitats (due in large part to the number of species available as colonizers) and that, here too, stochastic effects play a significant role in maintaining species richness. Work with mathematical models suggests that a fluctuating patchy environment can support a more highly dispersable species than a static patchy environment (Horn & MacArthur 1972, Levin 1974). If turnover is fast and the species pool is large, one can hypothesize that dispersability, shortened life cycles, and maintenance of cross-pollination despite low density populations have been important factors in the evolution of tropical wet lowland species.

Los bosques tropicales tienden a tener menos árboles grandes que los bosques de la zona templada e igualmente éstos viven menos tiem-

po. La mortalidad de árboles y los incrementos de reemplazo tienden a ser altos favoreciendo a las especies con alta habilidad reproductora. Como quiera que los sitios de reemplazo son imprevisibles, las interacciones competitivas decrecen y aún las especies poco competidoras tienen una oportunidad.

The species richness of the tropical lowland wet forest is especially difficult to comprehend when it has been argued that communities offer increasing resistance to invasion as they grow in richness (Terborgh 1973). The latter argument implies that competitive interactions make it progressively more difficult for new species to enter the habitat, whether by immigration, fusion, or adaptive speciation. Grime (1973) and Huston (1979) argue that the reduction of competitive displacement, primarily as the result of restriction on population growth, allows high species richness. Predation, disease, and the effects of minor unpredictable disturbance all serve to reduce the growth rates of competing species and counteract competitive displacement. This in turn makes successful immigration more likely and reduces the probability of extinction, allowing higher species levels than would be possible if competitive interactions were allowed to run their full course. In species rich tropical communities plant populations are often much more sparsely represented, and monospecific stands tend to be smaller and more patchily distributed than in temperate communities. These characteristics further reduce competitive interactions and the rate of competitive displacement (Huston 1979).

Connell (1978) argues that frequent disturbance and climatic change prevent the elimination of inferior competitors and are thus basic to supporting species richness in tropical rainforests and coral reefs. He believes that local assemblages in these biomes are continually changing and that it is this change through time that permits high species richness. However, the wide geographic range of many lowland rainforest tree species seems to argue against local or rapid change. It seems preferable to think of the lowland rainforest as a relatively stable, slowly evolving system of broad scale in which small-scale local disruption complements niche division and predation to support species richness (Hubbell 1979 and Huston 1979).

Where Did Costa Rica's Many Species Come From?

Of the historical factors contributing to Costa Rica's species richness, immigration is undoubtedly the most significant, with northern South America the single most important source area. For example, the Melastomataceae family, with a great many species in South America, is represented by about 275 species in Costa Rica (Almeda, personal communication), while Guatemala, much larger and further north, has only 155 species in this family. Moreover species richness may arise not only from immigration from distant source areas but also by the coalescence of formerly separate areas with distinct floras following climatic or sea level changes. Costa Rica has been more than a narrow bridge between two species-rich continental areas; it has also had an independent history as an archipelago with varying degrees of isolation (see Graham 1973, Malfait and Dinkleman 1972, Rosen 1975). In addition, the isolation as an archipelago undoubtedly resulted in speciation, some of which persists today as part of Costa Rica's endemic flora. Furthermore, speciation may continue as a consequence of disruptive selection on environmental gradients; a number of Costa Rican genera give evidence for such a process (Burger 1974).

En Costa Rica ha habido un lento pero continuo influjo de propágulos provenientes de otras regiones y ésto ha contribuido a aumentar la diversidad de especies.

As mentioned earlier, the species richness of Costa Rica does not require that Costa Rica itself has been stable for a long time. It does imply that Costa Rica is part of a broader area or province in which the larger species pool has suffered few reductions and has diversified over a long period of time. An excellent example of youthful areas with great species richness in an old biogeographic province are coral reefs known to be only 5000 years old in the western Indian Ocean (Taylor 1978). Here, as in the flowering plants, dispersability appears to be a key factor resulting in high migration rates essential to the maintenance of large provincial species-pools. Located on a narrow isthmus between two rich floristic provinces, one can imagine that little Costa Rica has been subjected to a slow but continuous influx of foreign propagules. The topographic and climatic heterogeneity of Costa Rica and the high local turnover rates have provided a fertile soil for this rain of immigrants and may be responsible for what is, in fact, a supersaturated flora.

Summary

Discussions of diversity have often centered on the imprecise concept of stability, a term perhaps best restricted to the dynamics of populations and communities (see papers in Dobben and Lowe-McConnell 1975). The words stability and instability have often been used to describe environmental patterns without distinguishing between severity/equability and predictability/unpredictability. Additional confusion has arisen from the fact that the scale of severity of unpredictability was not defined. In the previous pages large scale perturbations (instability) such as glaciations and long term drought and severe conditions such as extreme temperatures seemed adequate explanations of species-poor polar and desert environments. The lack of similar large scale or severe perturbations are an essential requirement for Costa Rica's species richness, both in ecological and evolutionary time. On the other hand, small scale unpredictable disruptions, such as landslides, flooding, storms and tree falls, are essential in promoting species richness on a local scale. Using the concept of disruption or instability in this way is not inconsistent because we were considering factors of very different magnitude. The minor disruptions, added to biological causes of rarefaction such as predation, reduce the likelihood of competitive exclusion and are responsible for a dynamic system, whether viewed at the scale of the local habitat or the broad ecosystem.

En Costa Rica la diversidad vegetal se fundamenta en su clima favorable, en sus suelos y en las condiciones generalmente estables, pero con muchos cambios pequeños e imprevisibles. Estos factores han favorecido la migración, especiación y la fusión de biotas de áreas cercanas.

In conclusion, Costa Rica's favorable climate in a setting of physiographic heterogeneity supports a great number and variety of biological interactions that are made less predictable by small scale disruption. These factors, together with migration, speciation, and fusion of areas once separate, have given rise to one of the world's richest biotas.

ADDENDUM Since this paper was first prepared for Brenesia (Burger 1980), important papers have been published which are integrated into the expanded bibliography. Luis Diego Gomez P. of the Museo Nacional de Costa Rica and the editors of Brenesia are thanked for allowing republishing of this article.

The importance of disruption in vegetation dynamics and in supporting species richness is being widely acknowledged in recent work (Denslow 1980, Foster 1981, Fox 1979, Garwood et al. 1979, Paine 1979, Pickett 1980, Raup 1981, Sousa 1979, Sprugel & Borman 1981, White 1979). The evolution of flowering plant diversity has been discussed by Burger (1981) and Stebbins (1981). Gilbert (1980) has described the importance of food-web complexity in maintaining tropical diversity. Huston (1980) continues to argue that diversity decreases with richness of resources, but he uses data from Costa Rica where the soils are generally fertile. Inger (1980) has suggested that the irregular but synchronous mass-flowering and fruiting by many species of dipterocarpaceous trees in Borneo accounts for the reduced species richness of forest-litter animals in Borneo as compared to Central America. Goldblatt (1978) and Kruger and Taylor (1979) have emphasized the importance of delta diversity in accounting for the exceptional richness of the flora of southermost Africa. I believe we will find regional differentiation in Nicaragua, Costa Rica and Panama to be an important contributing factor to the floristic diversity of isthmian mesoamerica.

Ashton, P.S. 1969. Speciation among tropical forest trees: some deductions in the light of recent evidence. Biol. J. Linn. Soc. 1:155–196.

Backer, C.A. & R.C.B. *van den Brink*. 1963. Flora of Java, vol. 1. Noordhoff, Netherlands. 1–648.

———. 1965. Flora of Java, vol. 2. Noordhoff, Netherlands. 1–641.

Baker, H.C. 1970. Evolution in the tropics. Biotrop. 2: 101–111.

Brunig, F.E. & H. *Kluge*. 1977. Comparison of the phytomass structure of equatorial "rainforest" in Central Amazonas, Brazil, and in Sarawak, Borneo. Gard. Bull. Singapore. 29: 81–101.

Bullock, J.A. 1967. The insect factor in plant ecology. J. Indian Botan. Soc. 46: 323–330.

Burger, W.C. 1974. Ecological differentiation in some congeneric species of Costa Rican flowering plants. Ann. Missouri Bot. Gard. 61: 297–306.

———. 1980. Why are there so many kinds of flowering plants in Costa Rica? Brenesia 17: 371–388.

———. 1981. Why are there so many kinds of flowering plants? Biosci. 31: 572, 577–581.

Buzas, M.A. 1972. Patterns of species diversity and their explanation. Taxon. 21: 275–286.

Caswell, H. 1978. Predator-mediated coexistence: a non equilibrium model. Amer. Nat. 112: 127–154.

Connell, J.H. 1971. On the role of natural enemies in preventing competitive exclusion in some marine animals and in rainforest trees. *In* P.J. den Boer & G.R. Gradwell, eds., Dynamics of Populations. Proc. Adv. Study Inst. 1970. Center for Agric. Publ., Wageningen, Netherlands. 298–312.

———. 1978. Diversity in tropical rainforests and coral reefs. Science. 199: 1302–1310.

Dayton, P.K. & R.R. *Hessler*. 1972. Role of biological disturbance in maintaining diversity in the deep sea. Deep Sea Res. 19: 199–208.

del Moral & L.A. *Standley*. 1979. Pollination of angiosperms in contrasting coniferous forests. Amer. J. Botan. 66: 26–35.

Denslow, J.J. 1980. Patterns of plant species diversity under different disturbance regimes. Oecologia. 46: 18–21.

Dobben, W.H. van & R.H. *Lowe-McConnell*. 1975. Unifying concepts in ecology. W. Junk, The Hague, Netherlands.

Feinsinger, P. 1978. Ecological interactions between plants and hummingbirds in a successional tropical community. Ecol. Monogr. 48: 269–287.

Fischer, A. 1960. Latitudinal variations in organic diversity. Evolution. 14: 64–81.

Foster, R.B. 1977. *Tachigalia versicolor* is a suicidal neotropical tree. Nature. 268: 624–626.

———. 1980. Heterogeneity and disturbance in tropical vegetation. Pp. 75–92. *In* M. Soule & B.A. Wilcox, eds. Conservation Biology: an evolutionary-ecological perspective. Sinauer Publ., Sunderland, Mass. 75–92.

Fox, J.F. 1979. Intermediate disturbance hypothesis. Science. 204: 1344–1345.

Garwood, N.C., D.P. *Janos* & N. *Brokaw*. 1979. Earthquake-caused landslides: a major disturbance of tropical forests. Science. 205: 997–999.

Gentry, A.H. 1976. Bignoniaceae of southern Central America: distribution and ecological specificity. Biotrop. 8: 117–131.

Gilbert, L.E. 1980. Food web organization and the conservation of neotropical diversity. Pp. 11–33. *In* M.E. Soule & B.A. Wilcox, eds., Conservation Biology: an evolutionary-ecological perspective. Sinauer Publ., Sunderland, MA.

Gillett, J.B. 1962. Pest pressure, an underestimated factor in evolution. *In* D. Nichols, ed., Taxonomy and Evolution. System. Assoc. Publ. No. 4: 37–46.

Goff, F.G. & P.H. *Zedler*. 1968. Structural gradient analysis of upland forests in the western Great Lakes area. Ecol. Monogr. 38: 65–86.

Goldblatt, P. 1978. An analysis of the flora of southern Africa: its characteristics, relationships and origins. Ann. Missouri Bot. Gard. 65: 369–436.

Gómez-Pompa, A. 1971. Posible papel de la vegetación secundaria en la evolución de la flora tropical. Biotrop. 3: 125–135.

Graham, A. 1973. Vegetation and Vegetational History of Northern Latin America. Elsevier, New York. 1–393.

Grijpma, P. 1976. Resistance of Meliaceae against the shoot borer *Hypsipyla* with particular reference to *Toona ciliata* .J. Roem. var. *australis*. *In* J. Burley & B.T. Styles, Tropical Trees, Variation, Breeding and Conservation. Linn. Soc. Symp. Ser. 2: 69–78.

Grime, J.P. 1973. Control of species density in herbaceous vegetation. Jour. Environ. Manage. 1: 151–167.

Grubb, P.J. 1977. The maintenance of species richness in plant communities. Biol. Rev. Cambridge Phil. Soc. 52: 107–145.

Harper, J.L. 1969. The role of predation in vegetational diversity. Brookhaven Symp. Biol. 22: 48–62.

Hartshorn, G.S. 1976. Tree species of La Selva. Tropical Science Center, San Jose, Costa Rica. 1–7.

———. 1978. Tree falls and tropical forest dynamics. Pp. 617–638. *In* P.B. Tomlinson & M.N. Zimmerman, eds., Tropical Trees as Living Systems. Cambridge Univ. Press, London.

Hepper, F.N. 1963. Flora of West Tropical Africa, 2nd ed. vol. 2. Crown Agents, London. 1–544.

———. 1964. Numerical analysis of the flora of West Tropical Africa 1. Kew. Bull. 19: 443–450.

Holdridge, L.R. 1976. Biological and genetical factors affecting exploration and conservation in Central America. *In* J. Burley & B.T. Styles, eds., Tropical Trees: Variation, Breeding and Conservation. Linn. Soc. Symp. Ser. No. 2: 199– 202.

Horn, H.S. & R.H. *MacArthur*. 1972. Competition among fugitive species in a harlequin environment. Ecology. 53: 749–752.

Hubbell, S.P. 1979. Tree dispersion, abundance, and diversity in a tropical dry forest. Science. 203: 1299–1309.

Huffaker, C.B. 1971. The phenomenon of predation and its role in nature. Pp. 327–343. *In* P.J. den Boer & G.R. Gradwell, eds., Dynamics of Populations. Proc. Adv. Study Inst. Dynamics Numbers Popul. 1970. Center for Agric. Publ., Wageningen, Netherlands.

Huston, M. 1979. A general hypothesis of species diversity. Amer. Nat. 113: 81–101.

———. 1980. Soil nutrients and tree species richness in Costa Rican forests. J. Biogeogr. 7: 147–157.

Inger, R.F. 1980. Densities of floor-dwelling frogs and lizards in lowland forests of southeast Asia and Central America. Amer. Nat. 115: 761–770.

Janzen, D.H. 1967. Why mountain passes are higher in the tropics. Amer. Nat. 101: 233–244.

———. 1970. Herbivores and the number of tree species in tropical forests. Amer. Nat. 104: 501–528.

———. 1973. Comments on host-specificity of herbivores and its relevance to species richness. Pp. 201–211. *In* V.H. Heywood, ed. Taxonomy and Ecology. Academic Press, New York.

Johnson, M.P. & P.H. *Raven*. 1970. Natural regulation of plant diversity. Evol. Biol. 4: 127–162.

Jones, N. 1976. Some biological factors influencing seed setting in *Triplochiton scleroxylon* K. Schum. *In* J. Burley & B.T. Styles eds., Tropical Trees: Variation, Breeding and Conservation. Linn. Soc. Symp. Ser. No. 2: 125–134.

Keay, R.W.J. 1958. Flora of West Tropical Africa, 2nd ed. vol. 1. Crown Agents, London.

Knight, D.H. 1975. A phytosociological analysis of species-rich tropical forest on Barro Colorado Island, Panama. Ecol. Monogr. 45: 259–284.

Kruger, F.J. & H.C. *Taylor*. 1979. Plant species diversity in Cape *fynbos*: gamma and delta diversity. Vegetatio. 41: 85–93.

Levin, S.A. 1974. Dispersion and population interactions. Amer. Nat. 108: 207–228.

——— & R.T. *Paine*. 1974. Disturbance, patch formation, and community structure. Proc. Nat. Acad. Sci. 71: 2744–2747.

Lloyd, M., R.F. *Inger* & F.W. *King*. 1968. On the diversity of reptile and amphibian species in a Bornean rainforest. Amer. Nat. 102: 497–516.

Louck, O.L. 1970. Evolution of diversity, efficiency and community stability. Amer. Zoolog. 10: 17–23.

MacArthur, R.H. 1969. Patterns of communities in the tropics. Biol. J. Linn. Soc. London. 1: 19–30.

Madison, M. 1977. Vascular epiphytes: their systematic occurrence and salient features. Selbyana. 2: 1–13.

Malfait, B.T. & M.G. *Dinkelman*. 1972. Circum-Caribbean tectonic and igneous activity and the evolution of the Caribbean plate. Geol. Soc. Amer. Bull. 83: 251–272.

Osman, R.W. 1977. The establishment and development of a marine epifaunal community. Ecol. Monogr. 47: 37–63.

——— & R.B. *Whitlatch*. 1978. Patterns of species diversity; fact or artifact? Paleobiol. 4: 41–54.

Paine, R.T. 1966. Food web complexity and species diversity. Amer. Nat. 100: 65–76.

———. 1974. Intertidal community structure. Experimental studies on the relationship between a dominant competitor and its principal predator. Oecologia. 15: 93–120.

———. 1979. Disaster, catastrophe and local persistence of the sea palm, *Postelsia palmaeformis*. Science. 205: 685–686.

Peet, R.K. 1978. Forest vegetation of the Colorado Front Range: patterns of species diversity. Vegetatio. 37: 65–78.

Pianka, E.R. 1966. Latitudinal gradients in species diversities; a review of concepts. Amer. Nat. 100: 33–46.

Pickett, S.T.A. 1980. Non-equilibrium coexistence of plants. Bull. Torrey Bot. Club. 197: 238–248.

Pielou, E.C. 1975. Ecological Diversity. Wiley, New York.

Poore, M.E.D. 1968. Studies in Malaysian rainforest. 1. The forest on triassic sediments in Jengka forest reserve. J. Ecol. 56: 143–196.

Prance, G.T., W.A. *Rodrigues* & M.F. *da Silva*. 1976. Inventário florestal de um hectare de mata de terra firme, km. 30 da Estrada Manaus-Itacoatiara. Acta Amazonica. 6: 9–35.

Raup, H.M. 1981. Physical disturbance in the life of plants. Pp. 39–52. *In* M.H. Nitecki, ed. Biotic Crises in Ecological and Evolutionary Time. Academic Press, New York.

Richards, P.W. 1973. Africa, the "odd man out." Pp. 21–26. *In* B.J. Meggers et al., eds., Tropical forest Ecosystems in Africa and South America: a Comparative Review. Smithsonian Instit. Press, Wash. DC.

Ricklefs, R.E. 1977. Environmental heterogeneity and plant species diversity: a hypothesis. Amer. Nat. 111: 376–381.

Rosen, D.E. 1975. A vicariance model of Caribbean biogeography. Syst. Zool. 24: 431–464.

Sale, P.F. 1977. Maintenance of high diversity in coral reef fish communities. Amer. Nat. 111: 337–359.

Sanders, H.L. 1968. Marine benthic diversity: a comparative study. Amer. Nat. 102: 243–282.

Sawyer, J.O. & A.A. *Lindsey*. 1971. Vegetation of the life zones in Costa Rica. Indiana Acad. Sci. Monogr. No. 2: 1–214.

Scott, N.J., Jr. 1976. The abundance and diversity of the herpetofaunas of tropical forest litter. Biotrop. 8: 41–58.

Shafi, M.I. & G.A. *Yarranton*. 1973. Diversity, floristic richness, and species eveness during a secondary (post fire) succession. Ecology. 54: 897–902.

Slack, N.G. 1977. Species diversity and community stucture in bryophytes: New York State studies. New York State Museum Bull. 428: 1–70.

Sousa, W.P. 1979. Disturbance in marine intertidal boulder fields: the non equilibrium maintenance of species diversity. Ecology: 60: 1225–1239.

Sprugel, D.G. & F.H. *Borman*. 1981. Natural disturbance and the steady state in high-altitude balsam fir forests. Science. 211: 390–393.

Standley, P.C. 1938. Flora of Costa Rica. Field Mus. Bot. Ser. 18: 1–1571.

Stebbins, G.L. 1981. Why are there so many species of flowering plants? Biosci. 31: 573–577.

Strong, D.R., Jr. 1977. Epiphyte loads, tree falls, and perennial forest disruption: a mechanism for maintaining higher species richness in the tropics without animals. J. Biogeogr. 4: 215–218.

Sutherland, J.P. 1974. Multiple stable points in natural communities. Amer. Nat. 108: 859–873.

Taylor, J.D. 1978. Faunal response to the instability of reef habitats: pleistocene molluscan assemblages of Aldabra atoll. Palaeontol. 21: 1–30.

Terborgh, J. 1973. On the notion of favorableness in plant ecology. Amer. Nat. 107: 481–502.

Valentine, J.W. 1969. Niche diversity and niche size pattern in marine fossils. J. Paleont. 43: 905–915.

Vandermeer, J.H., J. *Stout* & G. *Miller*. 1974. Growth rates of *Welfia georgii*, *Socrates durissima*, and *Iriartea gigantea* under various conditions in a natural rainforest in Costa Rica. Principes. 18: 148–154.

Van Valen, L. 1973. Body size and numbers of plants and animals. Evolution. 27: 27–35.

Webb, L.J., J.G. *Tracy* & W.T. *Williams*. 1972. Regeneration and pattern in the subtropical rainforest. J. Ecol. 60: 675–695.

White, P.S. 1979. Pattern, process and natural disturbance in vegetation. Bot. Rev. 45: 229–300.

White, T.C.R. 1978. The importance of a relative shortage of food in animal ecology. Oecologia. 33: 71–86.

Whitmore, T.C. 1975. Tropical Rainforests of the Far East. Clarendon Press, Oxford.

———. 1978. Gaps in the forest canopy. Pp. 639–658. *In* P.B. Tomlinson & M.H. Zimmerman, eds., Tropical trees as living systems. Cambridge Univ. Press, London.

Whittaker, R.H. 1969. Evolution of diversity in plant communities. Brookhaven Symp. Biol. 22: 178–196.

———. 1972. Evolution and measurement of species diversity. Taxon 21: 213–251.

———. 1977. Evolution of species diversity in land communities. Evol. Biol. 10: 1–67.

Wilson, J.W. 1974. Analytical zoogeography of North American mammals. Evolution. 28: 124–140.

Wilson, M.F. 1973. Tropical plant production and animals species diversity. Trop. Ecol. 14: 62–65.

Paleohistoria y Estado Actual de la Vegetación y la Flora de Colombia

Enrique **Forero**

With some 50,000 species, the flora of Colombia is one of the richest in the world. Colombia has an interesting geological history that includes the uplift of the Andes and past lowering of vegetation belts. This has led to a diversity of climates and soils and a wide floristic diversity. The country embraces four topographic regions; the Pacific coast, the Atlantic coast, the Andes, and the eastern lowlands. Colombia's ecological map recognizes 24 vegetation types which may be simplified into six basic types plus a miscellany of groups of limited expanse. Mangroves especially rich in species occur along the Pacific coast. Moist and wet forests are important lowland formations on the east and west of the Andes, with abundant lianas and impressive endemism. Evergreen mountain forests occur from 1000 to 3000 m where rainfall is 1000–5000 mm per year. There are fewer species than in the wet lowlands. Paramos are cold and humid regions occurring from 2800–5000 m. Plants here have adaptations to extreme conditions: small size, frequent rosettes, copious pubescence. Savanas occur on the eastern plains, along the Atlantic coast and inland, and they experience a marked dry season. Deserts occur on the Caribbean coast and elsewhere, with conspicuous cactus and deciduous shrubs and trees. Besides these vegetation types there are aquatic communities, beach vegetation, and successional communities.

Colombia está localizada en la esquina noroccidental de América del Sur; al oeste se encuentra el océano Pacífico, y al norte el océano Atlántico, siendo el único país del sub-continente con costas sobre dos mares. Los países limítrofes son Panamá al noroeste, Venezuela al oriente, Brasil al sur-este, Perú y Ecuador al sur. Su extensión total es de 1.138.914 km^2.

La gran diversidad de habitat que existen en Colombia se debe principalmente a su interesante historia geológica que ha producido una fisiografía muy diversa caracterizada por las grandes extensiones más o menos planas en el Oriente, los tres ramales de los Andes conocidos como Cordilleras Oriental, Central y Occidental, los profundos valles del Río Magdalena y del Río Cauca, las planicies costeras al norte y al oeste, y dos sierras aisladas de las cordilleras (la Sierra Nevada de Santa Marta y la Sierra de La Macarena).

Esa variada gama de ambientes hace difícil el estudio de la Flora, cuyo número de especies de plantas superiores ha sido calculado en 50.000 (Schultes, 1951), no existiendo siquiera una idea aproximada de la cantidad de hongos, musgos, hepáticas, etc. No es sorprendente, pues, que no exista una "Flora de Colombia". Aunque se cuenta con revisiones de varios grupos (Acanthaceae, Xyridaceae, Compositae-Astereae, Violaceae, etc.) y con algunas publicaciones regionales, el conocimiento de las plantas que crecen en el país está muy lejos de ser completo y aún se requiere un intenso trabajo.

Clima

En un país como Colombia, con una complicada topografía, los factores climáticos varían considerablemente. La gran diferencia en topografía, climas y características geológicas entre unas regiones y otras hace que los suelos sean también de distintas clases.

The complicated topography of Colombia leads

to varied climates and soil types, with some areas very fertile and well watered and others sterile deserts.

Suelos

Se encuentran suelos formados de arcillas terciarias, yesíferas y calcáreas en los desiertos de la Guajira, muy fértiles en el Valle del Cauca y en la altiplanicie de Bogotá, o arenosos o pedregosos en las partes más altas de las cordilleras cerca de las nieves perpetuas.

Evolución de la Flora y la Vegetación en el Pasado

La historia del clima, la flora y la vegetación del Norte de Sur América ha sido admirablemente documentada por los estudios palinológicos de van der Hammen y sus colaboradores (van der Hammen 1961, 1962, 1972a, 1972b, 1973, 1974; van der Hammen & González, 1965; van der Hammen et al. 1973; van Geel & van der Hammen, 1973). Las siguientes generalizaciones pueden hacerse con razonable seguridad teniendo como base los resultados por ellos obtenidos:

En el Cretáceo superior (Maestrichtiano) se han encontrado granos de polen de palmas del género *Psilamonocolpites* y de algunas dicotiledóneas primitivas, además de haberse reconocido la presencia de helechos. En aquella época el área era plana, con algunas colinas bajas localizadas en el lugar que hoy ocupa la Cordillera Central. Al comenzar el terciario, la flora se enriquece con nuevos elementos. Así, en el Paleoceno aparecieron varios géneros de palmas (entre ellos *Mauritia)*, bombacáceas, rubiáceas, myrtáceas, guttíferas, moráceas, etc. Además se formaron filas de colinas en la planicie que continuó cubierta en parte por arena y en parte por pantanos. Durante el Eoceno Inferior se encuentran cerca de las costas granos de polen de una especie de manglar, *Breviticolpites*, alternando con *Mauritia* y con *Psilamonocolpites*. Paralelamente ocurren algunos levantamientos aislados en el oriente. En el Eoceno Medio se presenta una vegetación abierta, posiblemente de sabana, de la que forman parte Gramíneas, Malváceas y el género *Ludwigia* (Onagraceae). En el Oligoceno aparece *Rhizophora* (Rhizophoraceae) en las costas y en los estuarios. En el Mioceno la sedimentación proveniente de las colinas aumentó, y aparecieron por primera vez las Compuestas. En el Plioceno, hace aproximadamente 7.000.000 de años, los manglares se encontraban tanto en el Pacífico como en el Atlántico, las colinas se elevaron un poco más, y se formaron los valles de los ríos Magdalena y Cauca. La flora de polen tropical es muy rica en esa época, y no existe polen de plantas de montañas altas. Hacia el oriente existían vegetación de sabana y de selva semejantes a las actuales.

The elevation of the three Andes ranges occured during the Pleistocene, 2 million years ago. At this time North and South America were joined by the isthmus of Panama.

Durante el Plioceno Superior y el comienzo del Pleistoceno hace unos 2.000.000 de años, se produjo el levantamiento definitivo de los tres ramales de los Andes y de otras áreas montañosas tales como la Sierra Nevada de Santa Marta, dando lugar a la aparición de condiciones ambientales completamente diferentes a las existentes hasta entonces. En esa misma época apareció la conexión de América del Norte con América del Sur a través del istmo de Panamá (Haffer, 1970).

El proceso de población de esos nuevos ambientes comenzó con la adaptación de elementos provenientes de las áreas circundantes, pero al mismo tiempo ocurrió la inmigración de especies vegetales de sur a norte y de norte a sur. Del sur llegaron entre otros los géneros *Weinmannia* (Cunoniaceae), *Gunnera* (Gunneraceae), y *Drimys* (Winteraceae), mientras del norte aparecen primero (en el Pleistoceno Inferior) *Myrica* (Myricaceae) y más tarde *Juglans* (Juglandaceae), *Alnus* (Betulaceae) y *Quercus* (Fagaceae). Los géneros *Alnus*, *Juglans* y *Myrica* existían en Panamá desde el Plioceno (Graham 1973).

En los páramos existen desde entonces numerosos géneros endémicos tales como la característica *Espeletia* (Compositae), *Aciachne* (Gramineae), *Distichia* (Juncaceae), *Puya* (Bromeliaceae), y *Rhizocephalum* (Campanulaceae), además de otros elementos como *Aragoa (Scrophulariaceae)*, *Polylepis* (Rosaceae), etc. Esta flora se enriqueció gradualmente con elementos antárticos tales como *Muehlenbeckia* (Polygonaceae), *Acaena* (Rosaceae), y *Azorella* (Umbelliferae), y holárticos como *Gentiana* (Gentianaceae), *Bartsia* (Scrophulariaceae), *Valeriana* (Valerianaceae), *Draba* (Cruciferae), *Hypericum* (Hypericaceae), y *Berberis* (Berberidaceae).

La ocurrencia en los Andes de marcados cambios climáticos durante el Pleistoceno ha sido reconocida ampliamente en años recientes, y puede resumirse en los siguientes hechos principales (Bartlett & Barghoorn 1973; van der Hammen 1974):

1. Las regiones tropicales y las regiones templadas

del norte fueron afectadas por los diversos climas del Pleistoceno. Tanto los trópicos como las regiones templadas sufrieron episodios climáticos glaciales (fríos) e interglaciales (más calientes).

2. Las temperaturas anuales promedio en partes altas de los Andes en el último Pleniglacial Superior deben haber sido entre 6° y 7° C (o quizá 8° C) más bajas que las actuales. Estas temperaturas son comparables a las calculadas para la misma época en Europa.

3. Los períodos glaciales en los Andes correspondieron aproximadamente con períodos de mayor precipitación (o pluviales), y los interglaciales fueron de menor precipitación (interpluviales), aunque ocurrieron fluctuaciones considerables en la cantidad de lluvia en los diversos períodos.

4. Tanto los glaciales y los interglaciales como las variaciones climáticas menores fueron contemporáneos con los ocurridos en regiones templadas.

5. El límite superior del bosque andino debió descender unos 1200–1500 m durante el glacial. Mientras el límite actual está algo por encima de los 3000 m, en esa época debió localizarse a 2000 m como promedio.

Climatic changes during the Pleistocene included; glacial periods for both tropical and temperate areas, temperate areas, temperatures approaching those of Europe at the same time, interglacial periods of less rainfall, and a descent of the tree line by some 1200–1500 m from the present 3000 plus m. The present lower paramo islands were all connected, but the upper paramos never were, leading to high endemism. The eastern plains alternated between savanna and scrub vegetation.

Durante los glaciales el área ocupada por la vegetación de páramo fue mucho mayor que en la actualidad (van der Hammen 1974, Simpson 1975). Muchas de las "islas" de páramo que existen hoy estuvieron unidas en el pasado. Sin embargo, los superpáramos nunca estuvieron en contacto. Esto explicaría el alto endemismo que existe en el superpáramo, especialmente de la Sierra Nevada del Cocuy.

La conexión directa entre los páramos permitió el intercambio de especies de un páramo a otro y la expansión del área de distribución de elementos provenientes de los bosques cercanos, y ofreció condiciones favorables para la inmigración de especies colonizadoras provenientes de lugares lejanos, (holárticas y antárticas), dada la mayor área disponible y la cercanía de unos parámos con otros. Durante los interglaciales las poblaciones permanecían aisladas y esta situación favorecía los procesos naturales de especiación (van der Hammen 1974; Simpson 1975).

En cuanto a las tierras tropicales bajas, la temperatura debió descender durante los períodos glaciales unos 3° C con relación a la actual. Existen datos palinológicos que indican que en los llanos orientales de Colombia y en la sabana de Rupununi en Guyana hubo alternación de períodos de matorrales dominados por *Byrsonima* (Malpighiaceae) y *Curatella* (Dilleniaceae), y sabanas abiertas con Gramíneas, Ciperáceas, y otros elementos herbáceos de sabana. El cambio de sabana hacia matorrales densos indica un clima más húmedo.

Los datos disponibles de una serie de muestras del territorio de Rondonia, en el Brasil (van der Hammen 1972; Absy & van der Hammen 1977), también demuestran la ocurrencia de cambios marcados en el clima y en la vegetación que corresponden con las variaciones climáticas reconocidas para el último período glacial. Areas que hoy están cubiertas por selva pluvial tropical contínua, tuvieron vegetación de sabana en algunas épocas. Estos últimos datos sirven de apoyo palinológico a la teoría de los refugios pleistocénicos (Absy 1979).

Fisiografía Actual

Los movimientos orogénicos del Terciario produjeron en el norte de América del Sur una topografía muy complicada. La descripción de la fisiografía actual de Colombia puede dividirse convenientemente en regiones que coinciden con los diversos tipos de vegetación. Estas regiones son:

Costa del Pacífico.—Es la continuación de la faja costera que viene desde el sur del continente. En Colombia es una franja estrecha e irregular que alcanza en algunos lugares hasta 100 km de ancho. Esta plataforma aparentemente plana es interrumpida por brazos de la cordillera occidental y por serranías aisladas separadas unas de otras por pequeños ríos. Las mayores alturas llegan a 1000 y raramente 1300 m. El Cerro Tacarcuna, en el norte, alcanza a 1900 m. (A. Gentry com. pers.).

The Pacific lowlands are below 100 m elevation and are interrupted by spurs of the western cordillera of 1000 or even 1900 m.

Costa Atlántica.—La región costera del Atlántico o planicie costera del Caribe se extiende desde el río Sinú hasta la Guajira. Es un área plana con pequeñas elevaciones esparcidas, generalmente semiárida y abierta, y a menudo pantanosa cerca de los ríos. La península de la Guajira es en su mayor extensión una región desértica. La serranía de Macuira, al norte, constituye una interesante excepción a esta regla general y presenta vegetación densa diferente de la de los desiertos circundantes.

The Atlantic lowlands are semiarid and open except for the Serrania de Macuira on the north.

Andes.—En Colombia el sistema montañoso de los Andes forma tres ramales o codilleras cuya altura promedio está entre 3200 y 3500 m sobre el nivel del mar. Estas cordilleras, los valles que las separan, otros valles inter-andinos, los numerosos picos de gran altura y las abruptas pendientes de sus laderas crean una gran diversidad de condiciones ambientales. Algunos de los puntos más altos, que presentan nieves perpetuas, son el Nevado del Huila con 5750 m, el del Tolima con 5620 m, y el del Cocuy con 5493 m. La Sierra Nevada de Santa Marta aparece como una isla en la parte norte del país, sin conexiones con las tres cordilleras. De ella forman parte los picos de mayor altura en todo el país, que son el Cristóbal Colón, con 5800 m y el Simón Bolívar con 5775 m.

The three Andes ranges have average elevation of 3200–3500 m. Their interspersed vallies and high peaks, ranging to 5800 m, create a great diversity of climates.

Tierras bajas al Oriente de los Andes.—Constituyen la continuación de la inmensa planicie que se extiende por Bolivia, Perú y Ecuador. La altura de esta región varía entre 100 y 250 m., pero puede alcanzar 1200 en la Sierra de la Macarena (Departamento del Meta). En la parte nor-oriental se localizan los llanos orientales y en la parte sur-oriental está la selva amazónica. La Sierra de La Macarena y otros lugares elevados son considerados como el extremo occidental de las "Tierras Altas de las Guayanas".

Vegetación y Flora

The eastern lowlands, 100–250 m in elevation, extend through Peru, Ecuador and Bolivia. In the northern part are the eastern plains, and in the south the Amazonian jungles.

La enorme diversidad de hábitat y microhábitat que resulta de las condiciones anotadas, crea condiciones favorables para el desarrollo de muchos y muy diversos tipos de vegetación. Las referencias de muchos autores a la vegetación de Colombia son, sin embargo, superficiales, sobre todo en lo referente a zonas altas. Las más conocidas tienen que ver principalmente con las selvas que cubren las tierras bajas en toda la América del Sur (Hueck 1972), o constituyen un bosquejo de las formaciones vegetales de todo el sub-continente (Eiten 1974), o, como en el caso del mapa publicado recientemente por Gentry (1977), se refieren a áreas más o menos amplias (en este caso particular a Centro América y el nor-oeste de Sur América). Cleef (1977, 1979) ha contribuído a solucionar esta situación, con sus estudios sobre la secuencia altitudinal de la vegetación de los páramos de la Cordillera Oriental y el análisis de las relaciones fitogeográficas de la flora de los páramos neotropicales.

Las principales fuentes de información disponibles sobre los tipos de vegetación que existen en Colombia son las publicaciones de Cuatrecasas (1958) y Espinal y Montenegro (1963), y la última edición del Mapa Ecológico y de su Memoria Explicativa (Inst. Geogr. 1977). Estos trabajos son básicos para cualquier estudio de la vegetación del país y en cada uno aparecen listas más o menos completas de los géneros y especies presentes en cada "tipo" o "formación".

Understanding the six basic vegetation types is important for any study of the country's plant life. They are determined by climate, soils, topography and historical factors.

De acuerdo con la Memoria Explicativa sobre el Mapa Ecológico, existen en Colombia 24 Zonas de Vida. En la descripción que sigue se emplea, no obstante, una clasificación más sencilla en la cual se estudian seis tipos de vegetación, y en un séptimo grupo se incluyen varios de distribución restringida.

Con el fin de comprender mejor el porqué de la riqueza de la Flora de Colombia es importante relacionar los factores climáticos, edáficos, topográficos e históricos antes mencionados. Por esa razón, es conveniente discutir los diversos tipos de vegetación, indicando en cada caso las familias, géneros y algunas de las especies representativas, y las variaciones más importantes dentro de cada tipo vegetacional.

Manglares

Los manglares ocurren en gran parte de la costa del Pacífico y en el extremo occidental de la del Atlántico. En el Atlántico la especie dominante es *Rhizophora mangle*, mientras que en el Pacífico es *Rhizophora brevistyla* acompañada de varias especies endémicas entre las que vale la pena mencionar el "piñuelo", *Pelliceria rhizophorae* (Theaceae), *Pavonia rhizophorae* (Malvaceae) y *Ardisia granatensis* (Myrsinaceae). También importantes son *Avicennia nitida* (Avicenniaceae) llamada "mangle iguanero" o "mangle negro", *Laguncularia racemosa* "mangle blanco", y *Conocarpus erectus* (Combretaceae) "mangle jeli", y *Acrostichum aureum* (Pteridaceae). Cuatrecasas (1958) describe asociaciones de diferentes especies controladas por el tipo de suelo, la salinidad y la distancia del océano.

The Atlantic coast has one dominant mangrove species, but along the Pacific there are others in several families, controlled by salinity, soil and distance from the ocean.

En primer lugar, en el frente marino, ocurren comunidades de *Rhizophora brevistyla*; a continuación, en lugares más altos aparece como dominante *Avicennia nitida*, y por último *Laguncularia racemosa* asociada con arbustos y árboles pequeños; en terrenos aún más elevados aparecen asociaciones de *Conocarpus erectus* con otras especies.

Selvas Húmedas y Pluviales

Las selvas húmedas y pluviales ocupan una importante porción de las tierras bajas al oriente de los Andes y en el occidente se presentan a lo largo de la costa del Pacífico continúandose en el Ecuador y en Panamá. En el norte de Colombia entran por el valle del Magdalena hasta los 5°, en La Dorada. Ocurren desde el nivel del mar hasta los 1000 m con temperaturas que varían entre 23 y 30° C y con precipitación anual entre 2.000 y 8.000 mm, llegando en algunos lugares del Chocó a 11.000 mm, distribuídos en la mayor parte del año (West 1957, cita un total anual de 15.058 mm en Quibdó durante 1939).

Las familias predominantes entre los árboles son Leguminosae, Annonaceae (géneros *Rollinia, Guatteria*), Lauraceae (*Nectandra, Ocotea*), Myristicaceae (*Virola, Dialyanthera*), Chrysobalanaceae (*Licania, Hirtella*), Euphorbiaceae (*Hevea brasiliensis, Mabea*), Bombacaceae (*Bombax, Ceiba*), Apocynaceae (*Couma*), Burseraceae (*Protium*), Lecythidaceae (*Eschweilera, Lecythis*), Moraceae (*Ficus, Brosimum*), Guttiferae (*Clusia, Tovomita*), Meliaceae (*Guarea, Trichilia*).

Arboles pequeños y arbustos comunes pertenecen principalmente a las Melastomataceae, Rubiaceae, y Piperaceae. Entre los géneros más comunes de palmas están *Astrocaryum, Euterpe, Iriartella, Socratea* y *Geonoma*.

Las lianas más abundantes son Bignoniaceae, Hippocrateaceae, Passifloraceae, Menispermaceae, Leguminosae, Dilleniaceae, Sapindaceae, y Malpighiaceae. Las epífitas son Orchidaceae, Bromeliaceae, Piperaceae y Gesneriaceae.

De acuerdo con la información botánica presentada por Prance (1974), en Colombia debieron existir cuatro refugios de selva durante el Pleistoceno, localizados en el Chocó, en el Río Nechí, en el área de la Sierra Nevada de Santa Marta y en el Catatumbo, mientras en el sur el refugio del Napo debió alcanzar parte de Nariño, Putumayo, Caquetá y Amazonas.

The wet lowland forests occupy important areas in several parts of Colombia. Rainfall ranges from 2,000–15,000 mm and temperatures are high. Lianas and epiphytes are abundant. During the Pleistocene, four refugia persisted in the Choco region, handing down a rich endemic flora.

El endemismo en Colombia no ha sido estimado cuantitativamente. Sin embargo, parece que la cantidad de taxones endémicos en el Chocó puede ser alta (Gentry 1982). Esta región era hasta hace poco tiempo una de las peor conocidas del Neotrópico (Prance 1977; Gentry 1978). Actualmente se adelanta un programa de colaboración entre el Herbario Nacional Colombiano y el Missouri Botanical Garden durante el cual, en un período de siete años, se ha triplicado el número de colecciones de la región.

Selvas Siempre Verdes de Altura

Las selvas perennifolias de altura, bosques subandinos y andinos o selvas húmedas de altura o de montaña. Aquí se incluyen tipos de vegetación que ocurren entre 1.000 y 3.000 m, con temperatura anual que varía entre los 6° y los 23° C, con lluvias que fluctúan entre 1.000 y 5.000 mm al año, y con neblina frecuente. Se presenta en las laderas de las tres cordilleras y en la Sierra Nevada de Santa Marta. Aquí la riqueza en número de especies disminuye con relación a las selvas de tierras bajas, pero la vegetación epífita es mucho más abundante y en algunos lugares cubre completamente los árboles formando "jardines colgantes" compuestos por helechos, musgos, Orquídeas, Aráceas, Piperáceas, Gesneriáceas, Rubiáceas y Bromeliáceas. Los helechos arborescentes son abundantes y se desarrollan sobre un suelo rico en humus y considerablemente húmedo. También aparecen varios géneros de palmas pero las trepadoras leñosas son menos frecuentes que en la selva pluvial.

Evergreen mountain forests occur between 1000–3000 m with 1000–5000 mm rainfall. Temperatures are somewhat cooler than in the lowlands, and there are fewer species. Epiphytes are more abundant, and tree ferns and palms are conspicuous. A few species from arctic and subarctic regions are present.

Hay ciertas variaciones en la composición de la vegetación a medida que se asciende por las montañas. Entre 1.000 y 2.400 m de altura los géneros más comunes son *Calliandra, Tara* e *Inga* (Leguminosae), las Melastomatáceas *Miconia* ("tuno") y *Tibouchina* ("sietecueros" o "mayos"), *Cecropia* y *Ficus* (Moraceae), *Clusia* y *Vismia* (Guttiferae), *Quercus* (Fagaceae), *Cordia* (Boraginaceae), *Saurauia* (Saurauiaceae), *Billia* (Hippocastanaceae) y *Tabebuia* (Bignoniaceae). Entre los 2.300 y los 2.500 m hay un área de transición en la cual aparecen los géneros *Weinmannia* (Cunoniaceae, "encenillo"), *Tovomita* y *Clusia* (Guttiferae), *Drimys* (Winteraceae, "canelo"), *Dendropanax* y *Oreopanax* (Araliaceae), *Matisia* (Bombacaceae), *Symplocos* (Symplocaceae), *Ilex* (Aquifoliaceae), *Ficus* (Moraceae), *Cedrela* (Meliaceae, "cedro"), *Palicourea* (Rubiaceae), las Melastomatáceas *Miconia* y *Topobea*, y la Ericaceae *Befaria*. Por encima de los 2.500 m los árboles son por lo general más pequeños, tienen hojas más pequeñas, y las epífitas herbáceas continúan siendo numerosas. Allí están representadas las familias Gesneriaceae, Begoniaceae, Lobeliaceae y Piperaceae; epífitas leñosas son algunas Ericaceae y hemiparásitas las Loranthaceae.

El número de especies arbóreas dominantes continúa decreciendo, y los géneros más abundantes son *Weinmannia* (Cunoniaceae), *Escallonia* (Escalloniaceae), *Hesperomeles* (Rosaceae), *Ilex* (Aquifoliaceae), y en algunos lugares *Polylepis* (Rosaceae, "colorados"). Un componente importante de las selvas siempre-verdes de altura es el género *Ceroxylon* (Palmae), el "arbol" nacional de Colombia. En los estratos inferiores son comunes las familias Solanaceae, Rubiaceae *(Psychotria)*, Piperaceae *(Piper)*, muchas Melastomataceae, Cyperaceae, Araceae, Acanthaceae y helechos.

Un hecho fitogeográfico interesante que ha sido mencionado por Hueck (1972) es que en las selvas siempre-verdes de altura del centro de Colombia se reunen especies de origen holártico como *Quercus humboldtii* (Fagaceae) y *Juglans neotropica* (Juglandaceae), con elementos subantárticos como *Podocarpus taxifolius* y *Podocarpus macrostachys* (Podocarpaceae).

Páramos

Los páramos son un tipo de vegetación que existe desde Venezuela a través de las tres cordilleras colombianas hasta el Ecuador como unidades separadas. Los páramos son regiones frías y húmedas que van desde los 2800 m hasta los 4700 o 5000 m, donde comienzan las nieves perpetuas de las altas montañas. La temperatura mínima puede estar por debajo de los 0° C, (−2°) y la máxima es de 12° C; la precipitación anual va desde los 590 mm hasta los 800 mm aproximadamente. La lluvia y la neblina son muy frecuentes. La temperatura diaria tiene amplias variaciones, pero los promedios mensuales y anuales no cambian mucho. No existen diferencias marcadas entre épocas secas y húmedas, y por esta razón los páramos no son comparables con otros tipos de vegetación que ocurren a grandes alturas en la zona templada; por consiguiente la designación de "alpinas" no puede ser empleada con propiedad.

Los suelos son de dos clases: suelos de arena y piedra en los que el humus prácticamente no existe, que ocurren cerca a las nieves perpetuas (en los superpáramos) y están cubiertos de escasa vegetación, y suelos ricos en humus, húmedos y muy ácidos (pH 3.8–4.5), que ocurren en el páramo propiamente dicho y en el subpáramo, están cubiertos de vegetación continua y presentan elementos leñosos y bosques en las áreas más protegidas.

Las plantas de los páramos están adaptadas a vivir en condiciones extremas de frío y sequía fisiológica. Algunas de sus adaptaciones son: tamaño pequeño de las partes vegetativas, frecuencia de plantas arrosetadas, almohadilladas y pegadas al suelo, reducción en el tamaño de las hojas y presencia de grandes cantidades de pubescencia (Forero 1970). Las fanerógamas epífitas son raras pero las criptógamas son abundantes.

Los páramos pueden dividirse convenientemente en tres fajas altitudinales (Cuatrecasas 1958) cuyos límites, no obstante, son difíciles de establecer en muchos casos. Estos niveles son el subpáramo, páramo propiamente dicho, y superpáramo. Cleef (1977) propuso subdivisiones adicionales basado en sus estudios de la vegetación de los páramos de la cordillera Oriental. Por una parte, reconoce diferencias considerables entre la ladera seca y la ladera húmeda de la cordillera, y por otra, establece que en cada una de las fajas altitudinales pueden reconocerse una zona baja y una zona alta.

Paramos are cold and humid regions lacking marked seasonality. In Colombia they ascend

From 2800 m to the snow line, 4700–5000 m. Soils tend to be either sandy or stony with little vegetation, or rich in acid humus with herbs or small woody growth. Different elevation belts—subparamo, paramo proper, and superparamo—have different floras as do the dry paramos and humid paramos.

Los subpáramos se encuentran entre los 2.800 y los 3.000 m aproximadamente. Los árboles por lo general no sobrepasan los 8 m. Algunas de las plantas más importantes son *Hypericum laricifolium* con amplia distribución desde Venezuela hasta el Ecuador, *Aragoa* (Scrophulariaceae), *Baccharis, Loricaria* y *Diplostephium* (Compositae), *Vallea* (Elaeocarpaceae), *Macleania, Cavendishia, Vaccinium* y *Pernettya* (Ericaceae), *Escallonia* (Escalloniaceae), *Symplocos* (Symplocaceae), *Miconia* (Melastomataceae), *Weinmannia* (Cunoniaceae), *Polylepis* y *Hesperomeles* (Rosaceae).

Según la reciente subdivisión propuesta por Cleef, puede presentarse una descripción algo más detallada de lo que ocurre en el subpáramo de la Cordillera Oriental:

En el subpáramo bajo del lado seco existen matorrales de Ericáceas *(Befaria, Macleania, Cavendishia),* Compuestas *(Senecio, Baccharis, Eupatorium y Diplostephium)* y Melastomatáceas *(Miconia, Bucquetia* y *Brachyotum*). Localmente, *Myrica parvifolia* (Myricaceae) constituye matorrales densos.

En el subpáramo alto aparece *Arcythophyllum nitidum* (Rubiaceae) asociado con *Calamagrostis effusa* (Gramineae) y otras plantas características del páramo propiamente dicho.

El subpáramo bajo del lado húmedo se caracteriza por la presencia de *Chusquea scandens, Neurolepis aristata* y *Aulonemia trianae* (Gramineae), que también están presentes en las selvas siempre-verdes de altura, mezclados con Compuestas *(Ageratina, Diplostephium, Baccharis),* Escalloniaceae *(Escallonia)* Loranthaceae *(Gaiadendron),* Melastomataceae *(Purpurella, Miconia),* Scrophulariaceae *(Aragoa)* y Ericaceae *(Gaultheria).*

El subpáramo alto en el lado húmedo tiene *Arcytophyllum nitidum,* asociado con *Swallenochloa* (Gramineae), numerosos musgos y hepáticas, Ericaceae *(Disterigma, Gaultheria),* Compuestas, etc.

El páramo propiamente dicho se localiza entre los 3000 y los 3800 m. Las gramíneas son dominantes, en particular *Calamagrostis* y *Festuca,* pero diversas especies, en muchos casos endémicas, de *Espeletia* ("frailejón"), determinan el aspecto característico del paisaje. Otros elementos comunes son las Ericaceae *(Vaccinium, Pernettya* y *Gaultheria),* Melastomaceae *(Tibouchina, Miconia,* y *Castratella),* Gentianaceae *(Halenia),* Valerianaceae y Rosaceae *(Polylepis).*

En la zona baja del páramo propiamente dicho en el lado seco de la cordillera Oriental hay dominio de *Calamagrostis effusa* mezclado con hierbas como *Castratella piloselloides* (Melastomataceae), *Rhynchospora paramorum* y *Oreobolus* (Cyperaceae).

En la zona alta *Calamagrostis* se abre parcialmente y aparecen *Jamesonia* (Polypodiaceae), *Paepalanthus* (Eriocaulaceae), *Hypericum* (Hypericaceae) y varias hepáticas y líquenes.

En la zona baja del páramo propiamente dicho, del lado húmedo, *Calamagrostis effusa* ocupa una extensión considerable, y *Swallenochloa* disminuye su cobertura. También aparecen *Oreobolus* y *Espeletia.*

El super-páramo ocurre sobre suelos arenosos y pedregosos, entre los 3800 y los 5000 m. La vegetación es escasa y está formada por especies de *Senecio* (Compositae), *Agrostis, Poa, Bromus* (Gramineae), *Lupinus* (Leguminosae) y *Cerastium floccosum* (Caryophyllaceae). Condiciones especiales locales más favorables permiten la presencia de *Gunnera magellanica* (Gunneraceae), *Werneria* (Compositae), algunas Cyperaceae y varias criptógamas.

En la zona baja del super-páramo en el lado seco de la cordillera Oriental existen matorrales de *Loricaria complanata, Senecio vaccinioides* y *Diplostephium.* Otros elementos son *Pernettya, Draba, Luzula* (Juncaceae), varias gramíneas *(Agrostis, Poa), Jamesonia* (Polypodiaceae), *Lycopodium* (Lycopodiaceae) y *Bartsia* (Scrophulariaceae).

En general los mismos elementos se presentan aunque mucho más esparcidos en la zona alta del super-páramo.

En la zona baja del super-páramo en el lado húmedo ocurre una faja de matorral enano de *Loricaria complanata* asociada con briófitas. También se encuentran otras compuestas *(Erigeron, Espeletia, Senecio),* Valerianaceae, Cruciferae *(Draba),* Caryophyllaceae *(Arenaria, Cerastium).* Un elemento característico es *Senecio niveo-aureus.*

La constitución de la zona alta del super-páramo en el lado húmedo es básicamente la misma descrita para la zona equivalente en el lado seco.

Sabanas

Las sabanas ocurren en los llanos orientales, en parte de la planicie costera del Atlántico, en el alto valle del Magdalena al sur de Honda y en algunas áreas del valle del Cauca. Este tipo de vegetación está restringido a regiones cuya altura sobre el nivel del mar no sobrepasa los 1000 m, con temperaturas anuales promedio de más de 24° C, y con precipitación de 1000 a 2000 mm por año. Una de las características más importantes y que ejerce mayor influencia sobre la vegetación es la existencia de una estación seca que puede durar cinco meses o más, y una estación húmeda en la cual las lluvias pueden ser muy fuertes ocasionando inundaciones en muchos lugares.

Savanas occur below 1000 m along the Atlantic coast and in some interior valleys. There is a dry season of five or more months and a rainy season with torrents and flooding. The grasslands are interrupted by gallery forests along watercourses that have distinctive species composition.

Las gramíneas más típicas en las partes altas de las sabanas son especies de *Andropogon, Axonopus, Paspalum, Aristida* y *Panicum*, siendo muy común la especie *Paspalum contractum*. Los arbustos más frecuentes pertenecen a las Melastomataceae, Labiatae, Leguminosae y Malvaceae. Los árboles más comunes son *Byrsonima crassifolia* (Malpighiaceae), *Curatella americana* (Dilleniaceae), *Palicourea rigida* (Rubiaceae), *Cochospermum vitifolium* (Cochlospermaceae) y *Bowdichia virgiliodes* (Leguminosae).

A lo largo de las corrientes de agua se encuentran los bosques de galería que interrumpen el paisaje sabanero. En éstos aparecen numerosas especies que no existen en las sabanas circundantes y cuyas afinidades están en las selvas pluviales. Algunas son *Hirtella elongata* (Chrysobalanaceae), varias Annonaceae *(Guatteria)*, Lecythidaceae, Leguminosae, etc. Un componente muy característico es la palma *Mauritia flexuosa* ("moriche").

En las partes bajas e inundables de la sabana ocurre una flora diferente a la de las partes más altas y bien drenadas; en esas partes bajas es dominante la gramínea *Trachypogon plumosus*.

En el Vaupés, Caquetá, Guainía y Meta ocurren las "sabanas casmófitas" cuya flora es diferente de las anteriores, y está formada principalmente por Velloziaceae, Rapateaceae, Ochnaceae, Lentibulariaceae, y géneros diferentes de Melastomataceae, Ericaceae y Xyridaceae.

Vegetación Xerofítica y Subxerofítica

Existen zonas desérticas en la costa del Caribe, desde el Río Sinú hasta la Guajira, y desérticas o semidesérticas en los valles de los Ríos Chicamocha y Dagua, y en los Departamentos de Huila y Tolima. Todas se localizan entre el nivel del mar y los 1000 m de altura, pero en el Chicamocha llegan a los 2200 m. La temperatura media anual es superior a los 24°C, y la precipitación de 100 hasta 1000 mm al año. Hay una época seca de seis meses o más en la que prácticamente no llueve.

La condición desértica de la parte norte de Colombia se debe a la influencia de los vientos alisios que soplan en dirección este-noreste durante todo el año sin encontrar barreras montañosas que los obliguen a condensar su humedad. Cuando esas barreras aparecen como en la Serranía de Macuira al norte de la Guajira la precipitación aumenta y la cobertura vegetal es diferente y más rica en especies (Sugden, 1976). La presencia de vegetación xerofítica en los valles inter-andinos puede explicarse por condiciones locales provocadas por la barrera que forman las montañas circundantes, la que no permite el paso de masas de aire húmedo.

The deserts of the Caribbean coast occur mostly below 1000 m. Rainfall ranges from 100–1000 mm per year and temperatures are high. Trees of cactus and deciduous legumes are conspicuous.

Diferentes clases de plantas presentan diversos tipos de adaptaciones a las condiciones desérticas o semi-desérticas en las que tienen que vivir; entre esas adaptaciones están la suculencia de los cactos, la pérdida de las hojas durante la época seca, la producción de espinas, la reducción en tamaño de las partes aéreas, etc.

Un árbol característico de regiones semiáridas es *Cavanillesia platanifolia* (Bombacaceae). Además, *Acacia* (Leguminosae) que es generalmente caducifolia; cactáceas columnares y especies del género *Randia* (Rubiaceae).

Cuando las condiciones de sequía no son extremas (precipitación anual entre 1000 y 1100 mm y 5–6 meses de sequía) los árboles pueden alcanzar 8 m o más de altura, aunque por lo general son pequeños y esparcidos. Muchos de los componentes de esta vegetación sub-xerofítica son espinosos y los más característicos son las leguminosas *Prosopis* ("trupillo"), *Caesalpinia coriaria* ("divi divi"), *Piptadenia* y *Acacia*, además de los cactos de los géneros *Lemaireocereus* y *Acanthocereus*.

En los bosques muy secos también llamados "cardonales" el género *Lemaireocereus* es uno de los elementos principales, y se encuentra mezclado con algunos de los géneros de Leguminosae ya mencionados.

En los enclaves xerofíticos de los Andes ocurren *Cercidium praecox* y *Prosopis juliflora* (Leguminosae), asociadas con cactáceas de los géneros *Cephalocereus, Lemaireocereus, Opuntia* y *Mamillaria,* además de *Dodonaea viscosa* (Sapindaceae), y *Lantana rugulosa* (Verbenaceae).

Otros Tipos de Vegetación

Además de los tipos de vegetación mencionados hasta ahora, existen otros de distribución estringida, tales como la vegetación de agua dulce con especies como las Pontederiaceae, *Eichhornia crassipes* y *Eichhornia azurea,* ("lirios de pantano" o "taruyas"), y *Victoria amazonica* (Nymphaeaceae) en las zonas bajas y *Azolla* (Azollaceae) e *Isoetes* (Isoetaceae) en las altas montañas. En reudales y saltos de agua ocurren especies de Podostemonaceae. En aguas marinas aparecen, además de las algas que son dominantes, algunas plantas superiores entre las que se destaca *Thalasia testudinum* (Hydrocharitaceae). En las playas y márgenes de ríos, lagos, etc. crecen gramíneas de los géneros *Paspalum, Hymenachne, Leptochloa, Panicum* y *Eragrostis,* algunas Cyperáceas, Araceae del género *Montrichardia,* y más tarde en el proceso de sucesión, *Cecropia* (Moraceae), *Calliandra, Albizzia* y *Pithecellobium* (Leguminosae).

Freshwater communities, salt water communities, and water's edge communities are also present as well as secondary vegetation types.

El proceso de colonización humana en todo el país produce condiciones para el desarrollo de una vegetación secundaria cuyas características aún no han sido estudiadas.

Conclusiones

La presentación anterior demuestra claramente que la flora de Colombia es una de las más ricas del área Neotropical debido a la gran variedad de condiciones ambientales existentes en la actualidad y a la interesante historia de acontecimientos ocurridos en el pasado que apenas ahora comienzan a ser explicados.

Como consecuencia de esa diversidad, el estudio de la flora está muy lejos de ser completo, no obstante los esfuerzos de colección, observaciones de campo y publicaciones realizados hasta ahora por botánicos colombianos y extranjeros. Debido al rápido proceso de destrucción al que está siendo sometida la vegetación, es urgente incrementar los programas de colección intensiva, los estudios ecológicos y otros proyectos semejantes en numerosas regiones del país.

The rapid destruction of vegetation of all types makes urgent an increase in field studies so that a balance can be found between conservation and acceptable use of vegetation resources.

Los resultados de este fundamental trabajo de campo pondrán a disposición de los especialistas materiales indispensables en la preparación de monografías, abrirán nuevos horizontes a jóvenes colombianos con inclinación científica, y podrán ser empleados para ofrecer recomendaciones a las entidades gubernamentales sobre uso apropiado de la tierra, buscando un balance entre conservación y aprovechamiento consciente de los recursos vegetales del país, y procurando poner fin a la destrucción indiscriminada e irresponsable de nuestra vegetación natural.

Absy, M.L. 1979. A palynological study of Holocene sediments in the Amazon Basin. Ph.D. diss. Univ. Amsterdam. Amsterdam.

——— & T. *van der Hammen.* 1977. Some palaeoecological data from Rondonia, southern part of the Amazon Basin. Acta Amazon. 6: 293–299.

Bartlett, A.S. & E.S. *Barghoorn.* 1973. Phytogeographic history of the Isthmus of Panama during the past 12,000 years (A history of vegetation, climate and sea-level change). Pp. 203–299. *En*: A. Graham, ed., Vegetation and Vegetational History in Northern Latin America. Elsevier. New York.

Cleef, A.M. 1977. Secuencia altitudinal de la vegetación de los páramos de la cordillera Oriental, Colombia. Simp. Int. Ecol. Tropic. Panamá.

———. 1979. The phytogeographical position of the Neotropical paramo flora with special reference to the Colombia Cordillera Oriental. Pp. 175–184. *En*: K. Larsen & L.B. Holm-Nielsen, eds., Tropical Botany. Aarhus.

Cuatrecasas, J. 1958. Aspectos de la vegetación natural de Colombia. Rev. Acad. Colomb. Cienc. Ex. Fis. Nat. 10: 221– 268.

Eiten, G. 1974. An outline of the vegetation of South America. Proc. Symp. Fifth Cong. Int'l. Primat. Soc. 529–545. Japan.

Espinal, S. & E. *Montenegro.* 1963. Formaciones vegetales de Colombia. Inst. Geogr. Agustín Codazzi, Bogotá.

Forero, E. 1970. Life in the chilly tropics: the vegetation and ecology of the Andean Paramos. Garden Journal (New York) 20: 84–88.

Gentry, A. 1977. Endangered plant species and habitats of Ecuador and Amazonian Peru. Pp. 136–149. *En*: G.T. Prance, ed., Extinction is forever. New York Botanical Garden, New York.

———. 1978. Floristic knowledge and needs in Pacific tropical America. Brittonia 30: 134–153.

———. 1982. Phytogeographic patterns in northwest South America and southern Central America as evidence for a Choco refuge. Pp. 112–136. *En*: G.T. Prance, ed., The Biological Model of Biological Diversification in the Tropics. Columbia Univ. Press, New York.

Graham, A. 1973. Arborescent temperate element in the northern Latin American biota. Pp. 302–314. *En*: A. Graham, ed., Vegetation and Vegetational History in Northern Latin America. Elsevier, New York.

Haffer, J. 1970. Geologic-climatic history and zoogeographic significance of the Uraba region in Northwestern Colombia. Caldasia 10: 603–636.

Hueck, K. 1972. As florestas da America do Sul. Ecología composiao e importancia económica. Trad. de H. Reichardt. Sao Paulo. Polígono, Ed. Univ., Brasilia.

Instituto Geográfico Agustín Codazzi. 1977. Zonas de Vida o Formaciones Vegetales de Colombia. Memoria explicativa sobre el mapa ecológico. Bogotá.

Prance, G.T. 1974. Phytogeographic support for the theory of Pleistocene forest refuges in the Amazon Basin. Acta Amazonica 3: 5–28.

———. 1977. Floristic inventory of the tropics: where do we stand? Ann. Missouri Bot. Gard. 64: 659–684.

Schultes, R.E. 1951. La riqueza de la flora Colombiana. Rev. Acad. Colomb. Cienc. Ex. Fis. Nat. 8: 230–242.

Simpson, B.B. 1975. Pleistocene changes in the flora of the high tropical Andes. Paleobiology 1(3): 273–294.

Sugden, A. 1976. Habitat Diversity in a Colombian Elfin Cloud Forest and Some Biological Implications. M.Sc. thesis. Oxford Univ.

van der Hammen, T. 1961. Late Cretaceous and Tertiary stratigraphy and tectogenesis of the Colombian Andes. Geologie Mijnb. 40: 181–188.

———. 1962. Palinología de la región de la "Laguna de los Bobos." Historia de su clima, vegetación y agricultura durante los últimos 5000 años. Rev. Acad. Colomb. Cienc. Ex. Fis. Nat. 11: 359–361.

———. 1972a. Changes in vegetation and climate in the Amazon Basin and surrounding areas during the Pleistocene. Geologie Mijnb. 51: 641–643.

———. 1972b. Historia de la vegetación y el medio ambiente del norte sudamericano. Mem. Simp. Primer Congr. Latino-Americano Bot.: 119–134. México, D.F.

———. 1973. The Quaternary of Colombia: Introduction to a research project and a series of publications. Palaeogeogr. Palaeoclimat. Palaeoecol. 14: 1–7.

———. 1974. The Pleistocene changes of vegetation and climate in tropical South America. Journ. of Biogeogr. 1: 3–26.

——— & E. *González*. 1965. A late glacial and Holocene pollen diagram from Cienaga del Visitador (Dept. Boyaca, Colombia). Leid. Geol. Meded. 32: 193–201.

———, J.H. *Werner* & H. *van Dommelen*. 1973. Palynological record of the upheaval of the northern Andes: a study of the Pliocene and Lower Quaternary of the Colombian eastern cordillera and the early evolution of its high-andean biota. Palaeogeogr., Palaeoclimat., Palaeoecol. 16: 1–24.

van Geel, B. & T. *van der Hammen*. 1973. Upper Quaternary vegetational and climatic sequence of the Fúquene area (eastern Cordillera, Colombia). Palaeogeogr., Palaeoclimat., Palaeoecol. 14: 9–92.

West, R.C. 1957. The Pacific lowlands of Colombia. Louisiana State Univ. Soc. Sci. Ser. 8: 1–278.

Contrasting Phytogeographic Patterns of Upland and Lowland Panamanian Plants

Alwyn H. Gentry

En la flora panameña, hay una clara dicotomía entre los grupos de las tierras altas de origen Laurasiano (norte) y los de las tierras bajas cuya afinidad está centrada en el Amazonas. Los grupos Laurasianos son predominantemente herbáceos, sin embargo también incluyen algunos árboles polinizados por el viento.

La flora de las tierras altas de Chiriquí y la del Cerro Tacarcuna cerca de la frontera Colombiana son bastante similares, sin embargo existe un número de especies disgregadas o disyuntas en estas áreas. Las plantas de origen sur (Gondwanalandia) predominan en la flora panameña, sin embargo no muestran en sus especies de árboles y lianas mucha evidencia de especiación por lo tanto éstas podrían ser de arribo reciente. Las plantas de altitud media de los Andes del Norte forman más del 35% de la flora y actualmente muestran señales de evolución activa, ya que muchos de los géneros incluyen muchas especies y muchas de éstas son locales.

An overview of distributional patterns of Panamanian plants shows a clear dichotomy between Laurasian-derived montane elements and South American-derived lowland ones. Among the essentially South American tropical groups, two surprisingly distinct kinds of phytogeographic patterns predominate—most groups have diversity centers either in the northern Andes or in Amazonia. A few Panamanian families show other patterns which do not fit into these broad categories: four families are characteristically centered in dry, mostly peripherally tropical areas, and three are clearly south temperate. There are ecological differences between these phytogeographic assemblages as well. Different trends in the pollination systems, dispersal biology, life form, evolutionary tendencies, and plant community diversity distinguish montane and lowland vegetations, as well as different lowland elements. These patterns, discussed in general terms for the entire neotropical flora in Gentry (1983), are also apparent in Panama.

Las plantas panameñas pueden dividirse en dos grupos, unas de origen montano Laurasiano y las otras de origen en las tierras bajas de América del Sur.

The main Panamanian data sets used in this analysis come from the Flora of Barro Colorado Island (Croat 1978) and a list of upland (above 1800 m) Chiriqui plant species extracted from the Flora of Panama and the Missouri Botanical Garden herbarium (in part compiled by W.G. D'Arcy), subdivided into three altitudinal categories: 1800–2000 m, 2000–3000 m, and 3000–4000 m. The upland Chiriqui list (Table 1) includes 1281 species, approximately the same number as the 1210 naturally occurring Barro Colorado Island (BCI) species, facilitating direct comparisons of the floristic composition of the two data sets. My own collections from Cerro Tacarcuna on the Colombia-Panama border (Fig. 1), the only other high elevation Panamanian forest, provide an additional data set. I have separated the Tacarcuna collections into those of the montane oak forest—generally over 1400 m on Cerro Mali and 1500 m on the main Tacarcuna massif (see Gentry, 1977) and those from the adjacent wet lowland and premontane forests—below 1400 m (see Table 2). Although only 313 non-fern species were collected in the Tacarcuna montane forest and 273 in the lowland forests, these data can be compared floristically with the other data sets by using relative percent of species belonging to various families.

Several of the points from Gentry (1983) are considered here with respect to the Panamanian flora.

TABLE 1
SPECIES PER FAMILY IN UPLAND PANAMA

	Tacarcuna (1400–1900 m) No. of Species	Chiriqui (1800–2000 m) No. of Species	Chiriqui (2000–3000 m) No. of Species	Chiriqui (3000–4000 m) No. of Species	Chiriqui (1800–4000 m) No. of Species
Acanthaceae	5	18	6	1	18
Amaranthaceae	1	4	3	1	4
Amaryllidaceae	0	3	3	1	4
Anacardiaceae	0	4	0	0	4
Annonaceae	1	2	1	0	3
Apocynaceae	2	7	1	0	7
Aquifoliaceae	0	2	1	0	3
Araliaceae	4	9	10	0	13
Araceae	16	20	9	0	20
Aristolochiaceae	0	0	1	0	1
Asclepiadaceae	0	6	4	0	8
Balanophoraceae	2	1	1	1	2
Balsaminaceae	0	1	0	0	1
Begoniaceae	2	12	9	1	14
Berberidaceae	0	0	1	1	1
Betulaceae	0	0	2	0	2
Bignoniaceae	4	4	1	0	4
Bixaceae	0	1	0	0	1
Bombacaceae	2	2	1	0	2
Boraginaceae	1	5	5	1	8
Bromeliaceae	11	13	9	1	16
Brunelliaceae	1	0	0	0	0
Burmanniaceae	1	1	1	1	1
Cactaceae	0	1	1	0	1
Campanulaceae	3	9	15	2	18
Cannaceae	0	1	0	0	1
Capparidaceae	1	3	1	0	3
Caprifoliaceae	0	4	5	1	5
Caricaceae	0	1	0	0	1
Caryophyllaceae	0	7	4	3	8
Celastraceae	0	4	3	1	5
Chenopodiaceae	0	1	0	0	1
Chloranthaceae	1	0	0	0	0
Clethraceae	0	1	1	0	1
Commelinaceae	2	5	2	1	5
Compositae	6	61	51	15	89
Convolvulaceae	0	7	2	1	8
Coriariaceae	0	1	1	1	1
Cornaceae	0	1	1	0	1
Crassulaceae	0	2	0	0	2
Cruciferae	0	4	1	0	4
Cucurbitaceae	2	4	2	0	5
Cunoniaceae	2	1	1	0	1
Cyclanthaceae	4	3	2	0	3
Cyperaceae	1	12	7	5	14
Dichapetalaceae	1	0	0	0	0
Dilleniaceae	1	3	2	0	3
Dioscoreaceae	2	1	2	0	2
Elaeocarpaceae	1	1	0	0	0
Ericaceae	14	23	23	5	28
Erythroxylaceae	0	1	0	0	1
Euphorbiaceae	7	20	1	1	20
Fagaceae	1	7	4	1	11
Flacourtiaceae	2	2	3	0	4
Garryaceae	0	0	0	1	1
Gentianaceae	1	3	5	2	8

TABLE 1 Continued

	Tacarcuna (1400–1900 m) No. of Species	Chiriqui (1800–2000 m) No. of Species	Chiriqui (2000–3000 m) No. of Species	Chiriqui (3000–4000 m) No. of Species	Chiriqui (1800–4000 m) No. of Species
Geraniaceae	1	1	1	1	2
Gramineae	3	23	17	10	33
Gesneriaceae	10	26	19	2	33
Guttiferae	5	4	3	0	5
Haloragaceae	0	1	1	0	1
Hippocastanaceae	1	2	1	0	2
Hippocrateaceae	0	1	1	0	2
Hydrophyllaceae	0	1	1	0	1
Hypericacaceae	0	3	2	1	3
Icacinaceae	1	1	1	0	2
Iridaceae	0	3	5	2	6
Juglandaceae	0	0	1	0	1
Labiatae	1	17	3	1	19
Lacistemaceae	0	1	1	0	2
Lauraceae	11	18	17	0	26
Leguminosae	3	36	10	2	40
Mimosoideae	(1)	(15)	(3)	(0)	(16)
Caesalpinoideae	(0)	(5)	(0)	(0)	(5)
Papilionoideae	(2)	(16)	(7)	(2)	(19)
Lentibulariaceae	1	3	1	0	4
Loasaceae	0	2	3	0	3
Liliaceae	1	1	1	1	1
Loganiaceae	1	3	1	1	3
Loranthaceae	4	15	11	0	21
Lythraceae	0	2	2	0	2
Magnoliaceae	1	1	1	0	1
Malpighiaceae	1	2	0	0	2
Malvaceae	1	15	4	0	16
Marantaceae	3	3	0	0	3
Marcgraviaceae	1	4	3	0	4
Melastomataceae	24	42	32	0	58
Meliaceae	2	3	0	0	3
Menispermaceae	1	1	1	0	2
Monimiaceae	7	3	5	0	7
Moraceae	1	5	0	0	5
Musaceae	1	3	2	0	3
Myrsinaceae	7	16	17	1	24
Myrtaceae	3	11	3	0	13
Nyctaginaceae	1	1	2	0	3
Ochnaceae	0	2	0	0	2
Olacaceae	0	1	1	0	1
Oleaceae	0	0	1	0	1
Onagraceae	0	7	5	1	10
Orchidaceae[1]	11	ca. 153	ca. 103	ca. 15	ca. 176
Orobanchaceae	0	1	1	0	1
Palmae	6	8	4	0	9
Papaveraceae	0	1	1	0	1
Passifloraceae	1	9	6	1	2
Phytolaccaceae	0	2	2	0	3
Piperaceae	14	16	26	0	36
Plantaginaceae	0	3	2	0	3
Podocarpaceae	1	0	0	0	0
Podostemaceae	0	1	1	0	2
Polemoniaceae	0	2	1	0	2
Polygalaceae	1	9	2	1	9
Portulacaceae	0	1	0	0	1
Primulaceae	0	1	0	0	1

TABLE 1 Continued

	Tacarcuna (1400–1900 m) No. of Species	Chiriqui (1800–2000 m) No. of Species	Chiriqui (2000–3000 m) No. of Species	Chiriqui (3000–4000 m) No. of Species	Chiriqui (1800–4000 m) No. of Species
Proteaceae	0	2	0	0	2
Ranunculaceae	0	5	4	0	6
Rhamnaceae	1	5	2	0	6
Rosaceae	0	9	6	1	13
Rubiaceae	33	69	49	5	87
Rutaceae	2	3	1	1	3
Sabiaceae	1	2	2	0	2
Salicaceae	0	1	0	0	1
Sapindaceae	2	7	0	0	7
Sapotaceae	1	1	1	0	1
Saxifragaceae	1	3	1	1	5
Scrophulariaceae	0	17	11	5	20
Simaroubaceae	0	2	0	0	2
Smilaceae	2	3	2	0	3
Solanaceae	10	45	39	2	55
Staphyleaceae	0	1	1	0	1
Styracaceae	1	1	1	0	1
Symplocaceae	0	2	2	0	3
Theaceae	5	2	2	0	2
Thymelaeaceae	1	0	0	0	0
Theophrastaceae	1	0	0	0	0
Tiliaceae	0	3	1	0	3
Tropaeolaceae	0	0	2	0	2
Turneraceae	0	1	0	0	1
Ulmaceae	0	3	2	0	3
Umbelliferae	1	6	5	2	9
Urticaceae	3	16	12	1	21
Valerianaceae	0	4	2	1	5
Verbenaceae	3	20	7	0	22
Violaceae	2	2	1	1	3
Vitaceae	1	2	2	0	3
Vochysiaceae	1	0	0	0	0
Winteraceae	1	0	1	0	1
Zingiberaceae	4	3	1	0	3
Total	313	1038	698	109	1281

[1]The Orchidaceae estimates for upland Chiriqui are compiled from the number of species reported for each altitudinal level in the Flora of Panama (116 total upland species; 101, 68, and 10 species at the three altitudinal levels) plus a percentage based on the assumption that half of the 120 additional Chiriqui species reported by Dressler (1982) occur above 1800 m and that they are apportioned among altitudinal belts in the same ratio as the species in the published Flora.

Laurasian Species

Fifty-six Panamanian plant families are predominantly Laurasian as defined by Gentry (1983: Table 1) on the basis of better representation in Guatemala (and Mexico) than in Panama (Fig. 2), greater prevalence in the floras of the United States or Eurasia than in South America, better representation in the West Indies, especially the Greater Antilles, than in South America, presence of greater generic diversity in Asia than in the neotropics, and generic disjunctions between Central America and Asia. While most of these families share most of these criteria as well as the ecological similarities noted below, some may have only a few of them.

Las 56 familias de origen Laurasiano están principalmente constituidas por árboles polini-

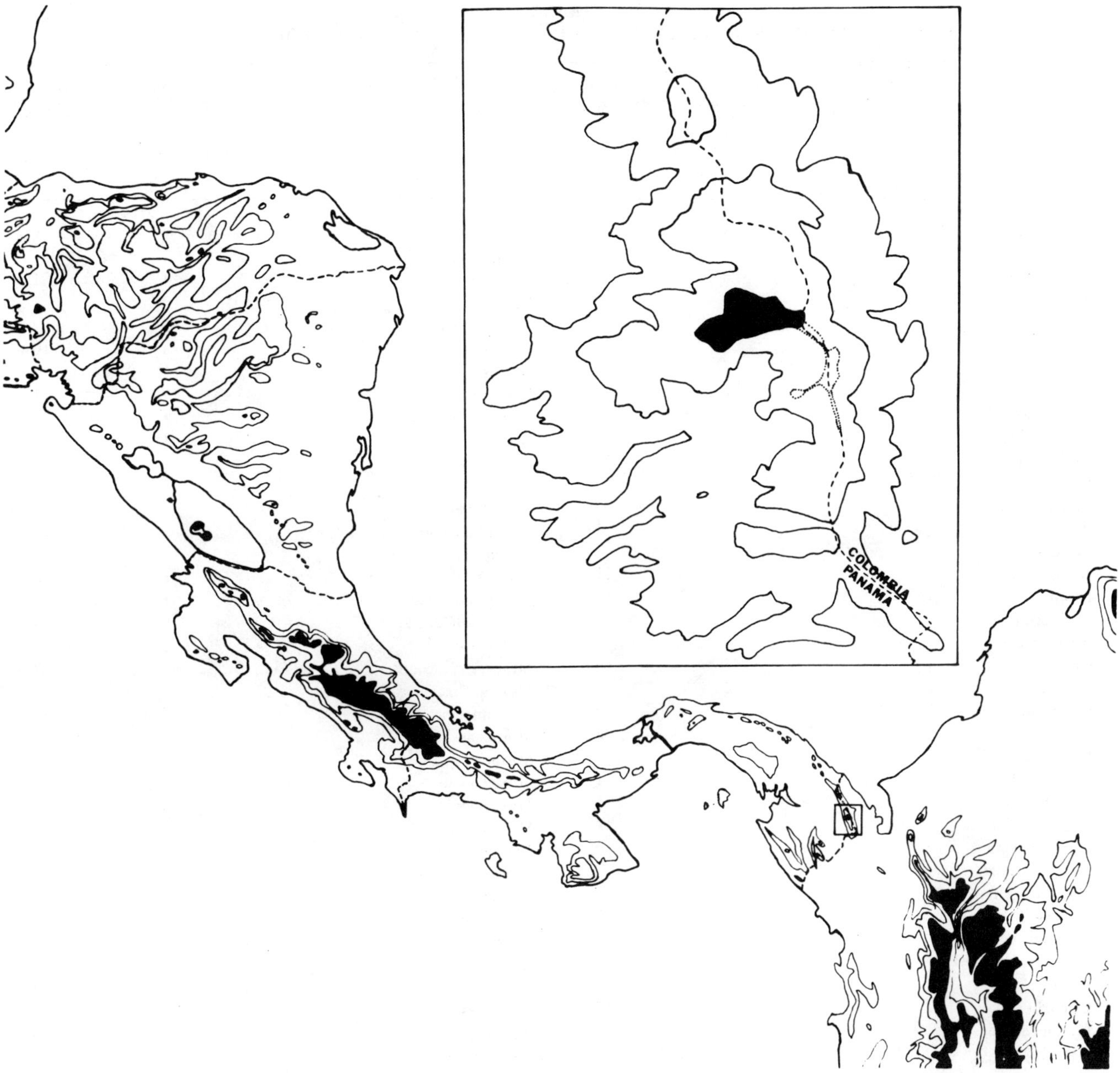

Fig. 1. Isthmian area uplands. Shaded areas are above the 1500 m contour line; boxed area around Cerro Tacarcuna is enlarged with the downward extension of the montane oak forest to 1400 m on the Cerro Mali ridge indicated by dotted line.

zados por el viento o son mayormente herbáceas.

Perhaps the more interesting shared aspects of the Laurasian groups are ecological. Most of them are atypical for the Panamanian flora as a whole in being wind-pollinated trees—Salicaceae, Myricaceae, Juglandaceae, Betulaceae, Fagaceae, Ulmaceae, Garryaceae or predominantly herbaceous—Caryophyllaceae, Ranunculaceae, Papaveraceae, Cruciferae, Droseraceae, Saxifragaceae, Rosaceae, Plumbaginaceae, Geraniaceae, Balsaminaceae, Umbelliferae, Primulaceae, Gentianaceae, Polemoniaceae, Hydrophyllaceae, Scrophulariaceae, Orobanchaceae, Plantaginaceae. There are no epiphytes in these groups except a single species of epiphytic cycad and very few lianas—one Ulmaceae, one genus of Ranunculaceae *(Clematis)*, one genus of Polemoniaceae

TABLE 2
SPECIES PER FAMILY IN TWO LOWLAND PANAMANIAN FLORAS

Family	Cerro Tacarcuna Lower Slopes (below 1400 m) No. of Species	Barro Colorado Island No. of Species
Acanthaceae	13	13
Alismataceae		1
Amaranthaceae	5	8
Amaryllidaceae	1	3
Anacardiaceae		5
Annonaceae	3	13
Apocynaceae	7	22
Araceae	11	43
Araliaceae	1	4
Aristolochiaceae		3
Asclepiadaceae	1	11
Begoniaceae	2	3
Bignoniaceae	15	28
Bixaceae		1
Bombacaceae	2	9
Boraginaceae		11
Bromeliaceae	3	18
Burmanniaceae		1
Burseraceae		6
Cactaceae	1	3
Campanulaceae	2	1
Capparidaceae		2
Caricaceae	1	2
Caryocaraceae	1	
Caryophyllaceae		1
Celastraceae		1
Ceratophyllaceae		1
Chrysobalanaceae		5
Combretaceae	1	7
Cochlospermaceae		1
Commelinaceae	2	7
Compositae		42
Convolvulaceae		10
Connaraceae		4
Cucurbitaceae	2	15
Cyclanthaceae		5
Cycadaceae	1	
Cyperaceae		28
Dilleniaceae	1	9
Dichapetalaceae	1	
Dioscoreaceae		6
Ebenaceae		1
Elaeocarpaceae	1	3
Ericaceae	2	
Erythroxylaceae	1	2
Euphorbiaceae	5	29
Flacourtiaceae	3	16
Gentianaceae		5
Gesneriaceae	4	10
Gramineae	3	77
Guttiferae	4	13
Haemodoraceae	1	1
Hippocastanaceae	1	
Hippocrateaceae	1	5
Humiriaceae		1
Hydrocharitaceae		2
Icacinaceae	1	
Iridaceae	1	1
Labiatae	1	5
Lacistemaceae	1	2
Lauraceae	4	10
Lecythidaceae		4
Leguminosae	7	111
Lentibulariaceae		2
Loasaceae	1	
Loganiaceae		6
Loranthaceae	4	7
Lythraceae		3
Malpighiaceae	2	20
Malvaceae		9
Marantaceae	2	12
Marcgraviaceae	1	2
Melastomataceae	10	34
Meliaceae	3	7
Menispermaceae	3	7
Menyanthaceae		1
Monimiaceae	8	2
Moraceae	5	33
Musaceae	3	8
Myristaceae	1	2
Myrsinaceae	3	5
Myrtaceae	1	13
Nyctaginaceae	1	3
Nymphaeaceae		2
Ochnaceae		2
Olacaceae	1	3
Onagraceae	1	5
Orchidaceae	11	89
Palmae	8	16
Passifloraceae	3	11
Phytolaccaceae		4
Piperaceae	23	31
Polygalaceae		3
Polygonaceae		9
Pontederiaceae		3
Proteaceae		1
Rafflesiaceae		1
Rhamnaceae		3
Rhizophoraceae	1	1
Rubiaceae	31	66
Rutaceae	1	4
Sapindaceae	5	26
Sapotaceae	1	6
Saxifragaceae		1
Scrophulariaceae		6
Simaroubaceae	1	3
Smilaceae	1	5
Solanaceae	7	24
Staphyleaceae		1
Sterculiaceae		8

TABLE 2 Continued

Family	Cerro Tacarcuna Lower Slopes (below 1400 m) No. of Species	Barro Colorado Island No. of Species
Theaceae		1
Theophrastaceae	ca. 3	
Tiliaceae		8
Tovariaceae	1	
Trigoniaceae	1	1
Turneraceae		1
Typhaceae		1
Ulmaceae	1	3
Umbelliferae	1	3
Urticaceae	3	5
Verbenaceae	3	7
Violaceae		3
Vitaceae		6
Vochysiaceae	1	1
Zingiberaceae	4	9
Total	273	1210

Fig. 2. Typical distributional pattern of a Laurasian taxon, *Rhamnus* (Johnston & Johnston 1978), showing number of species per country. In North America (and Eurasia) *Rhamnus* is found at low elevations but in Central America only in the uplands. In Panama it is restricted to upland Chiriqui and Cerro Tacarcuna (see Fig. 1). In South America *Rhamnus* is found only in the Andes except for a widespread specis which reaches subtropical southeastern Brazil.

(Cobaea), one section of *Hydrangea* (Saxifragaceae but often segregated). Most of these families are more prevalent in montane forests in Panama and many are restricted to them or at least have genera restricted to them.

Las familias Laurasianas están mejor representadas en las tierras altas, sin embargo la Vitaceae y la Aristolochiaceae son excepciones.

To document how much better represented these families are in montane than lowland habitats in Panama, we may compare the upland Chiriqui and BCI data sets. Forty of these families are better represented in upland Chiriqui (above 1800 m) than on BCI and eight are represented in neither set. Two Laurasian families have the same number of species in each area; one of these is Staphyleaceae with widespread *Turpinia occidentalis*, the only Panamanian species of the family, present in both upland Chiriqui and on BCI as well as in most other mesic Panamanian forests. Exceptions to the rule of the Laurasian families hardly entering the tropical lowlands are the two predominantly vine—as opposed to liana—families Vitaceae and Aristolochiaceae which are better represented in the BCI flora—and in Panama in general than in montane Chiriqui, just as they are elsewhere in the neo-tropics (Gentry 1983). The only other putatively Laurasian families to be better represented in the BCI flora than in upland Chiriqui are Boraginaceae (8 vs. 11 species) and Lythraceae (2 vs. 3 spp.) whose tentative classification as Laurasian are questionable, Typhaceae and Rafflesiaceae each with a single species on BCI and none in upland Chiriqui, and Gentianaceae (4 vs. 5 spp.) where the saprophytic genus *Voyria* has penetrated the tropical lowlands. Only 55 species (4.5%) of the 1210 seed plant species occurring naturally on BCI belong to putatively Laurasian families.

Las floras de Cerro Tacarcuna y de las tierras altas de Chiriquí son predominantemente Laurasianas, especialmente a mayores altitudes. Hay muchas especies disyuntas entre estas dos áreas.

Laurasian families are as predominant in montane Panama as they are poorly represented in the lowlands. Thus the dominant genus of the Chiriqui montane forests is *Quercus*, with such other Laurasian taxa as Theaceae, Magnoliaceae, and *Ulmus* also well represented. Even on isolated and less than 2000 m tall Cerro Tacarcuna, *Quercus* is the dominant canopy genus (Table 3).

The predominance of Laurasian elements in upland Panama is even more pronounced at higher altitudes. All of the families proportionately better represented above 3000 m altitude than in the rest of the upland Chiriqui flora are Laurasian except Ericaceae and the cosmopolitan Compositae and Gramineae (Table 4).

TABLE 3
COMPOSITION OF MONTANE FOREST ON CERRO TACARCUNA

Plants over 2.5 cm dbh. in a circular plot of 180 m^2. Elevation 1700 m.

	No. of Individuals	No. of large Trees* (over 20cm dbh)	Basal Area (in^2)
Euphorbiaceae, gen. nov.	5	1	155
Quercus humboldtiana	4	3	249
Guatteria allenii	3	0	26
Inga heterophylla	2	0	10
Miconia sp.	2	0	10
Guarea glabra	2	0	8
Phyllonema ruscifolia	1	0	7
Symplococarpon chiriquense	1	1	50
Drimys granadensis	1	1	113
Ardisia tacarcunana	1	0	3
Dendropanax gonatopodus	1	0	3
Dendropanax quercetti	1	0	1
Trichipteris williamsii	1	0	3
Cephaelis glabra	1	0	13
indet. (cf. Vochysia jefensis?)	1	1	254
Alchornea triplinervia	1	0	3
Sloanea zulianensis	1	1	254
Laplacea fruticosa	1	1	452

Source: Original data.
*Nine trees over 20 cm diam. were included; 3 of *Quercus humboldtiana,* 1 each of *Symplococarpon chiriquense, Drimys granadensis, Sloanea zulianensis, Laplacea fruticosa,* a new genus of Euphorbiaceae and an indet.

Ericaceae, too, was considered a basically Laurasian group by Raven and Axelrod (1974).

Although 1900 m tall Cerro Tacarcuna is very isolated, its flora is in most respects much more similar to that of upland Chiriqui than to that of the adjacent Darien lowlands. One striking manifestation of the floristic similarity of the two montane regions at opposite ends of Panama are the many species disjunct between them (Table 5). At least 152 of the montane Cerro Tacarcuna species are restricted to upland forests; 41 of them (25% of Cerro Tacarcuna's exclusively montane species) are otherwise known only from upland Chiriqui and adjacent Costa Rica. In contrast to the strong affinities between the Tacarcuna flora and that of other, geographically remote, upland areas, the overlap between the flora of the Cerro Tacarcuna montane forest and that of the adjacent lowland and pre-montane wet forests is strikingly low. Only 115 (41%) of the upland species also occur below 1400 m even though the two forest types are completely contiguous. The differences in composition between the upland and lowland Tacarcuna floras (Table 6) parallel the floristic differences between up-

TABLE 4
FAMILY DIFFERENCES: UPLAND AND LOWLAND CHIRIQUI

Families with more species in a 3000–4000 m subset of Chiriqui upland flora than in the flora of upland (over 1800 m) Chiriqui as whole (differing by 1% or more).

	3000–4000[1] meter subset	Upland[1] Chiriqui (over 1800 m)	Percent Difference
Caryophyllaceae	.028	.006	.022
Compositae	.138	.069	.069
Ericaceae	.046	.022	.024
Gentianaceae	.018	.006	.012
Gramineae	.092	.026	.066
Iridaceae	.018	.005	.013
Scrophulariaceae	.046	.016	.030
Umbelliferae	.018	.007	.011

[1]Percent of total species in family.

TABLE 5
DISTRIBUTIONAL PATTERNS FOR CERRO TACARCUNA MONTANE FOREST SPECIES
Totals: 313 species, 86 families. (354 species including ferns)

	No. of species
Distribution unknown (identification incomplete)	34
Also in lowlands	115 (41%)
Endemic to upland Tacaracuna	63 (23% of total; 39% of exclusively montane species)
Tacarcuna-Chiriqui disjuncts (not found elsewhere)	41 (15% of total; 25% of exclusively montane species)
Chiriqui/Costa Rica-Tacarcuna-Andes disjuncts (not found in lowlands)	13 (5% of total; 8% of exclusively montane species)
Tacarcuna-Andes disjuncts (not found elsewhere except one species also on Cerro Campana)	13 (5% of total; 8% of exclusively montane species)
Widespread montane species (Mexico to Andes but not found in lowlands in Panama)	13 (5% of total; 8% of exclusively montane species)

Costa Rica, Chiriqui, and Guayana (sometimes plus southern Andes) Trichipteris williamsii Podocarpus magnificus	Tacarcuna-Cerro Pirre Asplenium psilacrum Pitcairnia brittoniana
Tacarcuna-Cerro Jefe Eleagnia nitidifolia Conomorpha gentryi Columnea mira Vochysia jefensis	Guatemala-Tacarcuna Palicourea galeottiana Solandra brachycalyx
Tacarcuna-Jamaica Grammitis randallii	Venezuelan Coast Range Ormosia tovariensis

TABLE 6
FAMILY DIFFERENCES: CERRO TACARCUNA UPLANDS AND LOWLANDS

A. Families with noticeably more species in the lowlands around Cerro Tacarcuna than in the Tacarcuna montane forest (differing by at least 1%).

Family	Tacarcuna[1] lowlands	Upland[1] Tacarcuna (above 1400 m)	Percent Difference
Acanthaceae	.048	.016	.032
Amaranthaceae	.018	.003	.015
Apocynaceae	.026	.006	.020
Bignoniaceae	.055	.013	.042
Leguminosae	.026	.010	.016
Palmae	.029	.019	.010
Piperaceae	.084	.045	.039
Sapindaceae	.018	.006	.012

Laurasian families present but not also in montane forest: Ulmaceae.

Gondwanan families present but not also in montane forest: 16 families.

B. Families with noticeably more species in the Cerro Tacarcuna montane forest than in the lowlands around the base of Cerro Tacarcuna (differing by at least 1%).

Family	Tacarcuna[1] montane forest	Tacarcuna[1] lowlands	Percent Difference
Araceae	.051	.040	.011
Bromeliaceae	.035	.011	.024
Ericaceae	.045	.007	.038
Gesneriaceae	.032	.015	.017
Lauraceae	.035	.015	.020
Melastomataceae	.077	.037	.040
Myrsinaceae	.022	.011	.011
Rubiaceae	.105	.077	.028

Laurasian families present but not also in lowland forest: 14 families: Boraginaceae, Chloranthaceae, Fagaceae, Gentianaceae, Hippocastanaceae, Liliaceae, Magnoliaceae, Malvaceae, Rhamnaceae, Sabiaceae, Saxifragaceae, Styracaceae, Theaceae, Vitaceae.

Gondwanan families present but not also in montane forest: 11 families.

[1]Percent of total species in family.

Source: Original data.

land Chiriqui and BCI. The families noticeably better represented in the Cerro Tacarcuna montane flora are the same as the ones that are noticeably better represented in upland Chiriqui than on BCI except that families of predominantly weedy temperate herbs are missing, perhaps because of lack of suitable open habitats in the closed canopy Tacarcuna forest. The other exception to this clear pattern of floristically relatively homogeneous lowland floras is that a few of the families typical of low and midaltitude wet and premontane forests (Campanulaceae, Urticaceae, Orchidaceae) are better represented in the wet forest Tacarcuna lowlands than in the moist forest of BCI. Collecting artifacts account for some of the apparent differences: the two families (Araceae and Bromeliaceae) noticeably better represented in the upland than the lowland Tacarcuna data set but not the Chiriqui upland versus BCI data set are both epiphytes whose collection was greatly facilitated by the lower canopy of the elfin forest of much of the upper slopes of Cerro Tacarcuna.

Another question regarding the Cerro Tacarcuna montane flora is whether its parallels with the Chiriqui flora can suggest whether the ancestors of the Tacarcuna flora arrived by long distance dispersal or by migration across the intervening lowlands during some past cooler climatic phase. A disproportionate representation of bird-dispersed taxa, weedy taxa adapted for exozoochorous dispersal, or plants with easily long-distance dispersed dust-like seeds (ferns, orchids) would suggest a flora derived from long distance dispersal, but a balanced flora with a good representation of mammal-dispersed species would suggest that the Tacarcuna flora arrived via migration across the now inhospitable lowlands.

Table 7 compares the upland floras of Cerro Tacarcuna and Chiriqui. Eleven families are noticeably better represented on Tacarcuna than in Chiriqui. These are all either basically lowland "Amazonia-centered" taxa (e.g. Bignoniaceae) which spill over into the relatively low elevation Tacarcuna oak forest or represent typically wet or cloud forest and middle elevation "northern Andes-centered" taxa which are also well represented in upland Chiriqui. In Chiriqui their percent representation in the flora is reduced because of numerous species of groups like grasses and Compositae. All of these taxa are basically of Gondwanan origin, although the ecologically dominant woody Laurasian families of montane Chiriqui (Fagaceae, Theaceae, etc.) are also present and ecologically dominant on Cerro Tacarcuna. Since the Tacarcuna flora includes many exclusively montane, large-fruited, mammal-dispersed trees and lianas like *Quercus*, *Billia*, *Helicostylis*, *Dictyocaryum*, *Solandra*, and *Symplococarpus* its migration through the intervening lowlands seems favored. Most of the endemic species of upland Cerro Tacarcuna belong to the speciation-prone, predominantly lowland tropical, Northern-Andes-centered taxa—Araceae, Bromeliaceae, Campanulaceae, Ericaceae, Melastomataceae, Myrsinaceae, Piperaceae, Rubiaceae and probably represent subsequent speciation after the ecological isolation of the top of Tacarcuna through climatic change.

TABLE 7
FAMILY DIFFERENCES: UPLAND TACARCUNA AND UPLAND CHIRIQUI

	Tacarcuna (1400–1900 m)	Upland Chiriqui (1800–4000 m)	Percent Difference
A. Families with noticeably more species on upland Tacarcuna than in upland Chiriqui (differing by more than 1%). Figures = percent of total species in family.			
Araceae	.051	.016	.035
Bignoniaceae	.013	.003	.010
Bromeliaceae	.035	.012	.023
Cyclanthaceae	.013	.002	.011
Ericaceae	.045	.022	.023
Lauraceae	.035	.020	.015
Melastomataceae	.077	.045	.032
Monimiaceae	.022	.005	.017
Piperaceae	.045	.028	.017
Rubiaceae	.105	.068	.037
Zingiberaceae	.013	.002	.011
B. Families with noticeably fewer species on upland Tacarcuna than in upland Chiriqui (differing by more than 1%).			
Compositae	.019	.069	.050
Gramineae	.010	.026	.016
Leguminosae	.010	.031	.021
Orchidaceae	.035	.137	.102
Rosaceae	0	.010	.010
Scrophulariaceae	0	.016	.016
Solanaceae	.032	.043	.011

Once familias están mejor representadas en el Cerro Tacarcuna que en las tierras altas de Chiriquí, todas ellas son de afinidad localizada en tierras bajas amazónicas, en bosques nublados de los Andes del Norte o en altitudes medias.

There are few endemic Cerro Tacarcuna species in these Laurasian taxa. This relates to a general phenomenon of relatively little speciation in these groups (Gentry 1982a). Perhaps the prevalence of wind pollination in many woody Laurasian taxa is partially responsible for their evolutionary conservatism, since they thus lack the coevolutionary possiblities that are available for animal-pollinated plants. Another factor may be that populations of montane species which are isolated today would have been less so during glacial advances and their isolation may be too recent for extensive speciation. In any event, the evolutionary conservatism of these taxa relative to the northern Andes-centered taxa discussed below, which occupy many of the same habitats, is striking.

Gondwanan Taxa

Taxa of basically Gondwanaland families (see Raven & Axelrod 1974) overwhelmingly predominate in the Panamanian flora, with almost 5000 species constituting 84% of the documented flora. The trees and lianas that constitute the canopies of Panamanian lowland forests virtually all belong to Gondwana-derived families that have their neotropical centers of diversity in Amazonia (Gentry 1983). The epiphytes, palmettos, and understory shrubs that predominate numerically in the Panamanian flora have contrastingly extra-Amazonian distributional patterns. These two distributional patterns apparently reflect major differences in evolutionary mode and phytogeographic history.

Afinidades con Gondwanalandia (sur) son predominantes en la flora panameña.

Amazonia-centered plant families show little speciation or endemism in Panama. The 49 such families represented in Panama include only 26% of the Panamanian plant species despite their ecological dominance of the lowland forest canopy. These taxa generally decrease in diversity from east to west across Panama and northwestward through Central America. Examples (from Gentry 1982a) include Caryocaraceae which has 23 neotropical species with three reaching western Panama and adjacent Costa Rica and none north of Costa Rica. Humiriaceae has 31 species and eight genera in Amazonia, five species and three genera in Choco, and four species belonging to three genera in eastern Panama; only two species extend north to Costa Rica and none further north. Vochysiaceae has 182 neotropical species belonging to seven genera; only three genera reach the isthmian region with one species of *Qualea* in Choco and one species of *Erisma* and four of *Vochysia* in eastern Darien; only three species of *Vochysia*, including at least one which is exclusively Central American, reach western Panama and Costa Rica; north of Costa Rica there are probably only two widespread species, both reaching southern Mexico. There are ten species of Myristicaceae in eastern Panama but only five west of the Canal, four in Guatemala, and two in Mexico. These patterns are consistent with the interpretation that these families are recent arrivals in Central America where they probably arrived subsequent to establishment of an isthmian land connection about 3 million years ago, a conclusion supported by the palynological evidence (Graham 1976; Gentry 1983).

En Panamá hay poca especiación entre los grupos de afinidad Amazónica, lo que sugiere que han llegado al Istmo recientemente.

Canopy trees and lianas show little speciation or endemism in Panama or the rest of Central America, although taxonomic artifact has inflated the number of names in use in Central America making this pattern difficult to quantify (see Gentry 1978). Bignoniaceae, with all but one of the non-mammal-dispersed Panamanian species reaching Colombia, 89% Venezuela, and 72% at least to Amazonia, are probably typical; 75% of the non-mammal-dispersed Panamanian Bignoniaceae species reach northern Central America; no species is endemic to Panama. Thus the great majority of lowland Panamanian tree and liana species extend south into Amazonia and many range northward to Guatemala and southern Mexico as well. Overall only 15% of the neotropical species of these Amazonia-centered families reach Central America. Since these lowland canopy taxa have speciated little in Central America and decrease so markedly in diversity with distance from South America, Panama has a higher percentage of the total Central American species of canopy tree and liana taxa than do other Central American countries. Thus over a third (ca. 1405 species) of all Central American-Mexican species of the 49 Amazonia-centered families that reach Central America occur in Panama. This percentage can be expected to increase significantly as tree and liana species described from further north in Central America are critically compared with those of Panama: national boundaries have much less significance to the plants than they have had to the taxonomists who have studied them.

En Panamá están representadas más de un tercio de las especies de México y América Central las cuales están distribuidas en 49 familias de origen centrado en el área amazónica.

Strikingly different phytogeographic patterns are shown by the taxa of epiphytes, palmettos, and understory shrubs, which tend to have fundamentally extra-Amazonian distributions with diversity concentrated in the northern Andean region (Gentry 1983). Panama is a major center of evolutionary radiation for these "northern Andes-centered" families. These 34 families account for the majority (53%) of Panama's plant species and almost all of its endemics. About a quarter of all neotropical species of these taxa occur in Central America and 73% of these are endemic to Central America (Gentry 1983). Panama and Costa Rica are the richest Central American countries in representation of these taxa. Panama's 2910 documented species of northern Andes-centered plant families constitute over 40% of the total species of these groups for all of Central America plus Mexico. Many additional Panamanian species of these groups are still being discovered and these figures may be expected to increase.

Panamá es un centro principal de radiación evolutiva de epífitas, palmetos y arbustos del sotobosque, los cuales son afines a los del Andes del Norte.

These taxa (and thus Panamanian endemism in general) are concentrated in lowland and middle elevation wet and cloud forests. For example, almost all the Gondwanan families noticeably better represented in (wet forest) middle elevation Chiriqui than on (moist forest) BCI belong to these families—Campanulaceae, Ericaceae, Gesneriaceae, Melastomataceae, Myrsinaceae, Orchidaceae, Rubiaceae, Urticaceae. Although these families are well represented in middle elevation cloud forests, they are poorly represented at high elevations (compare Tables 1 and 7 with Table 4). Moreover they are equally well represented in wet lowland forests. A comparison of the BCI flora with that of wet forest Rio Palenque in coastal Ecuador (elevation 150 to 220 m.; Dodson & Gentry 1978) documents this point. Six of the eight families that are noticeably better represented at Rio Palenque are largely epiphytic and the other two families—Solanaceae and Urticaceae—are mostly understory subshrubs but also have a number of epiphytic species (Table 8).

The outstanding characteristic of northern Andes-centered families as a whole is the rapid, even explosive, speciation they appear to have undergone (Gentry 1982a, 1983). The same extremely dynamic speciation, often associated with pronounced local endemism, that is so characteristic of these taxa in northwestern South America, is clearly evident in Panama. One index of active speciation is provided by large genera. There are at least 68 genera with over 14 species in Panama (Table 9). Most (36) of these belong to northern Andes-centered families while only 14 large Panamanian genera belong to Amazonia-centered families (mostly Leguminosae). An additional five large genera belong to southern-Andes centered taxa and nine to cosmopolitan families which do not obviously fit any of the major patterns (Cyper-

TABLE 8
FAMILY DIFFERENCES: LOWLAND WET AND MOIST FORESTS
Families better represented at Rio Palenque, Ecuador than on BCI, Panama.

Family	Rio Palenque	Barro Colorado	Percent Difference
Araceae	.051	.036	.015
Bromeliaceae	.020	.012	.008
Gesneriaceae	.024	.008	.016
Moraceae	.038	.027	.011
Orchidaceae	.101	.074	.027
Piperaceae	.042	.026	.016
Solanaceae	.038	.020	.018
Urticaceae	.017	.004	.013

Source: Croat 1978; Dodson & Gentry 1978

aceae, Gramineae, Malvaceae, Verbenaceae). Only two large Panamanian genera (with a total of 33 species) belong to Laurasian families—*Cordia* (Boraginaceae) and *Hyptis* (Labiatae), and the familial phytogeographic classification of both of these is open to question.

Not only are there many more large genera of northern Andes-centered families in Panama but these genera are also more speciose. The 36 large Panamanian genera of northern Andes-centered families average 39 species per genus while the large Amazonia-centered, southern Andes-centered, Laurasian, and phytogeographically unclassified genera average, respectively, 22, 25, 16, and 22 species per genus. While I have no direct figures on how many Panamanian endemics different plant families have, many of the large Amazonia-centered and unclassified genera of Table 9 are composed primarily of wide-ranging weedy species with little speciation or endemism in Panama. In contrast the majority of species of most of the large northern Andes-centered taxa have restricted distributions. It is clear that rapid speciation and endemism in the Panamanian flora occur mostly in these northern Andes-centered taxa of epiphytes, palmettos, and understory shrubs.

Los elementos relacionados a los grupos de los Andes del Norte muestran especiación di-

TABLE 9
PANAMANIAN GENERA WITH MANY SPECIES

Largest Panamanian genera categorized by phytogeographical pattern of plant family[1]

		Northern-Andes Centered			
Laurasian					
Cordia	17	Anthurium	160	Faramea	17
Hyptis	16	Pleurothallus	137	Pilea	17
Total	33	Psychotria	97	Burmeistera	16
		Piper	95	Masdevallia	16
Amazonia-Centered		Epidendrum	90	Sobralia	16
Paullinia	37	Philodendron	75	Besleria	15
Cassia	34	Peperomia	72	Coussarea	15
Ipomoea	29	Maxillaria	52	Conostegia	14
Inga	28	Ardisia	50	Calathea	14
Pouteria	25	Miconia	36	Total	1385
Desmodium	24	Begonia	30		
Ficus	19	Oncidium	29		
Chamaedorea	19	Stelis	28		
Nectandra	18	Tillandsia	27	*Unassigned Families*	
Pithecellobium	17	Passiflora	27	Panicum	39
Zanthoxylum	16	Cavendishia	23	Paspalum	39
Machaerium	15	Palicourea	24	Cyperus	25
Serjania	15	Clidemia	21	Aegiphila	17
Ocotea	14	Cephaelis	21	Rhynchospora	16
Total	310	Phoradendron	20	Hibiscus	16
		Hoffmannia	20	Sida	16
Southern Andes-Centered		Ruellia	19	Scleria	15
Solanum	56	Drymonia	19	Pavonia	15
Eugenia	20	Encyclia	19		
Cestrum	20	Scaphyglottis	19		
Ludwigia	16	Aphelandra	18		
Mikania	15	Clusia	17		
Total	127				

[1]Numbers are number of Panamanian species, mostly taken directly from the Flora of Panama.

námica e incluyen muchos géneros grandes y muchas especies de distribución restringida.

Several other relatively minor distributional patterns are shown by a few Panamanian families. Southern Andes-centered taxa, mostly herbs and shrubs and often dry area specialists with amphitropical distributional centers (e.g. Cactaceae) are poorly represented in Panama. Some typically southern hemisphere trees like Cunoniaceae, Proteaceae, and Podocarpaceae show this distributional pattern, but most Panamanian representatives of southern Andes-centered taxa tend to be widespread weedy herbs.

A number of species of the eastern Panama uplands—especially Cerro Jefe and Cerro Tacarcuna—are apparently disjunct from the Guayana region, especially the Guayana Highlands area. In some cases a single disjunct species is involved; in others the nearest relative of an Isthmian species is in Guayana. Many of these species probably represent survivors of the old pre-Andean uplift flora of the islands which occupied today's isthmian region during the Miocene and earlier Pliocene. Prior to the Pliocene-Pleistocene uplift of the Andes, the Guiana shield would have constituted the nearest source of colonizers for these islands. A list of 19 such species was included in Gentry (1982a). Two recently discovered species, one in Panama and one in the Guiana area, add two species pairs to that list (Balslev & Mori, 1981)—Panamanian *Couratari scottmorii* Prance with Guianan *C. stellata* A.C. Smith and Guianan *Qualea amapaensis* Balslev and Mori with Panamanian *Q. cymulosa* Schery. It should be noted that the rest of the long list of Panamanian-Guiana species and species pairs presented by Balslev and Mori (1978) is quite irrelevant to the phenomenon discussed here and in Gentry (1982a) since the other species are not *limited* to these two regions but are much more wide-ranging.

Algunas especies panameñas de las tierras altas son especies disyuntas de las tierras altas de las Guayanas.

Mangroves are another group that shows an interesting pattern in Panama. The greater diversity of Pacific coast mangroves is especially pronounced along the Darien coast and in the Colombian Choco region where at least 14 mangrove-restricted species occur as compared to five on the Caribbean side. (Gentry 1982a). If *Rhizophora harrisonii* is a good species as suggested by A. Juncosa (pers. comm.), then one more species can be added to the list for both coasts. I interpret this phenomenon, which can only have developed subsequent to closing of the Panamanian isthmus 3 million years ago, as reflecting differential survival of formerly more widespread species in the wetter Choco area during dry periods of the Pleistocene.

Existen más especies de mangles a lo largo de la Costa Pacífica del Darién-Chocó que en cualquiera otra área de la región.

Conclusions

Several points can be noted in conclusion:

There is a fundamental dichotomy between Laurasia-derived and Gondwanaland-derived taxa as previously stated by Raven and Axelrod (1974) and others.

Laurasian taxa a) dominate in the uplands, b) are poorly represented in the lowlands where they tend to be restricted to specialized substrates, especially edaphically dry areas such as limestone outcrops, c) are typically wind-pollinated trees and temperate zone herbs such as Cruciferae, Ranunculaceae, Umbelliferae, and Scrophulariaceae, and d) have relatively few species in Panama (only about 6% of the Panamanian species but 30% of the families) and in the neotropics in general.

Gondwanan taxa, which make up the bulk (84%) of the isthmian flora, fall into two main groups: Amazonia-centered taxa which are all predominantly woody canopy trees and lianas, and extra-Amazonian taxa which have diversity centers either in the northern Andean region or subtropical South America and the southern Andes and are poorly represented in Amazonia.

Amazonia-centered Gondwanan taxa a) show little speciation or endemism in Panama (often contrary to the prevailing taxonomy), b) decrease in diversity from east to west across Panama and northward into Central America, but c) dominate the tropical lowland forests of Panama and the rest of Central and South America.

Northern Andes-centered taxa, which generally have strong secondary centers of radiation and much endemism in Panama, are mostly epiphytes, palmettos, and understory shrubs. Southern Andes-centered taxa, mostly herbs and shrubs but including such markedly primitive southern hemisphere trees as Cunoniaceae and Podocarpaceae, are poorly represented and frequently weedy in Panama.

A few other, relatively minor, distributional patterns are shown by several Panamanian families or subsidiary taxa: Guayana Highland disjunctions, beach and mangrove species—the latter much more diverse on the Pacific side of the Isthmus, and Choco-centered species and genera—mostly representing subsets of northern Andes-centered groups (see Gentry 1982a, 1983).

The Amazonia-centered taxa, which may be at ecological equilibrium (Gentry 1982b), account for nearly half of all neotropical plant species but only about a fourth of the Panamanian ones. The northern Andes-centered taxa seem to be in the process of explosive speciation both in northwest South America and in southern Central America. Contrary to the pattern shown by the Amazonia-centered taxa, there is no hint of ecological limits on their speciation, which may involve much coevolution with such pollinators as hummingbirds and bats (Gentry 1983). These groups account for over 53% of the Panamanian flora, and their unique presence in the neotropics may be largely responsible for the region's great diversity compared to tropical Africa or Australia.

Los taxa centrados en el Amazonas podrían estar en equilibrio ecológico, pero aquellos de los Andes del Norte parecen estar en proceso de evolución explosiva. Estos representan más del 35% de la flora panameña.

Acknowledgments

I thank the National Geographic Society for supporting the botanical exploration of Cerro Tacarcuna and W. G. D'Arcy for making available much of the Chiriqui data set.

Balslev, H. & S.A. *Mori*. 1981. *Qualea amapaensis* (Vochysiaceae), a new and phytogeographically interesting species from Brazil. Brittonia 33: 5–8.

Croat, T.B. 1978. Flora of Barro Colorado Island. Stanford Univ. Press, Stanford, CA.

Dodson, C. & A.H. *Gentry*. 1978. Flora of the Río Palenque Science Center. Selbyana 4: 1–628.

Dressler, R. 1982. Checklist of the Orchids of Panama. Missouri Bot. Gard. Monogr. Syst. Bot. 4: I-XXVI.

Gentry, A.H. 1977. Botanical exploration of Cerro Tacarcuna. Explorers Journ. 55: 40–45.

———. 1978. Floristic knowledge and needs in Pacific Tropical America. Brittonia 30: 134–153.

———. 1982a. Phytogeographic patterns as evidence for a Choco refuge. Pp. 112–136. *In* G. Prance, ed., Biological Diversification in the Tropics. Columbia Univ. Press, New York.

———. 1982b. Patterns of neotropical plant species diversity. Evol. Biol. 15: 1–84.

———. 1983. Neotropical floristic diversity: phytogeographical connections between Central and South America, Pleistocene climatic fluctuations, or an accident of the Andean orogeny? Ann. Missouri Bot. Gard. 69: 557–593.

Graham, A. 1976. Studies in neotropical paleobotany. 2. The Miocene communities of Veracruz, Mexico. Ann. Missouri Bot. Gard. 63: 787–842.

Johnston, M.C. & L.A. *Johnston*. 1978. *Rhamnus*. Flora Neotropica Monograph 20: 1–96.

Raven, P.H. & D.I. *Axelrod*. 1974. Angiosperm biogeography and past continental movements. Ann. Missouri Bot. Gard. 61: 539–673.

Vegetational Paleohistory Studies in Panama and Adjacent Central America

Alan Graham

La evolución cenozoica de la vegetación de América Central es poco conocida. Entre 159 tipos de microfósiles recobrados del Lago Gatún, 63 se identificaron. Estos incluyen Rhizophora, Utricularia, Allophyllus, Bursera, Hibiscus, Alchornea, Cedrela, Mortoniodendron, Cleyera, Echites, Roupala, Struthanthus, Alsophila, Cnemidaria y varias palmas y gramíneas. Esto refleja tipos de comunidades como manglares o bosques monsónicos, sin embargo comunidades de tierras altas, o secas a áridas no están representadas. La flora es primordialmente afin a la de Norte América, ya que presenta 8 grupos afines a los de Sur América y 39 de linaje norteamericano. Los resultados obtenidos mediante diferentes enfoques y métodos son consistentes.

Paleobotanical studies on the history of Central American vegetation began as early as 1888 with the publication of two papers by John Strong Newberry on Rhaetic (Upper Triassic) plants from Honduras. The specimens were collected in 1886 by a mining engineer named Charles Rolker from a site near San Juancinto, and from the collection Newberry described 10 genera and 14 species of fossil plants. Prior to these reports the only Triassic plants known for Latin America were 2000 miles to the north from deposits in Sonora, Mexico, and 2000 miles south from Peru (Newberry 1888b, p. 343).

In Panama early studies were made by Berry (1914, 1918). He described 17 species of Oligocene to Oligo-Miocene fossil plants collected by D. F. McDonald and M. I. Goldman from the Culebra Formation at sites 1/4 mile south of Empire Bridge, along the east wall of the Gaillard Cut just south of Canal Commission Station 1760, along the west wall of the Cut below the Miraflores Locks, and along the west wall opposite the Culebra Railroad Station. Other collections were made from the Cucaracha Formation along the Gaillard Cut; the Gatun Formation 7 miles northeast of Bejuco near Chame; the Bohio Formation at middle Bohio Bridge; the Caimito Formation 7 miles northeast of Bejuco, and from a section near the mouth of the Tonosi River in deposits considered Eocene in age.

During this same period there were papers by Berry (1921a) on plant fossils from Costa Rica, Stephenson and Berry (1929), on Upper Cretaceous plants *Araucarites* and *Dicksoniopsis* from Guatemala, and an anonymous article in 1886 and one by Humphrey (1916) on the San Juancinto (Honduras) flora. For Panama, Berry (1921b) published on *Iriartites vaughani*, a palm seed from the Miocene of the Canal Zone (Gatun Formation, 1 1/ miles northeast of Gatun overlooking the Gatun Locks); Berry (1928) on *Palmocarpon geonomoides* from Chiriqui Province; and Howe (1918) on some fossil Lithothamnieae algae from the Canal Zone. Thus, for the 50 year period from about 1888 to 1940, there was a total of 11 papers describing 40 genera and 42 species of fossil vascular plants for all of Central America.

More recently the pace of vegetational history studies in Central America has accelerated slightly, although detailed investigations on sizable fossil floras are still rare. For the period between 1940 and 1979, there were 37 published papers, a dissertation by Bartlett (1967), and a thesis by Krieger (1977). The first published palynological studies were by Tsukada (1966) on Quaternary sediments from Guatemala, Tsukada and Deevey (1962, 1967) on Quaternary lake deposits in Guatemala and El Salvador, and a brief summary (abstract) by Tsukada (1965). Other paleobotanical and paleopalynological papers include those for Costa Rica by Gomez-Pompa (1970, 1971, 1972, 1973; Kruckow and Gomez-Pompa 1974); El Salvador (Germeraad 1978a; Loeischert et al. 1975; von der Brelie and Teichmuller 1953); Guatemala

(Cowgill & Hutchinson 1966; Cowgill et al. 1966; Germeraad 1978b; Johnson & Kaska 1965; Johnson & Konishi 1960; Serrano 1948); Honduras (Delevoryas & Srivastava 1977; Maldonado-Koerdell 1953; Olsen & McGrew 1941); Nicaragua (Brown 1947; Mulleried 1948); and general papers by Berry (1942), Couper (1964), Graham (1972, 1973, 1974, 1977), Hunter (1977), and Muller & Caratini (1977). Recent papers on Panama plant fossils include Bartlett & Barghoorn (1973), Bartlett et al. (1969), and Graham (1979).

A pesar de que se han publicado aproximadamente 48 trabajos sobre paleohistoria de América Central, sólo 30 realmente describen las plantas fósiles provinientes de los depósitos del Cenozoico. Muchas de estas investigaciones sólo tratan pocos taxa y muchas de las determinaciones son erróneas. Información adicional puede proveer mejor conocimiento sobre las plantas del Cenozoico.

It is clear from this survey that little is known about the origin and history of Central American vegetation. Among the 48 published papers only 30 actually describe fossil vascular plants from Cenozoic deposits in Central America. Several of these treat only one to a few taxa, and experience has shown that many identifications in the earlier papers are likely erroneous. Recently, however, new sources of information are providing some insight into the physical and climatic changes in Central America during the Cenozoic and the influence of these changes on the compositon and distribution of Latin American biotas.

Caribbean Plate Tectonics

The Caribbean is one of the last major regions to be integrated into the global scheme of plate tectonics. Events between the present Canal Area and northwestern South America are especially poorly known, but a general picture of Caribbean tectonic history is emerging (Christofferson 1976; Emiliani et al. 1972; Freeland & Dietz 1971; Ladd 1976; Lonsdale 1978; Malfait & Dinkelman 1972; Mattson 1979; Molnar & Sykes 1969; Plafker 1976; Santamaría & Schubert 1974). Near the end of the Cretaceous North and South America were separated by some 3000 km, from about Oaxaca in southern Mexico to the Guyana Shield in northern South America (Freeland & Dietz 1971), but with some intervening islands (Dengo 1973). At this time an eastern segment of the Pacific plate (Farallon plate) protruded through this opening, with its eastern edge terminating along the Puerto Rican Trench and the Cayman Trough (Fig. 1).

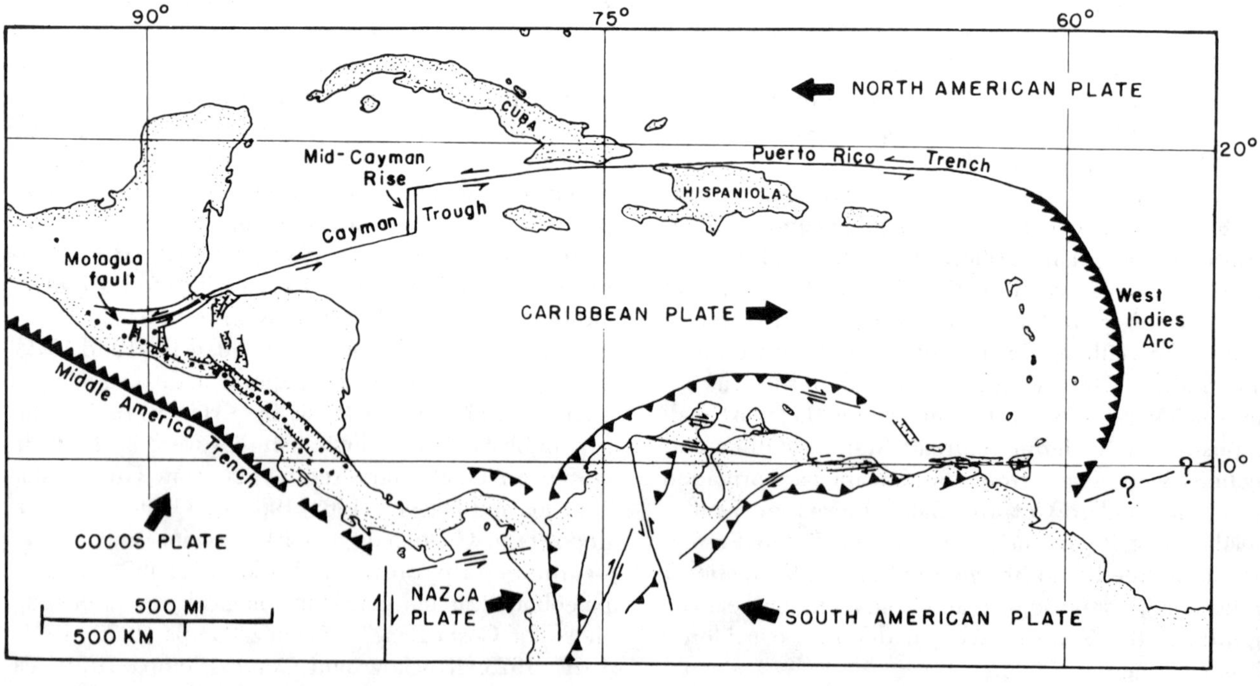

Fig. 1. Caribbean plate boundaries. Large arrows indicate relative plate movement directions. Black dots indicate major volcanos of the Middle American arc. (From G. Plafker, Science Vol. 193. 1976; 1201–1208, used by author's permission).

Aproximadamente al final del Cretáceo, Norte y Sur América estaban separadas por 3000 km y la placa del Pacífico (Farallón) se interpuso en ese lugar.

The history of the northern segment of this plate is more complex than suggested in earlier tectonic models. Mattson (1979) proposes that an initial event in establishing the northern boundary was development of a northward-dipping subduction zone; that is, the early Caribbean plate dipped under an oceanic portion of the North American plate along a subduction zone following the southern margin of the present Greater Antilles island arc. Lighter continental crust entering this zone from the south stopped subduction ca. 120–130 m.y. B.P. (Late Jurassic or Early Cretaceous). "Continued compression resulted in a new 'flipped' subduction zone dipping southward on the north side of the old island arc. The second episode of subduction began about 110 m.y. ago (Aptian/Albian to Santonian; Middle to Late Cretaceous), and formed most of the exposed island arc" (Mattson 1979, p. 293, Fig. 3). This second phase defined relationships currently observed along the northern plate and ended ca. 85 m.y. B.P. when buoyant Bahamas Bank material entered the new subduction zone from the north. Beginning ca. 45 m.y. B.P. (Middle Eocene) movement along the northern Caribbean plate boundary became lateral as the Caribbean plate moved eastward relative to the North American plate (Fig. 1) now overthrusting the North American plate along the Puerto Rican Trench (Fig. 1). Volcanism ceased in the Greater Antilles after Middle Eocene times and a new plate boundary was established along the Cayman Trough.

The northwest corner of the Caribbean plate contacts the North American plate along the Motagua fault system (Fig. 1). Movement along this plate boundary was responsible for the devastating Guatemalan earthquake of 4 February 1976 (Plafker 1976). Malfait and Dinkelman (1972) suggest this segment of the Caribbean plate is being pinned between the North American and Cocos plates while the remainder moves relatively eastward. Secondary fault systems reflecting regional extension are interpreted by Plafker (1976 p. 1206) as consistent with this view. As a result of these differential movements within the plate, Plafker (1976) believes separation may be taking place between the pinned northwestern segment of the Caribbean plate and the remaining eastward-moving section along the volcanic chain (Fig. 1). This would ultimately result in a marginal sea inland to the present volcanic chain, and the Gulf of Fonseca (Honduras/El Salvador) may represent an incipient stage in the development of this sea.

La historia de la placa de Norte América es compleja. Esta se unió a la placa del Caribe a lo largo del sistema de Fallas de Motagua; es aquí en donde ocurrió el terremoto de Guatemala en 1976. La placa del Caribe se une al sur por el noroeste de Sur América, alrededor de la costa oeste de este continente.

The northern margin of the plate continues around the Yucatan Platform and joins the Southern Mexico Trench following the west coast of Mexico. To the south it joins northwestern South America along a complex tectonic zone that continues southward and around the west coast of South America as the Cordilleran (Bolivar) Trench.

Ladd (1976) presents a summary of plate movements in the Caribbean during Mesozoic and Cenozoic times: "From Triassic to Early Cretaceous time, South America moved to the southeast away from North America. From Early Cretaceous to Late Cretaceous time, South America moved eastward with respect to North America. Southeastward motion of South America occurred again from Late Cretaceous to early Tertiary time followed by northward motion from early to late Tertiary. Tertiary compressonal structures on the northern and southern margins of the Caribbean can be attributed to Tertiary closure between North and South America . . . " (p. 969).

One effect of these movements of the North and South American plates on the protruding segment of the Farallon plate wedged between them was a decoupling of the segment to form the Caribbean plate. Edgar et al. (1971) suggest this occurred during the late Mesozoic, while Malfait and Dinkelman (1972) believe it was not completed until the end of the Paleocene. Beginning with the Eocene, and especially in the late Tertiary "when North and South America were converging, the Caribbean plate was moving independently eastward with respect to both North and South America as evidenced by the development of the Lesser Antilles island arc from Eocene to the present" (Ladd 1976, p. 975). This eastward movement of the Caribbean plate has been estimated from between 200 km since the Miocene (at an average rate of ca. 0.5 cm/yr; Malfait & Dinkelman 1972) to ca. 1000 km (at an average rate of ca. 2.2 cm/yr; Ladd 1976).

La placa del Caribe, formada a partir de la

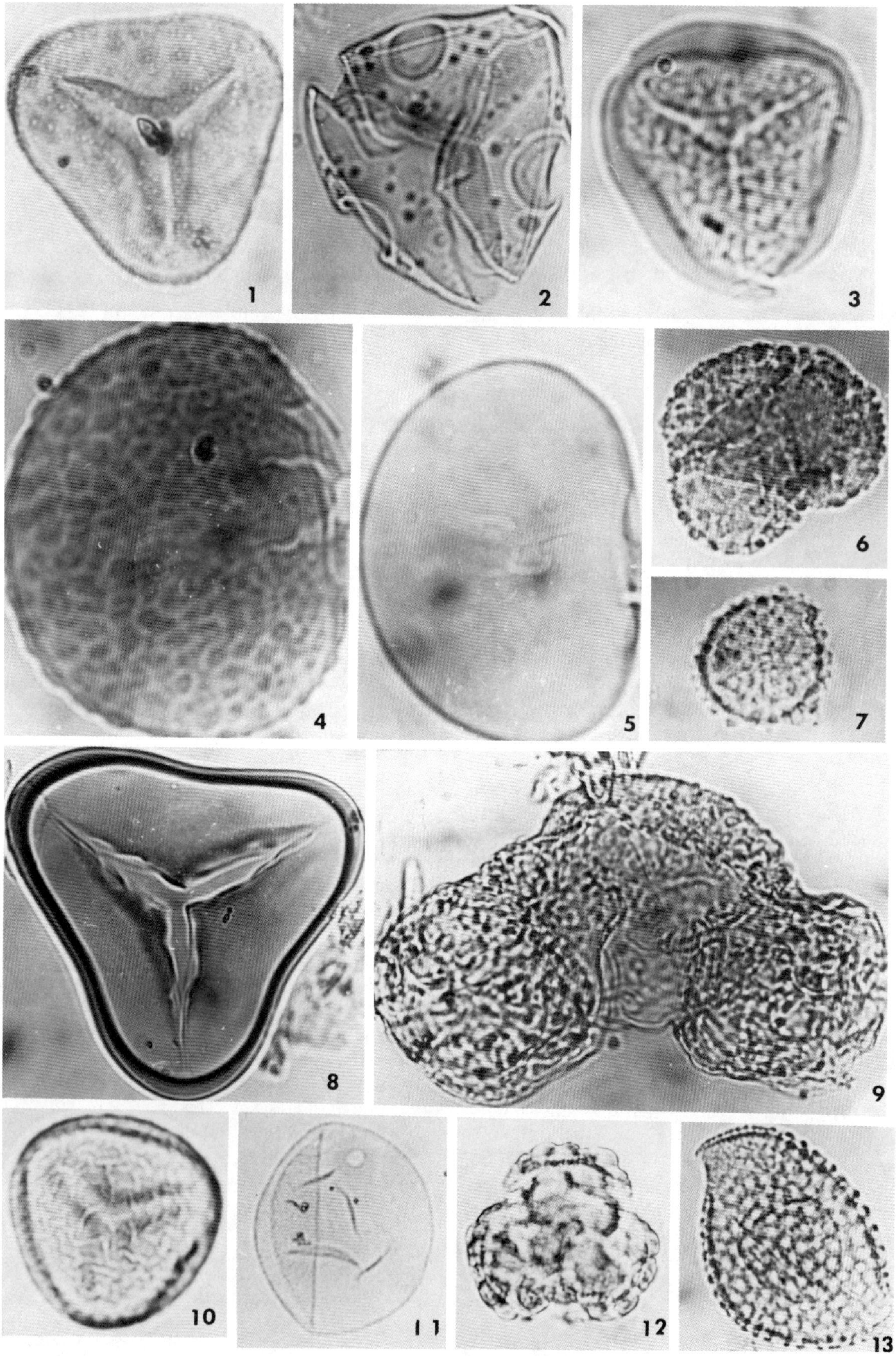
1
2
3
4
5
6
7
8
9
10
11
12
13

placa de Farallón, se ha movido hacia el este empujando las Antillas Menores hacia el este.

Concomitant with the relative motions of the North and South American plates and the eastward movement of the Caribbean plate was the extension of the older (Cretaceous-Paleocene) Southern Mexico Trench southward to form the Middle American Trench. This event is marked by the initiation of extensive andesitic volcanism during the Eocene. It effectively defined the western margin of the Caribbean plate and set up the classic physiographic relations observed at plate boundaries—a subduction zone with associated volcanic arc and trench where the eastern edge of the Farallon plate is underthrusting the Caribbean plate (Malfait & Dinkelman 1972, Figs. 2,3).

The final event affecting plates adjacent to the Caribbean plate was the development of the Cocos and Nazca plates impinging, respectively, on the Pacific coasts of southern Central America through the Panama Basin and on the northwestern coast of South America. The older Farallon plate broke apart during the Oligo-Miocene (22–27 m.y. B.P.; Hey 1977; Lonsdale 1978) to form the Cocos and Nazca plates (Malpelo Ridge on the Cocos plate, Carnegie Ridge on the Nazca plate; Lonsdale 1978, Fig. 1) which have interfered with the subduction of these plates. The ridges grew southeastward and eastward as the plates drifted over the hot spot. According to Lonsdale (1978), "With a plate convergence rate of 90 km/10^6 years we calculate that Carnegie Ridge entered the Ecuador Trench and enclosed the Panama basin in the late Pliocene, and began to underthrust the South American continent only during the Pleistocene". The effect of the ridge entering the subduction zone was a "strong recent tilting of Manabi basin, its uplift relative to the still submerged Gulf of Guayaquil, and the emergence of tablazados along parts of the Ecuadorian coast underlain by the Carnegie Ridge".

Los placas de Cocos y Nazca se movieron en el Pacífico, tocando la costa sur de América Central y la costa noroeste de América del Sur. La región caliente de las Galápagos influyó en este movimiento.

In a earlier review of Central American tectonic history Malfait and Dinkelman (1972, p. 254, Fig. 4) suggested a similar mechanism may have operated as the Malpelo Ridge entered the Central American Trench, essentially "zippering" the trench shut from east to west. According to Lonsdale (1978, pp. 2472–2473), however, there is a zone of normal oceanic crust between the Malpelo Ridge and the subduction zone; hence, tectonic forces set up by this mechanism were not involved in the final uplift of the Isthmian region.

Thus at least two tectonic forces have operated along the western edge of the Caribbean plate during the Cenozoic. One involved compression-expansion forces as the relative positions of North and South America changed (Ladd 1976), and as the Caribbean plate moved eastward. The second involved separation of the Caribbean plate from the eastern Pacific plate along the eastern coast of present-day Central America, with a resultant subduction zone and volcanic arc. The collective consequence of these factors has been progressive emergence of land surfaces eventually connecting the North and South American continents in late Pliocene time.

Fossil Faunas

Durante el Cenozoico, una de las fuerzas tectónicas involucró compresión y expansión a medida que Norte y Sur América cambiaron de posición y la placa del Caribe se movió hacia el este, mientras que otra incluyó la separación de la placa del Caribe a partir de la placa Pacífico este. Estos factores contribuyeron a que nuevas superficies de tierra emergieran.

Recent studies on the fossil faunas of Central America include those of Whitmore and Stewart (1965) on Miocene mammals from the Cucaracha Formation, and Woodring (1970, 1973) on Tertiary marine invertebrates. These data are supplemented by both older (Olson & McGrew 1941; Simpson 1950) and more recent studies (Brett-Surman 1979; Bonaparte 1979; Marshall, et al. 1979; Webb 1974, 1976) from adjacent North and South America. Much of this data

---

PLATE 1

Fig. 1–13. Pollen and Spores: 1, *Alsophila* 48 µm; 2, *Hemitelia* 52 µm; 3, Polypodiaceae 49 µm; 4, Monolete Spore type I 65 x 77 µm; 5, Monolete Spore type II 68 x 50 µm; 6, *Selaginella* type I 40 µm; 7, *Selaginella* type II 23 µm; 8, cf. *Alsophila/Dicksonia* 60 µm; 9, *Podocarpus* 90 µm; 10, *Lycopodium* 40 µm; 11, Gramineae 40 µm; 12, Palmae type I 36 µm; 13, Palmae type II 46 x 32 µm.

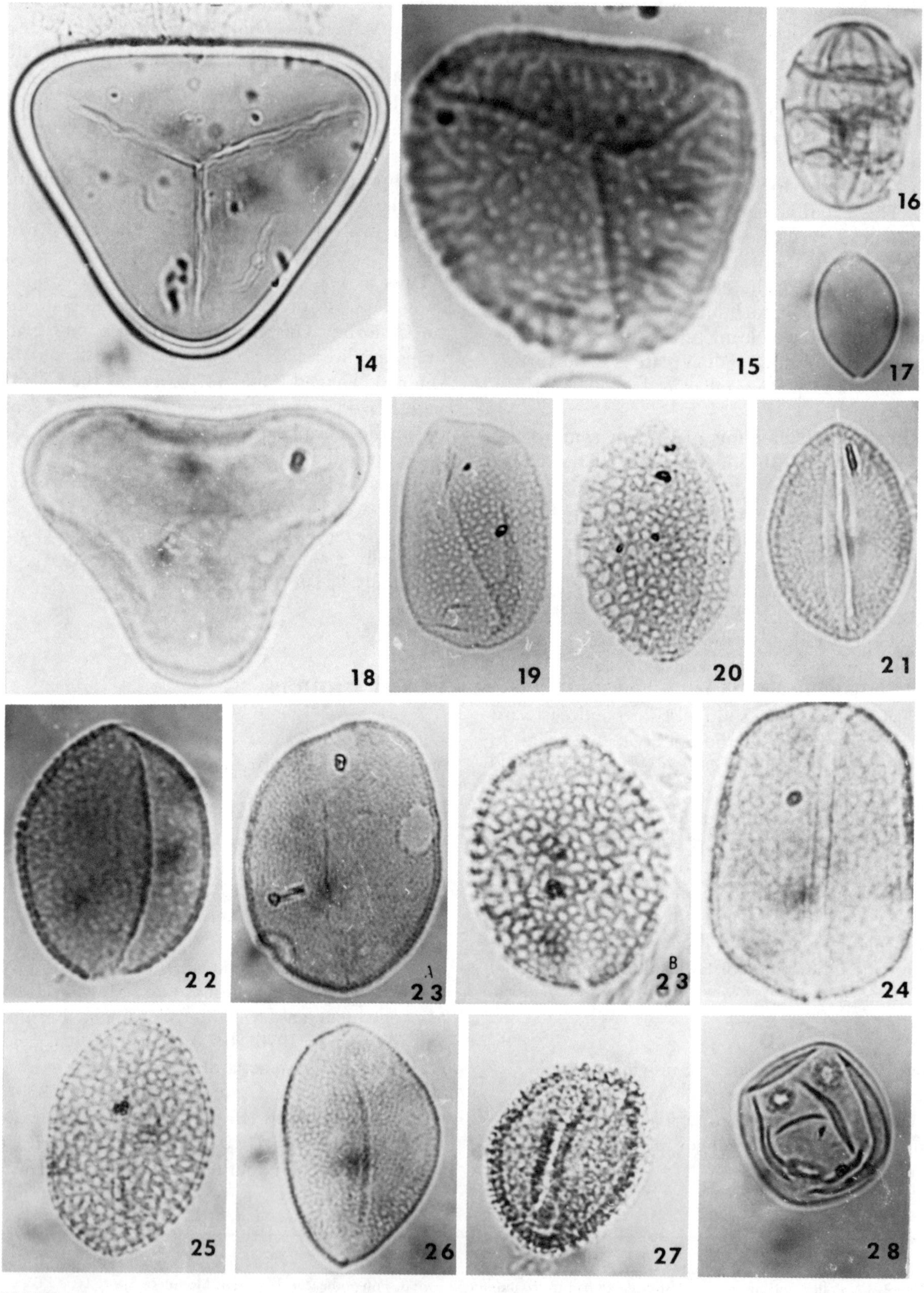

PLATE 2

Fig. 14–28. Pollen and Spores; 14, Trilete Spore type I 57 μm; 15, Trilete Spore type II 60 μm; 16, Fungal Spore Polyad 20 x 32 μm; 17, Basidiomycetes Spore 23 μm; 18, Trilete Spore type III 55 μm; 19, Palmae type III 27 x 29 μm; 20, Palmae type IV 40 x 27 μm; 21, Palmae type V 40 x 24 μm; 22, Palmae type VI 45 x 32 μm; 23a, Palmae type VII 50 x 35 μm; 23b, Palmae type VIII 45 x 36 μm; 24, Palmae type IX 54 x 26(31) μm; 25, Palmae type X 43 x 27 μm; 26, Palmae type XI 40 x 27 μ m; 27, Palmae type XII 36 x 27 μm; 28, *Echites* 27 μm.

has been reviewed by Duellman (1979), Raven and Axelrod (1974, 1975), and Vuilleumier (1971).

For the most part these studies present a consistent picture of Central American Tertiary zoogeography and one in accord with paleogeographic reconstructions based on plate tectonic data. For most of the Tertiary the larger terrestrial vertebrate faunas of North and South America were distinct, reflecting limited land surfaces available through present-day Central America. As progressively more complete connections developed throughout the Tertiary, interchanges became more frequent, and with the establishment of the Isthmian connection in latest Pliocene times virtually all physical barriers to migration were removed. At or near this time North American procyonids (raccoons and related groups) first appear in South America, ca. 6 m.y. B.P. South American ground sloths are first recorded in North America ca. 9.5–9.0 m.y. B.P. (Marshall et al. 1979); other late arrivals in North America include *Dasypus* (armadillo) and *Didelphis* (oppossum). Among plants, *Quercus, Ilex, Myrica,* and *Alnus* all initially appear in northern South America during Pliocene/Pleistocene times.

La evidencia de la fauna sugiere que durante la mayor parte del Terciario los vertebrados terrestres de mayor tamaño de Norte América, eran diferentes de los de América del Sur. Durante el Plioceno/Pleistoceno, plantas como Quercus, Ilex, Myrica y Alnus llegaron al norte de Sur América.

As with most developing models, new data periodically appear that are difficult to reconcile with the preponderance of existing evidence. Bonaparte (1979) has described a fossil carnosaur and sauropod fauna from Jurassic deposits in Patagonia. The carnosaur, *Piatnitzksaurus floresi,* is considered related to *Allosaurus* from the Upper Jurassic Morrison Formation of the United States. The sauropods are *Patagosaurus fariasi* and *Volkheimeria chubutensis,* and the former is considered comparable to *Cetiosaurus* and *Haplocanthosaurus* also of the Morrison Formation. Bonaparte (1979: 1378) concludes, "This record is the first indication that South America was populated during Callovian-Oxfordian time by carnosaur and sauropod assemblages. Paleogeographically, the new evidence supports the idea that during the Jurassic a terrestrial fauna interchange was possible between South America and other continents (for example, Africa and North America)".

Equally intriguing is the report by Brett-Surman (1979) of hadrosaurian dinosaurs common to North and South America in the Late Cretaceous. More recently (Brett-Surman, pers. comm. to Raven 1979) a new family of theropod dinosaurs was established that includes a species supposedly existing contemporaneously in Montana and Argentina. According to Brett-Surman (1979), "New evidence from vertebrate paleontology suggest that a substantial land connection between North and South America existed 85–100 m. y. ago which facilitated the passage of many terrestrial vertebrates between the continents, including large herbivorous dinosaurs" and "Dispersal between South America and Laurasia could have been by way of Africa or North America. Most workers believe that Africa was isolated from South America sufficiently early so as not to have been involved in the Upper Cretaceous dispersal of marsupials, so that dispersal must have occurred by way of North America. The two possible links with South America were through Panama or the Caribbean arc. Both hadrosaurs and ceratopsians are found in South America and so the route used must have been a major land connection at some time because both groups consist of large terrestrial vertebrates not subject to waif dispersal or rafting."

La evidencia fósil está en desacuerdo con la teoría que postula que Norte y Sur América estaban unidas durante el Jurásico y fin del Cretáceo.

At present most geophysical evidence and available data from Latin American fossil floras and faunas cannot be reconciled with a model involving extensive land connections between North and South America during the Jurassic and the Late Cretaceous. Nonetheless the reports cannot be taken lightly and emphasize the tentative nature of present concepts about Caribbean tectonics and biogeography.

Cenozoic Environments of Adjacent Regions

An important contribution to Latin America biogeography has been the refugium concept, particularly as applied to biotas of the Amazon Basin (Anderson 1979; Brown 1975; Brown et al. 1974, 1979; Haffer 1969, 1970, 1974; Prance 1973; Simpson & Haffer 1978; Vanzolini 1973, 1976; Vanzolini & Williams 1970). This model involves periodic restriction of the Amazon rain forest to physiographically controlled moist sites during times of decreased rainfall. The intervening areas are occupied by elements of the caatingas-cerrados vegetation now confined to the

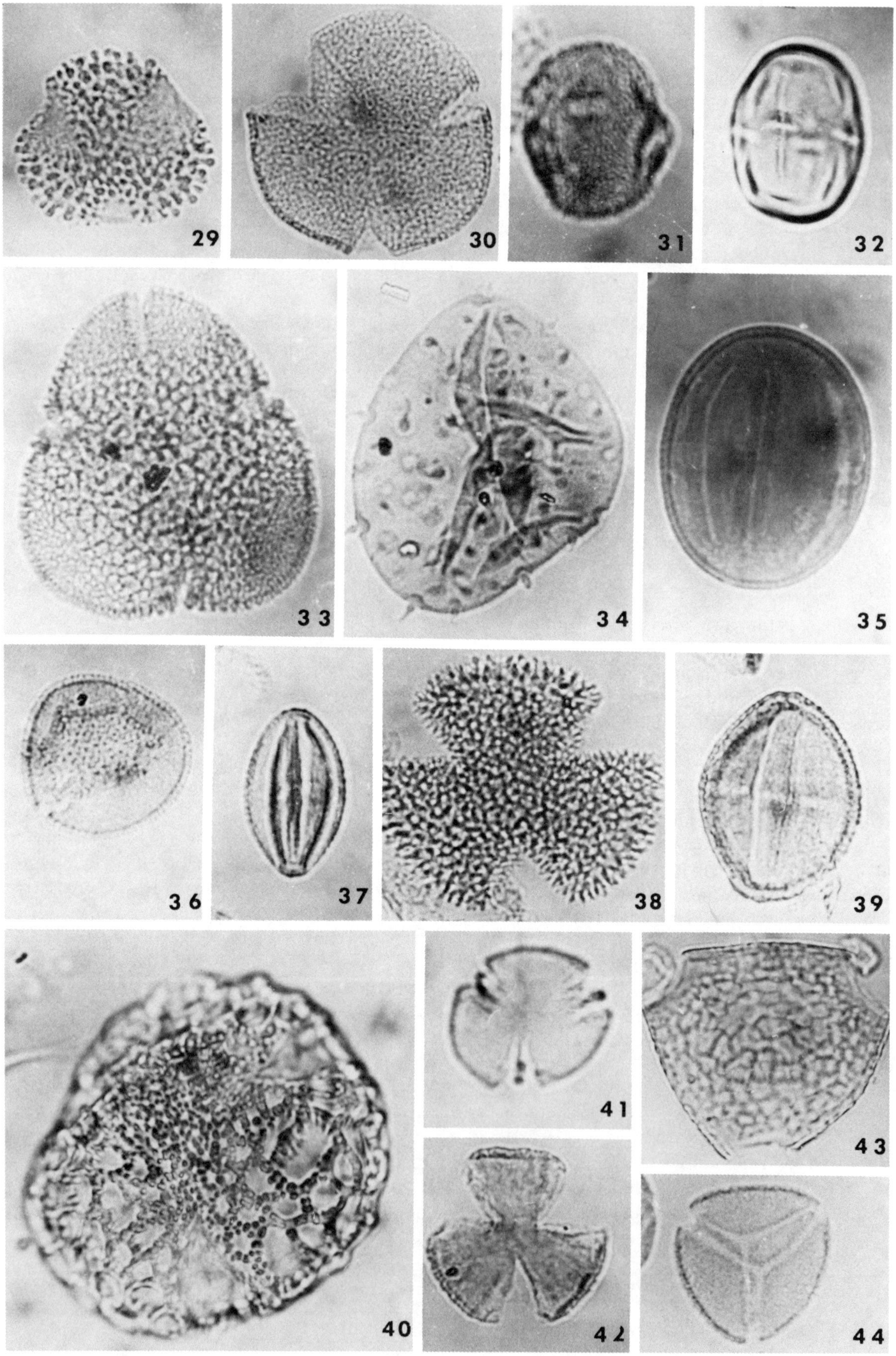
29
30
31
32
33
34
35
36
37
38
39
40
41
42
43
44

surrounding upland slopes. During periods of increased rainfall, as at present, the forest expands from these refugia and coalesces into the extensive lowland community that characterizes the Amazon Basin, while the caatingas-cerrados is again confined to the bordering margins of the basin.

El concepto de refugio, aplicado a la cuenca del Amazonas y a otras regiones, parece ser válido para América Central.

The initial basis for the refugium concept was the observation by Haffer (1969) that areas of high species diversity for Toucanidae birds coincided with areas likely to have remained moist during dry periods (at the base of mountain slopes, confluence of major rivers, etc.). Equally important were the palynological studies of van der Hammen and Gonzalez (1960) demonstrating that, at least in mid-altitude upland areas of Colombia, the climatic and vegetational changes required by the Haffer model were operative.

To the north studies in southeastern Veracruz, Mexico (Graham 1975, 1976), have shown that the rain forest was disrupted there during late Tertiary time. Lignite samples from the Miocene Paraje Solo Formation contain pollen of mangrove, lowland warm-oak, and bordering temperate (oak-*Liquidambar* vegetation, but little evidence of the present tropical rain forest. Collectively these studies demonstrate that plant communities both to the north and south of Central America were affected by changing Cenozoic environments.

In summary, ancillary sources of information are particularly important to vegetational history studies in Central America where direct paleobotanical information is limited. Plate tectonic data provide the physiographic framework for interpreting the history of Central American communities. Fossil faunas reflect paleoenvironmental conditions and provide a means of assessing migrations through Central America independent of plant distributions. Finally, results from biogeographic studies in adjacent North and South America set parameters for speculations on the history of Central American vegetation.

Studies on the Tertiary Communities of the Canal Zone

In the early 1950's a series of drillings was made in Gatun Lake to test the foundations for new constructions in the Canal Zone. These cores were stored in facilities of the Panama Canal Company, and in 1962 samples from along the cores were obtained by Elsa Barghoorn for studies on Quaternary vegetation and sea-level changes in Central America (Bartlett & Barghoorn 1973). At about the 162–foot level the lower portions of the cores penetrated into the Gatun Formation of middle Miocene age, and the sediments were found to contain abundant and well-preserved plant microfossils. During two subsequent field seasons (1963–64 and 1967) additional samples were obtained from the Eocene Gatuncillo Formation, and the Oligo-Miocene La Boca and Cucaracha Formations through the cooperation of Robert Stewart, Panama Canal Company. These samples form the basis for our current studies on Isthmian Tertiary vegetation, and some preliminary results are now available from the Miocene Gatun material.

At present 159 types of pollen and spores have been recovered from the Gatun Formation, and 63 are tentatively identified (Plates 1–5). Among the 47 genera and families listed in Table 1, 46 (95%) now occur in Panama, and 36 (ca 77%) grow on Barro Colorado Island adjacent to the collecting area (Croat 1978). Only *Glycydendron* (Euphorbiaceae) is not represented in Panama. It grows in the Amazon region of South America. These results are comparable to those from other sites in northern Latin America. Among the 110 identified plant microfossils from the Miocene Paraje Solo Formation of Veracruz, Mexico, 102 (92%) now occur in the state, and the remainder are found in adjacent Central America or the Greater Antilles. From the Oliogcene San Sebastian Formation of Puerto Rico, 34 of the 44 taxa identified (77%) now occur in the Greater Antilles.

Estudios de polen de la región de Gatún revelaron 159 diferentes tipos de polen. De los 47 géneros y familias identificados, 46 (95%) ocu-

PLATE 3

Fig. 29–44. Pollen and Spores; 29, *Ilex* 36 µm; 30, *Pseudobombax* 45 µm; 1, *Bursera* 32 x 27 µm; 32, cf. *Protium* 35 x 26 µm; 33, *Bombax* 60 µm; 34, Cucurbitaceae 60 x 50 µm; 35, *Bauhinia* 45 x 36 µm; 36, *Hedyosmum* 30 µm; 37, cf. *Margaritaria* 18 x 32 µm; 38, *Glycydendron* 50 µm; 39, *Sapium* 41 x 33 µm; 40, *Ochroma* 64 µm; 41, *Alchornea* 25 µm; 42, *Quercus* 31 µm; 43, *Erythrina* 40 µm; 44, *Struthanthus* 27 µm.

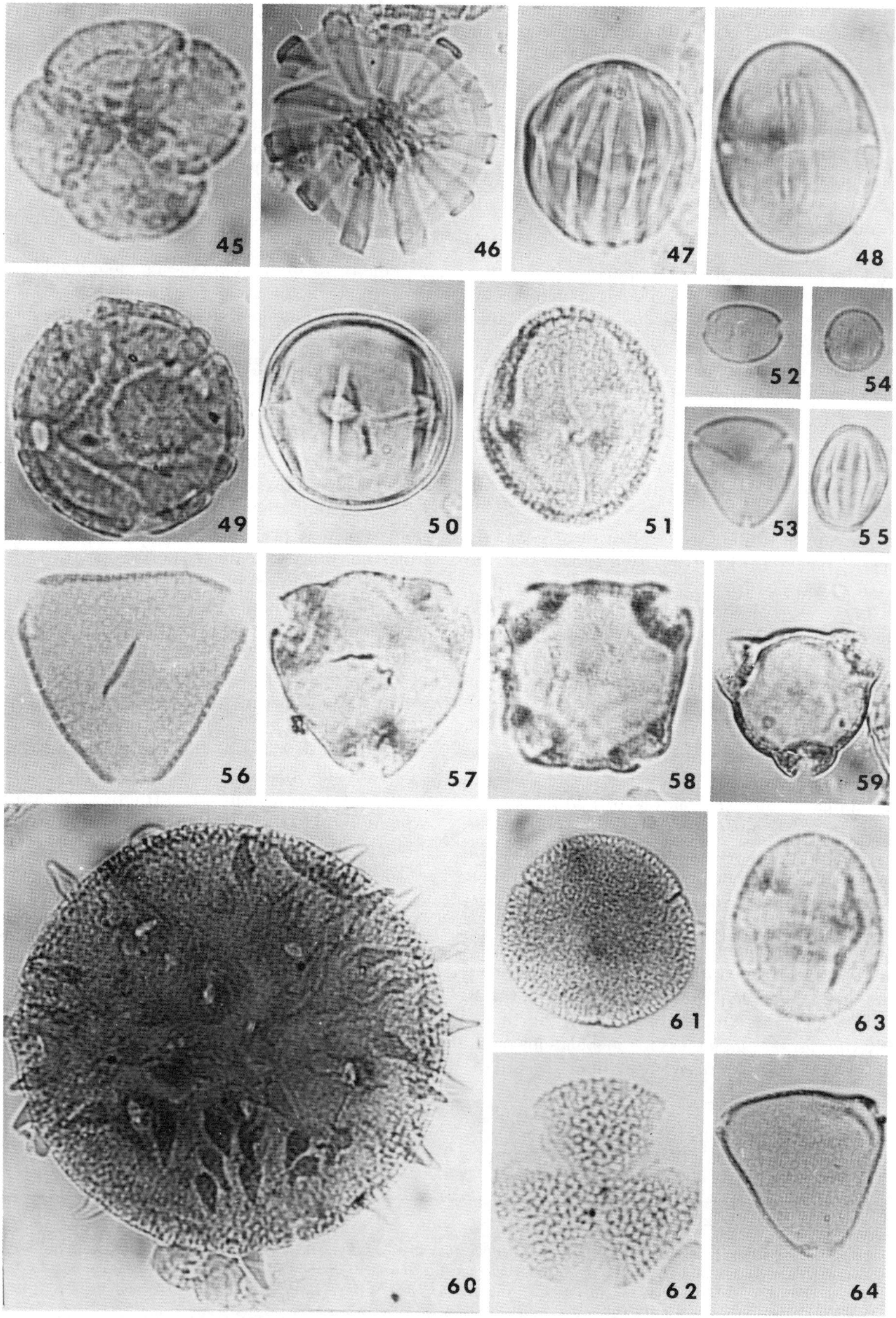
45
46
47
48
49
50
51
52
54
53
55
56
57
58
59
60
61
63
62
64

TABLE 1
LANT MICROFOSSILS FROM THE GATUN FORMATION
CANAL AREA, PANAMA
TENTATIVE IDENTIFICATIONS

yathaceae	*Chloranthaceae*	*Myrtaceae*
Alsophila	Hedyosmum	Eugenia/Myrcia
Cnemidaria	*Cucurbitaceae*	*Polygonaceae*
Hemitelia)	*Euphorbiaceae*	Coccoloba
ycopodiaceae	Alchornea	*Proteaceae*
Lycopodium	Glycydendron	Roupala
olypodiaceae	cf. Margaritaria	*Rhizophoraceae*
elaginellaceae	Sapium	Rhizophora
Selaginella	*Fagaceae*	*Rubiaceae*
odocarpaceae	Quercus	Faramea
Podocarpus	*Leguminosae*	Psychotria
ramineae	Bauhinia	Randia
almae	Erythrina	*Sapindaceae*
pocynaceae	cf. Macrolobium	Allophyllus
Echites	Schrankia	Cupania
quifoliaceae	*Lentibulariaceae*	Paullinia
Ilex	Utricularia	Serjania
ombacaceae	*Loranthaceae*	*Theaceae*
Bombax	Struthanthus	Cleyera
Ochroma	*Malpighiaceae*	*Tiliaceae*
Pseudobombax	*Malvaceae*	Mortoniodendron
urseraceae	Hibiscus	*Ulmaceae*
Bursera	*Meliaceae*	Trema
cf. Protium	Cedrela	
hloranthaceae	*Moraceae*	
Hedyosmum	cf. Ficus	

ren actualmente en Panamá y 36 (77%) crecen erca del sitio de colecta.

The data could mean environments at the time of lepostion were comparable to those of the present. Closer analysis reveals, however, that some major omponents of the modern vegetation were absent, while others were more prominent than would be expected from current floristic patterns. For example, the lowland tropical rain forest was poorly represented in the Veracruz material even though the sediments (lignites) were deposited under lowland environments. Rather, the temperate oak-*Liquidambar* forest, now growing in the 700–2000 m zone, was one of the principal communities represented in the microflora. This suggested climates were cooler near the end of the Miocene, resulting in disruption of the rain forest and lowering of the oak-*Liquidambar* ecotone. This view is supported by the presence of *Picea* in the fossil assemblage, since spruce in Mexico today is found only in the highlands to the north.

Such analyses are dependent upon arrangement of the identified fossil taxa into paleocommunities, and this has not yet been completed for the Gatun assemblage. Nonetheless, a few preliminary impressions of the paleoenvironments and paleophysiography are emerging. Among the microfossils now identified are elements from several modern plant communities of the Canal Zone and adjacent regions. The most abundant microfossil was *Rhizophora*, reflecting a well-developed mangrove community. Various aquatic to water-tolerant communities are represented by pollen of *Utricularia*, and by tree and shrub genera like *Allophyllus*, *Bursera*, *Coccoloba*, *Erythrina*, *Eugenia*, *Myrcia*, and *Hibiscus*. The Monsoon Forest included *Alchornea*, *Bauhinia*, *Cedrela*, *Cupania*, *Faramea*, *Ficus*, *Glycydendron* (South American), *Macrolobium*, *Margaritaria*, *Mortoniodendron*, *Paullinia* (liana), *Protium*, *Pseudobombax*, *Psychotria*, *Randia*, and *Sapium*. Understory herbs included *Lycopodium* and *Selaginella*, while ravines and other high-humidity sites supported the ferns *Alsophila* and *Cnemidaria (Hemitelia)*. Shrubs and trees of upland and more open and drier sites were *Cleyera*, *Echites* (liana), *Ilex*, *Hedysum*, *Quercus*, *Ochroma*, *Schrankia*, *Serjania*, and *Trema*. Wide-ranging genera occurring in several communities were *Roupala*, *Struthanthus* (a parasitic vine) and various palms and grasses.

Since most of these pollen and spores can be identified only to genus and the identifications are preliminary, precise reconstructions of paleophysiography and climates are not possible. It is apparent, however, that some modern communities are well-represented in the assemblage, while others are poorly represented or absent. Missing are high elevation communities (cloud forest, elfin forest) and those of dry to arid habitats (thorn forest, savannas). This is consistent with the concepts of Miocene paleophysiography based on plate tectonic data, which imply low relief during early stages in the development of the Isthmian landscape with few habitats available for up-

PLATE 4

Fig. 45–64. Pollen and Spores; 45, *Schrankia* 25 μm; 46, *Utricularia* 42 μm; 47, *Utricularia* 33 x 34 μm; 48, Meliaceae 42 x 36 μm; 49, *Malpighiaceae* 42 μm; 50, *Cedrela* 36 x 37 μm; 51, *Coccoloba* 36 x 42 μm; 52, Moraceae (cf. *Ficus*) 11 x 15 μm; 53 *Myrcia/Eugenia* 20 μm; 54, *Trema* 10 μm; 55, *Cleyera* 21 x 15 μm; 56, *Roupala* 40 μm; 57, *Faramea* type I 36 μm; 58, *Faramea* type II 36 μm; *Faramea* type III 27 μm; 60, *Hibiscus* 72 μm; 61, *Psychotria* type I 40 μm; 62, *Psychotria* type II 36 μm; 63, *Rhizophora* type i 24 x 27 μm; 64 *Allophyllus* 28 μm.

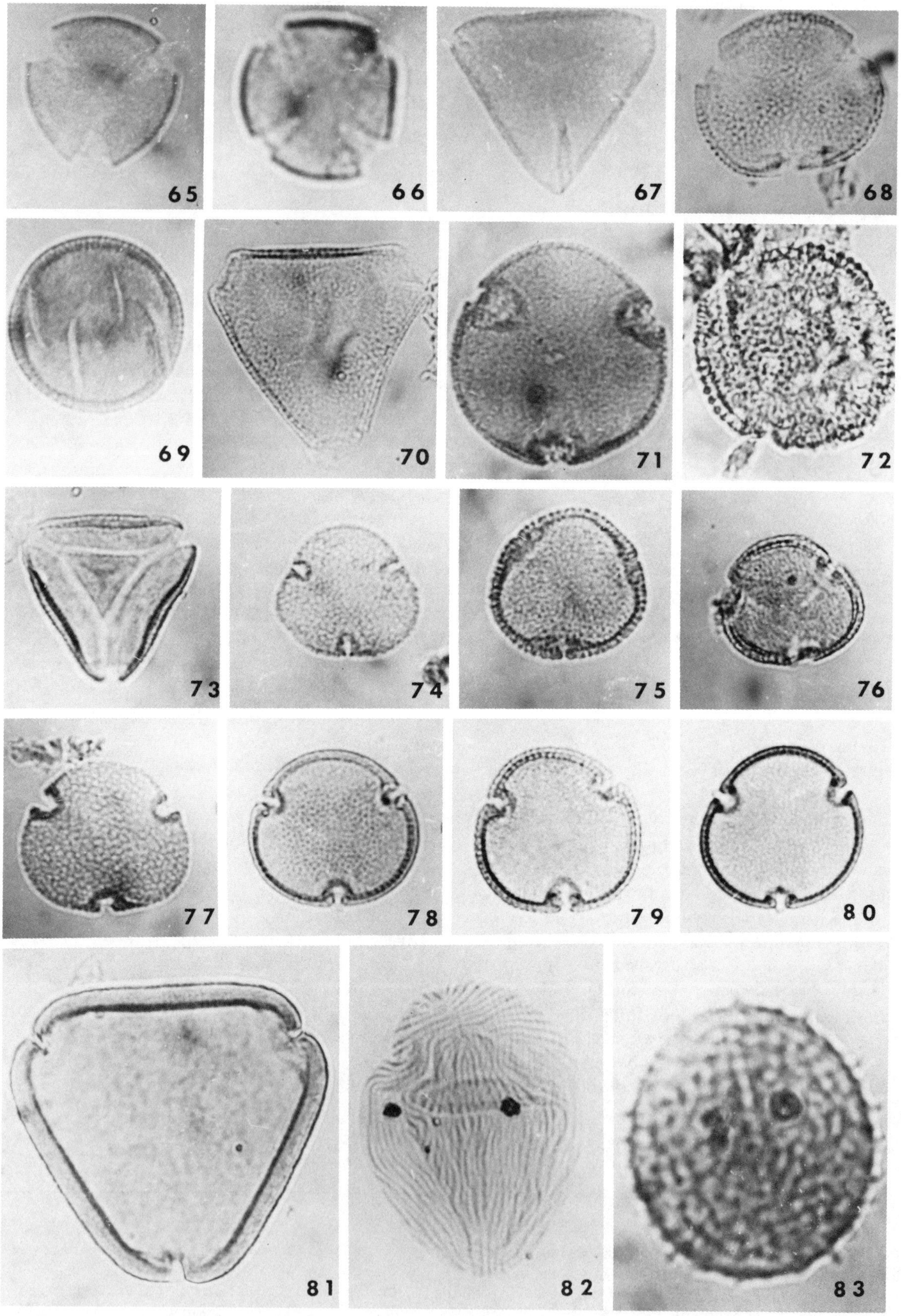
65
66
67
68
69
70
71
72
73
74
75
76
77
78
79
80
81
82
83

ınd communities. Seasonality in rainfall would likely ·e less pronounced, and differences in precipitation ·etween the Pacific and Atlantic coasts reduced, since ainfall would not be augmented by the present-day ɔw central cordillera.

:l polen sugiere que durante el Mioceno, las ierras eran bajas y no existían hábitat secos.

There are other impressions of the Miocene land-cape derived from these tentatively reconstructed ɔaleocommunities. Even without the upland and dry ıabitat communities, the flora seems diverse both in ypes of vegetation and genera represented. Also, lig-ıites, deposited along the margins of continental or nsular land masses, were forming at sites in the Canal Zone during at least Miocene (Gatun), Oligo-Miocene La Boca, Cucaracha, Culebra) and possibly Eocene Gatuncillo) times. If these islands were small and ephemeral, pollen representing mangroves, palms, grasses, and a few fast-growing trees and shrubs of ɔpen habitats would be expected. The presence of a nore diversified vegetation, including several climax orest tree genera and associated fresh-water swamp, riparian, herbaceous and shrubby understory genera ınd lianas, suggests extensive and enduring Miocene and surfaces in the region of Central Panama.

Fluctuations in Tertiary climates have been documented (see above) for areas both to the north ınd south of Central America. Yet the Miocene vegetation from the Canal Zone is remarkably similar to the modern vegetation in terms of generic composition. This apparent difference in climatic history is probably due to the contrasting physiography of the regions. Sites in Veracruz, Mexico and the Amazon Basin were from continental localities surrounded by highlands of substantial relief. Thus the effect of any climatic fluctuation on the vegetation would be heightened by physiographic diversity, and changes in the composition and distribution of the biotas more pronounced. By contrast, the Central American vegetation during the Tertiary existed under conditions of low relief and insular environments. Consequently any climatic fluctuations would likely be mollified with less effect on the biotas.

The continental affinities of genera and families found in the Gatun flora are important in estimating the extent of the Isthmian link and the nature of migrations taking place. Raven and Axelrod (1974, pp. 627–628) list plant families that probably spread from South to North America in Eocene time or subsequently. Among these only four were found among the microfossils of the Gatun Formation, and these included only four genera. The families are Chloranthaceae, the Mimosoideae of the Leguminosae, Malpighiaceae, and Proteaceae. By contrast, North American families number 14 in the Gatun fossil flora and include 25 genera. The families are Apocynaceae, Aquifoliaceae, Bombacaceae, Cucurbitaceae, Euphorbiaceae, the Caesalpinioideae of the Leguminosae, Meliaceae, Moraceae, Myrtaceae, Rubiaceae, Sapindaceae, Theaceae, Tiliaceae, and Ulmaceae. Clearly the Miocene vegetation in central Panama was composed of elements derived from the north, reflecting more continuous land surfaces than was available east and south through the present Darien. This is consistent with data from fossil vertebrates of the Oligo-Miocene Cucaracha Formation, Canal Zone (Whitmore & Stewart 1965). The affinities of this fauna were also distinctly northern. Further evidence that the discontinuity between North and South America was operative throughout the Miocene come from a variety of sources, including Caribbean Tertiary marine invertebrates. Woodring (1954, 1966) has shown that provincialization of these faunas into distinct Pacific and Caribbean communities took place only in latest Pliocene time. Further details on the timing of interchange for various terrestrial vertebrate groups are given by Marshall et al. (1979).

Few paleobotanical studies have been made in Central America, and the history of the vegetation is not well known, but for Panama this is likely to change in the near future. Fossil pollen and spore floras are known from the Eocene, Oligo-Miocene,

---

PLATE 5

Fig. 65–83. Pollen and Spores; 65 *Rhizophora* type I 27 μm; 66, *Rhizophora* type II 30 μm; 67, *Serjania* 35 μm; 68, *Randia* 28 x 34 μm; 69, *Psychotria* type III 32 μm; 70, *Paullinia* 36 μm; 71, Unknown I 41 μm; 72, Unknown II 38 μm; 73, *Cupania* 30 μm; 74, *Mortoniodendron* type I 25 μm; 75, *Mortoniodendron* type II 27 μm; 76, *Mortoniodendron* type II 23 μm; 77, *Mortoniodendron* type III 27 μm; 78, *Mortoniodendron membranaceum* (Modern Reference) surface view 32 μm; 79, *Mortoniodendron guatemalense* (Modern Reference) equatorial view 29 μm; 80, *Mortoniodendron membranaceum* (Modern Reference) equatorial views 32 μm; 81, Unknown III 54 μm; 82, Caesalpinioideae (cf. *Macrolobium*) 52 x 41 μm; 83, Unknown IV 51 x 42 μm.

and Miocene of the Canal Zone. Palynological investigations on the Quaternary of the Canal Zone have been completed (Bartlett & Barghoorn 1973). Extensive petrified wood material, now in the Hankins Collection at Harvard, is known from Ocu in far western Panama. These studies will provide more information on the history of vegetation in Panama than is available for any comparable area in Latin America. Fortunately contributions from several disciplines are available which broaden the base for interpreting the past history of Panamian and Central American vegetation. Prominent among these are the new *Flora of Barro Colorado Island* (Croat, 1978) and the completed *Flora of Panama*.

Anonymous. 1888. Fossil plants near San Juancito. Honduras Progr. 1: 3.

Andersson, L. 1979. Multi-layered distribution patterns and the hypothesis of rain forest refugia. Bot. Notiser 132: 185–190.

Bartlett, A. 1967. Palynological studies of Gatun Basin, Panama. Ph.D. diss., Harvard Univ., Cambridge.

——— & E.S. *Barghoorn*. 1973. Phytogeographic history of the Isthmus of Panama during the past 12,000 years (A history of vegetation, climate, and sea-level change). *In* A. Graham, ed., Vegetation and Vegetational History of Latin America. Elsevier, Amsterdam.

———, E.S. *Barghorn*, & R. *Berger*. 1969. Fossil maize from Panama. Science 165: 389–390.

Berry, E.W. 1914. Fossil plants in the Canal Zone. Science 38: 357.

———. 1918. The fossil higher plants from the Canal Zone. *In* Contributions to the Geology and Paleontology of the Canal Zone, Panama, and Geologically Related Areas in Central America and the West Indies. Bull. U.S. Natl. Mus. 103: 15–44.

———. 1921a. Tertiary fossil plants from Costa Rica. Proc. U.S. Natl. Mus. 59: 169–185.

———. 1921b. A palm nut from the Miocene of the Canal Zone. Proc. U.S. Natl. Mus. 59: 21–22.

———. 1928. A palm fruit from the Miocene of western Panama. Jour. Wash. Acad. Sci. 18: 455–457.

———. 1942. Mesozoic and Cenozoic plants of South America, Central America and the Antilles. 8th Amer. Sci. Congr. (Washington, D.C., 1940), Proc., Vol. 4(Geol. Sci.): 365–373.

Bonaparte, J.F. 1979. Dinosaurs: A Jurassic assemblage from Patagonia. Science 205: 1377–1378.

Brett-Surman, M.K. 1979. Phylogeny and palaeobiogeography of hadrosaurian dinosaurs. Nature 277: 560–562.

Brown, K.S. 1975. Geographical patterns of evolution in Neotropical Lepidoptera. Systematics and derivation of known and new Heliconiini (Nymphalidae: Nymphalinae). Jour. Entomol. 44: 201–242.

——— & Aziz Nacib *Ab'Saber*. 1979. Ice-age forest refuges and evolution in the Neotropics: Correlation of paleoclimatological, geomorphological and pedological data with modern biological endemism. Paleoclimas 5: 1–30.

——— et al. 1974. Quaternary refugia in tropical America: Evidence from race formation in *Heliconius* butterflies. Proc. Roy. Soc. London (B) 187: 369–378.

Brown, R.W. 1947. Fossil plants and human footprints in Nicaragua. Jour. Paleontol. 21: 38–40.

Christofferson, E. 1976. Colombian Basin magnetism and Caribbean plate tectonics. Geol. Soc. Amer. Bull. 87: 1255–1258.

Couper, R.A. 1964. Spore-pollen correlation of the Cretaceous rocks of the northern and southern hemispheres. Soc. Econ. Paleontol. Min., Spec. Publ. 11: 131–142.

Cowgill, U.M. & G.E. *Hutchinson*. 1966. La Aguda de la Santa Ana Vieja: The history of a pond in Guatemala. Arch. Hydrobiol. 62: 335–372.

——— et al. 1966. The history of Laguna de Petenxil, a small lake in northern Guatemala. Mem. Conn. Acad. Arts Sci. 17: 1–126.

Croat, T.B. 1978. Flora of Barro Colorado Island. Stanford Univ. Press, Stanford CA.

Delevoryas, T. & S.C. *Srivastava*. 1977. A Jurassic flora from central Honduras. Abstract, Bot. Soc. Amer. Misc. Publ. 154: 35.

Dengo, G. 1973. Estructura geologica, historia tectonica y morfologia de America Central. Centro Regional de Ayuda Tecnica, Mexico, D.F.

Duellman, W.E., ed. 1979. The South American Herpetofauna: Its origin, evolution, and dispersal. Mus. Nat. Hist., Univ. Kansas, Monogr. No. 7. 487 pp.

Edgar, N.T. et al. 1971. Seismic refraction and reflection in the Caribbean Sea. Amer. Assoc. Petrol. Geol. Bull. 55: 833–870.

Emiliani, C. et al. 1972. Neogene sedimentation on the Blake Plateau and the emergence of the Central American Isthmus. Palaeogeogr., Palaeoclimatol., Palaeoecol. 11: 1–10.

Freeland, G.L. & R.S. Dietz. 1971. Plate tectonic evolution of the Caribbean-Gulf of Mexico region. Nature 232: 20–23.

Germeraad, J.H. 1978a. Contribution to the palynology of El Salvador (C.A.)—A progress report. Mimeogr.

———. 1978b. Contribution to the palynology of Guatemala (C.A.)—A progress report. Mimeogr.

Gomez-Pompa, L.D. 1970. A first report on fossil fern-like Pteropsida from Costa Rica. Rev. Biol. Trop. 16: 255–258.

———. 1971. *Palmacites berryanum*, a new palm fossil from the Costa Rican Tertiary. Rev. Biol. Trop. 19: 121–132.

———. 1972. *Karatophyllum bromelioides* L.D. Gomez (Bromeliaceae), nov. gen. et sp., del Terciario Medio de Costa Rica. Rev. Biol. Trop. 20: 221–229.

———. 1973. *Criptonemiales* calcareas fosiles en las calizas tertiarias de Patarrá, Costa Rica. Rev. Biol. Trop. 21: 107–110.

Graham, A. 1972. Some aspects of Tertiary vegetational history about the Caribbean Basin. Mem. 1 Congr. Latinoam. Bot., Mexico, D.F., pp. 97–117.

———. 1973. Cenozoic history of temperate vegetation about the Caribbean Basin. Proc. III Internat. Palynol. Conf., Novosibirsk, U.S.S.R., Palynology of Cenophytic. Nauka, Moscow, pp. 93–96.

———. 1974. Tertiary history of vegetation about the Caribbean Basin. Abstract, Geosci. Man 9: 74.

———. 1975. Late Cenozoic evolution of tropical lowland vegetation in Veracruz, Mexico. Evolution 29: 723–735.

———. 1976. Studies in Neotropical paleobotany. II. The Miocene communities of Veracruz, Mexico. Ann. Missouri Bot. Gard. 63: 787–842.

———. 1977. New records of *Pelliceria* (Theaceae/Pelliceriaceae) in the Tertiary of the Caribbean. Biotropica 9: 48–52.

———. 1979. *Mortoniodendron* (Tiliaceae) and *Sphaeropteris/Trichipteris* (Cyatheaceae) in Cenozoic deposits of the Gulf-Caribbean region. Ann. Missouri Bot. Gard. 66: 572–576.

Haffer, J. 1969. Speciation in Amazonian forest birds. Science 165: 131–137.

———. 1970. Geologic-climatic history and zoogeographic significance of the Urara region in northwestern Colombia. Caldasia 10: 603–636.

———. 1974. Avian speciation in tropical South America. Publ. Nuttall Ornithological Club No. 14, Cambridge, MA.

Hey, R. 1977. Tectonic evolution of the Cocos-Nazca spreading center. Geol. Soc. Amer. Bull. 88: 1404–1420.

Howe, M.A. 1918. Contributions to the geology and paleontology of the Canal Zone, Panama. On some fossil and recent Lithothamnieae of the Panama Canal Zone. Bull. U.S. Natl. Mus. 103: 1–13.

Humphrey, E.W. 1916. *Sphenozamites rogersianus* Fontaine; an addition to the Rhaetic flora of San Juancito, Honduras. J. N.Y. Bot. Gard. 17: 56–58.

Hunter, V.F. 1977. The Tertiary geologic evolution of the southern Caribbean area and the palynological record. Abstract, Palynology 1: 174.

Johnson, J.H. & H.V. *Kaska*. 1965. Fossil algae from Guatemala. Prof. Contrib. Colo. Sch. Mines 1: 1–152.

——— & K. *Konishi*. 1960. An interesting late Cretaceous alga from Guatemala. Jour. Paleontol. 34: 1099–1105.

Krieger, L.J. 1977. Palynology of the Miocene Gatun Formation, Panama. Master's Thesis, Kent State Univ., Kent, OH.

Kruckow, T. & L.D. *Gomez-Pompa*. 1974. Notes on the palaeoecology of the fossil algae of Costa Rica. I. Brenesia 4: 23–29.

Ladd, J.W. 1976. Relative motion of South America with respect to North America and Caribbean tectonics. Geol. Soc. Amer. Bull. 87: 969–976.

Loeischert, W. et al. 1975. Die Plio-Pleistozäne Flora aus dem Sisimico-Tal, El Salvador. Ein Beitrag zur Frage der Kontinuität tropischer Regenwälder im Quartär. Geol. Jahrb. 13: 97–191.

Lonsdale, P. 1978. Ecuadorian subduction system. Amer. Assoc. Petrol. Geol. Bull. 62: 2454–2477.

Maldonado-Koerdell, M. 1953. Plantas del Retico Liasico y otros fosiles Triasicos de Honduras. Ciencia (Mexico) 12: 294–296.

Malfait, B. & M. *Dinkelman*. 1972. Circum-Caribbean tectonic and igneous activity and the evolution of the Caribbean plate. Geol. Soc. Amer. Bull. 83: 251–272.

Marshall, L.G. et al. 1979. Calibration of the great American interchange. Science 204: 272–279.

Mattson, P.H. 1979. Subduction, bouyant braking, flipping, and strike-slip faulting in the northern Caribbean. Jour. Geol. 87: 293–304.

Molnar, P. & L. *Sykes*. 1969. Tectonics of the Caribbean and Middle America regions from focal mechanisms and seismicity. Geol. Soc. Amer. Bull. 80: 1639–1684.

Muller, J. & C. *Caratini*. 1977. Pollen of *Rhizophora* (Rhizophoraceae) as a guide fossil. Pollen et Spores 19: 361–389.

Mulleried, F.K.G. 1948. Las facies de fauna y flora del Mesozoico en el N.O. de la America Central (Tehuantepec-Nicaragua). Internat. Geol. Congr., Great Britain, Titles and Abstracts.

Newberry, J.S. 1888a. Triassic plants from Honduras. Trans. N.Y. Acad. Sci. 7: 113–115.

———. 1888b. Rhaetic plants from Honduras. Amer. Jour. Sci. 136: 342–351.

Olson, E.C. & P.O. *McGrew*. 1941. Mammalian fauna from the Pliocene of Honduras. Geol. Soc. Amer. Bull. 52: 1219–1244.

Plafker, G. 1976. Tectonic aspects of the Guatemala

earthquake of 4 February 1976. Science 193: 1201–1208.

Prance, G. 1973. Phytogeographic support for the theory of Pleistocene forest refuges in the Amazon Basin, based on evidence from distribution patterns in Caryocaraceae, Chrysobalanaceae, Dichapetalaceae and Lecythidaceae. Acta Amazonica 3: 5–28.

Raven, P.H. & D. *Axelrod*. 1974. Angiosperm biogeography and past continental movements. Ann. Missouri Bot. Gard. 61: 539–673.

———. 1975. History of the flora and fauna of Latin America. Amer. Sci. 63: 420–429.

Santamaria, F. & C. *Schubert*. 1974. Geochemistry and geochronology of the southern Caribbean-northern Venezuela plate boundary. Geol. Soc. Amer. Bull. 85: 1085–1098.

Serrano, M. 1948. Introducción al estudio de la fitopaleontología Guatemalteca. Univ. San Carlos, Guatemala, Vasc. 1.

Simpson, B. & J. *Haffer*. 1978. Speciation patterns in the Amazonian forest biota. Ann. Rev. Ecol. Syst. 9: 497–518.

Simpson, G.G. 1950. History of the fauna of Latin America. Amer. Sci. 38: 261–389.

Stephenson, L.W. & E.W. *Berry*. 1929. Marine shells in association with land plants in the Upper Cretaceous of Guatemala. Jour. Paleon. 3: 157–162.

Tsukada, M. 1965. Late Pleistocene palynology in Central America. Rep. 7th. Congr. Internat. Quaternary Assoc., Boulder, CO.

———. 1966. The pollen sequence. *In*: Cowgill, U.M. et al., The History of Laguna de Petenxil, a small lake in northern Guatemala. Mem. Conn. Acad. Arts Sci. 17: 63–66.

——— & E.S. *Deevey*. 1962. Quaternary palynology in Central America. I. Guatemalan area. Internat. Conf. Paly., 1st, Tucson. Abstract, Pollen et Spores 4: 385.

———. 1967. Pollen analyses from four lakes in the southern Maya area of Guatemala and El Salvador. *In*: E.J. Cushing & H.E. Wright, eds., Quaternary Paleoecology. Yale Univ. Press, New Haven. Pp. 303–331.

van der Hammen, T. & E. *Gonzalez*. 1960. Upper Pleistocene and Holocene climate and vegetation of the "Sabana de Bogota" (Colombia, South America). Leidse Geol. Meded. 25: 261–315.

Vanzolini, P.O. 1973. Paleoclimates, relief, and species multiplication in equatorial forests. *In*: B.J. Meggers et al., eds., Tropical Forest Ecosystems in Africa and South America: A Comparative Review. Smithsonian Inst. Press, Washington, D.C. Pp. 255–258.

———. 1976. On the lizards of a cerrado-caatinga contact: evolutionary and zoogeographical implications (Sauria). Papeis Avulsos Zool. 29: 111–119.

——— & E.E. *Williams*. 1970. South American anoles: the geographic differentiation and evolution of the *Anolis chrysolopis* species group (Sauria: Iguanidae). Arq. Zool. S. Paulo 19: 1–240.

von der Brelie, G. & M. *Teichmuller*. 1953. Beitrage zur Geologie El Salvador. 3. Mikroskopische Beobachtungen an Mangrove-Sedimenten aus El Salvador. Neues Jahrb. Geol. Palaeontol. 6: 244–251.

Vuilleumier, B. 1971. Pleistocene changes in the fauna and flora of South America. Science 173: 771–780.

Webb, D.S. ed. 1974. The Pleistocene Mammals of Florida. Univ. Florida, Gainesville.

———. 1976. Mammalian faunal dynamics of the great American interchange. Paleobiology 2: 220–234.

Whitmore, F.C. & R.H. Stewart. 1965. Miocene mammals and Central American seaways. Science 148: 180–185.

Woodring, W.P. 1954. Caribbean land and sea through the ages. Bull. Geol. Soc. Amer. 65: 719–732.

———. 1966. The Panama land bridge as a sea barrier. Trans. Amer. Phil. Soc. 110: 425–433.

———. 1970. Geology and paleontology of Canal Zone and adjoining parts of Panama. Description of Tertiary mollusks (Gastropods: Eulimidae, Marginellidae to Helminthoglyptidae). U.S. Geol. Surv. Prof. Paper 306–D, pp. 299–452.

———. 1973. Geology and paleontology of Canal Zone and adjoining parts of Panama. Description of Tertiary mollusks (Additions of gastropods, scaphopods, pelecypods: Nuculidae to Malleidae). U.S. Geol. Surv. Prof. Paper 306, pp. 453–539.

Plant Study Techniques

Many kinds of natural history study involve col-cting materials from the wild and then examination f them in the laboratory.

The observation and collection of biological sam-les in tropical forests is often difficult because the rganisms are inaccessible from the ground. Sug-en's walkway is a method that is simple in principle nd very rewarding in use.

The history of past collecting activity is important) present day scientists, for by judging the work of arlier biologists in light of the knowledge and con-itions of their day compared with what is known oday, improvements can often be made in their nalysis. Dwyer's history of plant collecting in Pan-ma describes how many scientists have sought out nd collected the diversity of Panama's plant life in ne wild.

Collecting history can also illustrate the differ-ences in the extent of collecting in different parts of the world and point out areas that need attention. Forero describes plant collecting in the Choco region of neighboring Colombia, and it is clear that recent pioneering studies leave generations of work to be done. The region of Costa Rica-Panama-Choco may be florisitically the richest in the New World.

The next two papers are examples of different ways that materials provided by collectors can elucidate relationships and trends within selected groups of plants. Keating's anatomical study of a group of *Fuchsia* species shows how such work may be used to evaluate evolutionary trends in these plants. Rockwood examines the seeds taken from a large number of collections and finds there are relationships between their sizes and features of the environment where the plants grew.

Técnicas Para Estudiar Las Plantas

Diversos estudios de historia natural involucran la ecolección de materiales silvestres y el examen osterior de éstos en el laboratorio. Este capítulo resenta ejemplos de métodos y resultados y éstos nuestran algo de las actividades colectivas en la erspectiva histórica.

La observación y recolección de muestras en los osques tropicales es frecuentemente difícil debido la inaccesibilidad de los organismos desde el uelo. El puente aéreo de Sugden, en principio es encillo y su uso brinda considerables recom-ensas.

La historia de la recolección de plantas en el asado es importante para los científicos de hoy día, porque al juzgar el trabajo de los biólogos de ntes en relación al conocimiento y condiciones de esos días comparados a lo que se conoce en el resente, se puede perfeccionar su análisis. Dwyer en su estudio sobre la historia de la recolección de lantas en Panamá, describe cómo numerosos otánicos han buscado y colectado plantas tratando de captar la diversidad florística de la vida silvestre de Panamá.

La historia de recolecta de plantas también puede ilustrar la diferencia en la extensión del trabajo realizado en diversas partes del mundo y resaltar las regiones que precisan mayor atención. Forero describe y analiza el proceso de colecta de plantas en la región del Chocó en Colombia y concluye a partir de estudios pioneros recientes, que aun queda mucho trabajo por realizar especialmente por las generaciones futuras. La región de Costa Rica-Panamá-Chocó, quizás florísticamente la más rica de los neotrópicos, está sometida a una fuerte presión de deforestación.

La investigación de Keating y la de Rockwood constituyen ejemplos de cómo el material provisto por los colectores puede ilucidar relaciones y tendencias dentro de grupos selectos de plantas. El estudio anatómico de Keating realizado en especies de *Fuchsia* demuestra cómo evaluar las tendencias evolutivas en esas plantas. Rockwood examina semillas obtenidas de numerosas colecciones y determina las relaciones entre sus tamaños y las características de medio ambiente en donde esas plantas crecen.

The History of Plant Collecting in Panama

John D. Dwyer

Panamá posee una extensa y fructífera historia en relación a colecta de plantas caracterizada primero por la actividad de botánicos europeos y luego por la de botánicos de Norteamérica. A pesar de la intensa actividad botánica durante la última década, Panamá es un lugar que aún ofrece muchas recompensas a los colectores debido a que constantemente se descubren especies nuevas.

About 750 botanists or naturalists have collected plants in Panama for permanent deposit in herbaria. This may seem to be an exaggeration, considering that my current unpublished list (see also Dwyer, 1964, 1968) includes the names of only 450 collectors, but many new collectors can be found by examining herbarium specimens and in the literature. The list includes the names of all who have collected without regard to the number of collections.

Throughout the world there are about one half million numbers—not including duplicate material—of Panamanian plants. Of these about a quarter of a million are deposited in the Missouri Botanical Garden. In the last decade and a half five botanists from the Missouri Botanical Garden have collected approximately 50,000 numbers, these for the most part being gathered in non-team fashion (Table 1).

En Panamá, a partir de 1700 aproximadamente 750 botánicos han colectado cerca de medio millón de plantas y la mitad de éstas están depositadas en el Jardín Botánico de Missouri, Estados Unidos.

Starting at the beginning in 1700 James Wallace collected several herbarium specimens in the ill-fated Scotch Colony on the Atlantic Coast of the old Province of Darien. The subsequent 280 years of plant collecting can be divided into 3 phases:

1. 1700–1914
2. 1915–1957
3. 1958–1981

The first phase runs from the period of the Scotch Colony in Darien to the opening of the Panama Canal. The second period is marked initially by a lull period during and immediately after World War I, followed by some field work and considerable herbarium study, this to be again interrupted by a war, World War II. The third period, the golden period, extends from 1958, the year when William L. Stern and Kenton L. Chambers made an exciting collecting trip into Panama, up to the present time when herbarium material is being collected on a year-round basis.

El primer período de colección de plantas fue dominado por europeos, reemplazados, más tarde, por los norteamericanos.

A glance at Table 2 shows that in the first period (1700–1914) a shift from domination of plant collecting by Europeans to an eventual takeover by North Americans is obvious. Some collections made in this

TABLE 1
COLLECTORS WITH LARGEST NUMBERS OF COLLECTIONS FROM PANAMA

*Croat, T.	23,000	*Dwyer, J.	4,500
Standley, P.	7,500	*Hammel, B.	4,500
*Gentry, A.	7,000	Correa, M.	4,500
Duke, J.	6,000	*Dressler, R.	4,000
*Mori, S.	6,000	*Lewis et al.	3,700
*Folsom, J.	5,000	*Allen, P.	3,600
*Nee, M.	5,000	Pittier, H.	3,600
Tyson, E.	5,000	*von Wedel, H.	3,000
*D'Arcy, W.	5,000		

*Indicates individual was a staff member of, student at or professional collector for the Missouri Botanical Garden.

TABLE 2
EARLY COLLECTORS OF PANAMANIAN PLANTS

Wallace, J.	Scottish	1700
Jussieu, Jos. De	French	17??
Millar, R.	English	1734
Nee, L.	French	1789
Haenke, T.	Bohemian	1790–1791
Billberg, J.	Swedish	1826
Dahlin, E.	Finnish	1826–1831
Cuming, H.	English	1830–1831
Hinds, B.	English	1837 etc.
Barclay, G.	English	1837 etc.
Sinclair, A.	English	1837 etc.
Lobb, W.	English	1843
Seemann, B.	German	1846:1849
Warscewicz, J.	Polish	1848; 1850
Behr, H.	German	1848
Fendler, A.	American	1848–1849
Ball, J.	Irish	1852
Halsted, M.	American	1850–1854
Kuntze, O.	German	1874
Duchassaing, E.	French	1849–1851
Hayes, S.	American	before 1863
Roezl, B.	Czech	1869
Hart, J.	English	1885
Cowell, J.	American	1905
Howe, M.	American	1909–1910
Williams, R. S.	American	1908
Pittier, H.	Swiss	1910–1911; 1914–1915
Powell, C.	American	1907–1927
Ostenfeld, C.	Dane	1921–1922
Bro. Gervais	Panamanian	1912
Bro. Celestine	Panamanian	1912
Steven, F. L.	American	1924
MacBride, J.	American	1918
Rose, J.	American	1918
Maxon, W.	American	1911; 1923
Hitchcock, A. A.	American	1911
Killip, E. P.	American	1917–1918; 1922
Piper, C. V.	American	1923
Goldman, F. A.	American	189?-19??
Christophersen, E.	Norwegian	19??

period may never have been cited as exsiccatae, for example, those of Joseph Jussieu, Robert Millar, and B. Roezl. Noteworthy are the collections of Berthold Seemann, who wrote the first flora of Panama (Seemann 1854) as well as those of Henri Pittier, numbering over 1,850, drawn from one end of Panama to the other (Dwyer 1973). Several Christian Brothers living in Panama City collected several hundred numbers beginning in 1912. Interestingly, almost all the Americans who collected in the first period were from the eastern United States.

The second period (1915–1957) obviously overlaps somewhat with the first, making it the most difficult of the three to bring into proper perspective. Paul Standley's collection of 7,500 numbers in 5 months—end of 1923 and beginning of 1924, and one month in 1925—served as a basis for his Flora of the Canal Zone (Standley 1928) and Flora of Barro Colorado Island (Standley 1927). His collections, mostly unicates, were confined to the Canal Area and the vicinity of Panama City.

El segundo período se inicia con el comienzo de la Flora de Panamá y está representado por colectores como Greenman, Standley y Woodson.

As early as 1922 Jesse M. Greenman, Curator of the Herbarium at the Garden, made a few collections of phanerogams in Panama, and in 1923 the Missouri Botanical Garden was attracted to Panama by the living orchid collection of a retired postman, C. W. Powell, who lived in Balboa. To house this collection, received as a gift, the Garden maintained a tropical station at Ancon Hill, Canal Zone, which survived until 1939.

Beginning in 1934, up to the beginning of World War II, several botanists, functioning mostly in team fashion, collected throughout the isthmus, especially Paul Allen, Carrol Dodge, A. A. Hunter, Russell Seibert, Robert Schery, Julian Steyermark, and Robert Woodson, Jr. I estimate that their collections in Panama amount to fewer than 10,000 numbers, but this figure would be considerably higher if the vast majority of the 5,000 specimens collected by Woodson, Allen, and Seibert in 1947 were not destroyed by fire. In this second period the greatest exhibition of collecting was by C. V. Piper of the U.S. National Herbarium who, in 1923, collected 1,000 numbers in 10 days!

From 1948 until his death in 1963 Woodson believed that Panama had been adequately collected, and that the Flora of Panama could be written successfully by relying almost exclusively on the collections in the herbarium cases of the Garden. In light of the experience of the last two decades, this was a great mistake in judgement, although the lack of money for field work and the pressure of his academic responsibilities have to be considered. As we shall see in the third and last phase, the numbers of novelties described from Panama in the past decade, especially in the Ericaceae, Araceae, and Rubiaceae, strongly support such criticism (Table 3).

El tercer período, el moderno, constituye una revitalización de la Flora de Panamá y un aumento de la actividad de los botánicos de Missouri. Muchos otros comenzaron a colectar en

TABLE 3
NEW AND RECENTLY DESCRIBED SPECIES
FLORA OF PANAMA FAMILIES (1970–1980)

Year	Number of Families	Total Species	% New or Recently Described Species
1970	5	91	1
1975	11	413	13
1976	9	207	10
1978	8	217	32
1980	1	415	23

Panamá, incluyendo a Correa, a la vez que comenzó a operar el herbario de la Universidad de Panamá .

In 1949 Ivan Johnston's The Botany of San Jose Island appeared based essentially on a study of his own collections and those of C. O. Erlanson and J. Harlow. In this period the collections of George P. Cooper and George Slater amounted to about 450 numbers from the little explored province of Bocas del Toro, the higher reaches of which remain little explored today.

In 1958, the field trip of William Stern and Kenton Chambers, both then faculty members at the Yale School of Forestry, initiated the third stage of plant collecting in the isthmus, reopening the door of Panama as a land of promise for the systematics of flowering plants and ferns (Stern, 1960). The following year, Stern, Chambers, John Ebinger, and I collected more than 1,200 numbers, principally in Darien. In 1960, James Duke, under the auspices of the U.S. Army Research Office, initiated a decade of collecting in Panama. In 1963 Dr. Edwin Tyson, then of Florida State University, shifted much of his research in zoology to the systematics of the angiosperms. His collections, amounting to 5,000 numbers, as well as many made by Mireya D. Correa A., were housed in the U.S. Army Tropic Test Center, Canal Zone (Dwyer, 1967), which later became Summit Herbarium (Croat, 1971).

Soon after Woodson died in 1963, Walter H. Lewis became Director of the Herbarium of the Missouri Botanical Garden. He immediately set to work getting grant support for botanical exploration in Panama. Table 4 shows that in 1966 and 1967 13 Garden staff members and students made collecting trips in Panama. At about this time Robert L. Dressler became resident botanist with the Smithsonian Tropical Research Institute, at first living on Barro Colorado Island. His own collections and personal studies, particularly on orchids, coupled with his willingness to assist all of those botanizing in Panama, has been of inestimable value over the past 17 years.

One of those profiting from the grant support generated by Walter Lewis was Thomas B. Croat. His accomplishments in collecting during the past 14 years in Panama defy description. Statistics indicate (Table 1) that he tripled the number of plants collected by Paul Standley. Croat's total from the isthmus is more than 23,000 numbers. (See Croat, 1978).

A glance at Table 1, which shows those who collected more than 3,000 numbers in Panama, is revealing. With the exception of Walter Lewis, this list scores only those who collected alone. Thus some collectors on the list may also have assisted in team efforts. I have no statistics for the numbers collected by Sydney McDaniel, now in Panama, although they are several thousand.

In 1969, in the Canal Zone, the Missouri Botanical Garden established a field station with a house trailer and a modest herbarium (Croat, 1972). This allows pre-doctoral or post-doctoral students to pursue their research in systematics, as well as to collect. The Garden's resident collectors in Panama, who were also curators of Summit Herbarium, are shown in Table 5.

In 1973 the Missouri Botanical Garden in Panama and the Smithsonian Tropical Research Institute com-

TABLE 4
COLLECTORS ACCOMPANYING WALTER LEWIS ON FIELD TRIPS TO PANAMA (1966–1969)

*Baker, K.	*Duke, J.	*Macbryde, B.
*Blackwell, W.	*Dwyer, J.	*Oliver, R.
*Burch, D.	*Elias, T.	*Robertson, K.
Correa, M.	Escobar, N.	*Robyns, A.
*Croat, T.	*Hawker, J.	*Verhoek, S.
*Crosby, M.	Little, C.O.	Zaborowski, A.
Dressler, R.		

*Staff or student at Missouri Botanical Garden.

TABLE 5
RESIDENT COLLECTORS, SUMMIT HERBARIUM

Antonio, Thomas M.	May 1979–July 1980
Croat, Thomas B.	March 1970–August 1971
	February 1976–September 1976
Folsom, James P.	January 1977–January 1978
Gentry, Alwyn H.	August 1971–October1972
Hammel, Barry E.	January 1978–May 1979
Kennedy, Helen	September 1972–May 1973
Mori, Scott A.	September 1974–August 1975
Nee, Michael	August 1973–June 1974
Sytsma, Kenneth J.	August 1980–August 1981
Witherspoon, John T.	September 1975–November 1975

bined their herbarium resources in a restored portion of the former Tivoli Hotel in Balboa, Canal Zone (Croat & Crosby, 1974).

That the Missouri Botanical Garden dominated plant collecting in Panama from the 1940's onward is obvious. However when one considers the extensive collecting done by Edwin Tyson in the 1960's, as well as by Sydney McDaniel and Kurt Blum, sent into Panama originally by Robert Godfrey of Florida State University, as well as the extensive collecting of James Duke over the 1960–1970 decade, under the aegis of the U.S. Army and Battelle Memorial Institute, as well as the eight botanical expeditions of Robert L. Wilbur of Duke University and his students: Frank Almeda, James Luteyn, Richard Weaver, Robin Foster and others, a more fair perspective of collecting in Panama emerges. Wilbur himself has collected some 1,500 numbers in Panama.

Special tribute is due Mireya Correa, Novencido Escobar, Octavio Sousa, and others of Panama University for building up, over the past decade, the herbarium at Panama University, now numbering over 12,000 sheets. The major collector of plants held in this herbarium was Correa, who joined the University faculty as Curator in 1968 following a two year collecting effort with the U. S. Army Tropic Test Center.

Today, in Central America, the most exciting country for plant collecting is Panama. Evidence of this is the 45 new species of Rubiaceae which are awaiting description at the Missouri Botanical Garden, these having been encountered since publication of the Rubiaceae of Panama with a total of 415 species in 1980 (Dwyer, 1980). Table 3 shows the numbers of recently described species in parts of the Flora of Panama. There are certain areas of Panama, for example, the Fortuna Dam area, where a single day of intensive collecting will usually yield one or more species of flowering plants new to science. Obviously collectors are penetrating such areas because they and others before them have searched them out, often under great hardship. Also, new roads have been opened in the republic in recent years and local landing strips and helicopter pads have been established in remote areas, often related to industrial or national projects. The collector is aware that pristine collecting grounds of today may fall prey to tomorrow's ax of destruction. Deference is due the Government of Panama for granting collecting permits to qualified collectors. Many institutions reciprocate by forwarding identified duplicate specimens to the Herbarium of the Universidad de Panama, the national repository.

Actualmente, Panamá aún es el país que en América Central constituye el mayor estímulo para el botánico, ya que a medida que los colectores alcanzan lugares antes inaccesibles encuentran especies nuevas de plantas.

In conclusion, it is appropriate to pinpoint certain geographical areas of Panama which are "rich and recent" from the viewpoint of botanical collecting, at the same time acknowledging the pioneer efforts of the botanists. Such a listing is never complete. Only those localities botanized in realistic fashion for the first time between 1967 and the present are considered. Certain relatively new collecting areas, such as Cerro Jefe (Prov. Panama from 1965) and Cerro Pirre (Prov. Darien from 1960) have not been included as their "discovery" predated 1967. Table 6 indicates

TABLE 6
IMPORTANT NEW BOTANICAL COLLECTING SITES

1967	James Duke and Joseph Kirkbride, Jr. botanized from Chiriquí Grande (Prov. Bocas del Toro) to Caldera (Prov. Chiriquí).
1967	Walter Lewis, Thomas Croat, and Jon Hawker made important collections at Río Concepción (Prov. Veraguas). In the same year Will Blackwell and Mireya Correa made important collections at Santa Catalina (Prov. Bocas del Toro).
1971	Thomas Croat collected from Guácimo (Prov. Colón) to Miguel de la Borda (Prov. Colón), using a canoe.
1972	Al Gentry collected through the Darien with the British Trans-Americas Expedition.
1973	Thomas Croat, Ronald Liesner and Philip Busey botanized in the tropical wet forest on the Burica Peninsula at Bartolo Limite-Rabo de Puerco area (Prov. Chiriquí), utilizing a vehicle. Today the road is overgrown.
1975	Scott Mori was the first to collect at Cerro Colorado (Prov. Chiriquí), close to the Continental Divide, at an elevation of 1,200–1,500 m. Cerro Colorado is the site of a copper mine easily accessible by road. Mori, together with Alwyn Gentry explored botanically the summit of Cerro Tacacaruna, elev 1,975 m, on the Panama-Colombia border.
1976	Mireya Correa with others collected several times at the Fortuna Dam site and nearby.
1976	Thomas Croat collected in the Azuero Peninsula walking for 6 hours to the headwaters of the Río Pedregal (Prov. Los Santos) from Jobero in the same province, southwest of Tonosí.
1977	James Folsom began an era of rewarding collecting on the divide above El Cope (Prov. Coclé).
1978	Robert Dressler pioneered in collecting from La Pintada (Prov. Coclé) to Coclecito (Prov. Coclé).
1978	Barry Hammel collected at two important sites: Cerro Bruja (Prov. Colón) and the headwaters of the Río Indio and Río Boqueron (Prov. Panamá), and from Gualaca (Prov. Chiriquí) to the Fortuna Dam site, proceeding to Cerro Hornito (Prov. Chiriquí).
1979	Barry Hammel collected at Cerro Pate Macho (Prov. Chiriquí, the Río Culebra (Prov. Chiriquí), as well as along the Río Concepción (Prov. Veraguas, especially at a one-man operated gold mine).
1980	Mireya Correa collected in Bocas del Toro in areas which were to become part of the Changuinola-Teribe dam complex.

ow the various provinces of Panama have benefited rom the attentions of different collectors.

There is, of course, a limit to the number of species n Panama, as well as choice collecting sites. It is strange, however, that despite almost 300 years of collecting, the yield of novelties in our day is so great. Since the opening of the Panama Canal in 1914, more endemic species have been described from the sthmus than between 1700 and 1913. The richness of the flora is a reflection of the soil and climate of this great tropical country. It is no less a tribute to the efforts of the botanists who have been so richly rewarded.

Croat, T.B. 1971. Summit Garden, Canal Zone. Taxon 20: 769–772.

———. 1972. The Missouri Botanical Garden in Panama. Missouri Bot. Gard. Bull. 60: 19–22.

———. 1978. Flora of Barro Colorado Island. Stanford University Press.

———. & M.R. Crosby. 1974. Systematics program at the Missouri Botanical Garden. Taxon 23: 444–446.

Dwyer, J.D. 1964. Panama, plant collection and the Missouri Botanical Garden. Ann. Missouri Bot. Gard. 51: 109–117.

———. 1967. A new herbarium in the Canal Zone. Taxon 16: 158–159.

———. 1968. A list of localities botanized in Panama. Phytologia 16: 467–486.

———. 1973. Henri Pittier's botanical activity in Panama. Taxon 22: 557–576.

———. 1980. Rubiaceae *In* R.E. Woodson et al., Flora of Panama. Ann Missouri Bot. Gard. 67: 1–522.

Johnston, I.M. 1949. The botany of San José Island. Sargentia 8: 1–306.

Seemann, B.C. 1854. The Botany of the H.M.S. Herald (Flora of the Isthmus of Panama). London.

Standley, P.C. 1927. The flora of Barro Colorado Island. Smithsonian Misc. Collect. 78: 1–32.

———. 1928. Flora of the Panama Canal Zone. Contr. U.S. Natl. Herb. 27: 1–416.

Stern, W.L. 1960. Plant collecting in the land of Balboa. Gard. J. New York Bot. Gard. 10: 44–48, 64.

Estado Actual del Conocimiento de la Vegetación y la Flora del Chocó

Enrique Forero

With one of the richest floras in the world, there have been few plant collections from the Choco. The province has five main regions which correspond well with the seven Holdridge life zones represented. Al Gentry and I have conducted a plant survey in the Choco and have collected about 12,000 numbers. The largest families are the Rubiaceae and Melastomataceae. The largest genera are Piper and Anthurium.

Historia de las Exploraciones Botánicas

El Departamento del Chocó fue hasta hace pocos años una de las regiones menos conocidas del mundo desde el punto de vista botánico. Los cálculos adelantados por Schultes (1951) indicaban que en el Chocó se encuentra una buena parte de las 50.000 especies vegetales que posee Colombia. Sin embargo, expediciones realizadas por botánicos durante muchos años daban una idea muy limitada de su flora. Además de algunas colecciones de José Celestino Mutis y de José Jerónimo Triana en el siglo pasado, los datos disponibles para épocas más recientes se pueden resumir así: W.A. Archer visitó el Chocó hace más de 40 años, E.P. Killip realizó 3 expediciones al Chocó y coleccionó unos 1.000 ejemplares; Alvaro Fernández-Pérez, en una expedición a lo largo de la costa reunió unas 200 colecciones; Rafael Homero-Castañeda, en 3 expediciones coleccionó cerca de 500 números, principalmente en el área del Quibdó y a lo largo del río Atrato; Jesús M. Idrobo y James Duke coleccionaron 700 números en la región del Darién; H.P. Fuchs, Zanella y Torres, en el río Baudó, aproximadamente 700 números; Hernando García-Barriga trabajó en la región de Quidbó, y finalmente, David Lellinger y Elías de la Sota recorrieron el Departamento de sur a norte estudiando helechos, (Lellinger & de la Sota 1972) con un total de 1.000 colecciones.

Principal collectors in the Choco in past times were Mutis and Triana in the past century, and in this century Archer, Killip, Fernández-Pérez, Romero-Castañada, Idrobo and Duke, Fuchs, Zanella and Torres, García-Barriga, and Lellinger and de la Sota.

De las cifras anteriores se deduce que en un lapso de 40 años se reunieron solamente unas 4.000 colecciones entre helechos y fanerógamas. Numerosos botánicos que han colaborado activamente con la organización Flora Neotrópica en la preparación de monografías, habían expresado su preocupación por la carencia de información relativa de esta región. Prance (1977) consideró al Chocó como "una de las áreas más importantes y más interesantes de los trópicos para futuras colecciones." Gentry (1978) indicó que el Chocó era el área más pobremente conocida de toda la costa del Pacífica de América Tropical. En consecuencia, cualquier intento por establecer relaciones fitogeográficas entre ésta y otras regiones, lo mismo que la solución de numerosos problemas taxonómicos se tornaban notablemente difíciles.

Up until 1972 only about 4,000 collections had been made in the Chocó.

Fue con base en todas estas consideraciones que Al Gentry y yo resolvimos en 1973 afrontar el desafío de buscar una solución a esa situación y desde entonces hemos estado trabajando intensamente en la realización de un inventario de la Flora del Chocó, con financiación de la Universidad Nacional y de COLCIENCIAS en Colombia y del Missouri Botanical Garden y de la National Science Foundation en los Estados Unidos. Mi colega Roberto Jaramillo Wejía, del Instituto de Ciencias Naturales, ha colaborado decididamente en el trabajo de campo. En 19 expediciones hemos coleccionado más de 12.000 números. Además de los ejemplares depositados en MO y en COL hemos procurado distribuir duplicados para determinación a numerosos especialistas en todo el mundo, y en canje a instituciones colombianas y de otros países latinoamericanos, de los Estados Unidos y de Europa. Cuando pensamos que 12.000 números de colecciones se traducen en cerca de 50.000 ejemplares de herbario si se tienen en cuenta los 2, 3, 4 y hasta 10 duplicados de cada uno, podemos tener una idea de la magnitud de la obra. En solo 3 meses de 1979—enero, febrero y marzo—reunimos 3.300 colecciones, y en la última expedición realizada hace escasamente un mes, el aporte fue de 1.128 nuevas colecciones. Esto quiere decir que en 4 meses igualamos el número de colecciones realizadas en 40 años, y en 7 años (con algunas considerables interrupciones) hemos triplicado la cantidad de información disponible, en forma de ejemplares de herbario, sobre la Flora del Chocó.

In 1973 Al Gentry and I began a program to explore the Choco funded by the Universidad Nacional with COLCIENCIAS of Colombia and Missouri Botanical Garden with NSF of the United States. So far some 12.000 numbers have been collected yielding about 50,000 duplicates.

La investigación nos ha llevado prácticamente a todos los rincones del Departamento, desde Acandí sobre la costa del Caribe hasta la región del Río Calima y la desembocadura del Río San Juan en el Océano Pacífico; hemos coleccionado en el Río Tolo, el Cerro Tacarcuna, Arquía, Unguía, en el Parque Nacional de los Catíos, en el Río Truandó, Río Sucio, Pavarandó, Bahía Solano, Alto del Buey, Río Munguidó, en la región de Quibdó, Carretera Quibdó-Medellín, Itsmina, Lloró, Tadó, Carretera Panamericana, Andagoya, Río San Juan, Río Tamaná, área de San José del Palmar, Cerro Torrá, Noanamá, Río Fujiadó, Palestina, Docordó, y en las bocas del San Juan. Así mismo, las 19 expediciones nos han permitido cubrir los 12 meses del año.

Colombia estaba sufriendo en los últimos años la carencia de una nueva generación de jóvenes entusiastas por la "ciencia amable." En la investigación botánica del Chocó han participado activamente 8 estudiantes de botánica de la Universidad Nacional, y dos de ellos, Luis Eduardo Forero y Henry León están concluyendo actualmente sus trabajos de tesis sobre temas importantes. Luis Eduardo Forero permaneció varios meses con la comunidad Cuna de Arquía, en el norte, y con una comunidad Waunana del Río Pichimá, en el sur. Su estudio etnobotánico es uno de los poquísimos efectuados en el País. Henry León está preparando la Flora fanerogámica del Parque Nacional Los Catíos, situado entre el Río Atrato y la Frontera con Panamá.

Finalmente, los especialistas han comenzado a encontrar respuestas a muchos interrogantes, y las relaciones fitogeográficas se han comenzado a esclarecer. Thomas Croat había reconocido hasta hace un año cerca de 25 especies nuevas de *Anthurium* y varias más en otros géneros, Jim Luteyn ya está publicando descripciones de varias Ericaceae nuevas, John Wurdack ha encontrado un buen número de Melastomatáceas hasta ahora desconocidas para la ciencia, lo mismo que Lundell en Myrsinaceae; Maas, por otra parte, ha podido aclarar dudas taxonómicas, y así sucesivamente.

At last specialists have begun to find answers to long standing questions. Investigators include Thomas Croat (Anthurium), Jim Luteyn (Ericaceae), John Wurdack (Melastomataceae), Cyrus Lundell (Myrsinaceae), and Paul Maas (Monocotyledons).

Descripción del Area

El Departamento del Chocó tiene una extensión de 47.205 km^2, y su población actual es de 210.000 habitantes con una densidad de 4 habitantes por km^2. El Departamento se entiende entre los 4° 10′N y los 8° 10′N y entre los 76° y 78° de longitud Oeste del meridiano de Greenwich. Constituye la esquina noroccidental de América del Sur y es la única sección del país y del sub-continente con costas sobre los océanos Atlántico y Pacífico.

En la superficie del Departamento se reconocen las siguientes regiones principales:

1. Las tierras bajas que se extienden de sur a norte, compuestas por los valles de los Ríos Atrato y San Juan; el valle del Atrato presenta una amplia zona cenagosa en el norte.
2. La planicie costera que es muy angosta—entre 5 y 25 km de ancho—e incluye las regiones del bajo Baudó y del bajo San Juan.
3. A ambos lados de los Ríos San Juan y Atrato existen extensiones considerables de pequeños montículos cuyas alturas oscilan entre 100 y 300 m, es, decir que la llanura central no es completamente plana.
4. Las tierras altas, que están representadas al oriente por las estribaciones de la Cordillera

)ccidental y por grandes masas montañosas más o ıenos aisladas tales como el cerro Torrá en la hoya lel San Juan; al occidente las tierras altas están repreentadas por la serranía de Baudó y por la serranía del)arién. El punto más alto en la primera es el Alto del 3uey con cerca de 1.800 m y en la segunda el cerro 'acarcuna con algo más de 1.900 m, este último en la rontera Colombo-Panameña.

5. Una característica importante del departamento a constituyen las hoyas de los Ríos Baudó, San Juan 180 km de longitud) y Atrato (300 km de longitud).)esde el punto de vista geológico, y de acuerdo con 'an der Hammen (1961) y Haffer (1970), gran parte de as tierras bajas del Chocó se elevaron por encima del ıivel del mar hace unos 2.000.000 de años, durante el 'lioceno superior y el comienzo del Pleistoceno, es lecir paralelamente con el levantamiento definitivo le los 3 ramales colombianos de los Andes y de otras ıreas montañosas, y con el establecimiento definitivo le la conexión entre América del Sur y América del Norte.

The Department of Chocó has five main regions: 1) lowlands ranging from south to north formed by the Atrato and San Juan river valleys. 2) a narrow coastal plain. 3) a central plain with hills (100–300 m) at the sides of the Atrato and San Juan valleys. 4) highlands of the eastern Cordillera. 5) valleys of the rivers Baudo, San Juan and Atrato.

Una de las características más sobresalientes del Departamento es su pluviosidad, ya que su parte central es considerada como probablemente el lugar más húmedo del mundo. West (1957) indica que la estación de Debundscha en el Monte Camerún (Nigeria), a de Cherrapungi en Khasi Hills, India y la del Monte Waialeale en las islas Kauai en Hawai comparten esa característica, con promedios anuales de).850, 10.500 y 11.400 mm respectivamente. Gentry (1982) cita los dos últimos con promedios de 11.430 y 11.684 mm respectivamente y los compara con una cifra de 11.770 mm para Tutunendo en el Chocó.

The province is perhaps the rainiest area in the world with about 11,770 mm (464 inches) per year at Tutunendo.

Los valores máximos reportados para el Chocó aparecen en West (1957), quien da 15.058 mm como total anual en 1939 en Quibdó, y 13.108 mm en un promedio de 3 años en la estación de Lloró (al sur de Quibdó, sobre el Río Atrato). El mismo autor señala el promedio para Quibdó como 10.545 mm por año. Bustos & Ramírez (1977) ofrecen valores medios multianuales de precipitación para el período 1951–1973 según los cuales el área mas lluviosa se localiza en La Vuelta, al sur de Quibdó y sobre el Río Andágueda (posiblemente la misma estación que West llama "Lloró"), con 9.100 mm de promedio anual después de 31 años con datos directos. Le siguen Cértegui, un poco al sur, sobre la carretera Quibdó-Itsmina, y Pie de Pato, con 8.300 mm anuales.

Existe, pues, un enclave de altísima pluviosidad en la región de Quibdó. La misma ciudad capital del Departamento, Quibdó, recibe entre 7.200 y 7.600 mm anuales. Hacia el norte, a lo largo del Río Atrato, la abundancia de las lluvias decrece gradualmente. Así en Tagachi, una localidad sobre el Río Atrato, no muy lejos de Quibdó, el promedio anual desciende a 6.500 mm (una diferencia aproximada de 1.000 mm); en Riosucio caen 3.700 mm, en Sautatá 3.200 y en Turbo, sobre el golfo de Urabá, 2.400 mm o menos. Hacia el sur, en cambio, las condiciones no varían considerablemente en el Valle del San Juan, formando una franja continua con promedios anuales entre 6.000 y 7.500 mm, la cual se extiende aún hasta el departamento del Valle, en el puerto de Buenaventura.

Al norte de Río Sucio pueden reconocerse dos períodos durante el año: uno seco desde diciembre hasta marzo, y uno húmedo el resto del año. Estos mismos períodos son levemente discernibles en el bajo San Juan, en las localidades de Palestina y la Misión. En el resto del Departamento es prácticamente imposible reconocer épocas "secas" de épocas "lluviosas", ya que llueve prácticamente todo el año en forma más o menos constante.

North of the River Sucio there is a dry season from December to March and it is wet the rest of the year.

Ya en las estribaciones de la Cordillera Occidental, a medida que se asciende disminuye la precipitación. En el área del Río Torito (cerca a San José del Palmar) localizada a 700 m se ha registrado un promedio de 5.000 mm anuales en un lapso de 4 años; en El Carmen de Atrato, a cerca de 1.800 m, el promedio decrece a 2.700 mm anuales.

Cloud cover restricts the temperature at Quibdo to a mean of 28° C (82° F).

La temperatura media mensual raramente excede los 28° C (West 1957), con máximas de 41° C y mínimas de 15° C en Quibdó. Debido a la presencia de nubes la temperatura no alcanza niveles excesivamente altos. En algunos lugares, como por ejemplo en el límite entre los departamentos del Valle y Chocó, al sur de San José del Palmar, a 2.000 m, la neblina y la lluvia son constantes y la temperatura desciende considerablemente, hasta unos 10°-12° C aún en las horas del medio día.

Vegetación y Flora

Las especies vegetales de los trópicos por lo general tienen distribución geográfica amplia, pero su distribución ecológica es restringida y sus límites ecológicos están muy bien correlacionados con las zonas de vida del sistema Holdridge (Gentry 1977, 1978, 1982). Si se tiene en cuenta que cada zona de vida tiene un grupo de especies que le es característico, entonces entre más zonas de vida se presenten en una región dada, mayor será su número global de especies. La comparación que Gentry estableció ilustra muy bien este punto. Guatemala, Costa Rica y Panamá tienen 12 zonas de vida cada una y sus floras se calcula que cuentan con 8.000–9.000 especies. Nicaragua entre tanto, presenta 9 zonas de vida y una flora de cerca de 5.000 especies al igual que Honduras con 8 zonas de vida; finalmente El Salvador y Belize tienen 4 zonas de vida y floras que pueden estar entre 2.500 y 3.000 especies para cada país.

In the Choco the distribution of plants is well correlated with the Holdridge life zones.

En el Chocó, de acuerdo con la más reciente edición del mapa Ecológico de Colombia (Inst. Geogr. 1977), ocurren 8 zonas de vida. Si se tienen en cuenta los cálculos anteriores su flora debería contar con unas 5000 especies. No obstante, la presencia de áreas de gran pluviosidad mucho más ricas en especies que otras regiones, sumada a otros factores, permite suponer que en el Chocó pueden existir más de 8.000 especies. Las zonas de vida presentes en el Departamento del Chocó son las siguientes:

1. El bosque pluvial tropical (bp-T) que se localiza hacia el centro del Departamento, entre 4°45' y 6°20'.

2. El bosque pluvial pre-montano (bp-PM) cubre quizá el área más extensa dentro del Departamento, y se extiende a lo largo de la Serranía del Darién y algunas otras áreas más o menos aisladas sobre la frontera Colombo-Panameña, a lo largo del curso medio del Río Atrato desde Quibdó hacia el norte, en parte de la Serranía del Baudó,a lo largo de la vertiente occidental de la cordillera occidental, en el bajo Baudó, y cubre casi todo el valle del San Juan desde Andagoya hasta la desembocadura en el Pacífico. Esta zona de vida se continúa en el Departamento del Valle.

3. Bosque muy húmedo tropical (bmh-T). Se extiende a lo largo de la costa del Pacífico y de la frontera Colombo-Panameña; forma además una franja en dirección este-oeste más o menos a la altura del Río Truandó la cual se continúa en Antioquia. Algunos enclaves de bosque muy húmedo tropical se encuentran en la región sur-oriental del departamento, en el Río Hábita y en la parte alta del Río Sipi.

4. Bosque muy húmedo pre-montano (bmh-PM), ocurre en una franja angosta en el norte y en un área restringida en los alrededores de El Carmen de Atrato.

5. El bosque tropical (bh-T), cubre la parte baja y cenagosa del Río Atrato y se extiende a lo largo de la costa alrededor del Golfo de Urabá; un pequeño enclave se presenta sobre la costa del Pacífico en la frontera con Panamá.

6. El bosque húmedo pluvial montano bajo (bp-MB) se presenta a lo largo del flanco occidental de la Cordillera Oriental al sur del Carmen de Atrato y en los alrededores hacía el sur de San José del Palmar.

7. El bosque muy húmedo montano bajo (bmh-MB) ocurre en angostas franjas a lo largo de los límites entre el Chocó y Antioquia arriba de El Carmen de Atrato, y arriba de San José del Palmar a lo largo del límite entre el Chocó y los departamentos del Valle y Risaralda en las estribaciones del Cerro Tatamá.

8. Bosque pluvial montano (bp-M), del cual se presentan dos enclaves en el sur oriente del Departamento, uno al norte de San José del Palmar y el otro en el Cerro Torrá, al oeste de San José del Palmar.

The following life zones are represented: 1) Tropical Rain Forest in the center of the Department. 2) Pre-montane Rain Forest is the most extensive. 3) Tropical Wet Forest occurs along the Pacific coast and the Panamanian border. 4) Pre-Montane Wet Forest forms a narrow strip in the north. 5) Tropical Moist Forest covers the low swampy area of the Atrato. 6) Lower Montane Rain Forest occurs along the flanks of the eastern cordillera. 7) Lower Montane Wet Forest occurs in narrow belts along the Chocó-Antioquia border above El Carmen de Atrato. 8) Wet Montane Forest occurs in two patches in the northwest and west of the Department.

La suma de las colecciones provenientes del Departamento del Chocó, al incluir tanto las obtenidas dentro del estudio de Forero y Gentry como las de olectores anteriores, alcanza a un total de 16.000 úmeros. Desafortunadamente, en el momento de reparar esta contribución solo se tienen determinaciones de unas 8.000. Con esta base, se pueden tar para el Chocó 147 familias de angiospermas que epresentan aproximadamente el 75% de todas las milias presentes en el territorio colombiano. Es interesante anotar que Prance (1978) citó 161 familias ara Amazonia, con un área mucho mayor. En Panamá existen 191 familias. Hay además en el Chocó 3 milias de Gimnospermas, 6(?) familias de helechos, ıás Lycopodiaceae, Equisetaceae y Selaginellaceae. lo existe siquiera un dato aproximado del número de milias, géneros o especies de los grupos inferiores, unque recientemente K.P. Dumont y María Margara Pulido han coleccionado hongos en algunas áreas omo San José del Palmar, la carretera Panamericana, Quibdó y la carretera Quibdó-Medellín. Con 950 éneros reconocidos hasta hoy, el número de especies e angiospermas alcanza a 2.600; a éstas se agregan 53 éneros y cerca de 200 especies de helechos y otros rupos para un gran total de 160 families, 1.003 éneros y 2.800 especies.

Even with only half the collections available identified, the Flora of the Chocó includes 950 genera and 2600 species of angiosperms plus 53 genera and 200 species of ferns and fern allies or a total of 160 families, 1,003 genera and 2,800 species of vascular plants.

Gentry calcula que la flora del Chocó puede ser imilar en tamaño a la de Panamá, quizá con más de 3.000 especies. Un dato interesante que sirve para lustrar la riqueza de la flora chocoana es el siguiente: en 50 hectáreas de terreno localizadas entre 630 y 850 n sobre el nivel del mar cerca de San José del Palmar se coleccionaron 85 familias de angiospermas o sea alrededor del 60% de las conocidas para el Chocó.

Las 10 familias mejor representadas hasta ahora en el Departamento son las siguientes: *Rubiaceae*, con 205 especies en 62 géneros; *Melastomataceae*, con 160 especies en 26 géneros; *Leguminosae* (131/53); *Piperaceae* (126/5); *Compositae* (87/49); *Araceae* (72/10); *Solanaceae* (66/13); *Gesneriaceae* (62/15); *Bignoniaceae* (61/33), y *Orchidaceae* (59/37). Estas familias representan el 37% del total de la flora. Es de anotar que las Orquídeas constituyen uno de los grupos en los cuales se ha determinado quizá menos del 50% de los ejemplares coleccionados; su posición entre las 10 mayores familias, aún en estas condiciones, da una idea de la riqueza de la flora orquideológica del Chocó; sin embargo, también es posible que las orquídeas no sean tan diversas como en algunas otras regiones, pues la alta pluviosidad puede hacer difícil su establecimiento.

The ten best represented families are the Rubiaceae with 205 species in 62 genera, the Melastomataceae (160/26), Leguminosae (131/53), Piperaceae (126/15), Compositae (87/49), Araceae (72/10), Solanaceae (66/13), Gesneriaceae (62/15), Bignoniaceae (61/33), and Orchidaceae (59/37).

Puesto que no se cuenta aún con una lista completa de las especies no es posible todavía caracterizar los diversos tipos de vegetación empleando las especies dominantes en cada caso. En consecuencia, debo limitarme a ofrecer un resumen de los géneros y especies más comunes en el Departamento, basándome en los datos obtenidos hasta ahora.

El género *Piper* (Piperaceae) presenta el mayor número de especies registradas hasta hoy, 68, de las cuales *Piper aduncum* L., *P. augustum* Rudge y *P. brachypodon* (Benth.) C.DC. son comunes en todo el Departamento. *Peperomia*, por otra parte, cuenta con 37 especies; *P. ciliaris* C.DC. es una especie común en la región central del Chocó, en áreas relativamente cercanas a Quibdó, no habiendo sido coleccionada hasta ahora en el norte ni en el sur del Departamento. El género *Trianopiper* es endémico de la región fitogeográfica del Chocó; de 17 especies, 1 alcanza la parte oriental de Panamá y varias llegan al noroeste del Ecuador (Trelease & Yuncker 1950; Gentry 1982). *Anthurium* (Araceae) está representado por 54 especies de las cuales cerca de 25 parecen ser nuevas para la ciencia.

Piper is the largest genus with 68 species. Anthurium is represented by 54, of which about 25 are new to science.

Del género *Psychotria* (Rubiaceae) también existen 54 especies, siendo *P. acuminata* Benth. y *P. macrophylla* R. & P. las de más amplia distribución en el Departamento. Otros géneros representados en la flora son *Cephaelis* (14 especies), *Faramea* (10 especies) y *Palicourea* (10 especies).

Dentro de las Melastomatáceas, el género *Miconia* aparece con 49 especies, de las cuales *Miconia nervosa* (Sm.) Tr. ocurre prácticamente en todo el territorio chocoano. También *Clidemia*, con 29 especies, es un componente importante de la flora. Otras Melastomatáceas de amplia distribución son *Tococa acuminata* Benth. y *Tibouchina longifolia* (Vahl) Baill.

Solanum arboreum Dun. es la especie de más amplia distribución dentro de las 35 conocidas para el género en el Chocó. *Witheringia solanacea* L'Her. también se encuentra en todo el territorio; *Cyphomandra hartwegii* (Miers) Dun. ha sido coleccionada desde Quibdó hasta el parque de los Catíos, en el norte.

Cavendishia (Ericaceae) ocupa el octavo lugar con 25 especies, algunas de las cuales han sido reconocidas por Luteyn como nuevas para la ciencia.

De *Calathea* (Marantaceae) se han coleccionado 22 especies; la de más amplia distribución es *Calathea marantifolia* Standley, mientras *C. elegans* Kennedy se conoce únicamente de la región de Quibdó y el Alto del Buey. El género *Heliconia* (Musaceae) cuenta hasta el presente con 20 especies.

La familia Gesneriaceae está representada entre otras por *Alloplectus panamensis* Morton, cuya distribución abarca desde Quibdó hacia el sur hasta las bocas del Río San Juan; *Besleria barclayi* L. Skog en el Chocó Central (area Quibdó-Bahia Solano), *Chrysothemis friedrichstaliana* (Hanst.) Moore de amplia distribución en el departamento, y *Columnea parviflora* Morton en el centro y sur. *Cremosperma* es citado por Gentry (1982) como un género prácticamente endémico del Chocó, pues de las 18 especies que lo componen 15 ocurren allí, 1 en el Valle del Magdalena y 2 en Panamá.

A lo largo de los ríos es llamativa *Pentaclethra macroloba* (Willd.) Kuntze (Leguminosae-Mimosoideae). La familia Leguminosae está representada además por 14 especies de *Inga*, por lo menos 7 de *Desmodium*, 9 de *Cassia*, varias *Pithecelobium*, etc.

De las compuestas se encuentran comúnmente *Clibadium asperum* DC. y otras malezas. Otro género importante es *Mikania*, que incluye 7 especies.

En Bignoniaceae la especie mejor representada en nuestras colecciones es *Jacaranda hesperia* Dugand, cuya distribución hasta ahora aparece en el centro y el sur del Departamento. Otros géneros importantes son *Arrabidaea* con 7 especies, *Schlegelia* con 6 y *Tabebuia* con 5 especies.

Schlegelia es un género de distribución amplia pero con gran concentración en el Chocó, pues 9 de sus 18 especies ocurren allí, mientras que las restantes están distribuídas así: 5 en Amazonia, 3 en las islas del Caribe, 2 en las Guayanas, una en el sur-este del Brasil y 3 en Centro América (Gentry 1979a, 1979b).

Aunque las Annonáceas no aparecen como una de las familias más comunes en el Chocó, poco a poco hemos ido reuniendo numerosas colecciones. *Guatteria cargadero*, por ejemplo, se ha coleccionado repetidamente, principalmente en las regiones centro y norte. Se conocen hasta ahora 13 géneros y 36 especies de Annonaceae.

Dentro de las Apocynaceae, una familia que presenta 22 géneros y 40 especies, son comunes *Himatanthus articulata* (Vahl) Woods. en el centro y norte, mientras *Prestonia portobellensis* (Beurl.) Woods. y *Stemmadenia grandiflora* están ampliamente distribuidas en todo el Departamento.

Entre las familias con especies arbóreas no mencionadas aún sobresalen las Moraceae con 18 géneros y 44 especies de las cuales 18 pertenecen a *Ficus; las Bombacaceae con 11 géneros y 20 especies, 8 de las cuales son Quararibea; las Boraginaceae tienen 3 géneros y 15 especies; Cordia panamensis* Riley es común; las Guttiferae incluyen 11 géneros y 24 especies; de *Clusia* se conocen 7.

The Moraceae has 44 species in 18 genera, the Bombacaceae has 20 species in 11 genera, and the Guttiferae have 24 species in 11 genera.

Existen 25 géneros y 41 especies de palmas; de *Geonoma* ocurren 9 especies en el Chocó.

También dentro de las Monocotiledóneas, las Zingiberaceae ofrecen interés particular. Aunque hasta el presente sólo se conocen 23 especies en 4 géneros, quedé sorprendido ante su proliferación en el area cercana a San José del Palmar que estudié recientemente. Es este el lugar donde he encontrado mayor concentración de especies de *Renealmia* (calculo 5 especies), y de individuos reunidos en grupos. En general el area visitada aparece muy rica en familias que no son muy comunes en otras regiones del Chocó, tales como Zingiberaceae, Musaceae, Marantaceae y Araliaceae.

Las expresiones de asombro de los especialistas al estudiar las recientes colecciones provenientes del Chocó, debido al gran número de especies que les son desconocidas, indican que el área debe ser una de las de mayor endemismo en el mundo; es posible que un 20% de las especies sean endémicas.

Gentry and I expect to publish an annotated list of the plants of the Choco in the near future.

El trabajo que Al Gentry y yo hemos venido realizando no ha concluído. Nuestra esperanza es la de publicar a la mayor brevedad posible una "Lista anotada de las plantas del Chocó" que servirá de base para estudios posteriores de profundización y permitirá establecer comparaciones con floras de regiones vecinas como la de Panamá, para responder definitivamente a numerosos interrogantes fitogeográficos.

Bustos B., N. & A. *Ramírez* R. 1977. Estudio preliminar de precipitaciones y caudales medios en la vertiente del Pacífico. Tesis. Univ. Jorge Tadeo Lozano. Bogotá.

Forero P., L.E. 1980. Etnobotánica de las comunidades indígenas Cuna y Waunana (Chocó, Colombia). Cespedisia 9(33–34):115–301.

Gentry, A. 1977. Endangered plant species and habitats of Ecuador and Amazonian Peru. Pp. 136–149. *En*: G.T. Prance, ed., Extinction is Forever. The Status of Threatened and Endangered Plants of the Americas. New York Botanical Garden.

———. 1979. Distribution patterns of neotropical Bignoniaceae: some phytogeographical implications. Pp. 339–354 *En*: K. Larsen & L. Holm-Nielsen, eds., Tropical Botany. Academic Press, London.

———. 1978. Floristic knowledge and needs in Pacific Tropical America. Brittonia 30: 134–153.

———. 1982. Phytogeographic patterns as evidence for a Chocó refugium. Pp. 112–136. *En*: G.T. Prance, ed., Biological Diversification in the Tropics. Columbia Univ. Press. New York.

Haffer, J. 1970. Geologic-climatic history and zoogeographic significance of the Urabá region in northwestern Colombia. Caldasia 10: 603–636.

Instituto Geográfico Agustin Codazzi. 1977. Zonas de vida o formaciones vegetales de Colombia. Memoria explicativa sobre el mapa ecológico. Bogotá.

Lellinger, D. & E. *de la Sota*. 1972. Collecting ferns in the Choco. Amer. Fern Journ. 62: 1–8.

Prance, G.T. 1977. Floristic inventory of the tropics: where do we stand? Ann. Missouri Bot. Gard. 64: 659–684.

———. 1978. The origin and evolution of the Amazon flora. Interciencia 3: 207–222.

Schultes, R.E. 1951. La riqueza de la flora colombiana. Rev. Acad. Colomb. Cienc. Ex. Fis. Nat. 8: 230–242.

Trelease, W. & T.G. *Yuncker*. 1950. The Piperaceae of Northern South America. Univ. Illinois Press, Urbana.

van der Hammen, T. 1961. Late cretaceous and tertiary stratigraphy and tectogenesis of the Colombian Andes. Geologie & Mijnb. 40: 181–188.

West, R.C. 1957. The Pacific lowlands of Colombia. Louisiana State Univ. Sci. Ser. No. 8: 1–278.

Anatomy of *Fuchsia* section *Jimenezia* from the Flora of Panama

Richard C. **Keating**

En Panamá, la Onagraceae está representada por sólo cuatro de sus 17 géneros, los cuales son más característicos de las regiones templadas. Una especie de Onagraceae recién descrita de Panamá y Costa Rica es inesperadamente primitiva en una familia que tiene su mejor representación en la zona templada. Los dos especímenes disponibles tenían una combinación de características primitivas tales como rafidios compactos, células que los contienen con membranas acromáticas, venas medias bien desarrolladas, fibras y un diente con hidatodo tipo rosoide. Los hidatodos pueden haber evolucionado para efectuar funciones pesticidas. Esta combinación de características generales presentes en Fuchsia jimenezii del sur de América Central es inesperada y quizás tiene importantes implicaciones.

The Onagraceae, a flowering plant family in the order Myrtales, is well represented in the Americas and in high elevations and latitudes worldwide. Of the 17 genera only *Oenothera*, *Lopezia*, *Fuchsia*, and *Ludwigia* are recorded from Panama. In terms of comparative vegetative anatomy, the Panamanian species of *Oenothera* are quite specialized and one must look to western North America for the more generalized members of that group. *Lopezia* is restricted to Mexico and Central America while *Fuchsia* and *Ludwigia* both are centered in South America, the probable center of origin of the family (Boufford et al. 1978). Using morphological and floral anatomical criteria, *Fuchsia* and *Ludwigia* are clearly the most generalized or primitive members of the family (Carlquist 1975), but only recently has there been any suggestion that the most generalized species of the family might be found in Central America. Breedlove et al. (1982) recently described *Fuchsia jimenezii*, known only from Costa Rica and Panama, and placed it in the monotypic sect. *Jimenezia*. On the basis of recently completed comparative studies of the vegetative anatomy of the Onagraceae (Keating 1982), the leaves of this species appear to have the most generalized structure of any of the 650 species now included in the Onagraceae. *Fuchsia jimenezii* has a number of features in common with *F. arborescens* Sims of sect. *Schufia* and, in fact, was included in that species by Munz in the Flora of Panama (1959). The two species have several features of leaf histology that are striking for Onagraceae. The large midribs are deeply rounded abaxially and contain deep horseshoe-shaped midveins. The abaxial phloem is surrounded by fibers which in the Onagraceae are reported only in these species of *Fuchsia*, a few *Ludwigia* species, and in both species of the arborescent genus *Hauya* (Keating 1982).

Cuatro géneros de Onagraceae ocurren en Panamá, Oenothera, Lopezia, Fuchsia, y Ludwigia. Fuchsia y Ludwigia están centradas en América del Sur, mientras que Lopezia está restringida a México y América Central y Oenothera está distribuida tanto en América del Norte, como en América del Sur.

Fuchsia jimenezii

Two specimens were available for study. The first, *Wilbur and Luteyn 18566* Costa Rica, (MO) from Heredia, has within the center of the midvein a number of patches of intraxylary phloem capped with large irregular patches of fibers. In all other species of Onagraceae, adaxial patches of fibers are totally absent. The other specimen, *Haber et al. 434* (MO), Costa Rica, from Monteverde, lacks the internal fibers but does contain the internal phloem common in

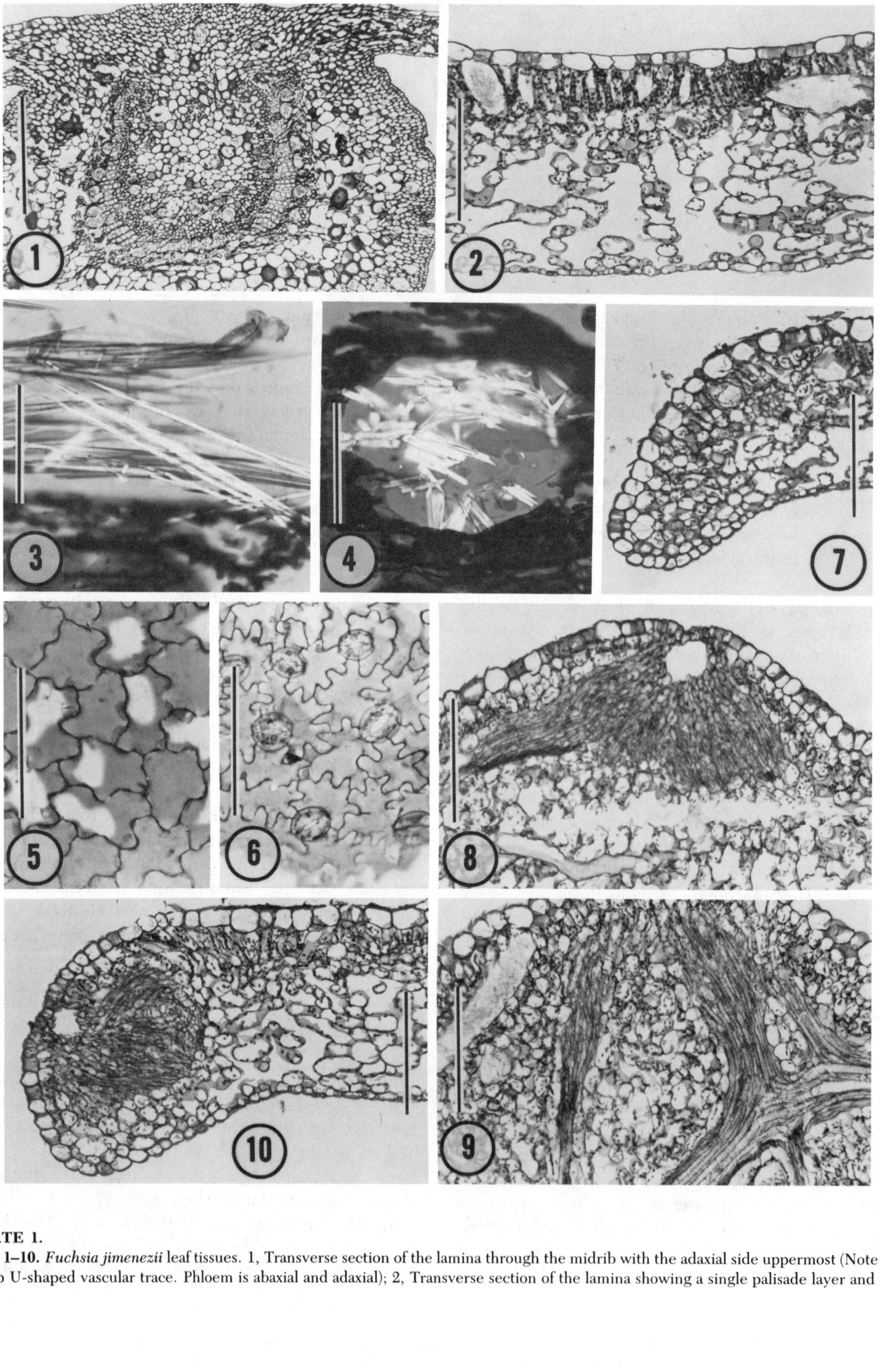

PLATE 1.

Fig. 1–10. *Fuchsia jimenezii* leaf tissues. 1, Transverse section of the lamina through the midrib with the adaxial side uppermost (Note the deep U-shaped vascular trace. Phloem is abaxial and adaxial); 2, Transverse section of the lamina showing a single palisade layer and two

Fuchsia species (Fig. 1). The leaves examined are dorsiventral with one layer of palisade cells (Fig. 2), abaxial stomata and large idioblastic cells containing either fine (Fig. 3) or coarse (Fig. 4) raphides. Epidermal cells seen paradermally have undulate margins adaxially (Fig. 5) and are deeply sinuous abaxially (Fig. 6).

The leaf margins in *F. jimenezii*, normally slightly recurved (Fig. 7), show the best developed hydathodal teeth yet described for the family. The conspicuous "glandular teeth" were noted by Breedlove et al. (1982). Anatomically, the glandular tooth is a hydathode comprised of a round lacuna opening to the outside surface of the leaf by one or more stomata (Fig. 8). Surrounding the lacuna internally is a glandular epithem which is in turn connected to a large plexus of vascular tissue. As is typical of this tooth type which Hickey (1980) has called the rosoid tooth, the plexus is resolved into three to four vascular bundles (Fig. 9) which connect to a marginal vein as in most leaves of the Onagraceae (Hickey 1981). In cross section, the hydathodal lacuna is not centered on the edge of the leaf margin, but is submarginal on the adaxial surface (Fig. 10). Also, as seen in both cross and paradermal sections, the hydathode has a circular lacuna unlike those of in other Onagraceae examined (Keating 1982). In other species of *Fuchsia*, as well as in species of *Epilobium* and *Calylophus* where these teeth are also well documented (Keating 1982), the glandular epithem is smaller; the hydathodal lacunae are more irregular in shape and marginal.

Las hojas de Fuchsia jimenezii tienen bien desarrollados los dientes con hidatodos de tipo rosoide en donde tres o cuatro fascículos vasculares están conectados a un plexo y a la vena marginal.

Discussion

On the basis of correlations with characters from wood anatomy, floral morphology, pollen morphology, flavonoid chemistry, and geographic information, Keating (1982) concluded that leaves of Onagraceae have undergone a general reduction in structural complexity and that this reduction has proceeded independently in most genera.

En muchos géneros de Onagraceae las hojas han sufrido una reducción general en su complejidad.

On the basis of several leaf anatomical criteria (Keating 1982), we conclude that *Fuchsia jimenezii* has probably the most generalized constellation of characters reported for the family. These include the presence of coarse raphides as well as raphide-bearing cells without stainable membranes or deposits of amorphous substances. The well-developed midveins and presence of fibers illustrate an elaborate, and therefore probably generalized, pattern. At present, interpretation of the hydathodal tooth is more problematical. This "rosoid" tooth structure has been reported from only three orders; in the woody Saxifragales, it has been reported in four genera (Stern 1974, 1978; Styer & Stern 1979a, 1979b). Hickey (1980, 1981) reported it in several genera of Onagraceae and in the Lythraceae, and Lersten (pers. comm.) noted the same structure in the leaf tip of *Punica granatum* (Punicaceae). In each of these cases the structure of the hydathode with its converging veins, position of the epithem and lacuna, looks convincingly monophyletic with the other known examples.

As a working hypothesis, the rosoid hydathodal tooth might be used as a marker for one phyletic series extending from the basal stocks of the Saxifragales, the Rosales and the Myrtales. This is compatible with the recent classifications of Cronquist (1981), Dahlgren (1975), Takhtajan (1980), and Thorne (1976). It also links the Onagraceae firmly to the Myrtales, although the family on all available evidence remains among the most isolated within the order.

The rosoid teeth of the woody Saxifragales and those previously reported in the Onagraceae (Keating 1982) are all quite similar in the irregular shape of the

raphide idioblasts in the palisade area; 3, Fine raphids in the mesophyll under polarized light; 4, Coarse raphids cut transversely under polarized light; 5, Paradermal view of adaxial epidermal cells showing undulating anticlinal walls; 6, Paradermal view of abaxial epidermal cells showing deeply sinuous anticlinal walls and anomocytic stomata; 7, Leaf margin transverse section showing slightly recurved, non-glandular portion of the margin; 8, Paradermal view of a leaf tooth showing hydathodal foramen and extensive subtending epithem tissue; 9, Paradermal section adaxial and slightly proximal to the hydathodeal foramen in Fig. 8 showing three converging veins; 10, Transverse section of lamina showing spherical foramen and subtending epithem. Scale lines: Fig. 1 = 0.5 mm; Figs. 2, 7, 8, 9, 10 = 0.2 mm; Figs. 3, 4 = 0.05 mm; Figs. 5, 6 = 0.10 mm.

lacunae, the presence of three veins leading to the tooth and in the generally triangular shape of the epithem which abuts the lacuna proximally in a flat boundary. Structural elaborations found in *F. jimenezii* include the four veins leading to an extensive epithem into which the nearly spherical lacuna is proximally half embedded. The leaf tip described by Lersten (pers. comm.) for *Punica granatum* has an enlarged epidermal area over the lacuna with numerous stomata visible suggesting an obvious hydathodal function in the nearly mature leaf.

El diente tipo rosoide podría ser un marcador de una serie evolutiva a partir de Saxifragales, Rosales y Myrtales. A pesar de que la Onagraceae es una familia aislada, ésto también la relaciona con las Myrtales.

From a functional standpoint, the hydathodes of *F. jimenezii* may have elaborated in response to the need for guttation for secretions for pesticidal functions. Leaves of *Epilobium ciliatum*, described in relation to turion growth (Keating et al. 1982), showed completely developed tooth hydathodes on leaves 2 mm long. Other than this, no developmental studies of hydathodes in this family have been done. Recognition of a morphocline of the rosoid tooth in the Saxifragales-Rosales-Myrtales must await a screening of genera and species.

Studies by Breedlove et al. (1981), including biogeography and floral biology, concluded that sect. *Jimenezia* is the sister group of the generally more specialized sect. *Encliandra*. It appears to be more primitive that other *Fuchsia* sections on the basis of its scandent, sub-shrub habit, simple inflorescence structure and hermaphroditic flowers. The feature "glabrous throughout" is noted by Breedlove et al. to be remarkable for the genus but its status is difficult to evaluate at this time. The peculiar combination of characters discovered in *F. jimenezii* and its occurrence in southern Central America has implications that must be considered carefully in further studies of the genus.

Fuchsia jimenezii parece ser más primitiva que las otras Fuchsia con base a su hábito de subarbusto trepador, estructura de las inflorescencias simple y flores hermafroditas.

Boufford, D., P.H. *Raven* & J.E. *Averett*. 1978. Glycoflavones in *Circacea*. Biochem. Syst. Ecol. 6: 59–60.

Breedlove, D.E., P.E. *Berry* & P.H. *Raven*. 1982. The Mexican and Central American species of *Fuchsia* (Onagraceae) except for sect. *Encliandra*. Ann. Missouri Bot. Gard. 69: 209–234.

Carlquist, S. 1975. Wood anatomy of Onagraceae with notes on alternative modes of photosynthate movement in dicotyledon woods. Ann. Missouri Bot. Gard. 62: 386–424.

Cronquist, A. 1981. An integrated system of classification of flowering plants. Columbia Univ. Press, New York.

Dahlgren, R.M.T. 1980. A revised system of classification of angiosperms. Bot. J. Linn. Soc. London. 80: 91–124.

Hickey, L.J. 1980. Leaf architecture of Onagraceae. Abstracts. Second Internat. Cong. Syst. Evol. Biol.

———. 1981. Leaf architecture of the Myrtales. Abstracts. 13th Internat. Bot. Cong. Sydney, Australia.

Keating, R.C. 1982. The evolution and systematics of Onagraceae: leaf anatomy. Ann. Missouri Bot. Gard. 69: 770–803.

———, P.C. *Hoch* & P.H. *Raven*. 1982. Perennation in *Epilobium* (Onagraceae) and its relation to classification and ecology. Syst. Bot. 7: 379–404.

Munz, P.A. 1959. Onagraceae. *In*: R.W. Woodson & R.W. Schery, eds. Flora of Panama. Part. 7, Fasc. 4. Ann. Missouri Bot. Gard. 46: 195–221.

Stern, W.L. 1974. Comparative anatomy and systematics of woody Saxifragaceae. *Escallonia*. Bot. J. Linn. Soc. 68: 1–20.

———. 1978. Comparative anatomy and systematics of woody Saxifragaceae. *Hydrangea*. Bot. J. Linn. Soc. 76: 83– 113.

Styer, C.H. & W.L. *Stern*. 1979a. Comparative anatomy and systematics of woody Saxifragaceae. *Philadelphus*. Bot. J. Linn. Soc. 79: 267–289.

———. 1979b. Comparative anatomy and systematics of woody Saxifragaceae. *Deutzia*. Bot. J. Linn. Soc. 79: 291–319.

Takhtajan, A.L. 1980. Outline of the classification of flowering plants (Magnoliophyta). Bot. Rev. 46: 225–359.

Thorne, R. 1976. A phylogenetic classification of the Angiospermae. Evol. Biol. 9: 35–106.

Seed Size and Plant Habit in Seven Families From Panama and Costa Rica

Larry L. Rockwood

Las semillas de plantas distribuidas en 7 familias se midieron y pesaron. Los datos se analizaron en relación a formas de vida, elevación y zonas de vida. Los resultados se compararon con aquellos de la zona templada disponibles en la literatura. Los árboles presentaron semillas más grandes que las epífitas, los arbustos o las hierbas. Las semillas de las epífitas fueron mayores que las de los arbustos y las hierbas, mientras que las semillas de los últimos no demostraron diferencias significativas (estadísticamente). No se logró obtener medidas significativas que indicaran que el tamaño de las semillas disminuye en relación a la elevación, tal como ocurre en la zona templada.

The variety of life history patterns found in nature shows that reproductive success is usually the result of a series of compromises among conflicting selective pressures. In higher plants seed size is a good example. Since seeds are units of dormancy, dispersal and multiplication, there is no single type or size which is entirely compatible with all of these functions. As the unit of dispersal, small seeds are advantageous since for a given energetic investment, they are more numerous and more easily dispersed (Harper et al. 1970, Baker 1972, Harper 1977). Small seeds have other advantages. They often germinate more quickly than large seeds and thus quickly occupy newly available sites. This can lead to competitive success (Ross & Harper 1972, Rabinowitz 1978). Small seeds may also suffer lower predation rates than large seeds (Janzen 1969, Jakhmola & Singh 1971). Large seeds, however, have more nutritional reserves. This leads to increased seedling vigor (McKell 1972, Kneebone 1972, Smith et al. 1973) and competitive advantages when seeds are grown at high densities (Harper & Clatworthy 1963, Harper et al. 1970, Capinera 1979). Seed size is especially important to the survivorship of seedlings under competitive or environmental stress (Harper et al. 1970). Larger seeds, with their large embryo and food reserves, make it possible for a seedling to grow and survive longer in an environment starved of such resources as light, moisture, or minerals (Grime & Jeffery 1965, Harper & Obeid 1967, Twamley 1967, Cook 1972, Harper 1977).

Seed size is heritable (Fehr & Weber 1968, Draper & Wilsie 1965), is powerfully affected by natural selection (Harper 1977) and is the least plastic (most constant) plant organ (Harper et al. 1970). There are numerous examples of seed size variation related to intraspecific and interspecific density (Harper et al. 1970, Reed & Stevenson 1973, Snell 1976), environmental factors such as moisture availability (Baker 1972, Cook 1972, Singh 1972, Schimpf 1977, Seavey et al. 1977, Sorenson & Miles 1978, Yeaton 1978), and other factors such as herbivory and predation (Harper et al. 1970, Capinera 1979). Nevertheless, seed size shows a high degree of constancy relative to the plasticity shown by other reproductive and vegetative features such as size and number of leaves, size of individual plant, development time, and number of seeds per plant. Thus intra-and interspecific patterns of seed size under different environmental conditions are of interest to biologists (Table 1) from both a practical and theoretical viewpoint.

El tamaño de la semilla tiende a ser constante y hereditario. Semillas pequeñas son ventajosas bajo ciertas condiciones competitivas, mientras que las más grandes son seleccionadas bajo otras.

TABLE 1
FACTORS AFFECTING SEED SIZE

Environmental Factor	Correlated with	Reference	Locality
Low moisture availability	Large seeds	Baker (1972)	California
		Singh (1972)	India
		Cook (1972)	California
		Schimpf (1977)	North America
		Seavey et al. (1977)	
		Sorensen & Miles (1978)	Oregon
		Yeaton (l978)	
Shade or low light availability	Large seeds	Salisbury (1942)	Britain
		Baker (1972)	California
		Carlquist (1966)	Pacific Islands
		Hupp & Rockwood (in prep.)	Virginia
High altitude	Small seeds	Baker (1972)	California
	Large seeds	Sorensen & Miles (1978)	Oregon
Short growing season	Small seeds	Baker (1972)	California
	Large seeds	Sorensen & Miles (1978)	Oregon
	No trend	Schimpf (1977)	North America
	Small seeds	Hupp (in prep.)	Virginia
Competition: little to great	Small to large seeds	Harper & Clatworthy (1963)	
		Harper et al. (1970)	Britain
		Stebbins (1976b)	France
		Williams et al. (1978)	Australia
Seedling vigor	Large seeds	McKell (1972)	
		Kneebone (1972)	
		Smith et al. (1973)	
Early germination	Small seeds	Rabinowitz (1978)	Missouri
		Ross & Harper (1972)	Britain
Dispersal ability	Small seeds	Harper et al. (1970)	
		Baker (1972)	
		Stebbins (1976b)	
		Rabinowitz (1978)	
		Yeaton (1978)	
High predation rate	Small seeds	Janzen (1969)	Central America
		Jakhmola & Singh (1971)	India
Islands/seacoasts	Large seeds	Cavers & Harper (1967)	Britain
		Carlquist (1966)	Pacific Islands
Saprophytes/ parasites	Very small small	Harper et al. (1970)	
		Harper (1977)	
Successional stage: early to late	Small to large seeds	Salisbury (1942)	Britain
		Yeaton (1978)	United States
		Hupp & Rockwood (in prep.)	Virginia
Annual herbs, perennials	Small to large seeds	Stevens (1932, 1957)	North America
Shrubs		Salisbury (1942)	Britain
Trees		Baker (1972)	California

Seedsizes over 10 orders of magnitude. Saprophytic and parasitic seeds are the smallest reported: 10^{-6}g. Herbs, grasses and weed seeds generally range from 10^{-5} to 10^{-2}g. Shrubs and tree seeds are large: 10^{-2} to 10^{4}g.

eed Size Patterns

Environmental and biological factors which have een correlated with seed size patterns are summa- zed in Table 1. The factor cited most often, and hich appears to be dominant as a determinant of ed size, is moisture availability. Seed weight tends increase with an increase in the probability that a edling will be exposed to drought. Several reports fer to intraspecific variation in seed weight or size lated to moisture availability. In each case larger eds are found in the drier environments (Singh 972, Cook 1972, Schimpf 1977, Sorensen & Miles 978). In other studies, species within a genus or mily inhabiting wet areas had smaller seeds than ose associated with drier sites. For example, Yeaton 978) found that pines of the subgenus *Haploxylon* ave smaller seeds than those of the subgenus *Di- loxylon*. *Haploxylon* seedlings are found on wetter tes less subject to environmental fluctuations than edlings of *Diploxylon*. Baker (1972) showed that ithin six genera of herbaceous California plants ean seed weight increased with moisture stress hen species found in hydric, mesic, and xeric sites ere compared. Baker also showed a strong correla- on between dry environments and large seeds for e California flora as a whole.

Other environmental factors which have been orrelated with seed size include shade (low light in- nsity), altitude, length of growing season, competi- on, and predation (Table 1). Salisbury (1942) and aker (1972) found that shade and large seeds were orrelated. This observation, however, is complicated y the general trend for an increase in seed size with uccessional stage and life form (Table 1). Thus, low ght conditions are found in late successional stages here trees dominate. The factors selecting for large eeds are difficult to identify, though they probably nclude competition for moisture and nutrients as ell as light. Certainly selection for shade tolerance is mportant in climax forest vegetation; large seed size ay be another factor heavily selected for in these nvironments.

Baker (1972) found a correlation between elevation nd decreased seed size in California. He examined eeds from sea level to over 3000 m in the Sierra evada range. Mean seed weights for whole com- unities decreased with elevation as did mean eights for species within a given genus (Baker 1972). aker also reexamined data from the work of Clausen, eck and Heisey (1940) on elevational ecotypes in *otentilla glandulosa*. He found a steady decrease in eed size with elevation (Clausen & Heisey 1958). aker (1972) found that larger seeds simply dropped out of high elevation communities. He hypothesized that with the short growing season there was not enough time to accumulate reserves needed for the production of large seeds. Supporting evidence comes from the general trend toward larger seeds as one moves from the Arctic to the tropics (Baker 1972) and from the work of Hupp (in prep.), who has found a positive relationship between length of growing season and seed size in grasses in Virginia. However, when moisture availability is combined with either of the above factors—elevation, length of growing season—it appears to be the dominant selective force. For example, Sorenson and Miles (1978) found that Douglas-fir seed weight went up with elevation and decreased growing season in Oregon. These same areas were also drier. Seed weight was thus positively correlated with xeric conditions as would be predicted from the previous paragraphs. Schimpf (1977) also found that moisture availability was more important as a selective force than length of growing season for pigweed, *Amaranthus retroflexus*. Both Schimpf (1977) and Singh (1972) noted that intraspecific seed weights are more closely related to long term moisture availability than to recent rainfall patterns.

En la zona templada, las semillas pequeñas están asociadas con elevación y las grandes con hábitats secos, sombra, duración de la estación de crecimiento y con los estados sucesionales avanzados.

For a recent review of the relationship between competitive ability and seed size see Capinera (1979). Predation may select for smaller seeds (Janzen 1969, Jakhmola & Singh 1971, Capinera 1979), but the data are not conclusive. More work on specific plant groups within given ecosystems such as that of Janzen (1969) is needed before generalizations can be made.

Dominant life forms in most temperate zone communities change during succession from primarily annual herbs and grasses to perennials, to woody shrubs, and finally to trees. Early successional plants have high reproductive efforts and high dispersibility, producing many small, easily dispersed seeds. Plants which occur late in succession are expected to have large seeds and to invest energy in persistent vegetative organs such as trunks and branches (Harper 1977). Even within a given genus, such as *Desmodium* in Virginia, species found in late stages of succession produce fewer but larger seeds than those species found in early stages (Hupp & Rockwood, in prep.).

Saprophytes and parasites have many extremely tiny seeds. They maximize dispersibility at the expense of nutrient reserves since they can gain their nutrition from the host plants immediately after germination (Harper 1977).

Las saprófitas y parásitas tienen semillas extremadamente pequeñas.

Seed size may also be correlated with dispersal type. Wind and mechanically dispersed seeds are usually small while water dispersed seeds—coconuts—are likely to be relatively large. Among seeds dispersed by animals, mammal dispersed seeds are likely to be larger when compared with those dispersed by birds and insects. Bat dispersed seeds in the tropics may resemble bird dispersed seeds in size. Tropical forest trees, many of which are animal dispersed, are commonly thought of as having large seeds as compared with temperate zone trees which are often wind dispersed. However, dispersal agents, fruit types and seed sizes have coevolved. Quantitative analyses are lacking in the literature.

Semillas dispersadas mecánicamente o por el viento son usualmente pequeñas, mientras que aquellas dispersadas mediante el agua pueden ser de mayor tamaño. Las semillas dispersadas por mamíferos generalmente son más grandes que las transportadas por aves o insectos. Las semillas dispersadas mediante murciélagos y aves pueden ser del mismo tamaño.

Seed Size Patterns in the Tropics

Most data on seed size and weight come from temperate locations (Table 1). Exceptions are from the Pacific islands (Carlquist 1966) and the Central American legumes of Janzen (1969). This paper examines seed size by life form and by elevation for plant families from Panama and Costa Rica, and the results are compared with predicted patterns and with data from the temperate zone (Salisbury 1942, Baker 1972).

Methods

Herbarium sheets of six plant families—Gesneriaceae, Araliaceae, Myrsinaceae, Campanulaceae, Bignoniaceae, Bromeliaceae, and Melastomataceae at the Smithsonian Institution (US) were examined individually for attached fruit with the following considerations: 1) Collections were from Costa Rica or Panama, to make use of life zone maps (Holdridge et al. 1971) available for these countries. 2) All herbarium sheets had fruit or seeds attached. 3) The herbarium label indicated where the plant was collected, in what type of habitat, and the life form of the plant. Fruits were prepared and seeds excised from the fruit under a dissecting microscope. Normally, lots of 100 seeds were weighed, but for very light seeds, lots of 200–500 seeds were used. Seeds were then weighed and mean seed weight calculated. Only some very light wind-dispersed seeds in the Bromeliaceae presented problems in obtaining mean seed weights, and this was not a problem when using the weight class system described below. Fruits and seeds were returned to the appropriate sheet when weighing was completed.

Arithmetic mean seed weight for a given life zone or for a given life form is influenced greatly by the presence or absence of an extremely light or heavy seeded species (Baker 1972). To minimize this problem, a series of seed weight classes on a logarithmetic scale was adapted from Baker (Table 2) (see also Salisbury 1942). The classes were narrow enough that clearly different environments produced significantly different mean seed weights, and different seed samples from the same taxon fell into the same weight class in all cases. When data were sufficient to compare intrageneric differences, raw weights were used instead of the classes. In Table 3, both class means and arithmetic means are presented, but the results are discussed in terms of class means. Conventional statistical methods were used in comparing mean seed weights as frequency of weight class plotted against class ordination produced a series of normal curves.

TABLE 2
SEED WEIGHT CLASSES
(from Baker 1972)

Seed Weight Class	Limits (in mg.)	
1	less than .003	
2	.003—	.009
3	.010—	.031
4	.032—	.099
5	.100—	.315
6	.316—	.999
7	1.000—	3.161
8	3.162—	9.999
9	10.000—	31.611
10	31.612—	99.999
11	100.000—	316.108
12	316.109—	999.999
13	1000.000—	3161.080
14	more than 3161.080	

TABLE 3
SEED WEIGHTS BY LIFE FORM FROM PANAMA AND COSTA RICA

Life Form	No of Species Examined		Class Mean	Arithmetic Mean (mg)
MELASTOMATACEAE				
Trees	16		5.25	0.251
Shrubs	64		3.20	0.048
Epiphytes	8		3.63	0.045
t-tests				
Trees vs. shrubs		p less than:	.001	0.001
Trees vs. epiphytes		p less than:	.001	0.01
Shrubs vs. epiphytes		not significant		
Miconia				
Trees	8		5.20	0.295
Shrubs	17		4.12	0.144
t-tests		p less than:	.02	.075
GESNERIACEAE				
Herbs	16		2.38	0.021
Shrubs	16		3.31	0.043
Epiphytes	13		4.15	0.090
t-tests Herbs vs. shrubs		p less than:	.05	n.s.
Herbs vs. epiphytes		p less than:	.001	.001
Shrubs vs. epiphytes		p less than:	.05	.05
ARALIACEAE				
Trees	15		7.93	6.330
MYRSINACEAE				
Trees	16		9.00	35.668
CAMPANULACEAE				
Herbs	8		2.88	0.018
Shrubs	6		3.67	0.091
t-test			n.s.	n.s.
BIGNONIACEAE				
Lianas	10		8.80	24.031
BROMELIACEAE				
Pitcarnia spp. (wind dispersed)	8		1.50	0.025
"Tank" epiphytes	34		5.85	1.761
RUBIACEAE				
Herbs	10		4.50	0.250
Shrubs	16		5.63	5.477
Small trees	10		5.40	7.569
Trees	12		5.00	22.702
Vines	4		3.50	0.080
All Plant Families Examined				
Trees and small trees	69		6.68[1]	
Epiphytes (excl. *Pitcarnia*)	55		5.13[2]	
Shrubs	102		3.63	
Herbs	34		3.12	
Miscellaneous				
Lianas (Bignoniaceae)	10		8.80	
Vines (Rubiaceae)	4		3.50	
Pitcairnia	8		1.50	

[1]Tree mean weight significantly larger than all other mean weights (p less than 0.001).
[2]Epiphyte mean weight significantly larger than shrub and herb mean weight (p less than 0.001).

Las semillas se pesaron y se midieron. Los pesos se clasificaron en grupos y los promedios de éstos se analizaron logarítmicamente.

Life Form

The results by life form were consistent. In the Melastomataceae, for example, tree seed weights are significantly larger than both epiphyte and shrub weights. There is no significant difference between shrub and epiphyte mean seed weights. Trees have larger seed weights within the genus *Miconia* (Table 3).

Las semillas de las Melastomataceae arbóreas fueron consistentemente más grandes que las de cualquier arbusto o hierba.

Data from the Gesneriaceae (Table 4) show that epiphytes have mean seed weights significantly larger than both shrubs and herbs. Shrub seed weights are significantly larger than herb seed weights when the weight class system is used, but not if raw seed weight data are used. The raw data show that epiphyte mean seed weight is twice that of shrub weight, and shrub mean weight is twice that of the herbs.

Las Gesneriaceae epífitas presentaron semillas más grandes que las de los arbustos o las hierbas.

All species examined in the Araliaceae and Myrsinaceae are trees. As expected the seeds were relatively heavy (Table 5). Within the Campanulaceae shrub seeds were heavier than herb seeds, but the difference is not statistically significant due to small sample sizes.

Los árboles de Araliaceae y de Myrsinaceae tenían semillas grandes. Los arbustos de Campanulaceae presentaron semillas ligeramente más grandes que las hierbas. Todos los bejucos de Bignoniaceae presentaron semillas grandes.

All data from the Bignoniaceae (Table 3) are from lianas. Again the seeds are large. Much more data are needed, but it appears that natural selection favors large seeds in plants such as lianas whose strategy is to "climb to sunlight."

All plants examined in the Bromeliaceae are epiphytic, except those of *Pitcairnia* which are terrestrial. Two types of seeds were present, wind-dispersed, very light seeds, and relatively large, bird-

TABLE 4
SEED WEIGHT BY ELEVATION
SEVEN TROPICAL FAMILIES

Family	No. of Species Examined	Class
Melastomataceae (0–2925 m)		
below 1500 m	73	3.77
above 1500 m	15	2.87
	(p less than .01)	
Gesneriaceae (0–2200 m)		
below 1500 m	40	3.32
above 1500 m	8	3.50
Araliaceae (30–2300 m)		
above 1500 m	7	7.70
below 1500 m	8	8.10
Myrsinaceae (0–2600 m)		
below 1500 m	9	8.67
above 1500 m	7	9.43
Campanulaceae (500–3000 m)		
below 1500 m	9	3.11
above 1500 m	8	3.25
Rubiaceae (0–3280 m)		
below 1500 m	33	5.54
above 1500 m	5	4.20
Bromeliaceae (0–2770 m)		
below 1500 m	26	5.00
above 1500 m	13	5.23

TABLE 5
PREDICTED SEED SIZES BY LIFE FORM
Sale of ascending size 1—6

Life Form	Predicted Seed Size
Protoepiphytes and wind dispersed epiphytes of other types	1
Herbaceous annuals and perennials	2
Vines	2
Shrubs	3
Lianas	4
Nest and bracket, tank, and hemi-epiphytes (non-wind dispersed)	4
Small trees	5
Large trees	6

See Richards (1966) for description of life forms.

dispersed seeds. All species within the genus *Pitcairnia* are wind-dispersed, and, as can be seen from Table 3, are extremely tiny. Other species of bromeliads, which are not wind-dispersed, have relatively large seeds. These species often begin in shady places and must be able to grow for a time in suboptimal conditions. Thus the large seeds give the seedlings a chance to establish themselves. Within the Bromeliaceae, there appear to be two adaptive strategies—very small, wind-dispersed seeds, and relatively large, bird-dispersed seeds.

En las Bromeliaceae las semillas dispersadas mediante el viento eran livianas, mientras que las dispersadas por aves eran grandes.

Table 3 summarizes the data on seed weight by life form for all species examined to date, regardless of family. It shows that temperate zone patterns generally hold true in the tropics. Thus, tree seeds are larger than shrub and herb seeds. However, shrub mean weight is not significantly larger than herb weight. Animal dispersed epiphyte seed weights, however, are significantly larger than either shrub or herb seed weights. Data from lianas, vines and wind-dispersed epiphytes are still too limited for generalizations.

Seed weight by elevation is summarized in Table 4, showing rank correlations between seed weight and elevation for each plant family and for subsets of plant families, and in no case was the correlation significantly different from zero. In addition, no significant seed size differences by elevation were found except in one plant family. In four plant families the mean class weight was larger for species collected above 1500 m. Thus a search for seed patterns by elevation has been unproductive.

Size by Life Form

Two questions can be asked: 1) are the data consistent with the temperate zone data (Table 1), and if so, can the data on life forms unique to the tropics be logically integrated into the temperate zone framework? 2) what seed size patterns can be anticipated for the many tropical life zones.

Richards (1966) pointed out that some of the smallest seeds known belong to epiphytic orchids, and he seemed to assume that small seeds would be characteristic of epiphytes. Yet preliminary data from the Gesneriaceae indicate that average seed weights from epiphytic species are larger than herbaceous or shrubby species from the same environment. The literature also suggests that many lianas have extraordinary life cycles (Strong & Ray 1975).

In Table 5, I assume that herbs, shrubs and trees follow temperate zone patterns. Herbaceous vines are predicted to have seed sizes comparable to herbaceous plants in general. Lianas, however, are expected to have relatively large seeds, equivalent to, or larger than, shrubs. Epiphyte seed size will depend on the plant family, dispersal type and epiphyte class (Richards 1966), but in general, epiphyte seeds are predicted to have a bimodal seed weight distribution. An epiphytic species may germinate on the ground, be associated with particular animal dispersal systems, or have mutualistic relationships with ants (Richards 1966, Janzen 1974, Kleinfeldt 1978). Under these conditions seeds are likely to be large. In other species, epiphyte seeds are wind dispersed or depend on sticky pulp which is deposited by birds on appropriate branches. In these species, seeds are predicted to be smaller than those of herbaceous plants.

Theoretically, small seed size should be important to epiphytes because of their need to be dispersed into the forest canopy. On the other hand, their position away from the ground makes the seedling susceptible to drought. As mentioned earlier, increased probability of drought usually selects for large seeds (Table 1). These conflicting selective pressures may well result in the bimodal seed sizes prediced in Table 5.

The actual results (Table 3) fit the predictions. The Bromeliaceae in particular show the expected bimodal distribution of seed sizes. Species of *Pitcairnia* which are wind-dispersed, have extremely tiny seeds of seed weight class 1. All other species surveyed have seeds of seed weight classes 5, 6 or larger. In general, epiphytes have seeds as large or larger than those of shrubs (Table 3). Few data are available for lianas, but the eight species of Bignoniaceous lianas had very large seeds.

Los resultados demuestran que los patrones de tamaños de semillas estudiados son comparables con aquellos patrones reportados en las zonas templadas; semillas dispersadas por el viento son pequeñas y las epífitas presentan semillas más grandes.

Size by Life Zone

Although there is surely a decrease in seed size with elevation, I have been unable to identify it within the selected plant families using this research method.

To address the second question—what seed size patterns can be anticipated for the many tropical life zones data from the temperate zone indicate that where several environmental factors interact, seed size is usually determined by moisture availability (Baker 1972, Schimpf 1977, Sorenson & Miles 1978). Elevation *per se* is probably unimportant, but productivity, as determined by length of growing season or other factors, has some influence. Intense competition for light or nutrients is also important, and correlations between seed size and shade reflect this competition. The dispersal agent is another factor to be considered. One prediction would be that at a given elevation, a drier life zone would favor larger seeds than a wetter life zone, this assuming that moisture availability is a primary seed size determinant as in the temperate zone.

Within a given humidity province—dry, moist, wet—seed size might be predicted not to change with elevation until the lower montane or montane life zones where seed sizes would be expected to decrease. This corresponds to the Terborgh hypothesis (1977) that above 1500 m, low temperature, low light levels, and poor soil conditions combine to lower plant productivity, and this is assumed to produce smaller seeds. Many have reasoned that plant productivity will not decrease at mid-elevational sites as compared with the lowlands. In fact Janzen (1973a,b), Janzen et al. (1976) and Terborgh (1977) have reasoned that net productivity may even increase at mid-elevations. If they are right, it is not surprising that I have been unable to find trends in seed size by elevation. However, even this line of reasoning would expect to find smaller seeds at elevations above 2000 m within a given humidity province. In future research an attempt to correlate seed size with life zone rather than simply by elevation may be a more fruitful line of inquiry.

Los datos obtenidos no apoyaron la hipótesis de que el tamaño de las semillas disminuye con la elevación.

Baker, H.G. 1972. Seed weight in relation to environmental conditions in California. Ecology 53: 997–1010.

Capinera, J. L. 1979. Qualitative variation in plants and insects: effect of propagule size on ecological plasticity. Amer. Nat. 114: 350–361.

Carlquist, S. 1966. The biota of long distance dispersal, Part 2, Loss of dispersibility in Pacific Compositae. Evolution 20: 30–48.

Cavers, P.B. & J.L. *Harper*. 1967. The comparative biology of closely related species living in the same area, Part 9, *Rumex:* The nature of adaptation to a sea-shore habitat. J. Ecol. 55: 73–82.

Clausen, J., D.D. *Keck* & W.M. *Heisey*. 1940. Experimental studies in the nature of species. I. Effects of varied environments on western North American plants. Carnegie Inst. Wash. Publ. 520, Washington, DC.

——— & W.M. *Heisey*. 1958. Experimental studies in the nature of species. IV. Genetic structure of ecological races. Carnegie Inst. Publ. 616, Washington, DC.

Cook, Alfred D. 1972. Polymorphic and continuous variation in the seeds of dove weed, *Eremocarpus setigerus* (Hook.) Benth. Amer. Midl. Nat. 87: 366–376.

Draper, A.D. & C.P. *Wilsie*. 1965. Recurrent selection for seed size in Birdsfoot Trefoil, *Lotus corniculatus* L. Crop. Sci. 5: 313–315.

Fehr, W.R. & C.R. *Weber*. 1968. Mass selection by seed size and specific gravity in soybean populations. Crop Sci. 8: 551–554.

Grime, J.P. & D.W. *Jeffery*. 1965. Seedling establishment in vertical gradients of sunlight. J. Ecol. 53: 611–642.

Harper, J. L. 1977. Population biology of plants. Academic Press, London.

——— & J.N. *Clatworthy*. 1963. The comparative biology of closely related species. VI. Analysis of the growth of *Trifolium repens* and *T. fragiferum* in pure and mixed populations. J. Exp. Bot. 14: 172–190.

——— & M. *Obeid*. 1967. Influence of seed size and depth of sowing on the establishment and growth of varieties of fiber and oil seed flax. Crop. Sci. 7: 527–532.

Harper, J.L., P.H. *Lovell* & K.G. *Moore*. 1970. The shapes and sizes of seeds. Ann. Rev. Ecol. Syst. 1: 327–356.

Holdridge, L.R., W.C. *Grenke*, W.H. *Hatheway*, T. *Liang* & J.A. *Tosi*, Jr. 1971. Forest environments in tropical life zones, a pilot study. Pergamon Press, New York.

Jakhmola, S.S. & L. *Singh*. 1971. Relationship of seed size in green gram, *Phaseolus aureus*, with the development and reproduction of pulse beetle. JNKVV (Jawaharlal Nehru Krishni Vidyalaya) Res. J. 5: 104–106.

Janzen, D.H. 1969. Seed-eaters versus seed size, number, toxicity & dispersal. Evolution 23: 1–27.

———. 1973a. Sweep samples of tropical foliage insects: description of study sites, with data on species abundances and size distributions. Ecology 54: 659–686.

———. 1973b. Sweep samples of tropical foliage insects: effects of seasons, vegetation types, elevation, time of day, and insularity. Ecology 54: 687–708.

———. 1974. Epiphytic myrmecophytes in Sarawak: mutualism through the feeding of plants by ants. Biotrop. 6: 237–259.

———, M. *Ataroff*, M. *Farnian*, S. *Reyes*, N. *Rincon*, A. *Soler*, P. *Soriano* & M. *Vera*. 1976. Changes in the arthropod community along an elevational transect in the Venezuelan Andes. Biotrop. 8: 193–203.

Kneebone, W.R. 1972. Breeding for seedling vigor. *In* V.B. Younger and C.M. McKeel, eds. The Biology and Utilization of Grasses. Academic Press, New York.

Kleinfeldt, Sally E. 1978. Ant-gardens: the interaction of *Condonanthe crassifolia* (Gesneriaceae) and *Crematogaster longispina* (Formicidae). Ecology 59: 449–456.

McKell, C.M. 1972. Seedling vigor and seedling establishment. *In* V.B. Younger & C.M. McKell, eds. The Biology and Utilization of Grasses. Academic Press, New York.

Rabinowitz, D. 1978. Abundance and diaspore weight in rare and common prairie grasses. Oecologia 37: 213– 219.

Reed, F.C. & S.N. *Stephenson*. 1973. Factors affecting seed number and size in burdock, *Arctium minus*. Mich. Acad. 5: 449–455.

Reeder, W.G. & S.E. *Riechert*. 1975. Vegetation change along an altitudinal gradient, Santa Cruz Island, Galapagos. Biotrop. 7: 162–175.

Richards, Paul. 1966. The Tropical Rain Forest. Cambridge Univers. Press, Cambridge.

Ross, M.A. & J.L. *Harper*. 1972. Occupation of biological space during seedling establishment. J. Ecol. 60: 77–88.

Salisbury, E.J. 1942. The reproductive capacity of plants. G. Bell & Sons. London.

Schimper, A.F.W. 1888. Die epiphytische Vegetation Amerikas. Bot. Mitt. Trop.

Schimpf, David J. 1977. Seed weight of *Amaranthus retroflexus* in relation to moisture and length of growing season. Ecology 58: 450–453.

Aerial Walkways in Rain Forest Canopies, Construction and Use in Panama, Papua New Guinea, and Sulawesi

Andrew M. Sugden

Los puentes áereos, bastante mejorados, son un magnífico medio de acceso al dosel de los bosques tropicales. Se han probado en Panamá, Nueva Guinea e Indonesia. Son seguros, simples de construir y causan mínimo perjuicio al bosque. Los puentes áereos proveen áreas disponibles a través del estrato superior y estos transectos permiten una variedad plural de repetidas observaciones durante varios años. Son excelentes para efectuar estudios de polinización, de fenología foliar, de animales herbívoros, de química de las hojas, y para colectar insectos y murciélagos. Los puentes aéreos son un avance notable sobre otros métodos de acceso al estrato superior del bosque.

The need for studies of the biology of the upper canopy of tropical rain forest requires little justification. The primary production of a tropical rainforest is concentrated in its upper layers, and many of the selective forces affecting the trees and their associated flora and fauna operate high above the ground. As rainforest biology expands from a largely descriptive and taxonomic science into an ecological discipline with opportunities for experiment and observation, our knowledge of the functioning of forest canopies is still limited by the inaccessibility of the tree crowns.

Las observaciones biológicas, los experimentos y las colecciones en el dosel del bosque siempre han estado limitados como consecuencia de su poca accesibilidad.

Most methods of access to the canopy—the structurally heterogenous upper layers of the forest—employed hitherto have tended to limit observation, experiment and collection to a relatively confined space. Towers and tree platforms, though often excellent for studying larger animals, especially birds, for vertical trapping of insects (Sutton & Hudson, 1980), and for basic records of forest phenology (Hingston, 1932; McClure, 1966) are of limited use for most botanical studies and for investigations of foraging by smaller animals, especially insects. Because of the diversity and heterogeneity of rain forest canopies, towers and tree platforms cannot usually give statistically valid samples—particularly of trees (Perry, 1978). Systems involving ropes suspended in spider's web fashion between emergent trees are a distinct improvement, being both cheap and versatile (Perry, 1980), but they are inconvenient for long-term repeated observations and experiments by teams.

Here we report on the aerial walkway—a technique which provides access to sizeable areas of canopy. Its basic design is that of a suspension bridge hung between the boles or branches of large emergent trees. Aerial walkways are not new; they have been used previously for projects in Malaysia (Muul & Liat, 1970), Zaire (Dieterlen 1978) and Panama (N. Smythe, pers. comm.), but have found few adherents, possibly due to limited publicity and scepticism from the scientific community.

Los puentes áereos sobre el dosel del bosque son seguros, fáciles de construir y de costo razonable. Estos proveen acceso a áreas razonables y permiten el estudio de este estrato del bosque.

Our own conclusion is that aerial walkways represent a significant advance in investigating canopy biology, and we publish this to encourage others to build walkways, especially in field stations and forests where there already exist good communications and a good taxonomic base. Contrary to some impressions, walkways are safe to use, they are not difficult to build and maintain, nor do they damage the habitat. Their cost need not be prohibitive.

Our studies entailed the construction and use of sets of walkways in three lowland tropical rainforest sites: at Puerto Escoces, San Blas, Panama; at Buso, Morobe Province, Papua New Guinea, and at Morowali, east central Sulawesi (Indonesia). At each site a set of two (Sulawesi) or three (Panama and Papua New Guinea) walkway spans were built, and used for short-term investigations of canopy biology. The building technique (see Appendix) was with minor modifications the same in each site, and a variety of different materials were tried out. Except for one of the spans at Buso, the walkways were dismantled after 2–4 months of use. The objectives were to assess the suitability of different materials for long-term studies, and to determine the feasibility of walkways for canopy biology.

Puentes áereos se han construido en San Blas en la República de Panamá, en Indonesia y en Sulawesi. El objetivo primordial ha sido el de probar su factibilidad.

The principal features of the spans in the three sites are given in Table 1, and Plates 1–4 illustrate some of the differences between the sites and the details of construction. The construction sequence and the recommended materials are listed in the Appendix.

TABLE 1
COMPARATIVE FEATURES OF THE WALKWAY SPANS ON THE THREE SITES

Site	Panama			Papua-New -Guinea			Sulawesi	
Span no.	1	2	3	1	2	3	1	2
Length (m)	25	42	54	25	24	37	41	70
Height above ground at ends (m)	25	27	27	31	36	33	28	28
No. trees directly accessible	6	5	9	8	5	7	7	7
No. trees indirectly accessible	5	3	8	12	3	19	15	—

Constraints and Requirements

The general aim of walkway construction is to position a set of spans in the canopy, usually 25–35 m above ground, with minimum damage to the forest. The life of the walkway should be long—at least three years in order to follow the fluctuations and structural changes expected in the forest. Most important, the walkway should be safe.

The first requirement is careful siting. Biologists of different disciplines may have different ideas about the ideal situation for their investigations—ornithologists may require a clear view, while a phytochemist may wish to have at hand a large and varied a sample of foliage. To be cost-effective, the final positioning must be a compromise between the members of a research team.

Healthy trees should be selected for the walkways. These should be emergents, or at least well above the average height of mature trees in the forest. Trees already burdened with epiphytes, lianas and stranglers should not be used, nor should trees which are stag-headed or show other signs of decay. This applies to the trees supporting the walkway, and also to trees in the walkway route.

En su ubicación en el dosel deben emplearse árboles emergentes sin numerosas epífitas y, adicionalmente, estos árboles no deben mostrar signos de descomposición.

The second requirement is that construction should inflict minimum damage on the supporting trees and the forest, and that the walkway should not excessively weaken the supporting trees. To climb the trees themselves we used pitons—a standard rock-climber's method. While satisfactory for short-term construction, this is not recommended because of the injury inflicted on the tree. A better method is that of Perry (1978) using a crossbow to fire a thin monofilament line over a branch followed by a thicker braided line and a climbing rope which can be anchored and ascended with jumars (ascenders) and a sit-harness. This may not give immediate access to the desired part of the tree because of obstruction by lower branches, but this can be remedied by further climbing. A climb of this kind can be accomplished safely given a little previous practice (Perry 1978).

Damage to the supporting tree is minimized by using lashes of thick rope or polyester tape at every point of attachment. The use of nails or wood screws should be avoided. Abrasion caused by tensioned

PLATE 1.

Figs. 1 and 2. Walkway spans in Panama (top) and in Papua New Guinea (bottom), showing access to canopy. The main cable materials for these two spans (polypropylene rope in Panama and 5 mm steel wire rope in Papua New Guinea) have been superseded by polyester tape in Sulawesi (See Table 1 and Figs. 3, 4). The wire mesh and Dexion flooring (see Appendix) was unsatisfactory because of rapid corrosion and buckling.

ropes and tapes can be reduced by padding which also reduces frictional damage to the ropes themselves.

Damage to the canopy intervening between the two supporting trees is minimized mainly by careful siting. In Panama, only two branches which impeded construction had to be removed. Given more time, the obstructing trees might be moved out of the path of the walkway during construction by attachment of tensioned guy ropes. In the Panamanian site, most of the crowns at walkway level (25 m) exhibited elastic properties in moderate winds which would have allowed bending; larger trees are not so tractable. Siting the walkway just above the surface of a relatively even canopy is the best solution; during subsequent years tree crowns may grow to envelop the walkway. The problem of obstruction will always, however, preclude the construction of walkways through very thick canopies.

La construcción de estos puentes no causan mucho daño a la vegetación. Los árboles pueden subirse utilizando algunos implementos como arcos, aparejos para trepar y otro equipo pertinente. Las sogas gruesas y de polyester sirven para fijar el puente y causan poco daño en el dosel del bosque.

The problem of tension on the supporting trees, causing weakening of the tree over a period of time, cannot be accurately assessed or resolved completely. A 50 m span weighs about 500 kg, and the directional force this exerts on the two supporting trees is greatly increased by the tension in the walkway. The directional force on the supporting trees may be reduced by applying an opposite force with tensioned guy rope(s) attached to the base of a nearby tree. Different tree species have differences in architecture and overall strength, and where known these should be taken into account. The locality should be flat, sheltered, and not subject to frequent high winds, cyclones or earthquakes. Frictional damage at points of attachment and the motion of branches caused by personnel on the walkway can be reduced by constant maintenance and controlled movement. The motion of a person walking fast or heavily on the walkway can be felt distinctly by an observer in a supporting tree.

Los árboles de soporte podrían encorvarse con el tiempo y para evitar tal cosa se tiemplan con retenes.

The third requirement is durability and replaceability of materials used in the walkway itself. We were able to test a number of different materials, and though none of the walkways was in place long enough to assess its potential lifespan, some recommendations can be made concerning the best materials (see Table 1, and Appendix). The design allows any part of the walkway to be replaced with signs of deterioration, so that in effect the life of the walkway is determined only by the survival of the supporting trees.

Safety, Ascent and Descent

Safety, during both construction and use, cannot be stressed too much. To fall 30 m from a walkway could easily be fatal. The walkway was designed for safety, and serious accidents will not occur if all personnel obey a set of simple rules. Members of construction and scientific teams always wear webbing sit-harnesses attached at all times with jumars or karabiners to safety lines independent of the walkway. Vertigo is not usually a problem, as the ground is rarely visible. Heavy rain can be hazardous. The flooring of the walkway can be slippery, although the danger is reduced by the reasonably flat floor of the bridge design, by wearing shoes with gripping soles and by using the safety lines. The noise of a heavy rain shower can prevent communication between the canopy and the ground, which is a problem during construction when there are members of a team both operating aloft and on the ground. But if there is always an independent means of descent the construction team can operate securely. Handrails on the walkway (see Appendix) provide additional support in rain and wind.

Although a walkway of this type can in theory support weights over one ton, no more than three or four people should be on a span at once. The motion set up by an extra person can be inconvenient to anyone making observations, and the presence of more peo-

PLATE 2.

Figs. 3 and 4. Walkway spans in Sulawesi (left) and Papua New Guinea (right). The Sulawesi span uses polyester tape for the main cable, while the Papua New Guinea span uses 15 mm steel wire rope. The shackles used to attach the suspenders to the main cable (right) are unnecessary with polyester tape (left) as the loops slide easily along the cable. Also note on the Papua New Guinea span; sheltered tree platform, insect traps, and lateral guy ropes.

PLATE 3.
Figs. 5 and 6. Direct (top) and indirect (bottom) access from a span in Panama. Note the use of a safety line.

TABLE 2
FEATURES OF THE WALKWAY SPANS AT THE THREE SITES

Site	Panama			Papua-New-Guinea			Sulawesi	
Span no.	1	2	3	1	2	3	1	2
Method of Access	Counter-balance hoist	Ladder, and from span 1	Hoist, and from span 2	Ladder,	and hoist	spikes in tree, and hoist	Ladder	From span 1
Main cable Material	polypropylene	rope	5mm diam.	Polyprop. steelwire rope	15mm diam. rope	polyester steelwire rope	webbing tape	
Flooring Material	expanded mesh on	metal Dexion frame			'Dosp'			

PLATE 4.
Fig. 7. Pulling the walkway across with the aid of a Tirfor jack (foreground) while obstacles are moved aside from the path of the walkway (right middle distance).

ple could lead to obstruction if it were necessary to descend to the ground in haste.

La prevención de accidentes requiere precauciones tales como que los trabajadores empleen correas o tirantes de seguridad, que las suelas de sus botas de trabajo posean tracción y que no más de 3 o 4 personas permanezcan simultáneamente en ninguno de sus tramos.

Total collapse of the walkway is highly unlikely as long as regular maintenance checks are carried out. Barring the collapse of a supporting tree—a danger minimized by measures already described—this could only occur if the main suspension cable were to snap. This risk is minimized in the design of the bridge by spreading the weight of the flooring evenly along the suspension cable. Each section of the flooring is suspended independently (see Appendix) so the collapse of one will not affect the others. Careful maintenance and replacement of suspect parts will, in any case, reduce the risk.

Safety considerations demand that ascent and descent be simple and rapid. Three different methods were assessed. In Panama access to the walkway was mostly by means of counterbalanced hoist systems. These consisted of a pulley block attached above the platform at the end of a span bearing a looped length of climbing rope carrying a weight equivalent to that of a person. Though this provided easy ascent and descent, it was necessary to adjust the counterbalance for the individual weight of each person. This inconvenience is increased by the problem of repositioning the counterbalance to allow a second ascent which involves the cumbersome and risky maneuvering of detachable heavy weights on the tree platforms.

A better method of ascent, used at all three sites, is by a tensioned flexible ladder (see Appendix) or set of metal ladders strapped to the tree. Albeit slow and initially requiring considerable exertion, this is safe and reliable provided that a safety line is used with a jumar attachment.

The best and fastest method of descent, used in Papua New Guinea and Sulawesi, is by abseiling (rapelling) with a figure of eight (descender). So long as each person has an ascender and descender, ladders and abseiling do not limit the number of people aloft at any time, with the added advantage that there are no moving parts. However, the ladders in Sulawesi showed considerable wear after three months of use.

Las escaleras son seguras y confiables pero tienden a gastarse y por tal razón es preferible emplear sogas para el descenso.

Scientific equipment and construction materials are hauled up and down on separate hoist systems, using pulley blocks attached above the level of the platforms in order to avoid encumbering personnel during ascent and descent.

Las escaleras son seguras y confiables, pero tienden a gastarse y por tal razón es preferible emplear sogas para el descenso.

Scientific equipment and construction materials are hauled up and down on separate hoist systems, using pulley blocks attached above the level of the platforms in order to avoid encumbering personnel during ascent and descent.

Use

Our walkways were used by biologists of several disciplines with emphasis on the collection and observation of plants, insects and bats. Flower visiting and leaf damage by insects were studied in all three sites. These short-term investigations provided good evidence of the potential of aerial walkways for long-term studies. The full results are currently in preparation for publication as separate papers, and this appraisal merely summarizes the opinions of those who used the walkways in the three sites.

The framework for the studies of plant-animal interactions is provided mostly by the tree crowns accessible from the walkway and is essentially a cylindrical transect running through the canopy. Access can be designated as direct when the branches of a tree can be touched, or indirect by means of telescopic pruners, etc. The number of trees directly accessible by site (Table 1) reflected the siting policy. The maximum number of trees directly accessible per 10 m of walkway was four—from the short span in Panama (Fig. 8). The walkway thus provides larger samples than other methods of access. As well as providing a transect through the canopy, the walkway allows observation of phenomena beyond the reach of collecting equipment. For instance, flower-visiting animals can be seen with binoculars from vantage points on a walkway.

El puente de Panamá permitió acceso a 4 árboles en cada tramo de 10 m.

Botanically, the walkways were particularly valu-

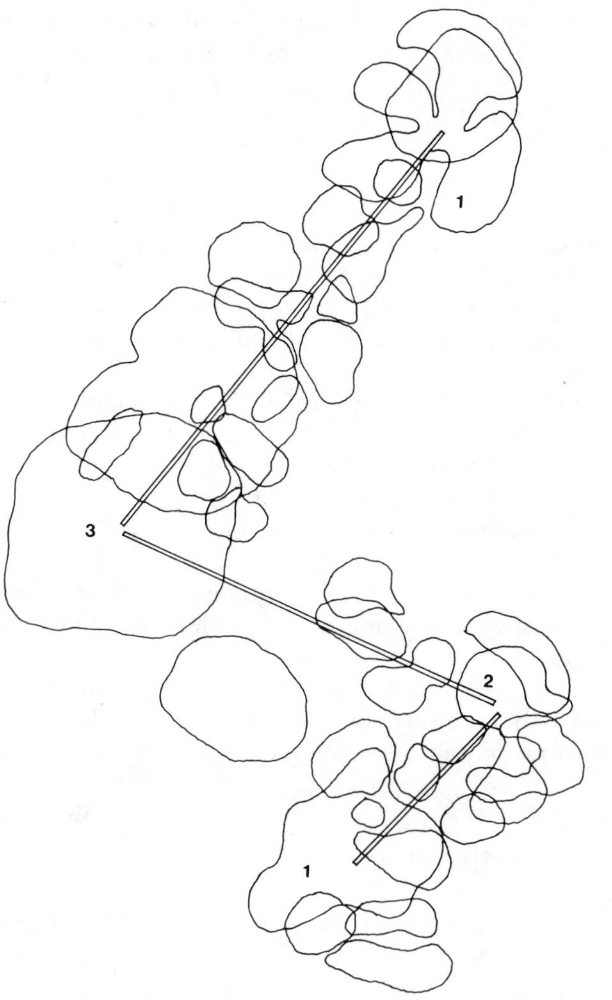

Fig. 8. Tree crowns accessible from the Panama spans. The supporting trees are 1(*Dussia* sp. 2) *Dialium guianense* 3) *Pterocarpus officinalis*.

able for studies of pollination and the effects of insect herbivores on foliage. Access to the flowers on the outside of tree crowns allows close observation of flower visitors, the sampling of nectar, the bagging of flowers and inflorescences, and accurate quantitative studies of reproductive success. Access to foliage permits quantitative monitoring of growth rates, leaf phenology and the progressive damage suffered by leaves during their life span. Sampling of different plant parts for chemical analysis is also facilitated. Examination of forest structure and the construction of profile diagrams is possible without resorting to the destructive techniques often employed. Long-term monitoring of crown competition and changes in forest structure can be envisaged. The advantages of walkways for botanical collecting are limited by the number of trees, vines and epiphytes accessible, but the information on the biology of canopy species, usually lacking on herbarium labels, is more than adequate compensation.

For collecting, entomologists can benefit more from walkways than botanists. Traps for flying insects can be placed in the crowns of different trees or in gaps in the canopy. Butterflies are mostly impossible to catch from walkways, except if visiting flowers. The capture of sessile insects and other arthropods is greatly facilitated—as with most botanical work, walkways obviate the need for destructive sampling techniques. A walkway also increases the scope for investigation of the foraging of pollinators and herbivores in relation to the local availability of resources.

Los puentes son de mayor utilidad para los entomólogos que para los botánicos, sin embargo permite buenos estudios de polinización, facilita estudios de las tasas de crecimiento y de la competencia al nivel de este estrato del bosque.

The only vertebrates studied in detail from the walkways were bats. In Papua New Guinea and Sulawesi, mist nets suspended from the walkways caught 196 (6 species) and 180 (8 species) bats on 463 and 213 m net nights respectively (B. Gaskell, pers. comm.). There is obvious potential for coordinating bat studies with observations on local distribution and adundance of resources—flying insects, fruit, nectar, etc.

Son excelentes para efectuar estudios de los murciélagos, pero en el caso de las aves no mejoran mayormente los métodos usuales.

Ornithologists concluded that walkways are not a significant improvement on well-positioned tree platforms for most observations of birds. Mist-netting for birds was unsuccessful. This may have been because of the disturbance caused by the system—a problem which might be less serious in a long-term study. We made few observations of reptiles and arboreal mammals, probably for the same reasons, although on one occasion in Sulawesi some lizards were watched for more than an hour on flowering branches of *Eugenia* sp. at a distance of 1 m. Long-term studies of arboreal vertebrates, possible once the animals have become accustomed to the walkway, are likely to be incompatible with regular use by botanists and entomologists.

In all of these studies, a major advantage of the walkway was the ease of manipulating large and unwieldy pieces of equipment—telescopic pruners, mist nets, insect traps, etc. The platforms could store collecting equipment and facilities for the immediate preservation of samples—for example, leaves for chemical analysis and in addition to purely biological studies, the walkway permits microclimatic measurements in different parts of the canopy.

Una de las principales ventajas de los puentes es que facilitan la manipulación de equipo difícil de manejar, de podadoras, de redes, de trampas para insectos y de otros aparatos.

Concluding Remarks

This account of the feasibility of walkway construction and the opportunities that walkways provide for research concludes that they represent a major advance on other techniques for access to the upper canopy. Ease of construction, safety, and access afforded to the canopy, strongly commend their use for long-term studies. We have touched the kinds of studies for which they are suitable and leave more specific uses to the imagination of the reader. We were constantly aware during our investigations of the range of studies potentially open to us but precluded by lack of time. Rainforest ecology would benefit from a great deal more attention to the canopy. Finally, there is no reason why a set of walkways spans should not be integrated with other methods of access such as Perry's (1980) 'spider's web' technique to provide even more access.

We welcome queries concerning construction techniques, sources of materials, and any other problems that are not fully answered here.

Acknowledgments

This project was made possible by the sponsors, organizers and members of Operation Drake. We are particularly grateful to M. Tindall and Package Control Ltd. for their interest in the project and for the supply of materials. Thanks are due to the Royal Engineers, in particular to R.S.M.H. Masley, for designing the walkway, and to Sgt. M. Christy, Sgt. L. Gallagher, Cpl. J. Rimmer and their construction teams, for building the spans. We valued the cooperation of the Universidad de Panama, The Smithsonian Tropical Research Institute (Panama), the Bulolo Forestry College and the Wau Ecology Institute (Papua New Guinea), and the Indonesian Institute of Sciences (LIPI). C. Ash, B. Gaskell, P.J. Hudson, A. Lack, D.J. Mabberley, C. Rees and M. Swaine made valuable contributions to the scientific work, and A. Lack, M. Swaine and F. R. Whatley made useful comments on the manuscript. The photographs in Figures 1–7 are by courtesy of Operation Drake, and the illustrations in Figures 8–3 are by M. Mitchell.

Dieterlen, Fl 1978. Zur Phänologie des aquatorialen Regenwaldes im Ost-Zaire (Kivu). Dissert. Bot. 47: 5–111.

Hingston, R.W.G. 1932. A Naturalist in the Guiana Forest. E. Arnold, London.

McClure, H.E. 1966. Flowering, fruiting and animals in the canopy of a tropical rain forest. Malayan Forest. 29: 182–203.

Muul, I., & L.B. *Liat*. 1970. Vertical zonation in a tropical rain forest in Malaysia: method of study. Science 169: 788– 789.

Perry, D.R. 1978. A method of access into the crowns of emergent and canopy trees. Biotrop. 10: 155–157.

———. 1980. An arboreal naturalist explores the rain forest's mysterious canopy. Smithsonian 11 (3): 42–53.

Sutton, S.L. & P.J. *Hudson*. 1980. The vertical distribution of small flying insects in the lowland rain forest of Zaire. Zool. J. Linn. Soc. 68: 111–123.

Appendix

It is essential that the construction team should include at least one member with previous engineering experience, and with an ability to improvise. Every walkway span presents its own individual problems of construction, and variations on the basic construction sequence, outlined below for a single span, are usually necessary. The construction team requires 2–5 members. Under good conditions, the following sequence can be completed in less than a week.

1) Select and climb emergent trees. It is preferable to climb a number of trees to survey the canopy before finally deciding on the route for the walkway. The trees should have a bole diameter of more than 50cm or equivalent stout branches at the level where the walkway is to be attached.
2) Position pulley system for hauling construction materials and equipment. The pulley block should be ca. 3 m above the level of the platform and walkway.
3) Position ladders and abseiling ropes, for ascent and descent. Ladders should be lashed to the tree, and tensioned using Load-Lok straps (see Materials) or similar.
4) Construct the basic platform using timber spars lashed to the tree. The design will depend on the

peculiarities of the tree. It is advisable to cover the floor with the same material as the walkway flooring for better purchase. Handrails and other improvements can be added after the walkway is complete.

5) Connect the two trees at platform level by firing a thin monofilament line with a heavy bolt from one tree to the other using a crossbow. Several shots are usually required before success.
6) Add a thicker cord to the monofilament line and draw it across and loop back to give two parallel lines. Follow this with the polyester main support cable so that its two ends are in the same tree. The looped end of the cable is attached to the other tree with a heavy snatch block.
7) Position the suspension tapes in the correct order on the cables (Fig. 9). It is useful to tie the end of each suspender to the next with a length of cord equivalent to the final distance between each suspender; this helps when pulling out the flooring sections after which the cords can be removed. (Fig. 10)
8) Secure and partly tension the main support cable. Tensioning is achieved with a Tirfor jack (see Materials) or similar attached to loops at one end of the cable (Fig. 11). Care should be taken to avoid abrasion at points of attachment to the tree.
9) Construct the flooring. Each length of floor is attached to a pair of suspenders with shackles (Fig. 12) and bolted to the next with fishplates. This is done on the platform, and as each piece is added the growing bridge is pulled across from the other side with a rope already connected between the trees (Figs. 7 and 13). It is helpful to use a Tirfor jack to pull the dridge across since the loops of the suspenders can become jammed on the main cables.
10) When the floor is complete, the main cables are tensioned to give, ideally, a sag of 1/50 of the length of the walkway. The ends of the floor are lashed to the platform or the tree. The tension will require periodic adjustments during the life of the walkway.

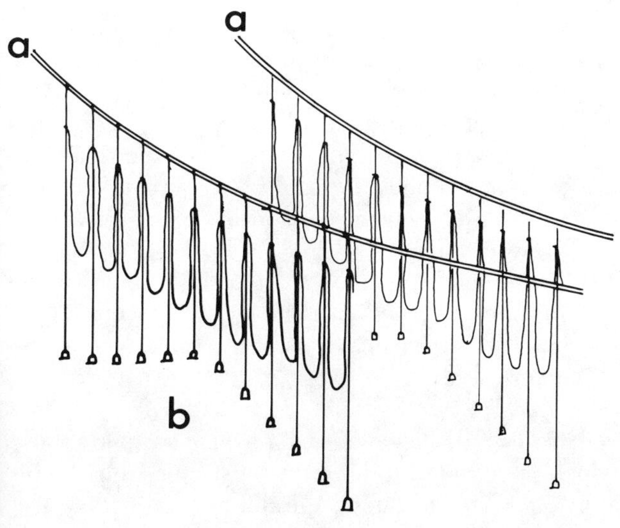

Fig. 9. The arrangement of suspenders (b) on the main cables (a) prior to building the floor. The suspenders are cut to individual lengths to accommodate the suspension bridge design.

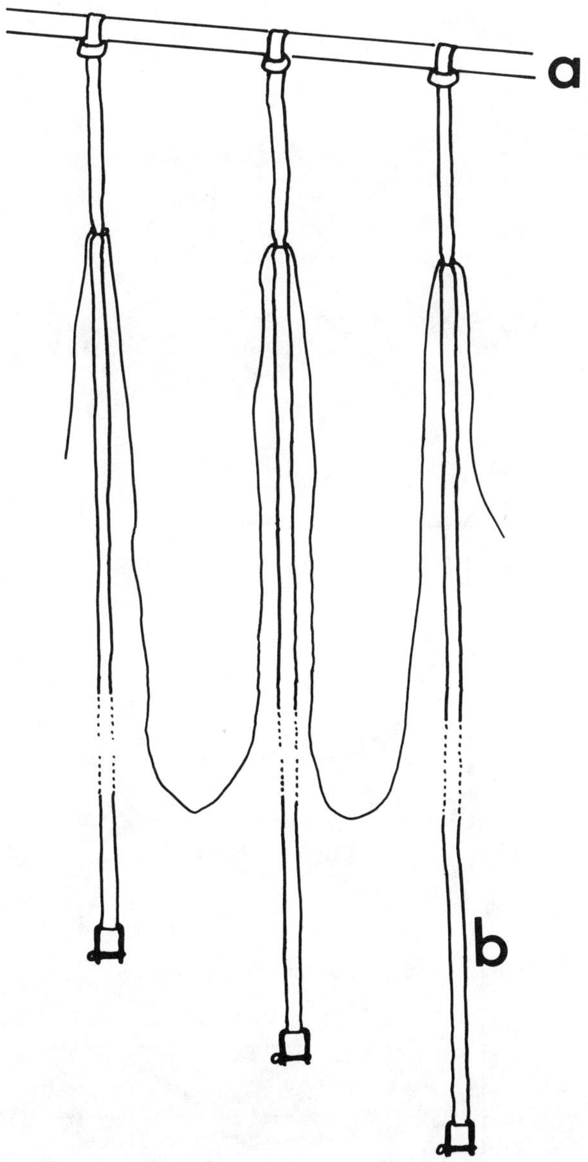

Fig. 10. The attachment of the suspenders (b) to the main cable (a) showing the cords tied between the suspenders to assist when pulling out the flooring sections.

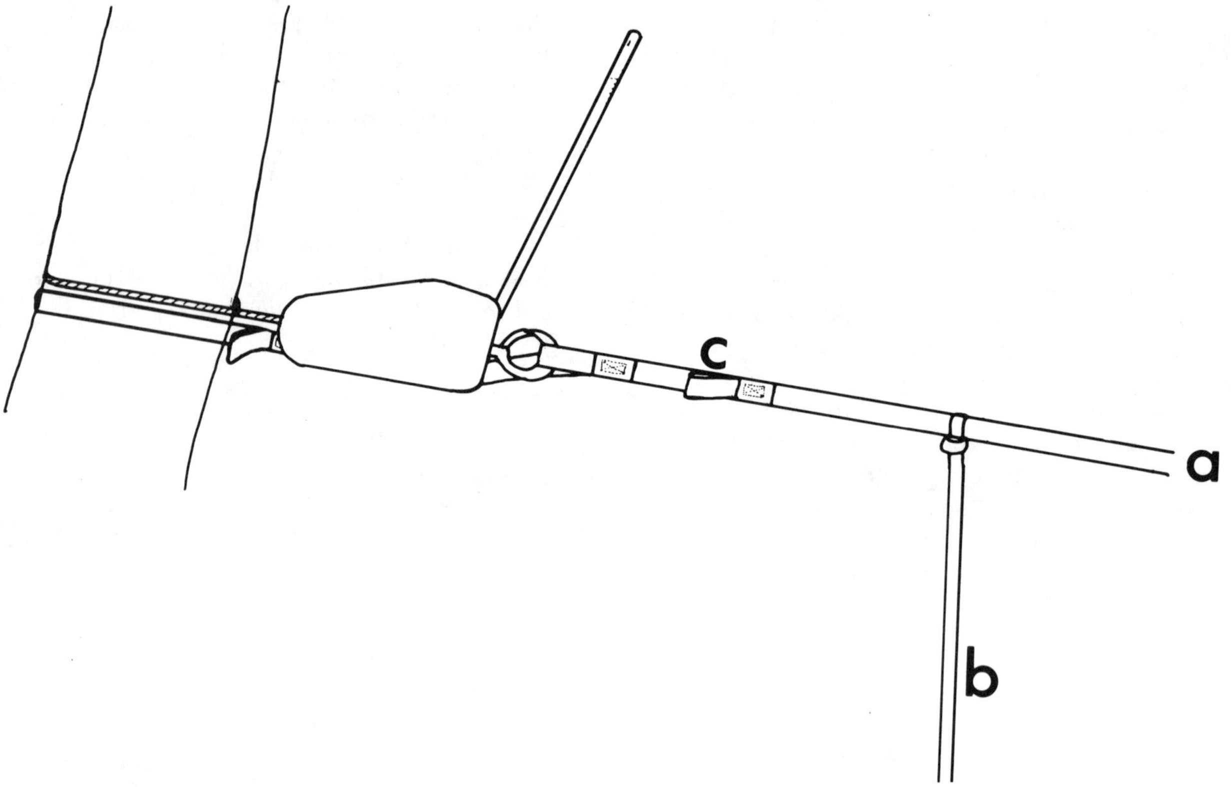

Fig. 11. Tensioning the main cable with a Tirfor jack. The loops (c) sewn in the main cable (a) are used for sequential attachment to the jack as tensioning proceeds. b = suspender.

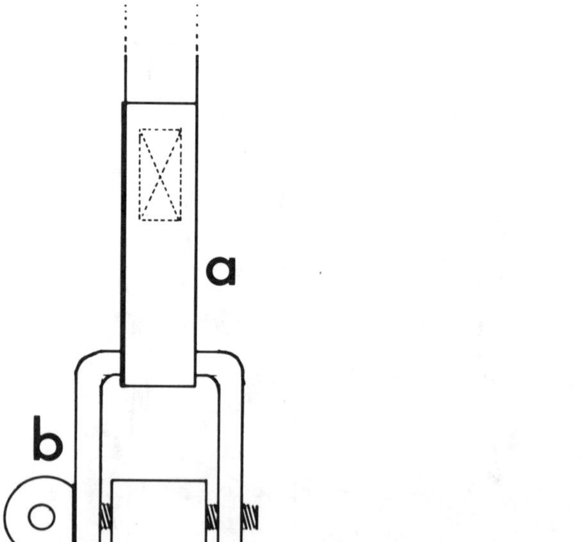

Fig. 12. Shackle attachment (b) of the suspender (a) to the floor (c).

11) Safety lines of steel wire rope are secured between the trees running above and completely independent of the walkway. There should be one safety line for each individual on the walkway.
12) A handrail of webbing tape is attached to the suspenders at waist level. An additional handrail, at shoulder level, can be useful at the ends of the span where the suspenders are longest.
13) Lateral guy ropes are added, from the ground to the supporting cables. These have the dual purpose of keeping the cables well apart (allowing freedom of movement) and of reducing sway in the walkway.
14) Tensioned guy ropes are attached between the supporting trees and the ground using Tirfor jacks and polyester tapes of the same type as the main cables.

The materials listed here, for a 50 m span, are those which were found to be most satisfactory during this study in terms of safety, durability and ease of use. Sources of supply in the U.K. are given where necessary. Some of the other materials tested are illustrated in Figs. 1–6 ; see also Table 1.

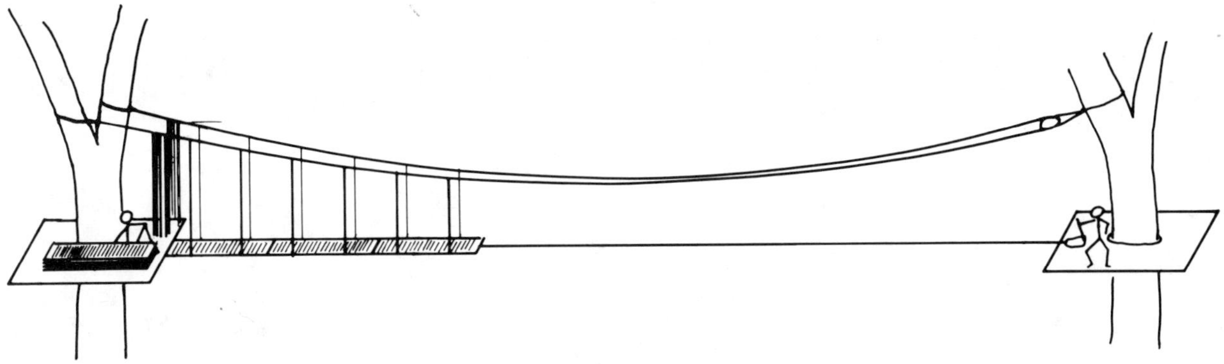

Fig. 13. Building the floor, which is pulled across with the aid of a Tirfor jack (right). Note the snatch-bolt attachment to the right-hand tree.

1) *Main support cables*. 120 m × 4.5 cm polyester webbing tape, 10,000 lb breaking strain, with loops sewn at one end for attachment to Tirfor Jack. Package Control Ltd., Crowthorne, United Kingdom.
2) *Suspenders*. 2.5 cm polyester webbing tapes, breaking strain 5,000 lb, with loops sewn at each end. 4 each of lengths 110, 125, 130, 135, 140, 145, 150, 155, 160, 165, 170, 175, 180, 185, 190, 195, 200, 210, 215 and 220 cm, supplied as a set by Package Control Ltd.
3) *Handrails*. 2.5 cm × 200 m (minimum) polyester webbing tape 5,000 lb breaking strain. Package Control Ltd.
4) *Flooring*. 1.8 m × 22.5 cm lengths of galvanized steel "Dosp" factory flooring. 28 lengths, extra lengths for platform flooring. Dexion Ltd., Wembley, United Kingdom.
5) *Fishplates and bolts*. for the flooring, supplied with the floor by Dexion Ltd.
6) *Shackles*. 80 plus 20 spare, minimum working load 5 cwt.
7) *Snatch block,* working load 5 tonnes minimum.
8) *Safety lines*. 3–5 mm diameter steel wire rope, minimum safe working load 5 cwt, 250 m length.
9) *Ladders*. Polyester tape with metal rungs, ca. 40 m per tree. Package Control Ltd.
10) *Load-Lok straps*. Ratchet straps and D-buckles for tensioning ladders. Package Control Ltd.
11) *Tirfor jacks*. 1 per span plus 1 or 2 extra. Tirfor Ltd., Halfway, Sheffield, U.K.
12) *Crossbow plus bolts*.
13) *Monofilament line*. At least 5 lengths of 100 m, to allow for breakages and loss due to tangling.
14) *Parachute cord*. 100 m, for drawing across after monofilament line.
15) *Lashing rope, and lateral guys*. 300 m × 15 mm diameter (minimum) sisal, hemp or non-stretch nylon.
16) *Main support guys*. 200 m × 4.5 cm polyester webbing tape. Package Control Ltd.
17) *Pulley blocks* plus rope, 2 per tree, for hauling equipment and construction materials.
18) *Climbing equipment*. 200 m × 11 mm nylon climbing rope. Webbing sit-harnesses (1 per person). Jumars (ascenders) (2 per person, plus extras). Figures-of-eight (descenders) (1 per person, plus extras). Karabiners (2 per person, plus extras). 2.5 cm × 1 m lengths of polyester webbing tape, with loops sewn at ends (2 per person, plus extras).
19) *Miscellaneous tools*. Pliers, spanners, etc.

Total cost, ca. $5,000. Note that some items are used repeatedly for different spans, so this figure overestimates the actual price of a walkway. Nevertheless, this will be offset by the purchase of spare parts. Checks on the security of lashings and bolts should be carried out twice weekly. Materials which show signs of wear should be replaced; the walkway is designed so that any piece, including the main support cable, can be replaced without disturbing the position of the other parts. Major overhauls should be undertaken by qualified personnel, but minor repairs and checks can be done by those using the walkway. Retensioning of ladders is frequently required.

PART II

THE INTERACTIONS—
LAS INTERACCIONES

Plants, Animals, and Habitats

A fundamental difference between tropical and temperate life is the degree to which life in the tropics interacts with other life. A prime requirement for an organism's success is for it to accommodate to the physical conditions of its living site. Physical conditions of the living site are partly determined by other organisms, and this is particularly true in the humid tropics. The individual must accommodate not only to the physical conditions provided by other organisms but also to the behavior and interactions of the other organisms, and these interactions are more complex in the humid tropics than elsewhere. This series of studies deals with examples of such interactions.

The euglossine bees are an interesting group of New World insects. The males collect substances from specific flowers, often traveling over long distances to find blossoms and mates. Ackerman examines the seasonality of these bees and of the flowers they feed on. In the dry season when nectar is more plentiful, small, short tongued bees predominate, but later larger bees become more numerous.

How do tropical plants know when to bloom? In lowland Panama, some species bloom all at once throughout a district. Sudden changes in moisture or temperature seem to regulate the timing. Augspurger evaluates these systems and finds more precision for flowering than for fruiting.

How do small birds select the fruits they eat? Denslow and Moermond watch plants and birds both in the wild and in captivity to see which fruits are eaten first. If shrubs place their fruits for maximum attractiveness to birds, do they also encourage feeding by rodents which do not disperse the seeds so well as the birds? The positioning of fruits for maximum dispersal is a tradeoff between several opposing factors.

Although leaves are present on most of the trees in central Panamanian forest all the time, these tropical forests actually undergo striking seasonal variations. Foster finds that different forests have differing periodicity of leaffall and reappearance which in turn affects different groups of animals, and that year to year changes caused by weather differences lead to an instability of animal numbers.

The forest is a dynamic system with new elements continually replacing old elements. One part of this replacement process is the removal of tree crowns by mechanical fracture and uprooting. Putz et al. examine this process, and they find that trees damaged in these ways sometimes resprout to join the canopy again.

One abundant and conspicuous element in the lowland tropical forest is the Heliconias, or wild bananas. In the neotropics they are almost exclusively pollinated by hummingbirds. The two papers by Kress and Stiles show how these two kinds of plants and birds have diversified, and both papers point out that the patterns are mutualistic. Groups of hummingbird species have evolved specialized structures and behaviors to take advantage of specialized forms of the flowers which are in turn specially constructed to encourage hummingbird visits and to discourage visits by other organisms.

Plantas, Animales y su Hábitat

Una de las diferencias básicas entre la vida en los trópicos y aquella de la zona templada es el grado de interacción de unos organismos con otros en los trópicos. Uno de los requerimentos primarios para el éxito de un organismo es el de acomodarse a las condiciones físicas del sitio que habita, el cual es parcialmente determinado por otros organismos. Esto es especialmente cierto en los trópicos húmedos. Asimismo los individuos deben adaptarse al comportamiento e interacciones procedentes de otros seres vivientes. Estas interacciones son más complejas en los trópicos húmedos que en cualquier otro lugar. La serie de estudios que se exponen en este capítulo versan sobre estas interacciones biológicas.

Las abejas euglosinas constituyen un grupo interesante de insectos de los Neotrópicos. Los machos recogen substancias de flores específicas y frecuentemente viajan vastas distancias para hallar las flores, al igual que a las parejas. Ackerman examina la variación anual en el número de abejas y de las flores de las cuales éstas se alimentan. En la estación seca, cuando abunda el néctar, predominan las abejas pequeñas de probóscide corta, mientras que las abejas de mayor tamaño son más numerosas más tarde en el año.

Cómo las plantas tropicales determinan cuándo florecer? En las tierras bajas de Panamá, algunas especies florecen simultáneamente en un área de cambios súbitos en humedad o temperatura; éstos parecen regular el momento de floración. Augspurger evalúa estos sistemas y demuestra que existe más precisión en la floración que en la fructificación.

Cómo las aves pequeñas seleccionan los frutos que comen? Denslow y Moermond observaron plantas y aves, tanto silvestres como en cautiverio, para determinar qué frutos comen primero. Están los frutos localizados en las plantas para atraer a los pájaros y en algunos otros casos a roedores los cuales no dispersan sus semillas tan eficientemente como las aves? La posición de los frutos en las plantas para su máxima dispersión es un balance entre varios factores opuestos.

A pesar de que los árboles de los bosques del centro de Panamá poseen hojas durante todo el año, en realidad esos bosques tropicales sufren cambios sorprendentes durante diferentes épocas del año. Foster demuestra que distintos bosques presentan periodicidad diferente en relación a la caída y reaparición de sus hojas, la cual afecta diversos grupos de animales. Año tras año, cambios causados por variación del clima provocan inestabilidad en el número de animales.

El bosque es un sistema dinámico en donde nuevos elementos constantemente reemplazan los viejos. Una parte de este proceso de reemplazo es la remoción de árboles mediante fractura mecánica o desarraigados del suelo. Putz et al. examinan este fenómeno y señalan que algunos de esos árboles retoñan y eventualmente se incorporan otra vez al dosel del bosque.

Un elemento abundante y conspicuo de los bosques tropicales de tierras bajas son las Heliconias, un género de monocotiledóneas bastante grande. En los Neotrópicos éstas son casi exclusivamente polinizadas por colibríes. El estudio de Kress al igual que el de Stiles muestran cómo dos tipos de aves y las plantas se han diversificado y ambos autores señalan que los patrones son mutualísticos. Grupos de especies de colibríes han desarrollado estructuras y comportamientos especializados para aprovechar las diversas formas de flores, las cuales a su vez están construidas para alentar las visitas de los colibríes y desalentar las de otros organismos.

Euglossine Bees and Their Nectar Hosts

James D. Ackerman

Los machos de abejas euglosinas y las flores que éstos visitan se estudiaron en la isla de Barro Colorado, localizada en las tierras bajas de Panamá. Se utilizaron como señuelos fragancias químicas que se ubicaron en el bosque y se efectuó un censo de las abejas que las visitaron. Las abejas estuvieron mejor representadas en especies y en número a partir de la mitad de la estación seca hasta la primera parte de la lluviosa (febrero a junio). Las abejas pequeñas con probóscide corta fueron las dominantes. Más tarde cuando el número de abejas fue menor, las abejas grandes con probóscide larga prevalecieron. La mayoría de los datos señalaron cómo plantas ciertas hospederas producen pocas flores muy especializadas, de acceso limitado por día durante un período largo. Estas florecieron primordialmente de junio a septiembre, mientras que el clímax de floración de hospederos menos especializados del dosel y subdosel del bosque ocurrió más temprano.

Durante la estación seca y comienzo de la lluviosa, abejas de todos los tamaños y con probóscide de diversa longitud fueron abundantes y se alimentaron principalmente de la masa de flores de la parte superior del bosque. Más tarde, cuando el néctar disponible disminuye, la abundancia de abejas decrece y las abejas grandes de probóscide larga prevalecieron.

The predominantly solitary euglossine bees (Hymenoptera: Apidae: Euglossini) are important pollinators of many low and mid-elevation neotropical plants. In central Panama, the tribe constitute 15–20% of the forest bee fauna (Roubik pers. comm. 1981). Some species are quite common. Females pollinate flowers as they search for nectar, pollen, and sometimes resin to construct and provision nests (Zucchi et al. 1969; Armbruster & Webster 1979). Males are not involved in nesting activities, but visit and pollinate the same nectar flowers as females. They also seek other flowers, most notably orchids, from which they obtain perfumes (Dodson et al. 1969). Both sexes are strong, rapid fliers and reputedly long-lived (Kimsey 1980; Dodson 1962, 1966; Janzen 1981; pers. obs.). Many of their food plants bloom for long periods and produce only a few flowers each day (Janzen 1971). Under natural, undisturbed conditions, female euglossine bees are thought to trapline widely spaced host plants (Janzen 1971, 1974). The traplines may cover several square kilometers (Janzen pers. comm. 1981). Males may be nomadic and vagabond (Dodson 1970) or exhibit foraging behavior reminiscent of traplining (Ackerman et al. 1982). Thus, both sexes may be important long-distance pollen vectors (Janzen 1971; Williams & Dodson 1972).

Abejas euglosinas se caracterizan por visitar plantas ampliamente distanciadas, las cuales florecen poco durante un período de tiempo largo. Sin embargo, en el centro de Panamá, tanto las abejas como las flores son periódicas.

In some habitats food plants with the long-term flowering may provide their pollinators with nonseasonal food resources so that severe fluctuations in bee abundance do not occur. This is the case for *Heliconius* butterflies in seasonal habitats of Trinidad (Ehrlich & Gilbert 1973). In central Panama, however, the male euglossine bee community and its nectar and orchid fragrance hosts are seasonal. There is no indication that the orchids have an important effect on bee seasonality.

Methods

Male euglossine bee populations in central Panama were monitored for over a year. The primary study site was within the tropical moist forest (Tosi 1971; Holdridge Life-Zone system) of Barro Colorado Island (BCI) at the Tower Clearing. The climate is distinctly seasonal with a dry period of 3–4 months beginning near mid-December. BCI's year is divided into four 13 week seasons: dry, early wet, mid wet, and late wet. Roughly, the dry season corresponds to the first quarter of the year, and the wet seasons cover the last 9 months, but the dry season actually falls 2 weeks earlier and later than this. A detailed description of the climate and vegetation is in Croat (1978).

The salient features of the censusing program (Ackerman 1983) are as follows. Every week from mid-February 1979 to mid-March 1980, and every four weeks thereafter through June 1980, I censused male euglossine bees which were attracted to 16 baits, each a different chemical applied to a paper blotter pad. All baits (cineole, benzyl acetate, vanillin, eugenol, methyl benzoate, methyl salicylate, methyl cinnamate, linalool, geraniol, skatole, 2–phenyl ethyl acetate, 2–phenyl ethanol, b-ionone, b-myrcene, d-carvone, benzyl benzoate) are naturally occurring compounds and most are constituents of the floral fragrances of the orchids pollinated by male euglossine bees (Hills et al. 1968, 1972; Dodson 1970; Givaudan Index 1961). During the mornings from 0830–1230 I presented the chemical attractants every 15 minutes and recorded all individuals and species at each pad.

Una serie de fragancias químicas, impregnadas en papeles, se utilizaron como señuelo en el bosque y las abejas que los visitaron se contaron y midieron de acuerdo con el tamaño del cuerpo, largo de la probóscide y edad.

I used two measures of the morphological complexion of the community: 1) body size, measuring the wing lengths of all species attracted to baits to the nearest mm. The length of a bee's wings is fixed and a function of the bee's size (Medler 1962; Cumber 1949; Michener & Lange, 1959). 2) The length of folded tongues to the nearest mm (after Ricklefs et al. 1969; Kimsey 1977). Euglossine bees are noted for their long tongues, and interspecific differences in some cases are considerable.

The age structure of the BCI community was monitored from November 1979 through June, 1980. Male bees attracted to the Tower Clearing baits were classified as either young or old depending on the size and number of wing nicks.

Seasonality of the Euglossine Bee Community

Nectar host plants of bees were determined by direct observation, literature records, and unpublished data provided by colleagues. The flowering seasonalities of these plants were taken from Gentry (1974) and Croat (1978) and substantiated by direct observation whenever possible. Seasonality was expressed as the range in blooming time.

Seasonality of Resources: Fragrance Hosts

Seasonal patterns of this male euglossine bee community of BCI were presented by Ackerman (1983). In one year, beginning March 1979, the 44 species of male euglossine bees attracted to the Tower Clearing baits included representatives of all four genera native to Panama: *Euglossa* (28 species), *Eufriesia* (9), *Eulaema* (5), *Exaerete* (2). Only 6 species known or expected to occur on BCI failed to come.

Se observaron 44 especies de abejas euglosinas distribuidas en 4 géneros y todas, excepto 6 de las especies esperadas, visitaron los señuelos.

Weekly bee abundance and species richness were highly correlated (Spearman's Rank Correlation Coefficient, Daniel 1978: r_s = .79, P less than .0001). Both were seasonally variable. Species richness was greatest (41) in the early wet and lowest (28) in the mid wet season. Bee abundance abruptly increased in the middle of the dry season, peaked in the early wet, then abruptly fell to low levels in the mid wet season (Fig. 1).

There are several body forms and size classes of male euglossine bees. *Exaerete* are large robust bees (20–29 mm long) with wasp-like bodies. *Euglossa* are small to medium sized bees (9–18 mm long) with a honey bee shape. *Eufriesia* and *Eulaema* are medium to large bees (13–30 mm long), conspicuously hairy, and bumblebee-like. Forewing length averages of these genera ranged from 7.1–23.1 mm. The majority of species had wings less than 10 mm long, and all of these were *Euglossa*. Smaller bees were dominant in the dry and early wet season, but their prevalence diminished in the mid and late wet seasons (Table 1).

Exaerete son abejas grandes con cuerpos simi-

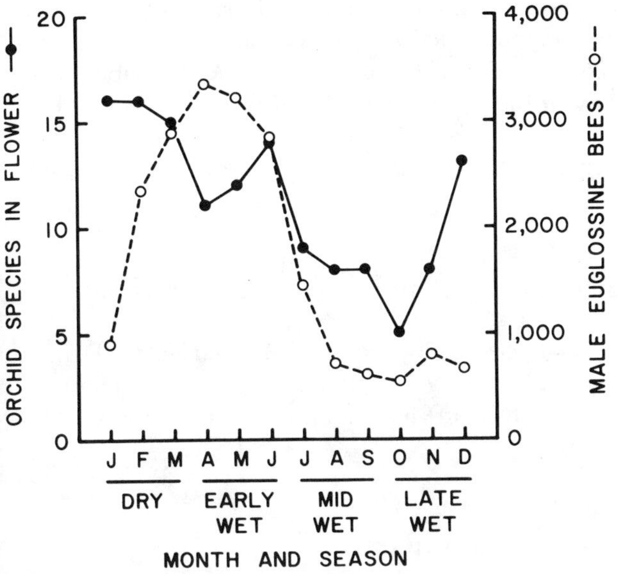

Fig. 1. Seasonality of male euglossine bee abundance and the number of orchid species in flower for 1979–1980.

lares a los de avispas; Euglossa, abejas más pequeñas parecidas en forma a aquellas que producen miel; Eufriesia y Eulaema son de tamaño mediano a grande con pelos y similares a los abejorros. La mayoría de las abejas poseían alas de menores de 10 mm y primordialmente eran especies del género Euglossa.

The average length of folded tongues ranged from 4.7–19.5 mm—within species variation was slight. Over half the species had tongues 5–8 mm long. *Eulaema* and *Exaerete* had long tongues (11.1–16.1 mm; 13.9–16.7 mm, respectively), and *Euglossa* and *Eufriesia* had a broad range of lengths (5.0–17.8 mm; 4.7–19.5 mm, respectively). The seasonal distribution of tongue lengths was similar to that of body size. The prevalence of short tongue bees is greatly reduced in the mid and late wet seasons (Table 2).

Thus, one may recognize two "bee seasons". From the middle of the dry season through the early wet season, male bees were abundant, and small and short-tongued species were dominant. For the remainder of the year, bee abundance was low, and long-tongued and large bodied species were more prevalent.

La cantidad y tipos de abejas varían periódicamente durante un año. Se notó que desde la mitad de la estación seca hasta la primera parte de la lluviosa predominan en gran cantidad las especies pequeñas de probóscide corta y durante el resto del año, cuando las abejas son menos abundantes, prevalecen las grandes de probóscide larga.

The baiting program included sampling at two mainland sites at low elevation 6 and 15 km from the Tower Clearing, and here the results mirrored the BCI patterns. However, bee communities in nearby upland areas, Cerro Campana, Cerro Jefe, and Santa Rita Ridge, (all 400–1000 m) are quite different (Ackerman 1983; Ricklefs et al. 1969; Roubik pers. comm. 1980).

Seasonality of female euglossine bees may be similar to that of males, but at present there is no chemical baiting technique for censusing female bees. The best method to assess male versus female abundances is to monitor visits to the flowers. Ackerman et al. (1982) observed 9 male and 8 female *E. imperialis* individuals nectar foraging at *Calathea latifolia* during the mid wet season. In the early wet season when males of this species were much more numerous at the Tower Clearing baits, 11 individually marked females and 9 individually marked males were

TABLE 1
SEASONAL ABUNDANCE OF BEES ON BCI

Body size classes as estimated by wing length

Wing Length Class (mm)	No. of Species	Season: Early Dry	Mid Wet	Late Wet	Wet
5.1-10.0	25	3798	5856	647	609
10.1–15.0	6	1437	2155	918	892
15.1–20.0	9	1214	1794	602	401
20.1–25.0	4	438	563	384	234

G test of independence (Sokal and Rohlf, 1969)
G equals 1466, P less than .005

TABLE 2
SEASONAL ABUNDANCE OF BEES ON BCI

Tongue length classes

Wing Length Class (mm)	No. of Species	Season: Early Dry	Mid Wet	Late Wet	Wet
4.0–7.9	25	3569	5752	682	483
8.0–12.9	8	1333	1793	463	486
13.0–19.9	12	1985	2823	1406	1167

G test of independence, G equals 1523, P less than .005

observed visiting *Stemmadenia grandiflora* (Ackerman & Montalvo unpubl.). For at least *E. imperialis*, one of the two most abundant species, the sex ratio appears roughly equivalent during these juxtaposed seasons with very different male bee activity levels.

Las hembras euglosinas pueden ser tan abundantes como los machos.

Circumstantial evidence suggests that male bee populations here are resident, and immigration is apparently rare. Only 6 of 35,668 bees censused at lowland sites had pollinariums of orchids not present in the area, and these carriers represented only 0.4% of the total pollinium laden bees. Furthermore, changes in bee abundance were accompanied by shifts in the age structure of the community. As bee abundance rose in the dry season, the proportion of young bees increased (Ackermann 1983). Thus, seasonal patterns appear related to population senescence and hatching.

Seasonality of Resources: Nectar Hosts

There are 59 observed or reported nectar hosts of euglossine bees on BCI and in surrounding forested areas (Table 3). Five families account for 80% of the

TABLE 3
NECTAR HOSTS OF EUGLOSSINE BEES BCI AND VICINITY

Flowering period, phenology type, and visitors[1].
APOCYNACEAE
Mandevilla villosa, all year, steady state: *Euglossa imperialis* (pers. obs.)
Mesechites trifida, all year, steady state: *Eg. tridentata* (Montalvo pers. comm.)
Odontadenia macrantha, Dec-Jun, cornucopia: *Eg. allo sticta*, *Eg. tridentata*, *Euglossa* sp. (pers. obs.)
Prestonia portobellensis, all year, steady state: *Euglossa* sp. (Montalvo pers. comm.)
Rhabdadenia biflora, all year, steady state: *Eulaema nigrita* (Steiner pers. comm.)
Stemmadenia grandiflora, all year, steady state: *Eg. imperialis*, *Eg. allosticta*, *El. leucopyga*, *El. cingulata*, *El. meriana*, *El. polychroma*, *Exaerete frontalis* (pers. obs.)
Thevetia ahouai, all year, steady state: *Ex. frontalis* (Kimsey 1979)
BIGNONIACEAE[2]
Adenocalymma arthropetiolatum, Oct-Nov, cornucopia: (Gentry 1974)
Adenocalymma apurense, Apr-Aug, cornucopia: (Gentry 1974)
Amphilophium paniculatum, May-Jan, steady state[3]: (Gentry 1974)
Anemopaegma chrysoleucum, all year, steady state: (Gentry 1974)
Arrabidaea candicans, Nov-Feb, cornucopia: (Gentry 1974)
Arrabidaea chica, Mar-Nov, cornucopia: *Euglossa spp. (Gentry 1974)*
Arrabidaea corallina, Feb-Apr, cornucopia: (Gentry 1974)
Arrabidaea patellifera, Jun-Nov, cornucopia: *El. polychroma*, *Eufriesia* sp., *Eg. variabilis* (Gentry 1974)
Callichlamys latifolia, all year, cornucopia: euglossines, 6 spp. (Gentry 1974)
Ceratophytum tetragonolobum, Apr-Aug, steady state: (Gentry 1974)
Jacaranda copaia, Feb-May, cornucopia: (Gentry 1974)
Macfadyena unguis-cati, Mar-Jun, big bang: (Gentry 1974)
Pachyptera kerere, all year, steady state: *Euglossa* sp. (Gentry 1974), *Eg. imperialis*, *El. leucopyga* (pers. obs.)
Paragonia pyramidata, all year, cornucopia: (Gentry 1974)
Pleonotoma variabilis, Feb-Apr, steady state: (Gentry 1974)
Stizophyllum riparium, Apr-Nov, steady state: *Eg.* aff. *igniventris* (Gentry 1974)
Tabebuia ochracea var. *neochrysantha*, Jan-Jun, big bang: *Euglossa* sp. (Gentry 1974)
Xylophragma seemannianum, Mar-May, cornucopia: *Euglossa* sp., *Eg. townsendi* (Gentry 1974)
EUPHORBIACEAE
Margaritaria nobilis, May-Jun: *Ex. smaragdina* (Heithaus 1979b)
GESNERIACEAE
Drymonia serrulata, Jun-Jan, steady state: *Eg. allosticta*, *Eg. dissimula*, *Eg. igniventris*, *Eg. townsendi*, *Eg. tridentata*, *Eg. variabilis*, *El. nigrita*, *Ex. frontalis* (Steiner pers. comm.)
LEGUMINOSAE
Andira inermis, Feb-May, Oct, cornucopia: *Euglossa* sp. (Frankie et al. 1976)
Centrosema pubescens, all year, steady state: *El. nigrita* (Steiner pers. comm.)
MALVACEAE
Hibiscus rosa-sinensis, all year, steady state: *El. nigrita* (Dodson 1966; pers. obs.), *Ex. frontalis* (pers. obs.; Kimsey 1979), *Ex. smaragdina* (Kimsey 1979)

TABLE 3 Continued

MARANTACEAE

Calathea insignis, Jun-Nov, steady state: *Eg. dissimula* (Dressler 1978), *Euglossa, Eulaema, Eufriesia* (Janzen 1971), *El. cingulata, El. polychroma, Eg. ignita, Ex. smaragdina* (Dodson 1966), *Eg. dodsoni, Eg. gorgonensis, Eg. hansoni* (Dodson 1966; Dressler pers. comm.), *Eg. allosticta* (Moure 1969), *El. nigrita, El. speciosa, Eg. cyanura, Eg. cybelia, Eg. igniventris, Eg. tridentata* (Dressler pers. comm.)

Calathea inocephala, Apr-Sep, steady state: *Eg. imperialis* (pers. obs.)

Calathea latifolia, Jul-Dec, steady state: *Eg. imperialis, El. nigrita, El. cingulata, El. meriana* (Ackerman et al. 1982; Kennedy 1978), *El. seabrai, Euglossa* sp. (Kennedy 1978), *El. leucopyga, Ex. frontalis, Ex. smaragdina* (Ackerman et al. 1982)

Calathea lutea, all year, steady state: *El. meriana, El. seabrai, Eg. imperialis* (Dressler 1978)

Calathea marantifolia, Jun-Oct, steady state: *El. polychroma, El. cingulata, El. meriana, El. seabrai, Eg. imperialis* (Kennedy 1978)

Ischnosiphon pruinosus, Jul-Oct, steady state: *Eg. ignita* (Dressler pers. comm.), *Eg. imperialis* (Ackerman et al. 1982; Croat 1978)

Stromanthe jacquinii, Jun-Dec, steady state: *Eg. cyanura* (pers. obs.)

MUSACEAE

Heliconia latispatha, May-Nov, steady state: *El. nigrita* (Steiner pers. comm.)

ORCHIDACEAE

Sarcoglottis acaulis, Mar-May, cornucopia: *El. cingulata* (pers. obs.)

RUBIACEAE

Alibertia edulis, Apr-Oct, steady state: *Eg. cyanura* (Steiner pers. comm.)

Faramea occidentalis, Mar-Jul, cornucopia: *El. cingulata* (pers. obs.)

Hamelia patens, all year, steady state: *Ex. smaragdina* (Kimsey 1979)

Isertia haenkeana, May-Nov, steady state: *Eg. cyanura, El. nigrita* (pers. obs.)

Manettia reclinata, Dec-Feb, steady state: *Euglossa* sp. (Montalvo pers. comm.)

Pentagonia macrophylla, Apr-Sep, steady state: euglossines (Dodson & Gentry 1978)

Psychotria acuminata, Mar-Jul, cornucopia: *Eg. cyanura* (pers. obs.)

Psychotria emetica, Mar-Aug, steady state: *Euglossa* sp., *Eg. cyanura* (pers. obs.)

Psychotria horizontalis, Mar-Aug, steady state: *El. nigrita* (pers. obs.)

Randia armata, Mar-Jun, big bang: euglossines (pers. obs.)

Sabicea villosa, all year, steady state: *Eg. championi, Eg. cyanura, Eg. dressleri, Eg. hansoni, Eg. hemichlora, Eg. heterosticta, Eg. mixta, Eg. sapphirina, Eg. villosiventris* (Dressler pers. comm.)

TILIACEAE

Apeiba tibourbou, all year, cornucopia (?): *El. polychroma* (Heithaus 1979b)

VERBENACEAE

Stachytarpheta jamaicensis, all year, steady state: *Euglossa* sp., *El. polychroma* (Heithaus 1979b)

VIOLACEAE

Hybanthus prunifolius, Feb-May, big bang: *Euglossa* sp. (pers. obs.)

ZINGIBERACEAE

Costus allenii, May-Nov, steady state: *Eg. imperialis* (Dressler pers. comm.), *Ex. frontalis* (Kimsey, 1979)

Costus guanaiensis, May-Nov, steady state: *El. meriana* (Dressler 1979), *Eg. allosticta, Eg. imperialis, Ex. frontalis, Ef. schmidtiana, Ef. surinamensis* (Dressler pers. comm.)

Costus laevis, Jun-Sep, steady state, *Ex. frontalis* (Kimsey 1979; Dressler pers. comm.), *Eg. imperialis, Ef. schmidtiana* (Dressler pers. comm.)

Costus villosissimus, Mar-Oct, steady state: *El. meriana* (Dodson 1966; Dressler 1979), *Ef. duckei, El. cingulata* (Dressler pers. comm.), *El. nigrita, Ef. surinamensis, Ef. schmidtiana, Eg. imperialis* (Dodson 1966; Dressler pers. comm.), *El. leucopyga, El. polycroma, Ef. ornata, Eg. gorgonensis, Eg. hansoni, Eg. ignita* (Dodson 1966), *Ex. smaragdina* (Dodson 1966; Kimsey 1979), *Ex. frontalis* (Kimsey 1979; Dressler pers. comm.), *Eg. turbinifex* (Dressler 1978), *Eg. allosticta* (Moure 1969; Dressler pers. comm.)

Dimerocostus strobilaceous, all year, steady state: *Ex. frontalis* (Kimsey 1979; Schemske pers. comm.), *El. meriana* (pers. obs.)

[1]Not all observations of visitors were made on BCI or in central Panama. *Eg.* = *Euglossa*, *El.* = *Eulaema*, *Ef.* = *Eufriesia*, *Ex.* = *Exaerete*.

[2]Where pollinator observations are not indicated, the form of the flower suggests pollination by euglossines.

[3]Steady state includes modified steady state phenology of Gentry's (1974) terminology.

species: Apocynaceae (6 species), Bignoniaceae (19), Marantaceae (7), Rubiaceae (10), and Zingiberaceae (5). Undoubtedly the list will grow. Most species bloom in the early and mid wet seasons, and primarily April-August (Fig. 2).

Two thirds of the nectar hosts bloom in the understory or are at least occasionally found there. Also, better than two thirds of the species have a steady state or modified steady state flowering strategy (terminology: Gentry 1974). That is, these plants produce few flowers at a time and bloom for extended periods ranging from several weeks to all year. Twenty-two of 58 nectar hosts are both steady state bloomers (including modified steady state) and flower in the understory.

Las abejas euglosinas visitaron por lo menos 59 especies para obtener néctar y el 80% de éstas están distribuidas en cinco familias: Apocynaceae (6 especies), Bignoniaceae (19), Marantaceae (7), Rubiaceae (10) y Zingiberaceae (5).

Peak blooming activity of steady state flowers occurs from June-September, and the fewest flower in December-March. Species with other flowering strategies bloom relatively early: March-May. Most of those species that flower in the upper forest strata are also in bloom at this time.

Dos tercios de las especies productoras de néctar florecen en el sotobosque y más de dos tercios producen pocas flores en cada floración durante un período largo.

Nectar host plant records of BCI are similar to records of hosts in other regions. The proportion of steady state flowers on BCI was not significantly different from that of all other nectar host records. The same is true for understory versus upper forest strata nectar hosts (Table 4).

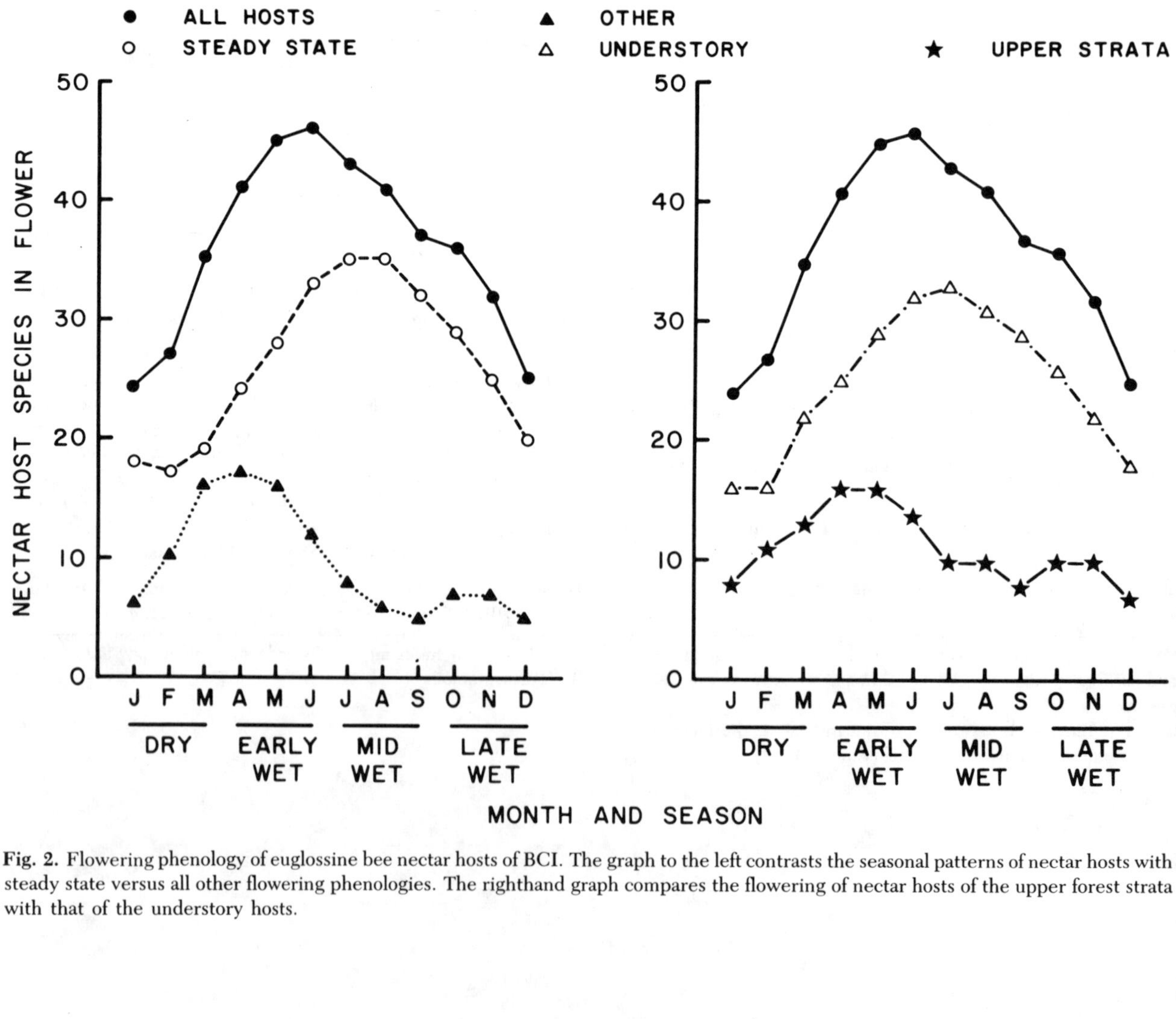

Fig. 2. Flowering phenology of euglossine bee nectar hosts of BCI. The graph to the left contrasts the seasonal patterns of nectar hosts with steady state versus all other flowering phenologies. The righthand graph compares the flowering of nectar hosts of the upper forest strata with that of the understory hosts.

TABLE 4
EUGLOSSINE BEE NECTAR HOST SPECIES

Flowering phenology type and inflorescence locations: BCI versus other areas

	Phenology Type[2]		Flower Location[3]	
	Steady State	Other	Understory	Other
BCI	37	21	37	22
Other localities[1]	52	20	46	26

[1]Includes nectar hosts that occur in Central and South America but not on BCI.
[2]chi squared = 1.1, P greater than 0.1
[3]chi squared = .019, P greater than 0.5

Discussion

The seasonal patterns of the andro-euglossophilous orchid flora and male euglossine bee community of lowland central Panama are reasonably concordant. Most orchid species bloom in the dry season, but peak activity of the bees is later. From this perspective, the overall phenological correlation is inexact. However, eclosion events are frequent in the dry season which may provide a flux of young bees not yet satiated with collected compounds. At peak bee activity, there is a greater proportion of older bees. These may have more discriminating needs and may visit orchids with complex floral fragrances less frequently.

The majority of euglossine bee nectar hosts on BCI have steady state flowering phenologies. The proportion these plants make of the total nectar hosts is no different than that found elsewhere. However, most of the steady state flowers bloom primarily in the mid wet season at a time when male bee abundance and species richness is lowest.

Steady state flowers are thought to be traplined (Janzen 1971, 1974), or at least foraged consistently day after day (Ackerman et al. 1982). If they are widely spaced as claimed, then their pollinators should possess extraordinary flight capabilities. Pyke (1978) and Levin (1978) suggest that the large bees are best suited for flying long distances and traplining plants. Certainly euglossine bees appear capable. A census of bees attracted to chemical baits placed at three sites on Lake Gatun offshore from BCI supports this view (Ackerman 1981). The furthest site was 1 km away from the bee source area on the island. The proportion of large bees (*Eulaema* and *Exaerete* to small bees (*Euglossa* recorded on the lake was significantly different from that of the island (lake, 112:18, Tower Clearing 341:1508).

The majority of euglossine bee nectar hosts have limited access flowers. For example, the long, narrow corolla tubes of most known BCI nectar hosts of the Marantaceae, Apocynaceae, Rubiaceae, and Zingiberaceae (Croat 1978) exclude most short tongue flower visiting wasps and bees, including many euglossines. These plants account for 69% of the total known nectar hosts with steady state flowering phenology, and they flower mostly in the mid wet season. It is not surprising that long tongue euglossine bees should be more prevalent at this time. Thus, the pecularities of the mid wet season nectar hosts; widely spaced plants, steady state flowering phenologies, and long corolla tubes, may account for the shift in the size make-up of the euglossine bee community.

La mayoría de las flores que las abejas euglosinas visitan para obtener néctar son de acceso limitado.

The flowering pattern of the classical euglossine bee nectar hosts, the understory steady state flowers, does not explain the seasonal fluctuations of male euglossine bee abundance. When the bees are most active during the middle of the dry season through the early wet season, the nectar hosts of the upper forest strata are at peak bloom (Fig. 2). These plants typically produce masses of flowers but generally do not bloom for much more than a month (Gentry 1974). Canopy and subcanopy flowers are generally inaccessible so our knowledge of euglossine bee nectar host plants may suffer from a strong observational bias.

Euglossine bees may frequently forage flowers of the upper forest strata, particularly in the dry and early wet seasons. During this time the bees may feed predominately on common, massive flowering, and relatively unspecialized bee flowers, (many Bignoniaceae, Gentry, 1974, 1976). In such situations traplining may be an inappropriate foraging behavior. In fact, many solitary bees foraging on mass flowering species tend to remain on a single tree (Frankie et al. 1976). Later in the year as the wet season progresses, flowering in the upper forest strata diminishes and the scattered understory nectar hosts become relatively more important resources (Fig. 2). Bees would then be more inclined to trapline.

Durante la parte media de la estación lluviosa la mayoría de las flores que producen néctar están ampliamente espaciadas y tienen los tubos de las corola largos. A la vez, las abejas de probóscide larga son más notorias.

This scenario is not without precedent. Heithaus (1979a) reported that in Costa Rica some bee and wasp species foraged in mass flowering trees in the dry season, but shifted to shrubs and small trees in the wet. This represented a switch from a concentrated resource to many dispersed ones. Heithaus suggested that the change in host plant choice is a response to plant phenology. Ackerman et al. (1982) predicted that the nectar foraging behavior of male euglossine bees may change with the seasons, and the patterns presented here support their hypothesis. The two bee seasons, high abundance with small bodied, short tongued bees dominant, and low abundance with large bodied, long tongued bees relatively more prevalent, appear to reflect changes in the dispersion, abundance, and type of nectar hosts as well as a possible change in foraging behavior.

Los patrones de variación de las abejas asociados a las estaciones y la estabilidad de los hospederos de néctar no coinciden exactamente y es muy probable que durante la parte final de la estación seca las abejas visiten especies que florecen densamente en el dosel del bosque, lo cual es difícil de observar. Más tarde se alimentan del néctar de flores de especies disponibles en el sotobosque.

Acknowledgments

I thank N.H. Williams, R.L. Dressler, and D.W. Roubik for discussions and for logistic and technical assistance. A.M. Montalvo prepared the illustrations and assisted in the field. D.R. Strong and the others mentioned criticized the manuscript. Robert L. Dressler, K. Steiner, and A.M. Montalvo provided numerous unpublished flower visitor data. The project was financed by a Smithsonian Institution predoctoral fellowship and a Smithsonian scholarly studies grant to D.W. Roubik.

Ackerman, J.D. 1981. The phenological relationships of male euglossine bees (Hymenoptera: Apidae) and their orchid fragrance hosts. Ph.D. diss. Florida State Univ., Tallahassee, FL.

Ackerman, J.D., M.R. *Mesler,* K.L. *Lu* & A.M. *Montalvo.* Food foraging behavior of maleEuglossini (Hymenoptera: Apidae): vagabonds or trapliners? Biotrop. 14: 241–248.

Armbruster, W.S. & G.L. *Webster.* 1979. Pollination of two species of *Dalechampia* (Euphorbiaceae) in Mexico by euglossine bees. Biotrop. 11: 278–283.

Buchmann, S.L. 1981. Preliminary anthecological observations on *Xiphidium caeruleum* Aubl. (Monocotyledonae, Haemodoraceae) in Panama. J. Kansas Entomol. Soc. 53: 685–699.

Croat, T.B. 1975. Phenological behavior of habit and habitat classes on Barro Colorado Island (Panama Canal Zone). Biotrop. 7: 270–277.

———. 1978. Flora of Barro Colorado Island. Stanford Univ. Press, Standford, CA.

———. 1980. Flowering behavior of the neotropical genus *Anthurium* (Araceae). Amer. J. Bot. 67: 888–904.

Cumber, R.A. 1949. The biology of bumblebees with special reference to the production of the worker caste. Trans. Roy. Ent. Soc.London. 100: 1–45.

Daniel, W.W. 1978. Applied nonparametric statistics. Houghton Mifflin Co., Boston.

Dodson, C.H. 1962. The importance of pollination of the orchids of tropical America. Amer. Orchid Soc. Bull. 31: 525–534, 641–649, 731–735.

———. 1965a. Studies in orchid pollination.The genus *Coryanthes.* Amer. Orchid Soc. Bull. 34: 680–687.

———. 1965b. Agentes de polínización y su influencia sobre la evolución en la familia Orquidácea. Univ. Nac. Amazonia Peruana, Inst. General de Invest.

———. 1966. Ethology of some bees of the tribe Euglossini (Hymenoptera: Apidae). J. Kansas Entomol. Soc. 39: 607–629.

———. 1967a. Studies in orchid pollination:the genus *Notylia.* Amer. Orchid Soc. Bull. 36: 209–214.

———. 1967b. Relationships between pollinators and orchid flowers. Atas do Simpósio sobre a Biota Amazonica 5: 1–72.

———. 1970. The role of chemical attractants in orchid pollination. Pp. 83–107. *In* K.L. Chambers, ed. Biochemical Coevolution. Oregon State Univ. Press, Corvallis, OR.

———. 1975. Coevolution of orchids and bees. Pp. 91–99. *In* L.E. Gilbert & P.H. Raven, eds. Coevolution of Animals and Plants. Univ. Texas Press, Austin, TX.

——— & A.H. *Gentry.* 1978. Flora of the Rio Palenque Science Center. Selbyana 4: 1–628.

———, R.L. *Dressler,* H.G. *Hills,* R.M. *Adams* & N.H. *Williams.* 1969. Biologically active compounds in orchid fragrances. Science 164: 1243–1249.

Dressler, R.L. 1967. Why do euglossine bees visit orchid flowers? Atas do Simpósio sobre a Biota Amazonica 5: 171–180.

———. 1968. Observations on orchids and euglossine bees in Panama and Costa Rica. Rev. Biol. Trop. 15: 143–183.

———. 1976. How to study orchid pollination without any orchids. P. 534–537. *In* K. Senghas, ed. Proceedings of the eighth world orchid conference. German Orchid Soc., Frankfurt, West Germany.

———. 1978. New species of *Euglossa* from Mexico and Central America. Rev. Biol. Trop. 26: 167–185.

———. 1979. *Eulaema bombiformis*, *E. meriana*, and muellerian mimicry in related species (Hymenoptera: Apidae). Biotrop. 11: 144–151.

———. 1980. Checklist of the orchids of Panama. *In* L.O. Williams & P.H. Allen. Orchids of Panama. Monogr. Syst. Bot., Missouri Bot. Gard. 4: I-XXVI.

Ehrlich, P.R. & L.E. *Gilbert*. 1973. Population structure and dynamics of the tropical butterfly *Heliconius ethilla*. Biotrop. 5: 69–82.

Frankie, G.W., P.A. *Opler* & K.S. *Bawa*. 1976. Foraging behavior of solitary bees: implications for outcrossing of a neotropical forest tree species. J. Ecol. 64: 1049–1057.

Gentry, A.H. 1974. Coevolutionary patterns in Central American Bignoniaceae. Ann. Missouri Bot. Gard. 61: 728–759.

———. 1976. Bignoniaceae of southern Central America: distribution and ecological specificity. Biotrop. 8: 117–131.

Givaudan Index. 1961. Specifications of synthetics and isolates for perfumery. 2nd ed. Givaudan-Delawanna, New York.

Heithaus, E.R. 1979a. Flower-feeding specialization in wild bee and wasp communities in seasonal neotropical habitats. Oecol. 42: 179–194.

———. 1979b. Flower visitation records and resource overlap of bees and wasps in northwest Costa Rica. Brenesia 16: 9–52.

Hills, H.G., N.H. *Williams* & C.H. *Dodson*. 1968. Identification of some orchid fragrance components. Amer. Orchid Soc. Bull. 37: 967–971.

———, 1972. Floral fragrances and isolating mechanisms in the genus *Catasetum* (Orchidaceae). Biotrop. 4: 61–76.

Janzen, D.H. 1971. Euglossine bees as long-distance pollinators of tropical plants. Science 171: 203–205.

———. 1974. The deflowering of Central America. Nat. Hist. 83(4): 48–53.

——— (in press). Bee arrival at two Costa Rican female *Catasetum* orchid inflorescences, and a hypothesis on euglossine population structure. Oikos.

Kennedy, H. 1978. Systematics and pollination of the "closed-flowered" species of *Calathea* (Marantaceae). Univ. California Publ. Bot. 71: 1–90, 20 pl.

Kimsey, L.S. 1977. New species of bees in the genera *Euplusia* and *Eufriesia* (Hymenoptera: Apidae, Euglossini). Pan-Pacific Entomol. 53: 8–18.

———. 1979. An illustrated key to the genus *Exaerete* with descriptions of male genitalia and biology (Hymenoptera: Euglossini, Apidae). J. Kansas Entomol. Soc. 52: 735–746.

———. 1980. The behavior of male orchid bees (Apidae, Hymenoptera, Insecta) and the question of leks. Animal Behav. 28: 996–1004.

Levin, D.A. 1978. Pollinator behavior and the breeding structure of plant populations. Pp. 133–150. *In* A.J. Richards, ed. The pollination of flowers by insects. Linnean Soc. Symp. Ser. 6. Academic Press, New York.

Medler, J.T. 1962. Morphometric studies on bumblebees. Ann. Entomol. Soc. Amer. 55: 212–218.

Michener, C.D. & R.B. *Lange*. 1959. Observations on the behavior of Brazilian halictid bees (Hymenoptera, Apoidea). IV. *Augochloropsis*, with notes on extralimital forms. Amer. Mus. Novitates 1924: 1–41.

Mouré, J.S. 1969. The central American species of Euglossa subgenus *Glossura* Cockerell,(Hymenoptera, Apidae). Rev. Bio. Trop. 15: 227–247.

Pyke, G.H. 1978. Optimal body size in bumblebees. Oecol. 34: 255–266.

Ricklefs, R.E., R.M. *Adams* & R.L. *Dressler*. 1969. Species diversity of *Euglossa* in Panama. Ecology 50: 713–716.

Sokal, R.R. & F.J. *Rohlf*. 1969. Biometry. W.H. Freeman and Co., San Francisco.

Tosi, J.A., Jr. 1971. Zonas de vida, una base ecológica para investigaciones silvícolas y inventariación forestal en la República de Panamá. Rome: Org. Nac. Un. Agric. y Aliment.

Wiehler, H. 1978. The genera *Episcia*, *Alsobia*, *Nautilocalyx*, and *Paradrymonia* (Gesneriaceae). Selbyana 5: 11–60.

Williams, N.H. 1982. The biology of orchids and euglossine bees. Pp. 119–171. *In* J. Arditti, ed. Orchid biology: reviews and perspectives, II. Cornell Univ. Press, Ithaca, NY.

——— & C.H. *Dodson*. 1972. Selective attraction of male euglossine bees to orchid floral fragrances and its importance in long distance pollen flow. Evolution 26: 84–95.

——— & R.L. *Dressler*. 1976. Euglossine pollination of *Spathiphyllum* (Araceae). Selbyana 1: 349–356.

Zucchi, R., J.M.F. *de Camargo* & S. F. *Sakagami*. 1969. Biological observations on a neotropical bee, *Eulaema nigrita*, with a review on the biology of Euglossinae (Hymenoptera: Apoidea). A comparative study. J. Fac. Sci., Hokkaido Univ., Ser. VI, 17: 271–380.

Flowering Synchrony of Neotropical Plants

Carol K. Augspurger

El sincronismo en la floración es importante para las especies tropicales y afecta el éxito reproductivo y la estructura de la población. La selección en relación al sincronismo puede actuar sobre el individuo a través de un número de mecanismos y el grado de sincronización varía con el estado reproductivo. El éxito del individuo es parcialmente determinado por su sincronización con el resto de su población.

Patterns of Flowering Synchrony

Early biologists visiting the tropics noted the highly "gregarious" (synchronous) flowering of certain species, especially those with individuals in flower for a very short time (Mueller 1882, Schimper 1903, Spruce 1908, Coster 1926, Holttum 1935). Spruce (1908) said of the Amazonian Myrtaceae: "They are remarkable for their simultaneous and ephemeral flowers. On a given day all the myrtles of a certain species, scattered throughout the forest, will be clad with showy fragrant flowers; on the following day nothing of flowers appears save withered remnants." Another well-documented example of synchronous flowering occurs in the pigeon orchid *Dendrobium crumenatum* that flowers for only 1–2 days but synchronously over a wide geographic area in the Far East (Holttum 1949, 1954).

On the other hand, Baker (1959) suggested that asynchronous flowering is common in the tropics and that the proportion of neighboring trees of the same species in flower at the same time may be extremely low. The first phenological study to provide systematic observations of synchrony for many different tropical species took place in Costa Rica where asynchrony occurred in 13% of the tree species and in 33% of the treelet and shrub species in a wet forest, and in 2% of the tree species in a dry forest (Frankie et al. 1974, Opler et al. 1980). A species with only infrequent deviants was not designated as asynchronous, and the category of asynchrony appears to have been assigned only if some individuals flowered at a completely different time than other members of the species. The sample size per species was small (2.5 individuals at wet site, 10.2 at dry site).

A more quantitative study indicated that synchrony is not a dichotomy—asynchronous versus synchronous—but rather is a continuum when comparisons are made among various species in a community (Augspurger 1983). There daily flower production was quantified for 20 individuals for each of 6 shrub species on Barro Colorado Island (BCI) (Fig. 1). The 6 species differed greatly in the mean length of an individual's flowering period and the mean number of flowers per individual (Table 1).

La sincronización en la floración ha sido señalada por varios científicos, sin embargo en muchas especies la floración es asincrónica. En la floración en los trópicos hay un patrón continuo entre sincronismo y asincronismo.

Synchrony of the day of onset of flowering was quantified as the standard deviation among 20 individuals. A decrease in the value of the standard deviation indicated greater synchrony. Peak day of flowering was not chosen for comparison because individuals in the 3 species with long flowering times did not have a readily defined peak day; their flower production was distributed evenly over time.

Synchrony of the onset of flowering increased as the mean length of flowering per individual decreased (Table 1). The day of onset of flowering was more

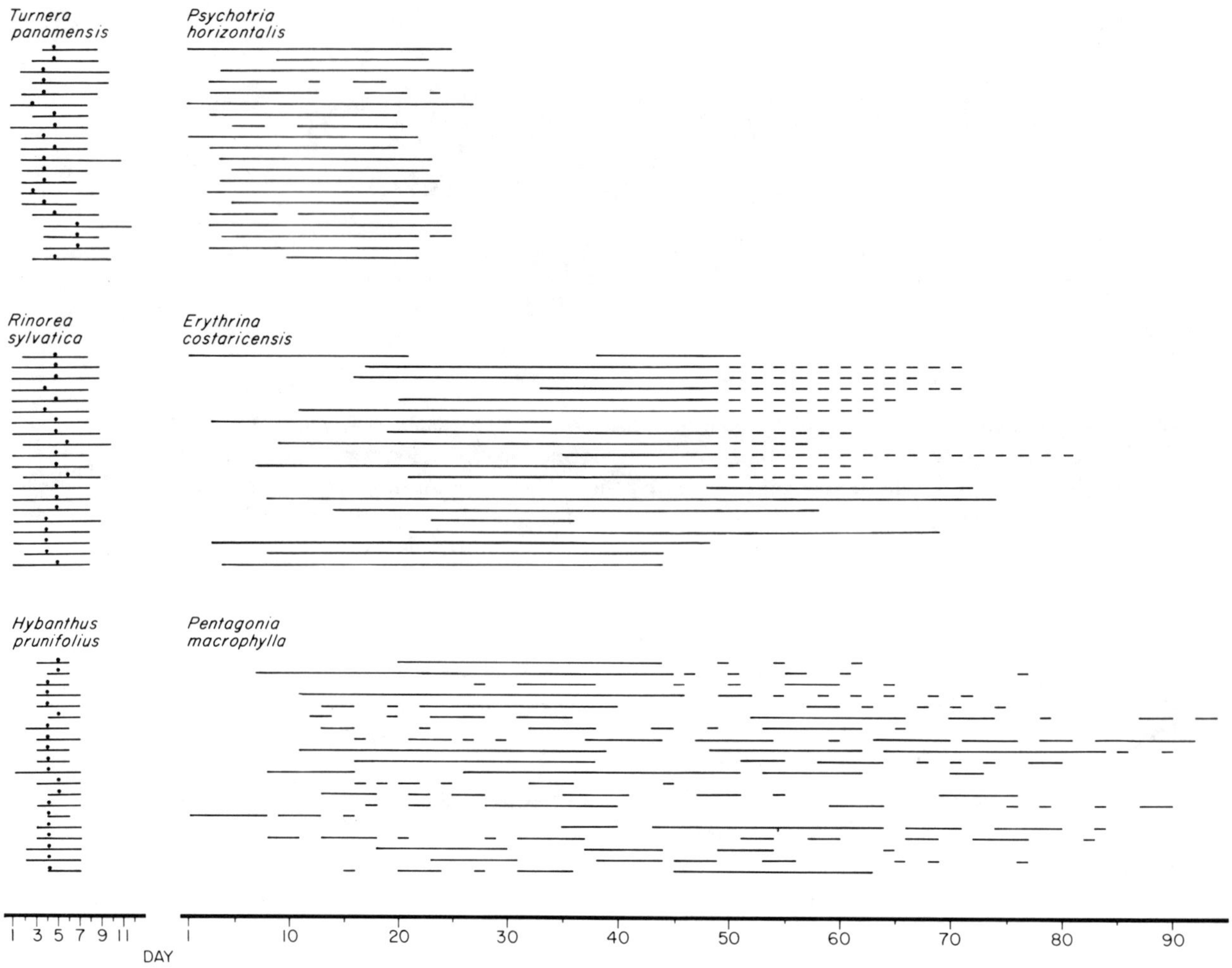

Fig. 1. Comparisons among 6 shrub species of timing of flowering of 20 individuals per species. Each line represents length of time an individual is in flower. Dot above line represents day with largest number of flowers (peak day). Flower number per day for the 3 species with longer flowering periods was fairly constant; therefore no peak day is noted for those species. A dotted line for *Erythrina costaricensis* equals period of estimation; only final day of flowering is known with certainty for those individuals.

TABLE 1
PHENOLOGICAL AND POLLINATION CHARACTERISTICS OF SIX SHRUB SPECIES DEPICTED IN FIGURE 1

	Hybanthus prunifolius	Turnera panamensis	Rinorea sylvatica	Psychotria horizontalis	Erythrina costaricensis	Pentagonia macrophylla
Mean no. of flowers per individual	226	499	802	2816	344	64
Mean no. of days per individual	3.7	7.3	8.1	21.0	44.0	59.0
Synchrony of onset of flowering[1]	1.02	0.97	0.31	2.14	11.82	16.16
Season of flowering	Dry	Dry	Dry	Dry-Wet	Wet	Wet
Compatibility and pollination system	Self-Comp. Needs Vistation	Incomp. Heterostyly	?	Incomp. Heterostyly	Incomp.	? Protandry
Pollinator	Social bee	Mixture	?	Small bees & wasps	Hummingbird	Hummingbird
Spacial density	High	Medium	High	Medium	Low	Low

[1]Synchrony calculated as 1 standard deviation around the mean day of onset for 20 individuals of each species.

synchronous in species with individuals that flower for a short time than in species with individuals that flower for a long time. Minor variability in synchrony occurred among populations of the same species at different sites and in different years (Table 2).

Importance of Flowering Synchrony

Flowering synchrony has important implications for both the individual plant and the population. For the individual, synchrony can affect both the quantity and the genetic qualities of its offspring. Flowering synchrony is a requirement for successful reproduction by an obligatory outbreeding individual—dioecious or self-incompatible. Synchrony can also influence the amount of outbreeding of a self-compatible individual and the number of mates with which it may breed.

Synchrony may therefore also influence mate selection if the female can differentially abort ovules on the basis of the genetic qualities of the male. As such, synchrony may influence the genetic quality of offspring of an individual.

Synchrony may also influence the quantity of offspring produced in one season and over an individual's lifetime. If synchrony with the population increases the ability of the individual to attract pollinators (see later), a synchronous individual may begin to reproduce at a smaller size or earlier age than an asynchronous individual. Synchrony then also influences the number of offspring produced in all subsequent reproductive episodes. The timing of flowering may also influence the individual's synchrony during its subsequent reproductive stages, such as fruit and ovule maturation, seed dispersal, and germination (see later).

TABLE 2
VARIABILITY OF SYNCHRONY IN FLOWERING AND SEED DISPERSAL BETWEEN SITES AND YEARS FOR *HYBANTHUS PRUNIFOLIUS*

	Synchrony of Onset[1]			
	1975	N	1976[2]	N
Flowering				
Site 1	1.69	10	0.56	16
Site 2	1.53	20	0.85	20
Site 3	1.18	20	0.91	20
Site 4	1.02	20	0.43	20
Mean	1.36		0.69	
Seed Dispersal[3]				
Site 3	3.01	20		

[1]Synchrony calculated as 1 standard deviation around the mean day of onset at each site.

[2]The rain triggering flower anthesis was much heavier in 1976 than in 1975.

[3]Synchrony of onset of seed dispersal is significantly lower than onset of flowering at this site ($F = 6.52$, p less than .0005).

Flowering synchrony influences the genetic structure of the population and delimits its effective size (Wright 1946). By defining the number of individuals with potential to exchange genes, synchrony can affect the probability of genetic drift and gene fixation and influence local differentation of populations and the rate of evolution.

La sincronización de la floración puede influir en la calidad genética de la descendencia al igual que en la cantidad de descendencia. Igualmente, influye en la sincronización de etapas subsiguientes, tales como la maduración de frutos y óvulos, dispersión y germinación de semillas.

At first glance it might appear that the less synchronous species (Table 1) have less potential for exchange of genes among individuals (i.e., for gene flow). However, synchrony in this analysis was quantified only for the day of onset of flowering and not for the entire flowering period. Even though species with long individual flowering periods are less synchronous in the day of onset, nevertheless each individual has a long overlap with other individuals when the entire flowering period is considered (Fig. 1). In contrast, in a species with a short flowering period, one day's deviation by an individual in the onset of its flowering results in a large loss of overlap over its flowering period (Fig. 1). The major conclusion is that species differing greatly in synchrony of onset of flowering may actually have quite similar overlap when the entire lengths of their flowering periods are considered. Selection for time of onset of flowering must be greatest in species with individuals with short flowering periods so there is overlap throughout the individual's flowering period. Until a better method of quantifying overlap throughout the flowering period is developed, the role of timing of flowering in influencing gene flow in a population cannot be accurately determined. The type of breeding system must also be considered in evaluating the potential for gene flow.

La sincronización del principio de la floración es mayor en especies con períodos de floración cortos.

Evolutionary Factors Influencing Synchrony

Ultimate (evolutionary) factors are those affecting the reproductive fitness of the individual. They explain the evolutionary basis for the timing of flowering of the individual plant. The most frequently mentioned hypothesis to explain the evolution of flowering synchrony is that it increases potential for outcrossing (Holttum 1954, Moore & Lewis 1965, Whitehead 1969, Stern & Roche 1974, Opler et al. 1976). This hypothesis has never been directly tested. Greater outcrossing is a possible result of synchrony, but outcrossing need not be a determining selective factor.

Alternative and supplementary hypotheses involving biotic factors leading to reproductive synchrony include 1) minimization of hybridization and pollen wastage (isolating mechanism) (Smith 1926, Grant 1949), 2) increasing foraging efficiency of pollinators that are (more) likely to remain constant to one species of plant (Heinrich & Raven 1972), 3) increasing pollinator attraction (Corner 1940, Beattie et al. 1973, Opler et al. 1976), and 4) lowering herbivory (predation) of an individual's seeds and fruits by the population satiating the herbivores (Smith 1970, Janzen 1971, Beattie et al. 1973). The last hypothesis is most relevant for those species attacked by herbivores during or shortly after the flowering stage when synchrony is still unchanged and for species with fruiting synchrony comparable to flowering synchrony (see later).

Se considera que la sincronización de la floración incrementa el cruzamiento, minimiza la hibridación y el desperdicio de polen, aumenta la eficacia de los polinizadores y la atracción a éstos, además de que disminuye la destrucción de semillas por herbívoros.

Limits to effectiveness of synchrony may occur if the amount of flowers produced becomes too large for the local pollinators to service. Pollinator satiation may then be a factor affecting the degree of synchrony among tropical trees with their massive floral displays.

These evolutionary arguments require a genetic basis to timing among individuals in a population. Numerous temperate studies indicate that timing of flowering is heritable (McMillan & Pagel 1958, Cooper 1960, Paterniani 1969, Harris 1970, Stern & Roche 1974, Hodgkin & Quinn 1978). In some species early- versus late-flowering of individuals can be explained by a single gene (McIntyre & Best 1978). Selection on variation of timing of a population can greatly affect synchrony of flowering in only a few generations (McNeilly & Antonovics 1968, Paterniani 1969). These genetic studies do not directly indicate which of the proximate factors that control timing of flowering are affected by selection. Data for non-domesticated tropical species on the genetics and timing of flowering are not yet available.

La sincronización de la floración podría ser heredada por mecanismos simples.

Ecological Factors Influencing Synchrony

Evolutionary factors explain why synchrony is advantageous to an individual, but how is such synchrony achieved by the individual? Proximate (ecological) factors relate to the mechanism an individual uses to ensure it flowers synchronously with its conspecifics. Selection can act on several aspects of the timing of an individual: 1) preparation for the flowering response, 2) nature of the environmental trigger initiating the response, and 3) the developmental changes involved in the response itself.

El proceso de selección de la sincronización puede actuar sobre el individuo mediante la preparación para florecer, mediante factores ambientales que inician la respuesta y a través de pasos de desarrollo asociados con la respuesta.

First, the preparation for the flowering response includes the induction of flower buds. Bud initiation may be followed by continuous development to flower anthesis, or it may be followed by a period of dormancy at some incomplete stage of development. Additional factors are then required to break dormancy and to complete development and flower anthesis.

Other things being equal, less variation in timing of flowering is expected if dormant buds are formed, especially if dormancy begins at a late stage of bud development. Then when the environmental trigger for anthesis occurs, less variation in development timing occurs and higher synchrony results.

In a phenological study of the shrub species, *Psychotria horizontalis* (Rubiaceae), on BCI, flowering synchrony was higher in a population that experienced bud dormancy than in one with no bud dormancy (Table 3). Normally this species responds to heavy rains, usually at the beginning of the wet

TABLE 3
ONSET OF FLOWERING SYNCHRONY IN *PSYCHOTRIA HORIZONTALIS* CHANGES WITH DORMANT BUDS PREFORMED PRIOR TO RAINS

	Onset of First Flowering		Onset of Second Flowering		
	Synchrony[1]	Range[2]	Synchrony	Range	n
ite 1[3]	3.09	13	1.00[4]	5	20
ite 2[5]	2.22	10	4.93	24	20

ynchrony calculated as 1 standard deviation around the mean day onset at each site.
Number of days encompassing the onset day of all individuals at ne site.
Plants at this site responded to the first heavy rain by forming ower buds; anthesis occurred in a few buds in all plants. Anthesis f most of these preformed buds occurred with the second rain.
ynchrony of second flowering at Site 1 (with preformed buds) was reater than first flowering of either Site 1 (no preformed buds) (F = 9.55, p less than .0005) or Site 2 (no preformed buds) (F = 4.93, less than .0005).
Plants at this site responded to the first heavy rain by forming ower buds; anthesis of all buds immediately followed. A few new ower buds were formed with the second rain.

eason, by undergoing an uninterrupted developnental sequence from bud induction through flower nthesis. However, in one year in one population on he plateau of the island, a heavy rain late in the dry eason induced bud development, but not much flowr anthesis. A second heavy rain at the beginning of he wet season was followed by highly synchronous owering in that one population. Other populations n other areas of the island flowered after the first ain, but with less synchrony.

La formación de yemas latentes permite mayor incronización de la floración. En la IBC, Psyhotria horizontalis presentó mayor sincronizaión en poblaciones que poseen yemas latentes. Esta especie normalmente manifiesta dearrollo ininterrumpido desde las yemas hasta as flores en respuesta a las lluvias fuertes.

Rutgers and Went (1915) and Seifriz (1923) noted hat tropical plants with high synchrony form dormant uds. In the study on BCI, the three species with the ighest synchrony form dormant buds; the three with he lowest synchrony do not normally do so (Augspurger 1982). Some species without dormant buds, however, also achieve high flowering synchrony, although he degree of synchrony was not quantified (Rees 1964).

Second, irrespective of the presence of preformed dormant buds, the type of environmental trigger affects the probability that other individuals are perceiving the trigger at a comparable time and are beginning or completing flower development and undergoing anthesis at the same time. In the tropics the most unambiguous environmental triggers with the sharpest, most rapid change in environmental conditions appear to be either a rain and change in soil moisture following a drought, or changes in temperature accompanying a tropical storm.

En los trópicos, los factores ambientales más definidos que inician la floración son la lluvia o los cambios de humedad después de una sequía o de temperatura después de una tormenta.

Based on the observation that temperate species known to use photoperiod for diverse developmental phases do not have a high synchrony of those events, photoperiod is less likely to be the cue used by species with such high synchrony of flowering. Photoperiod is known, however, to be involved in tropical species with less synchronous flowering (McClelland 1924, Njoku 1958) and with the initiation of vegetative growth (Stubblebine et al. 1978). Photoperiod may also control the time of floral bud initiation in synchronous species with buds undergoing dormancy and awaiting a second cue for flower anthesis (e.g., coffee, Piringer & Borthwick 1955).

En los trópicos, el fotoperíodo es el factor que influencia las especies con floración menos sincronizada y que puede determinar la iniciación de las yemas florales.

Rain interrupting an extended drought has been implicated as the trigger for many synchronous species in Guanacaste, Costa Rica (Opler et al. 1976, 1980) and on BCI (Augspurger 1982). The brief flowering response for each of these species occurs a consistent number of days after a rain. Repeated waves of synchronous flowering in the liana *Phrygonocydia corymbosa* (Bignoniaceae) occur in the wet season in Panama (Gentry 1974). The environmental trigger for this response is undetermined, although it may involve sudden large changes in temperature associated with tropical storms.

Experimental evidence of the environmental triggers for flowering of tropical species with high synchrony is limited. Coffee (*Coffea* sp.) and *Hybanthus prunifolius* both must experience a drought before they will respond to a rapid increase in soil moisture. The environmental change is perceived below ground; sprinkling of the foliage is not necessary to induce flowering in these two species (Alvim 1960,

1973, Augspurger 1982). A leaf water potential of -12 bars is required before the flower buds of *Coffea arabica* can break dormancy. Irrigation brings about the opening of the flowers by a rapid influx of water into the buds in response to the leaf water potential (Magalhaes & Angelocci 1976). In other tropical species with high synchrony of flowering, sharp drops in temperature (Holttum 1949, 1954) and simultaneous changes in temperature and moisture (Kerling 1941) have been experimentally demonstrated or implicated as the trigger.

Hybanthus prunifolius y café requieren de una sequía antes de que respondan a la humedad del suelo y a los cambios percibidos debajo del suelo.

Finally, after the trigger occurs, the amount of development required for flower anthesis influences the variability in timing of the response. Other things being equal, the smaller the flower, the faster the development, and the more advanced the stage of the dormant bud, the less the variation in the onset of flowering. Also, as discussed earlier, the synchrony of the flowering response within the individual influences the intensity of selection for synchrony with others. Generally, greater intra-individual synchrony was found to correlate with greater intra-population synchrony of the onset of flowering (Table 1).

An Experimental Study of Synchrony

An experimental study of *Hybanthus prunifolius* (Violaceae) on BCI (Augspurger 1981) illustrates the interaction between the evolutionary and ecological factors influencing synchrony. The shrub species used in the study occurs in high density on BCI. It forms dormant flower buds in the wet season. It requires a drought before it will respond by flowering 6–7 days after a heavy rain in the dry season (Augspurger 1982). These cue requirements parallel those of coffee (*Coffea* sp., Alvim 1960, 1973). *Hybanthus prunifolius* exhibits an extremely high synchrony of the onset of flowering (Fig. 1).

Different individuals of *Hybanthus prunifolius* were experimentally induced to flower in the forest by sprinkling them with water at different times following a drought during the dry season. A highly asynchronous population was created that was in flower for a 2–month period, although the population naturally is in flower for only 8–10 days. The length of time that any one individual was in flower was not significantly different between the two groups. Individuals in the induced asynchronous population were compared with those in the highly synchronous population for their ability to attract pollinators and to avoid seed predators.

En la IBC arbustos de Hybanthus prunifolius se indujeron a florecer asincrónicamente aplicándoles agua durante la estación seca. Estos individuos atrajeron menos polinizadores y sus semillas fueron mayor objeto de predación. Por consiguiente, la sincronización de la floración mejora la adaptabilidad de la población.

Individuals in the induced asynchronous population attracted fewer visits by the pollinator, a social bee. Flowering synchrony in *Hybanthus prunifolius* is followed by a high synchrony of fruit and seed development. The natural synchronous population satiated the seed predator but individuals in the induced asynchronous population incurred more seed predation. The net effect was that individuals synchronous with conspecifics experienced, on the average, a 10–fold increase over asynchronous individuals in reproductive output, as measured by the number of seeds dispersed. Clearly, the flowering synchrony of an individual with its conspecifics has the potential to greatly influence its overall fitness. Flowering in synchrony with the population had the biggest effect on the reproduction of individuals with the smallest number of flowers. Selection for synchrony therefore also interacts with the size and age of first reproduction.

In this species selection by pollinators and seed predators favors the individual in synchrony with its conspecifics. The individual achieves such synchrony in part by the use of a sharp and unambiguous cue, a heavy rain following a long drought. It is impossible to determine the evolutionary sequence by which these evolutionary and ecological factors affecting synchrony became intertwined. Selection of that particular cue may also relate to other evolutionary pressures, unrelated to synchrony, such as need for water for both flower and fruit. The biotic factors may then be of secondary or reinforcing importance in maintaining the present level of synchrony and the use of that particular cue.

Effects of Subsequent Reproductive Stages

The timing of an individual's later reproductive stages is not necessarily coupled to its timing of flowering. Selection may act independently on each stage of reproduction. Also, later developmental stages may use other environmental triggers to synchrony from previous stages. Hence high synchrony of flowering is neither a prerequisite for nor a predictor of high synchrony at later stages.

La sincronización de la fructificación no está en correlación con la sincronización de la floración y por eso la selección de la sincronización no es tan intensa.

In general, the longer the time between two reproductive stages—flowering and fruit ripening—the greater the probability that synchrony of the second stage will be lower than the first stage because the intervening stages are more likely to take variable times for different individuals. This prediction assumes that selection for synchrony is not acting independently on the second stage and that no additional trigger to increase synchrony initiates the second stage.

In general, selection for synchrony is likely to be more intense on the flowering stage than on fruit ripening. This is especially true if outcrossing, requiring a high degree of flowering synchrony, is of prime importance: an individual must be synchronous with conspecifics to achieve successful pollen transfer.

Comparable precision of synchrony has not been reported for fruit ripening. Seed dispersal does not require any movement between conspecifics. While attraction of animal dispersers may be enhanced by ripening of fruit synchronously, selection for precise synchrony is not expected. Also species using inanimate agents for dispersal would experience no selection for synchrony of fruit ripening with conspecifics, unless a post-dispersal seed predator is involved. Synchrony of earlier stages in fruit development may also be affected by pre-dispersal seed predators (Janzen 1971).

Synchrony of fruit ripening and seed dispersal is often more difficult to quantify than synchrony of flowering. This is especially true for those species with no clear indicator of when a fruit is fully mature and ready for dispersal. Exceptions to this include species that use inanimate means of dispersal, such as gravitational or explosive devices. In these species, time of dispersal coincides with fruit maturation and thus synchrony is quantifiable. Other examples are species that open the mature fruit for dispersal of arillate seeds. This action clearly demarcates the time of availability of mature seeds.

Synchrony of both flowering and seed dispersal has been quantified for *Hybanthus prunifolius* on BCI. Seed dispersal in that species is explosive and occurs about one month after flowering; it requires a progressive dehydration of the capsule valves. Its synchrony of dispersal is significantly lower than its highly synchronous flowering (Table 2).

En Hybanthus prunifolius la dispersión de semillas ocurre explosivamente un mes después de la floración.

Synchrony of fruit ripening of *Randia armata* (Rubiaceae), an understory tree on BCI, is markedly lower than its highly synchronous flowering (Augspurger, unpublished data). Flowering occurs after a heavy rain in the late dry season or the beginning of the wet season. Over several months the fruits develop slowly to full size; then they remain green and unripe for many months. Finally, up to 18 months after flowering, the fruits ripen to an orange color and are mammal-dispersed. Flowering of the population lasts about 10 days while fruit ripening and seed dispersal occur over a 2–month period.

En Randia armata la fructificación no está tan sincronizada como su floración y los frutos son dispersados por los mamíferos.

Seed germination may be synchronous as a carryover from synchronous flowering and fruiting. However, it is as likely to be asynchronous due to different selection pressures, such as unpredictability of environmental conditions required for seedling establishment. Likewise asynchronous flowering and fruiting may be followed by synchronous germination because of the use of a different environmental trigger for germination.

In *Hybanthus prunifolius* on BCI, seed germination requires a rain trigger; usually the seeds respond to the onset of the wet season. Even with this additional cue, however, synchrony of seed germination is markedly lower than either flowering or fruit ripening in this species. Germination occurs over a 4–5 month period, although 90% is completed within a 6–week period (Augspurger 1979).

La germinación de Hybanthus prunifolius responde a la lluvia, pero, aún así, está menos sincronizada que la floración.

Highly synchronous germination may have both positive and negative outcomes. Synchrony may produce density effects by increasing competition among a parent's offspring and with offspring of nearby conspecifics. It may also attract density-dependent herbivores and allow infected seedlings to pass fungal disease to adjacent plants (Burdon & Chilvers 1976, Augspurger, unpublished data). Alternatively, synchrony may produce such a high density of seedlings that the herbivores are satiated (Connell 1971).

Future Directions

The quantification of phenological events of neotropical plants at the population level is in a beginning stage. More such studies are needed, especially in different seasons and in different habitats with different selective forces and environmental triggers of phenological events. Also lacking is a rigorous measure of population overlap incorporating both the entire length of an individual's reproductive phase and variation in number of available reproductive units—flowers/fruits—over that time period. Such a measure would allow a more direct test of the relative potential for outbreeding in different species. A measurement of synchrony of either the onset or the peak day of flowering is insufficient for this use.

La cuantificación de fenómenos fenológicos presenta varias dificultades, sin embargo es necesario esclarecer el valor adaptativo de la variación de diferentes acontecimientos reproductivos.

Particularly lacking are studies comparing the relative synchrony of all reproductive stages of a species and the direction in which synchrony changes during the entire reproductive period. More information is also needed on environmental triggers available, which ones are used by different species, the number of triggers used by a species for its entire reproductive period, and the genetic basis for the timing of reproductive events. Finally, more experimental studies are needed, as they explore the adaptive value of the observed variability in synchrony among different reproductive events and different species.

Acknowledgments

This research was supported by a Smithsonian Institution Predoctoral Fellowship and a University of Michigan Predoctoral Fellowship. Field assistance from Tania Beliz, Paula Amann, and support from many BCI residents was indispensable. I thank M. Willson and B. Benner who provided comments.

Alvim, P. 1960. Moisture stress as a requirement for flowering coffee. Science 132: 354.

———. 1973. Factors affecting flowering of coffee. Pp. 193–202. *In*: A.M. Srb, ed., Genes, Enzymes, and Populations. Plenum Press, London.

Augspurger, C.K. 1979. Irregular rain cues and the germination and seedling survival of a Panamanian shrub (*Hybanthus prunifolius*. Oecol. 44: 53–59.

———. 1981. Reproductive synchrony of tropical plants: experimental studies on effects of pollinators and seed predators on *Hybanthus prunifolius* (Violaceae). Ecology 62: 775–788.

———. 1982. A cue for synchronous flowering. *In*: E.G. Leigh, Jr., A.S. Rand & D.M. Windsor, eds. The ecology of a neotropical forest: seasonal rhythms and longer-term changes. Smithsonian Instit. Press.

———. 1983. Phenology, flowering synchrony, and fruit set of six neotropical shrubs. Biotrop. 15: 257–267.

Baker, H.G. 1959. Reproductive methods as factors in speciation in flowering plants. Cold Springs Harbor Symp. Quant. Biol. 24: 177–199.

Beattie, A.J., D.E. *Breedlove* & P.R. *Ehrlich*. 1973. The ecology of the pollinators and predators of *Frasera speciosa*. Ecology 54: 81–91.

Burdon, J.J. & G.A. *Chilvers*. 1976. Epidemiology of *Pythium*-induced damping-off in mixed species seedling stands. Ann. Appl. Biol. 82: 233–240.

Connell, J.H. 1971. On the role of natural enemies in preventing competitive exclusion in some marine animals and in rain forest trees. Pp. 298–312. *In*: P.J. dem Boer & G.R. Gradwell, eds., Dynamics of populations. Centre for Agric. Publ. Docum., Wageningen.

Cooper, J.P. 1960. Selection for production characters in ryegrass. Proc. 8th Int. Grassland Cong. Pp. 41–44. *In*: R. Creed, ed., Ecological Genetics and Evolution. Blackwell Publ., Oxford.

Corner, E.J.H. 1940. Wayside trees of Malaya. Govt. Print. Office, Singapore.

Coster, C. 1926. Periodische Blüteerscheinungen in den Tropen. Ann. Bot. Gard. Buitenzorg 35: 125–162.

Frankie, G.W., H.G. *Baker* & P.A. *Opler*. 1974. Comparative phenological studies of trees in tropical wet and dry forest in the lowlands of Costa Rica. J. Ecol. 62: 881–919.

Gentry, A.H. 1974. Flowering phenology and diversity in tropical Bignoniaceae. Biotrop. 6: 64–68.

Grant, V. 1949. Pollination systems as isolating mechanisms in angiosperms. Evolution 3: 82–97.

Harris, W. 1970. Genecological aspects of flowering

and vegetative reproduction in *Rumex acetosella* L. New Zealand J. Bot. 8: 99–113.

Heinrich, B. & P.H. *Raven*. 1972. Energetics and pollination ecology. Science 176: 597–602.

Hodgkin, K.C. & J.A. *Quinn*. 1978. Environmental and genetic control of reproduction in *Danthonia caespitosa* populations. Austr. J. Bot. 26: 351–364.

Holttum, R.E. 1935. The flowering of Tembusu trees (*Tagraea fragrans*) in Singapore 1928–35. Gard. Bull. Straits Settlem. 8: 73–78.

———. 1949. Gregarious flowering of the terrestrial orchid *Bromheadia finlaysoniana*. Gard. Bull. Straits Settlem. 12: 295–302.

———. 1954. Plant Life in Malaya. Longman Group, London.

Janzen, D.H. 1971. Seed predation by animals. Ann. Rev. Ecol. Syst. 2: 465–492.

Kerling, L.C.P. 1941. The gregarious flowering of *Zephyranthes rosea* Lindl. Ann. Bot. Gard. Buitenzorg 51: 1–42.

Magalhaes, A.C. & L.R. *Angelocci*. 1976. Sudden alterations in water balance associated with flower bud opening in coffee plants. J. Hort. Sci. 51: 419–23.

McClelland, T.B. 1924. The photoperiodism of *Tephrosia candida*. J. Agric. Res. 28: 445–460.

McIntyre, G.I. & K.F. *Best*. 1978. Studies on the flowering of *Thlaspi arvense* L. IV. Genetic and ecological differences between early- and late-flowering strains. Bot. Gaz. 139: 190–195.

McMillan, C. & B.F. *Pagel*. 1958. Phenological variation within a population of *Symphoricarpos occidentalis*. Ecology 39: 766–770.

McNeilly, T. & J. *Antonovics*. 1968. Evolution in closely adjacent plant populations. IV. Barriers to gene flow. Heredity 23: 205–218.

Moore, D.M. & H. *Lewis*. 1965. The evolution of self-pollination in *Clarkia xantiana*. Evolution 19: 104–114.

Mueller, F. 1882. Bemerkungen zu Hildebrand's Abhandlung über die Lebensdauer und Vegetationsweise der Pflanzen. Bot. Jahrb. Syst. Bd. II. Pp. 391–394.

Njoku, E. 1958. The photoperiodic response of some Nigerian plants. West Afr. Sci. Assoc. 4: 99–111.

Opler, P.A., G.W. *Frankie* & H.G. *Baker*. 1976. Rainfall as a factor in the release, timing, and synchronization of anthesis by tropical trees and shrubs. J. Biogeog. 3: 231–236.

———, G.W. *Frankie* & H.G. *Baker*. 1980. Comparative phenological studies of treelet and shrub species in tropical wet and dry forests in the lowlands of Costa Rica. J. Ecol. 68: 167–189.

Paterniani, E. 1969. Selection for reproductive isolation between two populations of maize, *Zea mays* L. Evolution 23: 534–547.

Piringer, A.A. & *Borthwick*, H.A. 1955. Photoperiodic responses of coffee. Turrialba 5: 72–77.

Rees, A.R. 1964. The flowering behavior of *Clerodendrum incisum* in southern Nigeria. J. Ecol. 52: 9–17.

Rutgers, A.L. & F.A. *Went*. 1915. Periodische Erscheinungen beiden Blüten des *Dendrobium crumenatum* Lindl. Ann. Bot. Gard. Buitenzorg 29: 129–160.

Schimper, A.F.W. 1903. Plant-geography upon a Physiological Basis. Clarendon, Oxford.

Seifriz, W. 1923. The gregarious flowering of the orchid *Dendrobium crumenatum*. Amer. J. Bot. 10: 32–37.

Smith, C.C. 1970. The coevolution of pine squirrels (*Tamiasciurus*) and conifers. Ecol. Monog. 40: 349–371.

Smith, J.J. 1926. Ephemeral orchids. Ann. Gard. Bot. Buitenzorg 35: 55–70.

Spruce, R. 1908. Notes of a botanist on the Amazon and Andes. Vol. I. A.R. Wallace, ed. Johnson Repr. London.

Stern, K. & L. *Roche*, 1974. Genetics of forest ecosystems. Springer-Verlag, New York.

Stubblebine, W., J.H. *Langenheim*, & D. *Lincoln*. 1978. Vegetative response to photoperiod in the tropical leguminous tree *Hymenaea courbaril* L. Biotrop. 10: 18–29.

Whitehead, D.R. 1969. Wind pollination in the angiosperms: evolutionary and ecological considerations. Evolution 23: 28–36.

Wright, S. 1946. Isolation by distance under diverse systems of mating. Genetics 31: 39–59.

Fruit Display and Foraging Strategies of Small Frugivorous Birds

Julie Sloan **Denslow** and Timothy C. **Moermond**

La dispersión de semillas por aves es afectada por la estructura de la infrutescencia, la clase de fruto y los patrones de comportamiento de las aves. Algunas aves obtienen los frutos cuando están en vuelo y otras cuando se posan. Estudios de las aves en jaulas y de la dispersión de frutos en el campo, demuestran que las aves prefieren frutos accesibles, es decir prefieren los frutos que están cerca de las ramas en donde se posan. Los frutos menos accesibles son tomados por aves que son capaces de obtenerlos cuando están en vuelo o cuando se cuelgan de una rama.

Plants with fleshy fruits are an important component of Neotropical rainforest floras. Although mammals—especially bats, monkeys and rodents—also eat fruits and may disperse seeds, most of these plants show aspects of the bird dispersed seed syndrome described by Ridley (1930), van der Pijl (1969) and Snow (1971). Fruit characteristics include a fleshy nutrient-rich pulp; high visibility from combinations of red, black and white fruit coats and accessory structures; ripeness signaled by change in color, carbohydrate content and succulence; small sizes of fruits and seeds (or removable portions of fruits); protection of the seed from mechanical and chemical damage in the digestive tract of the birds; and semi-permanent attachment of the fruit to the stem, delaying fruit fall until removal by a frugivore. This convergence of fruit characteristics reflects traits that birds have in common: similar nutrient and caloric requirements from the fruits, a strong dependence on color vision for foraging, and generally small body sizes.

Las características de frutos dispersados por aves incluyen alta visibilidad, tamaño pequeño, protección contra daño cuando están en el aparato digestivo del ave y alto contenido energético.

The interactive system of birds and plants represents the integration of different strategies with somewhat divergent "goals." Plants should maximize the number of seeds dispersed for the energy invested in fruits and in associated display structures. Birds, on the other hand, should visit plants that yield the greatest return for foraging effort. We argue that energy (or nutrient) yield is increased in two general ways—through choice of fruits with high caloric (or nutrient) yield for the amount of non-digestible ballast and through the choice of accessible fruits which cost little in terms of time or energy to remove.

Thus the plant attracts seed dispersers both by offering caloric or nutrient reward in its fruit and by visual and spatial displays. Infructescence structure, an important component of spatial display, includes the sizes, flexibilities and angles of stems in the vicinity of the infructescence and the distance of the infructescence from stems suitable for use as perches by the foraging birds. The rates of fruit removal will depend on the cost (benefit balance associated with removing and eating the fruit in different species of birds. The success of the dispersal strategy may be measured in fruits removed from the vicinity of the parent.

Seed-dispersal strategies of plants producing fleshy, colorful, bird-eaten fruits have been assumed to extend between two extremes of a continuum: 1) large seeded, difficult-to-handle fruits that attract a few reliable fruit specialists by offering a nutrient rich pulp around the seed and 2) small, easily handled, berry-like fruits which attract many species of small, generalized dispersers by an abundance of relatively nutrient-poor fruits (Snow 1971, McKey 1975, Howe & Estabrook 1977).

Las estrategias de dispersión de semillas se ex-

tienden desde frutos con semillas grandes, difíciles de manipular, ricos en pulpa, que dependen de agentes dispersivos específicos hasta frutos abundantes, pequeños, pobres en nutrimentos, bayas fáciles de manipular, que atraen aves pequeñas en general, pero no específicas.

Within the second group, the subject of this paper, fruits vary in color, in color changes during ripening, in pulp texture and content, in seed sizes and numbers and in the positions in which they are displayed on the plant. Fruits may be carried at the tips of twigs (terminal), in old or new leaf axils (axillary), or on the trunks of trees and shrubs (cauliflorous). Even within one family, the variation in fruit display may be tremendous. Species of Melastomataceae found at La Selva, an Atlantic lowland rainforest in Costa Rica, differ in fruit color, ripening patterns, texture and placement (Plate 1). These are all shrubs or trees and produce small, succulent fruits with many seeds.

Small berry-like fruits can be handled by many species of birds, but differences in foraging abilities mean that not all birds can reach and remove all kinds of fruits. Birds are constrained by their anatomies to different kinds of foraging movements and plants displaying their fruits in different ways are therefore likely to be differentially available to different birds. The critical variables here are the structures of the infructescences and the ways in which small frugivorous birds forage within fruiting plants.

Aves pequeñas y frugívoras difieren en los modos de alimentarse de las infrutescencias estructuralmente diferentes.

If frugivorous birds differ in foraging behavior, such differences should be reflected in their diets. Snow and Snow (1971) described differences in foraging ecology within a group of tanagers and honeycreepers on Trinidad. They saw important differences in the behavior associated with insect eating but described broad overlap in the fruit portions of the diets. Behavioral and anatomical differences associated with the search and capture of different insects might also lead to use of different fruiting plants.

We re-analyzed a data matrix published by Snow and Snow (1971) which consisted of a list of the birds with numbers of feeding records at each fruit species. Our analysis is presented in the form of a polar ordination (Cottam et al., 1973) in which the birds are arranged spatially according to similarities in composition and relative abundances of different fruit species in their diets (Fig. 2). Birds falling close together on the ordination have similar diets and those widely separated have different diets. We can then plot a series of morphological attributes over the birds' positions to examine the relationship between diet and morphology. Body weight and culmen depth (from Hartman 1961 and Snow 1971) show a significant association with position of the bird in the ordination (Spearman Rank Correlation coefficients, n = 13, body weight: $R_s = 0.84$, p less than .05; culmen depth: $R_s = 0.88$, p less than .05). This analysis suggests that morphological differences, particularly those correlated with body size, are associated with differences in fruit choice, even within these similar sized birds. Diamond (1973) also described differences in fruit choices associated with body size in the fruit pigeons of New Guinea.

This analysis indicated that a study of the patterns and the constraints of foraging behavior on fruit choice might give insight into the structures of both the bird and plant communities. Our approach has been two-fold 1) under controlled aviary and field conditions we attempted to determine which components of infructescence structure are likely to be important in the bird's choices of fruits and how they affect birds foraging in different ways, and 2) we used detailed field observations of foraging behavior, fruit display and removal patterns in plants to examine conclusions drawn from experimental situations. The following field and aviary studies were carried out at the La Selva Biological Station in the Atlantic lowland of Costa Rica.

Tanto en el campo como en las jaulas ex-

PLATE 1.

Figs. 1–10. Fruit display in the family Melastomataceae. Color changes during ripening and approximate diameters of ripe fruits are given in parentheses. 1, *Leandra mexicana* (Naud.) Cogn. (green to red to black, 9 mm); 2, *Conostegia subcrustulata* (Beurl.) Triana (pink to black, 7 mm); 3, *Miconia simplex* Triana (red to black, 9 mm); 4, *Miconia nervosa* (Smith) Triana (white to light blue, 10 mm); 5, *Clidemia septuplinervia* Cogn. (dark blue to bright blue, 11 mm); 6, *Miconia affinis* DC. (white to light blue to black, 5 mm); 7, *Clidemia epiphytica* (Triana) Cogn. (green to black, 9 mm); 8, *Clidemia japurensis* DC. var. *heterobasis* (DC.) Wurdack (green to light blue, 9 mm); 9, *Ossaea robusta* (Triana) Cogn. (pale green to white, 5 mm); 10, *Henriettea tuberculosa* (Don. Sm.) L.O. Wms. (green to orange, 5 mm).

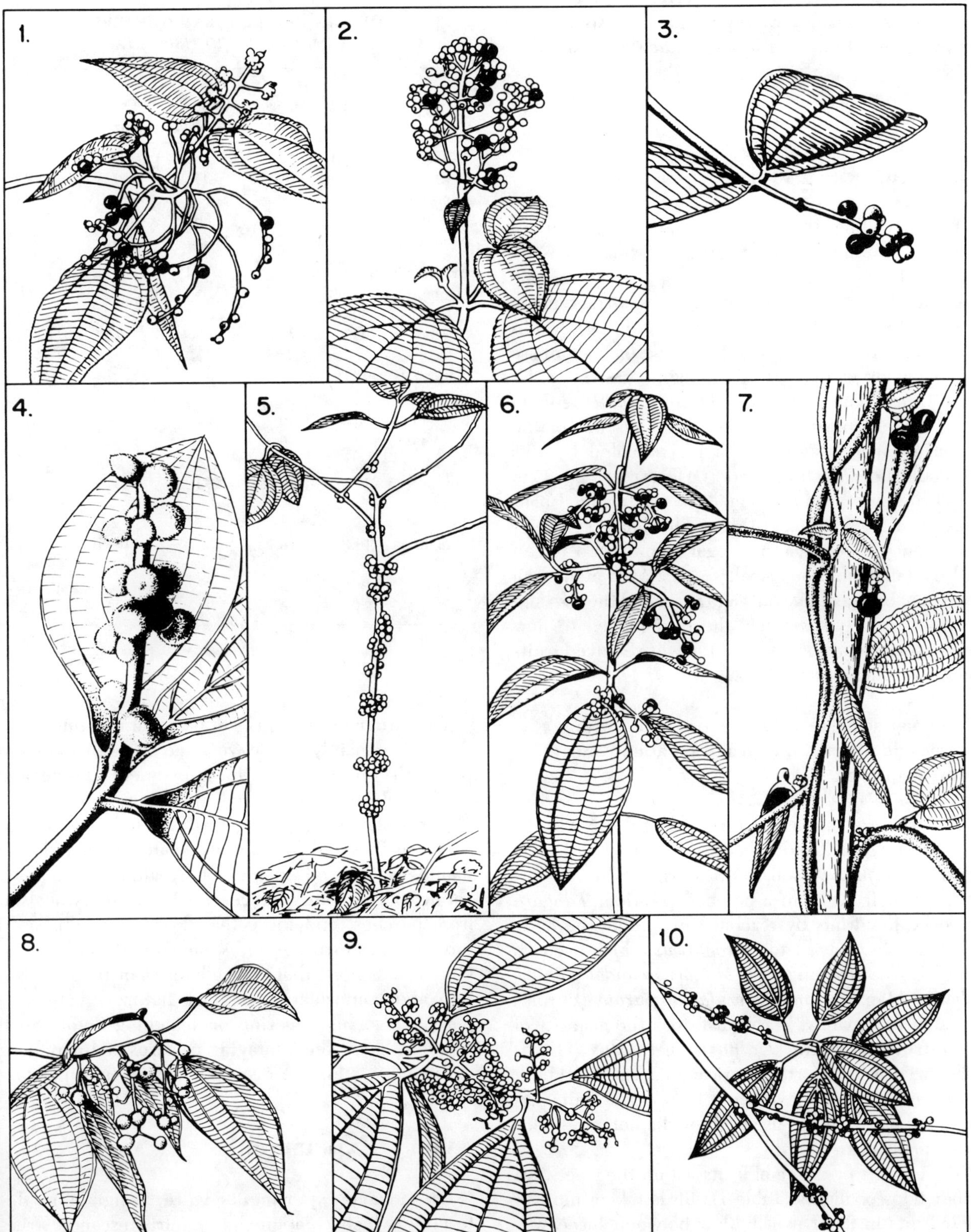
1.
2.
3.
4.
5.
6.
7.
8.
9.
10.

perimentales se analizaron las estructuras de las infrutescencias, para determinar los aspectos más importantes en la alimentación de las aves.

Field Observations of Foraging Behavior

To see whether there are important behavioral differences among birds foraging for small, easily-handled fruits we selected a tree species producing abundant crops of fruits. Such trees attract many species of birds (Eisenmann 1961, Land 1963, Willis 1966, Leck 1969, Howe 1977) allowing us to compare the behavior of many birds foraging under similar conditions. The data from *Miconia multispicata* Naud. (Melastomataceae) demonstrate some of the differences in foraging behavior revealed by this method. *Miconia multispicata* is a forest tree 20–35 m tall. Fruits are produced copiously on terminal panicles. The fruits are less than 1 cm diameter, blue-black, juicy, sweet (ca. 5% sugar in pulp and juice), and contain numerous small seeds. We watched all species of birds eating fruits and recorded the size and angle of every perch, how the bird moved to new perches (by flights or hops) and how it removed fruits (picks, reaches or snatches).

La alimentación se estudió en detalle en dos especies de Miconia que producen abundantes frutos pequeños.

The data are summarized in Table 1 for five birds. *Tangara larvata* (Golden-hooded Tanager) and *Pteroglossus torquatus* (Collared Aracari) pick the majority of their fruits from a perched position. *Tangara* removed few fruits by reaching (extending the body and legs), in contrast to *Pteroglossus* which used its long bill as a reaching tool. *Manacus candei* (White-collared Manakin) and *Querula purpurata* (Purple-throated Fruit Crow) in contrast, took the majority of their fruits in flight. *Catharus ustulatus* (Olive-backed Thrush, a North American migrant) used both flight and picking movements to remove fruits. As expected, large birds used larger diameter perches than small birds.

Movements in search of fruits within the tree were either hops or flights (Table 1). Birds taking most of their fruits in flight usually flew between successive perches in the tree, and birds that perched to remove fruits hopped between perches.

Algunas aves se alimentaron cuando en vuelo, mientras que otras cuando estaban posadas.

TABLE 1
FRUIT FORAGING BEHAVIOR OF DIFFERENT BIRDS IN *MICONIA MULTISPICATA*

	Tangara larvata	Manacus candei	Catharus ustulatus	Querula purpurata	Pteroglossus torquatus
FRUIT REMOVAL					
Pick	86 %	27	44	19	41
Reach	8				58
Snatch	6	73	56	81	1
Number	63	11	95	37	254
FRUIT SEARCH					
Hops	61 %	36	61	23	88
Flights	39	64	39	77	12
Number	308	11	168	13	413
PERCHES USED					
1/8 in.	22 %	20	5		
1/8–1/4 in.	56	40	24	24	52
1/4–3/4 in.	22	40	58	56	33
3/4 in.			12	20	14
Number	246	10	193	34	419
Body Weight (g)	18.8	18.2	35.1	103.0	226.0
Number of Birds	34	6	32	20	25

Thus frugivorous birds can be separated into those that forage mainly on the wing and those that depend primarily on their legs for movement. Large and small birds are represented in both groups.

For instance, perching birds include many tanagers, finches and the toucans; snatchers include the manakins, several species of flycatchers and most cotingas. Few birds forage equally on the wing and from perches, and those that do are medium sized birds such as thrushes and some cotingas and tityras. Our data suggest that tradeoffs exist in the distribution and conformation of musculature involved in perching versus hovering or snatching flight. Such tradeoffs likely affect foraging patterns and may have consequences for the dispersal strategies of plants.

Aviary Experiments

Our aviary experiments were designed to determine which components of infructescence structure were important to the birds in their choices of fruits. Three variables were of particular interest (1) the perch size (5 mm or 12 mm diam.), (2) whether the fruits were above or below the perch and (3) whether the fruits were close to the perch or at the limit of a bird's reach. To account for differences between the two sides of the cages we included a fourth variable—left and right. Combinations of the four variables were presented to the birds in the form of a series of choices between two infructescences of 16 fruits each. We could index the bird's choice by the number and order in which it removed the fruits. A replicated series of four such runs could then be analyzed to determine which of the four variables most influenced the choice (Box et al. 1978).

Results for eight species are included in Table 2. The single most important factor in the choices of all birds was proximity to the perch; all species chose fruits close to the perch in preference to those requiring reaching or flying (all fruits were within reaching distance of all birds). Although it would seem a trivial result, this pattern has important consequences. It says that minimizing energy spent (or perhaps risk of injury through falling) in removing fruit is a likely consideration in the bird's choice of fruit. Proximity to a sturdy perch was even important to the small, agile olive-backed euphonia *(Euphonia gouldi)* which was the only bird capable of landing directly on the wires of the infructescence and did so as soon as the closer fruits were removed. It also suggests that fruits displayed closest to suitable perches are likely to be removed before those at twig tips or pendant below the branches.

Las aves escogieron los frutos más cercanos a la percha.

A second factor, important to three bird species, was the position of the fruits above or below the perch

TABLE 2
FRUGIVOROUS BIRDS AND INFRUCTESCENCE POSITION
Aviary experiments on fruit preferences.

	Weight (g)	Near/ Far	Above/ Below	Ability to Reach -Down	Ability to Hover
PIPRIDAE					
Corpipo altera	12.5	Near	Above	—	x
Manacus candei	16.0	Near	Above	—	x
MUSCICAPIDAE: TURDINAE					
Catharus minimus	32.0	Near	Above	—	—
EMBERIZIDAE: THRAUPINAE					
Euphonia gouldi	12.5	Near		xx	x
Tangara larvata	17.5	Near		x	x
Ramphocelus passerini	26.5	Near		x	x
Thraupis palmarum	36.5	Near		xx	—
EMBERIZIDAE: CARDINALINAE					
Saltator maximus	45.0	Near		x	—

(Table 2). These species showed preferences for fruits above the perch and all three had difficulty hanging below a perch to reach fruits. The thrushes *(Catharus)* made numerous aborted attempts to reach the lower fruits. When the bird did hang below the perch it often had difficulty in righting itself, sometimes falling from the perch. Manakins *(Manacus* and *Pipra)* consistently removed less accessible fruits on the wing rather than reaching for them, but they preferred to hover up to a fruit rather than down. Within the confines of the cages, only the smaller birds were able to remove fruits in flight; among those, only the manakins with short, broad, rounded wings did so regularly. Although *Tangara* and especially the *Euphonia* showed little difficulty in taking fruits on the wing, they usually took all fruits possible by reaching before resorting to flight. All of the other species have been seen to take fruits on the wing in the field—usually by snatching or hovering. Their unwillingness to take fruit on the wing in the cages and the manakins' preference for the upper fruits suggest that maneuvering room is necessary for this foraging tactic. These preferences also suggest that fruits exposed away from the foliage are more likely to be taken on the wing than those within the plant. Such preferences for position have also been observed in nectarivorous bats (Faegri & van der Pijl 1966).

Como quiera que algunas aves tenían dificultades de colgarse de las perchas, prefirieron los frutos que estaban por encima de éstas.

Finally we noted that perch size was not an important factor to any of these birds. This is in contrast to field observations and to findings in other experiments. We suspect that perch size is more important in flexible or vertical perches than in the fixed horizontal perches used in these experiments.

El tamaño de la percha parece no ser importante.

In summary, the aviary experiments showed that the first choice of all birds tested was for fruits close to a perch. Behavioral differences became most evident when birds foraged for less accessible fruits. At this point a second component—whether fruits were above or below a perch—became important.

Patterns of Fruit Removal from *Urera baccifera*

If birds preferentially take accessible fruits over similar but less accessible fruits, what are the consequences for the seed dispersal of plants displaying fruits in different ways? A short study on the large nettle, *Urera baccifera*, is illustrative.

Urera baccifera (Urticaceae) is a large (2 m), second growth herb that produces white berry-like fruits highly preferred by many species of birds. The large axillary infructescences are produced along the long, upright, leafless stem and at the bases of the large upper leaves. Like many nettles, the leaves, stems and infructescence axes are covered with spines and urticating hairs that inflict a painful sting. The only places we have seen birds perch have been on the large petioles of the terminal group of leaves which are largely free of urticating hairs: they never perch directly on the infructescences. Consequently only fruits within about 15 cm of these petioles are available to small perching birds; the rest must be taken on the wing.

What are the consequences to the plant of this difference in availability of fruits? We counted ripe fruits, green fruits and fruit scars on infructescences that were within 15 cm of the petioles or other suitable perches and on those that were out of reach (Table 3). Less accessible fruits were removed more slowly than were those nearer perches which is consistent with our conclusions from the cage experiments. The lower fruits are more likely to be taken by birds that forage on the wing. *Urera baccifera* berries are so highly preferred that within a few weeks it is difficult to find ripe fruits on the plants. Although this might suggest that the position of fruit display is of little consequence, unremoved ripe fruits are subjected to loss and destruction by fungi, rodents, insects and rain. This loss is minimized by rapid removal of the fruits by seed dispersers.

Aves perchadas sólo en los pecíolos terminales de la ortiga Urera baccifera primero removieron los frutos cercanos.

TABLE 3
PERCENTAGES OF FRUIT OR SCARS REMAINING ON *URERA BACCIFERA*

	Ripe Fruits	Unripe Fruits	Fruit Scars	Total
Accessible Fruits (within 6 in. of perch)	3.0%	12.0%	85.0%	4171
Inaccessible Fruits (beyond 6 in. of perch)	12.2	11.8	76.0	2246

X^2 = 212.0, p less than .001, d.f. = 2

Plants producing fleshy fruits should disperse their seeds more rapidly if fruits are displayed in positions accessible to suitable perches than if they are displayed at the tips of narrow twigs or in pendant infructescences. Axillary infructescences should be more common than terminal ones among berry producing plants. This does not, however, appear to be the case. Table 4 summarizes fruit display data from dicotyledonous plants of Barro Colorado Island, Panama (Croat 1978). If all species are considered together, terminal infructescences are slightly more common among small fruited plants than are axillary infructescences. In four large families producing predominately small fleshy fruits, two (Moraceae and Solanaceae) produce mainly axillary and two (Melastomataceae and Rubiaceae) produce mainly terminal infructescences. Some species even produce poorly accessible axillary fruits on pendant peduncles (Fig. 1).

The fact that the overwhelming majority of these fruits or their accessory structures are red, black or blue (Table 4) suggests that they are most likely bird and not bat-dispersed species (van der Pijl 1957) for which one would expect fruits displayed away from the foliage. What then are the tradeoffs associated with infructescence structure? Why are so many small fruits carried in relatively inaccessible terminal positions and not axillary positions when our evidence suggests that the reverse should be true?

Factors Affecting Fruit Removal Rates

We reasoned from the aviary studies and from the observations on *Urera* that shrubs displaying fruits in axillary positions should show faster removal rates than those displaying fruits in terminal positions. Does accessibility influence fruit removal rates in the field as it does in the aviary? What are the important agents of fruit removal?

In our experiment (Denslow & Moermond 1982) we tied portions of infructescences with 10 to 20 fruits into axillary and terminal positions in existing shrubs and counted fruits removed or partially eaten at dawn and again at dusk. For each experiment we used 20 pairs of infructescences, all of the same kind and quality of fruit. We repeated the experiment many times in three different habitats—forest understory, forest gaps, and old fields—and in two different seasons—one of fruit abundance (Aug–Sept) and one of fruit scarcity (Dec–Jan).

The two most likely sources of fruit and seed destruction at La Selva are insects (especially orthopter-

TABLE 4
DISPLAY OF SMALL FLESHY FRUITS

Distribution of display characteristics of fleshy fruits less than 1 cm across on Barro Colorado Island, Panama (Data summarized from Croat 1978).

	Melastomataceae	Moraceae	Solanaceae	Rubiaceae	All Dicots
POSITION OF INFRUCTESCENCE					
Terminal	69.2	40.0	33.3	85.3	49.8%
Axillary	19.2	60.0	66.7	14.7	46.7
Cauliflorous	11.5	0.	0.	0.	3.2
COLOR					
Red-Yellow-Orange	0.	60.0	50.0	32.3	43.4
Black-Purple	80.8	0.	42.8	29.4	30.1
Blue	11.5	0.	0.	23.5	4.4
White	0.	0.	7.1	5.9	6.0
Green	0.	40.0	0.	0.	10.8
NUMBER OF SPECIES	35	31	23	79	249

ans) and arboreal rodents. We estimated the importance of the first by counting the number of fruits that had been partially eaten since the previous count. Observations at night had shown that long-horned grasshoppers (tettigoniids) were probably responsible for much of this damage. We estimated fruit loss to rodents by counting fruits removed from axillary infructescences at night. We assumed that all fruits removed during the day were taken by birds and that fruits removed from terminal infructescences during the night were taken by bats.

The data show clearly that the birds forage as previous experiments would predict, that is, "axillary" fruits are taken more rapidly than are "terminal" fruits. The highest diurnal removal rates are in the old fields and during the season of fruit scarcity (December to January). Insect damage was highest at night and in forest gaps and understory, but there was no consistent difference in herbivory on axillary and terminal infructescences nor in seasonal levels of damage. Nocturnal fruit removal rates, although lower than diurnal, were highest in the forest understory and in gaps. Removal from axillary infructescences—presumably due to rodents—formed a large component of this nocturnal frugivory. Other experiments with caged spiney rats *(Hoplomys gymnurus)* and pocket mice *(Heteromys desmarestianus)* showed that both species are to some extent arboreal *(Hoplomys* more than *Heteromys)* and that both eat fruits and destroy the seeds. Fruit loss to rodents may constitute an important counter pressure on the display of fruits along stems.

Cuando se amarraron frutos en posiciones terminales y axilares de arbustos, los axilares fueron tomados primero. La destrucción de los frutos a lo largo de los tallos puede aumentar por acción de roedores y otros animales.

Discussion and Conclusions

Fruit placement affects removal by both birds, the primary seed dispersers, and rodents which are more likely to destroy than to disperse seeds. Moreover rodents, being less agile than birds and therefore less able to reach far for fruits, may be even more sensitive to fruit placement. Rodents may require that fruits be closer to large, sturdy stems before the fruits can be easily removed.

La posición del fruto podría ser un balance entre los propósitos de la remoción de frutos y la disminución de su destrucción por roedores.

The effect of placement on the balance between seeds dispersed and seeds destroyed will depend on the fruit preferences of both rodents and birds. Rodent-preferred fruits—perhaps those with large seeds—displayed in poorly accessible positions will less likely be lost than accessible fruits. Less preferred fruits may be placed in positions more accessible to birds with less risk of loss to rodents.

Secondary compounds that protect fruit from insect damage (Herrera 1982) may also affect fruit choices of the birds. Like a rodent, a bird will reach farther for a desirable fruit than for a less desirable one (Moermond & Denslow 1983). Although secondary compounds may lower fruit palatability, more accessible placement may compensate for reduced preference.

Birds are also likely to be constrained by size, morphology, and foraging behavior such that fruits are not equally accessible to all species. Which birds take which fruits will be influenced by structural aspects of the display. Fruits in long, pendant panicles may be taken mainly by birds that can snatch fruits on the wing (Pipridae, Cotingidae, Tyrannidae) or by those that are small or agile enough to land directly on the infructescence (*Euphonia gouldi*, honeycreepers). Sessile fruits carried on a long flexible stem are likely to be taken by honeycreepers (*Dacnis cayana*, *Cyanerpes lucidus* that can hang on vertical stems rather than by the larger perching birds (*Ramphocelus passerini*, *Saltator maximus*) or by snatching birds (*Myiozetetes* spp., *Manacus candei*). Larger birds (Cotingidae, Trogonidae) will require larger perches to forage from and perhaps more nutritious fruits to repay foraging costs.

The factors entering into a bird's choice are diverse. The diet represents a balance between the costs of digesting, handling, and foraging for fruits on the one hand and the energetic and nutritional rewards gained on the other. Such a balance will vary from place to place, season to season, and perhaps habitat to habitat. As a result, close co-evolutionary relationships between individual bird and plant species are unlikely. Rather we see that dispersal patterns of small fruits are highly variable. A species may be dispersed primarily by one bird in one part of its range and another elsewhere, even though both dispersers are present in both regions (Howe 1977, Howe & Vande Kerckhove 1979). Fruits with similar sized seeds may be dispersed predominantly by a single fruit generalist (yellow-green vireo dispersing *Casearia nitida* in Costa Rica, Howe & Vande Kerckhove 1979) or by several species of generalists (*Guarea glabra* in Panama, Howe & De Steven 1979). Diverse assemblages of birds visit large crops of small fruits (Eisenmann 1961, Land 1963, Willis 1966, Leck 1969, Howe & Primak 1975, Howe 1979, Kantak 1977, McDiarmid et al., 1977, Howe & De Steven 1979, Howe & Vande Kerckhove 1979). Apparently many species of birds may exploit a super abundant food source which when less abundant would not be energetically rewarding. Similarly many species of birds not usually observed to sally for insects may sally for swarming termite reproductives (Eisenmann 1961).

También son importantes el tamaño de las semillas y de las plantas, el hábitat y las rutas de alimentación de las aves.

Neither is the bird-plant relationship entirely opportunistic. The interaction between dispersers and their food plants is a dynamic one involving choices between fruits of different qualities and accessibilities. Because the choices vary with season and habitat, the diets of the birds also vary. The dispersers nevertheless respond in predictable ways to particular combinations of variables.

Acknowledgments

These studies benefited from the logistical help of D. Stone, D.A. Clark, D.B. Clark, and the Organization for Tropical Studies; from conversations with D. Levey, G. Stiles, and M. Willson, from funding through the Zoology Department and Graduate School of the University of Wisconsin and NSF grant No. DEB7910991, and from advice from many researchers at La Selva Biological Station, Costa Rica. B. Wentworth helped design a synthetic diet to maintain caged frugivors.

Box, G.E.P., W.G. *Hunter* & J.S. *Hunter*. 1978. Statistics for Experimenters: an Introduction to Design, Data Analysis, and Model Building. John Wiley & Sons, New York.

Cottam, G., F.G. *Goff* & R.H. *Whittaker*. 1973. Wisconsin comparative ordination. Pp. 193–222. *In* R.H. Whittaker, ed. Ordination and Classification of Communities. Dr. W. Junk, The Hague.

Croat, T. 1978. Flora of Barro Colorado Island, Panama. Stanford Univ. Press, Stanford CA.

Denslow, J.S. & T.C. *Moermond* (1982). The effect of accessibility on rates of fruit removal from neotropical shrubs: an experimental study. Oecol. 54: 170–176.

Diamond, J.M. 1973. Distributional ecology of New Guinea birds. Science 179: 759–769.

Eisenmann, E. 1961. Favorite foods of neotropical birds: flying termites and *Cecropia* catkins. Auk 78: 636–637.

Faegri, K. & L. *van der Pijl*. 1966. The Principles of Pollination Ecology. Pergamon Press, New York.

Hartman, F.A. 1961. Locomotor mechanisms of birds. Smithsonian Misc. Col. 143(1).

Herrera, C.M. 1982. Defense of ripe fruits from pests: its significance in relation to plant-disperser interactions. Amer. Nat. 120: 218–241.

Howe, H.F. 1977. Bird activity and seed dispersal of a tropical wet forest tree. Ecology 58: 539–550.

——— & D. *De Steven*. 1979. Fruit production, migrant bird visitation, and seed dispersal of *Guarea glabra* in Panama. Oecol. 39: 185–196.

——— & G.F. *Estabrook*. 1977. On intraspecific

competition for avian dispersers in tropical trees. Amer. Natur. 111: 817–832.

——— & G.A. *Vande Kerckhove*. 1979. Seed dispersal by birds and the fecundity of a tropical tree. Ecology 60: 180–189.

Kantak, G.E. 1976. Partitioning of fruit resources by some birds of the Yucatan Peninsula, Mexico. M.S. thesis, Univ. Wisconsin-Madison.

Land, H.C. 1963. A tropical feeding tree. Wilson Bull. 75: 199–200.

Leck, C.F. 1969. Observations of birds exploiting a Central American fruit tree. Wilson Bull. 81: 264–269.

McDiarmid, R.W., R.E. *Ricklefs* & M.S. *Foster*. 1977. Dispersal of *Stemmadenia donnell-smithii* (Apocynaceae) by birds. Biotropica 9: 9–25.

McKey, D. 1975. The ecology of co-evolved seed dispersal systems. Pp. 159–191. *In* L.E. Gilbert & P.H. Raven, eds. Coevolution of Animals and Plants. Univ. Texas Press, Austin.

Moermond, T.C. & J.S. *Denslow*. (1983) Fruit choice in neotropical birds: effects of fruit type and accessibility on selectivity. J. Anim. Ecol. 52: 407–420.

Ridley, H.N. 1930. The Dispersal of Plants Throughout the World. L. Reeve, Ashford, England.

Snow, B.K. & D.W. *Snow*. 1971. The feeding ecology of tanagers and honeycreepers in Trinidad. Auk. 88: 291–322.

Snow, D. W. 1971. Evolutionary aspects of fruit eating by birds. Ibis 113: 194-202.

van der Pijl, L. 1957. The dispersal of plants by bats (Chiropterochory). Acta. Bot. Nederlandica 6: 291–315.

———. 1969. Principles of dispersal in higher plants. Springer-Verlag, New York.

Willis, E.O. 1966. Competitive exclusion of birds at fruiting trees in Western Colombia, Auk. 83: 479–480.

Plant Seasonality in the Forests of Panama

Robin B. Foster

La periodicidad en la pérdida y reaparición de hojas, en la floración y de fructificación son importantes para la fauna de Panamá. La pérdida y reaparición de hojas presentan diferente periodicidad en los distintos tipos de bosque y ésto influye en las diversas clases de animales. La periodicidad en la floración varia en relación a los grupos ecológicos y frecuentemente se asocia a los polinizadores. La periodicidad de la fructificación influye en los frugívoros. La periodicidad y las fluctuaciones de año a año hacen que los recursos vegetales disponibles a los animales sean inestables y éstos son vulnerables a cambios futuros tales como la restricción de las posibilidades de migración causada por la destrucción del hábitat.

Many people in Panama are aware that the forests are not constant throughout the year. The brief and brilliant flowering of *Tabebuia guayacan* in March and the numerous deciduous trees in the dry season are changes that can hardly be missed even by city dwellers. But many changes in the forests are subtle and are missed by all but the most observant of those who live in close contact with these forests. In addition the seasonality is different from place to place in the diverse climatic regions of the country. With the current massive destruction of Panama's forests the opportunities for further understanding forest seasonality are fading fast or already gone. It is only small comfort that the most poorly known forests—the wet forest of the Atlantic slope and the Darien frontier—are also the areas that will be last to be destroyed.

The information used here comes primarily from my own studies and those of my students on Barro Colorado Island (BCI). Barro Colorado Island, the largest island created by Gatun Lake of the Panama Canal Area, is situated in the intermediate 'moist' climate of Panama, and as such is roughly representative of what was the most prevalent forest type in the country (Tosi 1971). Information from several other studies is incorporated to predict what can be expected in other climatic regions of Panama for which we now have little information. For simplicity I distinguish only three climatic regions based on the map of Tosi (1971); dry forest, moist forest, and wet forest. In Panama the dry season runs between November and March, and November is the wettest month (Fig. 1).

Deciduousness, Leaf Renewal, and Litter Fall

There is often large variation between years in climate and plant response, and the patterns described are meant only as roughly average conditions.

Forest deciduousness and leaf renewal in Panama are not simple phenomena. In dry forests a few species with deep roots and leathery leaves remain leafy green during the dry season, —*Hymenaea courbaril.* The understory shrub *Jacquinia pungens* in Costa Rica is even deciduous throughout the rainy season (Janzen 1970). The loss of leaves might seem an adaptation that reduces plant transpiration as the water table drops during the dry season, and that may be true for many species. However, several species lose their leaves for variable amounts of time during the rainy season. *Cordia alliodora* is exceptional in this regard, becoming deciduous for several months following the start of the rains. In areas where the species is exceptionally common, one would rightfully conclude that the peak period of forest deciduousness was May through July, during the first part of the rainy season.

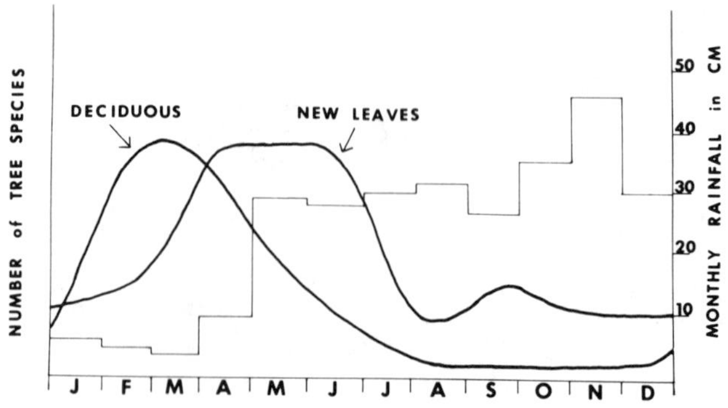

Fig. 1. Estimated seasonal pattern of deciduousness and leaf flushing for a tropical moist forest. Vertical bars represent mean monthly rainfall in Barro Colorado Island.

Los patrones de periodicidad en la pérdida de las hojas, en la floración y en la fructificación son importantes en la comunidad fáunica de los bosques de Panamá.

In moist forest many or even most of the lianas and shrubs flush new leaves throughout the dry season. But most surprising, and not adequately studied, are the many tree species that drop their leaves as the dry season progresses but manage to flush a full set of new leaves before the rains and when soil water availability is at a minimum.

Los patrones de reaparición de las hojas varían en las diferentes especies. En los bosques húmedos, la mayoría de las lianas producen hojas nuevas durante la estación seca, pero muchos árboles lo hacen casi al final de esta estación a pesar de que la humedad es baja.

The appearance of the forest—the proportion of trees or canopy area that is deciduous—is a function of the relative abundance of certain species. In moist forests, which are mainly 'semi-deciduous', usually only a small proportion of the tree species are deciduous. But these species may constitute a large fraction of the trees, especially in forest less than 150 years old. The Bayano Valley and the Darien Basin would look considerably less deciduous with the absence of a single species, the cuipo *(Cavanillesia platanifolia*. Furthermore, the loss of leaves and reflushing of leaves occurs at different times for each species, and hence the naked component of the vegetation may have very different species composition from month to month.

El cuipo (Cavanillesia platanifolia) es la causa de la aparencia decidua del Valle del Bayano y de los bosques asociados a las cuencas hidrográficas del Darién.

Seasonal trends in forest deciduousness and leaf flushing have not been studied carefully, nor is it clear how they should be measured. Fig. 1 presents a model for moist forest based on personal observation and data from selected trees in mostly young forest on BCI (Smithsonian Environmental Sciences Program). There is significant overlap in the deciduous and leaf flushing curves because many species flush leaves before arrival of the rains and many species drop leaves shortly after the rains before flushing anew. This overlap should be much greater in wet forests and the two curves much more separate in dry forests (see Frankie et al. 1974).

El clímax de pérdida de hojas se localiza muy temprano en la estación seca y continúa hasta el comienzo de la lluviosa. Este proceso prosigue y es errático durante el resto de la estación lluviosa.

The fall of leaves to the ground reaches a sharp peak early in the dry season (Fig. 2). Many of the trees 'thin out' at this time even if they do not become entirely deciduous. Leaf litter fall continues high during the rest of the dry season and into the early part of the rainy season. The progressive increase in leaf fall in October, November, and December before the dry season begins is surprising. The physiology of tropical leaf abscission remains to be studied. The highly erratic leaf fall in the late rainy season is associated with

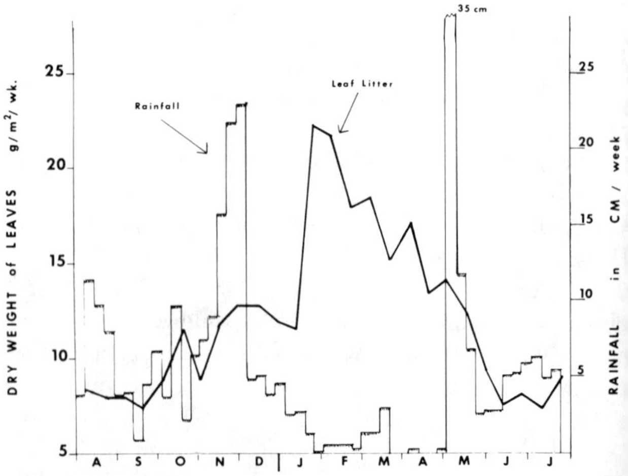

Fig. 2. Leaf liter fall during one year in moist forest on Barro Colorado Island. Weight based on 120 litter traps 1/12 m^2 sampling 1 km^2 of forest.

heavy rains following short droughts of several days, but this is also a common rainfall pattern throughout all of the rainy season.

Litter on the ground continues to build up during the dry season when populations of litter fauna are low and the decay process is slow. The quantity of litter on the ground reaches a maximum in the months before and after the start of the rains and a minimum in September and October. In wet forest there is probably much less buildup of litter since the populations of litter fauna are maintained at higher levels during this drier period.

La máxima acumulación de hojarasca sobre el suelo se encuentra justo antes e inmediatamente después del comienzo de la estación lluviosa. Probablemente hay menos acumulación en los bosques muy húmedos que en los húmedos.

Leaflessness increases the light, temperature, and air circulation near the ground and has an obvious impact on organisms in the understory. An abundance of leafless trees must also alter foraging and predator-avoidance behaviors among canopy animals.

The timing of new leaf emergence is of special importance to most generalist herbivorous animals, such as sloths *(Bradypus* and *Cholepis)*, howler monkeys *(Alouatta)*, iguanas, and leaf-cutter ants *(Atta)*, all of which seem to prefer the young leaves of most plant species. Leaf flush is indirectly important to insectivores since so many insects, even specialists, feed primarily on young leaves (Coley 1980, 1982). A community-wide flush of insects becomes available at the same time as leaf flush (Wolda 1978, 1980, Smythe 1982).

La sincronización de los brotes foliares es importante para los herbívoros que no son específicos, los cuales parecen preferir las hojas jóvenes. En este período de renuevo de hojas, los insectos son más abundantes.

Even the commonest members of the litter fauna, such as isopods, can hardly be found in the dry season. Their populations build up gradually following the start of the rains as they feed on the stockpile of litter. These populations apparently peak in June or July on BCI and decline sharply in the following months, perhaps with a secondary buildup at the end of the rainy season. The appearance of fungal fruiting bodies also seems to be greatest in the first part of the rainy season. Animals such as coatis *(Nasua)* and army ants *(Eciton, Labiolus)* which forage in the litter must be strongly affected by the seasonal fluctuations in these litter organisms.

Flowering

It is literally true that some plant species is in flower in a forest in Panama every day of the year, inconspicuous though it may be. Individual species have a wide range of flowering patterns ranging from abrupt short flowering periods to long intermittent flowering (Gentry 1974, Augspurger, this volume). More important is that one-half the year, February–July, has much more flowering than the other half, August–January (Fig. 3). In dry forests flowering is more clearly concentrated in the dry season; whereas in moist and wet forests an equally high proportion of species is in flower in the first few months of the rainy season (Daubenmire 1972, Frankie et al. 1974, Croat 1978, Foster 1982a).

En los bosques secos, la floración se concentra durante la estación seca. En los bosques húmedos y en los muy húmedos muchas especies florecen a principios de la estación lluviosa al igual que durante la seca.

Within the overall community pattern are various ecological groups, many having distinct patterns of their own. More of the understory and forest herb species flower in the early rainy season and more of the liana species flower in the dry season (Croat 1978, Opler et al. 1980). Most species pollinated by hummingbirds and bats flower in the late rainy season and dry season. It is unclear if these group differences are the result of availability of pollinators at different times of year, avoidance of flower and seed predators, different physiological requirements, or some combination of adaptations (Janzen 1967, Frankie et al. 1974, Foster 1982a).

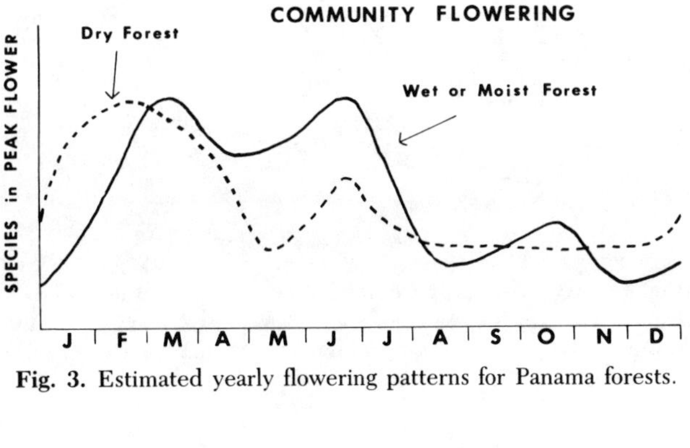

Fig. 3. Estimated yearly flowering patterns for Panama forests.

Muchos grupos ecológicos tienen patrones periódicos diferentes. Muchas especies del sotobosque florecen a principios de la estación lluviosa.

Many environmental factors or 'cues', alone or in various combination, can trigger flowering in the tropics. The most obvious and best studied cue is the onset of rains following a drought (Alvim 1960, 1978, Opler et al. 1976, Augspurger 1982). This is probably the most important cue for species flowering in early rainy season.

The seasonal availability of flowers must certainly affect the populations of numerous flower-visiting and flower-eating insects. It must also affect at least the behavior of hummingbirds and nectarivorous bats. But the availability of flowers may be of minor importance to the vertebrates which occasionally devour them with gusto.

Fruiting

Dry forests have a peak of fruiting toward the end of the dry season (Fig. 4). Wet forests have a peak of fruiting in the middle of the rainy season (Fig. 5). Moist forests, the most abundant in Panama, have two peaks—end of dry season and mid to late rainy season (Fig. 6). These patterns emerge from tropical forests in many parts of the world and should also apply to most of Panama. Tropical areas of the world with more complicated rainfall patterns—two dry seasons—probably do not fit these models.

Los bosques secos presentan el clímax de fructificación al final de la estación seca, sin embargo en los húmedos el clímax ocurre a mediados de la lluviosa.

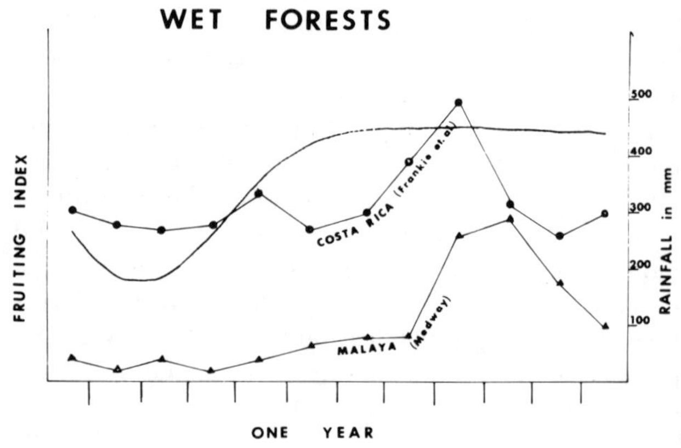

Fig. 5. Wet forest yearly fruiting patterns. Scale and measure for fruit activity are different for each study. Heavy solid line represents moist forest pattern of rainfall with dry season of intermediate length and severity.

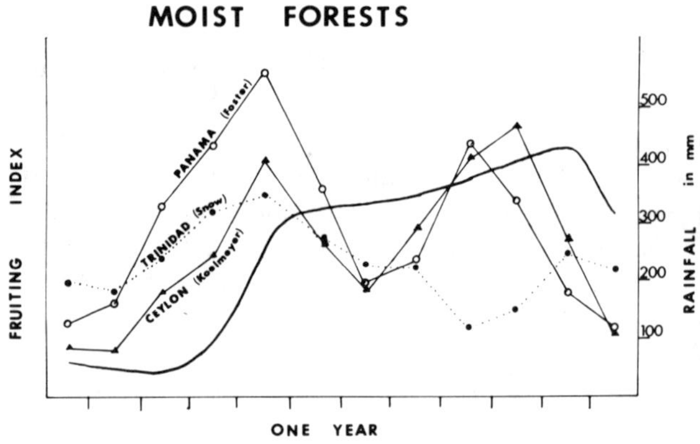

Fig. 6. Moist forest yearly fruiting patterns. Scale and measure for fruit activity are different for each study. Heavy solid line represents moist forest pattern of rainfall with dry season of intermediate length and severity.

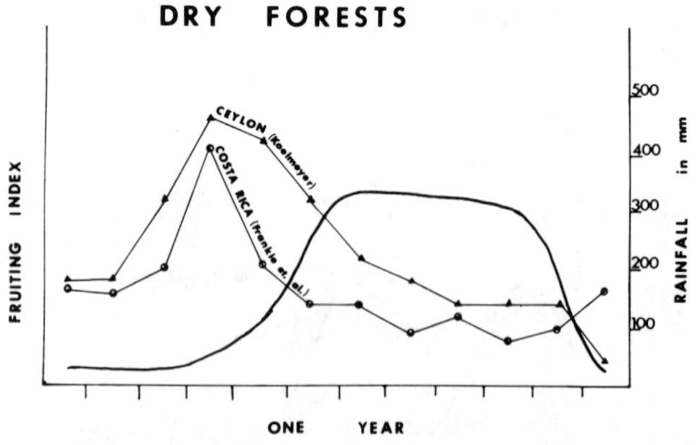

Fig. 4. Dry forest yearly fruiting patterns. Scale and measure for fruit activity are different for the two studies. Heavy solid line represents dry forest patterns of rainfall with long and severe dry season.

These general fruiting patterns are the product of many diverse individual species patterns. Species can fruit intermittently throughout the year (Fig. 7) or have two fruiting peaks each year (Fig. 8), but by far the majority of species, at least on BCI, have a more or less bell-shaped curve of fruiting once a year (Figs. 9, 10). The period of fruiting can be short (Fig. 9) or extended over several months (Fig. 10), and this is independent of whether the fruit is animal-dispersed or wind-dispersed. Species in the March–June fruiting peak tend to have a longer period of activity than those in the September-October peak. A greater pro-

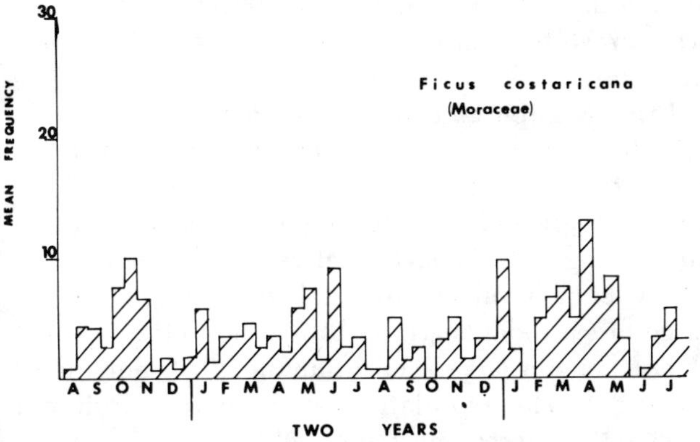

Fig. 7. Fruiting pattern of the strangler fig, *Ficus costaricana*, on Barro Colorado Island over a two year period. This kind of pattern is apparently characteristic of figs though there is sometimes predictable seasonality (Milton et al. 1982, Janzen 1979).

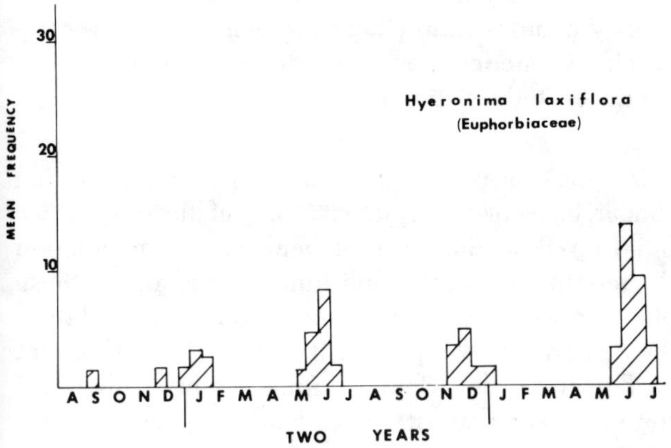

Fig. 8. Fruiting pattern of the tree *Hyeronima laxiflora* on Barro Colorado Island over a two year period. This same species also occurs in Guyana in a climatic regime of two dry and two rainy seasons.

portion of the species in March-June are wind-dispersed (Foster 1982a) and are more likely to be second-growth or 'pioneer' species. Shrubs are more likely to fruit late in the rainy season (Croat 1978, Opler et al. 1980, Foster 1982a).

El patrón de fructificación de cada especia varía y en la Isla Barro Colorado es independiente de los mecanismos de dispersión.

The duration of a species' fruiting is not necessarily a reflection of its flowering duration. *Randia armata*, for example, is in flower for only one week but may ripen fruit over a period of several months. Also the time from flowering to fruiting is not dependent on the size of the fruit and seed. *Gustavia superba* can develop very large fruits in one month, while *Alseis*

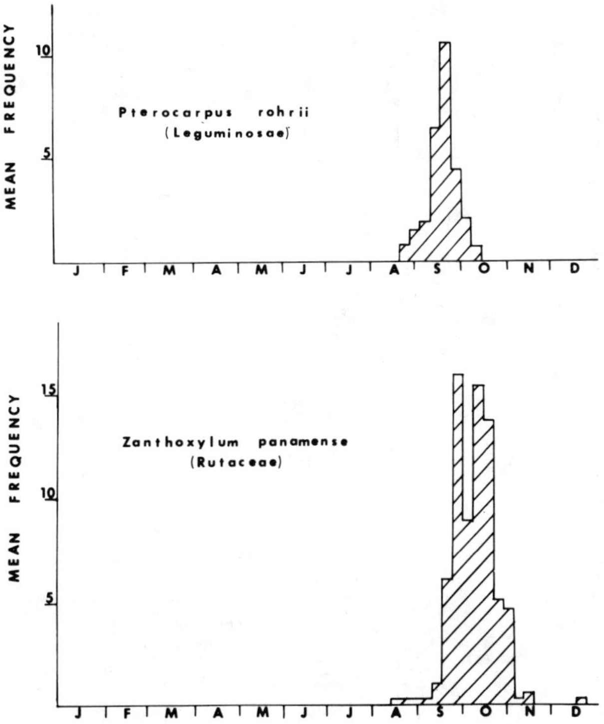

Fig. 9. Fruiting patterns of *Zanthoxylum panamense* (animal-dispersed) and *Pterocarpus rohrii* (wind-dispersed) during 1 year on Barro Colorado Island.

blackiana takes one year to develop tiny fruits. There are no data yet supporting the argument of Snow (1965) that competition forces species into different fruiting times. Certainly many species potentially competing for the same fruit dispersers do overlap considerably in their fruiting time. Wind-dispersed species show as much displacement and overlap as animal-dispersed species (Foster, unpublished data).

La duración de la floración y de la fructificación no están correlacionadas.

Species fruiting at the end of the dry season commonly germinate their seeds early in the rainy season (Fig. 11). Garwood (1982, 1983) has shown that the first seedlings to germinate in the rainy season have a definite growth advantage over those that germinate later when competing in treefall light-gaps. Species that germinate in early rainy season may also take advantage of the greater probability of treefalls and gap-openings in mid-rainy season (Brokaw 1982) (Fig. 11).

The fruiting peak in mid-to-late rainy season is more difficult to explain (Foster 1982a). One possibil-

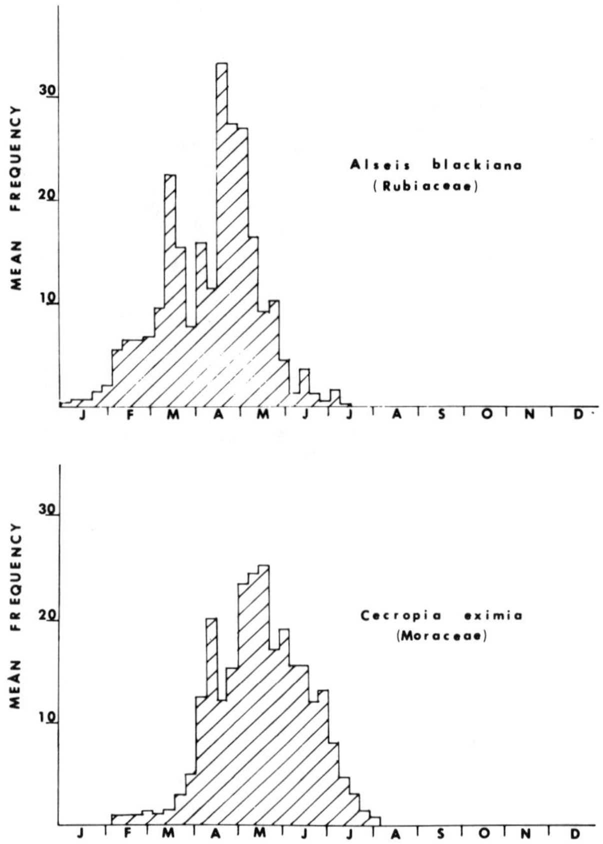

Fig. 10. Fruiting patterns of *Cecropia insignis* (animal-dispersed) and *Alseis blackiana* (wind-dispersed) during 1 year on Barro Colorado Island.

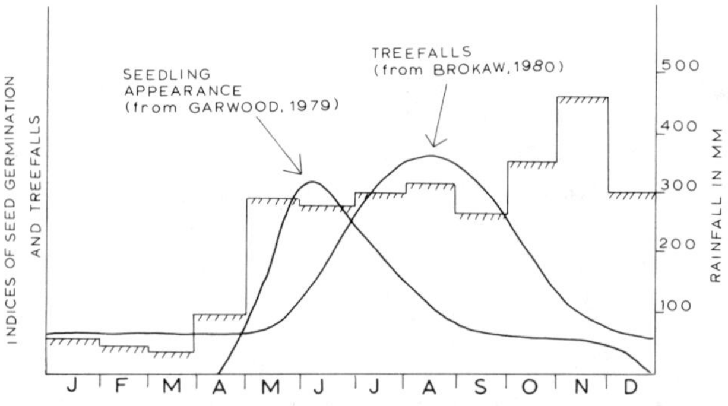

Fig. 11. Seasonal patterns of seedling emergence and frequency of treefalls (diagrammatic) on Barro Colorado Island. Bars are mean monthly rainfall for Barro Colorado Island.

ity is the availability of birds to disperse seeds, since this fruiting peak corresponds with the peak abundance of south-migrating birds in Panama (Karr 1976). In moist forest the seeds of most of these late-fruiting species remain dormant on the soil surface until the start of the following rainy season (Garwood 1982). In wet forest they are more likely to germinate immediately.

Fruit is important food to a wide variety of vertebrates though very few are exclusively frugivorous. Obligate frugivores, such as parrots and fruit bats, travel widely and the latter depend heavily on fruits such as figs which have relatively little seasonality. Many behavioral and physiological adaptations of large forest rodents are probably due to the seasonal fluctuations in large seeds (Smythe 1970, Smythe et al. 1982). The populations of numerous fruit-and seed-eating insects must also be highly susceptible to seasonal availability of fruit.

Los frutos son importantes para los vertebrados, a pesar de que sólo algunos de ellos son frugívoros. Los frugívoros obligados como loros y algunos murciélagos, viajan largas distancias y pueden depender de frutos que se producen todo el tiempo.

In moist forest we can now see there is a striking concurrence of plant activity in leaf flushing, litter build-up, flowering, fruiting, and seed germination in the months March through June. In wet forests these plant activities extend over more of the year and their fluctuations are dampened. But in dry forests, except for flowering, the effect on animals is probably much the same as for moist forest. Not every activity happens in March through June, but it is a time of abundant food for animals relative to the rest of the year.

If we add to this yearly seasonality the large fluctuations between years in the quantity of any of these plant activities (Foster 1982b), it is clear that for most of Panama the plant resources for animals are dramatically unstable. For many animals this instability forces migration or local extinction. As the forests of Panama become smaller and smaller there are fewer places to migrate to and fewer places from which recolonizers can come. The local extinctions are no longer temporary. With the disappearance of animals, the survival of plants pollinated and dispersed by them also becomes tenuous or impossible.

Debido a la periodicidad y a las grandes fluctuaciones de un año a otro, los recursos vegetales ofrecidos a los animales son muy inestables. Esto causa migración o extinción local. Como quiera que los bosques son cada vez más pequeños, hay menos áreas para migración y, consecuentemente, la extinción local puede ser permanente.

Summary

During the year most of the forests in Panama experience pronounced changes in the amount of leaf drop, new leaf emergence, flowering, fruiting, and seed germination. Much of this seasonality in plant activity is directly or indirectly attributable to the distinct dry season that affects most of the country. The forests in different climatic regions have seasonal patterns that are distinct from each other. These seasonal changes are of major importance in governing the population size and behavior of much of the fauna—and indirectly the populations of much of the flora as well. For most of Panama the period March through June is the time of greatest forest plant activity and greatest food availability for most animal species. The fluctuations are probably smallest in the wettest forests. The natural instability of these forests makes them all the more vulnerable to species extinctions as the forest around is devastated due to short-sighted human goals.

Alvim, P. 1960. Moisture stress as a requirement for flowering of coffee. Science 132: 354.

——— & R. *Alvim*. 1978. Relation of climate to growth periodicity in tropical trees. *In* P.B. Tomlinson & M.H. Zimmerman, eds. Tropical Trees as Living Systems. Cambridge Univ. Press.

Augspurger, C.K. 1982. A cue for synchronous flowering. *In* E.G. Leigh, Jr., A.S. Rand, & D.M. Windsor, eds. The ecology of a neotropical forest: seasonal rhythms and longer-term changes. Smithsonian Instit. Press, Washington, DC.

Brokaw, N.V. 1982. Treefalls: frequency, timing and consequences. *In* E.G. Leigh, Jr., A.S. Rand, & D.M. Windsor, eds. The ecology of a neotropical forest: seasonal rhythms and longer-term changes. Smithsonian Instit. Press, Washington, DC.

Coley, P.D. 1980. Effects of leaf age and plant life history patterns on herbivory. Nature 284: 545–546.

———. 1982. Rates of herbivory on different tropical trees. *In* E.G. Leigh, Jr., A.S. Rand, & D.M. Windsor, eds. The ecology of a neotropical forest: seasonal rhythms and longer-term changes. Smithsonian Instit. Press, Washington, DC.

Croat, T.B. 1978. Flora of Barro Colorado Island, Stanford Univ. Press, Stanford, CA.

Daubenmire, R. 1972. Phenology and other characteristics of tropical semideciduous forest in northwestern Costa Rica. J. Ecol. 60: 147–170.

Foster, R.B. 1982a. The seasonal rhythm in fruit fall on Barro Colorado Island. *In* E.G. Leigh, Jr., A.S. Rand, & D.M. Windsor, eds. The ecology of a neotropical forest: seasonal rhythms and longer-term changes. Smithsonian Instit. Press, Washington, DC.

———. 1982b. Famine on Barro Colorado Island. *In* E.G. Leigh, Jr., A.S. Rand, & D.M. Windsor, eds. The ecology of a neotropical forest: seasonal rhythms and longer-term changes. Smithsonian Instit. Press, Washington, DC.

Frankie, G.W., H.G. *Baker*, & P.A. *Opler*. 1974. Comparative phenological studies of trees in tropical wet and dry forests in the lowlands of Costa Rica. J. Ecol. 62: 881–919.

Garwood, N.C. 1982. Seasonal rhythm of seed germination in a semideciduous tropical forest. *In* E.G. Leigh, Jr., A.S. Rand, & D.M. Windsor, eds. The ecology of a neotropical forest: seasonal rhythms and longer-term changes. Smithsonian Instit. Press, Washington, DC.

———. 1983. Seed germination in a seasonal tropical forest in Panama: a community study. Ecological Monographs. (in press).

Gentry, A.H. 1974. Coevolutionary patterns in Central American Bignoniaceae. Ann. Missouri Bot. Gard. 61: 728–759.

Janzen, D.H. 1967. Synchronization of sexual reproduction of trees within the dry season in Central America. Evolution 23: 620–637.

———. 1970. *Jacquinia pungens*, a heliophile from the understory of tropical deciduous forest. Biotrop. 2: 112–119.

———. 1979. How to be a fig. Ann. Rev. Ecol. Syst. 10: 13–51.

Karr, J.R. 1976. On the relative abundance of migrants from the north temperate zone in tropical habitats. Wilson Bulletin 88: 433–458.

Opler, P.A., G.W. *Frankie*, & H.G. *Baker*. 1976. Rainfall as a factor in the release, timing and synchronization of anthesis by tropical trees and shrubs. J. Biogeog. 3: 231–236.

———. 1980. Comparative phenological studies of treelet and shrub species in tropical wet and dry forests in the lowlands of Costa Rica. J. Ecol. 68: 167–188.

Smythe, N. 1970. Relationships between fruiting seasons and seed dispersal methods in a neotropical forest. Amer. Nat. 104: 25–35.

———. 1982. The seasonal abundance of night-flying insects in a neotropical forest. *In* E.G. Leigh, Jr., A.S. Rand, & D.M. Windsor, eds. The ecology of a neotropical forest: seasonal rhythms and longer-term changes. Smithsonian Instit. Press, Washington, DC.

———. W.E. *Glanz*, & E.G. *Leigh*, Jr. 1982. Population regulation in some terrestrial frugivores. *In*

E.G. Leigh, Jr., A.S. Rand, & D.M. Windsor, eds. The ecology of a neotropical forest: seasonal rhythms and longer-term changes. Smithsonian Instit. Press, Washington, DC.

Snow, D.W. 1965. A possible selective factor in the evolution of fruiting seasons in tropical forest. Oikos 15: 274–281.

Tosi, J.A. 1971. Inventariación y demonstraciones forestales, Panamá. Zonas de vida, basado in la labor de J.A. Tosi. FAO, Rome.

Wolda, H. 1978. Seasonal fluctuations in rainfall, food and abundance of tropical insects. J. Anim. Ecol. 47: 369–381.

———. 1980. Seasonality of tropical insects. I. Leafhoppers (Homoptera) in Las Cumbres, Panama. J. Anim. Ecol. 49: 277–291.

Uprooting and Snapping of Trees on Barro Colorado Island

Francis E. **Putz**, Phyllis D. **Coley**, Karen **Lu**, Arlee **Montalvo**, and Annette **Aiello**

Un estudio de los árboles caídos en la Isla de Barro Colorado demostró que la mayoría se partieron, mientras que otros se arrancaron de raíz. Los contrafuertes de los árboles y la pendiente del terreno no influyeron en el tipo de caída de un árbol. El tamaño de la abertura en un bosque y la luz que incide en ella, al igual que el tipo de caída de un árbol, tienen implicaciones con relación a la secuencia de especies en la sucesión.

Después de su caída, muchos árboles partidos retoñaron y los árboles pequeños lo hicieron con mayor frecuencia que los grandes. Retoñar parece ser una vía mediante la cual los árboles de crecimiento rápido y madera débil compensan el hecho de que son propensos a partirse.

Tree diversity in tropical forests has long astonished naturalists, for example, Wallace (1878). Many hypotheses have been proposed to explain how this diversity is generated (for example, Federov 1966) and maintained (for example, Janzen 1970). Recently the role of disturbance in allowing coexistence of species has attracted much attention (Connell 1978; Hubbell 1979; Huston 1979; White 1979). It has been pointed out that cycles of disturbance and regrowth provide a wide variety of environmental conditions for which a multitude of species appear to be specialized (Grubb 1977). One major type of disturbance in forest communities is treefalls. The dynamics of treefall gaps have increasingly become the subject of study, particularly in tropical forests (Brokaw 1980, 1982; Denslow 1980; Hartshorn 1978 and 1980; Whitmore 1975 and 1978). In this paper we summarize our findings from Barro Colorado Island, Panama, concerning the way trees die and the effects of these different modes of tree death on species composition in treefall gaps. Specifically, we discuss the factors influencing whether a particular tree is snapped off or is uprooted, and possible effects of these contrasting types of treefalls on gap colonization by early successional (pioneer) tree species.

La caída de árboles en los bosques es un importante medio de disturbio de sus comunidades.

Barro Colorado Island (BCI) is a 15.6 km^2 island isolated from the mainland between 1910 and 1914 when the Chagres River was dammed to form Gatun Lake. Most of the island has a moderately rich and well drained soil classified as Frijoles Clay (Bennett 1929); small poorly drained areas have a gley soil. Annual rainfall between 1929 and 1975 averaged 265 cm (Rand & Rand 1979) with an eight month rainy season from late April until mid-December.

The older forest on BCI, with which this paper is concerned, is classified as moist tropical forest (Holdridge et al. 1971) and covers approximately one-half of the island area. Foster and Brokaw (1982) estimated that it has been at least 200 years since the last major disturbance to the old forest. The largest trees are 40–50 m tall and 3–4 m dbh (diameter at 1.3 m or above the buttresses). The average hectare of forest has 164 canopy trees (trees more than 20 cm dbh; Putz & Milton 1982) and between 50 and 100 species. Forest turnover rates due to fallen trees are calculated to be 114 years, based on a three-year census (Brokaw 1982). Another estimate of tree mortality rates based on five years of observation of two 1 ha plots in the old forest indicated that 1% of the canopy trees died per year (Putz & Milton 1982). Of these tree deaths, 52% were due to snapped trunks, 17% were uprooted, 14% died standing, 8% snapped at ground level, and the mode of death could not be determined for the remaining 9%.

TABLE 1
COMPARISONS OF UPROOTED AND SNAPPED TREES

Different samples sizes (N) reflect availability of data. Probabilities calculated using Student's t-test. For the 2 heteroscedastic variables (dbh and slope) we used the separate variance estimate (Nie et al. 1979) to calculate the t-statistics.

Variable	Uprooted		Snapped
dbh (cm)	N = 78		N = 217
	x = 44.6	P less than .05	x = 33.6
	s = 28.9		s = 23.9
Height (m)	N = 78		N = 217
	x = 21.6	P less than .01	x = 19.0
	s = 6.9		s = 7.4
Modulus of rapture	N = 63		N = 145
(kg cm^{-2})	x = 881.2	P less than .01	x = 641.9
	s = 346.6		s = 245.1
Modulus of elasticity	N = 63		N = 45
(10^3kg cm^{-2})	x = 148.8	P less than .01	x = 111.4
	s = 114.9		s = 84.0
Specific gravity	N = 74		N = 210
(dry weight green $volume^{-1}$)	x = 0.53	P less than .0001	x = 0.41
	s = 0.13		s = 0.12
Slope (degrees)	N = 78		N = 217
	x = 11.0	P = .16	x = 12.6
	s = 11.8		s = 13.7
Buttressing	N = 64		N = 78
	x = 2.5	P = .78	x = 2.2
	s = 1.1		s = 1.2
Height dbh^{-1}	N = 78		N = 217
	x = 58.2	P = .001	x = 68.1
	s = 22.0		s = 24.3

El promedio de árboles por hectárea en el dosel del bosque en Barro Colorado es de 164 y la tasa de cambio del bosque es de 114 años.

In addition to the rate at which trees die and the canopy is opened, the way trees die may be an important factor influencing forest dynamics and, in particular, the success of pioneer trees in treefall gaps. We monitored treefalls on BCI in order to determine frequencies of snapping and uprooting of trees, as well as possible causal factors. We determined the species of 310 recently fallen trees (greater than 10 cm dbh), and measured the dbh, height, basal (ground level) diameter measured along and perpendicular to the largest diameter, and the slope of the terrain on which the tree was growing. In the case of snapped trees, we measured the height to the snapping point. Our field measurements were compared with published data on the density, strength (modulus of rupture), and elasticity of the wood for each species of fallen tree.

The probability of uprooting as opposed to snapping for trees on BCI was correlated with their mechanical and architectural characteristics. Uprooted trees were significantly larger in dbh and had lower height/diameter ratios than snapped trees. Trees with strong and dense wood—trees with high modulus of rupture and high specific gravity wood—were more prone to uprooting than trees with weaker and less dense wood. Trees with a high modulus of elasticity, indicating stiff wood, also showed a significant tendency to uproot. We expected buttressed trees to be less likely to uproot than to snap because buttresses are thought to reduce the tensile forces acting on the roots (Henwood 1973). Contrary to our expectations, buttress development, expressed as the ratio of average basal diameter to dbh, was no different for snapped and uprooted trees. Terrain slope also had no effect on the probability of uprooting, even though soil depth on BCI generally decreases with increasing slope (Dietrich et al. 1982) and shallow-rooted trees seem to be prone to uprooting.

Los árboles arrancados de raíz presentaban un DAP mayor y una proporción entre altura y diámetro menor que aquellos partidos. Los contrafuertes de los árboles y la pendiente del terreno no influyeron en el tipo de caída de

éstos, es decir, si se arrancaron de raíz o si se partieron.

A high proportion of trees that snapped resprouted. Although sprouts from snapped trees are prone to heartrot (Hepting & Hedgecock 1937), they do have a positional advantage above seedlings and sprouts from uprooted trees. This would be particularly important for shade-intolerant pioneer species. We have observed that snapped trees survive for several years, occasionally flowering and setting fruit. Sprouting may therefore be one means by which weak-wooded trees partially compensate for their tendency to snap.

Muchos de los árboles que se parten, retoñan.

In addition to identifying several factors which influence the probability of uprooting or snapping for an individual tree, we determined the relative frequencies of different types of tree death in the forest of BCI. Of the 310 fallen trees greater than 10 cm dbh censused in this study, 25% uprooted, 70% snapped off between their lowest branch and the ground, and 5% snapped off at ground level. Only 15% of the snapped trees had heartrot or other signs of damage or weakness.

Un cuarto de los árboles caídos se arrancaron de raíz, mientras que tres cuartos se partieron.

The relative frequencies of treefall gaps created by uprooted as opposed to snapped trees may influence the composition of species able to regenerate in the forest as a whole. When a tree uproots it mixes the soil creating a pit and mound. In temperate forests, several plant species have been shown to specialize on these microsites (Hutnik 1952; Falinski 1978). Putz, working on BCI, found that pioneer tree species were more common on the pit and mound created by uprooted trees than elsewhere in treefall gaps. Several possible explanations for this pattern deserve mention. First, the pit and mound are likely to be free from overhead shade. Since pioneer trees, by definition, are shade intolerant, these sites would be desirable. In addition to being free from overhead shade, pits and mounds created by uprooted trees are at least initially clear of leaf litter. The exposure of mineral soil and the absence of the buffering effects of leaf litter may be important factors leading to the germination and establishment of pioneer trees. Seed germination in many pioneers is known to be stimulated by light and fluctuating temperatures (Sauer & Struik 1964; Bell 1970; Vazquez-Yanes 1974 and 1976), both of which would be enhanced by the removal of litter. Considering that the rain of seeds into treefall gaps seems to be light (pers. obs.) and that the population of buried viable seeds is quite high (742 seeds/m^2) in the top 10 cm of soil under the closed canopy of older forest, it may be that the exposure of buried seeds due to tree uprooting is a major factor leading to the concentration of pioneer trees on soil disturbed by uprooted trees. Therefore, because of both soil and light characteristics, gaps created by uprooted trees may provide better microsites for the establishment of pioneer tree species than those created by snapped trees.

Las aberturas en el bosque y los montículos de suelo producidos por los árboles que se arrancan de raíz, proveen un excelente hábitat para las especies pioneras de árboles.

Brokaw (1980, 1982) found that treefall gap size on BCI was smaller in young forest (approximately 70 years old) than in mature forest greater than 200 years old. He found that the density of pioneer trees on BCI increases significantly with gap size, and that pioneers are almost completely absent in gaps smaller than 150 m^2. The fact that secondary forest trees tend to be smaller than primary forest trees and hence open small gaps when they die, coupled with the observation that they are more likely to snap than to uproot, could contribute to the replacement of trees characteristic of young forest by more shade tolerant and longer lived primary forest species. This also helps to explain the persistence of pioneer trees in mature forests.

Las especies pioneras están asociadas con las aberturas producidas por la caída de árboles grandes en los bosques maduros, lo cual podría explicar su persistencia.

Acknowledgments

This research was supported by the Smithsonian Tropical Research Institute. We thank N.V. Brokaw and E.G. Leigh, Jr. for advice and encouragement.

Bell, C.R. 1970. Seed distribution & germination experiment. *In* H.T. Odum and R.F. Pigeon eds., A Tropical Rain Forest. U.S.A.E.C., Oak Ridge, TN.

Bennett, H.H. 1929. Soil reconnaissance of the Panama Canal Zone and contiguous territory. Techn. Bull. U.S.D.A. 94: 1–46.

Brokaw, N. 1980. Gap-phase Regeneration in a

Neotropical Forest. Ph.D. thesis. Univer. Chicago. 175 pp.

———. (1982). Treefalls: frequency, timing, and consequences. *In* E.G. Leigh, Jr., A.S. Rand & D.M. Windsor eds., The Ecology of a Tropical Forest: Seasonal Rhythms and Long-term Changes. Smithsonian Instit. Press, Washington, DC.

Connell, J.H. 1978. Diversity in tropical rain forests and coral reefs. Science 199: 1302–1310.

Denslow, J.S. 1980. Gap partitioning among tropical rainforest trees. Biotropica 12 (Suppl.): 47–55.

Dietrich, W, D.M. *Windsor* & T. *Dunne*. (1982). The physical environment and hydrology of Barro Colorado Island. *In* E.G. Leigh, Jr., A.S. Rand & D.M. Windsor eds., The Ecology of a Tropical Forest: Seasonal Rhythms and Long-term Changes. Smithsonian Instit. Press, Washington, DC.

Falinski, J.B. 1978. Uprooted trees, their distribution and influence in the primeval forest biotope. Vegetatio 38: 175–183.

Fedorov, A.A. 1966. The structure of the tropical rain forest and speciation in the humid tropics. J. Ecol. 54: 1–11.

Foster, R.B. & N.V. *Brokaw*. (1982). The vegetation of Barro Colorado Island. *In* E.G. Leigh, Jr., A.S. Rand & D.M. Windsor eds., The Ecology of a Tropical Forest: Seasonal Rhythms and Long-term Changes. Smithsonian Instit. Press, Washington, DC.

Grubb, P.J. 1977. The maintenance of species-richness in plant communities: the importance of the regeneration niche. Biol. Rev. 52: 107–145.

Hartshorn, G.S. 1978. Tree falls and tropical forest dynamics. pp. 617–630. *In* P.B. Tomlinson & M.H. Zimmermann. Tropical Trees as Living Systems. Cambridge Univ. Press, Cambridge.

———. 1980. Neotropical forest dynamics. Biotropica 12 (Suppl.): 23–30.

Henwood, K. 1973. A structural model of forces in buttressed tropical rain forest trees. Biotropica 5: 83–93.

Hepting, G.H. & G.G. *Hedgcock*. 1937. Decay in merchantable oak, yellow poplar, and basswood in the Appalachian Region. U.S. Dept. Agric. Tech. Bull. 570.

Holdridge, L.R., W.C. *Grenke*, W.H. *Hatheway*, T. *Liang* & J.A. *Tosi*, Jr. 1971. Forest Environments in Tropical Life Zones. Pergamon Press, New York.

Hubbell, S.P. 1979. Tree dispersion, abundance and diversity in a tropical dry forest. Science 203: 1299–1309.

Hutnik, R.J. 1952. Reproduction on windfalls in a northern hardwood stand. J. For. 50: 693–694.

Huston, M. 1979. A general hypothesis of species diversity. Am. Nat. 113: 81–101.

Janzen, D.H. 1970. Herbivores and the number of tree species in tropical forests. Am. Nat. 104: 501–528.

Putz, F.E. & K. *Milton*. 1982. Tree mortality rates on Barro Colorado Island. *In* The Ecology of a Tropical Forest: Seasonal Rhythms and Long-term Changes. E.G. Leigh, Jr., A.S. Rand & D.M. Windsor eds., Smithsonian Instit. Press, Washington, DC.

Rand, W.M. & A.S. *Rand*. 1979. The rainfall of Barro Colorado Island, Panama. Actas del IV Symposium International de Ecologia Tropical 1: 31–49, Panama.

Sauer, J. & G. *Struik*. 1964. A possible ecological relation between soil disturbance, light flash and seed germination. Ecology 45: 884–886.

Vazquez-Yanes, C. 1974. Studies on the germination of seeds of *Ochroma lagopus* Swartz. Turrialba 24: 176–179.

———. 1976. Seed dormancy and germination in secondary vegetation tropical plants: the role of light. Comp. Physiol. Ecol. 1: 30–34.

Wallace, A.R. 1878. Tropical Nature and Other Essays. Macmillan, London.

White, P.S. 1979. Pattern, process, and natural disturbance in vegetation. The Botanical Review 45: 229–299.

Whitmore, T.C. 1975. Tropical Rain Forests of the Far East. Clarendon Press, Oxford.

———. 1978. Gaps in the forest canopy. Pp. 639–655. *In* P.B. Tomlinson & M.H. Zimmermann eds., Tropical Trees as Living Systems. Cambridge University Press, Cambridge.

Pollination and Reproductive Biology of *Heliconia*

W. John Kress

Las Heliconias polinizadas por colibríes ermitaños presentan diferente morfología que aquellas polinizadas por colibríes no ermitaños. La mayoría de las especies de Heliconia son autocompatibles, pero la fidelidad del polinizador e inhibición en la superficie del estigma tienden a reducir la hibridación. Los híbridos son pocos y generalmente estériles.

Pollinators

Hummingbirds (family Trochilidae) are the primary and in most cases the exclusive pollinators of *Heliconia* L. (Heliconiaceae) in the Neotropics. Skutch (1933) made early observations in Central America on visitation of *Heliconia* by several hummingbird species. More recently Stiles (1975, 1978, 1979) has contributed greatly to our knowledge of the coevolution of *Heliconia* and their hummingbird pollinators. The majority of *Heliconia* species possesses brightly colored red and yellow inflorescences and long, tubular, diurnal, odorless flowers that produce copious nectar. All of these traits are characteristic of bird pollinated plants (Faegri & van der Pijl 1979). The habitat and phenological patterns of *Heliconia* are also strongly attuned to the life history and distribution of their associated hummingbird species (Stiles & Wolf 1979).

La mayoría de las Heliconias presentan inflorescencias rojas y amarillas brillantes. Las flores carecen de olor, poseen el tubo de la corola largo y son polinizadas por colibríes.

Species of *Heliconia* can be divided into two groups according to the type of hummingbird that visits the flowers (Linhart 1973; Stiles 1975, 1979). Plants visited by territorial, non-hermit hummingbirds (Trochilinae) grow in large, often monoclonal stands at forest edges or in open habitats, have short perianths and produce many flowers with copious nectar each day. Species that tend to form small clumps in closed forest habitats, have longer curved perianths, and produce few flowers per day with small amounts of dilute nectar, are generally visited by traplining hermit hummingbirds (Phaethorninae). These two basic groups of heliconias actually represent the ends of a continuum and many species are intermediate in various characters. Most species of *Heliconia* are visited primarily by hermits and most hermits are dependent upon *Heliconia* for their energy requirements. For this reason Stiles (1979) believes that hermit-visitation is the primitive pollination type in *Heliconia* and that territorial non-hermit pollination is a derived condition. All Central American species with pendent inflorescences are pollinated by hermits except for *H. mariae* J.D. Hook. (Stiles 1979; pers. obs.). This latter species often grows in large stands, produces many flowers per inflorescence per day, has a relatively straight, short perianth and is visited by the territorial crowned woodnymph *(Thalurania furcata)* in Costa Rica. It is one of the few heliconias with red flowers.

Heliconias visitadas por colibríes territoriales y no ermitaños crecen en grupos grandes y densos y en hábitat abiertos. Estos individuos presentan flores con perianto corto y muchas de ellas producen abundante néctar diaramente.

Las Heliconias visitadas por colibríes ermitaños crecen en grupos pequeños, en hábitat boscosos, poseen periantos largos y presentan menos flores, las cuales producen escaso néctar.

Although there is a diverse arthropod fauna that acts as nectar thieves or is associated with fluids that collect in the cincinnal bracts (Seifert 1975, 1982; Seifert & Seifert 1976; Stiles 1979; C. Baird pers. comm.), it is unlikely that any of these are predominant pollinators of *Heliconia* flowers. Dodson & Gentry (1978) suggest that *H. aemygdiana* Burle-Marx from Ecuador is visited by bats since hummingbirds were never seen visiting the flowers which are green and supposedly fragrant at night. Of the several distinctive disjunct species of *Heliconia* that are endemic to the South Pacific region some, especially those with pendent inflorescences, are visited by bats (Kress unpubl.). These species have inconspicuous green cincinnal bracts and green or white flowers that open in the late afternoon and last only until morning. I have observed numerous claw marks on the inflorescences of many of these species, and in one species, *H. solomonensis*, macroglossine bats pollinating the flowers. Sympatric species of the genus *Musa* possess a very similar inflorescence in aspect and color and are known to be pollinated by nocturnal Megachiropteran bats (Nur 1976; Start & Marshall 1976). Studies on pollination of Old World heliconias are currently in progress.

Algunas especies de Heliconia son polinizadas por murciélagos.

Breeding Systems

Physiological self-incompatibility is rare in Central American *Heliconia*, and in the few species where such systems have been found they are usually only partially effective in screening out self-pollen. In a detailed study of self-incompatibility in nine species of *Heliconia*, both pollen tube growth in styles and fruit set were monitored—only two species showed a significant decrease in fertilization after self-pollination (Table 1). One of these species, *H. mathiasiae* Daniels & Stiles, is completely self-incompatible, while *H. tortuosa* Griggs is partially self-incompatible. The remaining species tested were self-compatible although individuals within each species showed varying degrees of acceptance of self-pollen. The genetic control of self-incompatibility in *Heliconia* is unknown. Circumstantial evidence, such as binucleate pollen, suggests gametophytic determination of self-incompatibility. Autogamy is present (*H. colgantea* Daniels & Stiles, *H. umbrophila* Daniels & Stiles) in the genus. Most plants, however, require an external pollination vector in order to set full fruit. Separation of the stigma and anthers within any flower is not great and probably has little effect on promoting outcrossing. Furthermore, protandry or protogyny do not appear to play any part in the breeding systems of the species studied. Anthers dehisce before anthesis and stigmatic surfaces are receptive at flower opening.

La mayoría de las especies de Heliconia son autocompatibles. Sin embargo parece ser que la mayor parte de ellas requieren de un agente polinizador externo para producir fruto con éxito.

Pollinator foraging behavior may be the primary mechanism by which inbreeding is regulated in *Heli-*

TABLE 1
NATURAL HYBRIDS OF *HELICONIA* RECORDED FROM CENTRAL AMERICA

Parents	Country	Collection[1] or (literature citation)
curtispatha X pogonantha	Panama	Kress 80–1144, 80–1156
latispatha X bourgaeana	Mexico	Kress et al. 77–718
latispatha X imbricata	Costa Rica	Kress 76–528, 79–1098
mariae X pogonantha	Panama	Williams 686
	Costa Rica	Kress 76–627, 76–659, 77–795, 77–875
nutans X lankesteri	Panama	Kress & Clarkson 79–1089c
nutans X tortuosa	Costa Rica	Daniels & Stiles (1979)
sarapiquensis X imbricata	Costa Rica	Daniels & Stiles (1972)
secunda X clinophila	Costa Rica	Kress 77–777
stilesii X unnamed species	Costa Rica	Daniels & Stiles (1979)

[1]Kress specimens deposited at DUKE.

conia. In species visited by non-hermits, territoriality by the pollinators will severely limit pollen flow between individual plants and thereby lead to inbreeding. In contrast, pollen flow, and presumably the degree of outcrossing, will be significantly greater in species that are visited in a traplining fashing by hermits. These hypothesized patterns of pollen flow in *Heliconia* were demonstrated in a limited way by using histological dyes to track pollinator movement within and between plants of traplining and territorial species (Linhart 1973). Recently Ray, Stiles & Kress (unpubl.) showed that inter-territorial pollen flow is severely restricted in the non-hermit pollinated *H. imbricata* (Kuntze) Baker, although at various times during the morning territoriality breaks down and pollen is exchanged among territories.

Isolating Mechanisms and Hybridization

El flujo de polen, determinado por los colibríes, podría ser el regulador principal de la reproducción en Heliconia.

Plant/bird morphological specializations, foraging behavior and nectar requirements of the hummingbirds regulate pollen exchange between species (Stiles 1975, 1978). Interspecific variation in perianth morphology and position of anthers leads to deposition of pollen of different species in different places on a given type of hummingbird. In addition, variation in caloric parameters of nectar production and presentation influences the specificity of the type of pollinator that will visit a given species of *Heliconia* and promotes conspecific flower visitation. The conspicuous inflorescences that characterize the genus serve as visual cues to the pollinators and increase visitation constancy. Stiles (1975) has pointed out certain additional habitat and temporal specializations that may limit interspecific pollen flow.

La especialización morfológica y del néctar de las Heliconias, así como el comportamiento de los colibríes tienden a aumentar la fidelidad del agente polinizador, pero son insuficientes para prevenir intercambio de polen al nivel interespecífico.

In spite of the pollinator-related specializations that increase pollinator fidelity, sympatric species may exchange significant amounts of pollen. A preliminary study at Finca La Selva in Costa Rica in which fluorescent powders were used to track hummingbird movement among four species of *Heliconia* (two hermit and two non-hermit types) showed that ethological and morphological isolating mechanisms often are insufficient to prevent interspecific pollen exchange (Kress et al. 1979). It was concluded that pollinator infidelity was frequent, and a given species of hummingbird was not restricted to any single species of *Heliconia*. Non-hermit hummingbirds may forage on several different non-hermit species, whereas hermits favor hermit-type heliconias but may visit non-hermit types also.

Of all the mechanisms restricting gene exchange among species of *Heliconia*, physiological regulation at the stigmatic surface and in the style plays the most important role (Kress 1983b). An extensive controlled crossing program was performed in Costa Rica among seventeen sympatric and allopatric species with pendent and erect inflorescences, including types visited by hermit and non-hermit hummingbirds. Observation of pollen tube growth by fluorescence microscopy and seed set were used to determine crossability. It was found that foreign pollen may be inhibited at the stigmatic surface, within the stylar tissue or within the ovary. The site of inhibition is consistent for each pair of species and is dependent on the parentage and the direction of the cross. Cross-compatibility is rare, and very few interspecific combinations produce mature seed. In most cases of ovarian inhibition the techniques could not separate pre- and post-fertilization incompatibility.

La regulación fisiológica en la superficie estigmática es importante para prevenir la polinización por otra especie.

Specific crossability barriers have no obvious ecological correlates. The type of inhibition found in any combination of species is not associated with their geographical distribution; stigmatic, stylar and ovarian inhibition occur between allopatric as well as sympatric species. Species that might normally share the same individual pollinators or same type of pollinators (hermit vs. non-hermit) also fail to exhibit any particular type of interspecific inhibition.

The specificity of the interspecific crossability barriers is significant from a taxonomic standpoint, particularly for species with pendent inflorescences (Kress 1983b). In a study performed at Las Cruces Tropical Botanical Garden in Costa Rica, nine species with pendent inflorescences were crossed in nearly all reciprocal combinations. Each species was categorized by the ability of the style to accept foreign pollen tubes and the ability of the pollen to grow in foreign styles. The species can be grouped according to the main type of crossability barrier that characterizes the pistil and pollen of each. For example, in *H. pogonantha* Cufod. and *H. colgantea* foreign pollen is inhibited at the stigmatic surface, and pollen of these species is inhibited on the stigma of most foreign species. In contrast, although foreign pollen is nearly always inhibited at the stigma in *H. collinsiana* Griggs and *H nutans* Woods., the pollen of these species breaks the stigmatic barrier of most other species, and pollen tube growth is arrested in the foreign style. *Heliconia trichocarpa* possesses the least specific pistil and pollen inhibition barriers of the nine species tested. The same taxonomic groups that are characterized by specific crossability barriers are also supported by vegetative, floral and pollen characters. It appears, therefore, that possession of a specific crossability barrier reflects the phylogenetic history of the species in most cases. Some species, however, show disparities in morphological characters and interspecific inhibition sites. For example, *H. colgantea* that shares numerous derived characters with *H. trichocarpa*, possesses crossability barriers more similar to those found in *H. pogonantha*. Similarly, *H. mariae* is grouped with *H. curtispatha* and *H. stilesii* according to the pollen index but is more closely related to *H. pogonantha.*

Las especies pueden agruparse de acuerdo al principal tipo de barrera presente en la reproducción y estos grupos concuerdan con los grupos taxonómicos.

Natural hybrids are occasionally produced (Table 1), but the number is very small in proportion to the large number of sympatric associations. Most hybrids are found in areas of major ecological disturbance (often man-made). These hybrids may be the result of penetration of the strong physiological crossability barriers after repeated breakdowns of habitat and ethological isolating mechanisms. The few hybrids that have been found in the wild generally are pollen sterile, pistil sterile, or both (Stiles 1979), and some degree of natural selection against hybridization is to be expected.

Los híbridos naturales son poco frecuentes y generalmente son estériles.

Daniels, G.S. & F.G. *Stiles*. 1979. The *Heliconia* taxa of Costa Rica. Keys and descriptions. Brenesia 15(Supl.): 1–150.

Dodson, C.H. & A.H. *Gentry*. 1978. Heliconias (Musaceae) of the Rio Palenque Science Center, Ecuador. Selbyana 2: 291–299.

Faegri, K. & L. *van der Pijl*. 1979. The Principles of Pollination Ecology. 3rd ed., Pergamon Press, Oxford.

Kress, W.J. 1981. Systematics of Central American *Heliconia* with pendant inflorescences. J. Arnold Arboretum. 62:243–260.

———. 1983a. Self-incompatibility in Central American *Heliconia*. Evolution. 37: 735–744.

———. 1983b. Crossability barriers in neotropical *Heliconia*. Ann. Bot. 52: 131–147.

———, D. *Gill*, N. *Nadkarni*, R. *Rice*, J. *Snyder*, M. *Stephens* & D. *Wrobel*. 1979. Pollinator fidelity in sympatric members of the genus *Heliconia* at La Selva. O.T.S. Tropical Biology Course Book 79–3.

Linhart, Y.B. 1973. Ecological and behavioral determinants of pollen dispersal in hummingbird-pollinated *Heliconia*. Amer. Nat. 107: 511–523.

Nur, N. 1976. Studies on pollination in Musaceae. Ann. Bot. 40: 167–177.

Seifert, R.P. 1975. Clumps of *Heliconia* inflorescences as ecological islands. Ecology 56: 1416–1422.

———. 1982. Neotropical *Heliconia* insect communities. Quart. Rev. Biol. 57: 1–28.

——— & F.H. *Seifert*. 1976. Natural history of insects living in inflorescences of two species of *Heliconia*. J. New York Entom. Soc. 84: 233–242.

Skutch, A.F. 1933. The aquatic flowers of a terrestrial plant, *Heliconia bihai* L. Amer. J. Bot. 20: 535–544.

Start, A.N. & A.G. *Marshall*. 1976. Nectarivorous bats as pollinators of trees in West Malaysia. *In* J. Burley & B.T. Styles eds., Tropical Trees: Variation, Breeding and Conservation. Linn. Soc. Symp. Ser. No. 2, pp. 141– 150. Academic Press, London.

Stiles, F.G. 1975. Ecology, flowering phenology, and hummingbird pollination of some Costa Rican *Heliconia* species. Ecology 56: 285–301.

———. 1978. Ecological and evolutionary implications of bird pollination. Amer. Zool. 18: 715–727.

———. 1979. Notes on the natural history of *Heliconia* (Musaceae) in Costa Rica. Brenesia 15(Supl.): 151–180.

——— & L. L. *Wolf*. 1979. The ecology and evolution of lek mating behavior in the long-tailed hermit hummingbird. Amer. Ornithol. Union Monogr. 27: 1–78.

Aspectos de la Ecología, Distribución y Evolución del Género *Heliconia* en Costa Rica

F. Gary Stiles

The genus Heliconia includes 150–170 species of which about 40 occur in Costa Rica. They form about five groups based on habit, size, inflorescence type, corolla shape, and other characteristics which appear to be adapted to the habitats and the two major types of hummingbird pollinators. The primitive form may be adapted to medium sized light openings and pollination by hermit hummingbirds, while other forms represent departures from the original form.

El género *Heliconia* incluye aproximadamente 150–170 especies descritas actualmente (aunque es probable que queden todavía 50–100 especies por describir). Todas son plantas herbáceas, de tamaño moderadamente pequeño a bastante grande; casi todas poseen flores que son polinizadas por colibríes, y sus frutos son dispersados por aves frugívoras (Stiles 1979). Con excepción de unas pocas especies de Melanesia, todo el género (pertenece a la familia Musaceae) se encuentra en el Neotrópico (Smith 1968). Los orígenes de *Heliconia*, tanto como sus relaciones con los otros grupos que componen el orden Scitamineae, todavía son poco entendidos (Tomlinson 1961).

Aproximadamente 40 especies de *Heliconia* se encuentran en Costa Rica, donde su taxonomía e historia natural han sido estudiadas extensivamente en los últimos 10 años (Daniels & Stiles 1979; Stiles 1975, 1979, 1980). Por lo tanto, en la medida en que las especies costarricenses de *Heliconia* formen una buena representación de la diversidad morfológica, ecológica y taxonómica dentro del género, pueden considerarse una base o punto de partida ideal para intentar una reconstrucción de algunos procesos evolutivos dentro del género. Quizás pareciera algo arriesgado este procedimiento, puesto que el centro de diversificación del grupo está indudablemente en Suramérica, y que Costa Rica puede contar con no más de 15–20% del número total de especies del género. Sin embargo, los grupos principales de especies de *Heliconia* están bien representados en el país, y no hay ninguna otra área donde esté tan bien estudiada la historia natural del grupo. Por cierto, Panamá sería un lugar ideal para poner a prueba las teorías desarrolladas aquí, puesto que tiene una representación muy rica y diversa de *Heliconia*, que sólo en parte está compartida con Costa Rica.

Costa Rica is one of the best studied areas for Heliconia, although it has only about 40 species or 15–20% of the genus. For simplicity, the Costa Rican species are placed into five groups.

Los Grupos de Especies de *Heliconia* de Costa Rica

Para facilitar la discusión de las especies de *Heliconia* costarricenses, las he repartido en 5 grupos (algunos con subgrupos) (Cuadro 1). Este esquema representa una modificación y simplificación del que presenté antes (Stiles 1979), una divergencia marcada de la clasificación "tradicional" del género, como la de Smith (1968). No estoy proponiendo una nueva clasificación subgenérica en este momento, pues me faltan conocimientos de muchas especies fuera de Costa Rica. Sin embargo, el esquema propuesto aquí es mucho más conforme con lo que se sabe de las adaptaciones de las varias especies, y puede servir como la base o punto de partida para el desarrollo de una clasificación mejor.

A. Grupo *pogonantha*: plantas generalmente de tamaño grande a muy grande, con pseudotroncos masivos, hojas de textura dura; inflorescencias péndu-

CUADRO 1
GRUPOS Y CARACTERISTICAS ECOLOGICAS DE LAS ESPECIES DE *HELICONIA* DE COSTA RICA

Grupos y Especies	Habitat(s) de Maxima Abundancia[1]	Zonas de Vida[2]	Polinizador(s)[3] Principal(s)
I. Grupo pogonantha			
1. curtiospatha Petersen	2	Tmh, Pmh, Pp	Eep
2. longa (Griggs) Winkler	2	Tmh, Pmh, Pp	Eep
3. marginata (Griggs) Pittier	3	Tmh, Pmh	Eep
4. marie J.D. Hooker	3	Th, Tmh, Pmh	N
5. ramonensis Daniels & Stiles	2	Pmh	Eep
6. pogonantha Cufodontis	2	Th, Tmh, Pmh	Eep
7. vellerigera Poeppig	2	Tmh, Pmh, Pp	Eep
II. Grupo bihai			
1. wagneriana Peterson	3	Th, Tmh, Pmh	Ep
III. Grupo trichocarpa			
1. colgantea Daniels & Stiles	2	Pmh	Ep
2. nutans Woodson	1	Mmp	Ep
3. platystachys Baker	2	Th, Tmb, Pmb	Ep
4. secunda R.R. Smith	2	Pp, Mp	Ep
5. talamancana Daniels & Stiles	2	Tmh, Pmh	Ep
6. trichocarpa Daniels & Stiles	2	Tmh, Pmh, Pp	Ep
7. viridiflora Daniels & Stiles	1–2	Tmh, Pmh	Ep
IV. Grupo tortuosa			
A. Subgrupo tortuosa			
1. beckneri R.R. Smith	2	Pp, Mp	Ep
2. irrassa R.R. Smith	2	Th, Tmh, Pmh, Pp	Ep
3. latispatha Bentham	3	Ts, Th, Tmh, Ph, Pnh, Pp	N
4. monteverdensis Daniels & Stiles	1–2	Mp	N
5. sarapiquensis Daniels & Stiles	2	Tmh, Pmh,	N
6. tortuosa Griggs	2	Ph, Pmh, Pp, Mh, Mmh, Mp	Ep, N*
7. umbrophila Daniels & Stiles	1	Tmh, Pmh	Ep
8. vulcanicola Stiles	1–2	Mp	N
B. Subgrupo imbricata			
1. atropurpurea Daniels & Stiles	2	Pmh	Ep
2. imbricata (Kuntze) Baker	3	Tmh, Pmh, Th	N
3. lophocarpa Daniels & Stiles	1–2	Tmh	Ep
4. reticulata (Griggs) Winkler	1	Tmh, Pmh, Pp	Ep
C. Subgrupo lankesteri			
1. clinophila R.R. Smith	2	Pp, Mmp, Mp	Ep
2. lankesteri Standley	2	Mmh, Mp	Ep, N*
D. Subgrupo gracilis			
1. gracilis Daniels & Stiles	1–2	Pp	Ep
2. ignescens Daniels & Stiles	2	Pnh, Pp	Ep
3. H-40 sp.	2	Pmh	Ep
V. Grupo osaensis			
A. Subgrupo osaensis			
1. calatheaphyllus Daniels & Stiles	1	Tmh	Ep
2. deflexa Daniels & Stiles	1–2	Th, Tmh, Pmh	Ep
3. golfodulcensis Daniels & Stiles	1–2	Tmh	Ep
4. mathiasei Daniels & Stiles	2	Tmh, Pmh, Pp	Ep
5. longiflora R.R. Smith	2	Tmh, Pmh	Ep
6. osaensis Cufodontis	2	Th, Tmh	Ep
7. wilsonii Daniels & Stiles	2	Phh, Pp, Pmh	Ep
B. Subgrupo aurantiaca			
1. aurantiaca Ghiesbreght	1	Tmh, Pmh	Ep

1. Hábitats: 1 = sotobosque, aperturas pequeñas; 2 = aperturas medianas, crecimiento secundario veijo; 3 = aperturas grandes, crecimiento secundario joven.
2. Abreviaciones: T = tropical; P = premontano; M = Montano Bajo; h = humedo; mh = muy humedo; p = pluvial.
3. Abreviaciones: Ep = ermitaños como *Phaethornis*; Eep = ermitaños como *Eutoxeres* y *Phaethornis;* N = no ermitaños típicos; N* = no-ermitaños convergentes con los ermitaños (vea texto).

las, grandes, pesadas, duraderas; frutos oblongos, expuestos cuando maduros por una extensión rápida del pedicelo; épocas de floración largas, frecuentemente 6 meses o más. La especie que menos cabe aquí es *Heliconia marginata*, por ser mucho más pequeña en todas sus dimensiones que las demás (tal vez merece constituirse en un subgrupo aparte). Sin embargo, concuerda con la forma de la inflorescencia y del fruto, y en la manera de exposición de éstos, con el grupo de *pogonantha* y no con el grupo de *trichocarpa*. El tamaño pequeño de la planta puede ser una especialización relacionada con su hábitat preferido (pantanos, ríos) con sustrato sumergido e inestable. El otro cambio notable es la inclusión de *H. mariae*, que había sido incluida por Smith (1968) y otros en una "Serie Imbricatae" con *H. imbricata* y *H. reticulata*. Sin embargo, casi lo único que tiene en común *H. mariae* con *H. imbricata* y sus afines es la imbricación de las brácteas espatáceas. En todos los otros aspectos de su morfología y fenología, calza perfectamente con el grupo de *pogonantha*. Esta forma híbridos con *H. pogonantha*, pero con ninguna especie del subgrupo de *imbricata*.

Group pogonantha are large plants with heavy, hanging inflorescences.

B. Grupo *bihai*: inflorescencias erectas, dísticas, las brácteas profundas y especializadas para contener un "charquito" de agua; frutos redondos, suculentos, expuestos por la extensión rápida del pedicelo que los empujan hasta arriba de la superficie del agua; bordes de las hojas ondulantes en muchas especies. En Costa Rica sólo hay 1 especie. Es el grupo más diverso en la región del Caribe. Se ha considerado como una sección aparte dentro del género, pero hay varios puntos de semejanza con el grupo *tortuosa* (especialmente subgrupo *lankesteri*; la forma de exponer el fruto puede representar una especialización relacionada con las brácteas profundas y acuíferas. Debido a ésto puede que una separación tan grande entre los grupos *bihai* y *tortuosa* no se justifique.

Group bihai have erect, distichous inflorescences.

C. Grupo *tortuosa*: inflorescencias erectas, frutos más o menos redondos, expuestos por distorsión de las brácteas espatáceas y pudrición de las brácteas florales; épocas de floración generalmente cortas, aproximadamente 3 meses, y bien definidas; plantas pequeñas a moderadamente grandes, de hábito musoide. Este es el grupo más grande y variado de *Heliconia* en Costa Rica, y lo he dividido en cuatro subgrupos. Antes consideré algunos de éstos como grupos aparte, pero el grado de semejanza, e incluso la presencia de híbridos entre varios de ellos favorece el arreglo actual. Los subgrupos son:

Group tortuosa have erect inflorescences: subgroup tortuosa have rotate inflorescences; subgroup Imbricata have distichous inflorescences; subgroup Gracilis are smaller plants, and subgroup Lankesteri have "candelabra" type inflorescences.

1. Subgrupo *tortuosa*: inflorescencias rotáceas, brácteas espatáceas tocándose a distancia, pero no imbricadas.

2. Subgrupo *imbricata*: inflorescencias dísticas, brácteas espatáceas, imbricadas y cortas (la punta no aguda). Este grupo ha sido considerado como una serie aparte (con la inclusión de *H. mariae*, pero dada la frecuencia con que *H. imbricata* forma híbridos con ciertos miembros del subgrupo *tortuosa* y la dificultad de hacer divisiones muy tajantes entre ciertos subgrupos, creo que mejor merece ser subgrupo nada más, dentro del grupo *tortuosa*.

3. Subgrupo *gracilis*: plantas más pequeñas, inflorescencias dísticas, imbricadas basalmente, acuminadas en la punta; hábito musoide a semi-musoide. Este grupo fue "incertae sedis" en mi esquema anterior (Stiles 1979) pero creo que la morfología de *H. rodriguezii* definitivamente establece la relación con el grupo *tortuosa*.

4. Subgrupo *lankesteri:* inflorescencias dísticas, de tipo "candelabro" (Smith 1968), brácteas espatáceas acuminadas, tocándose por poca distancia. La morfología de este subgrupo muestra varias semejanzas con los subgrupos de *imbricata* y *gracilis*, además con el grupo *bihai*.

D. Grupo *trichocarpa*: plantas de tamaño pequeño a mediano, musoides; inflorescencias péndulas, rotáceas, de tamaño pequeño a mediano; brácteas espatáceas acuminadas; frutos redondos, expuestos por distorsión de las brácteas y pudrición de las brácteas florales; épocas de floración generalmente cortas y bien definidas. Antes los miembros de este grupo, más los integrantes del grupo *pogonantha* (menos *H. mariae* se consideraron como la sección *Pendulines*. Sin embargo, en muchos aspectos de su morfología y fenología, los miembros del grupo *trichocarpa* son muy parecidos a los del grupo (y subgrupo) *tortuosa*, no a los del grupo *pogonantha*. Además *H. nutans* forma híbridos con *H. tortuosa*; y especies como *H. secunda* y *H. viridiflora*, con inflorescencias muy contorsionadas, casi podrían pertenecer a cualquiera de los dos. Las especies del grupo *trichocarpa* son esencialmente como los miembros del grupo *tortuosa* pero con la inflorescencia péndula en vez de erecta; es más que nada por tradición que considero éste como un grupo aparte y no un subgrupo del grupo *tortuosa*.

Group trichocarpa are smaller banana-like plants with hanging or rotate inflorescences.

E. Grupo *osaensis*: plantas pequeñas a medianas, cannoides a semicannoides, inflorescencia pequeña, apical; frutos exertos desde la floración.

Group osaensis have small apical inflorescences and are canna-like: subgroup Osaensis have distichous inflorescences and truncate fruits; subgroup Aurantiaca are cannoid plants with rotate inflorescences.

1. Subgrupo *osaensis*: inflorescencias dísticas, frutos generalmente truncados. La taxonomía tradicional (Smith 1968) reconoció una Sección *Stenochlamys* para las especies con inflorescencias erectas y brácteas espatáceas más o menos angostas. Esta sección se dividió en dos series: *Cannoideae*, con especies cannoides y semi-cannoides, y *Distantes*, con especies musoides y semi-cannoides. Ciertas especies semi-cannoides muy relacionadas fueron puestas unas, en una serie y otras en la otra. En realidad es difícil o imposible hacer una división tajante entre cannoides y semi-cannoides: por ejemplo, la especie *H. wilsonii* (semi-cannoide) tiene una inflorescencia mucho más parecida a las de ciertas especies cannoides (e.g. *H. mathiasii* que a las de la mayoría de las especies semi-cannoides. Además el conjunto de cannoides y semi-cannoides muestra más semejanzas entre sí, que con los miembros de cualquier otro grupo. Sin embargo, las afinidades de este grupo probablemente son con el grupo *tortuosa*, especialmente el subgrupo *gracilis* que tiene inflorescencias algo parecidas y un hábito algo intermedio entre musoide y semi-cannoide ("semi-musoide").

2. Subgrupo *aurantiaca*: cannoide con inflorescencia rotácea, frutos redondos (una combinación única de carácteres que merece para esta especie un subgrupo aparte).

Distribución de *Heliconia* en Costa Rica

En Fig. 1 se presentan resultados de conteos para 102 localidades que he recorrido con suficiente tiempo y cuidado, lo que me induce a creer que mis cuentas son bastante completas. Prefiero basar este análisis en número de especies por localidad y no diversidad de especies, debido a que ésta requiere datos cuantitativos sobre abundancia de especies y mis observaciones en Finca La Selva (Stiles 1975) y otros lugares en Costa Rica me han convencido que tales datos no tienen gran validez, a menos que se trate de censos realizados sobre un período de varios años. La abundancia local de especies de *Heliconia* puede cambiar drásticamente dentro de pocos años debido a disturbios naturales o provocados por el hombre, o a cambios sucesionales de la vegetación.

Existen dos centros de diversidad del género *Heliconia* en Costa Rica, dentro de los cuales se puede encontrar de 8 a 11 especies por localidad (Fig. 1). El centro más extenso está a alturas de 100–700 m en la vertiente del Atlántico, en las faldas de las principales cordilleras del país (notablemente la Cordillera Central y la Cordillera de Talamanca). El número de especies disminuye a alturas mayores, y también hacia la costa. El segundo centro de diversidad de *Heliconia* está en los alrededores del Golfo Dulce y la Península de Osa en la zona Pacífico Sur. En esta zona, el número de especies por localidad disminuye hacia las montañas y en los valles interiores del General y del Coto Brus. Las dos regiones con pocas o ninguna especie de *Heliconia* son: el Pacífico seco (incluyendo el Valle Central), y las tierras altas del país.

The Atlantic slopes of 100–700 m and the Pacific neighborhoods of Golfo Dulce and the Osa Peninsula are the two centers of Heliconia diversity in Costa Rica.

La disminución del número de especies de *Heliconia* por localidad en altura está confirmado en más detalle en la Fig. 2. En las zonas húmedas del país, la relación entre números de especies y altura s.n.m. es negativa (especialmente arriba de los 500–700 m), linear y altamente significativa. Sólo una especie, *H. lankesteri*, alcanza los 2000 m y el género desaparece a alturas mayores de 2000–2200 m. En las zonas secas el número de especies presentes es baja, con sólo una especie, *H. latispatha*, presente en las tierras bajas de Guanacaste. En estas zonas casi no cambia el número de especies con la altura; y arriba de los 1500 m no hay diferencia entre zonas secas y húmedas en términos de diversidad de especies de *Heliconia*.

Los datos de la Fig. 2 dan indicio que el número de especies de *Heliconia* en una localidad podría reflejar una combinación de factores, tanto de temperatura como de humedad, sin excluir otros parámetros como suelos, etc. Además de estos factores, la zona geográfica del país puede tener cierta influencia en algunos casos. Las principales variables ecológicas son tem-

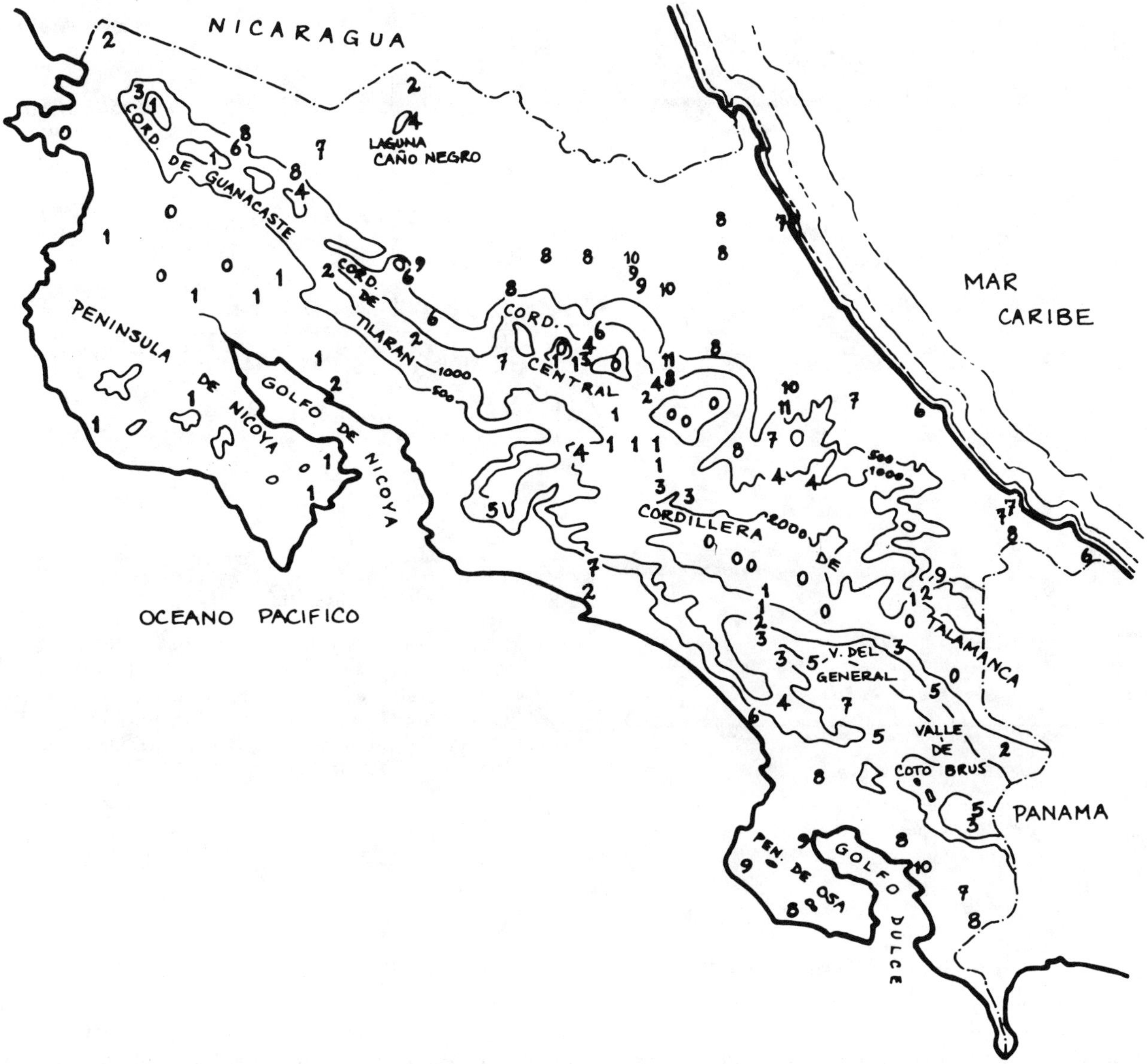

Fig. 1. Mapa de Costa Rica con 104 localidades ploteadas con sus respectivos numeros de especies de *Heliconia* encontrados en cada localidad. Elevaciones (lineas de contorno) en metros s.n.m.

peratura y precipitación. Calculé el número de especies de *Heliconia* en las localidades de cada zona de vida (Holdridge 1967, Tosi 1969). Los resultados (Fig. 3) son muy claros: las zonas de Bosque Tropical Muy Húmedo y Bosque Premontano Muy Húmedo son las más ricas en especies de *Heliconia*. El número de especies disminuye hacia las zonas más secas o—cosa algo inesperada—zonas más húmedas. Evidentemente las zonas de vida Pluviales son menos favorables para *Heliconia* que las zonas Muy Húmedas (la zona de Bosque Tropical Pluvial no está representada en el país). También hay una disminución fuerte en especies de *Heliconia* arriba de la línea crítica entre la faja Premontano y la de Montano Bajo; ninguna especie de *Heliconia* alcanza la faja de Bosque Montano.

The Premontane Wet Zone is the richest in species followed by the Tropical Wet Zone.

También en la Fig. 3 se da el número de especies de *Heliconia* que habita cada zona de vida. La zona más rica en especies es la del Bosque Premontano Muy Húmedo, seguido de cerca por la zona de Bosque Tropical Muy Húmedo. Igual que la primera, las zonas de Bosque Premontano Pluvial y Bosque Montano Bajo Pluvial cuentan con más especies de lo que

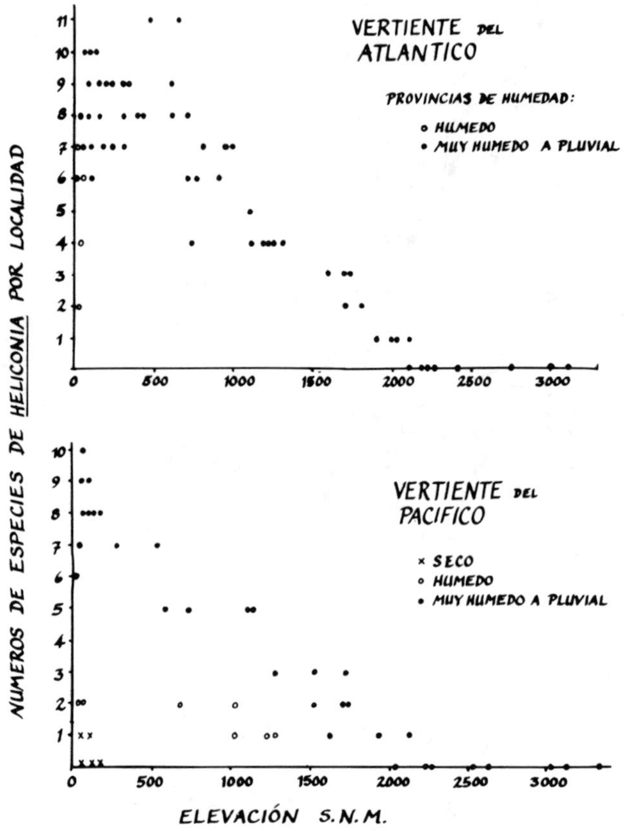

Fig. 2. Numeros de especies de *Heliconia* encontrados en 104 localidades de Costa Rica, según los vertientes, elevaciones, y provincias de humedad de estas.

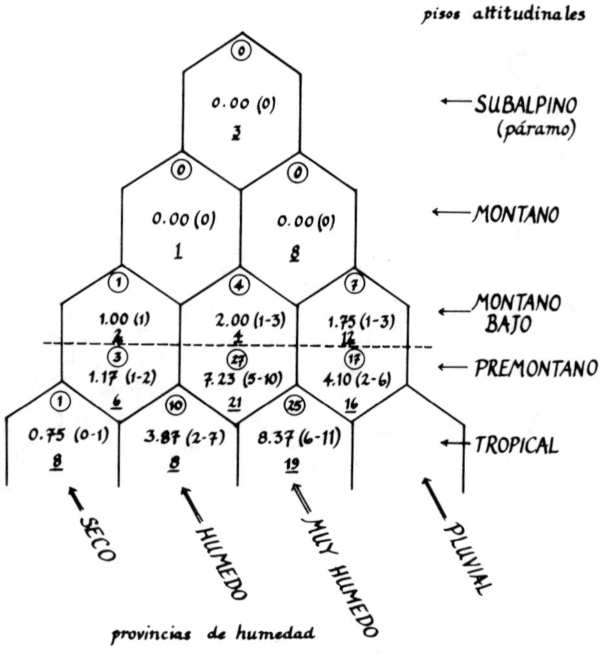

Fig. 3. Comparación de las floras de *Heliconia* de las diferentes zonas de vida (según Holdridge 1967) de Costa Rica. Los datos dados son: promedios y ámbitos de los numeros de especies por localidad dentro de la zona; numero total de especies upuntoda para la zona (en el circulo de arriba); y el numero de localidades observado para la zona (subrayado, abajo).

se esperaría a base de los números promedio de especies por localidad: es decir, hay mayores diferencias entre localidades en estas zonas. Esto se debe sin duda a la naturaleza discontinua de las montañas mismas; hay varias especies que sólo se encuentran en una o dos de las cuatro cordilleras principales del país. Todas las especies endémicas a zonas muy restringidas (*H. becknerii*, *H. monteverdensis*, *H. vulcanicola*, *H. ramonensis*, etc.) son especies de montaña. La diferencia entre las zonas Montano Bajo Muy Húmedo y Montano Bajo Pluvial, es que la primera tiene más especies por localidad y la segunda casi el doble del número de especies en total; probablemente se debe a lo poca extensa que es la primera en el país, y así el pequeño número de localidades representadas. La zona de vida más heterogénea, en términos de número de especies en distintas localidades, es la del Bosque Tropical Húmedo, y ésta refleja más que nada las posiciones geográficas de las localidades dentro del país. Por ejemplo, las dos localidades de esta zona en el sureste, Cahuita y Sixaola, tienen 6 y 7 especies, respectivamente. Los sitios en el Pacífico y Atlántico Norte del país tienen 2 a 4 especies cada uno, una reflección de contactos más extensos con las zonas más secas y relativamente pobres en especies de *Heliconia*. En el sureste, las montañas con zonas más ricas en *Heliconia*, se acercan mucho a la costa. Es interesante que Caño Negro, en la zona norte pero muy cercano a la transición del Bosque Tropical Muy Húmedo, tiene 4 especies mientras que los Chiles, en la misma zona de vida pero más lejos de esta transición, tiene sólo 2.

En general, una especie dada de *Heliconia* se encuentra representada en más de una zona de vida—la mayoría habitan 2 o 3 zonas de vida (Cuadro 1). De las 11 a 12 especies que están presentes en sólo 1 zona de vida, 8 son especies de altura (las fajas Premontano a Montano Bajo). Las dos zonas más parecidas en términos de *Heliconia* son precisamente las dos más ricas—el Tropical Muy Húmedo y el Premontano Muy Húmedo. La similitud de especies entre las dos (basada en el Cociente de Sorensen) es mucho más alta que entre cualquier otro par de zonas (Cuadro 2), y la gran mayoría de las especies de cada zona tam-

CUADRO 2

ESPECIES DE *HELICONIA* EN COMUN DE ZONAS DE VIDA

Numeros de especies en comun y cocientes de similitud

Zona de Vida	Zona de Zonas de vida en Costa Rica								
	Ts	Th	Tmh	Ph	Pmh	Pp	Mh	Mmh	Mp
Numero total de especies por zona de vida[1]	1	10	25	3	26	16	1	3	7
Ts		1	1	1	1	1	—	—	—
Th	.18		9	2	9	2	—	—	—
Tmh	.07	.51		1	21	8	—	—	—
Ph	.50	.31	.07		3	2	1	1	1
Pmh	.07	.50	.82	.21		9	1	1	1
Pp	.11	.15	.39	.21	.43		1	1	2
Mh	.00	.00	.00	.50	.07	.12		1	1
Mmh	.00	.00	.00	.33	.07	.11	.50		2
Mp	.00	.00	.00	.20	.06	.18	.25	.40	—
	Cocientes de Simulitud (C. de Sorensen)								

1. Abreviaciones: T = tropical; P = premontano; M = Montano Bajo; h = humedo; mh = muy humedo; p = pluvial.

bién ocurre en la otra (Cuadro 1). Ambas zonas comparten un grado de similitud de cerca de 0.5 con el Tropical Húmedo, y cerca de 0.4 con el Premontano Pluvial. Todas las otras zonas guardan menores grados de similitud entre ellas, y es notable que ninguna especie ocurre en las tres fajas altitudinales, así que la similitud entre las zonas de la faja Tropical y la del Montano Bajo es cero en todos los casos. Es más frecuente que una especie dada ocupe la misma provincia de humedad en dos fajas altitudinales, que encontrar una especie en dos o más provincias de humedad dentro de la misma faja.

Most Heliconia species inhabit 2 or 3 Life Zones.

La Ecología de *Heliconia:* Hábitat

La ecología de *Heliconia* ha sido descrita en términos generales (Stiles 1975, 1979). Dos aspectos particulares, los de hábitat y polinización de *Heliconia*, son los que mejor ilustran tendencias evolutivas dentro del género a este nivel.

Las plantas del género *Heliconia* representan un componente muy característico de la flora de los bosques húmedos a pluviales de Costa Rica. Sin embargo, sólo una minoría pequeña de las especies de *Heliconia* alcanzan su máxima abundancia dentro del bosque propiamente dicho. La gran mayoría del género está constituida por plantas del borde del bosque, o aperturas en el bosque, donde los rayos solares pueden alcanzar directamente el suelo. Existe una serie de tipos de éstas "aperturas de luz" dentro del bosque tropical, que difieren en sus tamaños y grados de permanencia, y aparentemente diferentes especies de *Heliconia* están especializadas para ocupar diferentes tipos de apertura. Se pueden distinguir tres tipos generales de hábitats asociados con el bosque tropical, basados en las condiciones lumínicas: el sotobosque de "fase madura" del bosque (Hartshorn 1980) y aperturas pequeñas y transitorias, como las causadas por la caída de un sólo árbol; aperturas intermedias, como a lo largo de quebradas y riachuelos dentro del bosque o cuando varios árboles se caen juntos; en estas aperturas el sol puede llegar directamente al suelo durante una parte del día solamente, y pueden durar varios años. El tipo de apertura más grande ocurre cuando la vegetación está removida casi en su totalidad de vez en cuando (por ejemplo en situaciones de inundaciones periódicas en los cauces de ríos grandes), dando lugar a áreas donde el sol pega directamente a la vegetación baja por todo o casi todo el día. Estas condiciones de pleno sol son muy parecidas a las que se obtienen en el crecimiento secundario joven que resulta de actividades humanas, pero una vez que la sucesión haya seguido por unos años más, resultando en la formación de un estrato arborescente, poco homogéneo, las condiciones se asemejan más a las de aperturas de luz del tipo intermedio. De esta forma incluyo "aperturas medianas" y "crecimiento secundario viejo" dentro del mismo renglón en la discusión que sigue.

The majority of species are plants of the forest margins or openings.

La gran mayoría de las especies de *Heliconia* de Costa Rica alcanzan su máxima abundancia en las aperturas del segundo tipo (incluyendo crecimiento secundario viejo), irrespectivo de sus afinidades taxonómicas o zonas de vida (Cuadro 3). Aproximadamente una cuarta parte de las especies son más que nada del sotobosque y una octava parte de aperturas grandes y crecimiento secundario joven. Sin embargo, es evidente que dentro de este marco general existen algunas tendencias y diferencias de interés, particularmente con respecto al número de especies de los dos hábitat más extremos en cuanto a condiciones lumínicas.

Most species are most abundant in openings of intermediate size.

CUADRO 3
ABUNDANCIA DE *HELICONIA* ESPECIES EN DIFERENTES HABITS

Factores	Habitats: Sotobosque y aperturas pequeñas	Aperturas medianas y crec. secundario viejo	Aperturas grandes y crec. secundario joven
Grupo Taxonómico			
pogonantha	0	6	2
trichocarpa	1	5	0
bihai	0	0	1
tortuosa	3	10	2
osaensis	5	5	0
Faja Altitudinal[1]			
Tropical	6	15	4
Premontano	6	22	4
Montano Bajo	2	7	0
Provincia de Humedad[1]			
Seco	1	0	0
Húmedo	3	2	3
Muy Húmedo	7	21	5
Pluvial	3	14	1

1. El numero de especies en estas categorias suma más de 40, debido a que unas especies son muy abundantes en más de una zona de vida.

Con respecto a afinidades taxonómicas, el grupo de *osaénsis* tiene más especies adaptadas al sotobosque que cualquier otro. Las especies cannoides y semicannoides generalmente son plantas de estatura y área foliar relativamente bajas, de tal forma que en ningún caso lograrían dominar áreas de vegetación de crecimiento secundario joven. En todos los grupos, las especies con más afinidades hacia el sotobosque son también las especies más pequeñas: *H. aurantiaca*, *H. reticulata*, *H. gracilis*, *H. nutans*, y *H. umbrophila* son buenos ejemplos. En cambio, ninguna especie del grupo de *pogonantha*, ya que todas (con la excepción de *H. marginata*) tienen plantas muy grandes, se ha adaptado al sotobosque. Las especies que se han adaptado a sobrevivir y competir en las aperturas más grandes y en crecimiento secundario joven, frecuentemente son plantas (brotes) grandes. *Heliconia mariae* es tal vez la especie con plantas más masivas, que a veces alcanzan 10 m de altura; *H. imbricata* es la especie más grande de su grupo también. Las otras especies más características de hábitat con pleno sol (*H. latispatha*, *H. wagneriana*, *H. marginata* alcanzan la dominancia vegetativa que les permiten sobrevivir en tales hábitat tan dinámicos, por medio del crecimiento rápido del clon, y no del gran tamaño de las plantas (en realidad brotes) individuales. Dentro de una sola especie, o comparando diferentes especies, el tamaño de los clones (en términos de números de plantas o brotes) aumenta con el grado de apertura del hábitat (Stiles 1975).

Within a single species, the size of the clone increases with the size of the forest opening.

Las proporciones de las especies que se encuentran más abundantes en los diferentes hábitat no varían mucho con la elevación excepto que en las zonas altas (la faja del Montano Bajo) no hay plantas adaptadas a crecimiento secundario joven (Cuadro 3). Esto tal vez se debe a que no hay ríos grandes a estas alturas; mas bien los sitios en que se elimina la vegetación naturalmente son áreas de derrumbes, en las cuales las plantas de *Heliconia* probablemente tienen cierta dificultad en establecerse debido a las condiciones de sequía frecuente (sin un río muy cerca) y erosión por aguaceros frecuentes. Con respecto a provincias de humedad, lo más llamativo es el aumento progresivo en la proporción de especies que alcanzan su máxima abundancia en el sotobosque, conforme uno pasa de zonas más húmedas hacia zonas más secas. Sin embargo, sólo una especie, *H. latispatha*, se encuentra en la zona del Bosque Tropical Seco. Siendo ésta la especie que más aprovecha los hábitat abiertos de pleno sol en zonas más húmedas, se puede sospechar que esta especie está preadaptada para ocupar zonas de menor humedad.

Only one species, H. latispatha, is found in Tropical Dry Forest.

Ecología de *Heliconia:* Polinización

Todas las especies de *Heliconia* de Costa Rica son polinizadas por los colibríes (familia Trochilidae), pero diferentes especies muestran especializaciones para polinización por diferentes tipos de colibríes. La familia de los colibríes está dividida en dos subfamilias: los Phaethorninae, que generalmente tienen picos largos y corvos, y colores apagados; y los Trochilinae, con colores y tipos de pico muy variados. Los Phaethorninae son los "ermitaños," así llamados porque son aves de bosque en su mayoría, y que forrajean solitariamente, a través de largas rutas en el sotobosque o a lo largo de las quebradas o los bordes del bosque. Los Trochilinae o "no ermitaños" tienen en su mayoría picos más cortos y rectos que los de los ermitaños. Sin embargo, como esta subfamilia incluye más del 90% de las aproximadamente 320 especies de colibríes, no es sorprendente que exista una gran variedad de tipos de picos entre los Trochilinae, desde cortos y rectos hasta largos y curvos, mostrando así convergencia hacia el pico típico de los ermitaños. Igual que en su morfología, las especies de no ermitaños son muy variadas en su comportamiento de forrajeo. Machos (y a veces hembras de muchas especies) defienden territorios en áreas ricas en flores; en el otro extremo hay especies que visitan flores esparcidas de la misma manera que lo hacen los ermitaños. Muchas de estas últimas también tienen picos largos y a veces curvos, de tal forma que son realmente convergentes con los ermitaños tanto en su morfología como en su ecología. Los ermitaños son típicos del sotobosque y de los bordes del bosque de áreas húmedas, de alturas bajas a medianas (en Costa Rica, por ejemplo, ningún ermitaño vive en alturas mayores de 1800–2000 m, ni tampoco en las zonas más secas del país). Las especies de no ermitaños más convergentes con los ermitaños, ocupan microhábitat o zonas de vida donde los ermitaños son escasos o ausentes—por ejemplo el dosel del bosque, o zonas de vida más altas.

All Costa Rican Heliconia species are pollinated by hummingbirds. The "hermit" species, Phaethorninae, generally have long curved beaks and drab colors, while the remainder, Trochilinae, have various kinds of beak and are often bright colored.

Con respecto a la polinización de *Heliconia*, las adaptaciones para polinización por los ermitaños vs. los no ermitaños son bastante distintas en la mayoría de los casos. Las flores de *Heliconia* están entre las más ricas en la producción de néctar (Stiles 1975 y datos no publicados; vea también Cruden et. al. 1981, Feinsiger 1976). Entonces son: la morfología de la flor y los patrones fenológicos, los que determinan para cuál tipo de colibrí la flor está adaptada en su polinización. Adaptaciones que favorecen las visitas de los no ermitaños territoriales son: flores con corolas cortas y rectas—que facilitan el acceso al néctar de estos colibríes de pico corto y recto—muchas flores producidas diariamente en cada inflorescencia, muchas inflorescencias disponibles simultáneamente en cada clon (que a su vez implica clones grandes de muchas plantas o brotes). Lo opuesto puede adaptar a a *Heliconia* para polinización por ermitaños: flores largas y curvas, clones pequeños y esparcidos o poca sincronización entre plantas de un clon en la producción de inflorescencias, pocas flores por inflorescencia (Stiles 1975, 1979).

Adaptations for hermit hummingbirds are large, curved flowers, small clones, and little synchronization of flowering, while for the non-hermits adaptations are short, straight corollas, many flowers produced daily on each inflorescence, and many inflorescences flowering at the same time.

Como en el caso de los hábitat en la polinización también hay un patrón predominante, exhibido por la gran mayoría de las especies de *Heliconia* de Costa Rica: polinización por ermitaños del género *Phaethornis* (Cuadros 1, 4). Las dos especies de *Phaethornis* implicadas en estos sistemas son: *P. superciliosus* de alturas de 0 a 500–700 m y *P. guy* de alturas entre 500 y 1800–2000 m. Existen dos variantes sobre este patrón predominante: un grupo de especies (la mayoría del grupo de *pogonantha* adaptadas para polinización tanto por *Phaethornis* como por *Eutoxeres aquila*, una especie de ermitaño con pico más curvo y de tamaño mayor; y tres especies polinizadas tanto por *Phaethornis* como por un no ermitaño con morfología y comportamiento muy parecidos, *Campylopterus hemileucurus*. Las especies del segundo grupo son las que alcanzan mayor altura (1800–2200 m), y *Campylopterus* efectivamente reemplaza a *Phaethornis guy* como polinizador en estas alturas. Las especies adaptadas para la polinización por *Eutoxeres* tienen flores más fuertemente curvas que las otras del género, aunque algunas son perfectamente explotables por *Phaethornis* y pueden representar los recursos florales más importantes para ciertas poblaciones de *Phaethornis* (Stiles y Wolf 1979).

Especíes polinizadas por no ermitaños típicos (es decir, no convergentes con los ermitaños, ni en morfología, ni en su comportamiento) ocurren en sólo dos grupos: *H. mariae* en el grupo de *pogonantha*, y varias especies en el grupo de *tortuosa* (Cuadro 4). Excepto para las especies de altura, *H. monteverdensis* y *H. vulcanicola*, todas las especies adaptadas para polinización por no ermitaños típicos alcanzan su máxima abundancia en hábitat abiertos, donde la luz solar es directa y la competencia para dominancia vegetativa es intensa (Cuadro 1). En estos hábitat, clones grandes son más frecuentes, y la floración simultánea de muchas inflorescencias facilita la atracción de no ermitaños territoriales. La ventaja principal de tener un polinizador territorial puede ser que una mayor proporción de las flores disponibles serían visitadas y polinizadas (Stiles 1978). Así las adaptaciones vegeta-

CUADRO 4
PATRONES DE POLINIZACION EN ESPECIES DE *HELICONIA* ESPECIES COSTARICENSES

Factores	Polonizador (Tipo de Colibrí) *Eutoxeres Phaethornis*	*Phaethornis* y afines	No Ermitaños Tipicos	No Ermitaños Convergentes con Ermitaños
A. Grupo de Especies				
pogonantha	5	1	1	0
bihai	0	1	0	0
tortuosa	0	12	5	2*
trichocarpa	0	7	0	1*
osaensis	0	8	0	0
TOTALES	5	29	6	3*
B. Zona de Vida[1]				
Tropical seca	0	0	1	0
Tropical Húmedo	0	5	3	0
Tropical Muy Húmedo	4	17	4	0
Premontano Húmedo	0	1	1	1
Premonano Muy Húmedo	5	18	4	1
Premontano Pluvial	3	11	1	2
Montano Bajo Húmedo	0	1	0	1
Montano Bajo Muy Húmedo	0	4	0	4
Montano Bajo Pluvial	0	5	2	3
C. Habitat de Maxima Abundancia				
Sotobosque-Aperturas Pequenas	0	9	0	1*
Aperturas Medianas, Crec. Sec. Viejo	5	18	3	2*
Aperturas Grandes, Crec. Sec. Joven	0	2	3	0

*Estas tres especies están polinizadas tanto por ermitaños como por no ermitaños que en su morfología y comportamiento son análogos al género *Phaethornis* (vea texto); así se incluyen en ambas columnas.

1. Una especie de *Heliconia* puede ocurrir en varias zonas de vida (vea Cuadro no. 1)

tivas relacionadas con ocupación exitosa de estos hábitat pueden facilitar la adaptación del sistema de polinización para no ermitaños territoriales—una alternativa no posible para la mayoría de las especies de hábitats más cerrados. Sólo tres especies de aperturas medianas muestran una especialización para polinización por no ermitaños típicos: las dos especies de altura, que ocupan una zona en que los ermitaños son poco abundantes y *H. sarapiquensis*, cuyo hábitat (aperturas de bosque pantanoso) puede ser el resultado de selección por aislamiento genético de la especie abundante *H. latispatha*, de áreas abiertas (Stiles 1975).

Only two groups of Heliconia are specialized for non hermit hummingbirds: H. mariae (group Pogonantha), and various species of group Tortuosa.

En las zonas de vida de máxima abundancia y diversidad de *Heliconia* (Tropical Muy Húmedo, Premontano Muy Húmedo, Premontano Pluvial), la polinizacioñ por ermitaños, especialmente *Phaethornis*, es el patrón predominante; una minoría de las especies están polinizadas por no ermitaños típicos. En las zonas más secas, la polinización por no ermitaños es progresivamente más importante; la única especie del Bosque Seco Tropical, *H. latispatha*, es polinizada por no ermitaños típicos (especialmente los del género *Amazilia*). En la faja altitudinal del Montano Bajo, cerca del límite superior de altura de los ermitaños, la polinización de no ermitaños también se vuelve más importante. En este caso se puede distinguir entre especies polinizadas por no ermitaños típicos (especialmente los del género *Lampornis*) y las polinizadas indistintamente por ermitaños (*Phaethornis guy* y no ermitaños parecidos en su morfología y comportamiento de forrajeo (principalmente *Campylopterus hemileucurus*)).

Patrones Evolutivos en el Género *Heliconia* en Costa Rica

A pesar de las grandes variaciones en los tamaños, formas y colores de las inflorescencias (Daniels y Stiles 1979), la gran mayoría de las especies de *Heliconia* de Costa Rica se conforman en un sólo patrón en términos de su distribución, hábitat, y polinización. Este patrón consiste en tener la abundancia centrada en aperturas medianas del bosque o en crecimiento secundario viejo, en las zonas de Bosque Tropical o Premontano Muy Húmedo y en ser polinizada por colibríes del género *Phaethornis*. Una mayoría de las especies de cada grupo bien representado en Costa Rica se conforma a este patrón, lo que me hace suponer que representa el patrón original o "primitivo", por lo menos en lo que se refiere a Costa Rica. Desde este punto de vista, la ocupación de otros hábitat, zonas de vida o polinizadores representaría una divergencia evolutiva de este patrón original.

The primitive Heliconias in Costa Rica may be those of medium sized openings of Tropical or Premontane Wet Forest and pollinated by hermit hummingbirds, while other habitats, life zones and pollinators may be an evolutionary divergence.

Diferentes grupos de especies muestran diferentes grados de divergencia en este sentido. Los grupos de *osaensis* y *trichocarpa* son muy conservadores en términos de polinizadores y muestran un grado limitado de radiación a otros hábitats y zonas de vida. Es difícil profundizar mucho en cuanto al grupo de *bihai*, porque sólo hay una especie en Costa Rica, *H. wagneriana*. Esta especie conforma al patrón básico en todo excepto que alcanza su máxima abundancia en hábitat soleados. Datos de otras partes (Snow 1974 en Trinidad) indican que muchas especies del grupo ocupan hábitat más sombreados.

El grupo de *pogonantha* muestra un mayor grado de divergencia del patrón básico, especialmente en la polinización. El patrón básico de este grupo aparentemente es el de polinización por *Eutoxeres* spp. más que de *Phaethornis* spp. Éste representa el único caso en que un grupo como tal se ha alejado del patrón básico del género, aún parcialmente. De este grupo, una especie se sale todavía más del patrón general: *H. mariae*, que es polinizada por no ermitaños típicos y que conjuntamente con este cambio ha ocupado hábitats más soleados que la mayoría de las especies del grupo. Si es correcto colocar *H. marginata* en el grupo de *pogonantha*, esta especie habría sufrido varias modificaciones bastante profundas con respecto a las otras especies del grupo, probablemente en relación con un cambio de hábitat hacia áreas de sustrato inestable y frecuentemente sumergido, a las orillas de ríos y pantanos, frecuentemente en pleno sol. Un aspecto de ésto sería la reducción en tamaño de la planta, tanto la parte vegetativa como la inflorescencia; otro sería el cambio hacia la polinización por *Phaethornis* y, tal vez más, por *Glaucis* spp. (que son los ermitaños más abundantes en estos ambientes) a expensas de la adaptación hacia *Eutoxeres*.

Group Pogonantha shows the greatest divergence from the basic system.

Dentro del marco general de *Heliconia* de Costa Rica, el grupo que muestra en mayor grado una verdadera radiación adaptada es el grupo de *tortuosa* (ésto se refleja en su división en subgrupos, varios de los cuales están ligados a cierta especialización, por ejemplo, el subgrupo de *lankesteri* y la ocupación de las alturas). Este es el grupo que más ha logrado colonizar las zonas de vida de las alturas y las zonas secas del país; es también el grupo dentro del cual existe la mayor variedad en cuanto a hábitat. Todos los sistemas de polinización presentes en el género en Costa Rica están presentes en este grupo excepto adaptación hacia *Eutoxeres* (ésto, debido a la curvatura extrema del pico del ave y de las flores correspondientes, prácticamente requiere que la inflorescencia sea colgante).

También podríamos enfocar el aspecto evolutivo de *Heliconia* en Costa Rica desde el punto de vista de las especies mismas. Hay varias especies de amplia distribución que muestran una variación geográfica marcada (a veces formalizada por el reconocimiento de subespecies y variedades), que pueden ser consideradas como especies relativamente antiguas. Buenos ejemplos serían: *H. pogonantha*, *H. irrasa*, *H. tortuosa*, y posiblemente, *H. osaensis*, *H. trichocarpa*, *H. lankesteri*, y *H. mathiasii*. Similares en este sentido son algunos pares de especies, ahora aisladas en diferentes vertientes del país, que obviamente están muy relacionadas, sin embargo, muestran un grado de diferenciación entre sí, que pone en tela de duda si son o no son aisladas reproductivamente. Ejemplos de tales pares son: *H. longa* y *H. curtispatha*, *H. mathiasii* y *H. golfodulcensis*. Estos pares tal vez serían mejor considerados como las semiespecies componentes de superespecies. Cabe señalar aquí también ciertas especies que son de amplia distribución y muy variables, pero que por lo menos una gran parte de esta variación no está organizada geográficamente, como *H. latispatha*.

Al otro extremo hay una serie de especies de distribuciones geográficas muy restringidas. Estas especies parecen ser de derivación relativamente reciente, o de una especie en particular de amplia distribución, o bien de la hibridización entre dos especies. En cualquiera de estos casos, la especiación implicaría una reorganización rápida del genomio para alcanzar el aislamiento reproductivo, como ha sido discutido ampliamente por Raven (1980). Las distribuciones restringidas de estas formas siguen lógicamente la manera de la especiación. Posibles ejemplos entre *Heliconia* son: *H. ramonensis, H. beckneri, H. calatheaphyllus, H. monteverdensis, H. vulcanicola*, y tal vez *H. secunda* y *H. viridiflora*. Hay que destacar que una distribución de este tipo puede ser difícil de distinguir de una distribución relictual, especialmente cuando hay dos formas locales parecidas pero aisladas geográficamente (por ejemplo, los casos de *H. monteverdensis* y *H. vulcanicola*, o *H. secunda* y *H. viridiflora*). En tales casos la existencia de un extenso hábitat apropiado, pero desocupado entre las dos formas, favorece la hipótesis de orígenes independientes de ambas tal vez de la misma especie "padre", o por eventos parecidos de hibridización (Stiles 1980). Sin embargo, es probable que cuestiones de esta naturaleza no se resuelvan satisfactoriamente sin la aplicación de nuevas técnicas citología, hibridización de ADN, etc.), que podrían generar los datos necesarios como para permitirnos salir del marco subjetivo en que estamos en esta fase todavía muy temprana, en el entendimiento de la evolución de este grupo de plantas.

Series of species of recent derivation are difficult to distinguish from relictual groups.

Perspectivas para el Estudio de *Heliconia* en Panamá

Me siento obligado a señalar algunas perspectivas que ofrece el estudio de *Heliconia* en Panamá. En primer lugar, hay que enfatizar que la taxonomia del género en Panamá necesita urgentemente de una revisión y actualización. El tratamiento del género en la "Flora de Panamá" es muy obsoleto, debido en parte a que éste se basó exclusivamente en especímenes de herbario.Tal como encontramos en Costa Rica, esos especímenes no forman una base adecuada para la clasificación de plantas tan grandes y carnosas: hay que basarse en las plantas vivas, utilizando material seco como testigo y suplemento (Daniels y Stiles 1979). Estudiadas de esta forma, las *Heliconias* de Panamá indudablemente van a rendir varias especies nuevas para la ciencia; pero más allá, podrán contribuir a un entendimiento mucho más amplio de la ecología y evolución del género en total. Varios problemas taxonómicos aún existentes entre las *Heliconia* de Costa Rica podrían ser iluminados por estudios en Panamá, y varias de las hipótesis taxonómicas y evolutivas expuestas aquí podŕian ser sometidas a la prueba, y modificadas o desechadas si el caso lo amerita. Pero a la par del estudio taxonómico tan necesario, quiero abogar por el estudio ecológico de las especies de *Heliconia* de Panamá. Creo que, por somero que sea, algún conocimiento sobre la ecología del género, incluyendo los polinizadores y dispersantes de frutos, los hábitat, etc, sería muy importante para entender a fondo la evolución de estas plantas tan bellas y fascinantes.

The taxonomy of Heliconia is in need of revision; the treatment in the Flora of Panama is quite obsolete and lacks input from study of living plants. A study of the group for Panama will illuminate problems of the genus in Costa Rica.

Agradecimientos

Este trabajo fue posible gracias a un beca proporcionada por el CONICIT y la Vicerrectoria de Investigación de la Universidad de Costa Rica (No. 07–03–028). Quiero agradecer a Rafael Campos por la ayuda en la redacción del español y a Virginia Zeledón por la ayuda en la mecanografía.

Cruden, R.W., S.M. *Hermann-Parker* & S. *Peterson*. 1981. Patterns of nectar production and plant-pollinator coevolution. *En* T.S. Elias & B. Bentley, eds.: The biology of nectaries. Columbia Univ. Press, New York.

Daniels, G.S. & F.G. *Stiles*. 1979. The *Heliconia* of Costa Rica: keys and descriptions. Brenesia 15 (supl): 1–150.

Feinsinger, P. 1976. Organization of a tropical guild of nectarivorous birds. Ecol. Monogr. 46: 275–291.

Hartshorn, G.S. 1980. Neotropical forest dynamics. Biotrop. 12 (supl): 23–30.

Holdridge, L.R. 1967. Life zone ecology. Trop. Sci. Center, San José.

Raven, P.H. 1980. Hybridization and the nature of species in higher plants. Can. Bot. Assoc. Bull 13 (supl): 3–10.

Smith, R.R. 1968. A taxonomic revision of the genus *Heliconia* in Middle America. Ph.D. diss. Univ. Florida, Gainesville.

Snow, B.K. 1974. Lek behavior and breeding of Guy's Hermit hummingbird *Phaethornis guy*. Ibis 116: 278–297.

Stiles, F.G. 1975. Ecology, flowering phenology, and hummingbird pollination of some Costa Rican *Heliconia* species. Ecol. 56: 285–301.

———. 1978. Ecological and evolutionary implications of bird pollination. Amer. Zool. 18: 715–727.

———. 1979. Notes on the natural history of *Heliconia* (Musaceae) in Costa Rica. Brenesia 15 (supl): 151–180.

———. 1980. Further data on the genus *Heliconia* (Musaceae) in northern Costa Rica. Brenesia 18: 127–134.

———. 1981. Geographical aspects of bird-flower coevolution, with particular reference to Central America. Ann. Missouri Bot. Gard 68: 323–351.

——— & L.L. *Wolf*. 1979. Ecology and evolution of lek mating behavior of the long-tailed hermit hummingbird, Ornithol. Monogr. 27: 1–78.

Tomlinson, P.B. 1961. Phylogeny of the Scitamineae: morphological and anatomical aspects. Evol. 16: 192–213.

Tosi, J.R. Jr. 1969. Mapa Ecológico de Costa Rica. Trop. Sci. Center, San José.

PART III

THE HUMAN ASPECT—
EL ASPECTO HUMANO

Plants That Affect Man

From the earliest times, mankind has used plants for curing and has also suffered from contacts with toxic plants. And although in modern medicine many curative substances are prepared from non-biological sources, plants are still an important source of new drugs, offering hope for some of the most persistent and serious illnesses. This group of papers shows that plants for human curing are assessed on a variety of criteria in many different places and at many levels of sophistication.

The paper by Arauz discusses how the Kuna healers undergo a long preparation to understand which plants should be used for healing and other purposes. Reading this paper against the papers by Dina and Joel Sherzer one realizes that medicinal and other plant usages are deeply intertwined with Kuna cultural attitudes and practices. In any society, much of day to day health practice is based on the folklore of mothers and neighbors. The Kuna recognize specialists in this knowledge who have an intimate knowledge of the forest and the many things that grow in it.

Relating to modern medical technology in contrast to this 'folk' technology, Duke describes a major program of the United States government to seek plants that may point to a cure for cancer. This scientific effort has screened many plants and come up with several species that warrant further study.

Plants toxic to man are usually toxic to animals too, and Escobar's study of the effect of poisonous plants on livestock highlights another way that the chemistry of plants can affect man.

The Lewises show how human responses to expectations from the plants in the environment are similar in widely separated countries. They note common medicinal uses of plants in Panama and Nepal.

Plantas que Afectan al Hombre

La humanidad, desde la antiguedad, ha utilizado las plantas para curar, pero también ha sufrido experiencias desagradables con las plantas tóxicas. A pesar de que en la medicina moderna muchas de las sustancias empleadas son preparadas a partir de recursos no biológicos, las plantas aún son importantes en la búsqueda de nuevas drogas que ofrezcan esperanzas en la curación de algunas enfermedades serias y persistentes que afectan al hombre. Este grupo de investigaciones muestra que las plantas utilizadas para curar al hombre están relacionadas a una variedad de criterios en diferentes lugares y a muchos niveles de sofisticación.

Araúz discute cómo los curanderos cunas son sometidos a una larga preparación para entender qué plantas deben usarse para sanar y cuáles para otros propósitos. En esta investigación al igual que las de Joel y Dina Sherzer, es notable el hecho de que tanto las plantas medicinales como las de otros usos están profundamente relacionadas con actitudes y prácticas culturales de los Cunas. En cualquier sociedad muchas de las actividades diarias en relación a la salud están basadas en prácticas folklóricas de madres y de vecinos. Los Cunas reconocidos especialistas en este aspecto, poseen un profundo conocimiento del bosque y de lo que crece en éste.

En relación a la tecnología médica moderna y en contraste con las técnicas folklóricas cunas, Duke describe el programa que está realizando el gobierno de Estados Unidos para buscar plantas que pueden apuntar hacia la cura del cáncer. En este esfuerzo científico se han analizado miles de plantas y de éstas se han escogido algunas especies con posibilidades para someterlas a estudios y análisis más detallados.

Las plantas tóxicas al hombre son usualmente tóxicas a los animales. El estudio de Escobar señala el efecto de las plantas venenosas para el ganado y constituye un ejemplo más de la forma en que la química de las plantas puede afectar.

Algunas veces, poblaciones ampliamente separadas en distancia utilizan plantas de las mismas familias o géneros con propósitos medicinales similares. Los Lewis presentan una revisión de los usos medicinales de plantas nativas de Panamá y del Nepal, estableciendo semejanzas en los usos locales de éstas.

Ethnobotánica Cuna

Reina Torres de Araúz[1]

Kuna society has profound magic and religious aspects, and this relates to the botanical pharmacopoea and to kinds of cultivation. Each plant has its own use, and this is derived from observation and a doctrine of signatures which derives properties from appearances. Kinds of crops and ownership, and many other features are unique to the Kuna system.

El concepto tradicional de etnobotánica ha enfatizado el interés en los estudios sobre las plantas cultivadas, producto pues del hacer cultural del hombre. De esa manera, al estudiar los conocimientos botánicos que existen en determinada población, el enfoque ha sido tradicionalmente extra cultural, ya que es como una sistemática y una ciencia propias de la llamada civilización occidental, como se ha procedido a establecer las distintas taxa, géneros, especies, variedades, etc.

Al estudiar los conocimientos botánicos de los indios Cuna, uno de los grupos indígenas más estudiados en Panamá se aplicó hasta ahora este mismo criterio, tal como podemos leerlo en los estudios iniciales hechos por la escuela antropológica sueca, con los nombres de Nordenskiold (1938), Henry Wassén (1949) y Nils Holmer (1951) a la cabeza y posteriormente con los estudios de Charles Bennett (1962) y James Duke (1975), de carácter ya especializado en identificación botánica.

La escuela antropológica sueca de 1927 en adelante remarcó el aspecto mágico religioso y socio cultural de sus investigaciones entre los Cunas, por lo que no puede hablarse propiamente de identificación de especies botánicas, ya que cuando ofrecen un nombre científico para determinada planta mencionada en textos curativos o mágicos, se utiliza generalmente nombres genéricos. Igualmente sucede en lo referente a nombramientos de productos agrícolas de uso común entre este grupo indígena.

The early Swedish school noted the religious-magic aspects of Kuna society, and Bennett gave an appreciable list of cultivated products used in cooking and medicine. The most complete ethnobotanical study was made by James Duke in 1966–68, providing the fullest list of cultivated and wild products. We worked together then, exchanging data and ideas.

Es en realidad Charles Bennett (1962) el primero que entra con un método etnoecológico, a analizar lo concerniente a los conocimientos y derivaciones de la botánica entre los Cunas de Bayano. En su ensayo: The Bayano Cuna Indians, Panama: An Ecological Study of Livelihood and Diet, este autor nos da una lista apreciable de productos cultivados, tanto dentro del campo de la culinaria como dentro del campo de las técnicas curativas.

Sin embargo, el estudio más completo realizado hasta el presente sobre este tema le correspondió a James Duke, que entre 1966 y 1968 tuvo oportunidad de realizar una intensa investigación botánica en la zona este de Panamá (Provincias de Panamá y Darién) que incluyó la recolección de muestras de especies cultivadas y silvestres utilizadas por los grupos humanos de esa región. Así, este científico le aplicó particular interés a la etnobotánica Cuna, ofreciéndonos la más extensa lista taxonómica de productos cultivados y silvestres, de uso culinario y mágico religioso, hasta ahora existente en los estudios especializados sobre esta cultura.

Tuvimos oportunidad de trabajar conjuntamente en el campo y, tanto en las aldeas de la selvática región

[1] The author died in February 1982: La autora murió en febrero 1982.

CUADRO 1
NOMBRES DE PLANTAS Y LOS USOS CUNA

Nombre Científico	Nombre Común (Español)	Usos
Musa paradisiaca	Plátano	Alimento
Musa sapientum	Guineo	Alimento
Phaseolus vulgaris	Habichuelas	Alimento
Phaseolus lunatus	Habas	Alimento
Cajanus cajan	Frijol de palo	Alimento
Guilielma gasipaes	Pixbae	Alimento
Cocos nucifera	Cocos	Alimento
Aechmea magdalenae	Pita	Alimento o Industrial
Anacardium occidentale	Marañón	Alimento
Annona reticulata	Anón	Alimento
Eryngium foetidum	Culantro	Condimento
Saccharum officinalis	Caña de azúcar	Para hacer bebidas embriagantes. Para dar sabor a las bebidas.
Theobroma cacao	Cacao	Para hacer chichas y dulces.
Dioscorea spp.	Ñame	Alimento
Manihot esculenta	Yuca	Alimento
Xanthosoma violaceum	Otoe	Alimento
Dioscorea trifida	Ñampí	Alimento
Cucurbita moschata	Zapallo	Alimento
Persea americana	Aquacate	Alimento
Mangifera indica	Mango	Alimento
Ananas comosus	Piña	Alimento
Carica papaya	Papaya	Alimento
Citrus sinensis	Naranjas	Alimento
Citrus paradisi	Toronja	Alimento
Citrus aurantium	Naranja agria	Alimento
Citrus medica	Toronjas o cidra	Alimento
Lycopersicon esculentum	Tomate	Alimento
Capsicum frutescens	Ají picante	Condimento
Bixa orellana	Achiote	Condimento
Citrus aurantifolia	Limón	Alimento
Annona muricata	Guanábana	Alimento
Artocarpus altilis	Fruta de pan	Alimento
Spondias purpurea	Ciruela	Alimento
Calocarpum mammosum	Mamey	Alimento
Chrysophyllum spp.	Caimito	Alimento
Citrullus vulgaris	Sandía	Alimento
Citrus aurantium	Naranja ácida	Alimento
Oryza sativa	Arroz	Alimento
Zea mays	Maíz	Alimento
Ipomoea batatas	Camote	Alimento
Psidium guajava	Guayaba	Alimento
Rheedia reticulata	Fruta de mono	Alimento
Annona cherimola	Chirimoya	Se toma el vino que produce y se coce con corozos
Annona purpurea	Toreta	Alimento
Annona squamosa	Anón	Alimento
Averrhoa bilimbi	Grosella	Alimento
Averrhoa carambola	Grosella	Alimento
Byrsonima crassifolia	Nance	Alimento
Cassia bicapsularis	Café de pobre	Infusiones
Cassia fistula	Cañafístula	Alimento
Hymenaea courbaril	Algarroba	Alimento
Chlorophora tinctoria	Mora	Alimento
Coffea arabica	Café	Estimulante
Tamarindus indica	Tamarindo	Alimento
Diospyros ebenaster	Zapote negro	Alimento
Eugenia malaccensis	Marañón curazao	Alimento
Genipa americana	Jagua	Alimento y para pintarse el cuerpo
Inga edulis	Guaba	Alimento
Mammea americana	Mamey de cartagena	Alimento
Melicoccus bijugatus	Mamón	Alimento
Passiflora quadrangularis	Granadilla	Alimento

de tierra firme, en las riberas del Chucunaque y del Bayano, como en el hábitat insular costero de San Blas, pudimos intercambiar datos sobre el uso de especies, nombres locales y la identificación botánica correspondiente (Vea Araúz 1970, 1975 y Araúz & González 1971).

Por supuesto, los datos en cuanto al uso y los nombres locales constituyeron el fuerte de nuestra contribución, y la identificación botánica fue el fuerte de la del Dr. Duke. Aún así, me complazco en admitir haber aprendido de sus observaciones propias muchos elementos etnográficos referentes a esta cultura.

Cuando el Dr. Duke, refiriéndose a la planta *nalup* (en Cuna) o *pixbae*, (nombre local panameño) o *Guilielma gasipaes* (nombre científico) dice "los Cuna deben ser excelentes taxónomos de las palmas," destacando el uso alimenticio y curativo de la misma; y cuando refiriéndose al *pachar* (en Cuna) o *pimienta*, (nombre local panameño), *Piper* spp. (nombre científico) dice: "para cada una de las docenas de especies de *Piper* en la selva los curanderos Cunas tienen un nombre y un uso distinto," está admitiendo la existencia de una taxonomía propia de esa cultura, sin entrar directamente en ella.

En realidad, se ha trabajado hasta ahora en la identificación, usos y derivaciones de las especies botánicas en esta cultura. Así se ha podido establecer el uso con fines alimenticios de un gran número de plantas cultivadas (Cuadro 1), como también la utilización con fines curativos y cosméticos de una serie de especies silvestres y cultivadas que conforman la rica farmacopea Cuna.

En lo concerniente a los productos alimenticios hay que distinguir los obtenidos mediante técnica agrícola tipo plantación y los que se obtienen mediante la técnica de "roza." Entre los primeros, es indudable que el cacao, el café y el género *Musa* constituyen los ejemplos clásicos. Este último, en razón de su sistema reproductivo—de tallos o "hijos" que prosperan una vez cortada la planta madre—representa el género de mayor reclamo alimenticio, tanto en la especie *Musa paradisiaca* (el plátano) o la especie *Musa sapientum* (guineos). El coco *(Cocos nucifera)* representa en San Blas la especie de mayor incremento tipo plantación.

En lo que se refiere a la agricultura de "roza," encontramos tanto la vegecultura (cultivo de especies vegetativas) como la semicultura (cultivo de especies por semillas). De las primeras, se destaca el cultivo de *Xanthosoma violaceum* (otoe) y de las Dioscoreas (ñame y ñampí), como, igualmente, de amplia difusión es la yuca o *Manihot*. El maíz *(Zea mays)* y el arroz *(Oryza sativa)* configuran las dos gramíneas de mayor cultivo dentro de la técnica de "roza," dándose comúnmente el pluricultivo o cultivos mixtos. En estos casos se combinan el cultivo de las especies del género *Musa* con la fruticultura. El aguacate *(Persea americana)*, los cítricos *(Citrus)* y los mangos *(Mangifera indica)* constituyen las frutas de mayor popularidad tanto por el gusto difundido como por la derivación económica, como es el caso del aguacate.

Cacao, coffee and coconuts represent a plantation type of cultivation, while taniers (Xanthosoma violaceum), yams (Dioscorea), cassava (Manihot) and rice (Oryza sativa) represent "roza" or subsistence crops.

En el Cuadro 1 presento una lista de las especies etnobotánicas más difundidas en esta cultura, tanto en el hábitat insular costero de San Blas como en la región selvática pluvial del Darién.

La botánica curativa merece un aparte, ya que de las prácticas de farmacopea Cuna, prácticamente todo el mundo vegetal es susceptible de uso, ya sea por vía de ingestión, baños, frotaciones y sahumerios.

In the Kuna pharmacopoea every plant has its use, be it by ingestion, for baths, for rubbing, or for incense. There is a magic and religious character in the use and harvest of plants.

Los *inatuledis* (curanderos botánicos) y los *neles* (shamanes) son poseedores de un rico conocimiento sobre la aplicación curativa de muchas plantas silvestres y cultivadas. Es en la vecina tierra firme, en el caso de San Blas o en la espesa selva que circunda la aldea en el caso de Bayano o Darién, donde se encuentran las múltiples especies que brinda la naturaleza o que el curandero cultiva o protege para su provecho. Hay una relación de carácter mágico religioso, tanto en la apreciación de las especies, como en su consecución y cosecha.

Se incluye una lista de especies identificadas botánicamente y el uso curativo correspondiente (Cuadro 2).

La Etnobotánica como Etnociencia

Hasta aquí nos hemos movido conceptualmente en el campo tradicional de la etnobotánica concebida como la ciencia de las plantas cultivadas por el hombre. Nos ubicaremos ahora dentro del ámbito cognoscitivo de la cultura Cuna, entrando propiamente dentro del campo de la Etnociencia. Entendemos, entonces, ahora Etnobotánica en un sentido reflejo, o sea la ciencia y sistemática que sobre el mundo bo-

CUADRO 2
PLANTAS CURATIVAS

Especie	Uso
Clystostoma isthmicum	Afecciones respiratorias
Piper spp.	Afecciones respiratorias
Spondias mombin	Afecciones respiratorias
Cyathula achyranthoides	Hemorragias
Drymaria cordata	Purgativo de la sangre
Erigeron bonariensis	Dolores de parto
Salvia occidentalis	Dolores de cabeza
Acrostichum aureum	Baños de infantes
Centrosema plumieri	Dolores de estómago
Cuphea sp.	Febrífugo
Dieffenbachia seguine	Resfriado
Dorstenia contrayerba	Resfriado
Erythrina berteroana	Enfermedades femeninas
Ficus citrifolia	Heridas
Jatropha curcas	Purgativos
Morinda sp.	Malaria
Neurolaena lobata	Dolores de estómago
Ocimum basilicum	Dolores de cabeza
Piper darienense	Mordeaduras de culebra y resfriados
Piper sp.	Expectorante
Pothomorphe peltata	Resfriados
Sabal allenii	Problemas femeninos
Selaginella sp.	Problemas femeninos
Siparuna guianensis	Mordidas de culebra y resfriados
Tussacia friedrichsthaliana	Mordeduras de culebra
Virola sebifera	Afecciones de la piel
Xiphidium caeruleum	Problemas femeninos
Zamia sp.	Purgante
Zanthoxylum microcarpum	Golpes
Ceiba pentandra	Problemas ginecológicos

tánico existe en esta cultura. Esta a su vez, forma parte de la Etnoecología, dándole igualmente al término el significado de la "perspectiva autóctona de las interrelaciones bióticas." Como ha dicho Catherine Fowler (1967), "Etnociencia y Etnoecología son ramas modernas de la ciencia, con objetivos y métodos en parte idénticos. La Etnociencia utiliza las técnicas de Lingüística descriptiva, Biología Sistemática y de Sicología para estudiar los modos congnoscitivos de pueblos de otras culturas". Dentro de este nuevo enfoque poco ha sido lo que se ha podido investigar hasta el momento. Sin embargo, es menester reconocer que Nordenskiold y Wassén autor y editor respectivamente, del clásico estudio "An Historical and Ethnological Survey of the Cuna Indians" al estudiar uno de los cantos curativos—específicamente el canto para curar el vómito de sangre y tuberculosis, con las raíces de un arbusto, presentan uno de los primeros intentos de detectar las claves taxonómicas utilizadas por los Cuna: "*batsal*" podría decirse que forma una familia de plantas. Tienen el mismo olor. Las variedades son diferenciadas por las flores y por la variación en la forma de la hoja. Las palmas no forman tales grupos. Entre las bijaos, *tagar*, muchas especies son diferenciables por el tamaño de las hojas y por las flores. Existe evidentemente, en la cultura Cuna un profundo conocimiento de las interrelaciones bióticas propias del medio que habitan, conocimiento basado en la observación práctica y experimentación a lo largo de su historia. Ese conocimiento, transmitido oralmente y en la experiencia diaria, de generación en generación le ha permitido un aprovechamiento equilibrado de los particulares nichos ecológicos donde se asientan, sin llegar a producir situaciones extremas de agotamiento de nutrientes, deforestación y eliminación de especies. La relación del Cuna agricultor con su medio circundante es íntima y respetuosa.

Based on a long history of observation and experimentation, the Kuna culture has a profound knowledge of the biotic relationships in the environment.

Incluso el nombramiento de los meses manifiesta una clara relación con los fenómenos ecológicos representados en el comportamiento cíclico de la flora y de la fauna. Fue un lingüista, Nils Holmer (1951), también de la escuela antropológica sueca, quien recogió de boca del jefe Cuna Olotebiliguiña, los nombres (Cuadro 3) que pretenden definir períodos equiparables a nuestros conceptos de meses (30 días calendarios, según el Calendario Gregoriano).

Pero resulta comprensible que si pretendemos proseguir dentro del enfoque de la etnoecología, tal como lo define la "nueva etnografía," así llamada, debemos persistir en la detección del conocimiento propio de la cultura Cuna en cuanto al calendario propio. Observadores tan antiguos como el anónimo caballero escocés que nos dejó el documento "Una carta que da una descripción del Istmo de Darién, donde la colonia escocesa está situada," reconocieron ya la existencia de este año lunar y así lo consignó este desconocido informante al afirmar: "ellos reconocen los años por la luna adjudicándole quince a cada año con lo cual hacen un ciclo entero como nuestros doce meses" (Anónimo, 1699).

Aunque indudablemente se ha producido una tentativa de ajuste con el calendario Gregoriano, es todavía posible detectar la referencia a las fases lunares que configuran el mes y que son las siguientes: *Ni sichi* (luna oscura); *Ni pipigwa* (luna pequeña o luna nueva); *Ni engusa* (luna creciente); *Ni talano* (luna

CUADRO 3
MESES CUNAS

1. *Yolanii* (mes de la estación seca, verano, aproximadamente enero)
2. *Arinni* (mes de la iguana, aproximadamente febrero)
3. *Tilanni* (mes del árbol *Tilla*, aproximadamente marzo)
4. *Olorii* (mes de los grillos aproximadamente abril)
5. *Yauknii* (mes de la tortuga, aproximadamente mayo)
6. *Masarii* (mes del bambú, aproximadamente junio)
7. *Punurii* (mes del árbol *Purur*, aproximadamente julio)
8. *Kikinii* o *Masarkorekenii* (mes del pájaro *Kiki* o del viento del sudeste o cuando el bambú se pone amarillo, aproximadamente agosto)
9. *Appinnii* (mes del árbol *Appin*, aproximadamente septembre)
10. *Kiplonii* (mes del pájaro *Kiplo*, aproximadamente octubre)
11. *Inanii* (mes de la medicina, aproximadamente noviembre)
12. *Partutnii* o *Asuuenii* o *Wainii* (mes del árbol *Partur* o del árbol de aguacate o mes del hombre blanco, aproximadamente diciembre)

Holmer, 1951; 128–129.

llena); *Ni orokuagusa* (luna menguante); *Ni yo bergwe* (aún no se terminó la luna); *Ni gwalagwen Ni nagwe* (casi pasó un mes de la luna). Esta información sobre fases lunares recogidas por personal de la Dirección Nacional del Patrimonio Histórico entre los Cunas de Bayano (Brizuela, 1971) pone en evidencia una detallada observación del ciclo lunar que tiene como finalidad aplicarlo a las labores agrícolas o de recolección de plantas consideradas como fármacos. También la observación de las estrellas *(níisguamar)* tiene relación con el proceso agrícola. A Venus llamada en Cuna *bugasu* o *bugsu*, se le atribuyen en el verano funciones de vigilancia de los cultivos y en el invierno, de protección en las jornadas de pesca.

The Kuna have names for the phases of the moon: Nii sichi (dark moon), Ni pipigwa (new moon), Ni engusa (waxing moon), etc. Each month has 24 days related to the agricultural cycle, and 15 days were added to the year which then matched the Gregorian cycle. The agricultural cycle begins in January and February with cleaning the lands, continuing to April with burning, and followed in May by planting of subsistence crops or renewal of bananas and plantains. This last coincides with the harvest of fruit crops like avocado and mango.

Cada mes lunar que tendría 24 días se aplica a un momento determinado del ciclo agrícola; por ejemplo, con la luna nueva se siembra el maíz lo cual suponen garantizará una mejor calidad de la cosecha. El ciclo agrícola se inicia en los meses de enero y febrero con la limpieza de los terrenos y prosigue hasta abril con las faenas de quema, adecuando el terreno para las labores de siembra que se inician en mayo con la siembra de las especies aplicadas al cultivo de "roza" o renuevo de plantación del plátano y guineo y que coinciden también con la cosecha de las especies de fruticultura como aguacate y mango.

Hasta el momento, las observaciones etnoecológicas realizadas no van más allá de la interrelación entre régimen de lluvias y ciclo de cultivos. Indudablemente, deberá existir una vasta área de conocimientos en lo relativo a la calidad de los suelos, ubicación de los sitios de labranza, y que todavía no han sido objeto de un estudio detallado. Como base de esta suposición, señalamos aquí las observaciones realizadas por nosotros anteriormente y que señalan la existencia de tres sitios de cultivo; la pequeña huerta en torno a la casa, en la cual se encuentran algunas pocas plantas vegetativas, algunas unidades del género *Musa*, ocasionales árboles frutales (papaya, naranja, totumo) y plantas de uso medicinal o cosméticos (ají, magueb, albahaca), un campo de cultivo en la ribera del río, o a corta distancia de la casa y donde se dan cultivos mixtos de vegecultura y semicultura; y un tercer tipo de campo de cultivo en zonas más alejadas de la habitación y que requieren la construcción de instalaciones temporales de vivienda, y donde se encuentran gran cantidad de árboles frutales, el tipo de agricultura casi de plantación que es el género *Musa*, el cacao, combinado casi siempre con algunas especies de vegecultura (otoe, ñame, ñampí, etc). Esto se pudo observar fehacientemente en la región de tierra firme habitada por los Cunas (Bayano y Darién). La situación en San Blas, con su particular hábitat insular costero, perfila una situación distinta por las limitaciones propias del medio. En efecto, pocos son los sembríos en torno a la casa ya que la delgada capa vegetal de las islas no permite mayores cultivos que la omnipresente palma de coco, ocasionales arbustos de algodón y raramente algún papayo *(Carica papaya)* que ha podido sembrarse trayendo tierra de la porción continental.

Three kinds of cultivation were distingushed: the small farm around the house with bananas, papaya, oranges, calabash, and medicinals; cultivated fields along the rivers not far from the house with mixed crops, and cultivation in areas further away requiring temporary dwellings and devoted to plantation crops such as bananas, cacao and accompanied by root crops such as taniers, yams, etc.

En relación a la diversidad de sitios de cultivo, el antropólogo Cuna Arnulfo Prestán ha señalado la contemporaneidad de hasta tres sitios de trabajo agrícola. Este sistema, que se basa en el aprovechamiento de las tierras según los años de barbecho (Cuadro 4).

Puede darse también la siembra en terrenos incultos por menos de cinco o siete años, si la necesidad lo señala (Prestán 1975).

Es justamente en San Blas donde ha podido realizarse un estudio sistemático en un campo semántico *sappi turba* (frutas). Este estudio, realizado por el lingüista Sherzer (Howe & Scherzer 1975) abre una amplia perspectiva al estudio de la sistemática taxonómica propia de la cultura Cuna. Es interesante destacar que la sitematización detectada por este investigador tiene un trasfondo económico innegable. En efecto, él distingue básicamente dos categorías de frutas o productos alimenticios:

1. *Ipet nikkat*, que significa "tiene dueño".

Son las especies botánicas de consumo humano que se siembran y que por lo tanto implican un esfuerzo laboral de connotación económica y social. Dentro de esta categoría hay una diversificación de seis tipos que definen especies que por su aplicación económica tienen claras prohibiciones para su usufructo. Así el frijol, el café, camote, calabazas, yucas, zapallo, pixbae, sandía, fruta de pan, coco, arroz, otoe, y ñame solamente pueden ser tocadas, y usufructuadas, por su dueño—o sea quien las sembró en sus tierras de cultivo—ya sea que se encuentren todavía en la planta o hayan caído al suelo.

Las otras variantes de esta categoría se refieren a productos de consumo diario, generalizado y de rápida putrefacción y permiten el ocasional usufructo por otro que no sea su dueño, siempre que se encuentren en el suelo, o muy maduros en la mata y requiere el aviso posterior al dueño.

Las variantes continúan hasta la variedad de frutas como el caimito, marañón, y ciruela, que, si bien tienen dueño, quien guste de ellas puede llevarse la cantidad que desee. Cabe destacar que estos productos son rápidamente perecederos.

2. *Ipet sulit*, que significa "no tienen dueño."

Esta categoría encierra una gran variedad de productos que no se siembran como ciertas clases de caimito, de guayaba silvestre, de mamey silvestre y la gustada *"igwa" (Dipterix panamensis)*. Esta segunda categoría no tiene subdivisiones encerrando bajo la denominación alterna de *"tios kat"* (que significan pertenecen a Dios) prácticamente todos los frutos silvestres de consumo humano, pero que tienen escasa derivación económica.

CUADRO 4
TIPOS DE CULTIVO CUNA

1. El cultivo en rastrojo *(Masargan saed)*
2. *Nagama saed* o cultivo en terreno de aproximadamente siete años de barbecho, se constata por el crecimiento do los árboles nuevos.
3. *Neg sergan saed* o cultivo en terrenos incultos por 13 o más años.

The Kuna distinguish crop property in two ways based in part on labor input: Ipet nikkat or having an owner—beans, coffee, calabash, etc., and Ipet sulit, having no owner—wild mamey, star apples, wild guava, and almost all wild fruit.

Otro campo semántico de la etnobotánica Cuna es el que se refiere al insondable mundo de la farmacopea botánica. Tiene el nombre de *ina* (medicinal). Ha merecido algunos esfuerzos investigativos que ha enfatizado más en su correlación con el mundo mágico religioso que con los aspectos científicos y económicos.

Pueden señalarse algunas categorías tales como: *uala* (maderas, cortezas); *ka* (hojas); *sarki* (bejuco). Como ya lo sugirió Nordenskiold (1938), las claves usadas en la clasificación pueden ser el olor, forma de las hojas, etc.

Another semantic area refers to the botanical pharmacopoea, having the name ina (medicine), and sometimes referring to categories such as uala (wood, bark), ka (leaves), or sarki (vines).

Alcibíades Smith, *inatuledi* de la Isla San Ignacio de Tupile, San Blas, hijo y nieto de *inatuledi* y cuya madre era partera *(coe gaedi)* es un reconocido curandero Cuna quien aprendió su oficio con dos *neles* y un *inatuledi*, durante un período de aprendizaje de más de treinta años. El cuenta ahora con 65 años de edad y se desempeña todavía como *inatuledi*. De él pudimos corroborar lo que habíamos observado en diversas temporadas de trabajo de campo en San Blas y Darién en relación a la existencia de una sistematización de los conocimientos botánicos. En efecto, este informante subrayó la metodología empleada por sus maestros que consistía en la correlación enfermedad-medicina *(ina)*. Se le enseñó a reconocer, observando las características del tallo, largo y grueso de las hojas, la existencia o no de espinas en la planta, todos los arbustos, árboles, plantas, trepadoras, etc. que servían para curar cada enfermedad. Esta enseñanza se realizaba en la selva.

Alcibiades Smith, a healer (inatuledi), is son and nephew of healers, and he learned his profession from two diviners and a healer over more than 30 years. He was taught in the forest to observe the morphological characteristics of the stem, size and thickness of the leaves, etc., of all the plants which could cure each illness.

Como he mencionado antes la diversificación de usos médicos de la flora, en bebidas, baños, ungüentos o frotaciones y sahumerios, exigen el discernimiento y aprendizaje de un gran número de plantas y sus respectivas variedades y las partes de las mismas que pueden ser usadas en uno u otro de los procedimientos de curación antes mencionados. Por ejemplo, las hojas de ciertas plantas pueden ser usadas y tomado el líquido, mientras que su corteza, macerada en agua expuesta al sol, se usa para baños.

El informante mencionado señala su propia experiencia de baños en los ojos con *sapi karta*, hojas que presentan la particularidad de manchas que semejan letras, lo que por concepto mágico imitativo se relaciona con el conocimiento y facultad de aprender. Es un amplio campo en el cual solamente un esfuerzo interdisciplinario podrá ayudar a conocerlo. Reiteramos aquí, que se impone un estudio más detallado y realizado conjuntamente por lingüistas, antropólogos y botánicos.

The informant described his experience with eye baths. These used sapi karta, leaves with markings resembling handwriting, which in magic concept indicate knowledge and ease of learning.

Pero una dimensión más se abre en relación al campo semántico *ina*. Esta es la del mundo mágico religioso dentro del cual se enmarca la farmacopea Cuna. No es posible olvidar que el concepto profundamente animista que en esta cultura se tiene del universo, se refleja en el mundo vegetal, particularmente en las especies aplicables al campo de acción que mayor relación presenta con lo espiritual y divino: *ina* (la medicina). De allí que, según hemos podido apreciar, en algunos breves sondeos realizados al respecto, una clave sistemática de *ina* es la mayor o menor relación de determinada especie vegetal con los espíritus que intervienen en el proceso curativo. Eso nos explica las actitudes de reverencia y recogimiento desplegadas por el *inatuledi* en una incursión de recolección de plantas curativas y que incluye siempre cantos mágicos que refieren a la relación entre determinada planta y el poblado mundo de los espíritus.

Todo parece indicar que el concepto Cuna del poder curativo de las plantas es profundamente místico. En efecto, se expone que son razones de carácter espiritual y mágico las que explican la capacidad curativa de las plantas. La mitología Cuna señala que fue *pat tummad* (Dios) quien, al principio de todo, dotó al mundo vegetal de fuerzas especiales para luchar contra las enfermedades. Así, el Dios creador designó al profeta y héroe cultural *dad-ibe*, con su séquito de siete hermanos y una hermana, para adjudicar a cada planta sus particularidades propiedades curativas. De allí que, por regla general, todo *inatuledi*, antes de cortar una planta curativa le recuerda en un canto especial que Dios le ha creado para que ayude a los Cuna o *tules* a curar las enfermedades. Este canto debe decirse necesariamente frente a la planta, aunque la premura exija hacerlo con rapidez, y luego en un canto más elaborado, diserta el *inatuledi*, ante la planta ya cortada qué enfermedad espera que pueda curar y cómo debe ser el proceso.

The Kuna concept of healing power is deeply mystical, and the spiritual and magic characteristics explain the plants' ability to cure. Therefore each healer before cutting a curative plant recalls in a special chant that God created it to aid Kunas to cure illness.

El mundo vegetal y la flora tienen tan estrecha relación con el mundo mágico religioso cuna que incluso, plantas como el cacao *(sía)* son consideradas *neles*. Tal es el caso también de plantas no identificadas aún como *kaubi*, *kateb*, *surub* o *mumut uala* que son usadas para fabricar los *suar mimi*, pequeñas figuras biomorfas (por lo general en forma humana) que se usan en las sesiones curativas. De estas plantas por lo general se usan también las hojas tiernas como medicina por ingestión y los tallos macerados para baños curativos. Es interesante la observación que nos hiciera Alcibíades Smith de que *sía* (el cacao) es un *nele* vegetal considerado femenino, el cual es el primero que envía el *nele* o *inatuledi* a luchar contra los *ponigana* o espíritu de la enfermedad, pues por su condición femenina tendrá mayor posibilidad de controlarlos o seducirlos.

Los conceptos imitativos como el mencionado en el uso de *sapi karta*, juegan un papel importante en la estructura de los conceptos de la medicina botánica Cuna. Igualmente, la dualidad sexual hombre-mujer *(Macheretome)* aparece en la conceptualización de los poderes intensos o atenuados, de un particular elemento botánico curativo. Estos conceptos parecen aplicarse también como elementos sistematizadores en la etnobotánica Cuna.

Creemos que no se ha hecho más que levantar el velo del rico acervo de la etnobotánica, en cuanto a etnociencia, entre los *tule* o Cuna.

Discusión

La necesidad de una investigación etnobotánica, en cuanto a Etnociencia, para la cultura Cuna, se impone cada vez más. Mientras el enfoque etnobotánico tradicional ha tenido una relación y derivación intercultural, el nuevo enfoque iniciado pretende tener una acción intracultural definida. Ello no implica de ninguna manera el reemplazo del uno por el otro. Mientras el uno nos permite la identificación botánica dentro de la taxonomía vigente en la ciencia occidental, como un vehículo de traducción y comprensión de un importante sector cultural, con las derivaciones que la Antropología Aplicada pone a disposición de toda empresa intercultural, el otro nos permitirá conocer la interrelación conceptual entre un determinado campo semántico y la economía y el mundo social y espiritual. Esto finalmente podría conducir a una más fiel comprensión de la cultura, puesto que es de su propia Etnociencia de donde emana el conocimiento que analizamos.

The ethnobotanical focus was traditionally on intercultural relations, but the new focus is intracultural. While the first allows botanical identification in terms of Western taxonomy as a means of transmitting knowledge, the other permits understanding of the interrelation between a given terminology and its economic, social, and spiritual spheres.

Anónimo. 1699. Una Carta que da una descripción del Istmo de Darién. Traducción por Reina Torres de Araúz 1970.

Araúz, R. Torres de, & R. *González.* 1971. Apuntes etnoecológicos sobre los Indios Cunas de Chucunaque y Bayano. Patrim. Histór. 1(1). Panamá.

——— 1975. Darién: Etnoecología de una Región Histórica. Panamá.

——— 1970. Human Ecology of Route 17 (Sasardí-Mortí) Región, Darién, Panamá. Panamá.

Bennett, C., Jr. 1962. The Bayano Cuna Indians, Panama: An Ecological Study of Livelihood and Diet. Ann. Assoc. Amer. Geog. 52:32–50.

Brizuela, A. 1971. Agricultura y Calendario Agrícola de los Cunas del Río Bayano, Panamá. *In* Act. II Simpos. Nac. Antrop., Arq. Etnohist. Panamá. Panamá.

Duke, J. 1975. Etnobotanical Observations on the Cuna Indians. Econ. Bot. 29: 278–293.

Fowler, C. 1967. Etnoecología. *In* D.L. Hardesty, Antrop. Ecol. Barcelona.

Holmer, Nils. 1951. Cuna Cherstomathy. Goteborg.

Howe, J. & J. *Sherzer,* 1975. Take and Tell: a practical classification from the San Blas Cuna. Amer. Ethnol. 2: 435–460.

Nordenskiold, E. 1938. An Historical and Ethnological Survey of the Cuna Indians. Goteborg.

Prestán, A. 1975. El uso de la chicha y la sociedad Cuna. I.I.I. Ediciones Especiales 72, México.

———. 1976. La enfermedad, la curación y la muerte en la sociedad Cuna. Rev. Nac. Cult. 2. Panamá.

Wassén, H. 1949. Contributions to Cuna Ethnography. Etnol. Stud. 16. Goteborg.

Neotropical Anticancer Plants

James A. **Duke**

El Instituto Nacional del Cáncer y el Departamento de Agricultura de Estados Unidos han investigado más de 10% (30,000) de las especies de plantas del mundo en la búsqueda de principios anticancerosos. Este programa ha durado aproximadamente 20 años. Actualmente la investigación es selectiva y la selección está apoyada en una serie de archivos de computadora, especialmente aquellos que producen floras tóxicas para los países bajo estudio. Aún puede haber 9000 especies con principios anticancerosos y de éstas, 45 pueden ser de mayor interés.

In twenty years, the cancer screening program of the National Cancer Institute (NCI) and the United States Department of Agriculture (USDA) has screened more than 10% (30,000) of the plant species of the world (conservatively 300,000 species). That 10% represents the easiest 10% to obtain. One plant species in ten will show some sort of anticancer activity. The often-quoted 10% activity in randomly collected species is actually too high, because about 1/2 to 2/3 of the time many of the same species will show up inactive. The active figure is closer to 5% at the extract level. In one summary report (Feb. 11, 1980), there were 4,870 active extracts out of 108,699 (covering nearly 6,500 genera). There is a distinction between anticancer activity—usually referring to *in vivo* activity—and cytotoxicity, where *in vitro* activity is usually high. Usually cytotoxicity is more harmful to the host than to the cancer. The World Health Organization (Spjut & Perdue 1976) notes that chances for anticancer activity (or cytotoxicity) are twice as good in medicinal species, thrice as good in anthelminthic plants, four times as good in fishpoison plants, and five times as good in arrow poison plants as in the average plant. Most arrow poisons, however, are either cardenolides (cardiotoxins) or curare-like producing respiratory paralysis, and none has shown useful in vivo activity. Out of the first 30,000 plants screened, only about 15, or one in 2,000, are still of major interest to NCI. Only one, the Madagascar Periwinkle, is a superstar, producing vincristine and vinblastine.

De 30,000 especies de plantas investigadas en relación a actividad anticancerosa, durante más de 20 años, sólo 15 aún son de interés para el Instituto Nacional del Cáncer.

During the first 16 years, 13,000 African samples were screened out of a tropical African flora of 30,000 species, while 29,000 samples were screened from Latin America out of a tropical American flora of 90,000 species (Prance, 1977) (Table 1). We look to the neotropics in our search for anticancer activity, hoping to engender a collaborative approach with neotropical countries interested in cooperating with our anticancer activities. The odds tell me that there are 9,000 species with anticancer activity or cytotoxicity (the one in ten), 45 species of major interest to NCI, and 3 superstars waiting for the world in the neotropical flora, if they do not fall victim to the axe of rural and industrial deforestation.

Until recently, we accepted anything from anywhere for our cancer procurement program. But after twenty years, many specimens coming in, either from our own collections or those of collaborators and contractors, were species that had already been screened. Duplication was approaching 60%. Today initial screening of a species costs about $150.00. We must refuse new collections of such things as (1) *Hagenia abyssinica* screened 65 times, (2) *Mangifera indica* screened 58 times, (3) *Trema orientalis* screened 51 times, (4) *Melia azedarach* screened 51 times, and (5) *Psidium cattleianum* screened 48 times.

Actualmente la investigación de nuevas especies debe ser selectiva.

Recently we combined and automated the list of

TABLE 1
CANCER EXTRACTS BY COUNTRY AND ACTIVITY (TO 1976)

Area	Plant Extracts to 1976	Confirmed Active	Percent
North America			
Canada	724	8	1.1
United States	30,021	874	2.9
SUBTOTAL	30,745	882	2.0
Mexico and Central America			
Belize	102	5	4.9
Canal Zone	195	2	1.0
Costa Rica	4,648	260	5.6
Mexico	3,800	175	4.6
Panama	2,218	64	2.9
SUBTOTAL	10,963	506	3.8
West Indies			
Puerto Rico	1,763	98	5.6
Other	191	3	1.6
SUBTOTAL	1,954	101	3.6
South America			
Brasil	5,701	103	1.8
Chile	743	23	3.1
Colombia	2,529	91	3.6
Peru	6,905	210	3.0
Uruguay	874	22	2.5
SUBTOTAL	16,752	449	2.8
Latin America	28,669	1,056	
Africa			
Ethiopia	1,024	85	8.3
Kenya	4,577	130	2.8
Nigeria	212	16	7.5
Rhodesia	358	4	1.1
S. Africa	3,301	61	1.9
Tanzania	3,141	102	3.2
SUBTOTAL	12,613	398	4.1

species which have been screened too many times (6+ file) and genera (GESOC file) which have been screened so often our sponsors at NCI deem these unproductive for further screening (Table 2). We term the combined list the SLOP (species low on priority). Calculations showed that the GESOC file eliminated nearly 20% of the species of the world (the easy 20%). A newcomer dropped off in any easily accessible forest in Panama or nearly anywhere, will be faced with a vegetation or flora, at least 50% of the biomass of which or 50% of the common genera or species of which have already been over-screened as far as NCI is concerned. The 6+ files and GESCO files will grow in proportion to the intensification of our efforts. If we aim for just such unscreened species, we will quickly reduce our "shopping list" for the cancer program to bare minima. A main objective in automation is to direct attention to the unscreened taxa and ignore the overscreened taxa.

Major activities now are aimed at converting the "shotgun" approach to the "shopping list" approach. This requires further automation. Our automated medicinal flora program is of such general interest that we expect to produce Medicinal Floras for several collaborating countries. In addition to Panama, I hope also to visit China, Egypt, Honduras, Korea, and Syria this year, where I have lain the groundwork for automated medicinal floras of each country.

Our Panama computer file indexes all the medicinal and poisonous attributes reported in the Isthmian Ethnobotanical Dictionary (Duke, 1972). The Gupta file indexes those reported by Gupta et al (1979). Our Mexico file indexes those medicinal and poisonous properties reported in the "Trees and Shrubs of Mexico" (Standley, 1920–6). Without a medicinal flora of Honduras, I can extract notes on Honduran uses of plants from nearby floras, such as Standley 1920–26, 1928, 1931, and Standley and Steyermark 1946) to make an"internal" Honduran Medicinal Flora, listing those plants that are used medicinally in Honduras by Hondurans. Once in the computer, a medicinal flora can be retrieved systematically, alphabetically by genus and species, or alphabetically by medicinal and poisonous attributes. An external Honduran Medicinal Flora lists medicinal uses outside Honduras of species that occur in Honduras but are not reportedly so used in Honduras. The creation of the external Honduran Medicinal Flora is automated also, once we have entered the correct nomenclature for the species reported from Honduras. The computer searches our files for all the medicinal and poisonous species attributes reported elsewhere. Because of ecological similarities in Mexico and Panama, it was quite simple to generate a quick external Medicinal Flora of Honduras. We simply asked the computer to

TABLE 2
FOLK "CANCER REMEDIES" OF THE NEOTROPICS

Genus (Country) Active extracts/Extracts tested

Abelmoschus (Bra) 0/0	Abuta (Bra) 12/42
*Abutilon (WI) 2/110	*Acacia (Ven) 25/622
*Acalypha (Gua) 7/198	Achillea (Mex) 1/83
Achras (WI) 0/1	*Acnistus (CR) 0/14
Adenanthera (Bra) 0/22	Adiantum (Mex) 0/12
Aeschynomene (Mex) 5/56	*Agave (Ven) 36/366

TABLE 2 Continued

Genus (Country) Active extracts/Extracts tested			
Ageratum (Ven) 0/27	*Allium (Peru) 2/91	Furcraea (Ven) 0/10	Gallesia (Bra) 0/15
*Aloe (Bra) 2/101	Alternanthera (Mex) 0/38	*Garcinia (Bra) 9/160	Genipa (Cuba) 2/27
Althaea (Peru) 1/25	Amaryllis (WI) 5/8	Gomphrena (Mex) 0/18	Gonolobus (Mex) 0/6
Ambrosia (WI) 5/83	Amphipterygium (Mex) 0/8	Gossypium (WI) 3/44	*Guarea (Bra) 6/201
Amyris (WI) 8/35	Anacardium (WI) 2/45	Gynerium (Ven) 0/14	Haematoxylon (Mex) 1/11
Anagyris (WI) 0/11	Ananas (Cuba) 2/15	Hancoria (Bra) 0/0	Harrisia (Hai) 0/0
Anemopsis (Mex) 0/12	Anethum (Peru) 0/13	*Helianthus (Ven) 2/139	Heliconia (Pan) 1/60
*Annona (Bra) 43/202	Anredera (Cuba) 0/5	Helicteres (WI) 1/18	*Heliotropium (Mex) 6/119
Apium (Cuba) 0/30	Arctium (Chile) 2/58	*Hernandia (?) 31/49	*Hibiscus (WI) 5/305
*Argemone (Mex) 4/52	Aristolochia (WI) 3/65	Hippomane (Cuba) 1/10	Hydrocotyle (Peru) 0/19
Arnica (Chile) 4/28	Arracacia (Mex) 0/1	Hydrolea (Ven) 0/6	Hymenocallis (WI) 3/15
Artocarpus (Bra) 2/86	Arum (WI) 0/13	Hymenophyllum (Ven) 0/3	Hyoscyamus (WI) 1/22
*Asclepias (Mex) 28/209	Astronium (Bra) 0/10	Hypoxis (WI) 2/9	*Hyptis (WI) 13/132
Avicennia (Cuba) 2/74	Ayenia (Mex) 0/3	Impatiens (Bra) 1/73	*Indigofera (WI) 3/179
Azorella (Peru) 0/1	*Baccharis (Mex) 10/233	Inula (Arg) 3/81	*Ipomoea (WI) 12/284
Basella (WI) 0/7	Bastardia (Ven) 0/1	Iresine (Nic) 0/23	Isocarpha (Ven) 0/1
Befaria (WI) 1/21	*Begonia (WI) 3/39	Isoloma (Mex) 0/1	*Jacaranda (Bra) 3/120
Beta (Mex) 1/30	Blighia (WI) 1/37	Jacquinia (Gua) 0/27	*Jasminum (WI) 0/124
*Bidens (Bra) 2/124	Bixa (Mex) 1/43	*Jatropha (Mex) 19/116	Jussiaea (Peru) 1/29
Bocconia (Mex) 8/31	Boerhaavia (Bra) 3/31	Kalanchoe (Bra) 16/60	Kallstroemia (Ven) 1/16
Boswellia (Peru 0/12	Bourreria (Mex) 1/11	Ketmia (WI) 0/0	Krameria (Mex) 1/22
Bowdichia (Ven) 0/1	Brosimum (Cuba) 0/58	Landenbergia (Ven) 0/25	Laetia (Ven) 2/24
Browallia (Ven) 0/2	*Buddleja (Mex) 5/165	*Lagenaria (Ven) 2/12	Lavatera (Peru) 0/19
Bursera (Mex) 20/98	*Byrsonima (Mex) 6/173	Lilium (Gua) 0/42	*Linum (Peru) 11/50
Byttneria (Peru) 1/22	*Caesalpinia (Mex) 7/168	Lippia (Mex) 3/109	Liquidambar (Mex) 0/41
Cakile (WI) 0/11	Calendula (Mex) 1/14	Lithospermum (Mex) 2/29	*Lonchocarpus (Bra) 7/159
*Calliandra (Mex) 6/108	Calonyction (WI) 0/7	Lophocereus (Mex) 5/28	Lophophora (Mex) 0/5
Canavalia (Mex) 3/46	Canna (Bra) 1/55	Luehea (Bra) 1/65	*Luffa (Ven) 2/32
*Capparis (WI) 9/174	Capsicum (Ven) 2/14	Lycopersicon (Ven) 0/16	Lycoseris (Ven) 0/0
Carica (Jam) 0/46	Carthamus (Ven) 2/14	Lythrum (Mex) 1/48	Malus (Chile) 1/29
Casimiroa (Mex) 1/20	*Cassia (Mex) 18/732	Mammillaria (Mex) 0/24	Mangifera (Bra) 2/68
Castilla (Mex) 3/10	Castilleja (Mex) 3/67	Manihot (Mex) 1/50	Marsdenia (Ecu) 1ˇLippia (Mex) 3/109
Catasetum (Bra) 0/4	Cayaponia (Bra) 3/17	Matricaria (Peru) 0/16	*Maytenus (Bra) 55/300
Cenchrus (Mex) 0/11	Centrosema (WI) 0/7	Melia (Col) 15/54	Melochia (Bra) 2/32
Cereus (WI) 1/15	*Cestrum (Arg) 9/162	Mentha (Ven) 1/55	Mentzelia (Mex) 0/45
Chelidonium (Gua) 1/10	*Chenopodium (Peru) 1/136	*Miconia (Ven) 19/715	*Mikania (Bra) 1/98
Chiococca (WI) 1/18	*Chrysanthemum (Ven) 1/90	*Mimosa (SA) 2/138	Mimusops (Bra) 2/68
Cinchona (Ven) 0/36	Cinnamomum (Cuba) 3/179	Mirabilis (WI) 4/71	*Momordica (Bra) 2/57
*Cissus (WI) 2/100	*Citrus (Chile) 2/165	Montanoa (Mex) 1/30	Montrichardia (FG) 0/7
*Coccoloba (Ven) 2/172	Cocos (Ven) 1/14	Moringa (WI) 2/37	Morisonia (WI) 0/4
Commelina (Mex) 1/41	Commicarpus (Mex) 0/8	Mucuna (Ven) 1/55	Musa (Peru) 0/23
Comocladia (WI) 2/5	Conyza (Mex) 3/81	Mutisia (Chile) 0/9	Myristica (Bra) 10/72
Copaifera (Bra) 0/27	*Cordia (WI) 8/385	Myroxylon (Pan) 1/19	Narcissus (Mex) 4/26
*Cornus (Mex) 10/168	Costus (Bra) 3/51	*Nectandra (Bra) 12/288	Neea (Ven) 3/77
Cotyledon (Mex) 1/5	Coutabea (WI) 0/2	Neomammillaria (WI) 0/4	*Nerium (Ven) 22/49
*Croton (Bra) 39/451	*Cucumis (WI) 5/45	Nesaea (Mex) 0/0	Neurolaena (WI) 0/10
*Cucurbita (Peru) 7/53	*Cupania (Bra) 5/134	Nicotiana (Mex) 2/74	Nopalea (WI) 1/4
Cuphea (Mex) 1/43	Cyrtopodium (Bra) 0/2	*Ocimum (WI) 3/77	*Ocotea (Bra) 8/433
Dahlia (Mex) 1/101	Dalea (Mex) 1/101	*Oenothera (Mex) 7/163	Olea (Mex) 3/199
Datura (Mex) 4/106	Davilla (Bra) 1/22	*Opuntia (WI) 2/177	Origanum (Ven) 1/26
Dendropanax (WI) 3/78	*Desmodium (Mex) 3/166	Pachycereus (Mex) 1/24	*Palicourea (Bra) 1/195
Didymaea (Mex) 0/0	Dieffenbachia (Cuba) 0/26	Papaver (Chile) 0/20	Parthenium (DR) 3/34
Dioscorea (Mex) 2/100	Diphysa (Gua) 2/19	Paullinia (Bra) 0/87	*Pedilanthus (WI) 3/17
Dodonaea (WI) 3/79	Dorstenia (Mex) 0/5	Pelargonium (Ecu) 0/59	Pereskia (Bra) 0/3
Drimys (Bra) 3/31	Drymaria (Mex) 1/17	*Persea (Peru) 7/140	Petiveria (DR) 0/7
Echeveria (Col) 1/11	*Elephantopus (Cuba) 6/40	Petroselinum (Mex) 0/11	Phaseolus (WI) 0/50
Enterolobium (Ven) 5/44	Equisetum (Gua) 0/84	Philadelphus (Mex) 0/25	Philodendron (Bra) 0/50
Erythroxylum (Peru) 5/117	*Eupatorium (Ven) 13/413	*Physalis (Mex) 11/95	*Phytolacca (Cuba) 1/136
Euphorbia (Bra) 48/641	Euxodus (Uru)	Pilea (Ven) 1/38	Pilocereus (Col) 0/0
Ficus (Peru) 45/1340	Funastrum (Col) 0/9	*Pinus (Ven) 25/544	*Piper (Bra) 10/483

TABLE 2 Continued

Genus (Country) Active extracts/Extracts tested			
Piptadenia (Ven) 0/33	Pistia (Bra) 1/8	Spathiphyllum (Mex) 0/3	Spilanthes (Bra) 0/16
*Pithecellobium (Bra) 9/210	*Plantago (Ven) 3/109	Spondias (Cuba) 1/70	Stachys (Peru) 2/75
Plumbago (WI) 0/33	Plumeria (WI) 2/65	Stachytarpheta (Ven) 0/62	Stellaria (Mex) 0/29
Polypodium (Hon) 2/63	Portulaca (WI) 0/20	Struthanthus (Bra) 2/37	Swietenia (Mex) 1/44
Pothomorphe (WI) 1/12	Prosopis (Mex) 6/73	*Tabernaemontana (WI) 33/215	Tabebuia (Bra) 6/116
*Protium (Ven) 6/164	*Prunus (Ven) 10/337	Tagetes (Cuba) 1/26	Talinum (Ven) 0/9
Pseusmagennetus (Ecu) 0/0	Psittacanthus (WI) 2/33	Tamarindus (Ven) 1/34	Taxodium (Mex) 3/48
*Psychotria (Bra) 17/476	Punica (Peru) 1/18	*Tecoma (Bra) 0/93	*Tephrosia (Bra) 19/107
Puya (Chile) 2/13	*Quercus (Mex) 11/432	*Terminalia (Mex) 26/334	Tetracera (Bra) 0/32
Ranunculus (Bra) 1/81	Renealmia (Bra) 0/25	Thalia (Bra) 0/5	*Thalictrum (Mex) 15/106
*Rhizophora (Bra) 6/80	*Rhus (Mex) 19/450	*Thevetia (Mex) 35/71	*Thuja (Cuba) 5/47
Ricinus (Bra) 5/45	Rivina (Mex) 0/6	Tillandsia (Bra) 2/87	Torrubia (Ven) 0/4
Rollinia (WI) 18/81	*Rosa (Peru) 3/118	*Tournefortia (Mex) 3/94	Tragia (Mex) 0/9
Rosmarinus (Peru) 0/5	*Rumex (Chile) 5/200	Tribulus (Ven) 0/25	Trichachne (WI) 0/8
Sabal (Mex) 0/20	Saccoglottis (Bra) 0/1	*Trichilia (Ven) 39/284	Trichocereus (Chile) 0/8
*Salix (Mex) 3/310	Samanea (Ven) 1/7	Triplaris (Peru) 0/47	Triticum (Peru) 0/7
*Sambucus (Chile) 5/150	*Sapium (Bra) 10/141	Trixis (Ven) 0/22	Tropaeolum (Peru) 1/16
Sauvagesia (WI) 0/1	Scheelea (Ven) 1/18	Tussilago (Chile) 0/20	Typha (Chile) 1/62
Schinus (Pan) 1/67	Scutellaria (WI) 2/40	Ullucus (Peru) 0/0	Urera (Pan) 3/69
Seguieria (Bra) 0/4	Semecarpus (WI) 5/18	Urtica (Mex) 1/47	Valeriana (Mex) 2/60
*Senecio (Mex) 16/474	Serjania (Mex) 3/57	Vanilla (Peru) 0/12	*Verbena (Mex) 0/143
Sesamum (WI) 0/15	Sesbania (Mex) 4/99	*Verbesina (Mex) 6/90	Viola (Peru) 2/63
Sickingia (Mex) 1/4	*Sida (WI) 1/111	*Vitis (Mex) 2/129	Waltheria (Bra) 1/39
Simmondsia (Mex) 1/16	*Smilax (Mex) 2/120	Xanthium (Bra) 1/72	Xanthosoma (WI) 0/6
Solandra (WI) 0/20	*Solanum (WI) 33/764	*Zanthoxylum (WI) 16/202	Zea (WI) 0/16
*Solidago (Bra) 7/190	Sonchus (Gua) 0/46	Zebrina (Mex) /7	Zephyranthes (Peru) 5/17

Extracted at the generic level from Hartwell, 1967–1971. (Plants Used Against Cancer—A Survey. Lloydia 30–34. In eleven parts.) The first or most frequent neotropical country in which the plant is reported as a folk remedy is reported in parentheses.

Arg = Argentina, Bra = Brazil, Col = Colombia, CR = Costa Rica, DR = Dominican Republic, Ecu = Ecuador, FG = French Guiana, Gua = Guatemala, Hai = Haiti, Hon = Honduras, Jam = Jamaica, Mex = Mexico, Nic = Nicaragua, Pan = Panama, SA = South America, Uru = Uruguay, Ven = Venezuela, WI = West Indies.

Genera with asterisks (*) are of no further interest for general collection.

list all the medicinal species we have reported from both Mexico and Panama in order to generate a first draft Medicinal Flora of Honduras.

Mediante la computarización del estudio ha sido posible crear floras tóxicas "externas" e "internas" para algunos países, los cuales, en muchos casos, ni siquiera tienen una flora medicinal publicada.

Another innovation completed with 1979 was the automation of Hartwell's (1967–71) "Plants Used Against Cancer." Hartwell combed literature up to about 1970 for folkloric remedies used against cancer and published a compilation of some 3,000 species. With this data now automated—with Dr. Hartwell's blessings—we can establish a "shopping list" according to how many entries a species or genus has in the Hartwell file (Table 2). We can find out how much more frequently folk remedies pan out in the cancer screen than non-medicinal species. Let me be quick to point out that the most popular folk cancer remedies in Latin America: matico, *Piper* sp.; calaguala, *Polypodium* sp.; and llanten, *Plantago* sp. and those I hear most about in the U.S.: aloe, *Aloe*; and creosote bush, *Larrea*, have not panned out in the NCI screen. And of the 14 plant-derived compounds cited by Suffness and Douros (1979) as reaching the Food and Drug Administration for approval, many occur in species listed in our Hartwell file.

The various automated files help us determine which species are more likely to produce results. When one is hauling out specimens via backpack, camelback, boat, rented truck or chartered plane, he must be selective. The logistical life lines for the cancer screen are long and expensive.

As I compile data for our various ethnomedical files, it is clear that much of the numerical weight may be due to bibliographic echoes going way back to Dioscorides. I have begun a new file to eliminate the bibliographic echo. In the new folk file, we include only the publications since 1970, and only files that are areally, ethnically and botanically specific, that is, including only data that clearly reflect local medicinal usage. In this file, independent similar usages by remote unconnected ethnic groups, would raise the probability that the usage is based on real medicinal attribute, be it anticancer or antifertility. For this file, I would use my specific Choco and Kuna papers (Duke, 1970, 1975) rather than my general Isthmian Ethnobotanical Dictionary (Duke, 1972).

Un archivo de computadora permite evaluar mejor los primeros reportes.

My interests in folk medicine, economic botany, phytochemistry, and ecology are neatly intertwined in cancer procurement activities. Prior to joining the cancer program, my long standing interest in herbs and medicinal plants had prompted me to prepare illustrated fact sheets on about 500 medicinal plants. These are being automated also, so that writeups for those medicinal plants which occur in Panama can be generated for our Panamanian Medicinal Flora. We now have writeups on over 200 Panamanian medicinal species, with notes on agronomy, ecology, chemistry, utilization, distribution, economics, as well as medicinal and other uses such as folk and real cancer activity.

Collaborating scientists who wish to have their flora screened for cancer activity are provided first drafts of the Medicinal Flora of their research area, based on checklists they provide. This is the working draft upon which he or she and our Laboratory can begin building. Then there is intensive effort to get vouchered three-pound collections of local medicinal species, with ethnobotanical notes on utilization and colloquial names. Vouchers are prepared with at least four objectives in mind; for positive identification of the species by specialists in case of significant biological activity, for label inclusion of ethnobotanical data, to provide data on the ecological amplitudes of medicinal plants (see Duke 1978, 1979a, 1979b), and to serve for color Xerox illustrations which may be converted readily into line drawings.

We believe that cooperating countries will benefit from our keen interest in screening genera and species that have not been screened at all, whether they are medicinal or poisonous or completely unused by local ethnic groups.

In the Cancer Symposium (Perdue & Hartwell 1976), Barclay, Perdue, and Spjut all touched on concepts useful in predicting anticancer activity. Using those concepts, especially the FOSI concept (families of special interest), I predicted (Duke 1979a) the activity of six species (*Brunsfelsia grandiflora*, *Pachypterys hymenaea*, *Picramnia cooperi*, *Prestonia rotundifolia*, *Stenosolen eggersii* and *Witheringia riparia*) collected during my first cancer-collecting trip to Ecuador. This points out both the strength and weakness of the FOSI approach; strong in that the activity was predictable, weak in that the activity was not really new, just the expected similar activity. At this point, we are more interested in moderate activity from new or unknown compounds than strong activity from old or well-known compounds or their close relatives. This reinforces our desire to study the unscreened families, genera, and species, in search of new activity, before these taxa fall to the axe.

Estamos más interesados en nuevas plantas que en los parientes cercanos de las plantas ya estudiadas.

NCI has a file (I call it the TEST file) showing all species and genera that have been screened, how many times, and what activity has emerged. Since NCI and USDA are both government agencies, all of our findings are in the public domain and shared by us with collaborating countries.

We are developing another file which automatically searches the local flora and identifies genera that have not been screened. We are building a GENERIC File which lists all the genera of the world, the approximate number of species in each genus, whether it is woody or herbaceous, or both, and where the genus occurs. Providing all our spellings are correct,—we rely heavily on Willis Dictionary (Shaw, 1973) for this generic file cross checking these two files against our FLORA File will tell us how many times each genus in our FLORA File has been screened and how much activity has been shown. NCI, EBL, and the cooperating country should share an enthusiasm for screening those genera and species identified as unscreened. Perhaps the cooperating country has a sleeping periwinkle which could become the cancer superstar of tomorrow, a basis for a new drug crop and industry for improvement of the agricultural and economic sectors of the cooperating country.

In exchange for vouchered three-pound specimens of medicinal or unscreened species, the collaborator from the cooperating country receives a first draft of a medicinal flora of the country, color Xerox copies of each voucher, and cancer-screening information, all to be used jointly in the preparation of a second draft of The Illustrated Medicinal Flora coauthored by the

Economic Botany Laboratory and the collaborating scientist(s).

Las personas que colaboren recibirán borradores de floras medicinales, copias a color de "vouchers" e información adicional para ayudar en la preparación de una flora medicinal ilustrada.

I hope we can discover the sleeping periwinkles of the neotropics before they have fallen to the "axe of progress".

Barclay, A.S. & R.E., *Perdue*, Jr. 1976. Distribution of anticancer activity in higher plants. Cancer Treatment Reports 60: 1081–1114.

Duke, J.A. 1970. Ethnobotanical observations on the Choco Indians. Econ. Bot. 24: 344–366.

———. 1972. Isthmian Ethnobotanical Dictionary. Harrod & Co, Baltimore.

———. 1975. Ethnobotanical observations on the Cuna Indians. Econ. Bot. 29: 278–293.

———. 1978. The quest for tolerant germplasm. Chap. 1, pp. 1–61 *In* ASA Special Symposium 32 "Crop Tolerance to Suboptimal Land Conditions" Am. Soc. Agron. Madison, WI.

———. 1979a. Silviculture and cancer. Pp. 15–19. *In* Tropical Silviculture Newsletter No. 2.

———. 1979b. Ecosystematic data on economic plants. Quart. J. Crude Drug Res. 17: 91–110.

Gupta, M.P., T.D., *Arias*, M. *Correa* & S.S. *Lamba*. 1979. Ethnopharmacognostic observations on Panamanian medicinal plants: Part I. Quart. J. Crude Drug Res. 17: 15–130.

Hartwell. J.L. 1967–71. Plants used against cancer: a survey. Lloydia.

Perdue, R.E., J. & J.L. *Hartwell*, eds. 1976. Proceedings of the 16th annual meeting of the Society for Economic Botany: "Plants and Cancer." Cancer Treatment Reports 60: 973–1214.

Prance, G.T. 1977. Floristic inventory of the tropics: where do we stand? Ann. Missouri Bot. Gard. 64: 659–684.

Shaw, H.K.A. 1973. J.C. Willis: A Dictionary of Flowering Plants and Ferns, 8th Ed. Cambridge. Univ. Press.

Spjut, R.W. & R.E., *Perdue*, Jr. 1976. Plant folklore: a tool for predicting sources of antitumor activity? Cancer Treatment Reports 60: 979–986.

Standley, P.C. 1920–6. Trees and shrubs of Mexico. Cont. U.S. Nat. Herb. 23:(1, 2, 3, 4, 5).

———. 1928. Flora of the Panama Canal Zone. Cont. U.S. Nat. Herb. 27: 1–416.

———. 1931. Flora of the Lancetilla Valley, Honduras. Field Columbian Mus. Nat. Bot. Series X. Publ. 283.

———. & J.A. Steyermark. 1946. Flora of Guatamala. Fieldiana: Bot. 26. Parts 1–6.

Suffness, M. & J. *Douros*. 1979. Drugs of plant origin. Chap. III, Vol. 16. *In* Methods in Cancer Research. Academic Press, New York.

Estudios Sobre la Flora Tóxica de Panamá

Novencido Escobar

Panama's toxic plants have been subject of several studies over the past decade. Numerous plants are toxic to livestock. The greatest number require open and sunny savannas. Some toxic plants are restricted in distribution.

Panamá posee una flora riquísima ya que es la conjunción de los elementos de los trópicos tanto de la América Central como de la América del Sur, así como también del Caribe. Constituye, por lo tanto, un centro espectacular de estudios para botánicos en todas sus ramas (Schultes 1972). Muchos investigadores han visitado nuestro país atraídos por la riqueza de su flora y, entre ellos, podemos citar a la pléyade de investigadores del Jardín Botánico de Missouri quienes se han dedicado en cuerpo y alma al estudio detallado de nuestra flora desde 1926. No obstante, queda mucho por hacer, ya que aún quedan regiones prácticamente vírgenes.

Uno de los aspectos parcialmente investigados es el de las plantas con características tóxicas. Ninguna importancia seria se le dió en el pasado a este tipo de estudios con excepción del breve opúsculo escrito por Paul Allen (1943) durante los años de la segunda guerra mundial.

In the past little attention was given to Panama's toxic plants.

Fue a partir de la fecha de la publicación de la Flora Tóxica de Panamá (Escobar, 1972) cuando el público se interesó de veras en estos tópicos.

La investigación se inició en 1963 exclusivamente en el grupo de las plantas superiores. No se ha realizado todavía ninguna investigación a fondo sobre la toxicidad de las algas, hongos y helechos, grupo muy vasto donde seguro hay una buena cantidad de plantas con características tóxicas. El grupo de las plantas ornamentales es el que mayor problema presenta para el hombre por ser un grupo doméstico de mucha accesibilidad para los humanos.

Toxic ornamental plants are a problem because of their accessibility.

Existe también un grupo numeroso de plantas tóxicas para los animales domésticos, en especial para el ganado vacuno.

Se han realizado algunas investigaciones por alumnos graduandos en sus tesis de graduación y el Laboratorio Especializado de Análisis (LEA) de la Universidad de Panamá investiga en este sentido, como también lo hace el Proyecto de Estudios Farmacognóstico de la Flora Panameña patrocinado por la Organización de Estados Americanos (OEA) y la Universidad de Panamá.

Por otro lado, la Fundación Internacional para las Ciencias de Suecia financia las investigaciones sobre las siguientes plantas: Lombricera, *Spigelia anthelmia* (Loganiaceae), Bellísima, *Antigonon leptopus* (Polygonaceae), Contragavilana, *Neurolaena lobata* (Compositae).

En Panamá, y ésto es general, las familias en donde más abundan las plantas tóxicas son las siguientes: Apocynaceae, Araceae, Euphorbiaceae, Leguminosae, Loganiaceae, Solanaceae y Sapindaceae.

In Panama, the families with the most abundant toxic plants are the Apocynaceae, Araceae, Euphorbiaceae, Leguminosae, Loganiaceae, Solanaceae and Sapindaceae.

Distribución Geográfica

La densidad de las plantas tóxicas en Panamá es mayor del lado del Pacífico que del Atlántico, es-

CUADRO 1
PLANTAS VENENOSAS COMUNES EN PANAMA

Abrus precatorius Peronillas ó Peonías	*Parte tóxica:* semillas negras y brillantes. *Principios venenosos:* abrina (Jequiritina). Fitoalbúmina aglutinante, está con el veneno de culebras. *Síntomas*: si se masca tan solo una semilla puede ser fatal. Se observa náuseas y vómitos, temblores, debilidad, cólicos, sudor y diarreas severas.
Allocasia macrorhiza Oreja de Elefante	*Parte tóxica*: hojas lanceoladas o lineares. *Principios venenosos:* oxalato de calcio. *Síntomas:* similares a los de la caña muda.
Blighia sapida Huevo vegetal, Seso vegetal, Akee	*Parte tóxica*: las paredes del fruto y de las semillas (los frutos inmaduros que no se han abierto estando en los árboles tienen venenos mortales). Un fruto inmaduro tiene una dosis letal de veneno que es soluble en agua hirviendo. *Principios venenosos*: hipoglicina. Hipoglicina A. La B es un dipéptico formado de hipoglicina A y ácido glutámico. *Síntomas:* la ingestión del fruto, especialmente por los niños, produce un síntoma conocido como enfermedad del vómito, que se caracteriza por una marcada hipoglicemia. Leucopenia, daño hepático. La tasa de mortalidad es alta, de 40–80%. *Tratamiento:* a) Medidas de urgencias: lavado gástrico. b) Antídoto: no se conoce ninguno. c) Tratamiento sintomático para controlar convulsiones.
Brugmansia candida Reina de la Noche Floripondio (ornamental)	*Parte tóxica:* todas las partes, especialmente las semillas. *Principios venenosos:* los alcaloides tropanos atropina, hiosciamina, o escopolamine. *Síntomas:* pulso rápido y débil, ceguera parcial, delirio, incoherencia, respiración lenta, baja temperatura, pulso rápido y débil, convulsiones o coma proseguidos de muerte. Puede causar dilatación de las pupilas si después de manipular las hojas hay un rociamiento en los ojos. *Tratamiento:* lavarse las manos después de su manipulación. Se recomienda el uso de ácido tánico, drogas colinérgicas (pilocarpina, fisostigmina), estimulantes respiratorios.
Cnidoscolus urens Ortiga brava, Pringamoza	*Parte tóxica:* los pelos duros de la planta. *Síntomas:* los pelos causan gran dolor al pinchar la piel. A menudo el dolor dura un día o más. En algunos casos hay fiebre alta. *Tratamiento:* el profesor Escobar recomienda amonia y $KMnO_4$.
Datura metel Floripondio (ornamental)	Véase *Brugmansia candida.*
Dieffenbachia picta Caña muda, Otoe de lagarto, Lluvia del plata	*Parte tóxica*: todas las partes incluyendo la savia. *Principios venenosos*: cristales de oxalato de calcio y proteína tóxica. *Síntomas:* la ingestión produce rápida irritación, enrojecimiento de la superficie bucal de la lengua y de los labios, salivación copiosa, y tumefacción edematosa, puede causar la muerte si la hinchazón o tumefacción bloquea el paso del aire a la garganta. El jugo puede causar irritación intensa en la piel. *Tratamiento:* sintomático, el dolor se alivia con hidrocloruro de meperidina; hidróxido de alumino o de magnesio es útil como demulcente y agente neutralizante.
Dieffenbachia seguine	Véase *D. picta.*
Euphorbia cotinifolia Barrabás	*Parte tóxica:* savia lechosa de las hojas y tallos; las semillas. *Síntomas:* dermatitis severa y reacciones alérgicas de la piel. *Tratamiento:* sintomático.
Euphorbia pulcherrima Flor de Pascua Flor de Navidad	*Parte tóxica:* Jugo de las hojas, tallos, flores o frutos, secos o verdes. *Principios venenosos:* en el acre lechoso de la savia. Puede ser resina. *Síntomas:* emesis intensa, dolores abdominales, diarrea, delirio. La savia causa dermatitis y ceguera temporal, si se frota en los ojos. Convulsiones. *Tratamiento:* se usa demulcentes, astringentes gástricos, estimulantes del SN y circulatorio. Administrar atropina, 2 mg por vía sub-cutánea; repetir según las circunstancias.
Hippomane mancinella Manzanillo	*Parte Tóxica:* la savia lechosa cáustica de todas las partes; frutos verdes. *Principios venenosos*: existen varios principios venenosos en el látex. El principio activo no se conoce con certeza todavía. En 1954 Schaffler y colaboradores lo llamaron brevifolina.

CUADRO 1 (Cont.)

	Síntomas: la savia lechosa es muy irritante. Causa inflamación severa y dermatitis, aunque como el caso del veneno de la hiedra, alguna personas son inmunes. *Tratamiento:* sintomático. Lavar con jabón y agua o alcohol.
Hura crepitans Tronador o Habillo	*Parte tóxica:* todas las partes látex. *Síntomas:* la savia lechosa del árbol causa inflamaciones de consideración e irritación cuando se pone en contacto con la piel. La semilla contiene un aceite que es un catártico drástico y produce un efecto peor que una sobredosis de aceite castor. Dos horas después de haber ingerido la semilla causa náusea, vómito y severo desarreglo gastrointenstinal.
Ipomoea crassicaulis	*Parte tóxica:* extractos de las raíces, tallo y hojas.
Lantana camara San Rafaelito, Pasarín, Hierba de Zorro	*Parte Tóxica:* granos. *Principios venenosos:* un triterpenoide policíclico llamado landadena. *Síntomas:* irritación gastrointestinal. Los envenanamientos severos producen una estremada debilidad muscular y colapso circulatorio, que generalmente resulta fatal. El síndrome se asemeja al envenenamiento con atropina.
Mangifera indica Mango	*Parte Tóxica:* la cáscara de la fruta y la savia del árbol. *Síntomas:* dermatitis, náusea, vómito y diarrea. *Tratamiento:* no ingerir la cáscara y evitar el contacto con la savia.
Nerium oleander Rosa de Lima	*Parte Tóxica:* todas las partes verdes o secas. *Principios venenosos:* glícosidos cardioactivos, oleandrina, neerina, oleandrosina. *Síntomas:* náuseas, depresión, midríasis, pulso débil e irregular, diarrea sanguinolenta, parálisis y muerte. *Tratamiente:* atropina, eméticos, lavado gástrico y se recomienda medicina.
Phytolacca rivinoides Calalú, Jaboncillo	*Parte tóxica:* raíces y hojas. El fruto es poco tóxico. *Principios venenosos:* puede ser una resina o saponina.
Ricinus communis Higuerilla Habras de Jequerití	*Parte tóxica:* todas las partes principalmente las semillas. *Principios venenosos*: ricina, una fitotoxina que produce aglutinación y habas de hemólisis de los eritrocitos a diluciones extremas. También dañina a todas las demás células. *Síntomas:* la ingestión de las partes de la planta no refinadas causa sensación de quemaduras en la boca y disturbios gastrointestinales. También daños renales. Dos semillas pueden causar un serio envenenamiento; ocho pueden ser fatales. *Tratamiento:* sintomático, lavados estomacales con solución salina caliente. Alcalinizar la orina administrando bicarbonato de sodio para impedir la precipitación de hemoglobina en los riñones.
Solanum tuberosum Papa	*Parte tóxica:* las manchas verdes "quemaduras por el sol" y los retoños de los tubérculos de papas, los tallos y hojas verdes. *Principios venenosos:* alcaloide esteroidal, solanina. *Síntomas:* piel húmeda, náusea, confusión mental, depresión respiratoria y cardíaca. Ha causado muertes.
Strychnos toxifera Canjuira, Fruta de murciélago	*Parte tóxica:* semillas y corteza. *Principios Venenosos*: toxiferina (alcaloides). *Síntomas:* causa parálisis de la unión neuromuscular de los nervios. El extracto crudo tiene un efecto tóxico en los vasos sanguíneos. El comer las semillas causa severas convulsiones, dilatación de las pupilas y termina en muerte por fallo del centro respiratorio.
Thevetia ahouai Huevo de tigre, Cojón de gato Lavaperro	*Parte tóxica:* la planta entera especialmente la nuez. *Principios venenosos:* glicósidos cardioactivos cerebrina, nerufolina, y la vetina, producen efectos parecidos a la digital. *Síntomas:* una o más nueces pueden ser fatal. Causa estimulación cardíaca, vértigo y vómito. El látex es irritante y causa dermatitis.
Thevetia peruviana Campanillo (ornamental)	Véase *T. ahouai.*

CUADRO 1 (Cont.)

Toxicodendron striatum Hinchador	*Parte tóxica: todas las partes al igual que el humo, que quema.* *Principios venenosos:* una oleorresina llamada urushiol que es 3–pentadecadienil catecol. Es un agente sensibilizador poderoso, ya que el contacto repetido parece aumentar la severidad de la reacción. *Síntomas:* produce una respuesta alergénica severa causando dermatitis al ponerse en contacto con ella, produciendo inflamación, ampollas y vesículas. Se observa formacioñ de grietas en la piel, se exuda un líquido y se forman escamas o una costra. Después de una exposición severa, ocurre irritación renal. En muertes producidas por envenamientos con hiedra venenosa y plantas relacionadas, los hallazgos patológicos incluyen lesiones renales y miocardíacas.

Source: Escobar 1972.

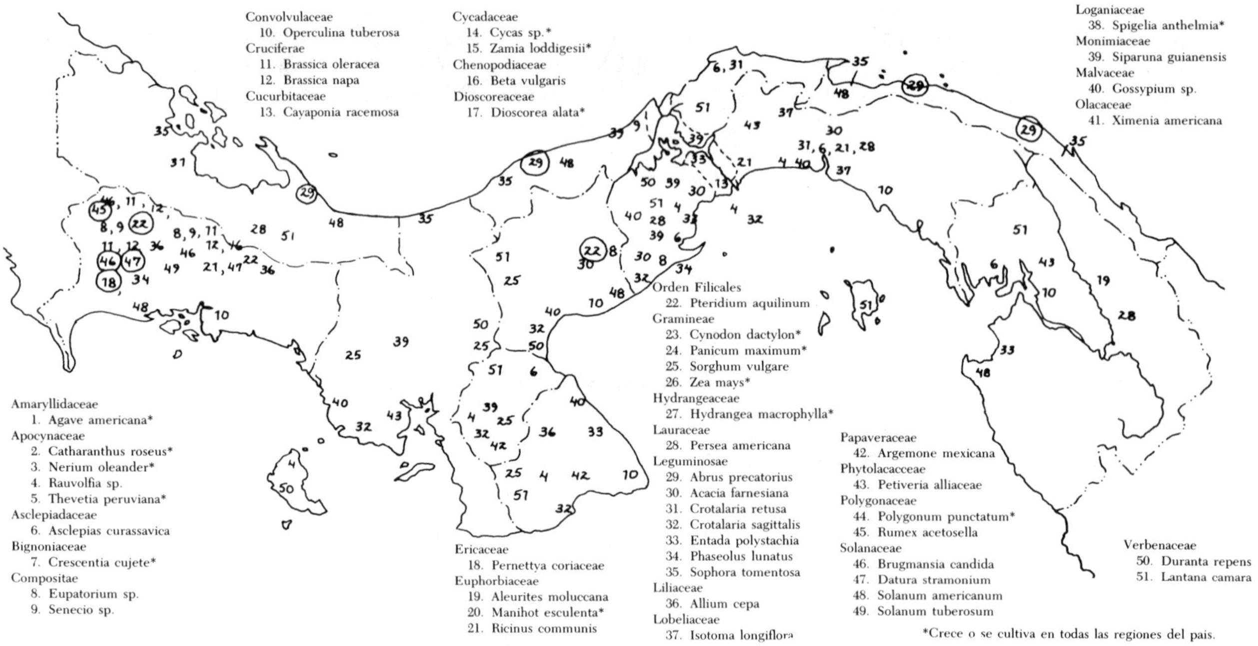

Fig. 1. Distribución Geográphica de las Plantas Tóxicas al Ganado en la Republica de Panamá.

pecialmente en las Provincias de Chiriquí, Los Santos, Herrera, Coclé y parte de la Provincia de Panamá. Pensamos que el ambiente agreste del Atlántico, selvas espesas como las de Bocas del Toro, parte de Colón, la parte continental de San Blas, el Este de la Provincia de Panamá y la Provincia del Darién no permiten la proliferación de este grupo de plantas, generalmente hierbas y arbustos que necesitan sabanas abiertas y soleadas (hay excepciones).

The density of toxic plants is greater on the Pacific than on the Atlantic side of the isthmus.

Algunas, como *Rumex*, (45) parecen estar confinadas a las tierras altas de Chiriquí. *Abrus precatorius* (29) crece en las costas de San Blas, Colón y Bocas del Toro. *Pernettya* (18) crece a 1200 m de elevación en los Llanos y Volcán (Chiriquí). *Brugmansia* (46) es abundante en Boquete y Cerro Punta en la Provincia de Chiriquí. Fig. 1 indica la distribución geográphica de las plantas tóxicas en la República de Panamá.

Ahora bien, crecen espontáneamente o se cultivan en toda el área geográfica los géneros *Asclepias, Catharanthus, Crescentia, Dioscorea, Hydrangea, Manihot, Nerium, Panicum, Polygonun, Ricinus, Spigelia, Thevetia*, por citar algunos. Presentamos Cuadro 1 de las plantas tóxicas más conspicuas de la flora Panameña con sus principales características fitoquímicas.

Allen, P.H. 1943. Poisonous and injurious plants of Panama. Am. Jour. Trop. Med. 23: 13, 33, 46.
Escobar, N. 1972. Flora Tóxica de Panamá. Editorial Universitaria. Univ. Panamá. Panamá.
Schultes, R.E. 1972. Prólogo a la Flora Tóxica de Panamá. Editorial Universitaria. Univ. Panamá. Panamá.

Efficacious Plants of the Neotropics

Walter H. Lewis and Memory P. F. Elvin-Lewis

Algunas veces, poblaciones ampliamente separadas utilizan plantas de las mismas familias o géneros para los mismos propósitos medicinales. Estas comparaciones pueden ser una magnífica recompensa en la obtención de información sobre la eficacia de diferentes remedios provinientes de plantas. Poblaciones en Panamá y Nepal presentan usos medicinales similares de algunas de sus plantas nativas. En estos usos locales de plantas con propósitos medicinales hay peligro de toxicidad.

It is appropriate to discuss efficacious plants of the neotropics emphasizing Panama because the epicenter of the New World tropics is Panama. This is not just a geographic concept, but extends to its rich indigenous flora within a small area associated with distinct native cultures, as well as European, African, and Asian invaders and immigrants from the 16th century onward. Many brought plants with them from abroad, and when they could not bring the plants they wanted, they adopted native ones that paralleled the appearance and efficacy of those at home, or they used new plants that had similar taste, texture, and composition to those previously used. This gives Panama an extremely diverse and cosmopolitan population that utilizes for health and other purposes a vastly greater vegetational resource and cultural base than found in most countries. For this reason, the people of Panama and their plants are extremely relevant for understanding worldwide human-plant interactions now, in the past, and also for the future.

La población de Panamá tradicionalmente considerada de origen cosmopolita adicionada a la rica flora proveen un magnífico apoyo histórico para estudios de las relaciones hombre-planta.

Plant Selections of Indigenous Populations

Plant Introductions in Cosmopolitan Medicine

Indigenous practitioners have long relied on the empirical method to distinguish between efficacious values of medicinal plants. Therefore, when similar plants (e.g., members of the same genus) have been used by disjunct populations to treat the same disease or symptom, it provides an initial clue to search for efficacious principles they may share (Lewis & Elvin-Lewis 1977, 1979). Should their potential be recognized in cosmopolitan medicine, they have been subject to rigorous examination before being adopted into modern pharmocopoiea. Of plants found in this category, no better example exists than that of curare, an extract from the menispermaceous *Chondrodendron* native to this area, but widely used to the southeast. This 20th century use of an arrow poison as a surgical muscle relaxant has paved the way for the development of modern synthetics based upon the alkaloid tubocurarine. Through their reversible relaxation of skeletal muscles, these substances permit certain types of constructive surgery that otherwise would have been difficult, if not impossible, to realize.

Chondrodendron se usó con Strychnos como veneno para las flechas, sin embargo en el este de América del Norte no se utilizaron plantas comparables ya que para tal fin existía el veneno proviniente de las culebras cascabeles.

As an arrow poison, *Chondrodendron* was used with *Strychnos* so that the rapid action of the *Chondrodendron* relaxant could be combined with the powerful convulsant and lethal actions of *Strychnos*. Today, the action of the *Chondrodendron* alkaloid is well-known, for it combines with acetylcholine receptor sites at neuromuscular junctions and effectively blocks all messages from the central nervous system.

Modern synthetic congeners also act as neurotransmission blocking agents with the same effect of reversibly paralyzing skeletal muscle.

Menispermum canadensis (moonseed) of eastern North America also possesses a similar isoquinoline alkaloid, known as dauricine, that has curare-like action. Recently an Osage Indian told us that the reason why plant extracts were probably not used as poisons by eastern native North Americans was that rattlesnake venom, dried at the end of an arrow tip, was widely utilized and thus the search for plant poisons was not a necessary alternative.

Potential Candidates

As we have discussed, an initial way to study potential medicinal candidates is to examine the species used to treat a disease or symptom by disjunct tribes or peoples. When plants are similar it is possible that related compounds exist having specific activities that may be of some therapeutic value. As an example, we have chosen one system, the gastrointestinal tract, to illustrate that certain plants selected empirically by the disjunct indigenous peoples of eastern Panama, the Choco and Kuna (Duke 1968), and those of Nepal (Suwal 1970) have common uses. We shall discuss the remedies of these disparate populations for the treatment of five disorders of the gastrointestinal tract.

Emetics: Three genera are used in common by Panamanian and Nepalese populations for the purpose of emesis. They are *Argemone* (Papaveraceae), *Entada* (Fabaceae), and *Momordica* (Cucurbitaceae), the species of the first two genera being used by both native Panamanians and the Nepalese, and *Momordica* by Jamaicans in Panama and the Nepalese. Both leaves and seeds are used and all possess known or suspected secondary metabolites that include saponins and alkaloids having emetic properties. For example, resin from the seeds and pericarp of *Momordica* contains a saponin glycoside that yields elatorin, as well as alkaloids that cause vomiting and diarrhea depending on the dose. Carefully monitored doses can elicit the action desired although serious poisoning may result if amounts are excessive.

Tanto en Panamá como en el Nepal, el hombre usa Argemone, Entada y Momordica como eméticos.

Purgatives: Six genera are commonly used for purging by both populations: *Euphorbia* (Euphorbiaceae), *Caesalpinia*, *Cassia*, and *Clitoria* (Fabaceae), *Aloe* (Liliaceae), and *Argemone* (Papaveraceae). Species of *Euphorbia* possess diterpenes known to elicit local gastrointestinal irritation, and among the legumes, *Cassia* of which six different species are utilized by these populations, is known to have anthraquinone glycosides in many plant parts. Their anthraquinones are also irritant purgatives, but they are absorbed from the small intestine and carried by the circulation to act on the parasympathetic component of the autonomic nervous system affecting the large intestine, thereby stimulating evacuation. These compounds act within six to eight hours following ingestion. *Aloe* species contain the anthraquinone emodine with parallel action to the *Cassia* anthraquinones except that they are more violent. In ancient English herbals, for example, they are specifically recommended for purging horses.

Ambas poblaciones también utilizan purgantes similares provinientes de Euphorbia, Caesalpinia, Cassia, Clitoria, Aloe y Argemone. Estos estimulan la evacuación en un período de 6 a 8 horas.

Antidiarrheals: To control diarrhea of various origins, several genera are commonly used in both cultures. These include species of *Terminalia* (Combretaceae), and *Acacia* and *Mucuna* (Fabaceae). Of these, species of *Terminalia* are known to have broad spectrum antibiotic activity. *Acacia* species contain astringent tannins and also tryptamines that through adrenergic effects are antagonistic to cholinergic activity, thus reducing muscle activity and other nerve-mediated bowel activities.

Ambas culturas además utilizan plantas similares para controlar la diarrea: Terminalia, Acacia y Mucuna.

Anthelmintics: To eleviate worm infestations both cultures use two species of *Carica* (Caricaceae), two species of *Chenopodium* (Chenopodiaceae), and the pomegranate *Punica granatum* (Punicaceae). Oil of wormwood from *Chenopodium* is widely used against roundworms and hookworms, although the therapeutic dose closely approximates minimum toxic levels and death may result from side effects (Lewis & Elvin-Lewis 1977). The action of *Carica* is presumably through its proteolytic enzymes and that of *Punica* by its isopelleturine alkaloid.

Hemorrhoids: To treat this disorder, both populations use two species of *Capsicum* (Solanaceae). Fruits of *C. frutescens* (Panama) and *C. annum* (Nepal) possess the oleoresin capsaican that probably acts as a counterirritant when administered topically.

Tanto en Panamá como en el Nepal se utilizan especies de Carica, Chenopodium y Punica granatum para tratar las lombrices. Igualmente usan Capsicum en el tratamiento de las hemorroides.

These examples illustrate instances of parallel use of similar plants for the same purpose by very different populations. They emphasize the value of the empirical method and the significance of examining on a world-wide basis similar plant selections in order to facilitate the search for substances of therapeutic value.

Plant Selections of Adventive Populations

Many peoples have moved to Panama and when they did so they found that the same plants were not always available to treat familiar disorders and for affecting hygiene. In time, however, they might choose new plants having similar properties that were as effective as those used originally. *Gouania lupuloides* (Rhamnaceae), native to the Caribbean basin and common in Panama (Nowicke 1971), is such a choice having been adopted by those of West African extraction because its properties closely resembled Old World species used as foamy chewing-sponges. The practice of using chewing-sponges to clean teeth rather than chewing-sticks is limited to certain tribes from southern Ghana and to adventive populations in the Caribbean area (Elvin-Lewis 1980), but in Ghana species of *Acacia* (Fabaceae) and *Terminalia* (Combretaceae) are employed in contrast to *Gouania* in the neotropics. Clearly the choice was not morphological but for other reasons including taste, texture, and astringency; these properties probably relating more to chemical and structural than to gross morphological similarities. The wood of all species is soft and fibrous. Salivation is promoted when *Acacia* and *Terminalia* are chewed because their tannins are extremely astringent, while *Gouania* like *Acacia*, possesses saponin glycosides that are naturally foamy. Selection was not because of taxonomic relationship based on morphological similarities, but rather as a consequence of functional parallels.

Cuando personas recién llegadas no encontraron las plantas requeridas, escogieron otras plantas con propiedades similares.

Another example would be the West African's use of *Garcinia kola* (Clusiaceae) as a teeth cleaning chewing-stick and the black American's use in the southeastern United States of *Cornus florida* (Cornaceae) for the same purpose. Both are similar in texture and flavor (bitter), but like *Gouania* and *Acacia* they are not morphologically related. When peoples select for functional similarities they may also be unknowingly choosing for fundamental chemical likenesses that could have important utility in cosmopolitan medicine.

Plant Selections that are Potentially Hazardous

In most cultures employment of plant products as therapeutic remedies follows very precise regimens based on long-term familiarity. However, when cross cultural contact occurs information regarding correct uses may not be transferred, and indiscriminate use might result in harmful rather than beneficial effects, and the real value of a remedy be lost in subsequent evaluations by both folk and cosmopolitan practices. This is particularly true when therapeutic and toxic doses closely approximate one another, or when the route of administration and the method of preparation are critical. Often herbals are written without concern for these important data and thus real or potential values of plants are diminished, lost, or their use may present unexpected hazards to the novice.

Phytolacca species are widely used for both food and medication, but they can also be dangerous. In rural parts of the southern United States young leaves of *P. americana* are frequently eaten as a part of salads and as a pot herbs after changing the water several times during cooking. Older leaves are also boiled, but greater care must be exercised in their use. Nevertheless, accidental poisonings occur even when extreme cautions are taken to reduce toxic properties, as evidenced by the recent illness of 20 persons in New Jersey who on the average of 3 hours after ingestion of cooked leaves experienced nausea, stomach cramps, vomiting, headache, dizziness, burning sensations in the stomach or mouth, and diarrhea (Callahan et al. 1981). Although most powdered or shaved root preparations of *P. americana* for use as herbal teas have been withdrawn from retail stores in the United States as a result of a reported poisoning associated with ingestion (Lewis & Smith 1979), some roots are still available and none should be ingested, for not only do they contain phytolaccigenin that causes gastroenteritis, often hemorrhagic, but also histamines that may result in life-threatening hypotension. The fruit and perhaps the roots possess powerful mitogens (PWM) that can indiscriminately stimulate the immune system producing an overabundance of white blood cells leading to death (Hsu 1979). Tinctures of the plant, however, were recognized as official in the U.S. Pharmacopoiea during the 19th century for use as antirheumatics, a practice that continues today in rural America where self-medication is more widely practiced and also in the practice of homeopathic medicine. In Africa this introduced American species is widely used medicinally (Watt & Breyer-Brandwijk 1962).

Cuando los usos de plantas se trasmiten de cultura a cultura, la información importante acerca de la dosificación puede perderse. En Phytolacca, por ejemplo, la dosis útil y la tóxica están muy cerca y han ocurrido algunos casos de envenenamiento.

All species of *Phytolacca* closely related to *P. americana* are probably toxic, though some are used as food, and some are potentially therapeutic with a close threshold between doses. Those grouped in the same subg. *Phytolacca* sect. *Phytolacca* (Nowicke 1968), particularly *P. rivinoides* and *P. icosandra*, whose young shoots are sometimes used as pot herbs by blacks in Panama, are nevertheless considered dangerous and ingestion of their roots or berries after a comparatively long interval produces symptoms of vomiting, convulsions, and in extreme cases death (Allen 1943). Pigs that accidently eat the roots of *P. rivinoides* may develop hemorrhagic gastroenteritis (Escobar 1972). In East Africa *P. dodecandra* is responsible for more poisonings than any other plant (Verdcourt & Trump 1969); roots are used in small quantities as a purgative, and against tapeworm, but many deaths to humans result from wrong dosages. Also from Africa are reports of fatal overdoses when using *P. heptandra* and *P. octandra* as medicinals (Watt & Breyer-Brandwijk 1962), and of severe poisonings on ingestion of *P. dioica* in Australia (Everist 1974). Clearly the toxicity of probably all species of *Phytolacca* must be weighed against the potential benefits either medically or as food before they are used.

A second instance would be the use of *Asclepias curassavica* for both food and medicine in Latin America. In Mexico the plant is sometimes used to flavor tortillas and throughout Central America it has been used medically for a variety of purposes (Morton 1977). Topically it is supposedly important for its astringent, hemostatic, and analgesic properties, and preparations are also ingested to induce vomiting and purging, and as a vermifuge. Its topical use is not without hazards, however, for cases of irritant dermatitis have been reported (Allen 1943), and should large amounts be ingested its emetic and purging properties suggest toxic principles that have proved fatal to cattle. One such compound is the cardiac glycoside calotropin that in Africa has been used as a lethal arrow poison and in the neotropics for suicides and for committing murder.

A third example is *Equisetum*, for its species are widely used for medicine even though toxicity is known. We observed *Equisetum* stems being sold in small bundles by street vendors in Panama City as Morton (1977) has reported for *E. bogotense*, *E. giganteum*, and their hybrid *E.* × *schaffneri*. She found that decoctions were popular diuretics and taken especially in cases of gonorrhea, yet various species have been linked to fatalities in grazing animals. *Equisetum* contains the alkaloid equisetine (a neurotoxin) and thiaminase that destroys vitamin B1 , so we agree wholeheartedly with Dr. Morton who states: "It is hard to believe that the people of Latin America can imbibe so much *Equisetum* with impunity. Surely its effect on the human system should be studied, especially in view of the current promotion of *E. hyemale* in the United States as a healthful tea."

Equisetum es tóxico, sin embargo es ampliamente utilizado como medicinal. Su uso excesivo puede ser fatal.

Plant Selections of Little or No Efficacy

Whereas the value of certain plants remains an enigma, with others the significance of reported efficacy is seriously doubted. We include in this category most panaceas and elixirs (perhaps not ginseng, Lewis & Elvin-Lewis 1977), plants used because of their shape and color (i.e., doctrine of signatures), and those recommended as aphrodisiacs and for snakebites.

Snakebite remedies, for example, abound where poisonous snakes are common, probably reflecting desperate attempts to treat the bites rather than an efficacy of the plant materials. Yet we should consider the possibility that washing or rubbing with onion or garlic, for instance, might be repugnant to some snakes and thus reduce the incidence of bites since no research has proven otherwise. Even so accidental contacts will be made, and there is no known topical or systemic remedy that will effectively neutralize snake venom activity. When venomous action is localized to the bite area, the activity of plant remedies may be somewhat effective, but in the absence of controlled experiments even this is speculative. A widely used plant in Panama is *Sansevieria* (Liliaceae) or the snake plant, a clear reference not to its value in treating snakebites, for no major secondary metabolite is known for the genus that could be significant, but rather to its leaf shape and thus to the doctrine of signatures. In fact no clinically documented evidence exists to corroborate any claims made for snakebite remedies in folk medicine.

En Panamá, Sansevieria representa la doctrina de los símbolos, ya que se denomina "planta culebra" en referencia a que sus hojas son similares a culebras. Esto ha causado que se utilice en tratamiento de las mordidas de culebras a pesar de que no se conoce que sea efectiva para tal fin.

Aphrodisiacs, those supposed stimulators of sexual drive (libido) and sexual performance (erection and orgasm), are another example of the doubtful value of plant products even though, like snakebite remedies, their use abounds in most cultures. Among the many plants selected as aphrodisiacs (Lewis & Elvin-Lewis 1977), many are nutritive in value, others contain volatile oils, and some possess compounds that affect the neuromuscular system, particularly by inducing a mild calming or tranquilizing effect and eliminating agitation or nervousness. The volatile oils undoubtedly stimulate respiration, raise blood pressure, and induce a feeling of well-being in some persons, but these substances may also be irritant and anesthetic and aid in obtaining an erection while delaying ejaculation. They are not aphrodisiacs *per se*.

Otro paralelismo entre Panamá y Nepal es el uso de Abrus precatorius como afrodísiaco. Es sumamente venenoso, si no se toma en dosis pequeñas.

Nevertheless, an interesting parallel exists between the natives of Panama and Nepal, for they both ingest the seeds of *Abrus precatorius* (Fabaceae) for sexual stimulation. The Indians of eastern Panama boil the seeds and roots, while the Nepalese eat the seeds (the means of seed preparation is not known). This is particularly important since the seeds of *Abrus* contain the highly toxic abrin that can be fatal on ingestion of only four raw seeds. Since *Abrus* was the only similar aphrodisiac among dozens of possible candidates, its use implies exciting possibilities if dose and pharmacological effects were better understood. Generally, the selection of aphrodisiacs tended to be more specific to a culture than other herbal preparations and thus remain the most exotic of the folk remedies.

Conclusion

Panamanian cultures are rich in medicinal plant folklore with an unrealized efficacious potential. Although most plant species have not been subject to pharmacological scrutiny, the discovery that active principles exist in related species utilized elsewhere suggest the rationale behind their use. With a completed *Flora of Panama* that provides the country with a floristic foundation found in few other regions of the neotropics, it is now possible to expand productive research in pharmaceutical botany. It must be emphasized, however, that unless care is taken to obtain as much information as soon as possible from herbal practitioners among both indigenous and adventive cultures, the medicinal wealth of the Panamanian flora may never be fully realized.

Los curanderos poseen la riqueza de la información sobre los usos medicinales de las plantas y ésta puede perderse a menos que se tomen las medidas adecuadas para preservarla.

Allen, P.H. 1943. Poisonous and injurious plants of Panama. Amer. J. Trop. Med. 23 (suppl.): 1–76.

Callahan, R., F. *Piccola*, K. *Gensheimer*, W.E. *Parkin*, J. *Prusakowski*, G. *Scheiber*, & S. *Henry*. 1981. Plant poisonings—New Jersey. Morbidity Mortality Weekly Rep. 30: 65–67.

Duke, J.A. 1968. Darien Ethnobotanical Dictionary. Battelle Memorial Institute, Columbus, OH.

Elvin-Lewis, M.P.F. 1980. Chewing-sponges for teeth cleaning. J. Prev. Dentistry 6: 75–80.

Everist, S.L. 1974. Poisonous Plants of Australia. Angus and Robertson, Sydney.

Escobar, N. 1972. Flora Tóxica de Panamá. Eupan, Panama.

Hsu, T.C. 1979. Human and Mammalian Cytogenetics: An Historical Perspecitve. Springer, New York.

Lewis, W.H. & M.P.F. *Elvin-Lewis*. 1977. Medical Botany: Plants Affecting Man's Health. Wiley Interscience, New York.

———. 1979. Systematic Botany and Medicine, pp. 24–31. *In* I. Hedberg, ed., Systematic Botany, Plant Utilization and Biosphere Conservation. Almqvist & Wiksell, Stockholm.

——— & P.R. *Smith*. 1979. Poke root herbal tea poisoning. J. Amer. Med. Assoc. 242: 2759–2760.

Morton, J.F. 1977. Some folk-medicine plants of Central American markets. Quart. J. Crude Drug Res. 15: 165–192.

Nowicke, J.W. 1968. Palynotoxonomic study of the Phytolaccaceae. Ann. Missouri Bot. Gard. 55: 294–364.

———. 1971. Rhamnaceae *In* R.E. Woodson et al. Flora of Panama. Ann. Missouri Bot. Gard. 58: 267–283.

Suwal, P.N. 1970. Medicinal Plants of Nepal. His Majesty's Government Press, Katmandu, Nepal.

Verdcourt, B. & E.C. *Trump*. 1969. Common Poisonous Plants of East Africa. Collins, London.

Watt, J.M. & M.G. *Breyer-Brandwijk*. 1962. Medicinal and Poisonous Plants of Southern and Eastern Africa, ed. 2. Livingstone, Edinburgh.

Indigenous Peoples

Panama is the home of many different groups of people. It is the indigenous people who show the most marked differences from social and philosophical patterns of European thought: they are also the most closely linked to the forests and natural landscape. Studies of indigenous inhabitants offer opportunity for testing the universality and validity of beliefs of the European tradition, and they also often show how different solutions are possible for similar problems within the same Isthmian environment. The Kuna have been the favorite group for study among the Indian groups in Panama, and by now serious studies of the Kuna tend to examine narrower questions than for the other groups. Four of the studies presented here examine different aspects of Kuna life, and when read together they give a picture of Kuna belief, lifestyle, and the evolution of the society over time. One study describes an activity of the Choco of eastern Panama's forests, and one study describes changes in the society of the Guaymi, the most populous Indian group in Panama.

Howe examines the bargain between succeeding generations that is struck by marriage within the Kuna family.

Moore investigates how Kuna government attempts to structure itself for relating to the central government of the land. Kuna language and social behavior is shaped in part by their perception of wild animals and plants and the magical features they possess.

Dina Sherzer discusses communication with natural spirits by magic specialists.

Joel Sherzer describes how objects have different names depending on the context within which they are viewed.

The paper by Sexton illustrates the simple but ingeneous solution of a Choco to a logistics problem. Both the problem and the solution will disappear as landscape alteration and acculturation of the Choco advance with time.

Young analyzes changes in Guaymi culture over time, giving an insight into the evolution of Guaymi life as it has experienced outside contact.

Each group discussed here is still culturally self sufficient but is inexorably, if reluctantly, destined to assimilate in major ways with the dominant culture of the isthmus. As this assimilation continues, will the indigenous groups become urbanites or campesinos with the same problems of economic displacement and attitudes to land clearing as other Panamanians? Or will some of their traditional wisdom and respect for the natural landscape prevail and permit wild plants and animals to coexist with man?

Población Indígena

Panamá alberga muchos grupos diferentes de gente, sin embargo son los indígenas los que muestran las diferencias más marcadas de los patrones sociales y filosóficos europeo. Asimismo, ellos constituyen el grupo que más estrechamente está ligado a los bosques y a las áreas naturales. El estudio de las poblaciones indígenas ofrece la oportunidad de probar la universalidad y validez de las creencias de la tradición europea y frecuentemente demuestra que soluciones diferentes son posibles para problemas similares dentro del mismo ambiente ístmico. En las investigaciones de los grupos indígenas de Panamá, los Cunas han sido los favoritos y actualmente los estudios más serios sobre este grupo tienden a examinar interrogantes más especificas que los realizados sobre cualquiera de los otros grupos. Cuatro de los trabajos, que se incluyen en este capítulo, examinan diversos aspectos de la vida Cuna; uno está relacionado con los Chocoes en los bosques del este de Panamá y otro describe los cambios que han ocurrido en la sociedad Guaymí, el grupo más populoso de Panamá.

Howe describe el pacto que se realiza entre

generaciones sucesivas de Cunas y que usualmente es sellado por un matrimonio dentro de la familia.

Moore explica cómo el gobierno de los Cunas intenta estructurarse en relación al gobierno central de Panamá. El lenguaje Cuna y el comportamiento social es configurado en parte por percepciones de animales y plantas silvestres y en las características mágicas que ellos poseen.

Dina Sherzer examina la comunicación de los especialistas en magia con los espíritus naturales.

Joel Sherzer describe cómo los objetos tienen diferentes nombres dependiendo del contexto dentro del cual se miren.

Cuando los trabajos anteriores son considerados en conjunto el resultado es un panorama claro de las creencias Cunas, de su estilo de vida y de la evolución de la sociedad en el tiempo.

Sexton ilustra la solución simple e ingeniosa que los Chocoes aplican a un problema de logística. Tanto el problema como la solución desaparecerán a medida que las áreas naturales sean alteradas y que la cultura Chocoe sea influenciada por la occidental.

Young examina los cambios que a través del tiempo ha experimentado la cultura Guaymí. El presenta información detallada de la evolución de la vida Guaymí al estar influida por corrientes externas.

Cada grupo de indígenas discutido aún se basta culturalmente pero inexorablemente, aunque renuentes, están destinados a asimilar, en forma general, aspectos de la cultura dominante del Istmo. A medida que la asimilación continúa sería interesante constatar si los indígenas constituirán un grupo urbano o campesino con los mismos problemas de desplazamiento económico y actitudes hacia la deforestación al igual que otros panameños o si prevalecerá su sabiduría tradicional en relación a áreas naturales que permitirá que plantas y animales silvestres coexistan con el hombre.

Marriage and Domestic Organization Among the San Blas Kuna

James Howe

El sistema de parentesco cuna está basado en la matrilocalidad, en la que la pareja casada vive en la casa de la madre de la novia y el novio trabaja para el suegro. El sistema ha sido modificado recientemente, pero, tradicionalmente, la separación de las jóvenes fue la clave que las hizo un objeto que motivó a los jóvenes al trabajo y dependencia de la generación de mayor edad. Los Cunas aún mantienen este ideal, pero las unidades familiares son más pequeñas e independientes y la separación de los solteros basada en el sexo ha disminuido.

Kinship plays a restricted but important role in the organization of social life among the San Blas Kuna of Panama. Kin ties do not dominate either politics or ritual, and except for secondary land-holding units found in some villages (Howe 1976), corporate kin groups outside the household are notable for their absence. In the domestic sphere however, and especially in production, consumption, co-residence, and child-rearing, kinship is as important in San Blas as anywhere else. Older sources on this topic include Nordenskiöld et al. (1938: 28–43), Puig (1948: 109–15), Stout (1947: 25–28), and Torres de Ianello (1957: 23–37). More recent work comes from Holloman (1969: 136–97, 1976), Howe (1974: 40–47, 1975, 1976, 1981), Prestán (1975: 61–106), Costello (1975: 61–108), Helms (1976), and Stier (1979: 212–22, 291–347).

This article brings together scattered data on household composition, and it elaborates on the development cycle of Kuna households, based in part on new material.

The matrilocal household is the key unit in the Kuna kinship system. The term matrilocal (or uxorilocal), usually refers to a rule of postmarital residence where a bridegroom goes to live with his wife in her parents' home (Murdock 1949: 16, 34). In Regina Holloman's examination of Kuna household dynamics (1969: 159–97, 1976), she analyzes the matrilocal household largely in terms of the relations between the male household head and his in-married sons-in-law, and all later analyses must take her basic insight into account. Her point of view fits with Oberg's (1955: 481) analysis of matrilocality in lowland South America as a device by which senior men can exploit the labor and support of junior men (Turner 1979: 157–64; Bohannan 1957, Kracke 1976). In extending Holloman's analysis, I focus on repetitive cyclical modifications to household form (Fortes 1958, Goody 1958) and on historical transformations usually considered under "social change." I begin with the early years of this century as described to me by elderly informants in the 1970's, after which I trace changes over the intervening years. Because parts of the past practice still apply today, I sometimes use the present tense.

The Way It Was

It would be foolish to assume that Kuna kinship in 1900 or 1920 was the same as it had been in 1800 or 1700. Cultures change continuously, and at least one element of the system as it was in 1900, private property in land, had appeared only a few years previously: the turn of the century however, is the period whose culture the Kuna now see as traditional, and the first for which we have fairly reliable information on kinship practices.

In those days it was expected that young children would help with at least a little work for the household. Indeed, when indicating the sex of a newborn child, the Kuna still refer to a male as a "fisherman" *(ua soet)* and a female as a "water filler" *(tii palet)*. Girls' work, which consisted largely of household tasks and trips to the river for water with their mothers and aunts, started earlier than did boys' work. The boys however, began fishing and accompanying their fathers and brothers to the forest at an early age, and by adolescence, both sexes were making significant contributions to the household economy. Neither sex attended school, since there were no schools to attend.

Tradicionalmente, las jóvenes cunas comienzan a cargar agua desde temprana edad, mientras que los jóvenes empiezan a pescar con sus padres ligeramente más tarde. Las muchachas se casan uno o dos años después de la pubertad, mientras que los jóvenes esperan entre 3 a 10 años después de ésta, para hacerlo. Algunos salen de la comunidad para obtener trabajo y otros permanecen en el área y trabajan en las tierras de los padres.

With adolescence, the life cycle of males and females moved out of phase. At puberty, a young maiden *(yaakwa)* underwent a series of rituals, and in most cases she married within a year or two afterwards. Youths *(sappinkana)* on the other hand, usually waited to marry, creating an age differential between spouses of anywhere from three to ten years.

What were these boys doing between the onset of puberty and the time they were thought ready for marriage? Some left San Blas for a year or more of wage labor, but they mostly stayed at home and worked in subsistence agriculture. In addition to working with their fathers on parental lands, they were expected to clear their own fields on previously unclaimed land, and to plant corn, rice, bananas, yams, and other crops, which met a considerable portion of their households' needs. Meanwhile the fathers of the young girls of the village were watching the boys and assessing their industriousness.

During all this time, the boys were carefully kept away from the young unmarried girls of the village. Girls spent most of their time secluded at home, and if they did go out in the streets, it was only with an escort and their faces completely covered with a headcloth. Boys were kept busy, during the morning and early afternoon working in the forest or fishing, and at night in the village meeting house or with a teacher learning ritual. In the hours between returning from the forest and going to the meeting hall, they were welcome as visitors only in the homes of kinsmen and family friends, and even there they were kept away from young girls and prevented from talking with them. Adolescent boys and young men were warned off from households that they visited too often or for the wrong reasons, and even in their parents' homes they were not made especially welcome. Instead, they were encouraged to spend the late afternoon together in a group, typically at a spot where the canoes were hauled up, talking and playing panpipes.

Las jóvenes solteras son apartadas y segregadas de los jóvenes, quienes están ocupados pescando o cultivando y quienes, además, no son bien recibidos en las casas donde están las jóvenes.

Marriage was arranged by the older generation. Typically, a girl's parents went secretly to the home of a youth to ask his parents for his hand. Potential fathers-in-law looked over the crop of young men, choosing husbands for their daughters in large part on the basis of how much they had cleared and planted in the forest.

Kuna weddings involve marriage by capture in which it is the groom who is caught. A number of authors have described Kuna marriage practices (Coope 1917: 167–69, Nordenskiöld et al. 1938: 28–30, Stout 1947: 37–38, Puig 1948: 109–13, Gassó 1948: 113–15, Torres de Ianello 1957: 23–25, Reverte 1967: 12–16, Prestán 1975: 91–94), but none has explicitly related wedding symbolism to the matrilocal complex.

The basic elements of "the dragging" *(kakaleti)*, as the marriage ritual is called, are the following:

1. A crew of young men ambush the unsuspecting groom, either in the streets or while he is drinking at a puberty ceremony, and carry him off to his bride's house, where the two are thrown in a hammock and swung back and forth. He flees, pretends to flee, or is carried back outside, and three times more he is hauled back inside and swung with his bride in the hammock. Today, when the couple have in fact chosen each other (see below) and the groom only pretends to resist, the dragging merely *symbolizes* the groom's reluctance and his subordinate position in his wife's household. In the past, when grooms had no advance warning and many fought back hard, the struggle *manifested* the young man's subordination and lack of choice: the symbolism, in other words was an inherent part of what it symbolized.

2. After the groom returns home from the dragging, his new father-in-law comes to fetch him, taking him forcefully by the arm, leading him back to his own house, and putting him in a hammock with his daughter. The couple spends the night together, immediately next to the hammock of the girl's parents, who make sure that the couple do not sleep. As with the dragging, these actions manifest the balance of power in the household. In some villages this second episode is repeated four times, with the couple stay-

ing together for a longer time each night. In others, they are brought together only once, after four nights of dragging and hammocking. Either way, in the past the repetition over several days gave the young man some choice in the matter, allowing him to flee the island for a while if he strongly objected to the match. Today, many islands have eliminated the repetitions, collapsing the whole ritual into a single night.

Las bodas son arregladas por los padres. Primeramente, el novio es "arrastrado" por un grupo de jóvenes hacia la casa de la novia para mecerse junto con ella en una hamaca. La oposición del novio simboliza y, en algunos casos, manifiesta su subordinación y la carencia de alternativas. Después de que el novio regresa a su casa, el nuevo suegro lo viene a buscar y lo pone en una hamaca con su hija durante la noche. La hamaca está ubicada al lado de la de los padres. Después de repetir el proceso varias veces, el joven se dirige al bosque a cortar madera la que compartirá con su suegro.

3. After the fourth night, the young man goes off to the forest, where he cuts a certain number of logs (12, 16 or 20) and a certain amount of kindling wood. A portion of the wood (one-half or two-thirds) is left behind in the forest and the rest is brought back to the island. In some variants of the ritual, father-in-law and son-in-law cut the wood together; in others, the son-in-law goes off alone, but on his return the father-in-law carries the wood from the canoe to the kitchen house. Contrary to some sources (Puig 1948: 112, Torres de Ianello 1957: 24), this task is not a serious test which the groom might conceivably fail. Rather it symbolizes and manifests the labors expected of the groom, along with the fact that he works, not alone, but in cooperation with his wife's father.

4. While the young man is away cutting wood, his wife visits his mother, bringing a sweet drink, and she repeats the visits for several mornings thereafter. The visits and gifts symbolize the somewhat more limited connection between the bride and her husband's family, and on at least some islands in the past her manner of dressing (in underskirt but not overskirt) symbolized her virginity. The groom returns repeatedly to his natal home after cutting the wood, each time being called back by his wife to eat and drink, and later in the day, the couple goes together to fetch his personal belongings.

5. That night the pair go to the village meeting hall, where several village leaders admonish them in turn on their rights and duties as a married couple (Salcedo 1980).

Mientras el novio está ausente, la novia visitará la suegra durante varias mañanas. El joven regresa a casa de su madre, pero su novia lo lleva a comer y tomar a su casa. La pareja, entonces, es advertida por los mayores de sus derechos y deberes. Finalmente, el joven se muda a la casa de la novia y comienza a trabajar para el suegro, a la vez que en sus propias tierras.

After marrying and moving into his wife's home, a young man began to work for her father. The head of the household would apportion its work, giving his sons-in-law opportunities to cultivate their own lands and their parents' as well as his own. On some days each man would work separately on a different task, one or two fishing for instance, one cutting bananas, and one gathering coconuts. On other days the father-in-law and his sons-in-law would work together, and on such occasions they followed a formal protocol: on a forest path, they would march in single file in order of seniority, the youngest first and the father-in-law bringing up the rear, and in a canoe they would seat themselves in the same order, with the father-in-law steering in the stern, unless he wished to rest, in which case he would sit in the middle and let the senior son-in-law steer.

Once their daughter had married, her parents no longer had exclusive control over access to her. Indeed, the formalized admonishments of newlyweds state that her parents should not interfere with their marital relations (Salcedo 1980). Her parents did however, continue to exercise influence over her actions, and in particular they could keep her from leaving the household under normal circumstances. Her father continued to buy her clothes and to contribute to her expenses and those of her children. Her father moreover, continued to be the first authority in household disputes. A marriage under such circumstances tends in its early years to be strained and brittle. Temporary separations, in which the son-in-law returned to his parental home after a quarrel, were and are common, and divorce moderately frequent. Here again, fathers-in-law exerted a great deal of influence. They could decide whether or not to take a son-in-law back after a separation, and occasionally they could even throw one out against a daughter's wishes.

Over the years of marriage, the relationship between father-in-law and son-in-law moved from one of dominance to relative equality. The son-in-law increasingly took the initiative in setting his own work, and although he continued to consult with his father-in-law, the latter issued fewer orders than before.

Eventually the headship of the house passed from father-in-law to the most senior of the sons-in-law still living with him, as the son-in-law's daughters married, the father-in-law began to retire from heavy labor, and the son-in-law cut the principal timbers in a periodic rebuilding of the house. Such transfers however, were often gradual and only tacitly acknowledged until the father-in-law reached an advanced age.

A través de los años el yerno gana mayor igualdad y a la postre la jefatura de la casa pasa del suegro al más antiguo de los yernos que viven con él. La transferencia de autoridad es gradual.

Meanwhile, if the father-in-law had multiple sons-in-law, and if the island had free land left for building, some of the sons-in-law might be given permission to found their own households when their daughters were ready to marry or the house was overly crowded or quarrelsome. (The Kuna say however, that fathers-in-law let daughters go much less readily then than now.) Most often the sons-in-law left, if they did, in order of seniority, eldest first, juniors later. At the death of the household head, the senior son-in-law continued on in his place, with his late father-in-law's widow as senior woman of the house. The other sons-in-law were free to leave, to found their own households or even to rejoin their parents if they wished. If these men got along well however, or if alternative houses or house sites were in short supply, they might stay together as a household of in-married brothers-in-law *(ampesukan)*. Either way, the cycle that began with the birth of the household head's first daughter had at this point completed itself.

The Way It Was

As noted at the beginning of this paper, the theory of Oberg (1955: 481) and Turner (1979: 157–64), which depicts matrilocality as a device by which senior males can exploit the labor and support of junior males through their control of junior females, works well for the Kuna. In more general terms, Kuna matrilocality makes sense within the perspective, often identified with Claude Lévi-Strauss (1969), which treats marriage as a form of exchange. Still more generally, the matrilocal complex exemplifies theories that treat almost all forms of social relationship and interaction as exchanges (Homans 1958, Blau 1964, Ekeh 1974, Heath 1976).

The Kuna have elaborated their own version of exchange theory. A household, they say, gives up sons in order to get sons-in-law. With a married daughter, parents exchange the cost and effort of raising her, their continuing support for her children and herself, and the money and time they expend on her puberty ceremonies and jewelry, in return for her continued presence in their household after she marries. If a married daughter and her husband try to move out of her parents' house without their blessing, someone will inevitably bring up her indebtedness for the cost of the puberty ceremonies, and if the young couple does get permission to break away, in many cases the village will force them to reimburse her parents.

A son-in-law, according to this indigenous exchange theory, receives a wife from her parents, in return for which he gives work, diffuse support of the household, his presence in it, and acknowledgment of her parents' senior status as household heads. In addition to the wife he receives, a son-in-law is thought to work for the property his wife will eventually inherit. Typically, a mature couple turns over its lands to its adult children over a long period of time, often beginning with one or two plots given out when a child marries, saving the final pieces of land to be shared out after they die. A son-in-law never becomes the legal owner of the lands his wife inherits, but he manages them for her as part of their combined marital estate, and their children will eventually inherit from both of them. In many cases a young man may be working on land that still belongs to his father or father-in-law but which has been promised to his wife or himself, and in such cases the Kuna say he is really working for himself as much as for them. In general, private landholding and delayed inheritance offer a strong means of control to the senior generation.

La matrilocalidad está basada en un complejo intercambio de valores, pero tiende a favorecer a las generaciones más antiguas.

Notice that although Kuna matrilocality is based on a complex of balanced exchanges, ultimately the system tilts heavily in favor of the senior generation. The giving and taking can be seen as balanced, but only if the legitimacy is conceded of one party's prior control of crucial resources and its right to set exchange values. In effect, many such systems are rigged ahead of time, a fact that exchange theorists do not always acknowledge.

Notice also that although the senior generation reaps the benefits of this arrangement when daughters marry, the system in the early years of this century depended on the management of *unmarried* males and females. All of the parental efforts—

secluding young girls, hiding their faces, protecting their virginity—were the artificial means by which the senior generation turned marriageable maidens into a scarce good, one they controlled. Thus, even today, the Kuna put heavy emphasis on a bride's virginity, and they work harder at preventing premarital sex than adultery: a married woman's adventures do not threaten the system to the same extent.

The senior generation realized a number of advantages from secluding young girls and forcing young unmarried men to compete for the attention of potential parents-in-law. By the time they married, young men were fully socialized into hard work and productivity. They came to marriage moreover, with producing fields. And by staggering the age at marriage of the sexes, parents were able to bring into their household a fully mature, adult male worker only 14 or 15 years after the birth of their first daughter, when she herself had barely passed puberty.

Anteriormente, durante la separación de las jóvenes solteras de los jóvenes, a las muchachas se les proveyó poca comida, ésta controlada por los mayores para elegir la productividad de los hombres.

This understanding of how the matrilocal system worked, once set out, must be modified in certain important respects. First, the analysis does not depend on the assumption that things worked perfectly. When elderly informants claim that in their youth it was parents who invariably arranged matches for their children, and that young people were much more diligent, obedient, respectful, and chaste in those days, one need not take every word literally. On the contrary, one may assume that teenaged boys occasionally caught glimpses of young girls, some boys worked less hard than they might have, couples sometimes chose each other, and that a few girls got pregnant first and married second. Nonetheless, the ideals now fondly remembered as fact were those that structured the system, those it was designed to perpetuate, and one may conclude that it approximated those ends better than it does now.

The analysis must further be qualified in terms of the relationship between father-in-law and son-in-law, which was and is a partnership as well as an exchange. The two men share a common interest in the welfare of the young man's wife and his children, who are the older man's grandchildren, and if the two cooperate as the Kuna believe they should, they are building a patrimony for the grandchildren when they become adults. And as noted before, with the passage of years, the dominance of the father-in-law gradually diminishes, leaving partnership as the primary element in their relationship.

At the turn of the century moreover, the Kuna were going through a period of rapid territorial expansion. As a people, they had only recently moved from the river valleys of the northern slope out onto the coast and islands of San Blas. Considerable amounts of unclaimed land could still be found near most villages, but agriculturalists were rapiding clearing and planting both the forest and uninhabited islands, thereafter alienating these plots of land as private property. By encouraging young men to begin agricultural work and land clearing at an early age, the matrilocal system benefitted them and their descendents just as much in the long run as it did their fathers-in-law. Given moreover, the care and farsightedness with which the Kuna generally plan ahead in such matters, one may conclude that the fathers-in-law had these ends in mind.

Al comienzo del siglo, tierras aún sin reclamar motivaron que los hombres limpiaran un pedazo de tierra. Esto beneficiaba a los jóvenes y a sus hijos, al igual que al suegro.

Similarly, my emphasis on the relationship between father-in-law and son-in-law should not obscure the fact that in terms of managing their estates and sons-in-law, senior men and women work together. Although possession of a house reverts to the senior woman of a household in the event of a divorce, her husband is otherwise most often formally identified as the *nekipet* "owner/master of the house," and under normal circumstances, the two hold both house and farm lands together as a couple. Just as a senior couple acts as a unit vis-á-vis their daughters and sons-in-law, so a junior couple comes to act as a unit vis-á-vis other couples and generations.

Relations among females also play a crucial part in household organization. More is involved than merely the indispensibility of their presence and their work. As Helms, (1976) points out, women, because they seldom travel and almost never work away from home, constitute the continuously resident membership of a household, in effect its stable core (Prestán 1975: 71). Female children moreover, are socialized continuously and gradually into their adult roles by working with their mothers, sisters, and aunts, and after marrying, they continue working with the same set of women.

Las relaciones entre las mujeres son cruciales, porque rara vez viajan o trabajan fuera de la casa y ellas representan el centro estable de la

casa, mientras que los hombres se pueden considerar los satélites.

Helms suggests that household males are satellites or adherents of a stable female core and that co-resident women are linked by consanguineal ties while in-married males relate to the household primarily as affines. Male interests, especially politics and economic production, she sees as fragmenting and individualistic, while females symbolize and embody household continuity and solidarity. Following Cohen (1969), she sees kinship in all societies as a set of symbols representing and embodying stability, continuity, and general values, in contrast with politics and economics, which are manipulative and hedonistic. Thus among the Kuna she finds, "highly cohesive core groups of consanguineous women which serve as a stabilizing symbolic foci" for their households (1976: 134, 137).

Helm's argument needs qualification. If women symbolize household solidarity, they do so only tacitly and in certain frames of reference. The Kuna talk a good deal about kin relations and make extensive use of kinship metaphors, but my notes at least, do not indicate any explicit symbolic use of the unity of household women. I found no trace moreover, of a lineal ideology or emphasis on matrilines, even tacit ones. It may be moreover, that a young man initially relates to his wife's household as an affine, but as soon as he and his wife have children, he not only relates to the children as a consanguine, as much as possible he phrases his relationship to his wife's parents in terms of the connection all of them share with the children. Often almost everyone in the household refers to the senior couple as grandmother *(muu* and grandfather *(tata)*. Household heads moreover, relate to *everyone* except their daughters' husbands as consanguines, and such senior men are in every sense full members of the household core and symbols of its continuity and permanence.

No se enfatizan las líneas femeninas y una pareja trata sus intereses e identidad como diferentes a los de otras parejas.

It is also doubtful whether the Kuna see female householders as more unifying and less divisive than males, either in symbol or in fact. Even in the first tenuous period of marriage, a couple treats its interests and identity as different and even opposed to those of other couples, at least in some contexts. Especially after she has children, a woman acts in terms of the separate interests of her nuclear family at least as much as her husband does. Sisters, aunts, and nieces quarrel fairly often about work, privileges, and children, and such quarrels are often cited as the reason why couples have been allowed to move out and form their own households. Even without quarrels, in the later stages of household development, sisters often go their separate ways.

Even the emphasis on household organization can obscure the importance of other key aspects of the kinship system. As noted above, males frequently return home to their parents' home, especially in the first years of marriage, and the ties between out-married sons and their parents continue strong all the way to the parents' death. I have not carried out the specialized fieldwork necessary to identify the dominant symbols (Schneider 1968) or the dominant dyads (Hsu 1971) in the Kuna kinship system, but I feel certain that such a study would point not to the mother-daughter relationship in particular or to ties between co-resident women, but rather to the connection between parents (especially mother) and children of *both* sexes (Prestán 1975: 61–106). Not only do these ties and the family units they demarcate endure after children have married and dispersed, but wider cognatic stocks descended from grandparents and great-grandparents are important in organizing some activities and in sentiment (Howe 1976). Throughout, kinship reckoning is thoroughly bilateral. Thus, the unit and principle given attention here, the matrilocal household, crosscuts others also crucially important, the parent-child dyad and the cognatic stock.

Change

What has happened to the Kuna matrilocal complex in the last 60 or 80 years? To hear older informants tell it, everything has fallen into disarray, and nothing is as it should be. To the outside observer it may seem as if things continue to work reasonably well. Nonetheless, a great deal *has* changed.

First of all, households have gotten smaller. In the past many houses held twenty or more inhabitants, with several married daughters, their husbands, and their children, all under one roof. Today average household size in villages for which information is available (see below) falls between eight and ten people, and average size for larger households is about twelve or thirteen. Households of more than fifteen people are statistically uncommon, and those with more than twenty are rare.

Durante los últimos 60 años los núcleos familiares se han hecho más pequeños y el número

de parejas casadas que viven solas ha aumentado.

In interpreting informants' claims, one must take into account, not only their nostalgia for their youth, but also the tendency to focus on the extremes and positive cases. Nonetheless, the persistence with which informants point to a severe reduction in household size gives credence to their claims, as do Mac Chapin's (pers. comm.) detailed investigations into the composition of specific past households on one island. Undoubtedly, households have grown smaller for the most part because household heads now let daughters and sons-in-law move out sooner than was customary in the past. Informants recall households with five, six, and seven resident sons-in-law (though mean size may have been smaller), while today households with more than two sons-in-law are rare.

Along with a reduction in household size has come an increase in the number of small nuclear family households centered on a single married couple (see below). The prevalence of nuclear family homes (separate households, with their own cooking arrangements, as opposed to huts or houses which function as satellites of extended family households) is variable. They seem to be most common on a few highly acculturated islands (Holloman 1969: 193–94, Costello 1975), while on a few others they seem to be almost unknown (Holloman 1976). Without a reliable demographic baseline, one cannot be sure how rare nuclear family households were in the past, but it seems certain they were much less common than they now are on many islands.

On most islands for which information is available, it is still almost unknown for a young man to get permission to set up a nuclear family household immediately upon marrying. On my primary field site, the average age for heads of nuclear family households in 1970 was forty. In all examined cases the married couple had previously resided in an extended family household. Within a few years moreover, many of these couples had acquired sons-in-law, converting nuclear family households into matrilocal ones. Even on Costello's acculturated field site, many nuclear family households represent only a stage preceded and followed by matrilocality (Costello 1975: 82, 91). Thus the increasing number of these households should probably be interpreted as a loosening of the domestic cycle rather than as a radical change.

Households are not only smaller than in the past, they are less autonomous. Although many daily frictions and problems continue to be resolved within the household, others are appealed to village gatherings for adjudication (Howe 1974). Especially in cases of serious discord, divorce, separation, reunion, and remarriage, village leaders intervene and preempt the household head's powers much more than in the past. They attempt to bolster his authority, by for instance reminding junior household members to heed him, and by getting his permission before allowing his sons-in-law to leave for wage labor. In general they work to reinforce and preserve the matrilocal system. On the other hand, they also enforce the rights of junior household members, and occasionally they will even allow a young couple to move out against the wife's father's wishes if it is decided that he has not been treating them properly. Even when adjudicators back him up, each intervention weakens his capability for independent action.

La familia es ahora menos autónoma y muchos de sus problemas apelan a las reuniones de las aldeas, para que éstos sean juzgados por la comunidad. Actualmente los líderes de las aldeas intervienen más en casos de seria discordia, divorcio, separaciones, reuniones y de parejas que se casan por segunda vez.

Even outside the context of household quarrels, the power of fathers-in-law has diminished. They still hold formal authority, and they still channel the labor of their sons-in-law, but the latter are more self-directed and less willing to listen, and the former are more hesitant about giving direct orders. The gradual change in their relationship over the course of the developmental cycle, from inequality to partnership, progresses more rapidly now than in the past, and even allowing for exaggeration, clearly "work has become more a matter of free choice" *(arpaet pur ittoleket kusa)*.

Fathers-in-law also feel that more of the burden of household work falls on them than should. In their view, sons-in-law are all too likely to go away for years at a time, leaving them with extra lands to manage and extra mouths to feed. "You forget the wife you have, you forget the children you have, . . . you lose a lot of money, and you come back without a penny."(from a speech at a local meeting). The degree to which older men are left supporting households single-handed and to which their sons-in-law do or do not send home remittances varies widely. Many household heads arrange matters so that at least one son-in-law or grown son is always at home, but in other cases, elderly men who according to Kuna norms should be taking it easy are still toiling unassisted.

Los suegros sienten que actualmente reciben menos ayuda de parte de sus yernos.

Households finally, dominate the economy less than before. Early in the century, a man's household and his father's household organized almost all of his productive effort, and indeed almost all of his labor. He worked at house-building and a few other tasks organized by the village, but otherwise he worked with kinsmen and housemates. Today, most men belong to voluntary societies, organized for such economic ends as cash-cropping coconuts and running stores and restaurants. Village communal labor moreover, now requires a great deal more time and effort than it did, and villages themselves now own coconut groves and stores.

Contrary to the views of the gloomiest informants, the system is not the verge of collapse. Children are reared and enculturated, coconuts are planted and harvested, people are housed and nourished quite well by third world standards, and the economy continues to expand. A relatively small amount of work by men and women meets household needs (Holloman 1976: 146), leaving considerable time free for other pursuits, whether they are migrant wage labor, village communal work, political and religious meetings, ritual, sports, dancing, or *mola*-sewing. Most people reside in matrilocal households, the senior generation is still in charge, and most household heads still have considerable influence over the labor of their sons-in-law. But there is no doubt that households are less important than they were, and that the balance of power within them has shifted. Household heads manage smaller domestic groups, exact less labor, have less say, and work longer into old age. Why?

El sistema aún está muy vigente y la mayoría de la gente vive todavía en el sistema de matrilocalidad, sin embargo la familia es menos importante y más pequeña.

The Causes of Change

Because the modifications in domestic organization traced here form part of a wider social transformation, a great many factors have affected the household. The expansion of village government and communal labor for instance, and the growth of voluntary societies, have taken away some of economic and social control functions that once belonged to the household. The ideals of personal choice, romantic love, and the independent monogamous family enter San Blas through a great many media, affecting the thinking of relatively traditional men and women as well as of highly educated and acculturated ones. Outside agencies sometimes oppose or discourage matrilocal residence.

Three factors however—schooling, migrant labor, and the disappearance of free land seem especially important. Most villages in San Blas now have elementary schools, attended by children of both sexes, and some have had schools for several decades. These schools do more than teach reading and writing. They change the attitude of young people towards their elders, and they make older people, most of whom are illiterate, less sure of their position. Schools keep young boys on-island when they would have been working in the forest and competing for the attention of a father-in-law. Although most boys still become agriculturalists, they start later in life. Most important perhaps, schools have put an end to the seclusion of unmarried Kuna girls. Boys now know what the village girls look like. Although efforts are still made to keep boys and girls apart as much as possible, it is nearly impossible to prevent them from forming romantic attachments, and for most marriages today, it is the couple who choose each other, not the parents. Indeed, in a number of cases, couples are forced to marry by the girl's pregnancy. To sum up, the exchange between father-in-law and son-in-law has altered radically, because parents no longer monopolize access to their marriageable daughters.

Los ideales de elección personal, el amor romántico y la familia independiente y monógama entraron a la cultura de San Blas a través de muchas vías. La enseñanza en las escuelas, las labores migratorias y la desaparición de tierra libre han sido importantes. Las escuelas han finalizado con la separación de las jóvenes cunas solteras. Las labores asalariadas han cambiado las relaciones entre suegros y yernos. La ausencia de tierras ha limitado el valor de la labor de los adolescentes.

Wage labor also changes the relationship between father-in-law and son-in-law. Boys now leave home for wage labor much more often than in the past (Stier 1979), and many spend as much of their adolescence in Panama and Colon as in San Blas. While at home, they work in the forest as before, but they are often marking time until their next trip away. After marrying, they are likely to make more trips outside San Blas, where they are largely outside the control of their fathers-in-law. Dutiful absentee sons-in-law who contribute to their household in cash and gifts, or who return home with major items such as beds, sewing machines, or outboard motors, thereby affect the relative proportion of their father-in-law's economic support vis-a-vis their own. And regardless of how dutiful they may be, they usually leave the job of feeding the household to the men left behind.

The lack of land still in the public domain also limits the value of adolescent labor. Today in most parts of San Blas, the unclaimed land, if there is any, lies so far from the nearest village as to be economically marginal. Young men may be given access to a fallow field or two by their fathers or fathers-in-law before or after marriage, fields which they are expected to plant to coconuts, but otherwise a young man's agricultural labor benefits himself a good deal less than it did when unclaimed land lay close at hand. The matrilocal system, in its configuration at the turn of the century, was adjusted to a period of rapid territorial expansion, and when the expansion ended, so did some aspects of the system.

La escasez de tierras conlleva a que la pareja joven sea más dependiente de las parcelas que heredan, tanto para la agricultura, como para el sitio en donde construir su vivienda.

On the other hand, some factors supporting traditional arrangements and the power of the senior generation have held strong or even grown stronger. The disappearance of unclaimed land makes young people doubly dependent on the parcels they will eventually inherit, except in the case of the small educated minority with salaried positions. And, as noted before, the senior generation's control of land tenure offers perhaps its strongest source of influence over the junior one.

Agricultural land is not the only kind in short supply. Most island villages in San Blas have used up all of the space available for building new houses, and a young man who breaks free from the household of his wife's father often has no place to go except back to his own father. In quite a few cases however, sons-in-law have been able to at least build satellite nuclear family sleeping huts alongside the main house, and in some cases several smaller independent houses have been constructed on the site previously occupied by a large matrilocal household. The land shortage can be doubled-edged however. A man eager enough to found his own household can sometimes create land by filling in shallow water with rubble, and house sites created in this way are often inherited in nontraditional ways (Costello 1975: 61–108).

Absentee wage labor, at the same time that it takes young men out from under the direct control of the senior generation, makes them dependent on that generation for care of the wives and children left behind, and as Holloman notes (1976: 145), this dependence strongly favors matrilocality and disfavors neolocality. Even if a neolocal household headed by a grass widow had sufficient cash resources to purchase all its food, it would be especially vulnerable to sickness and other crises, and indeed even to minor day-to-day difficulties. Kuna women do not like living without other adults, and most men are strongly opposed to leaving their wives alone. Not surprisingly, most heads of neolocal households either hold salaried employment in San Blas (see Swain 1978: 235–36), they have given up leaving San Blas for wage labor, or (in recent years) they take their families with them to the city. As Helms (1976, 1970), Casselberry and Valavanes (1976), and others (Miller 1967, Harris 1979: 97) have pointed out, matrilocality is adaptive whenever males make frequent extended trips from home, whether for war, trade, or wage labor, and there is every reason to believe that matrilocality served this function among the Kuna even in past centuries.

A las mujeres cunas no les gusta vivir sin otros adultos presentes y la mayoría de los hombres se oponen a dejar solas sus mujeres.

The advantages of matrilocality, in fact, hold even when husbands are at home. Nuclear family households forfeit the economies of scale in work and the opportunity to allocate and rotate tasks efficiently enjoyed by extended families, and they are more vulnerable to the periodic dislocations and crises caused by illness, economic difficulties, guests, travel by household members, and other events of daily life. Many nuclear family households call on their ex-housemates for help quite frequently, and Swain for one, has cast doubt on the supposed independence of many of the nuclear family households on her field site (Swain 1978: 235).

Perhaps less obvious in its bearing on the matrilocal complex is the inflation in the expense of sponsoring puberty ceremonies. Costs have increased over the years to the extent that the long four-day ceremony *(inna suit)*, which ideally every girl should have, costs her father at least several hundred and sometimes more than a thousand dollars (Prestán 1975: appendix). Kinsmen assist a girl's father during the ceremony, housemates help him in collecting fish and game ahead of time, and the village supplies a large part of the needed food and labor, but the financial burden of the ceremony, along with a considerable portion of the planning and provisioning, falls almost entirely on him alone. Meeting this obligation requires a good deal of advance planning, especially from men with several daughters. Such men often go away for wage labor as their daughters' puberty approaches, and some men now sponsor one of the required ceremonies for a daughter *before* puberty, if they have the money in hand.

Los cargos financieros de los ritos de pubertad recaen sobre el padre, quien algunas veces viaja para trabajar por un salario que le ayude a cubrir estos gastos.

Fathers are willing to undergo this major burden and to spend further amounts on gold jewelry and clothes for their daughters in part, I suggest because it strengthens their control over their daughters and ultimately over their sons-in-law, who must pay them back if they wish to break free in the early years of marriage. The greater the cost of the ceremonies, the greater the effort needed to break free. Some girls in recent years have had an abbreviated version of the full sequence of puberty ceremonies or even none at all. But for many families that still sponsor the normal sequence, the cost of the ceremonies continues as a means of control over their daughters.

TABLE 1
KUNA HOUSEHOLD TYPES, SITE 1—1967

	No. of Households
A.1 Prototype Matrilocal, Father-in-law Household: married couple in senior generation, at least one adult daughter & her husband	8
A.2 Mother-in-law Household: married couple plus wife's divorced or widowed mother	5
A.3 Brothers-in-law Household: at least two sisters & husbands but no surviving parents of sisters	3
A.4 Absentee Son-in-law Household: same as A.1 except that a son-in-law was absent at time censused	5
B.1 Common Kitchen Household, same as A.1 except that at least one daughter and husband use a separate sleeping hut	5
B.2 Common Kitchen Household with Patrilocality: same as A.1 in main sleeping house, son of senior couple & wife in separate sleeping hut	2
C.1 Independent Conjugal Household Nuclear Family	1

[a]In some cases includes new husband of senior woman.
Source: Holloman 1976: 140–141.

Household Composition

In the decade and a half since Holloman's pioneering fieldwork, census data on household composition have accumulated for villages in San Blas. Although different typologies, different enumeration and data-gathering methods, and probably different criteria for what constitutes a household, as well as other discrepancies make it impossible to compare results from different Kuna communities in detail, still, with some adjustment and reconciliation the data convey the range of variation and some central tendencies in Kuna household form.

In 1967, Holloman censused 29 out of 86 households on her primary field site, a large island of over 1000 inhabitants in central-eastern San Blas (Holloman 1976, 1969: 184–92). Her sample is judgmental rather than random, tapping the range of household types rather than indicating their frequency. Table 1 gives her results. In 1970, June Howe and I censused 79 out of 80 households on a medium-large island of more than 600 inhabitants in central-west San Blas. This previously unpublished data is reported in Table 2. The typology is compatible with Holloman's but more detailed. The principal difference between the two classifications is that I have not distinguished separate types according to whether the household uses one or several sleeping houses, or whether a household member was residing away from home at the time it was censused.

In 1967, Reina Torres de Araúz, late head of the Patrimonio Historico in Panama, and Raul González, anthropologist with that agency, censused a number of communities in eastern San Blas and the Darien interior. Processed according to the categories used in Table 2, this data, which they kindly made available, is presented in Table 3. Sites 3, 4, and 5 are communities in eastern San Blas of roughly 150, 250, and 450 inhabitants respectively, and number 6 is a small village in the Bayano River valley across the coastal mountain range from San Blas.

Margaret Swain, working on a Kuna island of well over 1500 inhabitants in central-eastern San Blas, was given access to household data for two different years, which had been collected for the village's own census. Swain used a simple lumping typology with just three types to classify the data. Complicated and marginal cases, which must have occurred, presumably were aggregated to the nearest major type. Her results are shown in Table 4 (Swain 1978: 235–36).

TABLE 2
KUNA HOUSEHOLD TYPES, SITE 2

	Number of Households	% of Households	% of Population	Average Number Members	Average Age Household Head[a]
Nuclear Family, Single Conjugal Unit	21	26.6	15.5	5.1	40.0
Unmodified nuclear family	12	15.2	8.5	4.9	39.4
Nuclear family with relatives or boarders	6	7.6	5.5	6.3	41.2
Quasi-nuclear family[b]	3	3.8	1.4	3.3	—
Matrilocal, intact conjugal units in two generations	25	31.6	34.4	9.5	53.1
Unmodified matrilocal	16	20.3	22.4	9.7	48.4
One son-in-law	12	15.2	14.4	8.6	47.1
Two sons-in-law	3	3.8	5.8	13.3	50.3
Three sons-in-law	1	1.3	1.7	12	53
Matrilocal with unmarried relatives or boaders	5	6.3	6.8	9.4	67.5
Modified matrilocal[c]	4	5.1	5.2	9.0	67.0
Matrilocal, divorced or widowed parent only in senior generation	7	8.9	9.0	8.9	46.8
Mother-in-law	6	7.6	7.1	8.2	39.7
One son-in-law	3	3.8	2.5	6.7	—
Two sons-in-law	3	3.8	4.6	10.7	—
Father-in-law	1	1.3	1.9	13	68
Matrilocal, conjugal units in two generations, individual or conjugal unit in third or fourth ascending generation	8	10.1	13.2	11.4	53.0
Divorced or widowed woman in third/fourth generation	4	5.1	6.8	11.8	51.5
Divorced or widowed man in third generation	1	1.3	1.3	9.0	45
Three generations of conjugal units	2	2.5	3.3	11.5	65
Three generations of conjugal units, widow in fourth	1	1.3	1.7	12	55
Brothers-in-law, at least two sisters & husbands but no surviving parents of sisters	2	2.5	3.5	12.0	—
Patrilocal, man & wife with his parents	2	2.5	2.0	7.0	53.5
Matri-partilocal, senior couple, daughter & husband, son & wife	3	3.8	6.1	14.0	66.3
Complex[d]	9	11.4	14.2	10.9	36.4
Miscellaneous but Uncomplicated[e]	2	2.5	2.3	8.0	61.0
TOTALS	79	100	100	8.8	47.9

[a]In all cases the age of the senior male present. Average age information is partly unavailable resulting in ommissions and adjustments to total.

[b]Two households of a couple with foster children or boarders; a third of a couple with grandchild.

[c]Households where: the wife in the junior couple is daughter only of the woman in the senior couple; she is grandaughter of the senior couple (no intervening generation present); she is granddaughter only of the woman in the senior couple; the son of the senior couple is married to their adoptive daughter.

[d]These households mostly follow death, divorce, or dislocations in other households, and take a variety of configurations.

[e] One of these is two married couples without co-resident children; the other is a matrilocal household with a divorced woman in the senior generation plus her adoptive parents.

Richard Costello (1975: 61–108, n.d.) made a thorough study of household organization on a medium-sized island of about 500 inhabitants in central-western San Blas. Costello was concerned with the relationship of domestic organization to the economic and political positions of junior and senior men and with the relative importance of jural norms and personal interests in determining household form. He classified his data primarily in terms of which statuses adult men of the village occupied in their households, but he provided enough information to determine the incidence of different types, and this reclassification is presented in Table 5.

Frances Stier censused an island of over 800 inhabitants in eastern San Blas during 1976 (Stier 1979: 216). Her classification of household types, which follows Laslett & Wall (1972: 31), facilitates some cross-cultural comparisons and was appropriate to her study of labor intensity, migration, and household size, but it discourages comparison with other villages in San Blas. Among other things her use of the word "extended" has a different and more specialized meaning. Table 6 reports her results.

In 1974 Lawrence (1977: 68–75) censused the same island studied by Holloman seven years previously. Because Holoman did not randomly sample house-

TABLE 3
KUNA HOUSEHOLD TYPES, SITES 3, 4, 5, & 6

	Number of Households			
	Site 3	Site 4	Site 5	Site 6
Nuclear Family, Single Conjugal Unit	6	23	19	3
Matrilocal, intact conjugal units in two generations	7	4	13	1
Matrilocal, divorced or widowed woman only in senior generation	1	2	6	2
Matrilocal, conjugal units in two generations, widow in third ascending generation	1	—	1	2
Patrilocal	2	1	—	—
Matri-patrilocal	1	—	1	—
Polygynous	1	—	—	—
Miscellaneous	—	1	—	1
Insufficient information[1]	—	2	1	1
TOTALS	19	33	41	10

[1]None of these are nuclear family households.

TABLE 4
HOUSEHOLD TYPES, SITE 7

	Number of Households	
	1970	1975
Nuclear Family	19	28
Matrilocal	148	137
Patrilocal	9	21
TOTALS	176	186

From Swain 1978: 235–236.

TABLE 5
KUNA HOUSEHOLD TYPES, SITE 8

	Number of Households
Nuclear Family, Single Conjugal Unit, Male-headed	34
Nuclear Family, Single Conjugal Unit, Female-headed	2
Extended Family, Single Dwelling[1]	23
Matrilocal, conjugal units in two generations	9
Matrilocal, widow only in senior generation	7
Matrilocal, three generations of conjugal units[2]	1
Brothers-in-law	1
Patrilocal, man & wife with man's parent(s)	3
Matri-patrilocal	1
Complex	1
Extended Family, Single Hearth, Multiple Dwellings	10
Matrilocal, conjugal units in two generations	7
Matri-patrilocal	2
Complex	1
TOTAL	69

From Costello 1975: 61–108

[1]In an unpublished paper delivered in 1975 Costello reclassified some of his data, dropping the category of female-headed nuclear family household and giving the total for extended family households as 25.

[2]It is unclear whether both spouses in the highest generation were still living.

holds, the two surveys cannot be compared closely except perhaps for the greater incidence of nuclear family households recorded by Lawrence (see below).

Varios estudios han provisto censos sobre las familias cunas.

TABLE 6
KUNA HOUSEHOLD TYPES, SITE 9

	Number of Households	Average Number of Members	Average Age Household Head
Single Person	1	1.00	60.00
Nuclear Family, Single Conjugal Unit	19	6.32	37.74
Extended Up from Conjugal Unit: married couple in one generation, divorced or widowed parent in senior generation (?usually the wife's mother)	7	7.86	50.14
Extended Down: conjugal family with unmarried individual(s) (other than couples own children) in descending generation	10	6.80	47.80
Extended Laterally: married couple plus unmarried individual(s) in same generation	3	7.00	32.00
Combinations of previous three	5	8.20	35.20
Multiple conjugal units in different generations	46	12.61	55.00
Other multiple families	5	12.20	56.00
Member of Secondary Conjugal Family Away[1]	4	10.45	57.25
TOTALS	100	9.90	49.00

From Stier 1979: 216.

[1]This category is unclear. Presumably many households in other categories had absent members when censused.

Of all the household types reported in these surveys, only the one most identified with acculturation and change, the single conjugal unit or nuclear family household, appears in every one. The reports vary widely, from 1% to 70% of all households, seemingly making it impossible to generalize about the incidence of this form. The actual range of variation for the majority of communities however, is probably much narrower. First of all, the results from sites 4 and 5, 45% and 70% nuclear family households respectively, are highly implausible, both because of the marked conservatism of both villages, and because of their proximity to site 9, which shows a much more plausible 19% nuclear family. The data for sites 3 through 6 were collected for the purpose of establishing demographic indices such as birth and death rates, not to determine household form. Although many people in these surveys were definitely recorded by household, it is likely that others were recorded by nuclear family instead, strongly inflating the reported cases of neolocality. Thus the totals for this category should be discounted for sites 4 and 5, and probably for sites 3 and 6 as well.

Among the remaining sites, number 8 shows by far the greatest incidence of nuclear family households. This makes sense, given that building sites on this highly acculturated island disappeared many decades ago except for those created on fill, and given that the national government strongly encouraged neolocality there as long as a half-century ago (Costello 1975).

Se presentaron más núcleos familiares en las islas altamente aculturales.

At the other extreme, Holloman (1967:184–92, 1976) reported only one nuclear family household on site 1 in 1967, not only in her judgmental sample of 79 but also among all 86 households on the island, and she paid particular attention to this point. The much greater number of such households recorded by Lawrence (1977:70) on the same site in 1974, 26 out of 116, probably derives from different procedures in the two surveys, from different criteria for distinguishing an independent nuclear family from one attached to an extended family household, and from a real increase in nuclear family households. Lawrence (1977:7–8) noted that an acculturated translator had hampered and distorted some of her census taking.

If sites 1 and 8 are treated as extremes, the remaining cases, sites 2, 7, 9, show a narrower range of variation, from 11 to 27%. It is not surprising that among the three sites, number 2, the only one with significant amounts of free land for building, would have the highest rate. One would expect that such households, because they tend by nature to be smaller than extended family households, would account for a lesser percentage of population than of total households, and on site 2, the only one for which this data is available, this prediction holds true: households with a single conjugal unit account for 27% of households but only 16% of the population. One can conclude that nuclear family households form a small but significant minority on most Kuna islands.

Patrilocal households, in which only a married son or sons are present, and matri-patrilocality, in which both a son and a son-in-law are present, should probably be merged for this discussion, especially since they are not always distinguished clearly in some surveys. For sites 1, 3, 4, and 8 one can state that (matri-) patrilocality does occur but not with what frequency. For site 2, 6% of households took this form, and on site 7, 5% in 1970 and 11% in 1975. Patrilocality and matri-patrilocality form a small but significant (and perhaps growing) minority of households on many islands.

The different forms and stages of matrilocality (including brothers-in-law households) account for a majority of cases on all but one of the sites for which the data are clear and reliable, namely numbers 1, 2, and 7. Despite the disparate classification used for site 9, table 6 strongly suggests that matrilocality predominates there as well. Only in site 8, a highly acculturated village, is matrilocality a minority residence form. Even among individuals currently in other kinds of household, most of them probably have in the past or will in the future reside matrilocally.

The data also show however, that one cannot infer the incidence of a certain household type from a jural rule of residence or even from a characterization of the modal or ideal type. The Kuna case does not offer much comfort to cross-cultural studies of postmarital residence, even though it fulfills the assumptions or hopes implicit in most such studies. Not only do the Kuna have a strong and seemingly inflexible rule of residence, but they also have in mind and strive for a certain ideal type, Holloman's "Father-in-law, Prototype Matrilocal." Assuming a close fit between rule and the incidence of the ideal type, one would hardly predict that on site 2, for instance, father-in-law households account for only 32% of the total, and that even combined with three-generation households, they still account for less than half the total. Still less could one predict or explain the presence of complex or miscellaneous forms, which make up almost 14% of households and 16.5% of the population.

Los cunas tienen no sólo reglas fuertes e inflexibles sobre la residencia matrilocal, sino

que ellos también procuran alcanzar cierto tipo ideal.

Future Research

The main body of this paper is confined largely to principal structural and dynamic elements of Kuna domestic organization. The Kuna matrilocal system is portrayed as a set of strategies—collective and institutional rather than individual—by which household heads attempt to control the behavior of others. The temporal and spatial scope of the description, spanning a whole society over many decades, has required a discussion in broad terms which glosses over complications and local variation. Obviously, this form of analysis cannot exhaust the significance of Kuna domestic organization, especially not in an article-length presentation, and a great deal remains to be done.

Four forms of analysis of Kuna domestic organization seem especially appropriate. The first is a systematic and explicit application of the decision-oriented approach to the domestic cycle (Goodenough 1955, Buchler & Selby 1968: 49–52, Quinn 1975, Fjellman 1976). Among other things, such an analysis could assess the relative weight of jural rules and self-interest, a question raised by Costello (1975: 61–108) in his critique of Holloman (1969, 1976). Second, formal and quantitative models could tell us much more about domestic cycles than we now know (Buchler & Selby 1968: 52–68, Fjellman 1977). Third, Stier's important recent work (1979), which examines agricultural production on one Kuna island in terms of variation in household composition and wealth, highlights the utility of paying more attention to intra-community variation (Yanagisako 1979: 175–76) and the role of households as productive units. Fourth and perhaps most important, several strikingly different programs of comparative and cross-cultural research have generated a useful set of hypotheses about the relationship between domestic forms—in particular matrilocal households and the wider polities in which they are found (Kloos 1963, Van Velzen & Van Wettering 1960, Ember & Ember 1971, Ember 1974, Divale 1974, Pasternak 1976: 42–57; cf. Maybury-Lewis 1979, Turner 1979; cf. Yanagisako 1979: 190–91). This topic requires further consideration as it applies to Kuna matrilocality.

Acknowledgments

This article is based on field work carried out in 1970, 1971, 1975, 1977, and 1978, which was supported by the National Institute of Mental Health, the National Endowment for the Humanities, the Social Science Research Council, the Old Dominion Foundation, and the M.I.T. School of Humanities and Social Sciences. Thanks go to Jean Jackson who read the manuscript.

Blau, P. 1964. Exchange and Power in Social Life. John Wiley, New York.

Bohannan, P. 1957. An alternative residence classification. Amer. Anthrop. 59: 126–31.

Buchler, I. & H. *Selby*. 1968. Kinship and Social Organization. Macmillan. New York.

Casselberry, S. & N. *Valavanes*. 1976. "Matrilocal" Greek peasants and a reconsideration of residence terminology. Amer. Ethnol. 3: 215–26.

Cohen, A. 1969. Political anthropology: the analysis of the symbolism of power relations. Man 4: 215–35.

Coope, A. 1917. Anna Coope, Sky Pilot of the San Blas Indians. American Tract Society, New York.

Costello, R. 1975. Political Economy and Private Interests in Rio Azucar. Univ. of California, Davis, Ph.D. diss. University Microfilms, Ann Arbor.

Divale, W. 1974. Migration, external warfare, and matrilocal residence. Behav. Sci. Res. 9: 75–133.

Ekeh, P. 1974. Social exchange theory: the two traditions. Harvard Univ. Press. Cambridge, MA.

Ember, C. 1974. An evaluation of alternative theories of matrilocal versus patrilocal residence. Behav. Sci. Res. 9: 135–49.

Ember, M. & C. *Ember*. 1971. The conditions favoring matrilocal versus patrilocal residence. Amer. Anthrop. 73: 571–94.

Fjellman, S. 1976. Talking about talking about residence: an Akamba case. Amer. Ethnol. 3: 671–82.

———. 1977. The Akamba domestic cycle as Markovian process. Amer. Ethnol. 4: 699–713.

Fortes, M. 1958. Introduction, Pp. 1–14. *In* The Development Cycle in Domestic Groups, J. Goody, ed., Cambridge Univ. Press.

———. 1969. Kinship and the Social Order. Aldine, Chicago.

Gassó, P.L. 1948. Notas del Padre Gassó, pp. 113–15. *In* Padre M. Puig, Los Indios Cunas de San Blas. Panama.

Goodenough, W. 1955. Residence rules. Southwest. J. Anthrop. 12: 22–37.

Goody, J., ed. 1958. The Development Cycle in Domestic Groups. Cambridge Univ. Press.

Harris, M. 1979. Cultural Materialism. Random House. New York.

Heath, A. 1976. Rational Choice and Social Exchange. Harvard Univ. Press, Cambridge, MA.

Helms, M. 1970. Matrilocality, social solidarity, and

culture contact: three case histories. Southwest. J. Anthrop. 26: 197–212.

———. 1976. Domestic organization in eastern Central America: The San Blas Cuna, Miskito, and Black Carib Compared. West. Canad. J. Anthrop. 6: 133–63.

Holloman, R. 1969. Developmental Change in San Blas. Northwestern Univ. Ph.D. diss. University Microfilms, Ann Arbor.

———. 1976. Cuna household types and the domestic cycle pp. 133–49. *In* M. Helms & F. Loveland, eds. Frontier Adaptations in Lower Central America. Institute Study Human Issues, Philadelphia.

Homans, G. 1958. Social behavior as exchange. Amer. J. Sociol. 63: 597–606.

Howe, J. 1974. Village Political Organization among the San Blas Cuna. Univ. Pennsylvania. Ph.D. diss. University Microfilms, Ann Arbor.

———. 1975. Notes on the environment and subsistence practices of the San Blas Cuna. Working Papers on Peoples and Cultures of Central America, no. 1.

———. 1976. Communal land tenure and the origin of descent groups among the San Blas Cuna, Pp. 151–63. *In* Frontier Adaptations in Lower Central America. Inst. Study Human Issues. Philadelphia.

———. 1979. The effects of writing on the Cuna political system. Ethnol. 18: 1–16.

———. 1981. The star girl's descent: a myth about men, women, matrilocality, and singing. J. Amer. Folklore. 94: 292–322.

Hsu, F. 1971. Kinship and Culture. Aldine, Chicago.

Kaplan, A. 1964. The Conduct of Inquiry. Chandler, San Francisco.

Kloos, P. 1963. Matrilocal residence and local endogamy: environmental knowledge or leadership. Amer. Anthrop. 65: 854–62.

Kracke W. 1976. Uxorilocality in patriliny: Kagwahiv filial separation. Ethos 4: 295–310.

Laslett, P. & R. *Wall*, eds. 1972. Household and Family in Past Time. Cambridge Univ. Press.

Leví-Strauss, C. 1969. The Elementary Structures of Kinship. Beacon, Boston.

Maybury-Lewis, D., ed. 1979. Dialectical societies: The Ge and Bororo of central Brazil. Harvard Univ. Press, Cambridge, MA.

Miller, E. 1967. Pentecostalism among the Argentine Toba. Univ. Pittsburgh. Ph.D. diss. (as cited in Holloman 1976).

Murdock, G.P. 1949. Social Structure. The Free Press, New York.

Nordenskiold, E., R.P. *Kantule* & S.H. *Wassén*, eds. 1938. An Historical and Ethnological Survey of the Cuna Indians. Comparative Ethnographical Studies No. 10. Goteborg.

Oberg, K. 1955. Types of social structure among the lowland tribes of South and Central America. Amer. Anthrop. 57: 472–87.

Pasternak, B. 1976. Introduction to Kinship and Social Organization. Prentice-Hall, Englewood Cliffs, NJ.

Prestán, A. 1975. El uso de la chicha y la sociedad kuna. Ediciones Especiales 72. Instituto Indigenista Interamericano, Mexico.

Puig, P.M.M. 1948. Los indios cunas de San Blas: su origen, tradición, costumbres, organización social, cultura, y religión. Panama.

Quinn, N. 1975. Decision models of social structure. Amer. Ethnol. 2: 19–45.

Reverte, M. 1967. El Matrimonio Entre los undios Cuna de San Blas. 2nd. ed. Panama.

Salcedo, G. 1980. Un consejo matrimonial, Pp. 64–72. *In* J. Howe, J. Sherzer & M. Chapin, eds. Cantos y oraciones del congreso cuna. Editorial Universitaria, Panama.

Schneider, D. 1968. American Kinship, a Cultural Account. Univ. Chicago Press, Chicago.

Stier, F. 1979. The Effect of Demographic Change on Agriculture in San Blas, Panama. Univ. Arizona Ph.D. diss. University Microfilms, Ann Arbor.

Stout, D. 1947. San Blas Cuna Acculturation: An Introduction. Viking Fund Publ. Anthrop. no. 9, New York.

Swain, M. 1978. Ailigandi Women: Continuity and Change in Cuna Female Identity. Univ. Washington Ph.D. diss. University Microfilms, Ann Arbor.

Torres, R. de Ianello. 1957. La Mujer Cuna de Panamá. Instituto Indigenista Interamericano. Mexico.

Turner, T. 1979. The Ge and Bororo societies as dialectical systems: a general model. Pp. 147–78. *In* D. Mabury-Lewis ed., Dialectical Societies: The Ge and Bororo of Central Brazil. Harvard Univ. Press.

Van Velzen, H.U.E. *Thoden* & W. *Van Wetering*. 1960. Residence, power groups, and intra-societal aggression. Intern. Arch. Ethnogr. 49: 169–200.

Yanagisako, S. 1979. Family and household: the analysis of domestic groups. Ann. Rev. Anthrop. 8: 161–205.

The Form and Context of the Kuna General Congress

Alexander Moore

Los Congresos Generales Cunas se reunen varias veces al año y duran aproximadamente tres días. Delegados procedentes de cada comunidad local se reunen para revisar los eventos e intenciones del Gobierno Nacional de Panamá. La discusión es activa y las decisiones se toman por consenso. El patrón de este Congreso es similar al de los Congresos locales, sin embargo carece del poder de la policía. Comparado con los congresos de los Cheyennes y los Hurones, la institución Cuna juega el mismo papel en dirigir las relaciones del grupo con elementos foráneos, sin embargo la composición de estos dos congresos difiere. Algunas veces los poderes del Congreso General Cuna han sido malentendidos por el Gobierno Nacional de Panamá.

The Natural History of the Kuna General Congress

This paper examines the general (tribal) congress of the Kuna Indians of San Blas District, Panama, as a natural phenomenon. It describes that congress as a patterned, repetitive, political event and institution as it was in the late 1970s. The methodology is inductive, holistic, and comparative. That is, from a sample of three congresses observed, and from many native statements about them, I set forth a procedural model of such congress. That model is then explicitly contrasted with its immediate precursor and most relevant cultural context, the Kuna local congress, of which the tribal one is a natural outgrowth. Then I compare the tribal congress with the councils of other tribal groups which have formed confederacies uniting communities, and tribes or sectors, into single entities, as the Kuna have done. Finally, since the tribal congress is of great importance to the continued vitality of Kuna culture and institutions, a few developmental comments are considered which may prove useful to Kuna and Panamanian policy-makers alike.

The conclusions of this inquiry further show that the Kuna general congress, while based on the local Kuna congress, differs from its prototype in the growing formalization of its procedures, which reflect the "Standing Orders" of Western parliamentary bodies. It resembles the prototypical local congress in its open character—very fluid delegations from local communities attend—and in its approach to problem of succession to chiefly office. Like local congresses, the tribal congress functions to resolve conflict. However, its external relations or diplomatic function—especially vis a vis the Panamanian nation-state—is of far greater importance and magnitude than in any local congress. The general congress resembles the local prototype—and a great many other tribal councils as well—in its custom of arriving at decisions by consensus, and not by majority vote. The exact manner of arriving at consensual decisions is worthy of a paper in itself, and better left to an effort subsequent to this one.

As used here the term "natural history method" is borrowed from Arensberg and Kimball (1965: 34–38), and I have explicated the method at length elsewhere (1978: 34–38). Here the The Kuna tribal congress is the object, or field, of study. Here the task is to come to some statement of its behavioral regularities and patterned spatial settings. We must ask what personnel perform what activities, on what schedules, and in what context? In this paper, however, the dimension of the repetitive activities shall provide a focus for all dimensions of description.

El método de historia natural es usado para

estudiar cómo las actividades repetidamente asociadas en los congresos cunas definen la evolución de éstos.

The district (*comarca*) of San Blas is an autonomous unit within the Republic of Panama. It is by law a tribal reserve and by charter a tribal confederacy of some 50 Indian communities comprising perhaps 30,000 individuals. Located on the northeast coast, the district includes some 150 miles of strategic Atlantic shoreline, ending—for customs and defense purposes—a few miles short of the Colombian border. The Kuna do not encourage outsiders to reside in San Blas; the few who do so permanently are mostly persons married to Kuna.

La comarca de San Blas es autónoma dentro de la República de Panamá. Hay un grupo de Caciques generales que presiden el Congreso General y adicionalmente un grupo de jefes, congresos y comisarios procedentes de cada comunidad. Las instituciones nacionales panameñas también están presentes. El Congreso General es un forum de interacción entre las intenciones nacionales y las de los Cunas.

Tribal institutions include a set of high chiefs (*caciques generales*) and a general congress over which they preside (examined here). In addition each community has its own chiefs, congress and a set of local unarmed constables. National Panamanian institutions in the district include several detachments of national guardsmen, staffed exclusively by Indians, together with national schools, clinics, and extension agencies of various sorts. All of these bureaus have been escalating their activities, installations, personnel, and budgets in recent decades. The general congress is the primary forum wherein the national agencies announce their intentions in San Blas. In addition the district elects three representatives to the Panamanian national assembly. Their proper relation to the tribal constituency has been debated in the tribal congress and was the issue which ultimately produced a tribal schism in 1976. Within this policy the congress is a key institution. Certainly the Indians consider it paramount, and never tire of asserting that paramountcy as declared in the tribal charter, la carta organica (*La Carta Orgánica*), written at a tribal congress in l945, unified the San Blas Kuna. For an English translation and discussion see Holloman 1969).

The natural history attempted here is akin to the models of Kuna speech events won from the ethnography of speaking by the Sherzers (J. & D. Sherzer 1972, J. Sherzer 1974). The methodology of the ethnography of speaking is close to mine. The difference is in focus. Rather than being interested in speech events as vehicles of language, and in the mutual interrelations of language as a system and speaking as a system, I am more interested in the speaking system as it related, ultimately, to the political process. Thus for me, mythological chants, oratory, verbal reports, admonishments, the verbal drawing up of written resolutions are all artifacts in the political process as the Kuna search to build consensus, that most remarkable feature of Kuna decision-making at all levels: local, regional, and tribal (district).

The general congress has been the subject of a cursory and critical examination by Regina Holloman, based on one congress she observed in 1967 (1969: 357–370). Councils in general have been the subject of attention from British social anthropologists. Richards and Kuper for example have edited a volume, *Councils in Action*, of papers on councils, defined as the "institutionalized process of joint discussion, as distinct from informal arguments" (1971: 1). A council is characterized by these criteria: rules of place or location, and by "a series of conventions governing the behavior of its members" (1971: 3). These include formal openings, certain fixed seating arrangements, a given order of speaking among the participants, and perhaps special phraseology of speech. Richards further observes that "groups become councils . . . by the adoption of conciliar behavior" (1971: 3). The conciliar behavior of the San Blas Kuna is precisely what we shall analyze here.

A Procedural Model of the Kuna General Congress

I and my research assistant, Kuna sociologist Eligio Alvarado, observed three Kuna General Congresses in 1977–1978. All three were characterized by a similar sequence of events. Certain events are mandatory, others are optional. Some are likely to follow others. That order, and the variability within it, is clear enough from our sample congresses.

Scheduling. Kuna congresses are by the tribal charter supposed to be held three times a year in sessions of no less than eight days. By charter the first high chief (*primer cacique general*) is to be the presiding officer. The first order of business is to hear reports from him and the other two high chiefs, followed by general discussion from all delegates present. Delegates are to comprise one chief and others from each community, in unspecified numbers, in practice selected by each local congress.

Los Congresos Generales Cunas deben reunirse teóricamente tres veces al año y durar aproximadamente 3 días.

In spite of the stipulation that congresses must be held at least three times a year: this is not necessarily the case each year. Moreover I am told they never last as long as eight days. Indeed the three we observed lasted only three days. The scheduling of congresses is quite variable and depends on the vagaries of political pressures, rather than on any fixed dates, or on any calendrical rites or observances, as is the case in gatherings of other tribes as discussed below.

Personnel. Although all San Blas communities are required by charter to send delegations to each general congress, in fact a few communities may fail to do so for any given congress. Any number of contingencies may delay or prevent a delegation from arriving. Each delegation is required by the charter to include local chiefs and such special delegates as are elected by the local congress. Not all the local chiefs of each community attend, but it is safe to say that one of them always does; one of his interpreters or official spokesmen (Sp. *vocero,* Kuna *arkar*) is almost certain to be in the delegation. The *carta orgánica* is extremely vague in its wording; and the Kuna are vague in their response. Apart from the mass of delegates there are the tribal high chiefs, the advance preparations committee, and such other officiers as are appointed during the course of preceedings, as we shall see.

Location. Congresses are rotated among the more populous communities of San Blas, those which have the population and resources to house and feed some 225 or more delegates for three days or more. The next host community volunteers for the honor toward the end of each congress. Lead time is needed to make preparations, which usually include repair, even rebuilding, of the local congress hall. The one frequent host community, Narganá, which does not have a traditional congress hall—that is, a large thatched building held aloft by three or more imposing king posts down the center (Moore 1981a)—hosts general congresses under the high tin roof of the school basketball court.

A estos Congresos asisten delegaciones de jefes locales de la comunidad, jefes de alto rango de la tribu y otros delegados especiales. Algunas veces, pueden asistir hasta 225 personas. Casi al final del Congreso, se escogerá el lugar para efectuar el próximo Congreso.

Procedures

Arrival and checking in. Delegates arrive both individually and in community delegations on the appointed day starting early in the morning. Some come on private canoes powered by outboard motors. Many come as groups on one or more of the commercial boats operated by Kuna cooperatives and plying San Blas waters. Still others come aboard boats belonging to Panamanian government agencies in San Blas. Arrangements for such boats have been made long in advance in consulatation with the local organizing committee, which is in constant touch with the first high chief and his secretary.

Upon arrival the usual course of action is to report directly to that organizing committee in session at tables in the local congress house. Indeed local constables are likely to be at the dock to escort the visitors and bear their belongings. They are certainly ready at the congress house to escort the delegates to their assigned host households, after they have checked in and signed or dictated their names on a roster. Some delegates may already have relatives or special friends in the host community with whom they prefer to stay; and they go directly there in that case. All guests are immediately shown their hammocks where they may sleep that night, and provided with towels and soap, in case they wish to bath after a sea journey to remove salt spray. In that case they may hand over their clothing to their hostess for washing. They are very likely to be offered a rich repast upon arriving, and they will certainly be offered something to drink, such as coffee or a fruit mead.

Al llegar al muelle, los asistentes son llevados al lugar de hospedaje y después de comer se movilizan en pequeños grupos a través de las calles.

After bathing and changing, the delegates are likely to go back to the congress house, and then immediately melt into small groups in the streets of the village. Others cluster in and around the houses where prominent leaders are staying. These locales are the "lobbies" for intense lobbying to take place if in fact the congress is slated to discuss some controversial issue.

Inaugurating the congress. It is proper to inaugurate a congress with a chant. That is, the chief of the host community may be asked—informally and seemingly in jest by some leading notable milling around with the crowd in the congress house—to chant a welcoming chant. This was done in this spontaneous manner in the Mamituppu general congress of June, 1977, while delegates were still milling around and checking in. It is customary ritual several nights a week in Kuna communities for their chiefs to sit in the hammocks hung in the center of the congress hall and chant from a repertory of myths and morality called *Pap ikar*, God's Way. These in turn are interpreted by less ranking officials, interpreters, *arkarkana* (Sp. *voceros*). Community sings are very formal, with women and children being summoned by the town constables to take their allotted places in the centermost rectangle of seats around the singing chiefs. One general congress we observed, at San Ignacio Tuppile in March, 1978, did hold such a singing congress in the afternoon, summoning the women of the town to listen to welcoming chants while the delegates were still checking in. Very few women attended; they were too busy attending to their guests with the very hospitality which the chants exhorted of them!

In any case, in the evening the milling around stops, and the delegates take seats spontaneously as they find them. Seating patterns are thus informal; although there is a tendency for lower-ranking members of the same delegation to sit together on back benches and for higher ranking men and leading district notables to sit "down front" closer to the action. There is nothing like the formality of Western parliaments, such as opposing parties seated facing each other (Great Britain), or the spectrum of parties arranged from Left to Right, in front of the podium (France). At the "extraordinary" general congress in Carti-Suituppu in November, 1977, a small bloc of nativistic leaders from the Carti sector, who generally dominated this congress, sat during the debates in a group of chairs literally confronting the head table. Seating, then, is flexible and indeterminate save for those appointed to the head table.

En la noche, el Congreso se inaugura mediante cantos de los jefes anfitriones y sentarse no es obligatorio, excepto para aquellos que se hallan en la mesa principal. Los cantos se emplean para disminuir e instar a la armonía. Entonces el primer jefe en jerarquía anuncia que se toma asiento en la mesa principal. Esta consiste de 2 secretarias, un fiscal y traductores para ayudar a los oficiales del Gobierno panameño.

At the start of the evening session, as delegates are settling into the seats they have found on the rows of hard wooden benches, the organizing committee may request leading visiting chiefs and interpreters to chant and interpret before starting the business of the meeting. This happens while the congress is still controlled from tables staffed by the local committee (rather like the "local arrangements committee of a convention in the United States). Inaugural sessions are held in the evening, leaving the daylight hours for the delegates to travel, arrive, and check in.

The appropriateness of a chant depends in part on the tensions and implicit conflicts present at the start of a gathering. If the congress has been called to deal with controversial issues then the organizing committee is certain to call for chants, which can instill harmony. The same congress in Mamituppu of June, 1977, was to resolve a most vexing tribal schism and dispute involving succession to the high chieftainship. Hours were spent in chants by leading cantors. In contrast the congress held at San Ignacio Tuppile in March of 1980 had no controversial issues to consider, and no formal inaugural chants were held other than the afternoon signing congress to edify the women of the town.

Installation of the head table. The organizing committee may well be recognized, especially if its chairman is also the chief of the host community. In one case this man rose to introduce his committee members. However, he must yield the floor to the first high chief for the next item of business, the announcement of the personnel of the head table *(mesa directiva)*. During the checking-in the first high chief has been receiving suggestions—in fact nominations for the high table—at this point he announces his decisions. Officers of the head table are literate younger Kuna, educated in Spanish—some in English as well. They consist of two secretaries, together with a correspondence secretary as well as a treasurer. Another is the *fiscal* or bailiff of the congress whose task is to allow speakers to take the floor in the order recognized by the first high chief, who presides. In addition there may be as many as three official interpreters at the head table, who must make official verbal translations if Panamanian officials are to speak and be spoken to.

At this point the members of the organizing committee leave the table to take seats in the room. If the table had been placed in the center of the hall, where the hammocks of the chiefs normally hang, then it is moved to one side, leaving that space as a forum for oratory, or allowing it to be filled with chairs if other space is lacking. The newly appointed officers of the head table stand, raise their right hands and are formally sworn in over a copy of the tribal charter held by the district judge who administers a western-style oath. They are joined at the high table by the high chiefs, and perhaps by the official councilors of the first high chief.

Reading of correspondence and of the agenda. One of the newly appointed secretaries, or the first high chief's personal secretary, then rises to read any letters addressed to the general congress. These are written in Spanish, and the secretary translates into Kuna as he goes along. The letters may be reports from bureaucrats unable to attend; or they may be anonymous denunciations of putative abuses. Then a secretary reads the agenda, which has been handed him by the first high chief, who has delegated its putting-together to influential educated, often city-based Kuna. The agenda is one of the most notable western features of the congress, and is directly traceable to Panamanian practice. The list of topics is then debated vociferously and even angrily from the floor. Indeed at the general congress held in Carti-Suituppu in November, 1977, announcing the agenda sparked hours of discussion. The delegates were expecting to meet face to face with Head of Government General Torrijos the following day and wanted to present a united front on all possible issues and grievances the tribe might have with the national government. Accordingly they arrived on agreement on all items to be presented and appointed one spokesman to address each issue the following day before the General. So important was the need to achieve unanimity before the government that the congress remained in session after the chants from 10:00 P.M. to 3:30 A.M. The regular session next day was to start at 7:00 A.M.

Cartas dirigidas al Congreso se leen desde la mesa principal, seguidas por la agenda, que se discute.

As it turned out it was the Vice President of the Republic who attended the congress in place of Torrijos. Moreover the Indians were unable to keep to their strict plan and limit speeches to one speaker for each grievance item. Each spokesman was seconded and interrupted so many times that together they never were able to address all the items on the agenda before the Vice President's departure on the following, or third day.

However, it is tacitly understood that items on the agenda are to be taken up in order of importance. Once pressing issues seem to have been resolved, delegates are impatient and less willing to hear lesser issues if it means staying beyond a third day.

The reports of the high chiefs and representatives. The inaugural session is in the evening and does not extend past the announcing of the agenda. Next morning the congress resumes with a roll call from the roster prepared by the organizing committee and handed over to the officials of the head table. Then the high chiefs, starting with the first high chief and continuing in order of rank, give reports (*owisoe*, Sp. *informes*) detailing in often very rambling fashion, their activities since the last general congress. The Panamanian government has incorporated San Blas into its new electoral system of 1974, and carved it into three electoral districts (*corregimientos*), which send representatives to the National Assembly. These three men may now make reports immediately after the high chiefs. This was the case in the Congress of March, 1978, which was devoted almost entirely to hearing reports from national bureaucracy. Such a session of reports can take all morning, in which case the delegates adjourn to take generous repasts in their host households.

Durante la mañana siguiente los tres cunas representantes en la Asamblea Nacional de Representantes presentan informes. En la tarde se discuten tópicos hasta llegar a un consenso.

Action items. The afternoon session is likely to take up action items, questions which need to be discussed and hopefully resolved. Items are known in Spanish as *temas*, in Kuna as *ikars* (the literal Kuna plural is *ikarkana*). These follow the order set forth in the agenda. The first high chief recognized a lead-off speaker for each one, usually a man known for his oratory. In neither of the two congresses (June and November 1977) which had many items on the agenda, did they finish considering all of them. Yet this "old business" was not carried over to the next congress.

Ideally an item is discussed until consensus is reached. Formal votes are avoided, although I have seen the local congress of Narganá take formal votes on technical and unimportant issues. Rather the sense of the meeting is understood when speakers stop addressing contrary opinions on an issue, or when a contrary opinion is angrily and massively shouted down. A good sign that consensus has been reached is a sudden rapid-fire repetition of a position by a number of speakers all saying almost the same thing. Tacitly, then, the matter is dropped so that the next item reaches the floor. While debate starts on the new issue, the secretaries of the head table begin to draw up a resolution (*resolución*), writing it in Spanish to express the sense of the agreement just reached.

Confirmation and swearing in of newly appointed village chiefs. When the meeting seems to have tired of action items, and this may well be in the evening session or in the morning session of the third day, newly appointed village chiefs, of whatever rank (Kuna villages have from three to six chiefs, ranked serially), are asked to center-stage, where they are presented to congress by name and village. Then the district judge administers a western-style oath over a copy of the tribal charter. Then the new chiefs sit in a prominent down front position and are formally admonished by leading notables and orators to the murmured approbation of the listeners.

Delivering admonishments. Admonishments are mandatory for the swearing in of new officials such as village chiefs. They may, however, appear at several points in a congress. The high chiefs may receive them, for example, after they have delivered their reports. Admonishments (*unaet*, Sp. *amonester*) are a formal speech event (Howe: 1974: 123–126) in which a venerable person exhorts the person admonished to right conduct. They are notable in rites of passage. Here we observed them prominently after the resolution of conflict between political opponents. The resolution of long-standing tribal difficulties in the congress of June 1977, for example, meant that several erstwhile enemies were called to sit center-stage, together, while orators one by one standing over them exhorted them to harmony, mutual agreement, and to uphold the general consensus.

Luego se confirman los nuevos jefes de los pueblos y se presentan las amonestaciones.

So dear is the admonishment to the Kuna, that when General Torrijos was scheduled to appear before the November 1977 congress, it was decided that a young woman, decked out in full Kuna regalia—literally girded in a breastplate of gold—would admonish him in Kuna, while a secretary would translate. The young woman chosen for the task spoke perfect Spanish, being a schoolteacher, but that fact was not apparent when she appeared to harangue the General's substitute, a surprised Vice President Gonzalez!

The reports of Panamanian government officials and of Kuna associations. After exhausting its "own" business, the congress then listens dutifully to the reports of representatives of the Panamanian national bureaucracy. In March 1978 at the congress at Tuppile there were no action items, only reports. No less than ten officials gave reports, five of these persons were Kuna.

In addition the congress hears reports from the representatives of Kuna organizations that supersede the boundaries of the local communities. These may be associations of educated youth. They are very likely to include reports from the labor organizations which represent the migrant Kuna laborers on United States installations and on the banana plantations of northwest Panama.

Después de las discusiones de temas pertinentes al Congreso, se presentan los informes del Gobierno de Panamá, al igual que aquellos de las organizaciones Pan-Cunas.

The presentation and approval of the texts of resolutions. Final business of a congress is to listen to the text of resolutions prepared by the secretaries of the high table, which are read in Spanish, interpreted and often debated and amended before reaching their final form. The legal status of these resolutions and their binding character, or not, upon those Kuna or Panamanian officials to which they are addressed is an emerging matter of great concern as the congress tests its mettle against the ever more-powerful and penetrating national bureaucracy.

Las resoluciones son preparadas por los secretarios, presentadas y discutidas. Casi inmediatamente se levanta la sesión. La mayoría de los delegados recogen sus pertenencias, toman los botes y regresan a sus comunidades.

Adjournment. When there are no more reports or resolutions—the congress seems to end by tacit agreement—there is a stir to leave. However, the final moments may be embellished by a burst of farewell oratory as one speaker after another takes the floor to praise the work just accomplished. This was true of the two congresses, of June and November 1977, which dealt with emotional and hotly felt issues, but not with that of March 1978, which dealt almost exclusively with reports. The formal adjournment is announced by the first high chief, and the crowd is up and milling toward the door. Most go immediately to gather their belongings and take to the boats for home.

Comparison with the Kuna local congress. The Kuna general or tribal congress is explicitly modelled on the local congress (see Howe 1974: 349–404). There can be no question that the local variant is the prior form. It is the focus of all Kuna communities, the locale of most public interactions. All we know of Kuna culture tells us that this was so in the last century as well. Kuna general congresses, in contrast, have always been sporadic and intermittent. They seem to have been extraordinary convocations of the tribe in response to foreign threats, called by influential chiefs of populous communities. With encroaching pressures of Western powers in the late

19th and early 20th centuries, the institution was activated more frequently (Moore 1980). It was finally regularized and put on a formal legal footing by the tribal charter of 1945.

Personnel. Conceptually local congresses include the entire population of a community. For ritual purposes, such convocations may indeed include everyone at singing congresses or the rite of exorcism (Howe 1976). Some few acculturated communities include adult women with the men in certain talking congresses called to resolve crises of leadership. Most of the time and in most other communities, such convocations are of the adult men only. Within this convocation there are several definite hierarchical orders of serially ranked personnel. First are the chiefs (*sailakana*) who number from three to six in any given community. Traditionally they are practitioners of the sacred chants, men learned in myth and lore. A few acculturated communities do not have such traditional chiefs; their presiding officers do not chant, but still hold the title of *saila*. Ranked below the chiefs are *arkarkana*, interpreters or spokesmen, who also form a serially ranked order from first to last, and number 15 or so. Traditionally they interpret the chants of the chiefs, but in the three communities without chanting they are in effect a group of aldermen. Ranking still below them is yet another lineup of men ranked from first to last, the constables or staffbearers (*sualipkana*). These act as heralds and occasionally administer corporal punishments sentenced by the chiefs in congress. In addition, most communities have a number of committees or task forces which perform public tasks, and which form rank orders of their own.

Los congresos locales anteceden los de las tribus e incluyen toda la población local, pero en este caso los jefes, los traductores y los aguaciles son los primeros en jerarquía.

As we have said, and James Howe points out (1974: 350), the local talking or business congress theoretically embraces the entire male public of the village as well as these officers. In fact talking meetings usually convene only these officials (and not necessarily all of these) and such members of the public as have business there or who care to attend. Younger men are infrequently there. The more senior and influential a man, the more likely he is to attend and voice his opinion in discussion and debate, and just to hang around during off hours. (Many local congress houses acquire an air of a senior men's clubhouse during the afternoon.)

Schedule. A local singing congress is held two or three times a week; a local talking congress perhaps three times, twice a week is minimal. The local congress house is used very heavily. The local congress meets on a schedule many hundreds of times more frequent than that of the general congress.

Congresos locales de canto y de conversación, también se llevan a cabo. Usualmente éstos se efectuan 2–3 veces por semana.

Local talking congress procedures. The procedures of a local talking meeting are very similar to those we have been considering. There is, as befitting so frequent and commonplace an event, much less formality about the inauguration of sessions, the announcement of an agenda and priority granted of action items. One activity of the local congress has no counterpart on the tribal level: the granting of travel permits. In highly acculturated Narganá, with its myriad links to the metropolis, persons travel beyond the community nearly every day, and the congress meets nightly just to grant permits. This "pre-business" session consists of the chiefs and town secretary sitting at a table to receive travel requests from individuals. The town work foreman and heads of the task forces are required to be on hand to attest to the individual's status vis-a-vis communal labor projects. Persons in arrears on labor contributions owe cash fines to the town treasury which must be settled before a travel permit by the first chief is granted. In other communities, the granting of permits is on a weekly or even monthly schedule, rather than nightly.

Los Congresos de conversación usualmente proveen los permisos para viajar.

In general *ikars*, action items, are taken up on an informal implicit agenda resulting from behind-the-scenes consultation between the first chief, his advisors, and leading notables, or any plaintiff. In contrast to the general congresses, many of these items concern resolution of disputes between two parties. These sometimes include disputants from other communities, in which case the first chief may be asked by the local congress to negotiate directly with the other community's chief and congress for a satisfactory settlement. Such inter-community disputes seldom, if ever, reach the general congress—a logical forum for resolution.

Votes on action items are almost never taken, and each matter is resolved by the sense of the meeting in much the same fashion as in a general congress. When agreement cannot be attained on a particularly divisive issue, a community may fission. Several communities in the past have literally split in two. Allegiance to Roman Catholic or Baptist churches, for example has been such a schismatic issue. In these cases one island may have two congress houses and two sets of officials; yet the households pertaining to each may be interspersed among each others' throughout the island.

Local and general congresses compared. As an outgrowth, the Kuna general congress is true to its prototype in many respects. Like that prototype, it is insistently open to all comers. Like the local congress, it considers almost anything to be within it's purview for reporting and discussing. However, it differs noticeably in it's timing and frequency. Modelled on a frequent, small-scale representative institution, the general congress is an infrequent, large-scale institution which tries to be equally as representative.

However, there are some very notable organizational differences between the two forms. The general congress, like the local prototype, has (currently) as many as six chiefs. During my field work the first high chief was not a traditional chief, but rather a man skilled in Panamanian ways and the ins and outs of the national bureaucracy. These high chiefs, unlike the local chiefs, have no body of their own tribal interpreters, *arkarkana;* nor does the tribal congress have it's own set of *sualipkana*, staff-bearer constables to enforce congress decisions. In this sense, the general congress only imperfectly replicates the organizational form of the local congress.

Los Congresos Generales y locales difieren en duración, frecuencia y organización. Sin embargo hay fuertes similitudes entre ambos.

Nonetheless there are strong echoes of local form in the general congress. The high chiefs do have appointed advisors (Sp. *asesores*) to advise them throughout the year. The staff of the head table do act as spokesmen for the high chief; and in this sense repeat the vocero function. Moreover the two committees of the general congress—the local organizing committee and the head table which play such prominent roles in its procedures—closely resemble the local task forces.

The lack of an effective police force or corps of constables is more than a glaring contrast; it is an omission with major political consequences, since the general congress has great difficulty enforcing decisions upon a recalcitrant local community. It must rely upon the moral suasion of its influence—invoking the norm of a consensual decision openly arrived upon—or negotiate with several local communities and their chiefs while invoking the national guard detachments stationed on the islands. It falls upon the high chiefs to undertake such negotiations.

El Congreso General tiene dificultades en ejecutar las decisiones.

The relation of procedures to the resolution of issues. I have deliberately not examined here the nature of the action items considered in the Kuna general congress, nor of their resolution, but it is important to note that some of the major issues considered have been reflections of procedural conditions. A major vague and undefined feature of Kuna politics is succession to the high chieftainship. The tribal charter is silent on the question. At the local level, traditionally chiefs are ousted and instated by consensus of the assembled adult public, being selected from among those eligible to the chieftainship by training. Even in those three communities which have formal, written rules of electing their chiefs every four years, chiefs are in fact often ousted and instated by the other method—popular acclamation. Succession is thus like any other issue and so treated in Kuna local and general congresses. This point, deeply embedded in Kuna political culture, was not at all understood by the Panamanian officials concerned during the schismatic tribal crisis of 1976–1977.

El Congreso General puede desconocer a los altos jerarcas en cualquier momento.

However, by successfully weathering that crisis the general congress proved its viability as an instrument of Kuna statescraft in times of increasing pressures from the national state. It resolved the crisis in typical Kuna fashion, by appointing the dissident high chiefs named to that post by a general congress a year earlier as auxiliary chiefs. Thus San Blas ended up with six high chiefs instead of three; something perfectly in keeping with the local model since Kuna communities have from three to six chiefs in any case.

Moreover that crisis was provoked in part by inadequate understanding of Kuna general congress procedures on the part of Panamanian officials. The issue which touched off the crisis was the government's desire to build a large scale tourist hotel in San Blas. IPAT, the agency concerned, several times communicated its plans to the general congress. IPAT made two mistakes. First it presented the plans in the context of reports, *informes*, rather than action items. In this context the report was received passively and sceptically by the delegates, who did not really believe that the government was going to spend millions of dollars to build an artificial island for the resort. Second, IPAT got permission to conduct feasibility studies, not from the general congress, but from the three high chiefs, not realizing that the congress both in tradition and by charter, can overrule the three high chiefs at any time. It would have been better to have had the question out early on in the form of an action item. (See Moore 1980 for an extended discussion of this crisis.)

Comparison with Councils of Other Tribal Confederacies

Rather arbitrarily, partly because the sources are relatively good, I compare the Kuna general congress with some other face-to-face representative councils of native confederacies. These are the Council of 44 of the Cheyenne (described in Hoebel 1954, 1960) and the Council of Peace Chiefs of the Huron Confederation (Trigger 1969). I discuss the Cheyenne political system at length elsewhere (Moore 1978: 145:154). Here, as far as the brief sources allow, we shall answer the empirical questions of these councils posed earlier of the Kuna general congress, the question of who, when, where and what.

The Council of 44 of the Cheyenne. The Cheyenne council was composed of 44 Peace Chiefs, elders (past 50 years), former warriors and perhaps even former War Chiefs (heads of military fraternities). Each was the head of an extended family; each had sponsored a major tribal feast in the past, and was thus a master of tribal ritual and dance. There were four Peace Chiefs for each of the ten bands of the Cheyenne nation. Each chief served a ten year term, and then nominated a successor. In addition four Sacred Chiefs presided over the council. These were selected by the council from among those men who had served a full term as Peace Chiefs; thus they were likely to be senior to the rest.

El Consejo de 44 Cheyennes difiere del Congreso General Cuna en la selección y duración de los jefes, en las responsabilidades rituales y en que poseen poder policial.

The Council of 44 met on a fixed yearly schedule, convening at the start of the summer camp when the ten bands of the Cheyenne nation came together in the spring. They remained in session during the summer.

The council had important ritual functions. They were custodians of the sacred arrows, symbols of the tribe, which had to be cleansed ritually in the "Arrow Renewal Ceremony" whenever a Cheyenne murdered a fellow Cheyenne. When the murderer returned from self-imposed exile (a cooling-off period) the council imposed his fine and ritual penance. The Council, then, was a court for internal disputes.

In addition the council debated and set tribal foreign policy each season. They decided which enemies the Cheyenne war parties might move against, and which nearby tribes would be friends. Within this framework small war parties of braves were allowed to go on the warpath. Major revenge expeditions after loss of life in battle were also policy matters for the council to decide. Such major campaigns were undertaken by the military fraternities.

For internal order the council had an effective police power in the seven military societies, one of which was appointed each year as the "Buffalo Police" to keep order in the annual summer ceremonies and to maintain silence and discipline when the hunters moved out later to stampede herds of buffalos (bison).

The Huron Confederacy. The Huron Indians of what is now the province of Ontario, Canada, formed a confederation of tribes, each itself divided into villages. The Confederacy Council was composed of Peace Chiefs, men who also formed the much smaller village councils. Each village had one chief per resident matrilineage. These succeeded to office upon the death of the incumbent, a senior relative. Each new chief was by that time an elder, a former warrior with many war honors, and a wealthy man with many trading partners. Consequently he had given many feasts in the past, and was man with many friends. Upon succeeding to the Peace Chieftanship he was invested in the office at the annual meeting of the confederate council. This investiture which was an elaborate feast and ritual took the form of the "return from the dead" of the deceased incumbent, whose chiefly name the successor took.

La Confederación de Hurones difiere del Congreso General Cuna en la relación y duración de los jefes. Sin embargo, es similar en que carece de poder policial y en el arbitraje de discusiones durante la Confederación.

At home the small council of village Peace Chiefs decided policy within the community and resolved disputes. The house of the senior Peace Chief was the village council house. Theoretically each tribe of Huron also formed a tribal council, but this met only intermittantly, in the tribal capital village, in its council house, also the home of its senior chief. This council was definitely activated only once every ten years to hold the collective reburial and Feast of the Dead for the entire tribe.

The confederate council, composed of all the Peace Chiefs from all the villages of each tribe, met regularly each year for two weeks in the spring in the council house of the confederate capital (and thus in the senior chief's house of the senior tribe). Here one prime item of business at the start was the investing of new chiefs with much feasting. They also settled disputes within the confederacy. They were also, as senior trading partners with many non-Huron allies, concerned with cultivating peaceful relations with certain foreign groups, the Athabascan to the north. Unlike the Cheyenne Council of 44, however, the Huron Confederate council lacked an effective police force.

Indeed in contrast there was a much more loosely defined confederate war council, comprised of brilliant senior warriors with many past war honors who met at the tribal capitals during the winter and at the confederate council during the spring. Its membership was variable and voluntary, composed of men who wanted to mount war expeditions that season. It is possible that the activities of the war council were loosely coordinated with those of the Peace Chiefs. In any event, the two had different poses. The Peace Chiefs were poised to maintain peace within the nation and cultivate a diplomacy of peace toward certain neighbors. The war council was posed to channel aggressive energies outside the tribe—against traditional enemies with whom there were always many blood revenge scores to even.

The council compared. The two tribal councils examined were poised to maintain internal and external peace. Their diplomacy was backed by the threat of impetuous war parties guided by council policy itself (with the Cheyenne) or by a separate war council. In either case much of their pose was external, and such is the case with the Kuna general congress today. It is the arena for the collective interface of San Blas Kuna with the outside.

However, beyond that similarity of function, there are striking differences of form. Although the Kuna make up their council with delegates from the constituent communities, these delegates are not permanent, nor is council office a chieftainship in itself, held for fixed terms or for life. These conditions immeasureably strenthened the authority and prestige of those earlier councils. The two councils examined and conducted major rituals—indeed rites of intensification symbolizing the entire nation. In contrast the Kuna general congress does not have special ritual of its own—its chanting is inaugural only. Moreover, as Holloman notes (1969: 340–341) regional singing congresses—one for each of the three blocs of communities—are taking that major collective ritual function away from the general congress which is becoming increasingly secularized. These regional singing congresses convene delegates from the participating communities exclusively for as long as five days of singing. (In fact in informal lobbying, there are also major places for articulating attitudes toward general congress issues beforehand.)

We do not have readily accessible data that would provide a procedural model of the sort I have detailed for the two other tribal councils. Like most such councils, however, these also operated by a rule of consensus. Like them, this consensus represented the internal unity of the group against foreign opponents.

Conclusions

The Kuna general congress has proved a remarkable instrument of native statescraft. It has maintained its viability in the face of the tendency of Panamanian officials to treat it as a sounding board rather than a deliberative and policy making body. Based on the local prototype, the tribal congress performs many of the local congress's functions of conflict resolution—especially of conflict between different communities—of succession to high office, and of presenting a common front to outsiders, in this case of the Panamanian nation-state. Unlike the tribal councils which we have examined, the congress is not so much a council of authoritative peace chiefs with direct ritual controls over blood-thirsty younger war chiefs and warriors, but rather is a convocation of ever-changing delegates from the constituent Kuna communities.

Los Consejos operan por consenso representando la unidad interna del grupo en contra de oponentes foráneos.

Several structural issues need to be addressed if the general congress is indeed to survive as a policy-making organ. The nature of its membership, timing, secretarial and support staff, and police powers shall all have to be addressed. It is likely that delegate status may become a more fixed and authoritative office. Were the Kuna to follow the examples I have discussed, delegate status would be an honorific, chiefly office in its own right, occupied by distinguished and learned men for fixed terms or for life. To follow Kuna methods, the general congress would select its own members. However, a "modern" electoral, representative model is sure to be discussed. The case of the Navajo tribal council demonstrates that the imposition of such an electoral model can sometimes be successful among native peoples (Shepardson 1963).

The procedures of the Kuna general congress reflect a gradual acculturation to the general or "standing orders" of Western parliaments. This development in turn must reflect the influence of Panamanian advisors, who quite unconsciously tell the Kuna how to conduct a proper meeting. "Standing orders" are best known in the United States in that widely consulted manual, *Robert's Rules of Order* (1971/1915). In spite of the adoption of standing order of business, the actual conduct of Kuna debates and the formulation of resolution after debates bears no relation whatever to western procedures of making motions, amending them and submitting them to vote. I have explored these differences elsewhere (Moore 1981b), but I have yet to fully explicate the Kuna process of forging consensus. What is most remarkable about the Kuna general congress, then, is its preservation of a Kuna style of decision making, and a viable democratic method at that, in the face of the well meaning attempts of Panamanian officials to impose their own developmental schemes upon the San Blas Kuna without true prior deliberation.

A pesar de que los Congresos Generales Cuna reflejan similitud con los cuerpos parlamentarios occidentales, la conducción de debates no presenta ningún parecido con los procedimientos occidentales de discutir proposiciones, etc. En el futuro el Congreso General mostrará más evolución estructural.

It is very likely that the Kuna shall, partly at Panamanian behest and partly following their own reformist impulses, carry out further structural and constitutional reforms of their tribal government over the decades to come. Hopefully any such structural and consitutional reforms shall be undertaken by the Kuna themselves acting through their congress, rather than imposed from above. In any case the Kuna are vociferous in upholding the merit of their highest native governmental body. That merit deserves careful strengthening.

Acknowledgments

This paper reports on field work in Panama from December 1976 to June 1978 supported by the National Science Foundation (Grant no. BNS76–11900).

Arensberg, C. & S. T. *Kimball*. 1965. Culture and Community. Harcourt, Brace & World, New York.

Hoebel, E. A. 1954. The Law of Primitive Man. Harvard Univ. Press, Cambridge, MA.

———. 1960. The Cheyennes: Indians of the Great Plains. Holt, Rinehart & Winston, New York.

Holloman, R. 1969. Developmental Change in San Blas. Doctoral diss., Northwestern Univ., Evanston, IL.

Howe, J. 1974. Village Political Organization Among the San Blas Cuna. Doctoral diss., anthropology. Univ. Pennsylvania. University Microfilms, Ann Arbor.

———. 1976. Smoking out the spirits: a Cuna exorcism, ritual and symbol in native Central America. P. Young & J. Howe, eds. Univ. Oregon Anthrop. Papers. 9: 69–76.

Moore, A. 1978. Cultural Anthropology. Harper & Row, New York.

———. 1980. Planners, tourists, and Indians: national planning, regional development projects and the San Blas Cuna. Practicing Anthrop. 2: 5–6, 19–20 (Summer).

———. 1981a. Basilicas and king posts: a proxemic and symbolic event analysis of competing public architecture among the San Blas Cuna. Amer. Ethnol. 8: 2: 259–277.

———. 1981b. Lore and life: Cuna Indian pageants, exorcism and diplomacy in the 20th Century. Ethnohist. 4 (Summer).

———. 1981c. Cuna Indian rules of order and Robert's Rules of Order: an exercise in comparative parliamentarianism. Paper read at annual meeting of Soc. for Appl. Anthrop., Edinburgh, Scotland.

Richards, A. & A. *Kuper*, eds. 1971. Councils in Action. Cambridge Univ. Press, Cambridge, MA.

Robert, General Henry M. 1971 (1915). Robert's Rules of Order Revised, with a Foreward by Henry M. Robert III. Morrow Quill, New York.

Shepardson, M. 1963. Navajo ways in government: a study in political process. Manasha, WI: Amer. Anthrop. Assoc. Mem. 96.

Sherzer, J. 1974. *Namakke, sunmakke, kormakke*: three types of Cuna speech event. *In* R. Bauman & J. Sherzer, eds., Explorations in the Ethnography of speaking. Cambridge Univ. Press, New York.

Sherzer, D. & J. *Sherzer*. 1972. Literature in San Blas: Discovering the Cuna *Ikala*. Semiotica 6(2): 182–199.

Trigger, B. 1969. The Huron. Holt, Rinehart, and Winston, New York.

Transport of Deer by Choco

Owen J. Sexton

Un indio chocoe puede cargar un venado recién cazado sobre su espalda, mediante bandas cortadas en los puntos adecuados de la piel del animal muerto y las cuales ajusta a su cabeza, hombro y brazo.

Information about the customs of the Choco Indians of the Darien is limited and scattered. This note describes how large game animals are transported by single individuals.

On 29 March, 1962 two other investigators and I accompanied a middle-aged Choco, Sr. Porras, from our encampment on the bank of the Rio Chucunaque to the fore-runner of the Pan-American Highway.

PLATE 1
Figs. 1 and 2. Choco making ready to transport a deer. 1, Passing a strip of bark through slit 1; 2, Placing the head strap in position in slits 1 and 3. The relationship between the several positions of the limbs are barely discernable at the left of the picture.

Our route paralleled the Rio Canclon (Canglon), a tributary of the Chucunaque. At that time most of the area was heavily forested.

Upon our return, Sr. Porras saw a deer. He fired at it twice with a 22 caliber rifle. The first shot missed, but the second one hit the doe in the jaw. The animal remained upright but stunned. I grabbed the doe by the hind legs and threw it to the ground. Sr. Porras obtained a dead, fallen branch and killed the deer by clubbing it over the head.

He then made preparations for transporting the carcass back to camp. He cut off, using a knife, the hind limbs at the ankle. He did the same with the limbs at the wrist except that the forelimb was not completely severed. Instead, he left intact a tendon running between the foot and the upper part of the limb. A slit was then made through the distal part of the remaining hind limbs, and the partially severed portion of the anterior limb (below the cut) was passed through the slit. Thus, the fore and hind limb were more-or-less bound together.

Sr. Porras then cut a slit through the skin at the back of the nape of the neck. The location of the slit was approximately above the pectoral girdle. A second slit was cut through the skin just above the base of the vertebral column. A third slit was made about the level of the sternum, and the fourth and final one was made on the posterior portion of the right thigh. Then Sr. Porras cut some strips of bark. One was tied to slits 1 and 2 (Fig. 1) and another to slits 3 and 4. These two anchored strips served as arm and shoulder straps. A shorter strip was tied to slits 1 and 3, and it served as a head strap (Fig. 2). The deer was carried on Sr. Porras' back with the deer's back to Porras' right and with the deer's head upper.

It is interesting to note that the deer was neither bled nor eviscerated in the field.

Plants and Animals in the Language of Kuna Magic[1]

Dina Sherzer

Los Cunas de San Blas creen en la existencia de un mundo de espíritus presentes en los animales comunes los cuales muestran características humanas: ellos poseen sus debilidades y su fortaleza, pueden causar el bien y el mal, pueden producir enfermedades y curaciones y les agrada que se les divierta utilizando lenguaje poético. Las técnicas poéticas usadas para obtener los resultados deseados son ilustradas por "Naipe Ikar" que es el canto del espíritu de la culebra.

Flora and fauna are intimately involved in the life of the San Blas Kuna who live from hunting, farming, and fishing in the nearby jungle and Atlantic Ocean. In addition to being resources of subsistence, plants and animals, especially plants, are central to Kuna curing and magical practices. A plant or an animal can be used for its medicinal properties, but in conjunction with this medicine, spirits of plants and animals must intervene for the desired effect to happen.

Plants and animals of the Kuna environment appear also in the *molas*, the blouses worn by women, which can represent toucans, iguanas, fish, turtles, or banana leaves, for example. Sculptured wooden items found in Kuna households such as walking sticks and small stools are also decorated or are in the shape of animals.

Kuna Magic: Beliefs and Practices

The Kuna believe in a supernatural world of spirits in which there are spirits of men, animals, plants, and objects. An exact isomorphism exists between this supernatural world and the real, concrete world, in the sense that the spirits, whether human, animal, plant, or objects, behave like the Kuna do in everyday life. They eat, drink, cook, fish, hunt, and smoke tobacco; they like poetry, stories, and tales; and they enjoy laughing and joking like the Kuna. This world of spirits, copresent with the world of men, is constantly interacting with it. Good spirits can intervene to cure a sickness, to help a hunter to catch an animal, to prevent a snake bite, to lead the soul of a dead person to the afterworld, or to help in difficult childbirth. Bad spirits exist too and are responsible for sickness, for difficult childbirth or for misfortunes. A good spirit can neutralize the action of a bad one.

El "Ikar Wisit" puede communicarse con los espíritus. El puede cantarle a la culebra, a la planta, a la pimienta o al espíritu de los árboles para que proteja y controle las culebras, para que ayude al cazador, para que cure la fiebre y el dolor de cabeza.

The *Ikar Wisit* is a specialist who knows how to communicate with the spirits in order to make them act. He is an individual who has learned appropriate magical incantation by memory. The *Ikar Wisit* does not have any special characteristics such as special birth, body marks, gifts of premonition, ability to predict the future; he is an agriculturist like any Kuna, who has memorized a certain number of magical texts word by word. Most magical texts used today in Kuna life have to do with animals and plants found in the local environment, jungle, island, and ocean, where the Kuna practice slash and burn agriculture, hunt, and fish. For instance *Pisep Ikar* is chanted to the spirit of the *pisep* plant, a kind of basil, in order to make a hunter successful, that is to make him attractive to animals so that he can catch them; *Masar Ikar* is chanted to the spirit of *masar*, a type of bamboo or cane which leads the soul of a dead person to the

[1]This paper is dedicated to Reina Torres de Araúz.

afterworld; *Kapur Ikar* is sung to the spirit of hot pepper plants to ward off fever; *Kurkin Ikar* is addressed to the spirits of trees to ward off headache; *Naipe Ikar* is addressed to the spirit of a snake to grab and control the actual snake; *Ukkunaipe Ikar* is addressed to the spirit of the rattlesnake, which is considered a protector against other snakes; *Pulu Ikar* and *Yata Ikar* are addressed to the spirits of wasps and bees to control them; *Yauk Ikar* is addressed to the spirit of the sea turtle so that sea turtles let themselves be caught; and *Inna Sopet Ikar* is chanted to the spirit of the sugar cane in order to make cane juice ferment properly for consumption at puberty festivals. Other aspects of magical practices involving plants and animals which are not treated in this paper need to be mentioned. The *Nuchu* or stick dolls made of wood and found in every Kuna household represent good spirits. They bear the name of the tree they are made of. The *Inatulet*—pharmaceutical specialist—goes to the jungle to find the plants he needs to cure several diseases. When he cuts them and before making his medicine he chants to their spirit to activate their benefic properties.

Several characteristics inherent to Kuna magical texts and to their context of performance make them efficacious. Although a specialist might recite or chant a text in the presence of a person, an object, or a plant, the addressees are not the persons or objects or plants of this world but one or several spirits in the spirit world. The text is worded in a variety of the Kuna language that the spirits understand. It is not everyday Kuna, not the type of Kuna that political leaders use, nor the type of Kuna performed in the puberty rites. It is a variety of Kuna with its own vocabulary and grammar and which is not intelligible to the non-initiated. The function of the magical text is to activate the spirits; it does so by telling them what to do. Therefore magical texts are narratives which describe situations that are supposed to exist and actions which are supposed to happen at the moment they are uttered. Several verbal strategies are deployed in the texts in order to make the spirits act. The text can say to the spirit "I advise you to do something", or it can describe a battle which is going on in the spirit world between good and bad spirits. This battle is never violent or destructive but rather it stages verbal games and trickster deeds in which cunning and intelligence play crucial roles.

La culebra es controlada, cuando el especialista conoce su origen y esencia.

The other way to neutralize the bad spirit is to tell him that his origin and essence are known to the specialist. Realizing that the specialist has that kind of knowledge about his private and intimate life causes the spirit to let itself be controlled. The chants do not give orders, do not contain hostile statements, or any aggressivity. Rather, a friendly relationship is sought in which the spirits let themselves be convinced by means of poetry, verbal virtuosity, and playfulness. The length of the texts is also to be considered, and the texts can be shortened or lengthened according to the intensity, virulence, or persistence of the evil. The texts are efficacious and work in the society because the people believe in their value. *Naipe Ikar* is one of the several magical texts addressed to the spirits of snakes in order to prevent their malefic power.

An Example of a Magical Text: Naipe Ikar

In order to understand its value it is necessary to discuss its importance in Kuna society and the contexts of its performances. Since it is a secret text, it is not possible to quote any Kuna expressions here. *Naipe Ikar* is chanted to make a snake crawl across the back of the hand of a specialist who then raises his hand and shows the snake hanging in the air. This chanting which results in a display of virtuosity and daring without any seemingly immediate benefit is in fact in a very important act. The Kuna live on islands where there are no snakes, but they farm in the jungle where they see dangerous snakes and where snakebites sometimes occur. It is also believed that some individuals are born characteristically prone to snake bites. This is known at birth, and the midwife informs the mother about this fact. Furthermore, not only the snakes themselves are dangerous, but their spirits are as well. Even if a person is not bitten he or she can become sick because of a snake spirit. Therefore the danger of snakes or of their spirits is omnipresent in Kuna life and needs to be attended to. *Naipe Ikar* is one of the several magical texts addressed to the spirits of snakes in order to prevent their malefic power.

Naipe Ikar can be performed in different circumstances and contexts, as can other magical texts. A specialist can chant it in the presence of a snake and thus raise it without being bitten. Its content and its incantatory quality overcome the spirit of the snake. This magical act shows to the spirit of the snake that the specialist knows how to control it and consequently the spirit of the snake will not act against humans. The specialist can chant this text at his home, in his hammock, in the absence of any snake. He might chant to keep in mind his text or to practice it. The specialist can chant this text in his home in order to teach a student. The specialist then lies in his hammock with no snake around and chants line by line, and the student repeats. During the puberty rites of young girls people like to listen to magical incantations, and it is possible for someone to pay the specialist to chant *Naipe Ikar* because he likes its content or the voice of the specialist. In each of these circumstances, whether the specialist chants in front of a snake to charm it, or in his own house for practice or teaching purposes, or for the enjoyment of others, the spirit of the snake hears this chanting and realizes that the specialist knows how to control it. Overtly and covertly this text is always active, always magical, since it is always heard by the spirit, and even if it is only thought by the specialist it has power.

Poetics and Magic

El "Naipe Ikar," se dirige a la culebra y el contenido, vocabulario, gramática, y patrones de sonido trabajan en conjunto.

In *Naipe Ikar* as well as in other Kuna magical texts, the content, the vocabulary, the grammar, and the sound patterns work together to bring about the wanted result. The content affects the spirit by what it says; the vocabulary, the grammar, and the sound patterns do so because of the incantatory effects they create. *Naipe Ikar* is a third person narrative which stages the meeting of two characters, a specialist and a snake, in a particular setting, a field in the jungle at mid-morning. The specialist is cutting brush in his field when suddenly he sees a snake. From this point on, the snake becomes a protagonist described as having body parts like humans and speaking. He is called by a name which indicates that he is a young male; his tongue is described as having the dark blue color of the *koka* leaves used for dye; he raises his chin, sticking out his tongue, saliva out of his mouth. The snake asks the specialist: "Do you know my origins?" to which the specialist answers yes. Then the snake moves his "hooks" up and down. The specialist calls on several plants to help him, to give him strength, to weaken the snake; and to protect his hand he covers it with a substance. Then follows a verbal show of power by the specialist who lists parts of the body of the snake; lips, chin, pupils, point of tongue, arrows, tie, flat head, and spinal cord and tells him that he knows the origins of all these parts. These are the words of the text and they are names of human parts and clothes because what is described is a scene in the world of spirits where animals are like men. The outcome is that the snake is overpowered and crawls on the back of the hand of the specialist without biting it. The climax occurs when the snake is lifted in the air by the specialist and hangs over the specialist's hand like a rope, says the text. In the last lines of the text the specialist becomes a merciful hero. The snake asks if he is going to be killed and the specialist answers "no, because we are friends now". This narrative contains no esoteric component whose symbolism has to be explained. Rather it is an extremely realistic text which means what it says. This realistic and even banal content, however, is expressed in an esoteric vocabulary, which is understood by the spirit of the snake. For instance the names given to the spirit of the snake and to the spirits of the plants are magical names, different from everyday names of the snake or of plants. The name of the snake does not have a particular meaning except that it indicates masculinity and has a traditional prefix *olo-* and suffix- *appi* which are characteristic of names in the language of magic. The names of the plants contain statements of medicinal properties that the specialist wants them to have (medicine which renders fangs weak, medicine which causes fangs to double up). Grammatically this text is characterized by a great number of special nominal and especially verbal affixes, by non-elided and long and full forms and by complete sentences. Another device is the massive use of parallelisms created by systematic repetitions. The same word such as the name of the snake and the word specialist reappear in consecutive lines or in several lines but without any regularity, forming an intricate network of phonic and formal repetitions. The same types of sounds repeated over and over create an incantatory effect, because of the regular rhythm they set up.

From the Kuna point of view the spirit of the snake understands the magical language of the text, appreciates the special vocabulary, enjoys the embellishments brought about by grammar, and likes the sounds that produce an incantatory effect. Since he is pleased he acts positively. What the spirit likes, and the Kuna appreciate these qualities too, is what we call poetry. Poetry is generated by special vocabulary, by grammatical manipulations and by sound patterns.

En la sociedad cuna analfabeta, gramática es poesía y poesía es magia.

In this oral text coming from a nonliterate society, grammar is poetry and poetry is magic. And Kuna verbal magical practices, summoning the intervention of the spirits of plants and animals in order to obtain positive results and effects for humans, constitute a mediation between men, nature, and the world of spirits.

What's in a Name in San Blas? Kuna Plant, Animal, and Human Names

Joel Sherzer[1]

En el lenguaje Cuna, muchos objetos y actividades, especialmente plantas y animales, poseen varios nombres diferentes. Los indios Cunas tienen los nombres comunes, los usados en ritos y ceremonias, los utilizados de noche, tales como eufemismos y apodos. Frecuentemente, objetos nuevos al mundo Cuna adquieren nombres Cunas adicionales a aquellos utilizados en Español.

It is useful to classify Kuna ritual-speech into two basic types. On the one hand, there is the language of political leaders, manifested in chants and speeches in the gathering or meeting house, performed for the people of a village, and dealing with myths, legends, tribal history, and, more immediately, political matters. On the other hand, there is the more secretive and esoteric language of curing, magic, and puberty, addressed to spirits and used to control plants and animals and to cause events to happen. One of the ways in which differences between everyday and ritual-ceremonial speech are conceived and labelled is in terms of a set of *kaya*, which means language, dialect, linguistic variety, or linguistic style. Thus in addition to *tule kaya* (the *tule*, people, or Kuna language), the Kuna colloquial language of everyday, ordinary conversation and interaction, there are three ritual and ceremonial languages—*sakla kaya*, the language of the *saklas* or political leaders, used in the gathering house; *suar mimmi kaya*, the spirit language; and *kantule kaya*, the language of the *kantule* or master of ceremonies at girls' puberty rites, which is addressed to and understood by a flute spirit which is central to these rites. The differences among *kaya* involve both grammar and vocabulary. But the most striking feature distinguishing linguistic varieties and the one that is most diagnostic for the Kuna themselves is vocabulary. Another way to conceive of this situation, and one that is clearly in keeping with Kuna social and cultural conceptions and considerations, is that many objects and activities, especially basic features of the Kuna ecology and important aspects of Kuna society and culture, have associated with them not one, but a complex of names, each of which is appropriate for use in a different, well-defined situation. Vocabulary and name differentiation is used to distinguish the four major Kuna language varieties, *tule kaya*, *sakla kaya*, *suar mimmi kaya*, and *kantule kaya*, as well as to make subcategorizations within them. There are several ways in which this works.

Muchos objetos y actividades poseen un nombre diferente en cada situación. Existe el lenguaje común y diario, el de los líderes políticos, el de magia, curación, pubertad y otros.

1. First, the neatest, categorical instances, the ones the Kuna themselves cite most often as examples. In these, the same meaning or concept is translated by a distinct vocabulary item in each or some of the different linguistic varieties (Table 1).

2. In some cases the differences between Kuna linguistic varieties involves the use of distinct prefixes and suffixes, with the same or different lexical stem (Table 2). The prefixes *olo-*, *ina-*, and *ikwa-* and the suffixes *-kinya*, *-nappi*, *-lele*, *-tili*, and *-liler* are used

TABLE 1
LEXICAL DIFFERENTIATION IN FOUR KUNA LINGUISTIC VARIETIES

English	Tule kaya	Sakla kaya
foreigner	*waka*	*tulepiitti*
		Suar mimmi kaya
woman	*ome*	*walepunkwa*
eye	*ipya*	*tala*
		Kantule kaya
woman	*ome*	*yai*
cup	*noka*	*kila*

[1]This paper is dedicated to Reina Torres de Araúz.

TABLE 2
PREFIXES AND SUFFIXES DISTINGUISHING KUNA LINGUISTIC VARIETIES

English	tule kaya	sakla kaya
Kuna person	*tule*	*olo-tule*
necktie	*korpatta*	*olo-korpatta*
		Suar mimmi kaya
hammock	*kachi*	*ipepo-kachi-pilli*
blouse, shirt	*mola*	*uu-mola*
		Kantule kaya
blouse	*mola*	*ilukka-mola*
beads	*wini*	*esa-wini-kachi*

with nouns in both the *sakla kaya* gathering house and *suar mimmi* curing variety of Kuna. They are involved in the formation of the *nuka sunnat* (true names of the objects labelled by the nouns. According to the Kuna world view, plants, animals, and objects have spirits, as do humans; some plants, animals and objects were once human. The *nuka sunnat* are their human names. Humans, especially women, also have *nuka sunnat.* Human *nuka sunnat* also make use of these same prefixes and suffixes. Now it is mainly older men who are publicly called by their traditional, true names, such as *Nipakkinya*, *Mantiwekinya*, and *Olowitinappi.* Since Spanish and English names are now much more common, at least in everyday usage, the more traditional names for both humans and animals are felt to be quite humorous out of their serious contexts of usage. In fact, some of the animal *nuka sunnat*, such as *olopiskaliler* for *tulup* (lobster), are used playfully and jokingly, like the *totoet* (play) names or nicknames and riddle names of animals.

In addition to this vocabulary differentiation marking the four major linguistic varieties, there is further elaboration and specialization within each. Such vocabulary elaboration and specialization is most characteristic of certain areas of Kuna vocabulary—ecology (especially animals and plants), celestial bodies, socio-political organization, and curing. Table 3 gives examples of lexical elaboration and differentiation of animal names. Notice the use of the prefixes *olo-*, *ina-*, and *mani-* and the suffixes *-lele*, *-liler*, and *-appi*, which indicate that these are not merely different words for tapir and peccary, but actually different names for them.

TABLE 3
ANIMAL NAMES IN *SAKLA KAYA* AND *SUAR MIMMI KAYA*

English	Tule kaya	Sakla kaya	Suar mimmi kaya
tapir	*moli*	*oloalikinyalilele*	*ekwilamakkatola*
		olohalikinyappi	*ekwirmakka*
		oloswikinyaliler	
collared	*wettar*	*olomurkikkaliler*	*manikwillosakpia*
peccary		*olomurkippilele*	*murkimanikwillosakpia*
		inamuikkaliler	
		olomoikinyappi	

Dentro de cada variedad lingüística aún existe más elaboración.

3. Figurative, symbolic, metaphorical, and poetic language is a major distinguisher between everyday Kuna and ritual-ceremonial speech. Viewed in terms of vocabulary and names, we can note that words in ritual, ceremonial varieties may be words from colloquial Kuna that take on a different meaning in ceremonial contexts. The effect is metaphorical and often poetic (see Table 4).

Figurative language is often organized according to entire sets of words. Thus in *suar mimmi kaya*, the linguistic variety used in curing ritual, elements of clothing are used to represent the parts of plants and trees (Table 5).

It is especially in gathering house chanting and speaking that figurative and metaphorical speech

TABLE 4
FIGURATIVE VOCABULARY: THE THREE RITUAL CEREMONIAL VARIETIES OF KUNA

English	Tule kaya	Sakla kaya
woman	*ome*	*tuttu* (flower)
village	*kwepur,nekkwepur, tuppu* (island)	*kalu* (stronghold, of animals, spirits)
thunder	*mala*	*pap kinki sakla* (god's rifle)
		Suar mimmi kaya
be born	*kwalluleke*	*akteke* (to land)
evil spirit, sickness	*poni*	*urwetule* (angry person)
		Kantule kaya
pieces of meat	*kallis*	*ipya suli* (no eyes)
drink	*kope*	*puklu sae* (to do like a drinking gourd)

TABLE 5
CLOTHING USED TO REPRESENT PARTS OF PLANTS AND TREES

English	Tule kaya	Suar mimmi kaya
seeds	*kwakwa*	*wini* (bracelet, anklet, beads)
flowers	*tuttu*	*kurkin* (hat)
leaves	*kakan*	*mola* (blouse)

abounds. In fact, one of the most characteristic markers of a good speaker is the use and development of metaphorical language. This not only pleases the audience which appreciates the oral poetry involved, but also convinces and wins it over to a point of view, by means of a poetic rhetoric. In gathering house discourse, metaphorical vocabulary sets and semantic fields are open to individual development and creative adaptation. Thus, not only does *tuttu* (flower) represent women, but the size and color or type of flower, or their place or way of growing, can be described in order to symbolize women and children of various types or the raising of children by women. The different poles used in the construction of a Kuna house, their size and location, are used to represent the different roles in Kuna village political organization, the central pole standing for *sakla* (village leader), secondary poles standing for *arkar (saklas'* spokesmen or seconds), and other poles standing for other officials. This symbolism can be further elaborated, developed, and creatively and individually adapted, so that a central pole which is rotten stands for a bad *sakla* who must be removed from office or a pole which had its rotten part cut away stands for a *sakla* who has been reinstated (Howe, 1977).

Kuna figurative, symbolic vocabulary is not then a static set of lexical replacements. Rather there is an active, dynamic system of relationships. When a Kuna *sakla* chants or speaks about the physical structure of a Kuna house, he is commenting both about the house itself and Kuna political organization. And as he develops the metaphor in personal ways, he throws new and often poetic light on his audience's understanding of house construction, political organization, and their interrelationship.

El vocabulario figurativo no es simplemente un juego estático de reemplazo de palabras, sino que está sujeto al simbolismo creativo.

5. There are cases in which certain areas of vocabulary or lexical taxonomies are more developed, elaborated, or complicated in ritual and ceremonial varieties of Kuna than in colloquial, everyday Kuna. Specialists in ritual and ceremony who are intimately familiar with the natural, cosmological, and spirit worlds and their interrelationships are most knowledgable with regard to associated vocabulary areas and semantic fields in their full lexical depth. A good example is the taxonomy of *kaa* and *kapur*, hot Spanish peppers, which are used in curing rituals. An elaborate taxonomy of hot peppers forms the basis of a major portion of *kapur ikar* (the way of the hot pepper), an incantation which is addressed to the spirits of *kapur* and is used in calming high fever. Table 6 is a small selection of members of this *kapur* taxonomy, in both colloquial *tule kaya* and the linguistic variety of curing ritual, *suar mimmi kaya*. The list includes naturally occurring forms, such as *kaa sankwa*, as well as forms which, while not existing in nature, are part of the *kaa* spiritual world and therefore the *kaa* spiritual lexical taxonomy, such as misty *kaa*.

In one version of *kapur ikar* which I have studied, 53 types and subtypes of hot pepper are listed. Analogous ritual and ceremonial taxonomies exist for many animal and plant species. The function of knowledge of and especially utterance of long and complex vocabulary taxonomies is magical power and control. Since the incantations in which these words, really names, are uttered are addressed to spirits, they demonstrate to the spirits the performer's intimate and complete knowledge of them, and thus enable him to control them and get them to do his bidding.

Vocabulary and Name Differentiation in Everyday Speech

Vocabulary specialization and elaboration has been focused on as an important characteristic of Kuna ritual and ceremonial speech. In everyday speech as well, it is quite common for a single item, object, or activity, to be labelled with different words, or to have several names.

1. There are words which hover between everyday and ritual-ceremonial speech which exist in colloquial Kuna, but are somewhat esoteric in that not everyone knows or uses them; they are more commonly known by older men acquainted with ritual and ceremony. Those who do not know such words use more ordinary words in their place. These more esoteric words are more likely to occur in ritual, ceremonial language than in everyday, colloquial speech. The common word for bathe is *ope*. Another word meaning to bathe, *apallukke*, is considered esoteric and is rarely

TABLE 6
SOME MEMBERS OF THE *KAA* (HOT PEPPER) LEXICAL TAXONOMY

English	Tule Kaya	Suar mimmi kaya
Spanish pepper	*kapur*	*nele pinaisepa nele*
red kapur	*kapur kinnit*	*nele yolina nele*
sankwa type kaa	*ka sankwa*	*sankwalinele*
misty kaa		*nele pookwar nele*

heard in everyday speech. It is used with the meaning bathe in both curing and puberty rites incantations.

Many examples involve Kuna words for objects or concepts for which there are also Kuna words of Spanish origin, now integrated into the Kuna vocabulary, but clearly loan words in origin. The most common way to say 'to walk about' is today *pasearsae*, from Spanish *pasear*. But the more escoteric and truly Kuna way of expressing this is *pippirmakke*, literally 'to turn about.' The words for numbers are another example. It is possible to count as high using the Kuna numeral system as using borrowed Spanish lexical items, but speakers typically use Spanish loans for higher numbers. Thus, 234 according to the Kuna system is *tulattar irpo kakka turkwen kakk ampe kakka pakke* (100 twice plus 20 plus 10 plus 4), while the Spanish way is *dos cientos treinta y cuatro*. But in gatherings and meetings, where traditional language is expected and rewarded, speakers are under greater pressure to use the esoteric and somewhat cumbersome Kuna numerals. Matters can be even more complicated. The colloquial word for necktie in Kuna is *korpatta* (from Spanish *corbata*) or *morsuit* (long cloth). A common way of saying or chanting necktie in the gathering house is *olokorpatta*, making use of the ceremonial prefix *olo-*, typical of the language of the gathering house. But, still another word, *morkechu*, is known to some Kuna traditionalists and used in the gathering house.

Acuñar palabras en Cuna para evitar el préstamo de palabras del Español, es un pasatiempo cuna.

The coining of Kuna words out of native vocabulary for new objects is an active, creative process not limited to traditional and ceremonial contexts. Such words demonstrate the inventiveness and playfulness of the Kuna in language use and their cleverness at using native vocabulary when desired in order to avoid loans. Kuna men who worked in the Canal Area invented Kuna terms for the different kinds of eggs that they served American soldiers. Eggs sunnyside up are *tuttukwa* (soft); easy over, *pinnakwa* (slow, easy); scrambled eggs are *ichi ichi* (all bumpy). Ketchup is called *san okinnoeketi* (meat reddenner). Napkins are called *kaya elieti* (mouth wiper). Carpenters have Kuna names for tools. Sandpaper is called *karta muru muru* (paper full of points); a screwdriver is called *kinki unket* (nail undoer). While words for American style eggs and tools are in fairly common usage, the Kuna also take pleasure in coining new words on the spot, for immediate and personal use, even though they may often go no further than a single verbal interaction. Thus one afternoon I came upon two elderly men making small purses to use either for gifts or sale. They conversed a bit about what to call them and then came up with *mani osopet* (that which causes money to be saved).

2. Euphemisms are another example of vocabulary proliferation in everyday speech. The Kuna use euphemisms in the area of body parts, sex, childbirth, and bodily functions. The area of conception and childbirth is a particularly tabu subject for the Kuna, intimately related to curing, magic, and control. The usual euphemism for the act of conception is *mimmi pakke* (to buy a child). Some euphemisms for pregnancy are *nus yoles* (a worm entered), *ipya kummas* (the eye got burned), and *sakkan pichisa* (the wing is broken). The infant itself is euphemistically called *sikkwi* (bird). It is interesting to note here the use of the animal word in these euphemisms—worm, bird, and bird part (wing). The current common word for baby, *koe*, also means deer, and may have been euphemistic in origin.

En el lenguaje hablado se usan diariamente muchos eufemismos para las partes del cuerpo, para el sexo y para los nacimientos de los niños.

Related to euphemisms are day and night names for various animals and plants (Table 7). While few in number today, there might have been more of them at one time. The day names are tabu at night and should be replaced with the night names; if not, there will be malevolent consequences, which themselves are sometimes an aspect of the night name, such as *kaya piri*.

3. Proliferation of lexical expressions for a single meaning can be found in the everyday verbal play and joking of which the Kuna are extremely fond, in which a single word is associated with a set of *totoet nuka*, humorous play names or nicknames by which a thing is called. This is most common for animal names (Table 8). The names play on a physical characteristic of the animal.

TABLE 7
DAY NAMES AND NIGHT NAMES

English	Tule kaya day name	night name
guava	*marya*	*kaya piri* (curved mouth)
snail	*salu*	*tios uwaya* (god's ears)
peach palm	*nalup*	*ikko turpa* (spine fruit)

TABLE 8
ANIMAL *TOTOET NUKA* (PLAY NAMES)

English	Tule kaya	Totoet nuka
land crab	*suka*	*pormo yarkan* (tin can back) *kapur ipya* (hot pepper eyes) *kampulet mali* (chair legs)
lobster	*tuluppa*	*ipya kallakallat* (eyes moving back and forth) *napkiar pepe* (thorny forehead)
agouti	*usu*	*muttu saya* (black tar ass) *korta mali* (tar feet)
octopus	*kikkir*	*morpattun mali* (button feet)

Totoet nuka have several functions. They can be used by themselves as a humorous focal point in conversations whose sole purpose is recreation and amusement. They can be used in *kwento* (stories) about animals to talk about the characters in the *kwento* and embellish the humorous aspects of the performance. And they can be used as part of a riddling game *(oeke*, on the wane in recent years, in which one individual names a *totoet nuka* and challenges his partner to name the animal.

A: *pete kapur ipya kwat* (you, the one with hot pepper eyes)
B: *suka* (crab)

Humans often receive playful nicknames, frequently based on animals, whom the Kuna view as similar to humans in many ways and domains. The human nicknames pick up on a comical, physical or personality similarity between an animal and a particular human. Some names are *sikkwi seret* (old bird), for an ex-village secretary (all of whom are often called *sikkwi* 'bird') who liked to sit around and talk a lot without having much to say; *ia korki* (uncle pelican), for a young man who looks like a pelican; and *kelu* (jack, a Caribbean fish) for an albino businessman, who like the *kelu* fish, was fatty, slippery, and white.

"Totoet nuka" son apodos utilizados en cuentos y juegos.

Conclusion: What's in a Name for the Kuna

What's in a name for the Kuna—plenty. As has been shown here, many objects, especially plants, animals, and other elements and aspects of the Kuna natural world have not one but many names, ranging from humorous play and nicknames through secret and magical names which are used in ritual and ceremony. In this discussion of Kuna vocabulary and names, I have drawn attention to relationship between humans and animals, that is; in addition to their ordinary name animals might have human names and in addition to their ordinary name humans might have animal names. These names, from ritual-ceremonial to everyday playful, humorous, and joking contexts, provide insight into Kuna perceptions of and conceptions of the animal world, and especially of human-animal relationships. The cosmological as well as ecological affinities between humans and animals are reflected in Kuna naming practices. Explanations of the different Kuna names, their structure, meaning, and function, take us into the heart of Kuna politics, medicine and curing, rhetoric, poetry, and humor.

Las relaciones ecológicas entre hombres y animales se reflejan en los nombres, los cuales son importantes en la política, en la medicina y curaciones, en la retórica, en la poesía y en el humor de los Cunas.

Acknowledgments

Data for this paper were collected in San Blas beginning in 1970. Research was supported by fellowships and grants from NSF, NIMH, the Guggenheim Foundation, and The University of Texas. Assistance and comments from Anselmo Urrutia Salazar, James Howe, and Dina Sherzer are gratefully acknowledged.

Howe, J. 1977. Caring the village: Cuna political metaphors, Pp. 132–163. *In* J. D. Sapir & J. C. Crocker, eds. The Social Use of Metaphor: Essays on the Anthropology of Rhetoric. Univ. Pennsylv. Press. Philadelphia.

Guaymi Socionatural Adaptations

Philip D. Young

Después de la llegada de los Europeos, la población guaymí disminuyó, pero la cultura continuó en áreas aisladas y así mismo prosiguió la agricultura de subsistencia con algunas modificaciones. La crianza de ganado por los europeos conllevó a restringir el área de las tribus y más tarde a que los guaymíes también criaran ganado. El salario por labores ejecutadas fue aumentado paulativamente y ésto ha tendido a romper el sistema de los guaymíes. La presión sobre los recursos terrestres ha contribuido a la degradación ambiental y ha resultado en la emigración y el aumento de los salarios pagados. Como quiera que el sistema no está adaptado a las corrientes de presiones externas, los cambios en la sociedad y la política de los guaymíes han comenzado.

The Guaymi today number over 50,000 and occupy approximately 6,900 km^2 of hilly and mountainous terrain in parts of the three westernmost provinces of Panama, Bocas del Toro, Chiriqui, and Veraguas. They face a host of social, economic, and ecological problems to which they must adapt if they are to survive as a culture and a people. Most of their current problems have become prominent only since the 1930s, but the roots to these are to be found much deeper in the past. The combined social and natural environmental context is referred to, following Bennett (1976), as a "socionatural" system, and Guaymi behaviors within it are called socionatural adaptations. A few words are required concerning some of the more obvious differences in the Guaymi natural environment.

The Guaymi territory straddles both sides of the continental divide. The Pacific watershed is characterized by a marked dry season, whereas the Atlantic or Caribbean slope has no true dry season. Rainfall is greater and forest cover substantially heavier on the Atlantic slopes in western Panama. Population is much heavier on the Pacific slopes, and this has contributed to a more advanced deterioration of soils and natural vegetational cover in Chiriqui and southern Veraguas than is evident in Bocas del Toro and northern Veraguas. These differences are reflected in Guaymi agricultural practices and crop preferences on opposite sides of the cordillera (Bort 1980).

Many variables make up a socionatural system, but the features of most interest here are adaptations to outside influences on the Guaymi socionatural system.

Western Panama Circa A.D. 1500

Early accounts agree that Panama at about 1500 A.D. was heavily populated, but the magnitude of the population is in doubt. Estimates range from 250,000 to two million, with 600,000 to 1,000,000 being the currently accepted range (Bennett 1968; 101; Young 1968; 43–45). If we assume that approximately 50% of the land area was then agriculturally usable, probably a generous estimate, the density (Bennett 1968; 37) based on a population of 600,000 would have been 15.2 persons per square km; and based on 1,000,000, 25.2. In western Panama, the Pacific slope was more heavily populated than the Atlantic (Colon 1959; Espinosa 1864, 1873; Oviedo 1853, 1969).

Extensive savannas and low second growth almost to the exclusion of forest cover are reported throughout western Panama in the early sixteenth century. By 1500 A.D. swidden farming had resulted in massive alterations of the natural vegetational cover, and soil deterioration from overuse of the land may have been considerable. Forests became reestablished throughout much of the area following the severe population decline immediately after European contact (Bennett 1968).

Early accounts and some archaeological data confirm the presence of endemic warfare among partially stratified rank societies throughout much of this area in the sixteenth century. The overall picture is of small-scale rank societies with varying degrees of stratification (Helms 1979) cultivating a variety of seed and root crops by swidden agricultural methods and competing by warfare and temporary alliances for scarce resources. The settlement pattern was of few nucleated villages located far apart (a day's walk or more), each surrounded by dispersed hamlets. Average population density was high, 15 persons per km^2 to use a conservative figure $(600{,}000/(79{,}378\ km^2 \times 0.5))$.

Archaeological evidence (Linares 1977) suggests intensifying resource exploitation and population expansion from about 500 A.D. which, by 1500, may have resulted in considerable environmental degradation, thus contributing to scarcity of resources and competition for these scarce resources.

The endemic warfare reported by early Spanish observers was a cultural response to control subsistence resources which, with expanding population, were becoming dangerously scarce. Vayda (1961, 1968) and Meggitt (1962) argue that endemic warfare may signal resource scarcity for an expanding population of swidden agriculturalists, and Linares (1977) makes a similar interpretation for western Panama. Helms (1979) and Linares (1977) think that some of the resources being competed for had symbolic connotations, but I contend that in Panama in 1500, the main contests were pragmatic in their goals and concerned access to food supplies. Ranking was giving way to stratification in a milieu of resource scarcity and escalating warfare.

En los alrededores de 1500, la gran población humana había creado extensas sabanas reemplazando los bosques del oeste de Panamá. Los núcleos de poblaciones estaban bastante distantes, cada uno rodeado por aldeas dispersas, sin embargo la densidad de la población era alta. Sociedades de menor jerarquía estaban involucradas en luchas endémicas como medio de competencia para obtener recursos.

Warfare and ranking were proving inadequate to ecological conditions on the isthmus in the 16th century, as the population had grown too large and the soil too exhausted. Different degrees of complexity had already made their appearance in the isthmian rank societies as adaptive responses (Fried 1967; Helms 1979). I conclude from the scant evidence that the ancestors of the Guaymi were in 1500 already perilously close to ecological disaster. European contact truncated the line of cultural evolution, and created new inputs which evoked new responses within the Guaymi socionatural system.

The Effects of Conquest

The period following the conquest of the isthmus was one of disruption and turmoil for native societies when many sociocultural patterns were terminated and others altered. We can safely assume that the population decline was enormous and its impact severe, resulting in total extinction of some groups and radical reduction in others, which in turn caused major social changes. The Guaymi population may have been reduced to as few as 4,000–5,000 at its nadir sometime in the late eighteenth century.

Swidden agriculture continued, but when pressure had been removed from the land the resource scarcity of the immediate pre-conquest period was alleviated. Forest cover became reestablished in much of western Panama, and two trends appeared. The first was assimilation into the developing *campesino* culture of western Panama. The other was of withdrawal and isolation, and it is this trend which insured the continuance of a distinctive Guaymi culture, although in a form altered from its predecessors.

Al llegar los Europeos, la población guaymí disminuyó y la agricultura mediante corte y quema continuó en los bosques, los cuales se restablecieron. La cultura guaymí prosiguió aunque algo modificada debido a que asimiló prácticas del campesino al igual que abandonó algunas de las suyas.

The earliest detailed historical account of the Guaymi is that of Fray Adrian de Ufeldre (1682), a Flemish priest who lived and worked among the Guaymi between 1622 and 1637. His account permits characterization of the Guaymi as a kin-based society of highly dispersed small hamlets, a subsistence system based on swidden agriculture, and a set of rituals validated by cosmological beliefs which permitted the differential accumulation of prestige. Some elements of ranking are still evident in Fray Adrian's time and, indeed, never completely disappear from Guaymi society. Kin group polities predominated. Warfare had apparently ceased.

The Guaymi of 1500 were a rank society with some elements of stratification, and they were participants in a general isthmian interaction sphere which in turn had connections to the east (Colombia) and west (Costa Rica and beyond). We may assess the sociocultural continuities and changes after 1500 as follows. Most of the social stratification had disappeared. Elements of ranking continued but in an essentially unstratified context. The isthmian and larger interaction spheres were thoroughly destroyed, leaving the Guaymi isolated from other native populations, encapsulated in western Panama as it were.

Slash-and-burn continued as the agricultural technology, now more efficient with metal tools. The only important additions to the pre-Columbian crop inventory were bananas and plantains (*Musa* spp.) which were introduced early, enthusiastically accepted, and quickly became dietary staples. Their cultivation presented no technological difficulties to the Guaymi who were already familiar with tree crops. *Musa* partially, but never completely, replaced native root crops. Rice did not become an important crop for the Guaymi until the 20th century, although it was introduced by the early Spaniards.

Los guaymíes restantes cortaron relaciones con otros grupos. Los bananos y las herramientas de metal fueron aceptados rápidamente, sin embargo los animales domésticos provinientes del Viejo Mundo tuvieron poca influencia.

Old World domestic animals were introduced, principally beef cattle, horses, pigs, and chickens. Their impact among the Guaymi during this early period was negligible because of their small numbers.

Later accounts suggest that Guaymi society as described by Fray Adrian persisted in a general way into the 20th century (Young 1970). Thus it may be inferred that by Fray Adrian's time a new socionatural system had emerged out of the chaotic conditions of the conquest period and that this system persisted over 300 years, keeping in mind that homeostatic systems are dynamic and that this does not preclude social adaptations to systemic variations, either internally or externally induced.

Perturbations of the System

From the mid-17th to the early 20th century, the Guaymi population remained small (estimated at only 16,000 in 1930), and no general scarcity of essential resources occurred. Scarcities of short duration, localized geographically, were absorbed by reciprocity between networks of kinsmen.

The expansion of cattle ranching in Chiriqui reduced Guaymi territory markedly in that province and produced at least three responses in addition to sporadic protests and constant ill-feelings. First, the Guaymi retreated even further into the mountains—were driven back is a more accurate statement—onto lands less able to bear the strain of swidden agriculture. Secondly, some Guaymi began to work during a part of the year on the cattle ranches. Thirdly, some Guaymi began to raise a few cattle of their own. Later, they began to sell a few.

Wage labor and cattle sales were initially *not* for supplementing subsistence resources, but rather adaptations to a desire for consumer goods. Guaymi involvement in the money market economy remained marginal, sporadic, and largely inconsequential throughout this period in that it did not interfere with the traditional subsistence economy or with established patterns of social interaction. Later on, it does.

La población guaymí permaneció pequeña hasta comienzo del siglo XX. En Chiriquí las fincas ganaderas contribuyeron a aislarlos mucho más. Algunos guaymíes comenzaron a trabajar tiempo parcial para los ganaderos, mientras que otros empezaron a criar su propio ganado. Un deseo de adquirir bienes ofrecidos al consumidor más que recursos para suplementar su subsistencia lo impulsó al mercado monetario.

Swidden agriculture remained the mainstay of Guaymi subsistence, and settlements remained small and highly dispersed. Gradual population increase produced an increasing number of dispersed hamlets, but prior to the 20th century, this apparently created no land shortage or shortening of the fallow cycle. Guaymi swidden agriculture was still partly of the pioneering type (Harris 1972). Not all of the land was occupied, and community fission and movement to unoccupied land was still possible. Some of my Guaymi friends insist that limited amounts of such unspoken-for land—usually mature forest (*kaw kri*)—were still to be had in the 1920s and 1930s.

During this period and later, the central governments subjected the Guaymi to their politico-legal systems. Such externally imposed politico-legal institutions seem to have had little impact on the Guaymi system until well into the 20th century. The Guaymi continued with their own internal regulatory system of kin group consensus and the arbitration of disputes by respected men of prestige and renown who had achieved their positions by traditional means (Young 1971, 1976; Young & Bort 1976, 1979).

El Gobierno Central tuvo poca influencia sobre el sistema Guaymí.

The principal features of the Guaymi socionatural

system that were established by the early 17th century persisted into the 20th. Population increase, cattle raising, wage labor, and other forms of involvement in a money market economy became features of the system between the 17th century and the 20th, but were quiescent with regard to ecological impact. However, they set the stage for accelerated change beginning in the 1930s and required, and still require, new adaptational responses from the Guaymi.

The Modern Era (1930s to the Present)

From the 1930s to the present, change in Guaymi society has markedly accelerated. Population has increased at more than double its pre-1930 rate. Land quality has deteriorated from overuse, and productivity has declined. Dependence on the external economy has become significant, and the traditional acephalous political system has begun to give way to a more centralized one. These and other factors have strained the Guaymi socionatural system and have provoked a variety of responses from the Guaymi as they attempt to adapt to a rapidly changing situation. The higher level of interaction between the two systems, one a traditional rank society with a primitive mode of production and a subsistence economy based on reciprocity, and the other a nation-state with an industrial mode of production and a consumer economy based on the principle of profit maximization, has served to highlight the inherent contradictions between the two. The adaptation to the influences of the alien capitalist system on their own precapitalist system has not been easy for the Guaymi, nor has it been entirely successful.

During the modern era, the Guaymi population has increased substantially, from slightly more than 16,000 in 1930 to just under 45,000 in 1970. The average annual rate of increase between 1930 and 1970 was 2.8%, that between 1960 and 1970, 2.4%. If a projection is based on the former rate, the estimated 1980 population is 59,000. The latter rate gives a more conservative estimate of 56,600.

If we accept 6,900 km^2 as the approximate area of Guaymi territory and generously allow that 50% of this area can be farmed, then the population density in 1980 is 16.4 persons per km^2 using the conservative population estimate. This is a high density for swidden agriculturalists. The situation is probably worse than implied considering the poor quality of the lands occupied by the Guaymi, and because it is not likely that 50% of Guaymi lands are truly suitable for cultivation.

A partir de 1930 la población guaymí, 16.000 en ese entonces, aumentó hasta alcanzar aproximadamente 56.600 en 1980. En ese año la densidad de la población era de 16.4 personas por km^2, densidad bastante alta tomando en cuenta que los guaymíes utilizan agricultura de corte y quema.

A consequence of increased pressure on the land has been a shortening of the swidden fallow period. In 1965 some Guaymi fields in Chiriqui were being recultivated after fallow periods as short as six years. At that time my Guaymi friends considered 12 years to be an adequate fallow period and complained of the need to recultivate some areas after fewer than 12 years, recognizing that yields were lower when they did so. At the time, I accepted rather uncritically their estimate of 12 years as a sufficient period for recovery of second growth and regeneration of the soil (Young 1971; 74–81). I now believe that a 25 year fallow period is more realistic for the Guaymi area and that any lesser period results in environmental degradation (Croat 1972). The Guaymi lamented their inability to maintain the length of fallow that had been general in their fathers' time, which was already insufficient. In other words, by the 1930s the Guaymi were already being forced to overuse their lands, and their swidden agricultural practices were beginning to cause substantial environmental degradation, the evidence of which is before us today throughout Chiriqui and southern Veraguas. Environmental degradation caused by too short a fallow cycle results in lowered yields requiring larger areas to sustain the same total yield, which in turn results in still further deterioration of the environment (Weiner 1972; Bort 1980). In the Guaymi case, the problem is exacerbated by the ruggedness of the terrain farmed and the heavy rainfall, conditions which are highly conducive to erosion once the vegetational cover has been removed. Much erosion has already taken place.

El aumento de la presión por tierra ha acortado la duración de los períodos en que la tierra no se cultiva. Anteriormente, estos períodos eran de 25 y 12 años, y actualmente han disminuido a 6 años. En los alrededores del año 1930, los guaymíes estaban sobre-utilizando sus tierras causando sustancial degradación ambiental, ésta empeorada por el terreno escabroso.

Guaymi adaptations to this developing crisis have been classic. Permanent out-migration has been a major response, but the magnitude of this is not known.

The apparent decline in the population growth rate between 1960 and 1970 may be largely due to out-migration. Other responses have been economic. Many have attempted to derive a part of family needs through wage labor and other involvement in Panama's market economy, notably the sale of cattle. Wage labor and cattle raising were established patterns before this scarcity existed as responses to desire for consumer goods. Unfortunately, these responses are not yet adequate to reestablish balance in the system.

Las respuestas clásicas han sido la emigración, la consecución de salarios a cambio de labores y la ganadería.

In the 1940s, if not before, the Guaymi began to convert some of their lands in Chiriqui into permanent pasture to engage in what, for them, was the lucrative sale of cattle. Guaymi cattle declined in number between 1950 and 1960, reputedly due to a severe "screw-worm" fly *(torsalo)* infestation, but the cattle population appears to have more than recovered by 1970 (Censos Nacionales de Panamá de 1970). This temporary decline only depressed the overall economic situation of the Guaymi; insufficient time passed for any significant upgrading of the environment (Bort 1980).

Increasing resource scarcity and a continuing demand for consumer goods has stimulated greater dependence on wage labor. To the extent that wage labor substitutes for subsistence agriculture, it might be seen as relieving some of the strain on the natural environment and therefore as a successful adaptive strategy. Realistically, there is considerable doubt that any significant reduction of pressure on the land occurs. Only part of the cash earned is spent for subsistence goods, and wage labor has not even balanced the increased need produced by the single factor of population growth.

Further, the disruptive influence of wage labor on traditional Guaymi socioeconomic practices greatly outweighs, in a systemic view, any reduction of environmental strain that may occur. Wage labor conflicts with subsistence agriculture because most of the wage labor available to the Guaymi is also agricultural and therefore peak labor needs in the two sectors overlap considerably. Land shortages resulting from population increase and decreasing yields are aggravated further by the return after prolonged absences of wage laborers who are desirous of reestablishing their land claims. Tension, prolonged disputes, the disruption of amicable relations among kinsmen, and sometimes fissioning of the kin group are the usual results. In a traditional system geared to the distribution of limited surpluses in socially patterned ways to those in need, wage laborers do not produce their share and are often chronically in need. Cash earnings have not been incorporated into the distributional system beyond the household level. Thus, the flow of aid becomes increasingly non-reciprocal. "Under these circumstances, the claims of wage laborers to a share of the harvests of their kinsmen become disruptive" (Young & Bort 1979; 85). These examples show that wage labor is not a viable response within a system that is already under great strain from other sources even though it clearly represents a successful strategy for some individuals and families. It is one of those responses that, rather than serving to reestablish balance in the existing system, is more apt to lead to a transformation of the system into something different. Wage labor is disruptive in other ways as well.

El salario por labores ha roto muchas de las tradiciones guaymíes y en algunos períodos compite con su necesidades de subsistencia. Los resultados son tensiones, disputas y división de los grupos. Las ganancias obtenidas en dinero no circularon en el sistema, tal como es el caso de las cosechas y ésto añadió tensiones al grupo.

In the past, the Guaymi had a series of rituals that, as adaptive responses, had two important features (Young 1976, 1980; Young & Bort 1976). First, these rituals served as the only avenue to prestige and renown, and substantial subsistence resources were required for sponsorship of them. Kin groups supported their aspiring members in these ventures. The result was widespread motivation to produce beyond subsistence requirements. Rituals were held in years of abundance and foregone in times of scarcity when production for ritual was used to meet subsistence needs.

Second, a core feature of ritual was the ritual sibling *(etebali)* relationship. Linkages created by this relationship extended those based on consanguinity and affinity. This expanded the network of social relationships within which distribution of subsistence resources took place.

Together, these two features of Guaymi ritual were ecologically adaptive. They contributed to equitable distribution of resources among a dispersed population occupying an environment subject to unpredictable, but localized, short-term productive fluctuations. When population increased, swidden cycles shortened and when generalized scarcity appeared in the subsistence sector, these rituals became less frequent and also less important with respect to their adaptive functions. Increasing involvement in a cash economy opened up new avenues to prestige thereby further weakening the tradition of the rituals.

The decline of ritual reached a low point in the early 1960s when rituals came to be seen by some as one of the *causes* of resource scarcity and were banned by the leadership of the then new Guaymi religion, *Mama-Chi*, (Young 1971, 1980). When rituals reappeared after a sharp decline in the influence of this religious movement, they did so stripped of their ecological adaptive functions and most of their prestige value. There is some evidence, however, that rituals have taken on new functions as symbols of ethnic identity. With increasing pressure from outside to become integrated into the larger society and to relinquish their identity as Guaymi, something which many do not wish to do, ethnic identity has become more important to the Guaymi.

En el pasado los rituales guaymíes influyeron y motivaron grupos a producir por encima de sus necesidades inmediatas. Los guaymíes también establecieron relaciones extra-familiares y éstas les permitieron la distribución directa de sus recursos. Los rituales han declinado especialmente bajo la nueva religión guaymí Mama-Chi y actualmente estos rituales se consideran símbolos de unidad étnica.

The traditional Guaymi productive system required relatively low levels of cooperative effort. A sociopolitical system based on periodic cooperative effort among kinsmen, ownership and control of access to land by kin groups, and distribution of products through networks of kinsmen served well the needs of their non-monetary economy. This

> segmentary arrangement served to dissipate and render largely ineffective . . . external attempts to control Guaymi affairs . . . With no individual or group possessing the authority to speak for all or the power to impose and enforce decisions on even a significant segment of the population, attempts at outside control tended to be ineffective beyond the individuals or small groups upon which they were directly imposed. The most noticeable and consequential flaw in this arrangement was an inability to prevent the piecemeal usurpation of their lands by outsiders (Sarsanedas 1978) (Young & Bort 1979; 79).

While this segmentary system with its ethic of mutual aid still functions and, indeed, may be essential in the subsistence sector (Sahlins 1977; 220–1), it is ineffective in coping with a cash economy or the other numerous external pressures impinging upon the Guaymi in the modern era.

El sistema de producción tradicional guaymí requiere de poco esfuerzo cooperativo y los guaymíes no están unidos políticamente. El sistema tradicional no se adapta a las presiones externas.

The response of the Guaymi to the serious disjunction between their traditional way of life and the changes brought about by involvement in the outside world took dramatic form in a nativistic religious movement that flourished during the 1960s and advocated among other things an isolationist policy (Young 1978). Concurrent with this religious reaction to the perception of crisis, some Guaymi were exploring the viability of controlling outside influences by political means, realizing full well that this type of response would require a break with the isolationist views of the religious leaders. Dissatisfaction with the tangible results of appeals to the supernatural, along with greater systemic disruption precipitated by mineral exploration activities at Cerro Colorado in the Guaymi territory, resulted in the ascendency and clear dominance of political responses as coping behavior in the 1970s.

A description of the competing internal political forces in Guaymi society has been presented elsewhere (Young & Bort 1979). Important to note here is that political action is the most recent form of reaction of the Guaymi to systemic imbalance produced by outside pressures. However, the Guaymi remain far from unified politically. Guaymi polity may still be characterized as highly segmented. The kin group remains the most important locus of power in Guaymi society, making and implementing decisions and bringing pressures to bear on individuals for conformity. But many changes have occurred in the past few years, and many more are to be anticipated as the Guaymi adapt to unrelenting external forces. In the present context, the traditional sociopolitical system is becoming increasingly maladaptive.

What Does the Future Hold?

At the beginning of the 1980s, several trends are evident. First, the Guaymi population will continue to increase, probably at about its present rate (2.4%

annually). Second, permanent outmigration will continue but this is not likely to be sufficient to relieve pressure on the land base. Third, the continuance of the practice of swidden agriculture with fallow periods too short for recovery of the soil will contribute to environmental degradation, a process already much advanced in some parts of the Guaymi territory. Fourth, increasing resource scarcity resulting from population growth and environmental deterioration will force those Guaymi who do not permanently outmigrate either to farm larger areas to compensate for declining yields, which will hasten further the process of environmental degradation, or to seek alternatives in the outside economic system to meet subsistence needs. This last will result in an even greater weakening of the traditional system that many would not choose willingly if they had any real control over the situation. Fifth, the discovery of vast mineral deposits in the Guaymi area—most notably at Cerro Colorado—has greatly increased the economic interest of the government in the area and has brought much attention to the Guaymi. The Guaymi are now under much greater pressure than ever before to become "integrated." Parenthetically, environmental degradation due to agricultural overuse is inexorably slow compared to the amount and rapidity of environmental deterioration likely should exploitation of mineral deposits begin in the Guaymi area.

A pesar de la emigración, la población guaymí continuará aumentando y los períodos de descanso de la tierra serán aún más cortos lo cual no permitirá la recuperación del suelo. Los guaymíes tendrán que labrar más áreas aumentando así el detrimento del ambiente. El desarrollo de las minas de cobre en Cerro Colorado presionará a estos indios a integrarse a la economía monetaria.

The intensification of external pressures on a system already showing symptoms of strain from resource scarcity will require adaptations involving more effective articulation with external polities. The low level of intra-societal cooperation represented by the traditional kin based polities of the Guaymi will not be able to meet this challenge. And to the extent that more centralized forms of political organization successfully respond to this challenge, the influence of kin groups over their individual members will be lessened. Outside agencies already treat the Guaymi *as if* they were organizationally unified, and this will clearly influence the direction of change in Guaymi society. To the extent that Guaymi begin to identify primarily as part of a Guaymi collectivity and only secondarily as members of a kin group, the dominance of kin group polities will be weakened and the way opened for greater concentration of political power at supra kin group levels.

The Guaymi system already shows the initial stages of concentration of political power in the present system of provincial chiefs, clearly modeled after that of the Kuna and, equally clearly in the Guaymi case, deriving much of its power and legitimacy from external sources. The provincial chief form of organization came into being in the late 1960s, and its internal power base and claims to legitimacy are, to put it charitably, weak. Nevertheless, it is referred to in every Panamanian government publication that I have seen as the "traditional" political system of the Guaymi. This is, of course, myth, not historical fact, but it is a myth that, for better or worse, the Guaymi may ultimately be forced to accept because they must deal with the Government and the Government does not perceive the Guaymi reality.

La concentración de poder político sobre los jefes provinciales proviene de fuentes externas y es endeble. Sin embargo, éste es el punto de vista del gobierno sobre el sistema "tradicional" guaymí.

There are a great many features of the twentieth century context that I have not mentioned: for example, bilingualism, literacy, schools, health care facilities, roads, projects of directed change, self-help projects at the community level guided by national and international agencies, and the efforts and programs of various church groups; and of immense importance in the 1970s and 1980s, the Cerro Colorado mine project. Guaymi responses to these factors, and for that matter to those I have discussed, have not been uniform. They can perhaps best be characterized as a wide range of alternative responses, not all of which are viable.

Internal and external pressures on the Guaymi system are both leading in the same direction—toward massive transformation of a socionatural system that endured for over 300 years. I mourn its passing, chiefly because valuable lessons to be learned from this and similar traditional socionatural systems have not been learned by modern man.

Bennett, C.F. 1968. Human Influences on the Zoogeography of Panama. Univ. Calif. Press, Berkeley, CA.

Bennett, J.W. 1976. The Ecological Transition: Cultural Anthropology and Human Adaptation. Pergamon Press, Oxford.

Bort, J.R. 1980. Ecology and Subsistence on Opposite Sides of the Talamancan Range, *In* O. Linares & T. Ranere, eds., Adaptive Radiations in Prehistoric Panama. Peabody Museum Publications, Cambridge.

Censos Nacionales de Panama. 1970. Censo Agropecuario, vol. 2, Producción Pecuario. Panamá.

Colón, Fernando. 1959. The Life of the Admiral Christopher Columbus by His Son, Ferdinand. Trans. & annot. Benjamin Keen. Rutgers, New Brunswick, NJ.

Croat, T.B. 1972. The role of overpopulation and agricultural methods in the destruction of tropical ecosystems. Biosci. 22: 465–67.

Espinosa, Gaspar de. 1864. Relación hecha por Gaspar de Espinosa, alcalde mayor de Castilla de Oro, dada á Pedrarias de Avila, lugar Teniente General de aquellas provincias, de todo lo que sucedió en la entrada que hizo en ellas, de órden de Pedrarias (1516). Colección de Documentos Inéditos relativos al descubrimiento, conquista y organización de las antiguas posesiones españolas de América y Oceanía, sacados de los archivos del reino y muy especialmente del de indias. Tomo II, Pp. 467–522. *In* Comp. Joaquin F. Pacheco et al. Imprenta Española Torija, Madrid.

———. 1873. Relación e proceso quel licenciado Gaspar Despinosa, Alcalde Mayor, hizo en el viaje que por mandado del muy magnifico Sr. Pedŕarias de Avila, Teniente General en estos Reynos de Castilla del Oro por sus Altezas, fue desde esta Ciudad de Panamá a las provincias de Paris e Natá, e a las otras provincias comarcanas. (Año de 1519). Colección de Documentos Inéditos relativos al descubrimiento, conquista y organización de las antiguas posesiones españolas de América y Oceanía, sacados de los archivos del reino y muy especialmente del de indias. Tomo XX, Pp. 5–119. *In* Comp. Joaquín F. Pacheco et al. Imprenta del Hospicio, Madrid.

Fried, M. 1967. The Evolution of political society: An Essay in Political Anthropology. Random House, New York.

———, M. *Harris* & R. *Murphy,* eds. 1968. War, the Anthropology of Armed Conflict and Aggression. Natural History Press, New York.

Helms, M.W. 1979. Ancient Panama: Chiefs in Search of Power. University of Texas Press, Austin.

Linares, O.F. 1977. Ecology and the Arts in Ancient Panama: On the Development of Social Rank and Symbolism in the Central Provinces. Studies in Pre-Columbian Art and Archaeology, No. 17. Dumbarton Oaks, Washington, DC.

Meggitt, M.J. 1962. Growth and decline of agnatic descent groups among the Mae Enga of the New Guinea highlands. *Ethnol.* 1: 158–165.

Oviedo y *Valdes, Gonzalo Fernández.* 1853. Historia general y natural de las Indias. Vol. 3 (part II, vol. 2). Real Academia Historia, Madrid.

———. 1969. De la Natural Historia de las Indias. A facsimile edition issued in honor of Sterling A. Stoudemire. (A reproduction of the 1st ed., orig. publ. in Toledo on Feb. 15, 1526) Univ. N. Carolina Press, Chapel Hill.

Sahlins, M.D. 1972. Stone Age Economics: Production, Exchange, and Politics in Small Tribal Societies. Aldine-Atherton. Chicago.

———. 1977. Culture and Environment: The Study of Cultural Ecology. Pp. 215–231. *In* S. Tax & L.G. Freeman, eds., Horizons of Anthropology, Aldine. Chicago.

Sarsanedas, J. 1978. Tierra para el Guaymí: La Explotación de las Tierras Guaymíes en Chiriquí. Ed. Centro Capacit. Soc. Panama.

Ufeldre, Fray Adrian de. 1682. Conquista de la Provincia del Guaymí. . . . Tesoros verdaderos de las Indias, Vol. 3, Chap. 1. N.A. Tunassio, Rome.

Vayda, A.P. 1961. Expansion and warfare among swidden agriculturalists. Amer. Anthropol. 63: 346–358.

———. 1968. Hypotheses about functions of war, Pp. 85–91. *In* M. Fried, M. Harris & R. Murphy, eds., War, the Anthropology of Armed Conflict and Aggression. Natural History Press, New York.

Weiner, J.S. 1972. Tropical ecology and population structure, Pp. 393–410. *In* G.A. Harrison & A.J. Boyce, eds., The Structure of Human Populations. Clarendon, Oxford.

Young, P.D. 1968. The Ngawbe: Social and Economic Organization of the Western Guaymí of Panama, Ph.D. diss., Dept. Anthrop. Univ. Illinois, Urbana; Univer. Microfilms. Ann Arbor, MI.

———. 1970. Notes on the ethnohistorical evidence for structural continuity in Guaymí society, Ethnohist. 17: 11–29.

———. 1971. Ngawbe: Tradition and Change Among the Western Guaymí of Panama, Illinois Stud. Anthrop., No. 7. Univ. Illinois Press, Urbana.

———. 1976. The expression of harmony and discord in a Guaymí ritual: The symbolic meaning of some aspects of the balseria, Pp. 37–53. *In* M. Helms & F. Loveland, eds., Frontier Adaptations in Lower Central America, Philadelphia: Instit. Study Human Issues, Inc.

———. 1978a. La trayectoria de una religión: el movimiento de Mama Chi entre los Guaymíes y sus consecuencias sociales. La Antiqua 11: 45–75.

———. 1978b. Los rituales Guaymíes: Perspectivas simbólicas y culturales. Rev. Patrimonio Histórico 2: 7–38. Inst. Nac. Cultura, Panamá.

——— & J.R. *Bort*. 1976. Edabáli: The ritual sibling relationship among the western Guaymí. *In* P. Young & J. Howe, eds., ritual and symbol in native Central America. Univ. Oregon Anthrop. Papers, 9: 77–90. Univ. of Oregon Press, Eugene OR.

———. 1979. The politicization of the Guaymí, J. Steward Anthrop. Soc. 11(1): 73–110.

Modern Man and The Landscape

This chapter is devoted to activities of the dominant isthmian culture of today, the damage it inflicts on the environment, and pleas and plans for corrective measures. This culture operates on the natural environment in many ways, from simple farming and gathering to the complex operations of large-scale business. The first paper describes an activity having little effect on the environment. The following papers describe how human activity is impelled to remove the forests and some of the consequences of this. Finally, some pleas and proposals for corrective action are presented.

Joly's paper illustrates human activity at a local rural level, describing how one group of Panamanians has learned to exploit a natural resource in an unusual way.

Jaén provides a history of human activity in Panama since the European arrival. He describes how two features have shaped Panamanian settlement patterns and economic progress: the importance of cattle in the rural economy, and the development of the transisthmian route which earlier provided a corridor of settlement and now supports internationally oriented employment.

McKay details growth of the cattle industry in previous centuries, and he suggests that since most suitable lands were saturated by cattle by 1950, there is now opportunity for planned development.

Heckadon describes the process whereby the rural poor pioneer new lands before joining the poor of the cities. During this process, the forest is extracted of its economic value before entering a degraded, expulsive phase when peasants are forced to move elsewhere.

Deforestation leads to many changes in the environment, one of which is in local if not regional climates, and Read presents evidence of how changes in forest cover have altered wind and moisture patterns in Panama.

Deforestation is not the only negative aspect of increasing human populations. Méndez notes that restraint is needed in the hunting of wild animals.

In Central America forests are important as a source of fuelwood. Reyna points out ways that good management of forests could greatly reduce dependence on imported petroleum for heating and cooking.

MacFarland and Morales survey forestry and wildlife problems at a regional level, emphasising how training for environmental management is essential for survival of wildlife resources in the future.

Raven outlines deforestation as a global problem, pointing out that Panama, with one of the most diverse biotas in the world, is in a position to encourage understanding of uses for tropical resources. Raven makes a strong plea for rational utilization of this resource before it disappears.

Díaz, as spokesman for the Panamanian government, presents the hope of restoring degraded lands to productivity by bettering the lot and improving the abilities of the rural poor. He suggests some further steps that are needed.

In spite of great efforts to protect natural landscapes, it is difficult to avoid pessimism as to the future of tropical forests.

Hombre Moderno y Areas Naturales

Este capítulo está dedicado a las actividades de la cultura ístmica dominante en el presente, al daño que ésta inflige al ambiente y a los planes para aplicar medidas correctivas. Esta cultura opera sobre el ambiente natural de diferentes formas, desde cultivos simples y de recolección hasta operaciones complejas de negocios en gran escala. El primero de los trabajos describe una actividad que tiene poco efecto sobre el ambiente, mientras que otros discuten cómo la actividad del hombre los ha impulsado a remover los bosques y algunas de las consecuencias de esta acción. Finalmente, algunos argumentos y propuestas para iniciar medidas correctivas.

El estudio de Joly ilustra la actividad humana en un área rural y describe como un grupo de panameños ha aprendido a explotar un recurso natural en una forma poco usual.

Jaén provee la historia de la actividad humana en Panamá desde la llegada de los europeos. El señala cómo dos rasgos han moldeado los patrones de los establecimientos poblacionales y el progreso económico: la importancia del ganado en la economía rural y el desarrollo de una ruta transístmica que proveyó un corredor de establecimientos poblacionales que hoy día sostiene una serie de empleos de orientación internacional.

McKay detalla el crecimiento de la industria ganadera en los siglos pasados y sugiere que como quiera que las tierras más propicias estaban saturadas con ganado alrededor de 1950, en el presente, hay oportunidad de desarrollo planificado.

Heckadon describe el proceso de cómo las personas rurales y pobres se movilizaron primero como pioneros de nuevas tierras y luego se convirtieron en los pobres de las ciudades. Durante este proceso, el valor económico del bosque es extraido antes de ser degradado y luego la fase expulsiva comienza cuando los campesinos se ven forzados a mudarse a otro área.

La deforestación conlleva a cambios del ambiente, uno de los cuales es la alteración de climas locales y en algunos casos regionales. Read presenta evidencia de que los cambios en la cubierta boscosa del centro de Panamá han alterado los patrones de viento y humedad.

Deforestación no es el único aspecto negativo relacionado al aumento poblacional humano. Méndez enfatiza la necesidad de control en la cacería de animales silvestres.

En América Central, los bosques son importantes como fuente de combustible. Reyna señala que el buen manejo de los bosques podría reducir la dependencia de petróleo importado utilizado en diversas formas, sin conllevar a la destrucción de los bosques.

MacFarland & Morales revisan los problemas de silvicultura y de la vida silvestre en la región, enfatizando la necesidad de entrenamiento en el manejo ambiental el cual es necesario para la supervivencia de los recursos de vida silvestre en el futuro.

Raven esboza la deforestación como problema global, señalando que Panamá, con una de las biotas más diversas del mundo, está en posición de fomentar el conocimiento de la utilización de los recursos tropicales. El argumenta enfáticamente en favor del uso racional de este recurso, antes de que desaparezca.

Díaz, representante del gobierno, reconoce los problemas de deforestación y de los humildes. El ofrece esperanzas en el sentido de que tierras degradadas se conviertan en productivas para mejorar las posibilidades económicas, en especial la de los campesinos. Díaz sugiere algunos de los pasos que se necesitan.

A pesar de los esfuerzos que se realizan, es difícil evitar el pesimismo sobre el futuro de los bosques tropicales.

El Uso de *Piper auritum* para Alimentar y Atrapar Peces por los Naturales del Río Indio, Panamá

Luz Graciela Joly

This is a study of the efficiency of some of the fishing techniques used in the Indio and Uracillo Rivers of Panama, especially the use of leaves of Piper auritum as bait in "chiqueros" which are traps or weirs made of stakes.

El manjar mas ordinario de los indios é a que ellos tienen grande afición, son los pescados de los ríos é de la mar: é son muy diestros en las pesquerias é artificios de que usan, para los tomar; . . . lo hacen con varas delgadas é domables é quales convienen para ello, é con cuerdas é volantines é con redes de algodón é muy bien hechas, lo mas continuamente. Y tambien con corrales é atajos hechos a mano de estacadas . . . (Oviedo 1944, II:120)

Este relato de Oviedo sobre las técnicas de pesca de los indios a principios del Siglo XVI tiene importancia contemporánea para las técnicas de pesca usadas por los *naturales* del Río Indio, en la región de la Costa Abajo, en el centro septentrional de Panamá, entre los 9° 20′N, 8° 40′S y los 80° 00′E, 81° 00′O (Fig. 1). Los *naturales* es un término de referencia geográfica que identifica a los habitantes nativos o autóctonos de la región, de ascendencia hispano-indígena, y que habitan los cauces de arriba de los ríos de la Costa Abajo. El vocablo es usado por los habitantes de las desembocaduras de los ríos que dan al Mar Caribe, quienes son de ascendencia afro-colonial-hispana, a su vez identificados geográficamente por los *naturales* como los playeros, o sea gente de la playa (Joly 1979).

Los naturales son cholos campesinos que se dedican a actividades agropecuarias, que incluyen la cría de ganado vacuno y porcino y de aves (gallinas, patos y pavos). Sin embargo, estos animales de cría son principalmente para negociar en el sistema de mercadeo. Por ello, la "presa" más común, como dicen los naturales, o "el manjar más ordinario," como relata Oviedo, es el pescado de río. Y parafraseando a Oviedo, los naturales son muy diestros en la pesca y en las técnicas que usan para pescar.

The natives of Rio Indio on the Costa Abajo, Cocle Province, raise cattle for cash, and they are also skilled in fishing.

Este estudio describe, compara y evalua la eficiencia de algunas de estas técnicas de pescar. Destaca, en particular, el uso de la hoja del hinojo sabalero *(Piper auritum* Plate 1, Fig. 1) como cebo en chiqueros, que son "corrales o atajos hechos a mano de estacadas" como los describiera Oviedo.

Métodos

Las técnicas de pesca de los naturales fueron observadas personalmente, usando el método antropológico de participación observante. Las observaciones se complementaron con entrevistas del personal que participa en la actividad. Estas observaciones se hicieron entre los naturales de Boca de Uracillo, provincia de Coclé, durante el período de residencia en esta localidad de enero a junio de 1979. Esta convivencia formó parte de la investigación antropológica del sistema de mercadeo y adaptación ecológica de los habitantes del Río Indio y Boca de Uracillo, efectuada de agosto de 1978 a febrero de 1980.

The native fishing techniques were observed during a period of residence from January to June 1979.

Las observaciones del chiquero, la red y la máscara-foco-y-chuzo se hicieron en abril de 1979, durante la

PLATE 1.

Fig. 1. Hojas de "hinojo sabalero" (*P. auritum*) mostrando el seno basal los cordoncillos. **Fig. 2.** "Chiquero" o "corral" para atrapa sabalo pipon (*Brycon chagrensis*). **Fig. 3.** Un charco con un "comedero" para "engreir" al sabalo pipon con un haz de hinojo salalero amarrado con un cordel a un árbol en la orilla. **Fig. 4.** Haz de *P. auritum* de 0.5 kg para colocar dentro de un "chiquero." **Fig. 5.** Amarrando el haz de *P. auritum* a un palo dentro del "chiquero." La puerta esta levatada y abierta bajo el agua.

Semana Santa y días subsiguientes, en los que aumenta la demanda de peces para el consumo de comidas tradicionales de esta ocasión. El chiquero y la red se practican sólo durante el verano, cuando baja el nivel del río, disminuye la turbiedad y aumenta la visibilidad en el agua.

Observations of the trapping were made at Holy Week time when the demand for fish was greater.

Se hizo sólo una observación de estas tres técnicas e indudablemente se necesitaría más observaciones para obtener mejores datos cuantitativos. Sin embargo, las cantidades que se anotaron en estas observaciones dan una idea de la relativa eficiencia de cada técnica.

La pesca con cuerda y anzuelo se observó en junio de 1979, cuando las lluvias se habían reanudado. El mes de junio es uno de los más severos en el calendario tradicional agrícola del campesino y se refleja en el dicho "estamos pasando junio", refiriéndose a la escasez de alimentos, particularmente arroz y maíz, cuyas siembras se cosechan en julio y agosto.

Cada uno de los peces obtenidos durante las observaciones fue pesado, medido y fotografiado. También se fotografiaron los eventos. El contenido estomacal de un sábalo pipón (*Brycon chagrensis*) atrapado en el chiquero fue analizado en el campo, fotografiado y secado en periódicos al sol y al aire libre. Las muestras de peces fueron preservadas en el campo en una solución de formol al 10% en agua.

Samples of the fish Brycon chagrensis were preserved in alcohol, and specimens of the Piper auritum were dried and deposited at the University of Panama herbarium.

Muestras del hinojo sabalero (*Piper auritum*) se secaron en el campo dentro de periódicos mediante el sol y el aire libre. Además, muestras de las hojas fueron secadas en el Herbario de la Universidad de Panamá (PMA).

Las identificaciones de las especies de peces fueron hechas en el campo usando los textos de Meek y Hildebrand (1916) y Jordan et al. (1930) y de plantas con Standley (1928). Estas identificaciones fueron confirmadas por el Dr. Richard Goodyear, Ictiólogo, y la Prof. Mireya D. Correa A., Botánica, ambos de la Escuela de Biología, Universidad de Panamá.

Los análisis químicos de muestras secas y frescas de *Piper auritum* para determinar los alcaloides y los aceites esenciales se efectuaron en el Laboratorio Especializado de Análisis y por el Proyecto de Valoraciones Bio-Farmacéuticas, ambos de la Facultad de Ciencias Naturales y Farmacia de la Universidad de Panamá, bajo la dirección y sub-dirección del Dr. Tomás Arias y del Dr. Mahabir Gupta, respectivamente.

Sitio

Los sitios de observaciones están localizados en las secciones de los ríos Indio y Uracillo indicadas en la mapa (Fig. 1). Los sitios se identifican por su localización geográfica con referencia a Boca de Uracillo, por los topónimos comunes y por el nombre del "dueño del charco." Esto se refiere a la familia que vive en las inmediaciones de un charco hondo y quien tiene derecho primario de pesca en este charco. Esta primacía es sancionada por el respeto de los coterráneos quienes usualmente notifican o piden permiso a los dueños del charco para pescar en sus aguas y comparten el producto de la pesca con los dueños.

Boca de Uracillo se usa como punto central de referencia geográfica. El Uracillo es el principal tributario del Indio y el poblado de Boca de Uracillo se encuentra en la desembocadura del Uracillo donde éste se une al Río Indio. El lugar se localiza a los 8° 58′ de latitud y los 80° 11′ de longitud, a una altura de 20 m y a una distancia de 25 km terrestres o 60 km fluviales, contando los meandros del Indio, al sur, o sea río arriba de Boca de Río Indio.

In the geography of the Rio Indio, Boca de Uracillo is the gathering point for products and goods distribution for communities higher up the river, including Las Minas and other towns.

En la geografía económica del Río Indio, Boca de Uracillo es el "puerto", "muelle", o "embarcadero" y es un centro de recolección de productos y distribución de mercancías para las comunidades de "más arriba", al sur de Boca de Uracillo, en un área aproximada de 150 km^2, incluyendo Las Minas, Los Uveros, San Cristóbal, Coquillo, Las Marías y Cerro Miguel.

En el sitio del embarcadero, en la margen occidental de los ríos Indio y Uracillo, está localizado el núcleo básico de símbolos comunitarios que determinan la participación de los miembros de la comunidad de Boca de Uracillo. Estos símbolos son la iglesia Católica, la escuela pública primaria, las tiendas, el campo de bola suave, el centro de salud, el acueducto y recientemente la sala de baile. Dentro de

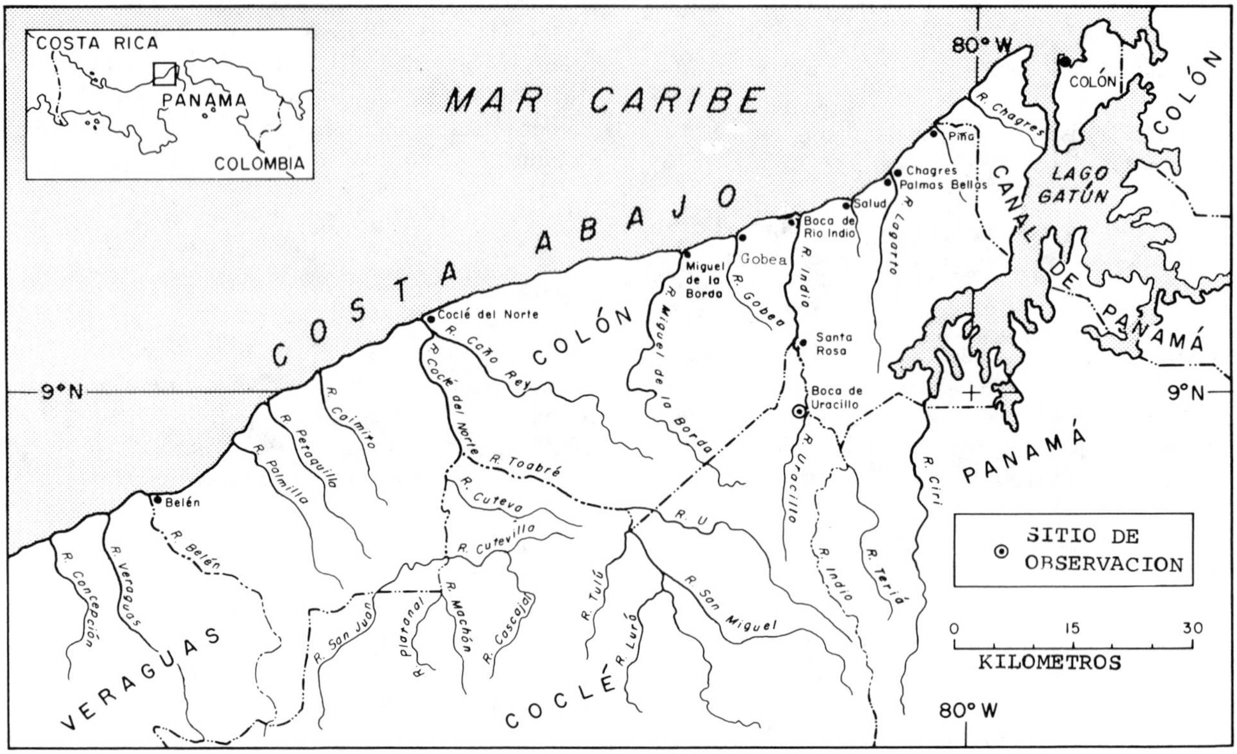

Fig. 1. Mapa de Panamá mostrando la región de la Costa Abajo y el sitio de observación de las ténicas de pesca en Boca de Uracillo.

este núcleo básico están radicadas las principales familias que controlan las actividades socioeconómicas de la localidad. El resto de las familias que se consideran miembros de la comunidad viven alejadas del núcleo básico, dispersas en una área aproximada de 30 km^2, aunque pueden tener viviendas temporales dentro del núcleo básico y que ocupan durante el año escolar o en las fiestas. En total, la comunidad está compuesta de 63 familias, con una población de 375 habitantes, de los cuales el 49% son de 15 años de edad.

Con excepción de una angosta galería de árboles en las riberas, compuesta en su mayoría por higuerones (*Ficus glabrata*) cuyas copas sombrean porciones de los ríos, las áreas aledañas a los ríos están desmontadas de vegetación arbórea y sembradas de maíz, yuca, café y pasto. En la banda occidental del Uracillo corre el "camino real" que conduce de Uracillo a Coquillo.

Except for a narrow gallery of fig trees, the area is denuded of trees and is planted in corn, cassava, coffee and pasture.

En los tramos descritos, el río Indio tiene un ancho aproximado de 10 a 12 m y el Uracillo de 8 a 10 m. Ambos ríos presentan una serie de corrientes, cascajales y charcos, cuya profundidad varía en el verano de 0.5 a 3 m en el Indio y de 0.5 a 2 m en el Uracillo. El subestrato varía de fango a cantos rodados y ocasionalmente lajas de piedra madre. Raíces, bejucos y troncos de vegetación terrestre, así como una capa de hojas descompuestas, cubren el fondo de los charcos y remansos. El agua de ambos ríos es típicamente clara pero se vuelve túrbida, por más de un día, después de fuertes lluvias.

Observaciones

En marzo de 1979, observé por primera vez un chiquero en el Guayabalito, a unos 13 km terrestres al sur, o río arriba, de la Boca del Río Indio. En abril de 1979, (Domingo de Gloria) tuve la oportunidad de observar el funcionamiento de un chiquero en El Barrero de Río Indio, a 1.5 km terrestre, o 5 km fluviales al sureste de la Boca de Uracillo. Este chiquero (Plate 1, Fig. 2) había sido principalmente construido por Juan Rivera, dueño del charco, en un plazo de unas seis horas.

El chiquero medía unos 2.5 m de largo por 2 m de ancho. Dentro del chiquero, el charco de agua tenía una profundidad de un metro, las estacas se construyeron de tallos de platanilla *(Heliconia mariae)* mientras que para la puerta se usó caña blanca *(Gynerium agitatum)*. La puerta estaba localizada hacia el charco, en el lado oriental del chiquero. Los tallos de platanilla y las cañas blancas medían unos 3 m de alto. Las platanillas estaban sumergidas un metro dentro del charco y amarradas con hilos sintéticos a un armazón de troncos de membrillo *(Gustavia superba)*. Dos troncos de membrillo, de unos 3 m de largo, enterrados en el barranco se extendían verticalmente sobre el charco, a una distancia de 2 m el uno del otro. Hacia el lado del charco, otro tronco los intersectaba formando un rectángulo de 3 m de largo por 2 m de ancho. Este armazón por fuera de las platanillas, sobresalía del agua. Sobre cada intersección de los troncos, fuera del agua, había amarrada una piedra de unos 25 kg. En el fondo dentro del chiquero, en la base de las platanillas, había piedras "aliñadas" para sostener firmemente los tallos en el fondo.

In April a weir 2.5 m wide by 2 m deep was built by the landowner in six hours. The sides of Heliconia mariae and the door of Gynerium agitatum were anchored by trunks of Gustavia superba.

Quince días antes de construir el chiquero, o sea más o menos desde mediados de marzo, los Rivera habían estado amarrando un haz de hinojo sabalero con una cuerda del tronco de un higuerón localizado cerca del lugar en donde se encontraría el "chiquero" (Plate 1, Fig. 3), para "engreir" o acostumbrar a los sábalos pipones a venir a comer el hinojo al pie del higuerón, cuyas hojas y frutos también come el sábalo. Luego de construido el chiquero, al atardecer de cada día, se amarraba a un palo dentro del chiquero, un haz de frescas y fragantes hojas del hinojo (Plate 1, Figs. 4, 5), dejando la puerta del chiquero abierta para que los sábalos entraran a comer las hojas dentro del chiquero.

La noche del Domingo del Ramos, 8 de abril, los Rivera atraparon 30 sábalos dentro del chiquero. Observé que durante el siguiente procedimiento durante el Domingo de Gloria, 15 de abril, atraparon 14 sábalos, que representaban un total de 10.2 kg y un promedio por pez de 0.73 kg de peso y 39.8 cm de largo.

Each day a bundle of fresh and fragrant hinojo salbalero leaves was tied to a tree in the weir. Then the night of 8 April, 30 sabalo fish were trapped in the weir, a total weight of 10.2 kg.

A las 5:45 pm. Pedro Mendoza fué a cortar unas hojas de hinojo sabalero en una "mancha" de esta planta en los potreros cerca de la casa de los padres de su mujer. Una "mancha" es un sitio donde crece un grupo de plantas de una misma especie. Pedro regresó a los 10 minutos con un haz de hojas, que pesaban 0.5 kg.

A las 6:00 pm. bajamos del "alto" donde está la casa en las faldas del Cerro del Barrero, por una ladera de unos 20 m hasta el "bajo" a orillas del río donde estaba el chiquero.

Después de revisar que los tallos del chiquero estuvieran firmes en su lugar y bien amarrados, Pedro sacó los restos del haz de hojas que habían en el chiquero desde el día anterior y colocó el nuevo haz de hojas las cuales amarró del palo el centro del chiquero ubicado en tendió un hilo que unos de los chiquillos amarraron a la ramita que Pedro había enterrado en el barranco antes de entrar al chiquero. Luego, con la ayuda de los chicos, templó bien la cuerda que levantaba la puerta de caña blanca y los chicos tensaron el amarre de la cuerda en el tronco de un árbol en la orilla, permaneciendo la cuerda en un ángulo inclinado del tronco a la puerta.

Para finalizar los preparativos, Pedro cogió una hoja de hinojo, la estrujó y partió en pedacitos que regó dentro del chiquero. Casi de inmediato se notó una agitación en el charco fuera del chiquero, señal indudable, según Pedro, de que los sábalos habían percibido las fragantes hojas frescas de hinojo, dentro del chiquero.

A las 9:00 pm. Pedro Mendoza y su cuñado Pedro Rivera, bajaron al chiquero para "aguaitar" los sábalos que entraban a comer el hinojo. Los dos hombres llevaron sus motetes, un foco, un chuzo y sus machetes. Estar "al aguaite" significaba alumbrar y vigilar la ramita enterrada en el barranco. Cuando los sábalos comenzaron a comer el hinojo, a eso de las 11:30 pm., el palo al cual estaba amarrado el hinojo comenzó a moverse, lo que hizo que la cuerda amarrada a la ramita se moviera y la ramita a su vez se moviera también. Al percibir este movimiento, Pedro Mendoza esperó unos minutos para que entrara la mayor cantidad de sábalos dentro del chiquero y luego soltó la cuerda de la puerta para que ésta cayera como una guillotina y encerrara a los sábalos dentro del chiquero. Pedro Rivera entonces entró al chiquero y sacó los sábalos con un motete, pasándole el motete a Pedro Mendoza quien permaneció en el barranco para recibir los peces y ponerlos en el otro motete que había aliñado con hojas de "fardo" *(Ischnosiphon leucophaeus)*. Aquellos sábalos que saltaban del motete, Pedro Rivera los arponeaba con el chuzo.

To trap the fish, fresh leaves of Piper auritum were placed in the weir and about 11.30 pm. when the fish were feeding the weir door was closed.

Los hombres regresaron a casa a la 1 am. y colgaron el motete con los sábalos de una viga de la casa, para que estuviera en alto fuera del alcance de perros y gatos. Cada pez se pesó y midió. En Uracillo examiné el contenido estomacal del sábalo que me dieron por mi participación en la pesca. Aparte de los fragmentos de hinojo, tenía hojas y semillas de higo *(Ficus glabrata)* y otra hoja de un bejuco que crece a orillas de los ríos, pero que no pude identificar. Una grasa blanca envolvía las vísceras. Tanto la grasa como las tripas después de limpiarlas son utilizadas por las mujeres para cocinar el arroz. La carne del sábalo, salada y ahumada, se mantiene grasosa y conserva el sabor del hinojo sabalero que es agradable al paladar y que ellos consideran como una ventaja de esta técnica. Esto sugiere que el *Brycon chagrensis* metaboliza e incorpora en sus tejidos grasosos algunas sustancias químicas del *Piper auritum*. Por esta razón, llevé muestras de la hoja para que fueran analizadas químicamente en el Laboratorio Especializado de Análisis y el Proyecto de Valoraciones Bio-Farmacéuticas.

Besides Piper auritum leaves, the fish stomachs contained leaves and seeds of Ficus glabrata. The flesh of the fish retained the fragrance of the Piper auritum leaves. The fish seems to metabolize some chemical substance from the leaves.

Los Rivera-Mendoza continuaron usando el chiquero una vez por semana durante el resto de abril y a principios de mayo, hasta que debido a las lluvias la creciente del río se lo llevó. El número de sábalos atrapados disminuyó gradualmente, pero se mantuvo en un promedio de diez peces atrapados semanalmente.

The weir continued in use, trapping an average of 10 fish per night once a week until rains raised the water level.

La eficiencia del chiquero, en términos del número de personal requerido, cantidad y peso de peces cogidos, promedio de peso y largo por pez, cantidad y peso por hombre y por casa puede determinarse de la Cuadro 1. Las técnicas de "redear" y de "montear" con máscara-foco-y-chuzo se describen brevemente a continuación.

El 12 de abril de 1979, Jueves Santo, un grupo de 28 hombres, mujeres, y niños de Boca de Uracillo, representando 17 casas unidas por parentesco, "redearon" el tramo del Río Uracillo denominado El Aguila, a 2 km terrestres de Boca de Uracillo. La red era de alambre gallinero, de 15 m de largo, fondeada con una "enhuevadura" de 300 piedras envueltas en sacos, cada piedra pesando un promedio de 0.5 kg. Diecinueve hombres sostenían la red, bajando con la corriente a lo largo de un tramo de un kilómetro en el río. Cuando se llegaba a un charco, se cercaba el mismo, reduciendo el círculo poco a poco. Hombres y niños con máscara y chuzo y un hombre con "arco y flecha" mataban los peces dentro del cercado y los tiraban a las orillas para que las mujeres y niños los ensartaran en bejucos. Aproveché esta oportunidad para pesar y medir los peces.

Durante 7.5 horas, desde las 9:00 am. hasta las 2:15 pm., se cogieron 60 peces, representando 6 especies, sumando un total de 14.5 kg. De éstos, 51 peces o sea el 85% eran sábalos pipones *(Brycon chagrensis)*, que pesaban un total de 12 kg., apenas unos 2.6 kg más que el peso total atrapado en el chiquero, sin embargo mediante la red se utilizó mucho más personal y tiempo. Además, el promedio por pez redeado y la distribución por casa eran menor usando la red que con el chiquero. Esto no representa mucho donde hay 8 o 10 personas por casa, pero la actividad, que se repitió el Sábado Santo (14 de abril), fue una manifestación de parentesco y de cooperación entre los dueños de la red, quienes vivían en la Quebrada El Silencio, afluente del Uracillo y los dueños de charcos en el tramo de El Aguila del Río Uracillo. Entre los dueños de charcos—todos emparentados por consanguinidad y afinidad—se encontraba Pedro Mendoza, el mismo que había participado en el chiquero de los padres y hermanos de su mujer. Conviene mencionar que el redear para la Semana Santa en los Ríos Uracillo e Indio también ha sido una práctica para los naturales de Las Marías, que está localizado río arriba de Boca de Uracillo (Martínez 1976).

Other species of fish were caught too.

Aunque los naturales dicen que los motores fuera de borda afectan la estructura del chiquero y por eso ya no se construyen tantos chiqueros a lo largo del Río Indio como hasta hace dos décadas, la eficiencia del uso de hinojo sabalero se puede comprobar por el hecho de que, sin construir el chiquero, algunos dueños de charco acostumbran en la actualidad poner en sus charcos hojas del hinojo sabalero todas las tardes, amarrando las hojas con un cordel a un árbol en

CUADRO 1
EFICIENCIA DE TECNICAS DE PESCA

Río	Técnica	Hombres	Casas	Horas	Especies	Cantidad	Kilos	Centimetros	Fecha
racillo	Red, Arco, Chuzo	23	17	7.5	5	60	14.5	1535.8	4.12.79
ndio	Chiquero con Hinojo sabalero	2	2	7.25	1	14	10.2	557.4	4.15.79
ndio	Máscara, Foco, y Chuzo	4	3	3.5	7	54	10.93	573.5	4.16.79

PROMEDIOS

Fecha	Técnica		Kilos	Centimetros
.12.79	Red, Arco, Chuzo	Por Pez:	.21	25.59
		Por Hombre:	.55	66.77
		Por Casa:	.74	90.34
.15.79	Chiquero con Hinojo Sabalero	Por Pez:	.73	39.81
		Por Hombre:	5.1	278.7
		Por Casa:	5.1	278.7
.16.79	Máscara, Foco y Chuzo	Por Pez:	.20	26.59
		Por Hombre:	2.73	359.08
		Por Casa:	3.64	478.77

a orilla. Con estos "comederos" dicen "engreir" a los peces, acostumbrándolos a venir a comer regularmente y de este modo asegurándose que haya una cantidad abundante de sábalos en el charco cuando se sale a "montear" de noche con máscara-foco-y-chuzo, cada 15 o 30 días.

Motorboats disturb the weirs, and not so many are built as in former times.

Este es el caso con la familia Martínez León, lueños del charco inmediatamente al norte (Río Abao) del embarcadero de Boca de Uracillo. Desde fines de enero hasta fines de abril de 1979, ellos colocaban regularmente al atardecer, un haz de hinojo sabalero en su charco, amarrándolo con un cordel a la raíz de un árbol en la orilla. En los potreros alrededor de su casa hay dos "manchas" de hinojo sabalero que, al igual que los Rivera de El Barrero, no "chapean" cuando "repican" los potreros.

Los resultados de la "montería" con máscara-foco-y-chuzo que hicieron dos de los hermanos Martínez León, junto con dos amigos, la noche del lunes 16 de abril de 1979, de las 8:30 pm. a las 12:00 pm. fueron de 44 peces y 10 camarones, sumando un total de 10.2 kg de peces y 0.73 kg de camarón. De los peces, el 66% eran sábalos pipones, que sumaban un total de 3.4 kg o sea, el 82% del peso total de peces cogidos. Este dato significó menos sábalos que los atrapados en el chiquero, a un promedio de peso y tamaño por pez menor que los del chiquero, usando el doble del personal en casi igual tiempo. Para los muchachos jóvenes, sin embaro, la máscara y el foco son instrumentos "modernos" y la técnica les proporciona la excitación de la cacería, por ello aplican el mismo vocablo que se usa para cazar animales del monte—"montear".

La pesca con cuerda y anzuelo es la técnica que se practica durante todo el año. Las mujeres pescan mientras lavan ropa en los ríos y quebradas y los niños lo hacen tan pronto salen de la escuela. De las cuatro observaciones de esta técnica los resultados no llegan siquiera a un kilogramo. De éstas observaciones, la que más pesó fue la de seis sábalos pipones que sumaban 0.65 kg y fueron cogidos por una mujer usando pizvá (*Guilielma utilis*) como cebo.

Fishing by hook and line is common but not so productive as the weirs using leaves of Piper auritum.

Conclusiones

Como los indios descritos por Oviedo a principios del Siglo XVI, los naturales de Boca de Uracillo hoy día son igualmente diestros en la pesca y en las técnicas que usan para pescar, y el "manjar más ordinario" continúa siendo los pescados de los ríos. Sin embargo, las observaciones hechas nos indican que las cantidades y pesos no son abundantes. Además, las técnicas que se usan para coger grandes cantidades de peces cuando aumenta la demanda sólo se pueden practicar durante los cuatro meses del verano, cuando el nivel del agua baja y los ríos están claros. En este aspecto, la situación es similar a lo que se ha reportado para los Guaymíes del occidente de Panamá, cuya pesca está confinada casi enteramente a la estación seca cuando los ríos y riachuelos están bajos y claros. Sin embargo, durante todos los meses del año, los naturales dependen de la cuerda y el anzuelo, cuyo rendimiento por período de pesca es menor que un kilogramo, como lo indican las observaciones. Esto no es mucho donde hay 8 o 10 personas por casa.

Ahora bien, de las técnicas de mayor rendimiento, el uso del hinojo sabalero *(Piper auritum)* para alimentar y atrapar el sábalo pipón *(Brycon chagrensis)* dentro del chiquero merece atención particular. Primero, entre todas las técnicas de alto rendimiento, el sábalo pipón es la especie más abundante. En el chiquero, ésta es la única especie que se atrapó. Es de suponer que tal vez existe una selección particular del sábalo pipón por las hojas de hinojo sabalero, cuyo nombre común en sí hace referencia al pez. Aunque se pudiera decir también que la técnica del chiquero mismo se usa particularmente para atrapar sábalos, ya que una estructura similar, denominada "corral" se fotografió en febrero de 1973 en el Río Tabasará, entre las provincias de Veraguas y Chiriquí, en la vertiente del Pacífico y se usaba para echar maíz y yuca para alimentar y atraer a sábalos que probablemente eran *Brycon striatulus* que es la especie de la vertiente del Pacífico similar al *Brycon chagrensis* de la vertiente del Atlántico (Pretto, comunicación personal). Una práctica alimenticia similar se ha reportado para atraer a los sábalos en Brazil, usando una preparación de "mandioca" (*Manihot* sp.) en el agua (Menezes 1969).

Aunque se atraparon menos sábalos pipones en el chiquero (14) que con la red (51) o con la máscara-foco-y-chuzo (44), el peso y largo promedio por sábalo del chiquero (0.73 kg; 39.81 cm) eran mayores que los de la red (0.25 kg; 26.7 cm) o que los de la máscara-foco-y-chuzo (0.29 kg; 29.6 cm). Es posible que los sábalos del chiquero representaran más adultos que juveniles, ya que se ha reportado una diferencia en los patrones alimenticios entre adultos y juveniles del *Brycon striatulus* en la cuenca del Río Chiriquí. Los juveniles son más carnívoros, mientras que los adultos son más herbívoros y por consiguiente tienen un intestino más largo que los juveniles para poder procesar mejor el material vegetativo y de menor valor alimenticio (Goodyear et al. 1977). El *Brycon chagrensis* ha sido caracterizado como una especie esencialmente herbívora de acuerdo con exámenes del contenido estomacal, y ésta es una tendencia general del género *Brycon* (Menezes 1969:220–221).

The weirs tend to catch mature fish, and they are better keyed to conservation than other techniques.

Si pudiéramos comprobar que los sábalos del chiquero son principalmente adultos y que los de las otras técnicas son principalmente juveniles, podríamos decir que el chiquero es una técnica más protectora para la especie. Los mismos naturales del Río Indio admiten libremente que el uso de la máscara, sobre todo, ha diezmado la población de peces en el río.

Otra ventaja del hinojo sabalero en chiqueros es que la hoja no es un veneno que mata a los peces indiscriminadamente, en grandes cantidades, como sucede con el uso de otras plantas en Centro y Sur América. La hoja del *Piper auritum* es un alimento atractivo para el *Brycon chagrensis*. Además, se puede asumir que este pez metaboliza los aceites de la hoja, ya que para los *naturales* los sábalos "engreídos" con esta hoja durante un período de 15 días retienen el sabor del hinojo, aún después de salado y ahumado el pez. Este sabor es agradable al paladar de naturales y playeros de la Costa Abajo, como si fuera un condimiento natural. En México, donde se conoce el *Piper auritum* como "hoja santa", se usa como condimiento para comidas, especialmente en el estado de Veracruz (Collera 1956 y comunicación personal). De comprobarse que en efecto el sabor de la hoja es absorbido por el pez, entonces el *Piper auritum* tiene un uso potencial en el alimento de peces usados en la acuicultura, tales como la *Tilapia* que es también herbívoro. Uno de los intereses en la acuicultura es el de hacer que el sabor de los peces criados sea agradable al consumidor (Pretto, comunicación personal). En este sentido el *Piper auritum* posee características comunes a otras plantas usadas como alimento de peces en la acuicultura. Estas características son hojas blandas, abundante superficie foliar y de fácil cultivo como las hojas de la yuca *(Manihot esculenta)*, bore *(Alocasia macrorhiza)*, chayamansa *(Cnidoscolus chayamansa)* (INDERENA 1979:39). Aunque los naturales del Río Indio no cultivan el hinojo sabalero, debido a su uso en la pesca y para infusiones medicinales, sí conservan las "manchas" de esta planta que se propaga libremente por las raíces corredoras o por dispersión de las semillas. Los cordoncillos con la semillas son más abundantes en esta especie durante el verano (Croat 1978:331). En Panamá, el *Piper auritum* se conoce de las áreas de vegetación tropical húmeda a lo largo de la vertiente Atlántica, en la franja canalera, en Chiriquí, Panamá y Darién; también en áreas de vegetación tropical seca en Coclé y de vegetación húmeda pre-montana en Colón, Chiriquí, Coclé y Panamá al igual que en vegetación tropical muy húmeda en Colón. También se encuentra desde México hasta Colombia y en las Antillas, desde el nivel del mar hasta los 1200 m de altura.

Conviene recomendar, también, que el sábalo pipón en sí merece mayor estudio como una especie de un potencial en la acuicultura. Experimentos en la cría de otras especies del género *Brycon* en la acuicultura se han realizado en Colombia (FAO 1971:3; Acero Sánchez 1979; INDERENA 1979).

Por todo lo anteriormente expuesto, considero que la técnica del chiquero usando el hinojo sabalero (*Piper auritum*) para alimentar sábalos pipones (*Brycon chagrensis*) tiene un uso potencial en la acuicultura, ya sea en el uso de la hoja o del pez, o de la estructura del chiquero como paso de transferencia al uso del estanque acuicultor. Son técnicas como el chiquero con el hinojo sabalero que son evidencias de la adaptabilidad ecológica del ser humano y que deben utilizarse en los programas de la "tecnología apropiada" (White et al. 1979).

Piper auritum is found throughout the lowlands of Panama and it ranges from Mexico to Colombia. The method of fishing described here merits major study for use in fish farming.

Agradecimientos

Agradezco a las siguentes personas por su asistencia y comentarios en la preparación de este informe: Tomás Arias, Laboratorio Especializado de Análisis, Universidad de Panamá; Julius Bird, American Museum of Natural History; Mireya D. Correa A., Escuela de Biología, Universidad de Panamá; Andre F. Clewell, Environmental Biology Division, Conservation Consultants Inc.; Richard Cooke, Arqueólogo; Robert Dressler, Smithsonian Tropical Research Institute; James Duke, Economic Botany Laboratory, U.S. Department of Agriculture; John Ewell, Dept. of Botany, University of Florida; Alwyn H. Gentry, Missouri Botanical Garden; Richard Goodyear, Escuela de Biología, Universidad de Panamá; Mahir Gupta, Proyecto de Valoraciones Bio-Farmaceuticas, O.E.A.(Universidad de Panama); Mary Helms, Dept. of Anthropology, University of North Carolina at Greensboro; Mary Power, Dept. of Zoology, University of Washington; Richard Pretto Malca, Dirección Nacional de Acuicultura, M.I.D.A.; Ira Rubinoff, Smithsonian Tropical Research Insitute; Philip Young, Dept. of Anthropology, University of Oregon; Elizabeth Wing, Florida State Museum.

Acero Sánchez, Alfredo 1979. Informe sobre la pesca continental y la piscicultura en Colombia. La Acuicultura en América Latina: Informes Nacionales, 3(159): 19–36.

Collera Zúñiga, O. 1956. Contribución al estudio del *Piper auritum*. Fac. Cien. Quím. Univ. Nac. México, México.

Croat, T.B. 1978. Flora of Barro Colorado Island. Stanford Univ. Press, Stanford, CA.

FAO (Food and Agriculture Organization) 1971. Breeding of *Brycon henni*. Aquaculture Bulletin 3: 3.

Goodyear, R., V. *Martínez* & J.B. *del Rosario*. 1977. Apéndice no. 4. Fauna Acuática. Pp. 265–334. *En* A.J. Adames, ed. Evaluación ambiental y efectos del Proyecto Hidroeléctrico Fortuna. Rev. Lotería nos. 254–255–256 (Abril, Mayo, Junio).

INDERENA (Instituto de Desarrollo de los Recursos Renovables, Colombia) 1979. Informe sobre la situación de la acuicultura en Colombia. La Acuicultura en América Latina: Informes Nacionales 3: 37–48.

Joly, L.G. 1979. Los que ya conquistaron el Atlántico: naturales y playeros de la Costa Abajo. Cat. Expos. Temporal conmemorando 3 Aniv. del Museo del Hombre Panameño. Panama. Imp. Nación. Inst. Nac. Cultura, Panamá.

Martínez, N. A. 1982. Breve reseña histórica de la comunidad de las Marías. Rev. Patrimonio Histórico 2: 61–78.

Meek, S.E. & S.F. *Hildebrand*. 1916. The fishes of the fresh waters of Panama. 10(15), Pub. 191, Zool. Ser. Field Mus. Nat. Hist., Chicago.

Menezes, N.A. 1969. The food of *Brycon* and three closely related genera of the Tribe Acestrorhynchini. Pap. Avulsos Zool. 22: 217–223. Sao Paulo.

Oviedo y Valdez, G.F. de. 1944. Historia General y Natural de las Indias. Islas y Tierra Firme del Mar Océano. Tomo 2. Prólogo de J. Natalicio González. Notas de José Amador de los Ríos. Ed. Guaranía, Asunción, Paraguay.

Standley, P.C. 1928. Flora of the Panama Canal Zone. Contr. U.S. Nat. Herbarium. Vol. 27. Smithsonian Instit. Washington, DC.

White, N., P. *Lernoux*, M. *Witaker* & E. *Ipsen*. 1979. Back to Basics. Newsweek 94(Aug. 3): 44–45.

Nuevos Hombres y Ganados y su Impacto en el Paisaje Geográfico Panameño entre 1500 y 1980

Omar Jaén Suárez

The European colonizers arriving in Panama early in the 16th century met a densely peopled landscape which had been influenced by man for many centuries. With the decimation of the native population, the landscape began to revert to forest, but the introduction of cattle and horses maintained the Pacific savannas in a modified open state. In the early years an occupation pattern was established which consisted of a north-south transit route across the Isthmus and an east-west axis of colonization based on cattle. Agriculture used newly introduced rice, yams and other crops. Late in the 19th century, the transisthmian route was enhanced by railroad and canal building, and many new immigrant groups were added to the scene. Plantation agriculture of bananas began, and new pasture grasses expanded the cattle industry. Revolutionary changes have taken place in the current century with completion of the canal and with large scale engineering modifications such as the Madden, Miraflores and Bayano reservoirs and the deforestation of their watersheds.

Este trabajo estará dedicado a plantear el problema del encuentro de los nuevos hombres y ganados, que desde el siglo XVI están llegando al Istmo de Panamá. Por ello, comenzamos por la descripción del paisaje geográfico que descubren los colonizadores del siglo XVI.

Luego, exponemos la explotación de los recursos y los fenómenos demográficos que se desarrollan después de la conquista y la primera colonización hispánica.

Después, analizamos la consolidación de las nuevas estructuras, que ocurre durante tres siglos, desde 1600 aproximadamente hasta 1900.

Finalmente, registramos los cambios revolucionarios del siglo XX, que han permitido el rápido crecimiento de las densidades humanas y la expansión de las fronteras de poblamiento y de explotación del territorio.

La aparición de una red de pueblos y ciudades y la creación de algunas estructuras de poblamiento que desafían a veces los siglos, de disposición del hábitat rural y de comportamientos migratorios, son también fenómenos relevantes en los cambios mayores del paisaje.

El Paisaje Geográfico hacia 1500

Diversos autores han adelantado cifras de población amerindia, en 1500, que se extiende entre 225.000 y 2.000.000 de habitantes, (Steward & Faron 1959; Oviedo 1959) aunque nosotros nos inclinamos más por un cuarto de millón de hombres que quizás lleguen hasta medio millón. En todo caso, ellos parecían encontrarse desigualmente distribuidos.

De acuerdo con la primera hipótesis, en las tierras altas veríamos (Bennett 1976) menos de 0.4 habitantes por km^2; quizás entre 0.4 y 1.0 habitante por km^2 en los piedemontes y la costa atlántica; y entre 4 y 6 habitantes por km^2 en los bajos valles aluviales y las llanuras litorales de la costa del Pacífico, en Chiriquí, la región central y las depresiones del Bayano, Tuira y Chucunaque. De todas maneras, el máximo de poblamiento pareciera concentrarse, en las amplias llanuras de la región central entre el Río Tabasará al Oeste y el Río Chame al Este (Cooke 1976).

Quizás un centenar de siglos de caza y recolección mediante las técnicas paleoindias que incluyen el uso del fuego, pero sobre todo los 1.500 años de agricultura formativa, con una población algo mayor, y luego los otros 1.500 años de la agricultura de gestos plenos con una población más abundante todavía y en aumento hasta principios del siglo XVI, han debido modificar profundamente el ambiente natural del territorio ocupado hoy por la República de Panamá.

There may have been a quarter or a half million inhabitants in 1500, and population was densest along the Pacific slope. Man's presence for centuries before had modified the natural scene.

Los testimonios brindados por los cronistas nos ofrecen, a principios del siglo XVI, un paisaje vegetal bastante variado, según la desigual naturaleza e intensidad de la ocupación humana. En la costa atlántica, la estrechísima planicie litoral cubierta a veces de maizales desde el Golfo de Urabá hasta por lo menos el área del Río Belén, se pone rápidamente en contacto con la selva tropical húmeda que sube el piedemonte, gana las próximas alturas de la Cordillera Central y se insinúa también por las bajas colinas del istmo central de Panamá hasta llegar al valle alto del Chagres y sus macizos de la Sierra Llorona y del Mamoní. El centro del Darién meridional, bastante poblado, conocía, por lo leído, la sucesión de plantíos, sub-bosques, matorrales y manglares, mientras que desde el valle medio del Bayano se abría la gran sabana antropógena. Desde Chepo ella se afirma con mayor vigor pasando por Panamá y, por el norte, alcanzaba las riberas del Chagres en el área de Cruces. Más adelante, ella se desarrolla también en los llanos de La Chorrera y Sajalices. Pero será a partir de Chame que la gran sabana se extiende sin interrupción pasando por las tierras de Chirú, Natá, Calobre, la peneplanicie de Veraguas central, Escoria y Guararé en el actual Azuero. Ocupa así esa sabana central, la región que corresponde hoy, a "grosso modo," al llamado clima tropical árido (Aw'gi de Köppen). Las tierras altas, por su parte, estarían ocupadas por una floresta vigorosa, desde las alturas de Veraguas hasta las de Chiriquí.

La costa norte fue visitada, en 1502, por el Descubridor de América, y " . . . el Almirante siguió navegando hasta que entró en Portobelo, al que puso este nombre por ser muy grande, hermoso y poblado, y tiene en torno mucha tierra cultivada . . . La región que rodea el puerto no es agreste, sino cultivada y llena de casas, distante una de otras un tiro de piedra o de ballesta . . . " (Bennett 1976) Luego, cerca de Nombre de Dios declara que "todos aquellos contornos e isletas estaban llenos de maizales . . . " Finalmente, al referirse a un área cercana al Río Belén al norte de Veraguas, menciona "seis leguas de maizales, que eran como campos de trigo" (cerca de 33 kilómetros). Sin embargo, este entusiasmo se matiza cuando, en 1508, se trata de colonizar el litoral caribe veragüense para explotar sus lavaderos de oro. Pedro Martir de Anglería nos dice que los colonos "en cierto valle de suelo fértil, que lo demás de la región es estéril, sembraron a usanza de su patria" (Castillero 1967). Detrás, la tupida selva era un obstáculo insalvable, aún en los mejores años de la minería de Veraguas, de 1559 a 1589. En el área del Golfo de Urabá, una densidad mayor del poblamiento y por lo tanto de plantíos agrícolas, atrae la colonización temprana, un poco más duradera, alrededor de Santa María la Antigua y Acla (Oviedo 1950).

El Darién, por su parte, conocía sabanas, sin duda extensas en el valle del Bayano, que en esa época semejaría la situación actual, resultado de los más recientes desmontes.

Desde Nombre de Dios hasta la ciudad de Panamá, el "camino asimismo es muy áspero y de muchas sierras y cumbres muy dobladas, y de muchos valles y ríos, y bravas montañas y espesísimas arboledas" (Oviedo 1950), mientras que "desde Panamá hasta el dicho río de Chagres hay cuatro leguas de muy buen camino, y que muy a placer le pueden andar carretas cargadas, porque aunque hay algunas subidas, son pequeñas, y tierra desocupada de arboleda, y llanos, y todo lo más de estas cuatro leguas es raso".

La región central parecía aún más abierta que las sabanas cerca de Chepo y la ciudad de Panamá. En 1517, "de Chame a la Provincia de Chirú hay ocho leguas de despoblado a la misma vía . . . Desta provincia a la de Natá hay cuatro leguas de despoblado. Todas estas tierras son finas y llanas y muy hermosa tierra, de muchos mantenimientos, de maíz y ages y melones diferentes de los de acá, y uvas, yuca y mucha pesquería en los ríos y en la mar y caza de venados" (Andagoya 1980).

Las tierras altas, desde Veraguas hasta Chiriquí, conocían una vegetación más tupida y, muy probablemente, densidades menores. Andagoya nos indica que desde Burica "y salidos desta provincia la vuelta de Panamá la tierra adentro, llegamos a una provincia de serranías, tierra fría, donde hallamos los montes de muy hermosas encinas cargadas de bellota" (Andagoya 1980). Se refería el autor a la palma pixbae, base de la extendida arboricultura practicada aún hoy por las poblaciones guaymíes (Linares 1977).

In the early 1500s the conquerors met intensive cultivation on the north coast and extensive savannas in the Pacific interior. But the uplands from Veraguas to Chiriqui probably had fewer people. The Isthmus was not the thick jungle that some people have thought.

El Istmo de Panamá no era, hacia 1500, la espesa selva que algunos han podido creer. La mayor parte del litoral pacífico del país estaba ocupado por una amplia sabana parcialmente antropógena, que retrocederá rápidamente en algunos lugares como el Darién y que evolucionará en otros, gracias a nuevas formas de uso del suelo implantadas por los conquistadores.

El territorio del Istmo era, antes de la conquista ibérica, un puente que unía las dos grandes masas terrestres del Continente, lo que se conocería después por la América del Norte y la del Sur. La intensidad de los movimientos migratorios o de paso parece aumentar cuando se advierten intercambios culturales marítimos, que sin duda acompañaban los comerciales (Linares 1977), entre las grandes culturas del Continente, las de Mesoamérica y las de las altas mesetas colombianas y peruanas.

Se han identificado 79 tribus en las cuales se encontraba distribuida la población panameña a la llegada de los cronistas hispanos (Mercado 1959), que constituían el mínimo de lo que podríamos llamar divisiones político-administrativas de la última etapa precolombina, con un promedio aproximado entre 1.500 y 3.000 habitantes cada una, dedicadas a una agricultura de corte y quema, complementada por la caza, la pesca y la arboricultura. Ellas cubrían, sin duda, los terrazgos más útiles del país.

At the time of Spanish conquest, the peoples were divided into many tribes without a unifying polity except in parts of the central lowlands. There was active trade between neighboring ecological areas.

Ningún poder superior parecía integrar esos cacicazgos y tribus a una organización que superase su propia autonomía y que cubriese todo el territorio ístmico o una buena parte de él. En el Darién los señores son pequeños y separados, según las descripciones de los cronistas hispanos. Sólo en la región de las sabanas centrales, parecía esbozarse un inicio de articulación política más sólida, extensa y compleja, con jerarquías territoriales y políticas organizadas por señores principales y vasallos que no superaba, sin embargo, las regiones del Chirú al Este y de Escoria al Oeste. Los cacicazgos de Natá y Parita, los más importantes, parecían controlar, a través de una decena de señores vasallos cada uno (Linares 1977), otras tantas zonas de medios naturales variados y complementarios. Quizás se establecían alianzas ofensivas y defensivas, pero nada nos sugiere la existencia de verdaderas confederaciones de cacicazgos—a pesar de la resitencia de Urracá en Veraguas—con propósitos más dilatados que los de un acontecimiento militar.

De todos modos, dos patrones diferentes de organización del hábitat parecen coexistir en el Istmo. Por una parte, el del Darién, caracterizado por la dispersión, sobre todo en los numerosos cursos de agua, en cuyas riberas se agrupan, si acaso, pocas viviendas. Mientras tanto, en el área de las actuales provincias centrales, también vemos aparecer concentraciones mayores, como las de Chirú, París y especialmente Natá, en donde Gaspar de Espinosa cuenta cerca de 1.500 habitantes que ocupaban entre 45 y 50 viviendas.

A pesar de este aspecto compartimentado del Istmo, existían, a la llegada de los conquistadores, activos flujos de intercambios comerciales entre zonas ecológicas complementarias, tanto en sentido Norte-Sur, como Este-Oeste, los más frecuentes. Esos flujos y en particular los últimos, más bien por mar, parecían realizar la función de paso transístmico de la época precolombina, entre las dos grandes masas continentales.

Creación de Nuevas Estructuras de Poblamiento y de Usos del Suelo: Siglo XVI

La llegada traumática de los españoles y los servidores africanos trastorna notablemente la estructura del poblamiento y de la organización del territorio panameño. Los amerindios son rápidamente diezmados y reemplazados, aunque parcialmente, por nuevos hombres llegados de Europa y Africa.

The arrival of the Spanish and their African servants was traumatic, for it spelled replacement, in part at least, of the Amerindians by newcomers from Europe and Africa and led to abandonment of much of the terrain.

En efecto, la guerra, las epidemias, las cabalgadas, los destierros de población indígena para las Antillas y sobre todo para el Perú (Góngora 1961) y el rechazo de la vida, como en otras partes de Mesoamérica, llevan a la casi completa desaparición de la población original. Primero, lo advertimos en el Darién y en particular la costa norte con Santa María la Antigua y

Acla que serán también definitivamente abandonadas en 1524; luego, desde la década de 1520, ellas surgen en las sabanas próximas de Panamá y de Natá, desde Chame hasta Pedasí. En el Darién, casi vacío, la sabana retrocede y la floresta ha prácticamente ocupado el territorio en la segunda mitad del siglo XVI. En 1559 se conquista Veraguas, se fundan algunos poblados de vida más bien efímera y se establece el régimen de la encomienda, originario de fuertes perturbaciones demográficas en esa gobernación. El resultado será una nueva estructura de población y de poblamiento en la cual los inmigrantes ocuparán un lugar relativamente confortable, aunque no logren alcanzar, a fines del siglo XVI, un décimo de la población amerindia de 1500.

Hacia 1575 se cuentan cerca de 800 blancos en el Istmo, más de 5.600 negros africanos ya sea esclavos, cimarrones y negros libres y entre 3.000 a 4.000 indios organizados en 5 ó 6 reducciones. Estos 10.000 hombres aproximadamente ocupan el territorio realmente sometido a la autoridad colonial y que cubre cerca de 1/4 del Istmo de Panamá. El resto estaría poblado por quizás otros 10.000 indígenas (Araúz 1595) que dispersos en las selvas y serranías practicaban la agricultura itinerante de corte y quema, complementada por la caza, pesca y recolección tradicionales.

Around 1575 the Isthmus was home to about 800 whites, 5,600 negro slaves, and 3–4,000 Indians organized in 5 or 6 missions (reducciones) as well as about 10,000 other Indians living outside the colonial authority. From about 1540, a north-south transit axis was established, and from the end of the century, an incipient net of nuclear population concentrations was formed throughout the Pacific lowlands west of Chepo.

Después del establecimiento de la función de tránsito, desde la década de 1540, se diferencian con mayor nitidez las dos grandes regiones activas históricas: la transístmica y el interior rural, es decir, las sabanas que miran el Pacífico.

Así, se establece un eje urbano en dirección Norte-Sur con los puntos terminales en el Atlántico y el Pacífico y los sitios de relevo a lo largo de la ruta intermarina. Desde el siglo XVI queda constituida la estructura jerárquica urbana con la ciudad de Panamá como capital, de la cual dependen las otras sedes de gobernación: Portobelo y el Real de Santa María, además de la importante ciudad de Natá, subordinada directamente a Panamá. Las jerarquías espaciales se refinan con la fundación de otros poblados como Cruces, Chepo, La Chorrera y Chame, que dependen sin intermediarios de la jurisdicción de Panamá y que, aunque corresponden a espacios ecológicos semejantes, tienen funciones complementarias. En el interior rural, las autoridades propician una política de hábitat concentrado y dividen el territorio en dos grandes circunscripciones: la Alcaldía Mayor de Natá que cubre las actuales provincias de Coclé, Herrera y Los Santos, y la Gobernación de Veraguas que se extendía hasta la frontera con Costa Rica. En la primera se funda Natá (1522) y la Villa de Los Santos (1569), pueblos españoles, y Parita, Cubita y Olá, establecidos en 1558, además de Penonomé (1581), reducciones indígenas. El Ducado de Veragua, integrado a la Corona en 1558, será enseguida conquistado y convertido en Gobernación dependiente de la capital de la Audiencia de Panamá. Allí, las primeras fundaciones realizadas en el espacio ecológico sabanero complementario de los selváticos lavaderos auríferos en la vertiente atlántica, difícilmente superan el ocaso de la minería a fines del siglo XVI, salvo Santa Fé (1559). Después Montijo (1590), Remedios (1589) y Alanje (1591) se convertirán en los núcleos del hábitat concentrado de la frágil construcción hispánica, reforzada desde el siglo siguiente.

Los españoles tratan de introducir la técnica del arado y los barbechos europeos, pero con tan poco éxito que será el cultivo tradicional itinerante de corte y quema el que finalmente dominará la agricultura panameña desde el siglo XVI. El huerto familiar, de tipo más bien africano tropical, conocerá la asociación de tubérculos, especias, caña de azúcar, cítricos según la región, mangos y palmeras. El ñame, raíz de origen asiático, es introducido por los negros esclavos desde principios del siglo XVI, mientras que el guineo, la caña de azúcar y las aurincáceas llegan con los españoles por lo menos antes de 1533 . El mango es traído por los españoles desde Asia. El arroz hace su aparición en las crónicas en 1605, aunque debió estar presente en Panamá desde algún tiempo antes.

After unsuccessful attempts at European style farming, the Spanish adopted an itinerant slash and burn agriculture, importing yams, sugar cane, bananas, citrus, rice and mangos. But it was the cattle and horses that had the greatest influence on the landscape.

Pero será el ganado, tanto vacuno como caballar que tendrá la mayor incidencia en las modificaciones del paisaje geográfico. Traídos desde temprano al Istmo por los colonizadores hispánicos, ellos ocupan rápidamente amplios sectores de la geografía de Panamá y en particular las sabanas que no habían revertido enteramente al estado de la floresta. La pronta ocupación por parte de nuevos hombres y sobre todo de ganados, impidió la entera regresión de la sabana antropógena heredada del poblamiento precolombino.

Evolución de un Nuevo Paisaje Geográfico: 1600–1900

Desde principios del siglo XVII los elementos principales de la organización del espacio colonial han sido establecidos en la mayor parte del Istmo de Panamá. Los dos siglos siguientes estarán dedicados más bien al fortalecimiento de esa estructura incipiente. Pero, más que al número de hombres tocará a los ganados provocar modificaciones tempranas del medio natural panameño, por lo menos durante los siglos XVII y XVIII.

En 1631, después de un siglo de haberse iniciado un nuevo tipo de ocupación humana hispánica y africana en la vertiente del Pacífico, Diego Ruíz de Campos nos ofrece una minuciosa descripción del paisaje natural y humano de la región comprendida entre la ciudad de Panamá y la Punta de Burica, 550 kilómetros al Oeste (Ruíz de Campos 1892). Nos describe el autor unas islas como Naos, Taboga, Coiba y Parida, con un paisaje natural de selva tropical y sus escasos claros producidos por la ocupación humana, sobre todo en las cercanías de la ciudad de Panamá. También nos descubre Ruíz de Campos un paisaje alrededor de la ciudad capital, como "tierra llana y sin arboleda . . . sabanas i llanadas que sirven de pasto y comedero para el ganado." Los medios intertidales sometidos a las fuertes mareas del Pacífico en donde prosperan los manglares, además de algunas arboledas dispersas, completan la descripción de una región con una muy débil población humana. Los manglares, más bien estrechos, se presentaban en 1631 más allá del Río Grande de Natá, lo que nos hace pensar que los actuales y extensos manglares que se desarrollan desde la boca del Río Hato hasta la del Río Grande, son una creación más reciente, resultado del aluvionamiento producido por varios siglos de erosión intensa de las planicies costeras, causada por un pastoreo extensivo y descontrolado. Las sabanas, amplísimas, continuaban sin interrupción, salvo el ancho bosque-galería del Río Santa María (Escoria), hasta el valle del Río Oria, en el litoral sur de Azuero.

Luego, en el Golfo de Montijo rodeado de densas selvas tropicales, un paisaje de ría dominaba, con sus manglares caracteristicos, mencionados sólo en las bocas de los ríos Martín y Tabarabá (San Pablo), aunque la peneplanicie veragüense cerca de San Pedro del Montijo y La Atalaya contase con "gran suma de ganado vacuno i de cerda i muchas gallinas . . ." sobre la estrecha sabana edáfica. Después de mostrarnos un litoral montuoso, con densas selvas tropófilas y manglares en los numerosos cursos de agua que drenan la Península de Las Palmas y la Cordillera Oriental de Chiriquí, el autor se refiere al área de Chiriquí (Alanje) donde nos encontramos con "gran suma de ganado vacuno de que se saca mucho sebo para traer a Panamá." Los manglares no aparecen aquí tan evidentes y el cronista habla más bien de arboledas en las desembocaduras de los Ríos Chorcha y Chiriquí y sobre todo hacia el occidente, en el valle de los Ríos Chiriquí Viejo y Garanche, en el área de Puerto Armuelles. Parte de los amplios manglares chiricanos, tanto como los coclesanos y azuereños que conocemos hoy, podrían ser una creación posterior producto del aluvionamiento y también, posiblemente, de movimientos isostáticos más recientes del Istmo de Panamá.

Habría que determinar la causa que ha podido provocar la erosión tan intensa que, al cabo de varios siglos, dos a tres cuanto más, ha creado los inmensos depósitos aluviales intertidales en el Golfo de Parita y en el de Chiriquí, para sostener los amplios manglares que conocemos hoy, que tienen en algunos casos hasta 10 kilómetros de profundidad. Ese responsable podría ser en parte, muy verosímilmente, los ganados vacunos y caballares, presentes en la región desde la llegada de los españoles. Ellos también serían responsables del mantenimiento primero y luego de la ampliación de la sabana antropógena, en condiciones climáticas que favorecían el desarrollo de una vegetación mucho más densa, arbustiva o, lo más a menudo, selvática.

By 1631, the Pacific savannas and areas around Panama City were essentially stocked with cattle, but other areas had gone back to forest. By 1607 there were 110,000 head of cattle, in 1790, 193,000 and 187,000 in 1873.

En 1607 contamos cerca de 110.000 reses en el Istmo de Panamá. En 1790 llegamos a 193.000 cabezas de vacunos y caballares y 187.000 en 1873. Pero ellas se encuentran muy desigualmente distribuidas en el territorio panameño, de manera que esas densidades generales se matizan mucho según las regiones y las localidades. Así, en las regiones de Coclé y Azuero pareciera que se hubiese llegado, desde muy temprano, a un punto de saturación, a una densidad máxima teniendo en cuenta la inmutabilidad de las técnicas de cría y del material genético. La cría ganadera va modelando poco a poco un paisaje natural uniforme en lo que se refiere a la vegetación. Mientras tanto, las densidades de ocupación humana, de población agraria, van aumentando lentamente. La dispersión demográfica, sobre todo en el siglo XVIII, puebla las campiñas aisladas. A pesar de haberse mantenido durante tres siglos en la región central la cantidad de ganados bastante constante el crecimiento demográfico exige la ampliación del espacio agrícola de cada comunidad, es decir, del terruño pueblerino.

Además, el mecanismo de la ampliación del territorio bajo una ocupación humana más permanente se apoya en el sistema de la roza tropical. Así, la contigüidad de las rozas va creando una zona de vegetación secundaria que fácilmente se puede convertir en potreros. Tal como lo observamos en las franjas pioneras selváticas, en los piedemontes de la cadena central los ganados seguían, en la época colonial, los desmontes producidos para cultivar las rozas.

También, la práctica del cercado por alambre de acero, data en Panamá de apenas fines del siglo XIX (Arias 1915). Antes, el cercado de fagina, (Franco 1972) formado por troncos unidos, muy costoso en trabajo, era poco practicado, por lo que resultaba más prudente localizar las rozas a cierta distancia del límite superior de los ganados, pero no tan lejos que su cuidado se convirtiese en una labor excesivamente onerosa en tiempo y esfuerzos.

Hacia 1607, el Istmo de Panamá estaba ocupado por aproximadamente 25.000 habitantes, lo cual arroja una densidad muy baja. Sin embargo, la mitad de estos hombres se encontraban dispersos en casi las tres cuartas partes del territorio y escapaban a la autoridad colonial. Se trata, esencialmente, de algunos negros cimarrones y de indígenas de las serranías de Veraguas y Chiriquí y del inhóspito Darién. Así, pues, en 1607 sólo cerca de 12.000 almas se concentraban en la zona sometida a la autoridad hispánica, de las cuales 5.708 habitan la ciudad de Panamá, y 6.300 el resto de la región transístmica y las sabanas de las provincias centrales. Durante el siglo XVII la población aumenta para alcanzar aproximadamente 40.300 habitantes en 1691 y 85.000 un siglo después, en 1790. El crecimiento es imputable casi exclusivamente a las sabanas centrales y de Chiriquí que pasan de aproximadamente 5.000 habitantes hacia 1600 a 45.000 por 1790. En ellas se produce un crecimiento natural vigoroso y también la inmigración, a menudo por aculturación, de poblaciones indígenas en los piedemontes de la cordillera central, desde Veraguas hasta Chiriquí.

During 3 centuries of the colonial period, the Panamanian population increased slowly and irregularly to 85,000 persons in 1790, over half of whom were in the Pacific lowland savannas of central Panama and Chiriqui.

Mientras tanto, en la región transístmica la población pasa de apenas 6.640 habitantes censados en 1607 a 16.679 en 1788. La ciudad de Panamá contiene, en 1790, 7.824 almas, mientras que el resto se distribuye en sus sabanas cercanas, desde Chepo hasta Chame y en la ruta misma hasta Portobelo cuya aglomeración, en plena decadencia, si acaso llega a los 1.000 habitantes.

Durante un poco más de tres siglos de la época colonial, que desde el punto demográfico podríamos llevar hasta mediados del XIX, la dialéctica entre los hombres y el medio natural se resuelve, en muchos casos, por una alta mortalidad endémica. A ella se añaden las epidemias periódicas que asolan el país (Jaén 1979) particularmente en los momentos de arribaje de nuevas poblaciones en tránsito.

Estos hombres, en crecimiento constante aunque irregular, practican esencialmente, durante la época colonial, una agricultura de corte y quema que, siguiendo la evolución de los ganados, va subiendo lentamente por los piedemontes de la cordillera central y va ganando también, en las planicies, los bosques-galería de los principales cursos de agua.

Durante el siglo XIX la población del Istmo de Panamá continúa creciendo a un ritmo más acentuado, pasando de 97.000 habitantes en 1803 a 316.045 estimados en 1896.

In the 19th century, the population rose more rapidly, from 97,000 persons in 1803 to 316,045 in 1896.

También, a pesar del auge de la región transístmica desde la década de 1850, del crecimiento de la ciudad de Panamá y de la fundación de Colón en 1852, el aumento demográfico es el resultado del vigor de los esfuerzos campesinos. Las provincias centrales y las sabanas de Chiriquí pasan así, entre 1788 y 1896, de 44.380 habitantes a 191.650 almas.

Durante el siglo XVII el Darién continúa su despegue demográfico que, muy probablemente, acompañó también el retroceso de una floresta secundaria relativamente reciente. Las migraciones humanas provenientes del Atrato, principalmente de Cunas y también aunque en menor medida de Chocoes, hacen elevar la población del Darién a mediados del siglo XVII a cerca de 15.000 habitantes, los cuales llegarán hasta aproximadamente 20.000 un siglo después, hacia 1741 (Cuervo 1892), a pesar de las epidemias, en particular de viruelas, que prácticamente diezmaban a tribus enteras. La población continúa creciendo lentamente para llegar al máximo antes de fines del siglo XVIII con cerca de 25.000 habitantes, sobre todo de lengua Cuna, pero las campañas militares de 1784 a 1792 provocan, más bien indirectamente por contagio y enfermedades, un descenso rápido para alcanzar, a mediados del siglo XIX, el punto más bajo, de aproximadamente 7.100 habitantes en todo el Darién histórico y un avance de las selvas tropófila y umbrófila, particularmente en la costa atlántica, con el consiguiente desmejoramiento de las condiciones bioambientales que obliga a los Cunas a emigrar, desde la década de 1850 a las islas de coral (Jaén 1979). A partir de entonces, el repunte demográfico se intensifica para alcanzar cerca de 13.000 habitantes en todo el Darién histórico en 1911, de los cuales 7.255 se encontraban en el litoral e islas de San Blas.

Estas perturbaciones demográficas producen, naturalmente, un avance de la floresta en todas partes. Ya hacia 1700, los escoceses hablan de la sucesión de selvas cerradas y de amplias sabanas en la costa norte, cerca de la Bahía de Caledonia (Borland 1700). A fines del siglo XVIII, en 1789, Fernando Murillo declara que "la maleza de los montes es tanta, que no da lugar a nuestras tropas a penetrarla, a menos que no la vayan rozando primero a fuerza de hacha y machete" en las cuencas del Tuira y Chucunaque (Cuervo 1892).

La vertiente atlántica de las montañas de Chiriquí y las islas de Bocas del Toro, por su parte, nunca fueron enteramente sometidas a la autoridad colonial. Su población aborigen era estimada, en 1620, en cerca de 4.000 personas (Del Salto 1620) dispersas y dedicadas a la arboricultura del pixbae, a las rozas de yuca y maíz y a la caza y pesca (Gasteazoro 1980). Desde entonces aunque tenemos pocos datos de la evolución demográfica contamos con testimonios de un lento pero seguro movimiento de poblaciones hacia las más altas cumbres de la cordillera central y hacia la vertiente del Pacífico, en donde se integraban a la sabana, sobre todo en el siglo XVIII, por aculturación en la frontera de poblamiento administrada por la Iglesia Católica. El clima tropical húmedo con abundantes precipitaciones notadas en enero y febrero de 1787, sustentaba una densa selva tropófila que se iniciaba al norte del poblado de Santa Fe y ocupaba, sin interrupción, todo el litoral atlántico hasta las islas de la Laguna de Chiriquí (Cuervo 1892). Este territorio insular se reputaba poblado por cerca de 3.000 indígenas en 1757, que son diezmados por invasores mosquitos, al año siguiente, a los cuales habría que añadir quizás otros tantos aborígenes, situados en la tierra firme, en las cuatro tribus mencionadas por cronistas de la época. En la segunda mitad del siglo XVIII, comienzan a llegar algunos inmigrantes ingleses con sus esclavos de las islas del Caribe, quienes constituyen el núcleo de un poblamiento nuevo pero que demuestra un tímido crecimiento. El censo de 1843 menciona por primera vez a Bocas del Toro con 595 habitantes (Jaén 1979) quienes, a causa de una intensa inmigración provocada por el auge de las plantaciones bananeras, se convertirán en cerca de 11.250 en 1896 (Jaén 1979), entre los cuales contamos una buena parte de indígenas recientemente aculturados.

Population in Darien grew more slowly but was increased by immigration from the Atrato region of present day Colombia. The Atlantic slopes of Bocas del Toro were never entirely subjected to colonial authority and continued as a forested area of low human density. The population was destroyed by Moskito invaders in 1757 but augmented by English immigrants with slaves in the second half of the 18th century.

Las relaciones históricas de los siglos XVI y XVII nos hacen sugerir una hipótesis: el espacio agrario se ha duplicado por lo menos en el intervalo comprendido entre principios del siglo XVII y fines del siglo XIX en las provincias centrales. La vegetación de sabana sólo cubría aproximadamente la mitad del territorio actualmente ocupado por la sabana y el bosque seco tropical, o sea, la sabana heredada de la ocupación agraria indígena que mantenía una población varias veces superior que la del siglo XVI, con un sistema de producción basado en el maíz en particular y que desconocía la cría de ganados.

Para mediados del siglo XIX se ha calculado que sólo 17% aproximadamente del Istmo de Panamá no conservaba su vegetación natural, aproximadamente un poco más de 10,000 km^2 de suelos agrícolas útiles, más planos y arables, de los cuales un poco menos de la mitad son aluviones recientes, que encontramos hoy en la vertiente del Pacífico, desde el valle del Bayano en el Este hasta la Punta de Burica y unos 6.000 km^2 de suelos menos fértiles y de relieve más ondulado, tapizan los piedemontes de la cordillera central y los glacis de volcanes antiguos que sustentan fácilmente, en la vertiente del Pacífico, una vegetación de sabana herbácea o de parque tropical, con sus asociaciones más o menos densas de matorral xerófilo. Esos serían el escenario principal de los desmontes y de la ganadería panameña hasta por lo menos las primeras décadas del siglo XX.

El espacio agrario ganadero es ocupado de una manera desigual en las principales regiones de cría del interior del país a fines del siglo XIX: en 1873, la mayor parte del territorio de la región central exhibe la presencia de la ganadería. Ello es así en 70% de la vertiente del Pacífico, pero en gran parte las densidades son muy reducidas, muy localizadas en los bajos valles aluviales de la antigua jurisdicción de Natá, lo cual nos sugiere una ocupación ganadera más completa de esta región. Las densidades intermedias de 10 a 30 reses por 100 hectáreas se encuentran en 29% del espacio ganadero, en particular cubriendo las zonas de contacto de pequeños valles aluviales y las planicies de llanos secos, región de ocupación ganadera temprana.

During the 17th and 18th centuries the area occupied by agriculture approximately doubled but the area of savanna covered only about half the area now occupied by savanna and tropical dry forest. Occupation of the savannas was very unequal.

En Chiriquí, más de la mitad del territorio se encontraba fuera del área de la presencia de los ganados. Pero también las mayores densidades ganaderas se desarrollaban, en 1873, en las zonas de más antiguo poblamiento y explotación agraria, en las llanuras bajas y las planicies aluviales que rodean David, Alanje y San Lorenzo. Los otras densidades dominan ampliamente el 65% del territorio, ocupando sobre todo y aún tímidamente, el piedemonte del Volcán Barú y las planicies aluviales recientemente desmontadas al Este y al Oeste. En la segunda mitad del siglo XIX los ganados, siguiendo los cultivos, continúan una ascensión más rápida hacia las tierras altas.

La curva de nivel de los 500 metros de altitud es realmente superada en Chiriquí a fines del siglo XIX. Hasta entonces aproximadamente, se establece en ella una frontera inestable de poblamiento amenazada por los indios mosquitos del litoral caribe, cuyas últimas incursiones a lo largo de la Cordillera Central se señalan en 1805.

The 500 m level in Chiriqui was effectively intruded on at the end of the 19th century when cattle and coffee came at about the same time.

Al mismo tiempo, el piedemonte del Volcán Barú, en Bugaba, Boquerón y Gualaca será ocupado con seguridad por los ganados y los cultivos en la primera mitad del siglo XIX.

Este fenómeno de la colonización rural animado por familias de la comunidad vecina se advierte en los ejemplos más notorios de la secunda mitad del siglo XIX en los valles de Boquete en Chiriquí y Tonosí (Morales 1928) en Azuero. En el primero se establecen familias de la comunidad de Gualaca, Bugaba y hasta David y algunos inmigrantes extranjeros que se dedican al cultivo del cafeto, de legumbres y a la cría de ganados. Otros pioneros semejantes, animados de un espíritu de colonización moderna, también se hacen presentes en las tierras altas de Coclé, más bien a principios del siglo XX, en Penonomé, La Pintada y El Valle de Antón, que se convertirá rápidamente en un lugar de descanso para las burguesías capitalinas.

Después de mediados del siglo XIX se amplia la frontera pionera de Panamá bajo la modalidad de la plantación tropical.

Primero, en la región transístmica, la presencia del ferrocarril terminado en 1855 y del puerto de Colón fundado en 1852, favorece el establecimiento, desde 1874, de plantaciones de banano (Jaén 1979). Sin embargo, en la región no podrá desarrollarse tan ampliamente como en Bocas del Toro, en donde las condiciones ecológicas óptimas propician el surgimiento de la gran plantación comercial de bananos para la exportación desde la década de 1880, la cual no ha cesado de prosperar, salvo algunas dificultades por enfermedades en las décadas de 1920 y 1930.

También, a principios del siglo XX, las empresas bananeras ganan el litoral pacífico de Chiriquí, en la región de Puerto Armuelles, en donde se desarrollan grandes plantaciones de frutales para la exportación.

En la costa atlántica las plantaciones de cacao del siglo XVIII en el litoral de San Blas, cerca del Golfo de Urabá, establecidas por colonos franceses, son destruidas por los indios Cunas hacia 1754 (Cuervo 1892) y tendremos que esperar hasta fines del siglo XIX para ver prosperar plantaciones semejantes en la región de Bocas del Toro.

After the middle of the 19th century, the Panamanian frontier was expanded by tropical plantations. Cacao had been planted by French immigrants a century earlier who were soon eliminated, but bananas became a persistant crop, with plantations around Colon and later Bocas del Toro.

El resto de la agricultura de gran plantación comercial será la obra del siglo XX: caña de azúcar, principalmente, en regiones de antigua colonización hispánica, cerca de Panamá, y en las sabanas de Coclé, Azuero, Veraguas y Chiriquí.

En la región transístmica distinguimos dos zonas: la comprendida por la misma ruta, y la formada por sabanas alrededor de la ciudad de Panamá, desde Chepo hasta Chame.

En la primera zona, el corredor de la ruta sigue más

menos el curso del Río Chagres y sus riberas, en donde la población radicada, que no superaba, hasta 1850 por lo menos los 1.500 habitantes, ocupaba las vegas aluviales y en particular los principales sitios de relevo en Cruces, que data de 1527, y, Gorgona, fundado en 1667.

Algunas pequeñas ganaderías establecidas sobre enormes latifundios más bien simbólicos, cubiertos en realidad por la selva, completaban la ocupación del suelo de una región poco alterada por el hombre durante la época colonial.

Entre 1850 y 1880, el ferrocarril transístmico reactiva un poco más la región, pero aparte de la estrecha vía, del establecimiento de los poblados de la línea férrea y de la creación de algunas plantaciones de bananos para la exportación por el Puerto de Colón, la enorme selva tropical continúa dominando la región por lo menos hasta el poblado de Frijoles (Otis 1867). Desde allí hasta Panamá, las rozas campesinas y las sabanas herbáceas se suceden como en la época colonial. La población apenas si ha aumentado alcanzando cerca de 2.000 habitantes en 1851.

Con los trabajos del Canal de Panamá, entre 1880 y 1920, el desmonte es más extenso, pero localizado en la inmediata proximidad de la vía de agua y en el futuro gran lago artificial, el Gatún. La población se multiplica por diez en pocos años, alcanzando cerca de 20.000 habitantes en 1896, la mayoría de los cuales ocupan los poblados de los trabajadores del Canal.

In the transisthmian region, villages were established along the transist route, and the area around Panama from Chepo to Chame was settled early. The area was bolstered in 1850 by the railroad, and after 1880 by construction of the Canal.

También, cerca del futuro canal se extrae madera para la construcción, pero los claros son rápidamente cubiertos por la vegetación secundaria que, al cabo de dos o tres decenios evolucionó hacia la pluviselva y el resto de la cuenca hidrográfica del Chagres permanecía prácticamente intacta.

Al contrario, la región de sabanas al Norte (San Juan), al Este (Pacora) y al Oeste (Río Grande, La Chorrera, Sajalices) de la ciudad de Panamá, es colonizada desde muy temprano por burgueses de la capital quienes establecen una próspera ganadería, la más densa del país. En la segunda mitad del siglo XIX se produce más bien un descenso de la actividad, y sin duda un cierto retroceso de la sabana antropógena. Resulta que la población, en aumento, se concentra más bien en los sitios urbanos de Panamá y Colón y se dedica a las actividades transístmicas, importando del interior, es decir de las sabanas centrales y de Chiriquí, lo esencial del abastecimiento agropecuario. La preocupación por las condiciones ecológicas de los sitios de hábitat concentrado que se debían establecer en el Istmo de Panamá aparecen claramente expresadas desde principios del siglo XVI, en las instrucciones que la Corona Española da a Pedrarias Dávila en 1513. En ellas se advierte una auténtica estrategia de adaptación al medio natural, cuyo cumplimiento resultó, en el Istmo, en el éxito de la mayor parte de los asentamientos humanos, pueblos y aldeas que, en los siglos XVI, XVII y XVIII se fundaron y que forman la estructura de la red urbana actual del país. La Corona recomendaba a Pedrarias que "así en el lugar que agora esta fecho como en los que de nuevo se hicieren, se ha de mirar que sean en sitios sanos e no anegadizos . . . y que sean de buenas aguas e de buenos aires e cerca de montes e de buena tierra de labranzas e destas cosas las que mas pudiere tener" (Pedrarias 1513). Quizás, hubo muy pocas excepciones a esta regla, que motivos más importantes explican, como son Nombre de Dios y Portobelo. Ya conocemos la pesada hipoteca demográfica que ello significó, hasta la década de 1850, cuando la función de la última fue sustituida por Colón.

Las instrucciones de Pedrarias, son fielmente respetadas en los patrones de poblamiento en hábitat concentrado: los llamados pueblos de indios y los de los españoles. De costumbre se escoge una planicie bien aereada que domina, a 5 ó 10 metros de altura, el curso de agua, situado a no más de un kilómetro de distancia. Además, esos poblados se localizan, generalmente, en la zona de contacto de los llanos secos y los valles de aluviones recientes, en donde son más frecuentes los taludes levemente inclinados, propicios al cultivo del maíz, que teme el exceso de humedad. Ya sea en medio de la sabana, como es el caso habitual de los pueblos de españoles o en los valles entallados de la cordillera central para los pueblos de indios, la estrategia de adaptación ecológica se revela un rotundo éxito.

Early towns were often sited at the juncture of dry plains and alluvial valleys. During the colonial period small towns were established with interconnecting roads which led to the sea, and by the end of the 19th century an urban net of towns and villages looking to the Pacific was complete.

Durante el período colonial hasta fines del siglo XVIII los espacios agrarios del interior del país son organizados mediocremente por cuatro aglomeraciones: Coclé, por Natá y luego Penonomé; Azuero por Los Santos; Veraguas por Santiago; Chiriquí por Alanje. Se trata de pequeñas aglomeraciones situadas en el interior de las sabanas, con funciones político-administrativas dominantes y que actúan sobre todo como centros intermediarios ampliamente cerrados y autosuficientes económica y demográficamente.

Pero a partir del siglo XIX, otros pequeños poblados agrarios situados cerca de los sitios con facilidades portuarias rudimentarias van a sustituir a las viejas ciudades administrativas, como organizadoras del espacio geo-económico: Aguadulce en Coclé, Chitré en Azuero y David en Chiriquí demonstraron la mayor dinámica de crecimiento urbano del interior rural desde entonces. Hasta la década de 1920 por lo menos, cuando se construye la primera carretera nacional, ellos funcionaron como puertos de mar, algunos kilómetros tierra adentro, cerca de los esteros y marismas del litoral pacífico, que aseguraban el cabotaje. En ellos convergía una esquelética red de caminos que drenaba cada región y la conectaba al naciente sistema de mercado y también contenían las pequeñas destilerías, principales industrias de transformación del país en el siglo XIX y principios del XX.

Los dos patrones geográficos de poblamiento que notamos en Panamá datan, fundamentalmente, de los siglos XVII y XVIII: la red urbana de pueblos y aldeas de las sabanas que miran al Pacífico por una parte, y, por otra, la amplia dispersión demográfica rural que asegura una presencia humana difusa en vastas áreas del país y que alcanza prácticamente entre un cuarto y un tercio del territorio de Panamá a fines del siglo XIX.

La Revolución del Siglo XX

La continuación de los trabajos del Canal de Panamá, atrae al Istmo una masa de trabajadores antillanos, norteamericanos y europeos de la cuenca del mediterráneo que producirán importantes efectos ecológicos, más bien en la región del istmo central de Panamá.

Continuation of work on the Canal in the 20th century brought a mass of immigrant workers from the Antilles, North America and Europe, many of whom remained as permanent residents.

En sólo 10 años, de 1904 a 1914, los empleadores norteamericanos del canal traen al Istmo a más de 60.000 trabajadores, que se unirán a los 320.000 habitantes estimados hacia 1904. A ello tenemos que añadir un número considerable de hombres que vienen a aprovecharse del auge de las actividades transístmicas. De tal modo que en 1911, la cuarta parte de los habitantes del Istmo de Panamá (la República y la Zona del Canal) es extranjero, elevándose esta proporción hasta el 50% en la región metropolitana. Muchos de estos hombres vinieron con la intención de retornar a su país de origen, pero muchos también se quedan en el Istmo de modo tal que en 1920, 12% de su población nació en el extranjero y un tercio de los habitantes de la región metropolitana. Durante la Segunda Guerra Mundial, la inmigración es muy fuerte gracias a la llegada de desarraigados europeos, colombianos y centroamericanos, atraídos por el repentino auge económico provocado sobre todo por la presencia de cerca de 50.000 soldados norteamericanos acantonados temporalmente en las bases militares. Muchos extranjeros optan por permanecer definitivamente en el país de manera que en 1950, aún 11% de los 862.585 habitantes del Istmo de Panamá habían nacido en al exterior, elevándose la tasa a 21% en la región metropolitana, de los cuales la mitad residía, es cierto, en la Zona del Canal.

Por una parte, la urbanización se convierte en un fenómeno definitivamente dominante al elegir estos nuevos habitantes, en su mayoría, los centros urbanos de la región metropolitana. Por otra parte, su llegada, sobre todo a principios del siglo XX, impulsa a las autoridades sanitarias norteamericanas a una inmensa obra de saneamiento en el istmos central de Panamá, en el área del canal y en las ciudades de Panamá y Colón, para adaptar mejor el medio natural a las necesidades de las nuevas poblaciones, es decir, de la necesaria mano de obra del canal.

The immigrants mostly chose urban centers to settle in, and at the same time health measures cut the death rate in these areas leading to large growth. Growth was also great in rural areas; the central plains, Bocas del Toro, and Chiriqui. In the 1970's, also, a net of urban and rural water conduits permitted population growth.

Así, se vencen rápidamente la fiebre amarilla, la malaria, la disentería y la tuberculosis. La tasa de mortalidad entre los empleados de las obras canaleras pasa de 41.7 por mil en 1906 a 9.1 por mil en 1910, mientras que en el mismo período, en la ciudad de Panamá ella desciende de 44.8 por mil a 31.7 por mil para alcanzar, finalmente, 21.4 por mil en 1918. Esto se logró, en gran parte, gracias a modificaciones del hábitat de los principales vectores de las enfermedades paludeas y de la fiebre amarilla, los mosquitos ıófeles y aedes. La construcción en 1905 del primer :ueducto y alcantarillado urbanos, el drenaje de los antanos, el relleno de manglares y la fumigación onstante de las áreas cercanas a las concentraciones rbanas y a los sitios de las labores canaleras se dindió luego a lo largo y ancho del país: primero fue la egión de Bocas del Toro y sus plantaciones ananeras, de la década de 1920, y, luego, principalıente en las décadas de 1930–1940, el interior del aís y en particular las sabanas centrales y de Chiriuí. En la década de 1970, se completó esta labor con mejoramiento de la medicina preventiva mediante ıejores cuidados a la infancia y también la construcón de una importante red de acueductos urbanos y ırales.

Todos estos fenómenos han provocado un aumento onsiderable de población y, naturalmente, una levación de las densidades de 4 hombres por kilóıetro cuadrado en 1896 a 23 en 1980. Llegamos ahora n la región transístmica y sobre todo en las sabanas ntropógenas a densidades rurales que superan los 40 ombres por km^2. Hacia 1920, se calculaba que cerca e un tercio del Istmo de Panamá estaba habitado llueca 1977). Hoy, prácticamente más de la mitad el territorio conoce los efectos permanentes en la resencia del hombre, principalmente mediante la ıodalidad muy difundida de una intensa dispersión emográfica en las zonas rurales.

Ya desde fines del siglo XIX el paisaje vegetal de la abana comienza a transformarse con la llegada de astos nuevos y más nutritivos como la hierba "pará", raída del Brasil (Arias 1915) y la hierba "guinea", de s Estados Unidos, que se adaptan con más éxito a las escas vegas aluviales.

lready at the end of the last century, pastures vere improved with better African grasses, 'ara brought from Brasil, Guinea from the Jnited States and Faragua from Brasil. From 950 to 1970 the area of pasturage doubled and his now forms perhaps 20% of the country's rea.

Pero la verdadera revolución tiene lugar con la faragua" introducida desde el Brasil hacia 1914 y que ápidamente desplaza a las tradicionales hierbas de la abana, gracias a su mejor adaptación a los latosoles anameños y a la larga estación seca en gran parte de a vertiente del Pacífico, desde los alrededores de la iudad de Panamá hasta los confines de Chiriquí. Esta ıierba conquistadora ha llegado a tipificar hoy la egetación de la sabana panameña.

Después de la destrucción de más de 60% de la anadería del país causada por las guerras civiles de 1900 a 1902, el repunte de la misma se revela con gran rapidez y el crecimiento del hato ganadero ha sido sostenido y, desde la década de 1930, casi paralelo al de la población. Asimismo, la densidad ha evolucionado notablemente. También, la superficie bajo pastos naturales o sembrados se ha ampliado recientemente, duplicándose entre 1950 y 1970, para llegar, en esta última fecha a 1.140.795 hectáreas, es decir, el 15% de la superficie del país. Hoy los pastos quizás superen el 20% del territorio de Panamá.

Todas las regiones participan de esta explosión del crecimiento del hato ganadero y de la ampliación de la superficie bajo pastos y en particular las regiones de Azuero y Chiriquí, cuyo vigor las lleva a contener, en 1970, 60% de los ganados del país. Pero hasta en Darién, Colón y Bocas del Toro se señala, en 1970, la presencia de una ganadería incipiente que transforma en sabana permanente, los claros de la agricultura de corte y quema, ampliamente difundida en tales regiones periféricas y de la intensa explotación maderera.

Sin embargo, a pesar del vigor del crecimiento demográfico y del aumento considerable del hato ganadero, la mayor parte del país permanecía, hasta hace poco, ocupado por bosques y densas selvas tropicales, que retroceden rápidamente bajo el impacto reciente de las franjas pioneras.

In 1950 only about 15% of the area was under human influence but this now exceeds 40%, the increase coming at the expense of the forests. Occupation intensified also during this time as modern tractors supplanted the machete. Itinerant agriculture is less important now because of regulations favoring land titles and more fencing.

En 1950, sólo el 15% del territorio panameño se encontraba bajo alguna forma de explotación humana, ya sea bajo cultivo, pastos naturales y sembrados o en descanso, cifra que casi se duplica en el término de 20 años para alcanzar, en 1970, el 27%. Si tenemos en cuenta que el 11% de los suelos del país son estériles, podemos estimar que las 2.098.062 hectáreas bajo explotación en 1970 representan más de un tercio de la tierra ocupable. Quizás hoy superemos el 40%.

Este cuadro de disponibilidad de bosques tropicales para talar y de las llamadas tierras vírgenes explica la presencia de las principales franjas pioneras que existen actualmente en el Istmo y que continúan modificando, con gran vigor y rapidez, el ambiente natural.

Pero al mismo tiempo que el espacio ocupado por el hombre se amplía horizontalmente, también regiones de antigua ocupación humana extensiva conocen una intensificación del uso del suelo mediante las modalidades de la agricultura comercial tecnificada que se extiende rápidamente en Panamá en los últimos decenios.

Aparte de la gran agricultura de plantación para la exportación que se desarrolla a fines del siglo XIX en algunas pequeñas regiones del país, dedicadas fundamentalmente a los bananos, en el siglo XX se realiza un gran esfuerzo de modernización de la explotación agraria, que impone modificaciones más permanentes del paisaje rural. Así, además de las grandes plantaciones de caña de azúcar para los ingenios, que se inician desde las primeras décadas del siglo y que se han ampliado considerablemente en la década de 1970, también se desarrolla la agricultura mecanizada de medianos y grandes productores, dedicados al cultivo del arroz, maíz, legumbres y pastos sembrados, que ocupa cerca de la mitad de la superficie bajo explotación, además de la tala comercial de bosques.

El machete tradicional y la candela estival, principales instrumentos técnicos de cultivo, están siendo sustituidos desde la década de 1940 por medios más modernos. En 1970 existían 2.693 tractores de algún tipo en Panamá, cantidad que sin duda se ha multiplicado. También los pastos sembrados ocupaban en esa fecha cerca de un millón de hectáreas, es decir, casi el 13% de la superficie del país y 84% de todos los pastos de la República.

El cultivo itinerante retrocede gracias a un ordenamiento más riguroso de la ocupación jurídica, favorecido por la Reforma Agraria y la titulación de las tierras, al mismo tiempo que el crecimiento del hato ganadero y la mayor densidad de los vacunos impone ampliamente el cercado que limita el libre movimiento de la agricultura tradicional trashumante.

En la región metropolitana, formada por las provincias de Panamá y Colón, los cambios ecológicos son aún más profundos y afectan también, como ya lo hemos dicho, la cuenca hidrográfica del canal. La terminación de la vía interoceánica introduce una alteración notable del paisaje geográfico de la región mediante el desvío del Río Chagres, la excavación del cauce del canal, la edificación de represas y esclusas y la creación del Lago Gatún de 423 km^2 y del lago Miraflores, mucho más pequeño. Luego más tarde, en 1936, del Lago Alajuela de 57 km^2. El área metropolitana de la ciudad de Panamá, incluyendo las áreas urbanas de la extinta Zona del Canal, pasa de 22,000 habitantes en 1905 a casi 700,000 hoy. Con la construcción del Canal de Panamá, se remodela enteramente toda el área urbanizada de los barrios de Balboa y sus alrededores, se crea, mediante relleno, Fuerte Amador y Albrook y se drenan los pantanos que quedan del Río Grande, cuyo cauce principal ocupa la nueva vía interoceánica. Hacia el Este, en 1915 nuevos rellenos crean el barrio de La Exposición. Más recientemente, extensas zonas de manglares al este de Chanis son recuperadas para integrarlas al tejido urbano y desde el valle de Pacora hasta Chepo el drenaje y los desmontes han permitido la creación de grandes plantaciones estatales de caña de azúcar. Con la reciente construcción de una represa y central hidroeléctrica, gran parte del valle del Bayano ha sido inundado creándose, en 1976–77, un lago de 260 km^2. La explotación ganadera y maderera se ha intensificado notablemente en la cuenca del Bayano.

In the metropolitan axis of Colon and Panama, the spread of urbanization has made profound changes, amplified by large water reservoirs; Madden Lake, Miraflores Lake and the Bayano Dam reservoir.

En el área de Colón suceden, aunque a una escala más modesta, cambios fundamentales del medio natural gracias a la urbanización, tanto en la isla de Manzanillo que continúa rellenándose como en el litoral cercano de Coco Solo y de Davis-Margarita-Gatún, en donde se producen grandes obras de relleno y drenaje de pantanos.

Más recientemente, alrededor de Puerto Pilón se desarrolla una urbanización acelerada. La población en el área metropolitana de Colón pasa así de 20,000 habitantes en 1911 a cerca de 90,000 hoy. El paisaje suburbano sigue también la ruta de la carretera transístmica, entre Panamá y Colón, inaugurada en 1944.

Las transformaciones más profundas del paisaje geográfico en el Istmo de Panamá durante el siglo XX se encuentran en la región metropolitana. Más aún que la extensión de los fenómenos urbanos, debemos mencionar aquí los efectos de la creación de tres grandes espejos de agua, (Gatún, Alajuela y Bayano) y de la intensidad de la deforestación que, en el período comprendido entre 1950 y 1980, alcanza a más de 70% de las dos principales cuencas hidrográficas, del Chagres y del Bayano, que cubren casi la mitad de la región metropolitana.

Conclusión

La lucha entre el hombre y el medio geográfico tropical dominado por los climas húmedos, que tiene lugar durante los cinco siglos más recientes de la historia de Panamá, se manifiesta mediante diversas esategias de adaptación ecológica realizadas por el uñado de nuevos hombres que, desde el siglo XVI, stán llegando al Istmo.

También los ganados vacunos y equinos in- oducidos desde el principio de la colonización hispá- ica, serán los aliados inapreciables en una empresa ifícil. La selva está presta a ganar el terreno que el lima le permite; los pocos hombres luchan desespe- adamente, durante varios siglos, por mantener una abana antropógena, frágil, heredada de los habi- antes precolombinos. Ella, como el pergamino, se ncoje en el vacío humano y bovino del Darién y hiriquí, pero luego, desde el siglo XIX en el Oeste y esde el XX en el Este, se amplia gracias a la llegada e nuevos pioneros y la introducción de la agresiva aragua.

The forest was ready at the conquest to regain he form that the climate would permit, but the ew newly arrived men struggled to keep avannas in the form inherited from the pre- olombian forerunners. Their efforts have een assisted in the current century by im- orted grasses, but nevertheless, large zones emain forested. Finally, metropolitan urban enters have a large effect on the country's cology and this is bolstered by more advanced echnology.

No obstante, aún permanecen amplias zonas como as cuencas del Tuira y del Chucunaque, que no han ecuperado las magnitudes demográficas pre- olombinas y el medio natural de sabana antropógena que en parte dominaba antes de la conquista. Pero, os fenómenos de franjas pioneras, con un vigor des- usado, están empujando la selva tropical en muchas regiones del país y en particular en los piedemontes de la Cordillera Central y las depresiones del Darién meridional.

Además, los fenómenos de una urbanización cre- ciente y cada vez más concentrada en la región metro- politana, la modernización de las técnicas agrarias y la creación de grandes lagos artificiales también for- talecen las modificaciones del medio geográfico en el Istmo de Panamá y hacen más permanentes las hue- llas del hombre en el paisaje natural. Las estrategias de adaptación ecológica continúan desarrollándose hoy con los instrumentos poderosos que ofrece una tecnología más avanzada.

Andagoya, P. de. 1980. *En* J.T. Medina. El Descubrimiento del Océano Pacífico. Tomo 2, doc. 11 de Cartas y Relaciones. (Citada por Gasteazoro 1980).

Arauza, Y., 1595. Carta de 20 enero de 1595, Colección Fernández. Tomo 5. Panamá.

Arias, R. 1915. La Ganadería Vacuna en Panamá. Panamá.

Bennett, C. F. 1976. Influencias Humanas en la Zoogeografía de Panamá. Editorial Universitaria, Panamá.

Borland, F. 1700. La Historia del Darién. Escocia.

Castillero, A.C. 1967. Estructuras Sociales y Económicas de Panamá, Veragua desde sus Orígenes Históricos, siglos XVI y XVII. Panamá.

Cooke, R. 1976. El hombre y la tierra en el Panamá prehistórico. Rev. Nac. Cultura. Vol. 2.

Cuervo, A.B. 1892. Colección de documentos inéditos sobre la geografía y la historia de Colombia. Tomo 2. "Relación hecha por Gaspar de Espinosa. . . ." Casa Editorial J.J. Pérez, Bogotá.

Del Salto. L. 1620.Carta del Gobernador de Veragua a S.M., Santiago, 21 de junio 1620. *En* colección Fernández, T.V.

Franco, J. 1792. Breve noticia o apuntes de los usos y costumbres de los habitantes del Istmo de Panamá y sus producciones. ed. O. Jaén S. Panamá 1980.

Gasteazoro, C.M. 1980. La Historia de Panamá en sus Textos. Editorial Universitaria, Panamá.

Gongora, M. 1961. Los grupos de conquistadores de Tierra Firme (1509–1530); fisonomía histórico social de un tipo de conquista. Santiago de Chile.

Illueca, J.B. 1977. Tendencias poblacionales y su impacto sobre el medio panameño 1800–1920. Rev. Nac. Cult. 7–8. Panamá.

Jaén, S. O., 1979. La Población del Istmo de Panamá del siglo XVI al siglo XX. Panama.

Linares, O.F. 1977. Ecology and the arts in ancient Panama: on the development of social rank and symbolism in the central provinces. Dumbarton Oaks, Trustees for Harvard University, Washington.

Mercado, E. 1959. El Hombre y la Tierra en Panamá (siglo XVI) según las primeras fuentes. Seminario de Estudios Americanistas, Madrid.

Morales, E.A. 1928. Ensayos, Documentos y Discursos. Colección Kiwanis, Editorial La Moderna, Panamá.

Otis, F. 1867. Historia del ferrocarril de Panamá. *En* C.M. Gasteazoro. La Historia de Panamá en sus Textos. Editorial Universitaria, Panamá. 1980.

Oviedo, G.Fernández de. 1950. Sumario de la Historia Natural de las Indias. Fondo de la Cultura Económica, México.

Oviedo, G. Fernández de. 1959. Historia General y Natural de Las Indias (reimpreso).

Pedrarias, P. 1513. Instrucción dada por el Rey a Pedrarias para Panamá, su viaje a la Provincia de

Castilla del Oro que iba a poblar Panamá . 1513. Valladolid. A.G.I. Patronato 17.

Ruiz de Campos, D. 1892. Relación sobre la costa Panameña en el Mar del Sur . . . , año de 1631. *En* A.B. Cuervo, Colección de documentos inéditos sobre la geografía y la historia de Colombia.

Steward, J. H. & L.C. *Faron,* 1959. Native Peoples of South America. McGraw-Hill, New York.

Ecología de la Industria Ganadera en Panamá

Alberto A. McKay

The cattle industry in Panama grew from the 16th century to the present with a temporary setback during the civil war at the beginning of the 20th century. The practice of enormous farms where cattle could be shifted seasonally to varied ecological regions ended as human population increased and new techniques and grasses were introduced. Now the cattle industry is on the decline except in a few areas, but it has had a great negative effect on the Panamanian landscape.

La ganadería es una actividad que desde principios el siglo XVI siempre ha estado íntimamente vincula-a a la historia del país y ha mantenido, por lo tanto, na poderosa influencia en la economía rural, en las structuras sociales e incluso, en los mecanismos de oder político. En lo que al medio natural se refiere, a sido uno de los más poderosos agentes de ransformación y ha afectado no sólo a la flora y a la una, sino también a los microclimas, a los regímenes idrográficos, a los suelos y al modelado del relieve.

Los datos estadísticos de diferentes épocas per-niten apreciar la magnitud de la actividad ganadera: n 1790, con una población de 63,000 habitantes, el aís poseía 193,000 cabezas de ganado vacuno; en 854, con unos 138,000 habitantes, las existencias anaderas llegaban a 368,000 reses.

'he numbers of cattle in Panama surpassed 68,000 head in 1854 but declined greatly dur-ng the War of a Thousand Days of 1899–1902. ince then the number grew to 1,374,000 in 977.

Luego de la decadencia de la segunda mitad del iglo XIX y especialmente de las enormes pérdidas de a guerra civil de 1899 a 1902, la ganadería ha pro-resado singularmente, pasando de 166,000 reses en 914 a 570,000 en 1950 y a 1,374,000 en 1977. Por eposar la ganadería panameña sobre sistemas ex-ensivos, el peso espacial de su crecimiento es con-iderable: en 1950, los pastos cubrían el 47.7% de las ierras en explotación agropecuaria y en 1970, 52.4%. ,as áreas rurales del país más pobladas, las de las ierras bajas de la vertiente del Pacífico, ofrecen paisajes desolados dominados por herbazales y por los alineamientos arbustivos que forman las cercas vivas. Debido a los pujantes embates de la colonización rural en el oriente del país y en la vertiente del Caribe, estos paisajes se reproducen con rapidez en regiones donde antes no existían.

Diferentes circunstancias han determinado que la ganadería sea la actividad que mayores proporciones de tierras bajo explotación ocupe. Unas de ellas es de raíz histórica y se puede encontrar en el escaso poblamiento del país que se mantuvo durante todo el período colonial, el siglo XIX y los principios del siglo XX. Ante las escasas posibilidades que ofrecía el medio tropical a hombres experimentados en la agricultura mediterránea, se optó por aprovechar los grandes espacios deforestados por la activa agricultura de los indígenas para desarrollar esta producción, la cual sólo requería de espacio abundante y posibilidades de mantener los pastos naturales mediante los incendios. Es así como al disminuir la población indígena en las tierras bajas del Pacífico con clima tropical de estación seca, los hatos ganaderos se multiplicaron rápidamente. Los medios limitantes a la expansión de las tierras ganaderas fueron las montañas, las tierras húmedas, las regiones de poblamiento indígena y aquellas con dificultades de acceso a mercados de consumo, de los cuales el más importante siempre fue la ciudad de Panamá y las otras zonas vinculadas al tránsito interoceánico. Pese a que las regiones con estación seca acentuada eran ideales para aplicar la técnica del fuego, indispensable para conservar a bajos costos y con poca mano de obra los pastos naturales, ésta ocasionaba la escasez de agua y alimentos durante parte del año, especialmente en la región de las Provincias Centrales. Para obviar este problema, se recurría cómodamente a la trashumancia. En la estación lluviosa, el ganado pastaba en las sabanas ubicadas en los interfluvios y en los piedemontes, en tanto que al llegar la estación seca, se le concentraba en los aluviones subirrigados de las llanuras y de los bajos fondos de los valles. También se conducía el ganado a tierras costeras cenagosas, próximas a esteros y a afloramientos de agua subterránea.

In previous centuries, a sparse human population with scant economic possibilities took advantage of the dry season in much of the country to use burning to keep down costs of raising cattle. Sparse population and immense farms bridging diverse ecology guaranteed efficiency of moving cattle seasonally. However, population increase led to fragmenting of farms toward the end of the last century.

La inmensidad de los latifundios compuestos de medios ecológicos diversos y la disponibilidad de abundantes tierras libres en ciertas regiones garantizaron la eficiencia de la trashumancia hasta los primeros años del siglo XX. De allí en adelante, las densidades de población aumentaron, los latifundios se fraccionaron en latifundios menores y las formas comunales de explotación de tierras y recursos desaparecieron casi completamente. Sin embargo, en este período avanzó una verdadera revolución técnica de la ganadería panameña que había tenido sus primeros brotes en los mediados del siglo pasado a partir de dos polos, uno ubicado en Chiriquí en las haciendas de José de Obaldía y otro en Panamá, en las de ciertos ganaderos de la capital.

La introducción de pastos sembrados tales como las hierbas pará y guinea aumentaron los rendimientos de los aluviones y las tierras anegadizas. Posteriormente, en la época de las exploraciones del canal francés, se comenzó a ensayar el alambre de púas, con lo cual se difundió el uso de las cercas y se dio paso a un mejor manejo de los hatos. Después de la independencia de Colombia, la ganadería se desarrolló rápidamente, recuperándose enteramente de los desastres de la guerra de los Mil Días, la cual había hecho desaparecer el ganado de muchas regiones.

Introduced grasses such as Para Grass, Guinea Grass and especially of Faragua together with the use of wire for fences and establishment of a coastal freight service were elements of progress for the cattle industry.

La nueva legislación que reconocía el derecho a usufructuar y luego a apropiarse de las tierras libres, el establecimiento de servicios costaneros de vapores, la constitución de los mercados de la Zona del Canal y la Región Metropolitana, fueron factores de progreso para la ganadería. Las dificultades de la trashumancia fueron vencidas con la adopción de nuevos pastos, adaptables a tierras de mala calidad y resistentes a la sequía de los cuales el más importante ha sido la hierba faragua. Posteriormente, se procedió al mejoramiento genético y a la mayor aplicación de la zootecnia.

En las nuevas condiciones técnicas del siglo XX, especialmente después de 1950, la ganadería no sólo ha saturado el espacio de las Provincias Centrales y Chiriquí ubicado dentro de los límites ecológicos y económicos del siglo XIX, sino que también ha incorporado nuevas tierras tales como las montañas de Azuero y el sur de Veraguas, las montañas medias y húmedas de Chiriquí, las tierras de la reserva indígena del Tabasará y grandes sectores de la Cordillera Central.

Por otra parte, ha recolonizado las inmediaciones de la ciudad de Panamá, en donde había estado en decadencia y, finalmente, ha penetrado, por medio de la colonización rural, la vertiente del Caribe, especialmente en Colón y el oriente del país (Chepo, Chimán y Darién).

By 1950, the cattle industry had saturated not only the Pacific lowlands but also the mountains of the Azuero, southern Veraguas, the vicinity of Panama, parts of the east of the country, and the Caribbean area near Colon.

En la actualidad, la ganadería no es, en nuestro país, un sector dinámico de la economía. A pesar de la elevación de los precios de la carne, el mercado de consumo se ha restringido debido a la reducción del poder adquisitivo de los consumidores, en tanto que han aumentado los costos de producción. Otro factor de estancamiento ha sido la política de incrementar el cultivo de plantas industriales, especialmente la caña de azúcar, al igual que la de exportaciones ganaderas excesivas hacia el extranjero. No obstante, debe tenerse en cuenta, también, que el sistema extensivo creado en los mediados del siglo XX tiene ya grandes limitaciones ecológicas y entre ellas podemos mencionar el deterioro del medio natural debido a las quemas anuales y a la degradación de los suelos, la acentuación de los estiajes en la estación seca, la fragilidad del sistema extensivo especialmente entre las explotaciones pequeñas y medianas durante los años de sequía acentuada, y, finalmente, las plagas que afectan hoy a la faragua.

At present, the cattle industry is growing only in the new production areas in Panama and Colon Provinces but is on the decline elsewhere. This has suspended the negative effect of the cattle industry on the environment, which provides opportunity for a planned development of the cattle industry for the future.

En el presente, las existencias ganaderas sólo aumentan en las nuevas regiones de producción de las provincias de Panamá y Colón, mientras que en las regiones de producción tradicional se nota una reducción de las existencias.

Es conveniente que la planificación panameña aproveche esta coyuntura especial del desarrollo ganadero del país para que se revitalice un sector económico y social arraigado en el medio, pero a través de una intensificación y un progreso técnico que suspenda el efecto negativo que la ganadería, hasta el momento, ha estado ejerciendo sobre nuestros recursos naturales renovables.

La Colonización Campesina de Bosques Tropicales en Panamá

Standley Heckadon Moreno

In Panama because of the expansion of cattle ranching, areas of tropical forest are being rapidly incorporated into the national economy at high ecological and social cost. Traditionally Los Santos province has been a region of small holders dependent on subsistence agriculture and cattle raising for cash. However, the demographic revolution and growing need for cash forces the Santeño to expand his activities and shrink the remaining forest. The extinction of forests followed by ecological deterioration negatively affects Santeño social organization and tends to phase out traditional institutions such as labor exchange and credit. These transformations contribute to a cycle that drives peasants from old frontiers of colonization to new ones.

Es muy probable que de aquí a veinte años, al cerrarse el siglo XX, los bosques tropicales de Panamá hayan desaparecido casi por completo. Pareciera entonces que hay algo de simbólica ironía en que los que hoy nos hemos dado cita aquí para celebrar la culminación de la magna obra "La Flora de Panamá", estemos participando sin darnos cuenta, en las tempranas exequias del rico y complejo legado ecológico del Istmo que nosotros los panameños no hemos sabido apreciar ni resguardar con celo.

It is likely that the forests of Panama will have almost completely disappeared by the turn of the century.

La destrucción de los bosques tropicales se debe, ante todo, a un proceso de colonización basado en la ganadería extensiva para el mercado interno. La ganadería se está expandiendo muy rápidamente: en 1950 el país tenía 570.000 cabezas de ganado vacuno y 550.000 hectáreas de pastos (Censos Nacionales de 1950), mientras que para 1970 el número de reses había subido a 1.2 millones y la superficie en potreros abarcaba 1.1 millones de hectáreas (Censos Nacionales de 1970).

Forest destruction is due largely to colonization based on the rapidly expanding cattle industry.

El objeto de esta ponencia es analizar a los grupos físicamente más asociados a la colonización de los bosques, es decir, los campesinos "interioranos" de las densamente pobladas planicies occidentales de la vertiente del Pacífico pertenecientes a las provincias de Coclé, Herrera, Los Santos, Veraguas y Chiriquí. Al mismo tiempo esta ponencia presentará el caso-estudio del más dinámico de todos los grupos de colonizadores campesinos, los "santeños", de la provincia de Los Santos.

La colonización campesina es uno de los fenómenos sociales más significativos del Panamá contemporáneo. Este proceso es cuantitativa y cualitativamente distinto a los movimientos de colonización del pasado. Las diferencias radican no sólo en el gran número de familias que hoy se movilizan sino, también, en las grandes distancias que recorren para llegar a las fronteras de colonización, la orientación de este campesinado hacia la ganadería extensiva para fines comerciales y la apropiación privada de la tierra.

El desplazamiento de los campesinos hacia las fronteras selváticas, como ha señalado McKay (1971), es la otra cara de la moneda del éxodo rural hacia los grandes centros urbanos. Ambas corrientes migratorias son de origen reciente y reflejan los profundos cambios ocurridos en la sociedad rural panameña en el presente siglo. Por tanto en la primera parte de este ensayo bosquejaremos, a muy grandes rasgos, algunas de las características más sobresalientes de la sociedad campesina "tradicional" que existió hasta las primeras décadas de este siglo. Aquí enfatizaremos el sistema de producción primitivo y la cosmogonía o visión del mundo que compartían los miembros de esta sociedad rural. Luego trataremos de establecer algunas de las causas de las transformaciones que han afectado a esta sociedad campesina y que han dado por resultado, entre otras cosas, al proceso colonizador que aquí nos preocupa.

Peasant colonization of the forest frontiers is concurrent with the rural exodus towards the cities. Both are of recent origin. The most evident colonizing group is the Santeños from Los Santos province.

En segundo lugar, se identificarán las instituciones sociales y el tipo de organización que tiene la sociedad santeña y que le permite a las familias campesinas establecerse en regiones distantes que, en muchos casos, son ecológicamente distintas de sus sitios de origen. Uno de los objetivos que se persigue con este punto es restar fuerzas al enraizado mito de que la colonización campesina es un fenómeno "espontáneo", cuando la realidad es todo lo contrario.

The regions colonized are often ecologically different from the Santeños' place of origin.

Finalmente abordaremos el tema de la relación entre los recursos naturales y el sistema de producción campesina basado en la agricultura de roza para el autoconsumo y la ganadería extensiva para la comercialización. Ambas actividades, tecnológicamente rudimentarias, hacen uso extensivo de los recursos naturales y pueden mantenerse en equilibrio con el medio ambiente sólo mientras se mantengan bajas las densidades demográficas y la participación en el mercado. Sin embargo, veremos como, bajo las nuevas condiciones en las que opera este sistema productivo, se obliga a los campesinos a intensificar ambas actividades, provocando un rápido y extenso deterioro de los recursos naturales, particularmente del bosque. En muchos casos la destrucción de los bosques desata un proceso de degradación ecológica que ejerce efectos negativos sobre la organización social, lo que eventualmente contribuye a la continua expulsión de los campesinos de viejos a nuevos frentes de colonización.

The destruction of forests has negative effects on social organization which contribute to new displacement of peasants towards the frontiers.

La expansión de la ganadería extensiva está provocando una transformación radical en los sistemas naturales del Istmo. Distintos investigadores, Bennett (1976), Mckay (1973), Heckadon (1978) y más recientemente Rubinoff (1979) han señalado las graves implicaciones ecológicas y socioeconómicas que estas alteraciones profundas del medio ambiente pueden acarrear para el país. Desafortunadamente, en algunos círculos a nivel de decisión, tanto gubernamentales como privados, prevalece aún una ideología desarrollista ecológicamente miope que ve los bosques tropicales como símbolo de subdesarrollo y un recurso que en su estado natural no aporta beneficio económico al país. Plantea esta ideología la sustitución de los bosques y la utilización del suelo para otros usos. A esta sustitución de la floresta se le llama, eufemísticamente, "la integración de la selva a la economía nacional." Con el fin de lograr la integración de nuestras últimas regiones boscosas se han venido tomando medidas que han cristalizado en dos amplias estrategias de desarrollo regional conocida una como "la conquista del Atlántico" y la otra como "la conquista del Darién."

El caso-estudio de los frentes de colonización de los santeños que aquí presentaremos indica con claridad que el sistema de producción extensivo que los campesinos reproducen en los frentes selváticos contiene muy serias limitaciones que lo descalifican como el vehículo más óptimo para lograr mejor uso de los recursos naturales. Simplemente no hay desarrollo con la ganadería extensiva. Estimular la expansión de esta actividad sería un error histórico de nuestra generación y sobre todo de quienes están en posición de ejercer la toma de decisiones.

With extensive cattle raising there is no economic development.

Antes de proceder a bosquejar la naturaleza de la sociedad campesina que precedió las grandes transformaciones sería de utilidad describir el marco geográfico de la provincia para así visualizar algunas de las limitaciones que el medio ha impuesto a la acción del hombre.

Algunas Características Geográficas de Los Santos

Los Santos, con 3.300 km^2, ocupa el margen sur oriental de la península de Azuero. En términos generales, la provincia es una región seca donde la precipitación es errática y los suelos pobres. Históricamente, un alto grado de incertidumbre ha rodeado las actividades productivas del hombre.

Los Santos, which occupies the southeastern side of the Azuero Peninsula, is famous for its droughts. Its rainfall is punctuated by a long and unpredictable dry season, and soils are poor.

El clima de la región es caliente y seco, registrándose los promedios de precipitación más bajos del país (entre 1.000 y 1.500 mm anuales). El verano es prolongado (de mediados de diciembre a fines de abril). En la temporada de lluvias la precipitación está mal distribuida. Todos los años hay un período de sequía que interrumpe la continuidad de las lluvias. A este período seco se le llama "veranillo de San Juan," y generalmente ocasiona fuertes pérdidas agrícolas, pues los cultivos se siembran en mayo al iniciarse la estación lluviosa.

Los Santos es notoria por sus sequías. Es difícil obtener información climatológica del pasado, ya que las estaciones meteorológicas son muy recientes. Sin embargo, fuentes históricas locales revelan, por ejemplo, que en un período de quince años (1930–1945) (El Eco Herrerano), escogidos al azar, se suscitaron sequías en 1930–1931, 1939 y 1945. Todas estas sequías provocaron la pérdida casi total de las cosechas y una alta mortalidad en el ganado. Donde las sequías son frecuentes la ganadería disfruta de una ventaja comparativa sobre la agricultura. Los suelos son pobres, generalmente latosoles y litosoles, bajos en los nutrimentos críticos y altos en acidez. Tan pronto la capa vegetal es destruida, los suelos se deterioran rápidamente por la acción de los elementos naturales, agravados por la acción cultural del hombre. En conclusión, si los sistemas tropicales del Istmo son frágiles, los de Azuero están entre los más frágiles.

La Sociedad Campesina Tradicional: Sus Transformaciones y los Orígenes de la Colonización Contemporánea

La sociedad campesina santeña comenzó a tomar forma desde mediados del siglo XVI como resultado del intenso proceso de mestizaje étnico y cultural entre españoles, indígenas y negros (Castillero 1971). Esta sociedad evolucionó muy lentamente durante el largo período colonial y la época departmental de unión a Colombia (siglo XIX). Manteniéndose en esencia como una formación compuesta por pequeños productores campesinos independientes que dependían de la ganadería primitiva y la agricultura de roza bajo un sistema comunitario de tenencia de la tierra, dentro del contexto de una mínima participación en el mercado y bajas densidades demográficas.

Formation of the Santeño society dates from the 16th century. It evolved slowly as a society of small independent farmers relying on primitive cattle raising and slash and burn cultivation under a communal land holding system. There was minimal market economy participation.

Cuál era la naturaleza de esta sociedad agraria que existió hasta las primeras décadas del siglo XX? En esta sociedad la meta central del campesino era la producción para la subsistencia y no para el mercado. Las familias vivían muy dispersas, dependiendo básicamente de la fuerza de trabajo que las propias familias podían proveer. En Los Santos, ni la esclavitud, ni el trabajo asalariado tuvieron papel importante en el proceso de producción. La gente se ganaba la vida practicando un equilibrio entre la agricultura de roza y la ganadería primitiva. En la ganadería primitiva el ganado criollo se criaba en soltura alimentándose de los pastos naturales de las sabanas cuya disponibilidad dependía a su vez de las lluvias, este factor ecológico impuso desde muy temprano un límite a la cantidad de ganado que podía existir en la región. Las cercas se utilizaban muy poco, al igual que las medicinas y los suplementos nutricionales. La mortalidad del ganado, por lo tanto, era muy elevada. Dada las dificultades de transporte del ganado al principal mercado del país (la ciudad de Panamá), los santeños desarrollaron industrias caseras como la fabricación del cebo, la elaboración de carne seca y artículos rústicos de cuero (sillas de montar, zurrones de carga etc.).

Dentro de la economía campesina las industrias caseras eran muy importante. De estas industrias rudimentarias la más importante eran las dedicadas al procesamiento de la caña de azúcar, que se sembraba en los bajos o terrazas aluviales de los ríos. En la provincia se encontraban centenares de trapiches, movidos por fuerza animal o humana, que molían la caña para producir panela y miel. La miel a su vez era vendida a los alambiques de los pueblos que la utilizaban para hacer aguardiente y alcohol. En Los Santos a fines de siglo habían 42 alambiques que daban empleo a gran cantidad de personas (Gaceta de Panamá 1894).

La agricultura y la ganadería estaban íntimamente vinculadas a un complejo sistema de derechos de acceso a las tierras comunales de propiedad de los pueblos o municipios desde el siglo XVII. Aunque en teoría la tierra pertenecía a toda la comunidad, distintos tipos de derechos posesorios ya habían tomado forma para el siglo XIX. La evolución hacia la propiedad privada era más avanzada en el caso de terrenos aluviales fértiles situados a los márgenes de ríos, los cuales permitían un uso más intensivo de la tierra y en donde también se concentraba la población. Los derechos individuales eran más débiles en el caso de los "montes comunes", o sea los bosques o rastrojos que por fuerza de la tradición sólo podían ser utilizados para los cultivos anuales de roza y para obtener materiales de construcción (un ejemplo típico fué el Colmón o Común de Macaracas). Nadie podía cercar estos montes o sembrar cultivos permanentes. En una posición intermedia se encontraban las sabanas para pastoreo que estaban divididas en diferentes "sitios de pastar," y sólo los hatos de los vecinos cercanos a tales sitios podían hacer uso de estos terrenos (Heckadon 1979).

Theoretically land was communal, but individual ownership evolved in the most fertile plots, while tradition held individual rights in abeyance in the forests and mountains.

Aunque el sistema tradicional se caracterizaba por la orientación de la actividad familiar hacia la autosubsistencia, las familias campesinas estaban vinculadas a las estructuras económicas más amplias del interior y de la ciudad de Panamá. Los campesinos santeños participaban en estos mercados como productores, así como de consumidores, aunque en pequeña escala. Entre las distintas regiones había un sistema de trueque fuertemente establecido. En el caso de Los Santos los productos más utilizados en estos intercambios eran sal, tabaco y café, productos que se originaban en los distintos micro-climas de la región. Para obtener artículos manufacturados esenciales y generalmente importados tales como herramientas y vestimenta, los campesinos vendían sus pequeños excedentes a los pequeños comerciantes de los pueblos quienes eran el vínculo principal entre los campesinos y la región metropolitana. Estos comerciantes a su vez dependían fuertemente de los grandes mercaderes de la ciudad de Panamá para crédito y mercancías.

In the subsistence economy, funds for purchase of tools were obtained by trading excess salt, tobacco and coffee through small merchants connected by credit to Panama City.

Otra característica distintiva de la sociedad tradicional era el uso dado a los excedentes económicos, al surplus, ya que la gente en vez de invertirlo en aumentar la capacidad productiva muchas veces lo utilizaban en adquirir joyas para adornar las vistosas polleras de las mujeres o participando en el extenso calendario festivo típico de las comunidades de Azuero.

Dentro de esta sociedad la religión tenía un sitio fundamental y era la Iglesia la que proveía las respuestas a las eternas interrogantes sobre el papel del hombre en esta vida, sobre el bien y el mal. Los campesinos se identificaban con su comunidad en virtud de que pertenecían a la misma parroquia la cual era una unidad religiosa más que administrativa. Hoy día aún se puede oír a los campesinos más viejos identificar su región de origen refiriéndose a la parroquia donde fueron bautizados, "Yo soy de la Parroquia de Macaracas."

Para el individuo era de suma importancia asegurar el paso del ánima al más allá y la gente daba por tanto muchas donaciones a la Iglesia. El liderazgo político también estaba rodeado de fuertes connotaciones religiosas. Dentro de la comunidad, el liderazgo estaba en manos de los "patriarcas", generalmente hombres de avanzada edad, no necesariamente ricos pero muy respetados por sus cualidades morales y su espíritu comunitario (Heckadon, 1976 ined.).

Finalmente la sociedad campesina santeña tenía una orientacion hacia el igualitarismo, lo que se reflejaba, por ejemplo, en el democrático sistema de las reglas de la herencia que estaba regido por el principio de "todos por igual." Los bienes de la familia se partían por igual entre todos los descendientes, a diferencia de otras provincias donde pareciése existir una tendencia a beneficiar al mayor (Heckadon 1977). A diferencia de hoy, la alta tasa de mortalidad aseguraba que pocos individuos sobrevivieran lo suficiente para participar en la partición de los bienes familiares.

The church and the community elders provided leadership, and there was an egalitarian attitude, expressed in inheritance laws giving each heir an equal share.

Por ser una sociedad compuesta de pequeños campesinos independientes que producían utilizando la mano de obra familiar y, ante todo, para el autoconsumo, la sociedad santeña permaneció por largo tiempo, como una sociedad generalmente homogénea sin grandes diferencias sociales entre ricos y pobres.

En el siglo XX la sociedad tradicional que hemos intentado esbozar hasta aquí comenzó a cambiar a un ritmo sin precedentes. Los cambios críticos están vinculados directamente a los procesos que se suscitaron con la independencia política de Panamá de Colombia (1903), pero sobre todo, con la construcción del canal interoceánico. La presencia del canal generó un crecimiento de la economía comercial y que fué acompañado de un intenso proceso de urbanización en las ciudades terminales de Panamá y Colón y la formación, por vez primera, de un significativo mercado interno para los productos agropecuarios del interior.

In the twentieth century, with independence from Colombia and the building of the Canal, a significant internal agricultural market emerged. Communal lands disappeared and fencing of lands left many peasants without land, especially in the most densely populated areas.

No contamos aquí con el tiempo, ni el espacio para discutir con mayores detalles los procesos más críticos que han alterado la sociedad campesina santeña en este siglo, sólo señalaremos algunos de ellos. Uno de los más importantes fué la eliminación del sistema de tierras comunales al expropiarlas el gobierno de la nueva república y luego facilitar su venta a particulares y compañías, tanto nacionales como extranjeras. Así se suscitaron los notorios "encerramientos" que dejaron sin tierra a muchos campesinos, sobre todo en las áreas más densamente pobladas de la provincia.

En segundo lugar, tenemos la rápida expansión de la demanda interna de carne de ganado vacuno que cambiaría la naturaleza de la industria de una actividad primitiva a una más tecnificada y orientada hacia la comercialización. Cesó el libre pastoreo y comenzó a generalizarse el sistema de "potreros" de pastos (artificiales) cercados con alambre de púas. La introducción de pastos como la faragua permitió a la ganadería salirse de su hábitat tradicional, las sabanas más secas, y penetrar en otras regiones. El ganado criollo fué reemplazado poco a poco por nuevas variedades genéticas como el Cebú. Paulatinamente se propagó el uso de vacunas y otras técnicas que redujeron la alta mortalidad de los animales. En pocas décadas en Los Santos, como en todo Panamá, ocurrió una vertiginosa expansión en el número de reses.

The number of cattle increased, bolstered by introduction of the pasture grass, faragua, by introduction of Zebu cattle, and use of vaccination.

Un tercer factor de cambio fue la desintegración de muchas de las industrias caseras y, entre ellas, la de la caña. En los Santos los alambiques desaparecieron con el surgimiento de las modernas destilerías que monopolizaron la destilación de licores.

Por último tenemos a uno de los mas importantes de los factores de cambio: la penetración de las instituciones nacionales al agro las cuales afectarían todas las facetas de la vida campesina. La construcción de caminos y carreteras comenzó a fracturar el aislamiento geográfico de muchas regiones. Se estableció por primera vez el sistema educativo nacional y la escuela se convirtió en uno de los principales focos de modernización de las actitudes de la población rural. En esta proyección de la escuela hacia la comunidad jugaron un papel importante los Clubes de Padres de Familias.

Construction of roads and national education helped disrupt geographic isolation. Population of Los Santos doubled between 1920 and 1960.

Una de las innovaciones críticas fue la introducción de la medicina y sanidad moderna por parte del estado, lo cual, en pocas décadas, eliminó muchas enfermedades que en antaño había diezmado la población rural.

Con el rápido decaímiento de la mortalidad, se suscitó una fenomenal explosión demográfica. En Los Santos, en un período de treinta años (1920–1950) la población se duplicó de 30 mil a 60 mil. Este salto demográfico, sin precedentes históricos, intensificó enormemente la presión sobre los recursos naturales, en particular las áreas de bosques donde eran posibles los cultivos de roza. Por otra parte, dado el sistema de herencia santeña, el mejoramiento de las condiciones sanitarias y la prolongación de la longevidad significarían que un mayor número de individuos sobrevivirían para reclamar las partes de los bienes familiares, lo cual llevaría al mayor fraccionamiento de la propiedad en porciones demasiado pequeñas para ser económicamente útiles.

La compleja interacción de los procesos señalados arriba han contribuído entonces a la intensa emigración de los campesinos santeños hacia las grandes ciudades de la región metropolitana y hacia fronteras de colonización. Un ejemplo de la intensidad de la diáspora santeña es que entre 1940 y 1960 cerca del 30% de la población salió de la provincia (Censos Nacionales de 1960). Durante la II Guerra Mundial miles de santeños emigraron hacia la región metropolitana atraídos por las oportunidades de empleo que surgieron producto de los gastos militares norteamericanos para defensa del canal. Después de la guerra, al disminuir el dinamismo de la economía comercial de Panamá, la migración campesina hacia las fronteras selváticas cobró mayor velocidad que se ha mantenido hasta el presente.

But between 1940 and 1960, 30% of the population left the province, mainly toward metropo-

litan areas but more recently towards forest frontiers.

Como ha señalado Mckay (1973), primero los santeños, colonizaron las últimas fronteras internas de la Provincia de Los Santos (Tonosí y Macaracas). Aproximadamente de 1950 en adelante los campesinos comenzaron a penetrar a provincias vecinas como Veraguas y Coclé. Más recientemente se han aventurado a regiones más distantes situadas al norte, en lo largo de la costa del Caribe y al este hacia las Provincias de Panamá y Darién.

La Organización y Etapas de la Colonizacion "Espontánea"

Desde el punto de vista del campesino la colonización no tiene nada de espontánea. Por el contrario, es una riesgosa operación que exige una planificación minuciosa de las actividades familiares, la adquisición de información sobre las características del nuevo sitio, la obtención de recursos financieros para trasladarse y asentarse, como también iniciar o reactivar una amplia red de relaciones sociales.

For the peasant, pioneering is a risky task demanding careful family preparation and relying on a broad social framework.

Establecer cuáles son las características comunes a todos los múltiples frentes es una tarea difícil, porque la colonización está ocurriendo en muchos sitios a lo largo de vastas áreas del país. Esta sección, por lo tanto, es un intento de sintetizar cuáles son algunos de los procesos que tienden a repetirse entre las múltiples experiencias colonizadoras que nuestro campesinado ha llevado y está llevando a cabo.

Pareciera como si el proceso de colonización de los bosques húmedos podría dividirse en dos etapas, una que llamaremos extractiva y la otra expulsiva (Foweraker 1977).

Durante la etapa extractiva la frontera se caracteriza por la abundancia de recursos naturales y la libre disponibilidad de tierras. La caza y la pesca abundan y forman una parte importante de la alimentación de las familias inmigrantes. Desde el punto de vista demográfico la fase extractiva es el período más dinámico del proceso de colonización, durante el cual un gran número de familias campesinas penetran a una región específica. En el caso del Distrito de Tonosí, por ejemplo, la población se duplicó en diez años de 4.170 habitantes al iniciarse la migración en 1950, a más de 8.000 habitantes en 1960, sobrepasando, finalmente, los 10.000 habitantes en 1970. Durante el período extractivo la población de Tonosí aumentó en 120%. Otro ejemplo de la velocidad con que se pueblan los frentes de colonización es el caso de Nuevo Tonosí en el Distrito de Portobelo en la costa del Caribe, frente que se abrió a la penetración de los campesinos interioranos con la construcción de la carretera costera que unió la ciudad de Colón con el histórico Puerto de Portobelo en 1970. Ya para mediados de 1975, De La Rosa y Varela (1977), encontraron en el área 537 inmigrantes, divididos en 103 familias, el 70% de las cuales venían del Distrito de Tonosí en la Provincia de Los Santos. De aquí el nombre de la nueva colonia, pues es otra costumbre de los santeños el de rebautizar los lugares que colonizan con el nombre de sus sitios de origen. Más recientemente, en el Darién, la apertura de la carretera interamericana ha atraído una ola humana sobre ésta otra poco poblada región. Recientemente, Cabrita (1980) estableció que están entrando al área entre 6 a 7 familias diariamente.

During the extractive phase, game and fish are important for food, access roads are built, population grows rapidly, and farming expands. Deforestation is rapid.

En todos los frentes de colonización ganadera se suscita el fenómeno que al rápido aumento de la población que se da en la etapa extractiva le sigue un proceso opuesto de expulsión de la gente que comienza a emigrar hacia nuevos frentes. En el caso del distrito de Tonosí, entre 1970 y 1977 la población decayó en un 10% (Shearer 1978). Actualmente el frente de Nuevo Tonosí también ha pasado a la etapa expulsiva y muchas familias están emigrando hacia Nombre de Dios, situado más al este en la misma costa, como también están dirigiendose hacia el Darién. En la Cuenca del Canal de Panamá también se nota hoy la intensa emigración de los colonos que deforestaron el área en el período entre 1950 a 1970, particularmente en los distritos de La Chorrera y Capira.

Otro fenómeno típico de la etapa extractiva es la rápida expansión de la frontera agrícola por el veloz surgimiento de nuevas fincas. En el caso de Tonosí, por ejemplo, el número de fincas aumentó tres veces entre 1950 y 1960, de 900 explotaciones a 2.461. En el mismo período la superficie de tierra en fincas se expandió seis veces de 14.000 hectáreas a 80.000. En el área de Darién, Cabrita (1980) ha estimado que la deforestación ha alcanzado ya 11.200 hectáreas "con una velocidad anual de 2.500 hectáreas". Por el contrario, en la fase expulsiva se suscita un proceso por el cual aumenta el número de fincas pero por fragmentación de las explotaciones existentes, dándose también un acaparamiento de la tierra en manos de un menor número de propietarios, es decir un proceso de concentración de la tierra en unidades mayores.

During the expulsive phase, people start to emigrate to new districts, and land holdings are concentrated in fewer hands.

La orientación del colono santeño hacia la ganadería es muy evidente. Volviendo al caso de Tonosí, en 1950 había en el área 12.000 reses, mientras que para 1970 su número ascendió a cerca de 70.000 lo que representa un aumento del 400%. La densidad de vacunos durante estos veinte años pasó de nueve cabezas por km^2 a 48 cabezas (Heckadon 1978).

Claramente los santeños son colonizadores muy dinámicos y efectivos y gran parte de este éxito es atribuible a las redes o sistemas de relaciones sociales y de amistad de las cuales dependen los campesinos que surgen por ser vecinos en los sitios de origen. El rol de estas redes es crucial en los diferentes pasos que debe tomar una familia cuando sale de sus regiones para establecerse en una frontera lejana.

El primer paso es preparativo, se trata de conseguir información sobre la nueva región a partir de familias y amigos ya establecidos allá. Luego sigue un corto viaje de reconocimiento por parte del padre, quien deja los niños con los abuelos o parientes, pero generalmente se lleva a su esposa o compañera. La mujer ejerce gran influencia al momento de decidir, si la familia se traslada o no. Pocas veces un campesino migrará, si la mujer no puede o no quiere seguirlo, sobre todo si hay niños pequeños de por medio. A las fronteras tampoco emigran los viejos.

Otra de las etapas distintivas de la migración es la obtención de capital y el traslado de la familia a la frontera. La colonización es cara y el éxito incierto. La magnitud de los riesgos y los costos actúan como filtros económicos que inicialmente eliminan los estratos más pobres del campesinado los cuales tienden a arribar a la frontera cuando los costos de transporte se han abaratado, usualmente con la construcción de carreteras. El capital lo obtienen vendiendo sus propiedades en sus sitios de origen, usualmente la tierra, pero no el ganado que muchas veces llevan a la frontera como semilla del nuevo hato. Los campesinos en mejor posición económica están en posición de dejar una reserva estratégica, en caso de fracasar en su intento de establecerse en una nueva región.

Capital for colonization is obtained by selling assets in the place of origin. Probability of failure is high.

La gente es muy consciente que hasta los planes mejor trazados pueden fracasar, sea por enfermedad, pérdida de las cosechas o del ganado. Por tanto cuando parten para la frontera dicen que van a "tratar", a probar suerte, pues es su creencia que mucho de lo que ocurre en la vida del hombre gira en torno al destino pre-establecido del hombre. Entre los colonos el porcentaje de fracasos es elevado, pero difícil de establecer. Green y Paredes (1969), estimaron que en el caso de los campesinos del distrito de Macaracas que se fueron a la colonización de Chepo, en la década del 60, más del 30% fracasaron viéndose obligados a retornar en peores condiciones que cuando partieron.

La movilización es siempre a inicios del verano lo que da tiempo a las familias de preparar tierra para la primera cosecha y construir la vivienda antes de que se inicie la temporada de lluvias. Cuando un grupo de familias son los primeros en "abrir" un frente donde no hay facilidades de ninguna clase son usualmente los hombres los que arriban primero seguidos un tiempo después, por las mujeres y niños. Si el frente ya ha sido abierto, entonces la familia se traslada toda a la vez y al llegar a la frontera hacen uso de las facilidades de los que arribaron primero.

Las familias campesinas no se establecen en cualquier sitio, sino en lugares donde tienen parientes y amigos. Los sitios preferidos para asentarse es cerca de los ríos o quebradas. La típica comunidad santeña en la frontera está compuesta por un número reducido de viviendas, físicamente dispersas, pero a la vez estos hogares están unidos por una multiplicidad de lazos sociales que forman una red funcional. El plano que sigue ilustra el fenómeno de las redes sociales entre el núcleo original de colonos santeños que abrieron el frente de Nuevo Tonosí cerca a Portobelo en la costa del Caribe. El plano establece las relaciones entre uno de estos campesinos (Bucho) y sus vecinos más cercanos.

Los derechos sobre la tierra se adquieren, ya sea por virtud de ser el primero en asentarse sobre la tierra, si ésta es nacional, o por compra al dueño anterior. Cuando, en los inicios de la colonización, un pedazo de tierra se adquiere por compra, se le llama "comprar la entrada", pues esta parcela original le permitirá al dueño encerrar tierras nacionales colindantes.

La construcción de carreteras ocasiona marcados cambios en el proceso colonizatorio y la naturaleza de la sociedad fronteriza. La carretera facilita la penetración de instituciones nacionales (p.e. el sistema educativo, etc.) y también, la entrada de nuevos grupos sociales a la región. Específicamente los estratos más pobres del campesinado (minifundistas y campesinos sin tierra), como también de los grandes ganaderos de los pueblos y ciudades que son productores que trabajan a gran escala utilizando el sistema bancario. Los grandes productores desempeñan varias funciones dentro de la ganadería, pues además de criar su propio ganado se dedican a la ceba, es decir a comprar novillos a los productores más pequeños a fin de engordarlos hasta que estén listos para el mercado, como también hay muchos que son matarifes. Es precisamente en la ceba y la matanza donde se obtienen las mayores ganancias dentro de la ganadería.

Construction of roads facilitates entry of national institutions such as schools and also of new social groups, including laborers and large scale operators.

Al aumentar el valor de la tierra se genera un intenso mercado de tierras, o como le dice la gente de "rastrojos" o parcelas de selva que han sido desmontadas y cultivadas brevemente. En todos los frentes de colonización de Panamá se puede encontrar un cierto número de campesinos que se dedican a especular con la tierra. Buscan ser entre los primeros en llegar a un nuevo frente y poder así adquirir una buena cantidad de tierra sea sin costo alguno si son nacionales o por compra si ya hay dueño. En la etapa extractiva la tierra se compra barata, entre 20 y 50 balboas la hectárea. Tan pronto la tierra se valoriza con la construcción de caminos, estos campesinos venden sus fincas que usualmente consisten, en una pequeña parcela de terreno desmontado y una sección mayor aún cubierta de bosques. Con el dinero así adquirido se dirigen de inmediato a otra región selvática donde pueden adquirir una cantidad mayor de tierra. Hay muchos casos de familias campesinas oriundas de Los Santos que han realizado esta operación especulativa en tres, cuatro y hasta cinco frentes distintos.

As the extractive phase proceeds and roads are built land values increase, and some colonists sell out to seek new lands to repeat the colonization process elsewhere.

La crisis de la agricultura de roza es la más seria amenaza a la estabilidad de la sociedad campesina de la frontera que surge como resultados de la desaparición de los bosques. Sin bosques los campesinos no pueden sembrar los granos básicos de la dieta familiar: arroz, maíz y yuca. La crisis de la agricultura de subsistencia desencadena toda una serie de cambios socio-económicos. Desaparecen instituciones tradicionales de obtener mano de obra como son las "juntas" como también instituciones de crédito como las "medias". Para los hogares que se han quedado sin "montes" para hacer sus cultivos se hace necesario conseguir dinero en efectivo para comprar estos productos y ésto usualmente conlleva la necesidad de salir a vender su mano de obra como peones.

Por otra parte la ganadería extensiva, que los campesinos practican para comercializar y obtener dinero, es de tan baja productividad que no es capaz, por si sola, de sostener la economía familiar. Debido a las técnicas utilizadas, el sobre pastoreo se suscita con suma frecuencia y con ello, la degradación, los potreros y el suelo que se ve afectado por la erosión. En el sector oeste de la cuenca del canal, por ejemplo, el 75% de los potreros de los campesinos santeños están fuertemente erosionados.

Extensive grazing practiced by settlers for cash income is of low productivity and cannot alone sustain the family but leads to degradation of pastures and soils, and then to the expulsive phase.

En la etapa expulsiva se suscitan cambios importantes en la tenencia de la tierra y la estructura de clases. Se observa por ejemplo, una mayor concentración de la tierra en manos de un número menor de propietarios, pero a la vez un rápido aumento en el número de minifundios y el surgimiento de un número mayor de campesinos sin tierras. Muchos campesinos consideran que ya no les es económicamente viable seguir en el área que han colonizado y optan por vender sus fincas ya hechas potreros para marcharse a nuevas fronteras de colonización en donde, lo más probable es que el ciclo que los ha expulsado de aquí los vuelva a expulsar también del frente a donde se dirigen.

A continuación trataremos de establecer la relación que existe entre el sistema de producción de los campesinos santeños y los recursos naturales. Como se verá, la naturaleza destructiva de esta relación es donde yace la inestabilidad de la colonización santeña.

Los Recursos Naturales y el Sistema Extensivo de Producción Santeño

El campesino santeño se gana la vida practicando una mezcla de agricultura de roza que le provee con los granos básicos de la dieta familiar (arroz, maíz y frijoles) y la ganadería extensiva su principal fuente de ingresos en efectivo. Ambas actividades requieren o hacen un uso extensivo de los recursos naturales pues dependen de una tecnología muy sencilla que se ha mantenido, salvo por algunas innovaciones en el caso de la ganadería, básicamente inalterada.

The Santeño practices mixed farming with very simple traditional technology, but his socio-economic conditions have changed markedly.

Mientras que tecnológicamente el sistema de producción ha evolucionado muy lentamente, las condiciones socio económicas bajo las cuales el sistema opera han cambiado radicalmente. El campesino responde a estas nuevas condiciones expandiendo simultáneamente, dentro de su pequeña propiedad, tanto sus actividades de subsistencia como las comerciales lo que en cortos años causa una rápida deteriorización de los limitados recursos naturales con que cuenta la finca familiar. Al disminuir los recursos naturales se desequilibra el sistema de producción.

Cuáles son las condiciones que provocan que el sistema de producción rompa rápidamente el equilibrio que debe guardar con los recursos naturales? Básicamente dos: la revolución demográfica y la creciente demanda de dinero en efectivo. En la frontera la intensa migración contribuye a comprimir el tiempo de crecimiento demográfico con el resultado que en pocos años superficies muy extensas de bosques deben ser derribadas para los cultivos de roza. La segunda condición es la continua y creciente necesidad de dinero en efectivo que confronta el campesinado según la economía rural se vuelve más monetizada y dado a que la agricultura de roza no deja casi excedentes vendibles, es la ganadería extensiva la manera más segura de obtener efectivo. La necesidad de "la plata" es el resultado de cambios objetivos y subjetivos que se han suscitado en el campesinado en las últimas dos generaciones. El efectivo es necesario para pagar por los artículos que el productor tradicionalmente ha considerado indispensable tales como herramientas, vestidos, etc. Sin embargo, la gente del campo ha ido adquiriendo nuevas necesidades sin cesar y ahora se considera necesario toda una nueva gama de artículos manufacturados (p.e. radios, relojes etc.) y de servicios. Entre los servicios tenemos la educación. Los campesinos, aún en las regiones más distantes, están dispuestos a realizar todo tipo de sacrificios a fin de enviar sus hijos a la escuela, pues la educación es vista como un pasaporte a una vida mejor. Pocos son los padres que quieren que sus hijos sigan ganándose la vida en el campo, como tiende a decir la mayoría "la agricultura es una vida muy dura."

Two factors upsetting past equilibrium are a demographic revolution and a growing need for cash, the second resulting from increasing monetization of the rural economy.

Veamos ahora con mayor detalle cómo funcionan en la vida diaria los dos pilares o sistemas de producción que sostienen la economía de la familia campesina. En cuanto al sistema de la agricultura de roza, sus características son bien conocidas: PRIMERO, requiere la existencia de bosques, pues sin ellos este tipo de agricultura no existiría, así como la agricultura mecanizada es inconcebible sin fertilizantes inorgánicos. Una familia de cinco personas utiliza, aproximadamente, de una a dos hectáreas de bosques anualmente para sembrar sus granos básicos, o como le dicen ellos, "la comida". Cuánta comida necesita sembrar una familia? Normalmente siembra una lata de semilla de arroz y otra de maíz. SEGUNDO, para poder obtener su comida utilizan herramientas muy simples que no han cambiado en generaciones tales como: hachas, machetes, coas y el fuego. TERCERO, al usar esta tecnología tan elemental el campesino tiene que invertir una gran cantidad de mano de obra. Por ejemplo: cultivar una hectárea de bosque primario puede llevar entre 80 y 100 días hombre y la mitad de esta cifra, si es monte secundario o rastrojo. Generalmente casi ninguna familia se basta por sí sola para suplir este volumen de trabajo y utiliza, entonces, un complejo sistema de instituciones sociales que le permiten al campesino obtener mano de obra de otras familias vecinas sin necesidad de pagar en dinero. En el caso de los santeños, la formas institucionalizadas de adquirir trabajadores dentro de la comunidad, son conocidas como la "junta" y la "peonada." La complejidad de estas instituciones no será analizada aquí. Finalmente, si bien el sistema de roza es ineficiente desde el punto de vista del mercado, pues pocas veces reporta excedentes de importancia para la venta, para el campesino tiene la ventaja que involucra menos riesgos que si utilizada otro sistema de producción más tecnificado. Mientras él se aferra a su sistema tradicional no tiene que utilizar insumos o tecnología cara que usualmente hay que pagar con dinero obtenido mediante crédito. En el sistema de roza, si la cosecha fracasa el productor no tendrá necesidad de hacer pagos en efectivo a ningún banco o temer la pérdida de su tierra debido a una hipoteca.

The peasant farming system requires forests

for cheap land, a few simple tools, and a heavy input of personal and demanded labor, but with this system financial risks are minimized.

Ecológicamente la agricultura de roza es un sistema racional de uso del suelo, siempre y cuando la población del área sea poca y por consiguiente sean abundantes los recursos naturales por la baja presión que hay sobre ellos. Por qué requiere tanto bosque el agricultor de roza? En nuestro medio esto obedece a que la pobreza del suelo y el problema de controlar la maleza hacen que en pocos años los rendimientos agrícolas sean anti-económicos y la parcela tenga que ser abandonada.

En Panamá los suelos de la mayoría de las regiones de bosques tropicales húmedos que están siendo colonizados son suelos pobres, de baja fertilidad. Al mismo tiempo los granos básicos que siembra el campesino son muy exigentes de los nutrimentos críticos. De aquí que en los frentes de colonización como los de la costa del Caribe las parcelas desmontadas para cultivos sólo pueden ser utilizadas por un año. El ciclo normal es arroz en la "primera coa," luego maíz y frijol en la "segunda coa." En ciertas regiones, como el área de Portobelo, la tierra es tan pobre en nutrientes que es factible hacer una sola cosecha, generalmente maíz.

Because soils in many forest regions of Panama are poor and the crops sown by the settlers are demanding of critical nutrients, lands cleared for cultivation can be planted for only one year.

Además de la baja fertilidad de los suelos el campesino que depende del sistema de roza tiene que enfrentarse al serio problema de las malezas. Hay ciertos sectores, como en los valles aluviales de ciertos ríos en Darién, en donde los suelos son de buena fertilidad y permiten que se cultiven por dos y hasta tres años seguidos. Sin embargo, una regla de la agricultura de roza es que a mayor el número de años que se utilice una parcela más grave será el problema de controlar las malezas, por tanto mayor la cantidad de mano de obra que hay que invertir en "deshierbar" los cultivos. En "tierra fresca" de bosques recién desmontados y trabajada por vez primera, es común realizar una deshierba, para el segundo año es necesario hacer dos y hasta tres limpiezas de la maleza para que los cultivos rindan fruto.

The more years a parcel is cultivated, the greater the problem of pests.

Terminado el ciclo agrícola, que como vimos puede durar de uno a tres años, la parcela de tierra debe ser dejada descansar de siete a diez años, a fin de que la vegetación crezca nuevamente convirtiéndose en rastrojo que al ser quemado proveerá nuevamente de nutrientes al suelo. Hasta hace una o dos generaciones atrás era una práctica común de los campesinos santeños dejar que los rastrojos crecieran para que no se agotaran los bosques en donde hacen los cultivos de roza. Hoy día el campesino ha perdido esta sabia práctica y lo que busca es transformar el bosque a pasto al plazo más breve. En la cuenca del Canal de Panamá, por ejemplo, muchos campesinos riegan la semilla de hierba aún antes que la primera cosecha agrícola se haya recogido. A qué se debe este pequeño, pero trascendental cambio en el uso de la tierra por parte del campesinado de Azuero?

Cuando la tierra "se cansa", es decir al dejar una parcela de ser rentable para los cultivos de roza, el campesino se enfrenta a dos alternativas; por un lado puede dejar la tierra en descanso para que se convierta en rastrojo y al cabo de 7 años a 10 años tumbarlo y quemarlo nuevamente para producir los granos para autoconsumo de la casa. Por otra parte, tiene la alternativa de "regarle paja" para hacer potrero y criar ganado. Estos animales a los dos años, ya tendrán suficiente peso para ser mercadeados. Entre las necesidades del autoconsumo y la comercialización gana la segunda alternativa pues generalmente el campesino opta por usar la tierra para la cría de ganado con la esperanza de que si le va bien con la ganadería obtendrá suficiente como para comprar los granos básicos que antes producía. Ante la creciente necesidad de contar con dinero en efectivo la alternativa ganadera es, desde el punto de vista económico, la alternativa más factible y menos riesgosa.

Although the farmer might prefer to fallow the land for 7-10 years, cash demands require immediate conversion of the land into pasture.

El hecho de que el campesino se incline por la ganadería no obedece exclusivamente a que ésta genera ingresos en efectivo a plazo más corto o tiene además otras ventajas sobre la agricultura en general, sea ésta de roza o de cultivos permanentes. En todas las regiones de colonización los sistemas de transporte son muy pobres y resulta muy caro y agotador llevar los productos al mercado. En donde los caminos son poco menos que intransitables durante los largos meses de la temporada lluviosa, el ganado, por su propia movilidad, ahorra mucho tiempo y energía al momento de la comercialización. No así con los productos agrícolas que muchas veces se pierden por las dificultades de transporte.

Otro factor que influye, en que el campesino se incline por la ganadería, es la forma en que opera el crédito agropecuario en Panamá. Las instituciones que más prestan al campesinado, como el Banco de Desarrollo Agropecuario, en general discriminan en contra de la agricultura de roza al tiempo que tienen muchos programas crediticios para facilitar la ampliación de los hatos ganaderos. Los bancos prefieren financiar la cría de ganado, porque ésta es más rentable a corto plazo, requiere menos supervisión y gastos en papeleo que la agricultura. Como ejemplo típico tenemos el caso de la cuenca del canal. En los últimos cinco años entre el Banco de Desarrollo Agropecuario y el Banco Nacional de Panamá han prestado en el Sector Oeste, unos dos millones de balboas a productores campesinos. De esta suma el 98% fueron préstamos ganaderos.

Panamanian banks generally prefer to finance cattle rearing over cultivation.

Es esta compleja red de factores que llevan al campesino a sustituir constantemente los bosques en potreros sin dejar tierras en descanso para la agricultura. De aquí que en todos los frentes de colonización, en un corto número de años, extensas regiones de bosques tropicales se conviertan en monótonos potreros.

Qué pasa cuando el campesino ya no tiene bosques dentro de su propia finca en donde sembrar sus cultivos de roza? Ante esta disyuntiva, los campesinos generalmente toman dos tipos de medidas que no son mutuamente exclusivas. Una de estas acciones es salir a vender su mano de obra, asalariarse como peones, para conseguir el dinero en efectivo con el cual comprar el arroz, el maíz y los frijoles, los cuales su finca ya no produce. La otra medida es tratar de obtener de otros campesinos "matas de monte" o sea parcelas de bosques, utilizando para ello el antiguo sistema de "las medias."

Las medias es una institución económica muy antigua de la cultura campesina santeña. Se utiliza tanto en la agricultura, y ganadería, como en la cría de aves y animales de corral. Esencialmente consiste en un arreglo entre dos productores que deciden compartir los riesgos y costos de la producción.

En el caso de la agricultura de roza los términos de estos contratos verbales varían mucho. El campesino dueño de la parcela de bosque se le llama el que "da las medias", mientras que el que la solicita se le llama el "coge las medias." Aquí hay que notar que las categorías lingüisticas utilizadas sirven para mantener la pretensión de que ambos productores son socialmente iguales, aunque en la práctica las diferencias pueden ser muy marcadas. Pero si ambos participantes son de estratos sociales similares el que "da las medias" derriba el bosque y lo quema, mientras el que "recibe las medias" lo siembra, lo limpia y cosecha y luego ambos parten por mitad la producción. Si la diferencia social es muy grande el que cede la parcela usualmente no contribuye con su mano de obra, mas deja que el otro se quede con toda la producción, pero exigiendo que deje la parcela regada con semilla de pasto tan pronto recoge la cosecha.

En última instancia la existencia del sistema de medias depende en que los cultivos rindan lo suficiente para que ambas partes obtengan una cantidad de granos adecuada para alimentar sus familias el año entero. Pero cuando decae la fertilidad del suelo y la productividad, las medias se convierten en un sistema impracticable y comienza a desaparecer. Las medias a su vez es remplazado por otro arreglo conocido como las "tercias" o a "tres partes una." En esta modalidad el que "da" el monte no invierte mano de obra, mientras el que las "coge" ejecuta todas las tareas del cultivo y a su vez le dará la tercera parte de la cosecha al propietario del terreno. Cuando el resquebrajamiento de la agricultura de roza esta bien avanzado por la desaparición de los bosques, el sistema de "tercias" es remplazado por el alquiler de la tierra pagado en efectivo.

When new lands are lacking, a system of sharecropping may keep peasants working existing lands until the fertility is reduced to such a degree that even such contracts fall into disuse.

Finalmente llega un momento en los territorios colonizados en que los bosques se agotan por completo. Cuando este factor se convierte en un recurso inconseguible la crisis de la agricultura de subsistencia es general. Para los campesinos la crisis de la agricultura de subsistencia plantea un serio dilema. La finca familiar pierde la capacidad no sólo de autoalimentarse, sino de dar empleo a sus miembros, al tiempo que las oportunidades de trabajo en la región escasean, porque la ganadería expulsa mano de obra.

When new land is completely exhausted and the family farm cannot feed or employ its members, cattle raising expels manpower from the land.

Por razones que no vienen el caso discutir aquí, de índole cultural y económico, la mayoría de los campesinos no están en condiciones o no se atreven a introducir nuevos métodos de cultivo que les supla los granos básicos que antes obtenían con el sistema de roza. Esto significa que los productores se ven obligados a depender más de sus actividades ganaderas para proveerse de un ingreso adecuado para así mantener sus necesidades alimenticias mínimas. Para la mayoría de los campesinos, sin embargo, la baja productividad de la ganadería extensiva, se convertirá en un obstáculo casi insalvable en su meta de mantener la estabilidad de la finca familiar y por tanto del hogar.

Para concluir trataremos de dar respuesta a la pregunta, cuáles son las limitaciones del sistema productivo de la ganadería extensiva que practican los campesinos santeños?

Ya que los campesinos dividen sus actividades ganaderas entre lo que ellos llaman las "cosas del potrero" y las "cosas del ganado", nosotros nos atendremos a esta categorización al hacer el análisis de este sistema extensivo de producción. En los últimos 30 años, no obstante la introducción de alguna innovaciones tecnológicas, la ganadería santeña (y en general la panameña) ha aumentado poco su productividad. La capacidad de carga de los potreros se ha mantenido alrededor de una res por hectárea. Durante este mismo lapso se han visto pocas mejoras en el manejo de los pastos que tienden a ser las mismas especies introducidas hace muchas décadas atrás, tales como la faragua, indiana, etc. El volumen de pasto fluctúa vertiginosamente con las estaciones. En el período lluvioso la disponibilidad de hierba es adecuada, no así en los meses de verano cuando el ganado enflaquece considerablemente. En este período hay que tomar en cuenta que pocos son los campesinos que suplen el ganado con complementos alimenticios tales como sales minerales y vitaminas. En general el campesino santeño no subdivide sus potreros en un número adecuado de parcelas o "mangas," con el resultado de que el ganado tienda a deteriorar más el pasto existente. En la ganadería extensiva la tarea más cara es limpiar la maleza de los potreros lo que se hace con machete y en verano quemando los potreros con fuego. La quema anual de los potreros, como bien ha señalado Miranda (1974), sirve para reducir la fertilidad de los suelos, que como se ha señalado en varias ocasiones, tienden a ser pobres en nutrimentos. En pocos años el contenido nutricional de los pastos es muy bajo, sobre todo en contenido proteínico.

In spite of some technical innovations over the past 30 years, the Santeño cattle industry has not improved productivity, still requiring about one hectare per animal. The heavy labor of removing weeds is eased by burning, which reduces soil fertility.

Una de las prácticas más comunes y dañinas es la tendencia de los productores a mantener en los potreros más ganado que la capacidad del pasto disponible. El exceso de reses lleva al sobrepastoreo con el resultado que las malezas que el ganado no come se convierten en la vegetación dominante. En terrenos de pendientes marcadas y muy quebrados, el sobrepastoreo y la quema producen la erosión del suelo y en pocos años se van formando enormes cárcavas u hondonadas. Cuando las malezas o la erosión afectan la mayor parte de la explotación, los potreros "se pierden" y son abandonados. Tal es el caso que hoy se ve en toda la parte suroeste del Lago Gatún.

Overgrazing continues until erosion, weeds and soil infertility lead to pasture abandonment, but most peasants cannot employ better methods, nor can they buy more land.

Como es de esperarse, estas prácticas contribuyen a que los ganados demoren de cuatro a cinco años para alcanzar su máximo peso de mercado. La mayoría de los productores, sin embargo, no están en condiciones de esperar tanto tiempo y venden sus novillos cuando estos tienen uno o dos años. Estos animales los compran generalmente los ganaderos más grandes con mayor capital y disponibilidad de potreros quienes los engordan hasta que estén listos para el mercado. En esa discusión hay que mencionar, de paso, que el sistema de mercadeo del ganado, en términos generales, opera en contra del pequeño productor y en beneficio de los ganaderos más grandes y de los matarifes.

Cuando las fuerzas del mercado obligan al campesino a intensificar su producción pecuaria, él usualmente responde intentando expandir el área de pastos y no introduciendo más tecnología. En la frontera las posibilidades de aumentar la superficie de las fincas son buenas durante la fase extractiva, cuando hay tierras libres o privadas para la compra a bajos precios. En la fase expulsiva adquirir tierras es difícil pues se ha tornado en un factor escaso y caro, tanto para comprar como para alquilar. La mayoría de los campesinos no están en condiciones de comprar tierras adicionales dado los altos precios.

El pequeño productor santeño podría aumentar su productividad, mejorando su tecnología, pero ésto implicaría entrar al sistema de crédito bancario comercial. En general, el campesino es cauteloso para solicitar créditos. Primero, porque ésto significaría hipotecar sus bienes, lo cual es un riesgo que se sopesa seriamente. En segundo lugar, aunque es difícil medir su influencia, se da la fuerza de los valores culturales del catolicismo popular que conceptúa que pagar intereses es moralmente injusto.

Sintetizando, podríamos decir que por su baja productividad la ganadería extensiva no es capaz por si sola de satisfacer las crecientes demandas de dinero en efectivo que confronta el campesino en nuestra sociedad. Este productor al no tener donde sembrar sus granos básicos y ante las dificultades económicas de ampliar la superficie de su finca para extender la ganadería, se encuentra administrando una empresa que ya, económicamente, no es viable. Cuando el campesino, imposibilitado de ganarse la vida dependiendo sólo de la ganadería extensiva se ve "estrecho," opta entonces por vender sus bienes de producción. Con el capital obtenido busca una nueva frontera en donde adquirir una finca más grande de donde, a los pocos años, el proceso de expulsión se repetirá nuevamente.

Because of its low productivity cattle raising cannot alone support the financial needs of the peasant, so he must eventually sell out to find new areas to pioneer. During the past decade social and political tensions have been reduced because of availability of open frontiers. What will happen in 20 years when all the forests have disappeared and our farmers are left without frontiers for colonization?

Quisiera terminar este ponencia con la interrogante ecológica que tiene serias repercusiones políticas. Durante la última generación las tensiones sociales de nuestro agro se han mantenido a un bajo nivel debido a que el país ha tenido fronteras abiertas a donde el campesinado ha podido emigrar. Qué ocurrirá, de aquí a veinte años, cuando hayan desaparecido todos los bosques tropicales y nuestros campesinos se queden sin fronteras de colonización?

Bennett, C. 1976. Influencias Humanas en la Zoogeografía de Panamá. Editorial Universitaria, Panamá.

Cabrita P., R. 1980. Posibilidades del desarrollo de un modelo agroforestal en el Darién. F.A.O., Panamá.

Castillero, A. 1971. La Villa de Los Santos y los orígenes históricos de Azuero. Minist. Educ., Panamá.

Censos Nacionales de 1950. Primer Censo Agropecuario. Contraloría General de la República. Vol. 1. Producción. Panamá.

———. 1960. Migración Interna. Contraloría General de la República. 8: 19.

———. 1970. Tercer Censo Agropecuario. Contraloría General de la República. Vol 4. Compendio General. Panamá.

De La Rosa, C. & E. *Varela*. 1977. Características socio-económicas del área de Nuevo Tonosí. Tesis Licenc., Escuela Trabajo Social, Univ. Panamá.

El Eco Herrerano. 1935–1945. Datos Climatológicos. Chitré.

Foweraker, J. 1977(ined.). The state and economic development; the impact of the periphery. Govt. Dept. Univ. Essex.

Gaceta de Panamá. 1894. Relación de los alambiques existentes en la Provincia de Los Santos. No. 722. Mayo 29. Panamá.

Green, J. & R.G. *de Paredes*. 1969. Panamanian District: a case study in the sociology of development. A.I.D., Panamá.

Heckadon, S. 1976 (ined.) Liderazgo y organización de la communidad. Pp. 82–97 *En* Tonosí. Informe Sociológico Preliminar. Minist. Planific. y Política Econom., Panamá.

———. 1977. Peasant systems and group farming models in Panama. *En* Int. Seminar Agrar. Reform, Inst. Innovation y ural Develop. Land Tenure Center, Jul. 14–22, Univ. Wisc.

———. 1978. Dinámica social de la cultura del potrero en Panamá. Trop. Ecol. 19: 209–217.

———. 1979. Santeño colonization of tropical forests in Panama; a peasant economy and natural resources. *En* Conference on the Development of Amazonia in Seven Countries. Univ. Cambridge.

McKay, Alberto. 1971. El éxodo rural en Panamá. Act. II Simp. Nac. Antropolog. & Ethnob. Panamá.

———. 1973. Colonización de tierras nuevas de Panamá. Rev. Panam. Antropol. 1: 35–55.

Miranda, O. 1974. Panamá, Producción de Ganado de Carne. F.A.O. Rome.

Rubinoff, I. 1979. Los problemas ambientales que confronta Panamá. Rev. Médica Panamá. 4: 21–33.

Shearer, E. Ánalisis, Comentarios y Sugerencias Preliminares Sobre el Censo de Tonosí. A.I.D., Panamá.

Interaction of Man and the Microclimate in a Tropical Forest of Panama

Robert G. Read

El clima tropical más que estable, es constantemente inestable. Es esta inestablidad la que permite que haya flujo de energía hacia otras áreas. Al cortar los bosques tropicales y crear grandes cuerpos de agua, el hombre altera los microclimas que lo rodean. En el área de Bayano en Panamá, los vientos son ahora más pronunciados, los ciclos diurnos de calentamiento y enfriamiento se han alterado, y la formación de neblina y de nubes ha cambiado.

The impression of a tropical climate is often one of stability in which year to year warmth is expected. It is often thought of as hot and humid with little change during the year except for a dry and rainy season. This is a false impression. The tropical atmosphere is constantly on the verge of instability and requires severe thunderstorms to exhaust excesses of heat which have accumulated over a period of time. Some 5,000 of these thunderstorms around the world in the wet tropics act as fire boxes to produce the heat energy, moisture, and kinetic energy that drives the world circulation. These thunder clouds act as vents or chimneys and exhaust the energy produced at the surface of the earth by the sun to high altitudes above the surface where it is carried northward or southward to latitudes less favored by the sun. As in a normal fire box in a dwelling, 60 to 80% of the energy is released through these vents and warms the atmosphere far from the source.

The humid tropical forest has emerged during the last decade as one of the most important natural resources of man. Man is often described as a bad steward of the land. With his deforestation he is accused of causing rapid siltation of important hydroelectric reservoirs and loss of natural water storage in large watersheds, impairing projects like the Panama Canal that depend on water for their operation, as well as lowering of quality of water used for human consumption. Man has done these things mostly in response to population pressure and ignorance. In Panama parks have been set aside for general use by the public, and there has been regulation of access to forested lands by check points at major entrances to these forests. An educational program to inform the public and the farmer on the value of these forests is actively being pursued by radio, television, and poster display. An attempt is being made to save or restore the forest.

In the last decade the world climate has undergone severe droughts and flooding in many parts of the world. Climatic excesses or deficits in the world water balance have been severe, and we enter the current decade in the same fashion. The main point of the discussion is the effects of flooding some 300 km^2 of forested land in the Bayano River Basin.

Deforestación y cambios climáticos durante la década pasada han estado muy relacionados. En Panamá, los cambios en la cuenca del Bayano se han discutido.

In the radiation budget of the earth the reflectivity of the surface becomes important because this is the part of the solar radiation which is lost directly to space. When the percentage of reflectivity increases the availability of heat, energy decreases. The forest in the Bayano area at Majé has a reflectivity of 0.20. That is, 20% of the direct and diffuse solar radiation is reflected to space. This figure is representative of tropical forests of Central and South America. The reflectivity of the cleared land has not been measured directly, but it is estimated to be 50% greater for green pastures and 100% greater for dry pasture—the reflectivity of dry pasture is 0.40 or twice as much as the forest.

En los bosques de Bayano, la cantidad de energía solar reflejada hacia la atmósfera es de .20,

mientras que en los pastizales verdes es de .30 y de .40 en los pastizales muy secos.

Most of the incoming radiation is used to evaporate water. About 90% is used in the production of latent heat, and 10% is used to produce sensible heat near the earth surface. An increase in reflectivity decreases the amount of evaporation into the atmosphere. That is, there is a decrease in the latent heat stored in the atmosphere as water vapor. There is also an increase in the surface temperature of the land and an increase in the wind speed. The whole air column from the surface to high altitude looses heat as a result of the decrease of latent heat in the air. This has a powerful climatogenic effect on the atmosphere globally.

A comparison with reflectivity in polar regions may illustrate this climatogenic factor. Sea ice has a reflectivity of 70% while open sea water surrounding the ice has a reflectivity of 10%. When the expansion of sea ice at -2°C replaces a water surface of 2°C, we are all aware of the global implications of this. Somewhat more difficult to imagine is the implications of the replacement of the tropical forest by the dry or green pasture. If we consider the radiation received in tropical regions to be at least 5 times greater than in polar regions, we can conceive of the two changes in reflectivity to be of the same order of magnitude. I could go on to speculate on consequences, but such speculation is weak unless it is quantified, measured, or a value is placed on the observations. This I have been able to do during the past 7 years in the tropical forest of the Bayano River Basin.

In September of 1972 a biometeorological study began of the effects of flooding 300 km^2 of tropical forest. A field station was established on a ridge overlooking the historical junction of the Maje and Bayano Rivers. With flooding, the field station occupied an island of 20 km^2 of untouched forest. The base line period prior to flooding was from 1972 through 1975. The flooding period occurred during the years 1976 and 1977, and the stabilization period began with the year 1978.

En una estación de campo se estudió la temperatura, precipitación durante 7 años. Los datos se obtuvieron en 5 niveles del bosque, antes, durante y despué de la inundación.

Measurements of temperature were taken at 5 different levels in the forest. Rain and evaporation were measured at canopy level and at the ground. Winds were measured in the free air above the canopy as well as the hourly temperature and humidity. All of these measurements are on file in the reference library of the Gorgas Memorial Laboratory along with detailed analyses of the moisture and radiation budget of this forested area.

Water is in excess of needs in the forest at Maje. An examination of Figure 1 indicates that there is a water deficit (the difference between evaporation and rainfall) until the middle of the calendar year on the forest floor and until July or even October at the canopy level; however, at the end of the year there is almost always a greater amount of rain than evaporation. Over the period 1972 through 1978 there was no significant change in annual rainfall that could be attributed to local deforestation.

En el bosque de los alrededores de la estación hay exceso de agua. El lago almacena calor durante el día, lo intercambia con la atmósfera en la noche.

The forest stores little heat over a twenty-four hour period of day and night. The lake or reservoir, however, stores heat in the upper layers by day and acts as a heat exchanger with the atmosphere at night. The heat loss from the lake was used in two ways. Sensible heat added to the air caused a nighttime thermal wind circulation. The winds at night become quite strong. Prior to the flooding, the wind at night was calm. In addition to the nighttime thermal winds, the air over the lake was warmed enough so that cooling to the point of condensation did not occur. Prior to the flooding, nighttime fogs covered the valley floor and persisted until 0800 to 0900 in the morning. With flooding the air temperature at all levels in the forest changed. At all five levels of measurement it became warmer by day and colder at night. Strong temperature inversions formed in the forest from the 16 m to 29 m levels.

The forest canopy functions in the radiation balance by providing shade, causing an entirely different climate in the forest. Climate is energy dependent, and air temperature and humidity depend on how much energy is available. The average daily radiation balance at canopy level is 133 watts per m^2, and at the forest floor it is 46 watts per m^2. This is a reduction of 65% in the energy available as latent and sensible heat. The persistent temperature inversion keeps the moisture in the air and in the ground as evaporation is inhibited. This is manifest in the patterns (Evaporation-Rain) shown in Fig. 1.

Daytime winds increased as the surrounding forest was cut and after the flooding. While visiting Barro Colorado Island, I had observed that the winds measured there were much stronger than those at Majé

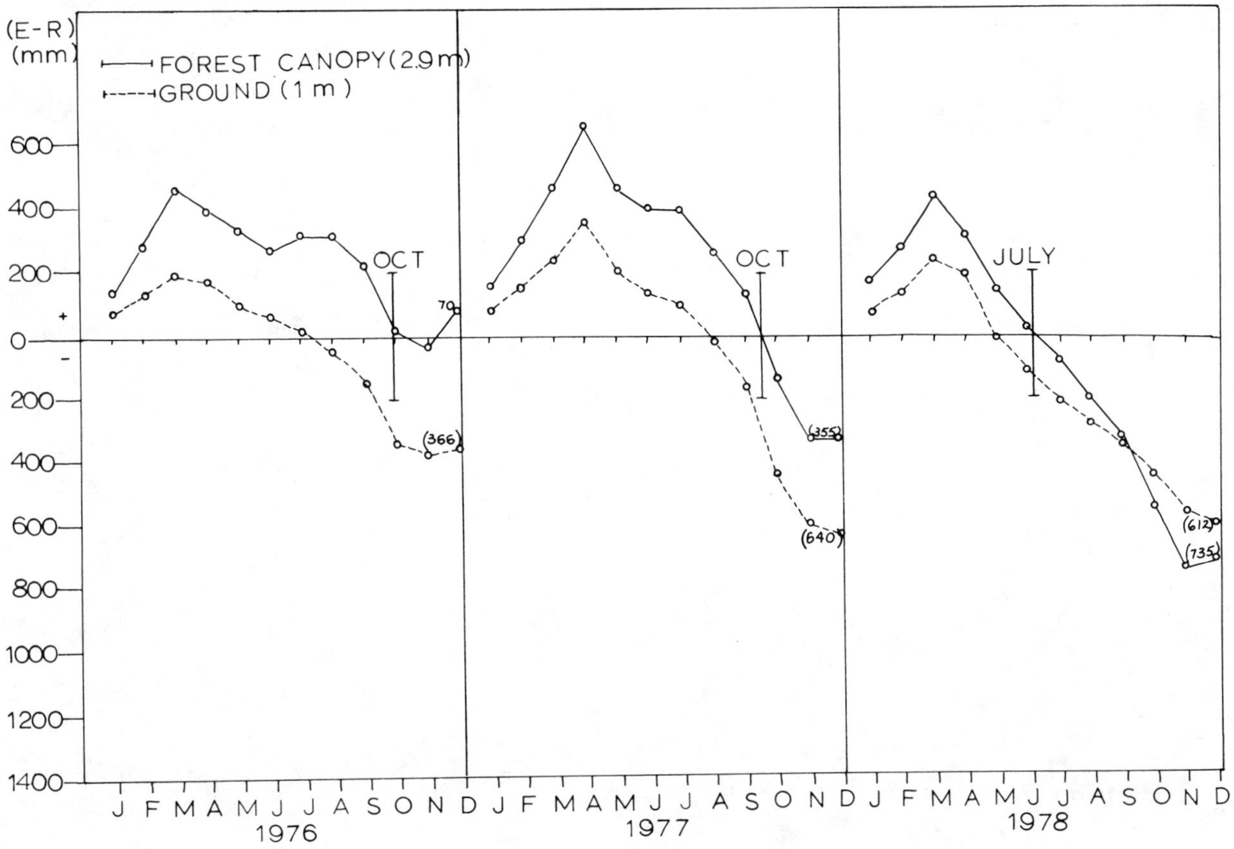

Fig. 1. Moisture balance (evaporation-rain) cumulative for year.

prior to the flooding in 1976. The acceleration of the daytime air flow at Majé after flooding was caused by the constant exchange of sensible heat from the lake to the atmosphere and the lack of the frictional drag of the previously forested area.

To conclude, there were several changes in the microclimate of the forest in the Bayano River Basin as a result of deforestation and the flooding of 300 km^2 of forested land:

a) Strong increase in the wind circulation by day and night.

b) Nighttime cooling and daytime warming of the forest from the ground level to the top of the canopy.

c) Change in the fog formation process from nighttime fog to low stratus cloud formation.

A medida que se cortó el bosque y se inundó el área de embalse los vientos aumentaron y la neblina disminuyó.

Local observation indicates there is an excess in the water needs of the forest from mid summer to the end of the year. The average daily radiation balance is 133 watts per m^2 outside the forest and 46 watts per m^2 at the forest floor.

La Conservación de la Fauna Silvestre en Panamá

Eustorgio Méndez

Serious environmental problems of the present day disturb the equilibrium of wild forest and fauna. Development of population centers and hydroelectric projects will continue, so it is necessary to examine the effects of this on the animals. Forest destruction and fire are prejudicial except in special situations. Excessive hunting, especially by sportsmen and for the market, is to be condemned. Introduction of exotic species has sometimes been disastrous, but some wild species, for example, the white tailed deer, may be suitable for domesticated rearing. A conservationist conscience for Panama is of urgent importance.

Los animales y las plantas constituyen recursos que conocemos como renovables por el hecho de que pueden reproducirse y mantenerse indefinidamente si son utilizados de una manera razonable. También el agua y los suelos son recursos naturales renovables, aunque en su reposición, por supuesto, intervienen factores de otra naturaleza. En cambio, aquellos recursos tales como los minerales y el petróleo se consideran como irrenovables, es decir, que no se pueden restablecer; por lo tanto, pueden explotarse hasta que se agotan totalmente.

Although many primitive cultures revered animals for artistic, medicinal and religious reasons and demonstrated conservationist tendencies, in our times, the idea that man must live in harmony with his environment was not taken seriously until there was evidence of serious environmental problems such as contamination of air, water and soils and the extinction of many valuable species of plants and animals.

Lamentablemente, existe en Panamá un desconocimiento casi general de la necesidad de conservar nuestros recursos naturales. La conservación implica el aprecio, valorización y ordenamiento de aquellos elementos que forman parte del ambiente de los seres humanos, es decir, el agua, el aire, el suelo, los minerales, plantas y animales, con el objeto de obtener un elevado nivel de vida para la humanidad. Estos nobles propósitos han inspirado una verdadera doctrina internacional que se conoce como el *CONSERVACIONISMO* y que constituye una de las causas más dignas y elevadas. A pesar de que en los tiempos modernos los principios de esta disciplina se definen con mejor claridad y, aparentemente, tienen mayor relieve e importancia, hay que reconocer las tendencias conservacionistas tienen su origen en épocas remotas y es posible que el hombre primitivo mostrara cierta preocupación por proteger algunas especies de animales poco abundantes que fueran de su interés como alimento o por otros motivos. La historia nos revela que tanto los babilonios, asirios, cartagineses, como egipcios, griegos y otros habitantes de antiguas civilizaciones tenían tanto interés en la fauna que domesticaron y utilizaron muchos animales salvajes. En América es notable la afición que mostraron los Incas, los Mayas y los Aztecas por la fauna y, se destaca el hecho de que el emperador Moctezuma era muy aficionado a los animales y tenía un hermoso zoológico en los jardines de su palacio. Otras civilizaciones indígenas americanas, así como las de otras regiones del mundo, han dado muestras de su aprecio e interés por los animales, los cuales no solamente eran cazados para utilizar su carne, la piel y otras estructuras, sino que también muchos de ellos eran venerados religiosamente y usados como motivos artísticos y medicinales.

Es sorprendente y llama la atención el hecho de que ciertas culturas primitivas han demostrado tendencias conservacionistas, que prácticamente han sido ignoradas por el hombre moderno. No ha sido sino ante la grave evidencia de los crecientes problemas ambientales, tales como la contaminación del aire, las aguas y los suelos, así como la extinción de muchas especies valiosas de plantas y animales, además de otros recursos naturales, cuando en nuestra época comienza a tomarse en serio el hecho de que el hombre debe vivir en armonía con su ambiente.

A medida que el país progresa y se desarrolla demográfica y técnicamente es necesario abrir nuevos caminos de penetración y se establecen núcleos de poblaciones agrícolas y ganaderas al mismo tiempo que se explotan los bosques, los ríos y otros recursos naturales. Por otra parte, la crisis de energía eléctrica que afecta la mayoría de las naciones del mundo se hace aguda en Panamá y nos vemos precisados a construir varias hidroeléctricas para contribuir a resolver ese problema e impulsar el desarrollo del país. Para la realización de esos objetivos es preciso transformar grandes extensiones del territorio nacional y producir notables e inevitables cambios ecológicos. Todas esas actividades repercuten en una forma desfavorable en nuestra fauna y flora produciendo un drástico desequilibrio de la naturaleza, que muchas veces elimina aquellos elementos más frágiles y vulnerables.

Progress and development of human settlements and technology are accompanied by alteration of forests, rivers and streams, and the construction of various hydroelectric projects. But all these activities unfavorably affect our fauna and flora and produce disequilibrium in natural conditions, leading to elimination of fragile and vulnerable elements.

Existe una profunda relación y una interdependencia entre los animales y su ambiente; por lo tanto, la destrucción de la vegetación no sólo priva a muchos de esos seres de alimentos constituídos por frutas, brotes, tallos, flores, hojas, semillas y diferentes invertebrados, especialmente muchos insectos y otros artrópodos, sino que también elimina la cubierta protectora que les sirve de abrigo y de refugio. Para los que viven primordial o exclusivamente en los árboles, significa la total eliminación del hábitat junto con sus nidos y crías. Es imposible percibir totalmente el inmenso daño que los incendios forestales ocasionan a la vida silvestre, pero sí es posible afirmar que constituyen una amenaza a la estabilidad ecológica de ciertas regiones. El fuego es nocivo tanto a los árboles como al suelo, a las fuentes de aguas, como quebradas, lagos y ríos, a los distintos tipos de animales y también a los seres humanos que, en forma directa o indirecta, pueden verse tarde o temprano afectados por sus efectos destructores. Sin embargo, es conveniente aclarar que, en situaciones especiales, el fuego puede ser utilizado como un medio efectivo para el manejo de cierto tipo de flora y fauna. Se ha observado que en algunas llanuras el fuego natural ha favorecido la evolución de ciertas hierbas contribuyendo por consiguiente al desarrollo de animales herbívoros, tanto grandes como pequeños y, consecuentemente, al progreso de algunos animales carnívoros o depredadores que se alimentan de aquellos. Las quemas realizadas en forma técnica y cuidadosa en algunos tipos de terrenos contribuyen al mejoramiento y al incremento de ciertas plantas. Naturalmente, deben ser efectuadas en terrenos previamente escogidos como aptos para una adecuada producción agrícola y que no son especialmente importantes desde el punto de vista de la fauna. Debemos aclarar que, aunque muchos de los animales que pudieran habitar estos terrenos pueden ser muy vulnerables a los efectos del fuego, existen otros que están facultados para sobrevivir por su adaptación a esa situación de peligro debido a su rápida movilidad o por tratarse de seres que usan refugios semisuperficiales o subterráneos.

Destruction of forest eliminates habitat for arboreal animals, and fire menaces the stability of forest life in many ways. However, there are special situations where fire under careful management is beneficial.

El uso indiscriminado y descontrolado de substancias tóxicas como los insecticidas y matamalezas contribuye también a la muerte de muchos animales valiosos y amenaza la salud de las personas.

El abuso de la cacería es otro de los factores que perjudica notablemente la estabilidad de la fauna silvestre y ha contribuído a la disminución de varias especies importantes de animales que antes abundaban en las selvas panameñas. El ejercicio de la caza en Panamá no está debidamente controlado, pero actualmente las autoridades de la Dirección General de Recursos Naturales Renovables de Ministerio de Desarrollo Agropecuario elaboran un proyecto de ley en el cual se establecen ciertas pautas para regular la caza, así como se declaran épocas de veda para las especies de mayor trascendencia nacional. El cazador de pocos recursos practica la caza de subsistencia, es decir, el campesino o el indio que mata animales para alimentarse él y su familia. En realidad, éste es el tipo de cazador que menos perjudica la fauna silvestre precisamente porque sus recursos de caza están limitados. El cazador deportista común caza por placer y generalmente tiene la tendencia a excederse en la matanza sin discriminar el sexo del animal y utilizando a veces métodos injustos para obtener las presas.

En ciertos casos se emplean lámparas eléctricas, con las que encandilan los animales en la noche y los convierten en fáciles blancos del tirador, o se permite que los perros de cacería se encarguen de acorralar y dar muerte al animal perseguido. El deportista de corazón es ya moderado, ya que antepone su destreza a la habilidad del animal para escapar del peligro haciendo gala de su instinto de conservación. Por lo general, esa persona despliega un interés (por no decir un aprecio) hacia los animales y tiene un mejor conocimiento de ellos respetando las hembras y los individuos jóvenes. Hay que reconocer que algunas personas que hoy se destacan como defensores de la fauna silvestre en Europa, en los Estados Unidos de América y otras partes del mundo, han sido notables cazadores deportistas.

Excessive hunting is a factor upsetting faunal stability, and a law is in the offing to regulate hunting in Panama. The campesino or Indian hunting for food causes minor damage because he has limited resources, but the sportsman often exceeds himself in killing without discriminating sex or age, and at times uses electric lamps or dogs which make his targets easier. In other places sportsmen have been important defenders of wildlife.

Existe otro grupo o categoría de personas que realizan la caza comercial de ciertas especies de animales para vender la carne a los mercados y hoteles, a pesar de que ya existen disposiciones legales que prohiben ese tipo de actividad. Es triste reconocer también que algunos cazadores eliminan ciertos animales simplemente por el placer de matarlos y sin el propósito de utilizarlos. Ni siquiera escapan a esta acción aquellas especies que resultan beneficiosas para el hombre porque sus hábitos favorecen la agricultura o la salud pública. Es preciso notar que las aves rapaces, ciertos reptiles y algunos mamíferos carnívoros como los gatos salvajes, las nutrias y otros semejantes, se alimentan principalmente de roedores que ocasionan tremendos daños a los cultivos. Aún los murciélagos, que son muy poco apreciados por los seres humanos, consumen muchos insectos manteniendo un equilibrio en sus poblaciones. Otros contribuyen a la repoblación de los bosques al dispersar semillas, al igual que a la polinización de las flores aquellos que se alimentan de néctar. Además, hay que considerar que el guano o excremento de los murciélagos es un excelente abono por la elevada cantidad de minerales que contiene.

El comercio con animales silvestres, ya sea en una escala grande o pequeña, es una actividad perjudicial que ha sido prohibida en muchos países y que debe ser condenada universalmente. Existen verdaderas compañías que operan en distintos países en vías de desarrollo (incluyendo a Panamá), las cuales se dedican a obtener, principalmente para su exportación al extranjero, distintos tipos de animales salvajes, desde ciertos insectos atractivos y otros animales inferiores, hasta reptiles, peces, anfibios, aves y mamíferos. Muchos de estos animales representan especies importantes desde un punto de vista científico o económico, ya sea por encontrarse casi extinguidos, por ser poco comunes, o tener una distribución muy limitada, o por otras razones importantes. Si bien es cierto que la fauna silvestre es una excelente fuente de proteínas que contribuye a solucionar el problema de alimentación a muchas personas, existe la necesidad de que se investiguen los factores de posibilidad de crianza y producción biológica de muchas de nuestras especies, tal como se está haciendo en otros países. Se ha observado que los animales silvestres están mejor adaptados a las condiciones climáticas y alimenticias locales que las especies domesticadas. Al mismo tiempo, son más resistentes a las enfermedades y a los parásitos del área, pues poseen una mejor capacidad de tolerancia a condiciones adversas del ambiente. Podemos citar como ejemplos, entre los varios animales silvestres con buenas perspectivas de explotación al venado cola blanca y al saíno, los cuales pueden criarse y reproducirse idealmente en áreas de poca extensión y con limitados recursos de alimentación. Ambos animales poseen buen rendimiento y una carne excelente y de gran valor nutritivo.

Commerce in wild animals for sale to meat markets and hotels or for other purposes has been prohibited in many countries and should be condemned universally. Export of showy insects and other animals such as reptiles, fish, birds and mammals often draws on species of limited distribution. It is necessary to investigate possibilities for rearing wild species in captivity. The white tailed deer presents good possibilities.

La introducción de ciertas especies exóticas en un país, muchas veces tiene consecuencias funestas para aquellas especies nativas que comparten los mismos hábitos alimenticios y otros requisitos ecológicos similares a los de los animales introducidos. Por otro lado, si la especie importada ya establecida, carece de competidores o de enemigos que podrían contribuir a mantener sus poblaciones dentro de un deseado equilibrio, entonces su número excesivo de individuos muchas veces se convierte en una gran calamidad que ocasiona perjuicios económicos y fáunicos, en vez de ser una fuente de beneficios. Además, existe siempre el peligro de que los parásitos y las enfermedades propias de los animales introducidos, encuentren una situación ideal para su diseminación en animales indígenos.

Existen clásicos ejemplos del fracaso de las importaciones de ciertas especies exóticas y de sus desastrosas consecuencias ecológicas. Tal vez entre los casos más notables podrían citarse el de la introducción del conejo europeo, *Oryctolagus cuniculi* en Australia y, el de la mangosta común de la India, *Herpestes mungo*, en la Isla de Hawai, en Jamaica y en otras islas antillanas.

Varios países han importado animales silvestres de pezuña, ya sea para aumentar las fuentes de proteína animal, con el fin de destinarlos al deporte de la caza, o con otros propósitos. No obstante, ni siquiera la introducción de estos ungulados silvestres de buena talla, ha ofrecido en todos los casos la seguridad de que dichos animales hayan podido mantenerse siempre dentro de los límites de las cercas de sus criaderos.

The introduction of exotic species can have disastrous results, as in the case of the European rabbit to Australia, and the Indian mongoose to Hawaii and the Antilles.

En Panamá se necesita con urgencia la creación de parques nacionales, refugios y reservas de vida silvestre donde los animales y las plantas prosperen idealmente y puedan ser admirados y estudiados. Es preciso, al mismo tiempo, que la cacería indiscriminada, la destrucción de los bosques y otras actividades perjudiciales a la vida silvestre sean debidamente controladas por las autoridades correspondientes. Ante todo, es muy importante inspirar una conciencia conservacionista en el ciudadano para que sea apreciado el valor científico, cultural y económico de nuestros animales silvestres y de otros recursos naturales y la necesidad de preservarlos e incrementarlos. Este objetivo puede lograrse mediante la enseñanza impartida en las escuelas primarias y secundarias, así como en las universidades, mediante los trabajos de divulgación, las conferencias, las películas especiales, los programas radiales y de televisión, además de los artículos publicados en la prensa.

Panama urgently needs national parks, refuges and reserves for wildlife. More important is the inspiration of a conservationist conscience in the citizenry, and this can be achieved by education in the primary and secondary schools and universities as well as by education of the general public.

Tanto los jardines zoológicos y botánicos como los museos de historia natural constituyen valiosas fuentes de conocimiento que están al alcance del público y contribuyen a estimular el interés y el aprecio por los animales. Un objetivo tan importante y tan patriótico, no puede ser realizado únicamente por las autoridades encargadas de promover la conservación y fomentar el uso adecuado de los recursos naturales. Para una obra de tal envergadura y tanta trascendencia nacional se necesita del concurso de aquellas organizaciones como el Ministerio de Educación, la Guardia Nacional, el Movimiento Scout y todas aquellas sociedades que despliegan una gran conciencia cívica y social.

El Combustible de la Foresta como Fuente de Energía para los Países Centroamericanos

Narciso Reyna R.

In Central America use of forests for fuelwood is mandated by the high cost of petroleum and the low personal wealth of the region, but with good management forest resources could lead to a significant long term benefit and avoidance of petroleum dependence. Several organizations such as CATIE, ITCO, CONICIT, and the National University of Costa Rica are working to this end.

El uso de los productos del bosque para cubrir las necesidades de combustible doméstico en la población preferentemente rural es una actividad ampliamente difundida en los países en vías de desarrollo. La práctica de esta actividad en la población tiene origen histórico, costumbrista, social y económico.

Dependencia Energética de Países Centroamericanos: Consecuencias y Alternativas

En la región Centroamericana países como Costa Rica, Belice, Guatemala, Honduras y Nicaragua consumen grandes cantidades de leña y carbón. Esta actividad sin el control debido está ocasionando graves daños al recurso bosque y obviamente, repercute en forma alarmante en el mantenimiento del equilibrio ecológico.

En la mayoría de los casos el uso actual de los bosques como fuente de energía calorífica, no sigue un ordenamiento técnico adecuado, debido particularmente a factores como falta de conocimiento por parte de los usuarios, escasez de materia prima, costo de extracción, preferencia por ciertas especies, etc.

Es frecuente observar la utilización de leña de madera fina con alto valor económico. En otros casos, la destrucción de la cobertura boscosa altera irremediablemente las condiciones físicas y bióticas del hábitat con las consecuencias ecológicas ya conocidas. Como ejemplo, cabe mencionar la destrucción de los bosques de mangles que provoca el desequilibrio del ecosistema y la pérdida irremediable de la fauna marina asociada a éstos, tales como crustáceos, peces y otros. En cuanto al suelo sin vegetación, éste se expone al constante embate de la erosión hídrica y eólica, causando la pérdida de la capa arable para finalmente aflorar la roca madre. Las prácticas agropecuarias en áreas inapropiadas incide en la alteración del hábitat y provoca generalmente la extinción de la flora natural. En relación a cierto tipo de fauna, se ha constatado que su existencia está relacionada directamente a la permanencia del bosque (Zambrana 1979). La fauna desaparece a medida que lo hace el bosque.

Por otro lado, sabemos que los países mencionados son particularmente importadores netos de petróleo. Según el análisis de economistas, en el año 1977 estos países destinaban el 10 por ciento del total de sus exportaciones para comprar petróleo y se estima que para el año 1985 esta tasa habrá aumentado al 27 por ciento. Otros consideran que la tendencia alcista seguirá un ritmo inusitado e imprevisible.

In 1977 Central American countries required about 10% of their exports to pay for petroleum products. In 1985 this will have reached 27%.

Por lo tanto las exportaciones de los artículos tradicionales que constituyen la base sobre la cual se sustenta la economía de estas naciones serán cada vez más insuficientes para hacer frente a los fuertes desembolsos por concepto de importación del petróleo y sus derivados. Como consecuencia de esta circunstancia los precios del kerosene ("canfin"), gas y luz han impactado negativamente en la población de escasos ingresos económicos ocasionando repercusiones graves en la dignidad humana e incidiendo en un clima de extrema inestabilidad política, cuyas perspectivas son difíciles de predecir.

The increase in petroleum cost has negatively affected these poor populations and has caused political instability in some countries.

Obviamente esta realidad requiere de la atención preferencial de los gobernantes de los países Centroamericanos, tendientes a diversificar las fuentes de energía para mantener o disminuir el incremento en el consumo de los hidrocarburos a través del tiempo.

Como alternativa viable a esta situación se plantea la producción de árboles maderables, mediante un aprovechamiento racional y sostenido, a fin de satisfacer la demanda de leña y carbón.

El recurso forestal puede generar una fuente inagotable de energía para satisfacer las necesidades humanas de consumo de leña y carbón, bajo técnicas racionales de producción y aprovechamiento forestal.

With good forest management, forest resources could be an inexhaustible source of energy, and this would have significant benefit to these poor populations and avoid total dependence on petroleum.

La alternativa planteada requiere de una política forestal bien definida. Las repercusiones de esta acción evidentemente tendrían profundo y positivo significado en los aspectos socio-económicos de un gran sector de la población de bajos ingresos salariales y en general en la economía de los países Centroamericanos, ya que evitaría una dependencia energética total del petróleo y derivados.

Establecimiento y Uso de la Foresta como Fuente de Energía Permenente

Producir energía permanente mediante el establecimiento de plantaciones destinadas al consumo de leña es técnicamente factible, aun con criterios evidentemente comerciales.

Las especies forestales a usarse deben tener como factores esenciales la máxima producción de biomasa en el período más corto, además poseer una conversión de calorías por unidad de peso seco relativamente elevada. En el primer caso se debe utilizar especies de rápido crecimiento que tengan la cualidad de fácil rebrote por cepas y en el segundo caso la densidad de la madera no debe ser muy baja.

The forest species to be used must produce the maximum biomass per unit of time and at the same time have a relatively high caloric conversion per unit of dry weight. Most species of Leucaena and Eucalyptus are thought to possess these qualities.

Existe evidencia que las cualidades anotadas pueden encontrarse en especies de los géneros *Eucalyptus* y *Leucaena*. Además, estas especies tienen la característica de ser esencialmente rústicas y adaptables a gran diversidad de suelos.

En la región Centroamericana existen ecosistemas que se caracterizan por ser sumamente frágiles. En muchos de ellos se asientan poblaciones humanas que en su gran mayoría están ávidas por disponer de leña y carbón para sus necesidades elementales de subsistencia. Las áreas de los ecosistemas mencionados comprenden preferentemente zonas de vida denominadas como bosque seco, bosque muy seco, estepa espinosa y monte espinoso, según el sistema de clasificación de Holdridge (1978). En estas zonas se pueden iniciar proyectos de reforestación para leña y carbón con grandes probabilidades de éxito.

Los organismos estatales y para-estatales del sector forestal, deben asumir el liderazgo desarrollando parcelas demostrativas tendientes a obtener datos en relación a crecimiento, rendimiento maderable, tratamientos silviculturales, espaciamiento más adecuado, factores fitosanitarios, calidad del sitio, posibilidad de introducir el sistema taungya como forma de reducir el costo de plantación y de obtener un valor agregado, determinar los costos de producción, analizar las propiedades tecnológicas, conocer el grado de aceptación por consumidor de la madera producida y otros datos que deben investigarse.

Proyecto de Leña en Costa Rica

La Universidad Nacional (UNA) en Costa Rica acorde con este realidad, está desarrollando el proyecto denominado: "Plantaciones forestales destinadas al consumo de leña". En la ejecución del proyecto, también se encuentran involucradas las instituciones como el CATIE, el ITCO y el CONICIT. Esta participación inter-institucional pone en evidencia la importancia que se está asignando a esta clase de programas.

The National University of Costa Rica has decided to give priority to the development of alternative energy sources from forests. Therefore, this institution together with CATIE, ITCO, and CONICIT is developing a project to produce wood entitled, "Forest Plantations for the Consumption of Wood." This is located in a dry forest at Canas in Guanacaste Province.

Asimismo, el Centro Agronómico Tropical de Investigación y Enseñanza (CATIE), conjuntamente con la Oficina Regional para Programas Centroamericanos, (ROCAP), han elaborado el proyecto denominado: "Producción de leña y carbón y desarrollo de fuentes alternas de energía".

CATIE in Turrialba is studying possible alternative sources of energy from forest resources.

Por lo tanto los organismos interesados deben establecer el contacto necessario con el CATIE, para solicitar asesoria a fin de posibilitar el desarrollo de la producción de leña.

Arnold, J.E.M. y *Jongma*, J. 1977. La leña y el carbón en los países en desarrollo. Unasylva. 29(118): 2–9. Roma.

Budowski, G. 1972. La conservación de la naturaleza como instrumento de desarrollo. México D.F. Instituto Mexicano de Recursos Naturales Renovables.

Burzell, L. 1979. Guía para el leño de *Leucaena leucocephala*, "Planta milagrosa del trópico," (Mimeo) Lowe Inc., Hawaii.

Candido, J.F. 1971. Comparacáo de especies de eucalipto para produáo de lenha, em Vicosa, Minas Gerais. Ceres (Brasil) 18(95): 53–59.

Centro Agronómico Tropical de Investigación y Enseñanza. 1979. Producción de leña y carbón. Resumen. (Mimeo) CATIE, Costa Rica.

Earl, D. 1975. Una fuente renovable de combustible. Unasylva. 27(110): 21–26.

Eckholm, E. 1975. The other energy crisis: firewood. Worldwatch paper no. 1. Worldwatch Institute, Washington, DC.

Holdridge, L.R. 1978. Ecología basada en zonas de vida. Trad. al español por Humberto Jiménez S. IICA, San José, Costa Rica.

International Consultation on ipil-ipil research. 1978. AID, Washington, D.C.

Reyna, N. 1979. La leña como fuente de energía con especial referencia a Perú. *En*: Simposio Internacional sobre las Ciencias Forestales y su Contribución al desarrollo de la América Tropical, Octubre 11–17. *CONICIT*, San José, Costa Rica.

Vaughan, C. 1978. Una metodología para determinar la distribución actual y datos sobre el estado actual de especies de la fauna silvestre, con énfasis en aquellas en vías de extinción. *En*: Morales, R. et al. Primera Reunión Regional Centroamericana sobre Vida Silvestre. Matagalpa, Nicaragua 1978. Actas. CATIE, Turrialba, Costa Rica.

Whitmore, J.L. 1979. Producción de madera en los neotrópicos por medio de plantaciones. Puerto Rico. Instituto de Dasonomía Tropical. IUFRO Boletin No. 4.

Zambrana, H. 1979. Producción de leña en las comunidades rurales. *En*: Simposio Internacional sobre las Ciencias Forestales y su Contribución al Desarrollo de la América Tropical, Octubre 11–17, 1979. (Mimeo) *CONICIT*, San José, Costa Rica.

Planificación y Manejo de los Recursos Silvestres en América Central: Estrategia para una Década Crítica

Craig **MacFarland**, Ramon **Martinez** and Roger **Morales**

Central America, globally defined, includes seven political entities: Panama, Costa Rica, Nicaragua, Honduras, El Salvador, Guatemala, and Belize. Because of its delicate natural resource situation and prevailing socioeconomic and political situation it is a critical area. The present decade will determine for Central America the possibility of balance with sane and sustained economic development based on conservation of natural resources. The situation has worsened in the last 20 years as we look at remaining forest resources and human poverty. Population expansion results in rapid deforestation, erosion, sedimentation, and impoverishment of soils, while rural population migrates to crowded cities or marginal lands.

Impressive advances have been made in conservation of natural resources over the past 10 years, especially the establishment of forest reserves, planning programs, and new environmental policies. The experience and understanding gained has been a significant advance. Major obstacles have been identified permitting adjustment of future activities.

La Región

Un entendimiento claro de las características individuales, de la diversidad y tendencias principales de los países de la región, en términos culturales, socioeconómicos y ecológicos es absolutamente básico para comprender el estudio de la conservación de los recursos naturales renovables y las medidas potenciales y necesarias para mejorar su manejo.

A clear understanding of the individuality of the countries in cultural, socioeconomic and ecological terms is basic to understand conservation of natural resources and possibilities for their management.

Diversidad y Características Socioeconómicas

Algunos detalles de estas características y tendencias están resumidos en el Cuadro 1.

Considerada su área territorial pequeña (541.090 km^2), América Central es una de las regiones más diversas del mundo.

Políticamente, la región está compuesta por seis estados independientes y una colonia autogobernada (Belice de Gran Bretaña). El Salvador y Belice son las entidades políticas más pequeñas, ya que territorialmente miden 21.000 y 23.000 km^2 respectivamente y Nicaragua la más grande, ya que su territorio es siete veces más grande que el del Salvador. La densidad de población tiene sus extremos en El Salvador y Belice, 203.7 habitantes por km^2 y 6.1 habitantes por km^2 respectivamente, mientras que en el resto de las entidades varía entre 16 y 60 habitantes por km^2.

The size and population density of the seven countries differ greatly.

Los elementos culturales principales (europeo, amerindio y africano) varían en la proporción de su mezcla de país a país, siendo en cinco de ellos el europeo y el amerindio los dominantes con el

CUADRO 1
CARACTERISTICAS GEOGRAFICAS, SOCIO-ECONOMICAS, CULTURALES Y ECOLOGICAS DE LOS PAISES DE AMERICA CENTRAL

Entidad Politica	Tamao 000's Km^2	Población 1977 o 1978	Densidad de Población[a] habit/km^2	Crecimiento Demográfico % (1970–77)	Población Est. Total 1985	Población % Urbana	Población % Indigenas
Panama	77.1	1.798	23.4	3.1	2.254	51.8	5.0
Costa Rica	50.9	2.044	39.5	2.9	2.961	43.7	.25
Nicaragua	148	2.346	16.6	3.6	3.347	54.9	5.0
Honduras	112.1	2.954	26.3	2.7	4.300	34.2	6.0
El Salvador	21.2	4.310	203.7	3.1	5.907	40.3	5.6
Guatemala	109.0	6.531	60.0	2.8	8.103	31.4	50.0
Belice	23.0	.140	6.1	2.8	n.d.	55.0	n.d.

Entidad Politica	% Alfabetismo 1975	Ingreso Anual por Habitante 000's US$[b]	% Area Terrestre Todavia Silvestre	Tasa de Deforestación (ha/año 1978
Panama	79.3	1.271	35	50.000
Costa Rica	95.7	1.048	33	60.000
Nicaragua	52.6	.857	45	40.000
Honduras	58.0	.514	40	60.000
El Salvador	63.0	.604	1	[c]
Guatemala	47.3	.898	30	75.000
Belice	90.0	n.d.	85	5.000

n.d. = no disponible
[a]1978
[b]1977
[c] casi no hay para deforestar
Fuentes de información: Almanaque Mundial. 1980. Editorial América, Panamá.; Belize Tourism Board, 1979; Agencias nacionales forestales en cada país, comunicaciones personales, 1978–79.

elemento africano como secundario y dominante principalmente en la costa atlántica. La población indígena tiene una representación alta solamente en Guatemala (aproximadamente el 50%). En cuatro de las otras seis entidades existen grupos indígenas que componen del 5% a 6% de la población total y solamente en Costa Rica éstos forman una minoría, el 1% de la población.

The indigenous population is low, ranging from 1–6%.

La economía de Costa Rica, Nicaragua, El Salvador y Guatemala dependen principalmente de las actividades agropecuarias, aunque en los otros tres países la agricultura y la ganadería también juegan un papel importante, y sus economías dependen, primordialmente, de otras actividades. Tal es el caso de Belice con la pesca, Honduras con la explotación forestal y Panamá con su comercio y actividades bancarias internacionales.

Cinco de los siete países tienen un alto porcentaje de su población (40% a 55%) en zonas urbanas, a pesar de que son países básicamente agrarios.

El idioma oficial y predominante en todos los países (excepto en Belice) es el español. Sin embargo en la costa atlántica de la región, el idioma inglés es común e igualmente es el idioma oficial de Belice.

Spanish is the official language of six countries but English is common on the Atlantic coast and is official in Belize.

Los recursos arqueológicos de la región son notables, especialmente los de origen Maya en Guatemala, Belice, Honduras y El Salvador. Los sitios arqueológicos en Panamá, Costa Rica y Nicaragua son menos espectaculares que los anteriores, pero de una gran relevancia dado su origen y confluencia de culturas.

The area is rich in archeological resources.

Ecológicamente la región es reconocida por su gran riqueza. En el área continental, el sistema de zonas de vida de Holdridge (Holdridge 1967) provee un método para expresar esa riqueza en forma adecuada. Todos los países de la región cuentan con un mapa ecológico según ese sistema (algunos de ellos actualmente bajo revisión).

La existencia de recursos naturales no renovables varía considerablemente en la región. A excepción de Guatemala, ningún otro país produce petróleo y las exploraciones hasta ahora indican que es dudoso que se encuentren reservas notables de este recurso. Casi todos los países tienen explotaciones minerales (oro, plata, cobre, etc.) en marcha o planificadas a escala mediana hasta relativamente grandes.

Natural resources are rich, but only Guatemala produces petroleum.

Es difícil estimar el uso potencial de la tierra a escala regional debido a la escasez de datos y a la falta de la recopilación y codificación de los existentes. Sin embargo, un estimado pareceria indicar que el mejor uso potencial sería bosques 60%, pastos 15%, cultivos 10% y otros 15%. Esto contrasta notablemente con el uso actual de bosques que es del 40%, pastos 30%, cultivos 20% y otros 10% (FAO 1976).

The optimal land use would be 60% forest, 15% pastures, 10% cultivation and other uses 10%.

Aunque no existe un sistema similar al de Holdridge para describir los recursos naturales marinos de la región, ciertas generalizaciones son necesarias. A diferencia de la situación en los mares templados, los recursos marinos en la región con ciertas excepciones como el atún, están concentrados en las zonas costeras del Pacífico y del Caribe, especialmente en éste último. Estudios sinópticos recientes de recopilación de una vasta cantidad de información y su mapificación bajo un sistema de traslape para indicar concentración de recursos, han demostrado que algunas de las áreas más importantes en recursos marinos (hábitats y especies) y de importancia para procesos ecológicos básicos en todo el Caribe, están en el Golfo de Honduras, costas de Belice, Guatemala y Honduras (Ray 1979, IUCN 1979). La costa caribeña de Panamá también tiene notables concentraciones de esos recursos, pero de menor importancia que los del Golfo de Honduras. La costa de Costa Rica es muy pobre en comparación con las áreas antes mencionadas. Manifiestamente, los recursos marinos del Pacífico son menos notables, por lo menos en términos de su explotación potencial aunque están, en general, más sobreexplotados que los de las costas caribeñas, debido a causas históricas de desarrollo en la región.

Marine resources, except for tuna, are concentrated along the coasts of the Gulf of Honduras. The Caribbean coast of Panama is also rich.

Desde el punto de vista de las economías nacionales, solamente Belice depende fuertemente de la explotación pesquera como elemento principal de su economía y de proteína para su población. El sector pesquero relativamente es mucho menos importante en los otros seis países. Sin embargo, en cuatro de los otros cinco países con costa caribeña, exceptuando a Costa Rica, se están planificando explotaciones mucho más grandes sin considerar adecuadamente la variable ecológica.

Fishing is of national importance only for Belize, but other countries are planning greater usage.

Tendencias y Obstáculos a la Conservación

Es un concepto común de que la causa básica de la degradación de los recursos naturales y de los problemas socioeconómicos en los países en desarrollo es el crecimiento no controlado de la población humana. Sin duda es un factor que contribuye, pero es necesario ponerle en perspectiva con otros factores de igual o quizás más importancia como el consumo de los recursos naturales de los países en desarrollo por parte de los países industrializados, la tendencia y uso de la tierra y la destrucción de los recursos marinos. Estos no son factores independientes, los tres son muy importantes en Centroamérica. Las densidades de población en la región, con excepción de El Salvador, no son alarmantes, pero si lo son dos aspectos no tan obvios. Primero, en todos los países (excepto Costa Rica), la tasa de crecimiento de la población (Cuadro 1) está incrementando o es estable, pero alta. Segundo, la mayoría de los terrenos silvestres de la región todavía no ocupados son áreas de uso potencial con excepción de áreas relativamente pequeñas únicamente "forestales", es decir, áreas que servirán mejor para la producción hídrica, conservación de recursos genéticos, explotación sostenida de madera, vida silvestre, etc.

The belief that population growth is the basic problem for underdeveloped countries must be considered along with the consumpion of their natural resources by industrialized countries, their land use, and destruction of marine resources. Population density in Central America is high but not alarming. Most forested lands not yet occupied have potential for use.

En realidad mucha de la tierra centroamericana, 30% a 40% de vocación forestal, ya fué deforestada y está bajo usos inapropiados y no sostenibles, transformándose éstos en los famosos "terrenos marginales" ocupados por una numerosa población rural pobre sin alternativas. Esto nos lleva al caso general del segundo factor, la tenencia y utilización inadecuada de la tierra. Este problema es crónico y de diferente intensidad en toda la región, en donde la concentración de la mayoría de las tierras ocupadas para fines agropecuarios está en manos de una relativamente pequeña proporción de la población. Además, la tendencia en la región es de que tales oligarquías, sean pequeñas o medianas, están extendiendo su control, no reduciéndolo. Así tenemos el caso de que grandes extensiones de tierras más aptas para otros usos y para satisfacer las necesidades de muchas familias, están siendo utilizadas para la ganadería o monocultivos, por pocos terratenientes, principalmente para la exportación de materia prima, no elaborada o semielaborada.

30–40% of natural production is exported, and these foreign demands are supported by international finance which favors large projects for short term gain and delivery of raw materials.

El tercer factor es la magnitud del consumo de los recursos de los países en Centroamérica por parte de los países industrializados. A nivel mundial, los países industrializados, con 20% de la población, están consumiendo 80% de los recursos comercializados en los mercados mundiales procedentes principalmente de los países en desarrollo (Myers 1979).

Es difícil obtener buenos datos sobre la exportación total de los recursos naturales explotados en Centroamérica, pero especulativamente se podría decir, probablemente entre 40% a 60% de aquellos recursos son consumidos en los países industrializados. Tales demandas son generadas y efectuadas principalmente por corporaciones multinacionales y bilaterales, y apoyadas por las agencias e instituciones internacionales y bilaterales de ayuda financiera. Histórica y tradicionalmente tales instituciones han financiado proyectos grandes, diseñados para obtener ganancias económicas altas a corto plazo, con la finalidad de exportar materia prima cruda y en pocos casos productos semielaborados a las naciones industrializadas.

Lo más preocupante, es que tales tendencias continuan e inclusive se incrementan en el caso de la mayoría de esas corporaciones e instituciones financieras. Pero, aún hay causa para optimismo limitado: tres de ellas (solamente tres!) US-AID, el Banco Mundial y el BID (Banco Interamericano de Desarrollo) han adoptado nuevas políticas ambientales en los últimos años, inicialmente requiriendo análisis de impacto ambiental para cualquier proyecto, y recientemente, progresando al punto de comenzar a financiar proyectos dirigidos primordialmente a la conservación de los recursos naturales, especialmente los renovables (Stein & Johnson 1979).

Three agencies, US-AID, the World Bank, and the BID have recently adopted new environmental guidelines.

Uno de los ejemplos más espectaculares es la demanda de carne de res barata para el mercado de comida rápida ("fast food") en Norteamérica (Myers 1979). Extensiones enormes de bosques han sido deforestadas en la región y convertidas en zonas de pastoreo para el ganado de carne. Desde 1950 el área total de zonas de pastoreo y el número de ganado de carne en Centroamérica se han duplicado. Cabe mencionar que este aumento en la producción de carne no ha contribuído casi nada a mejorar la disponibilidad de proteína en las dietas de los centroamericanos y más bien, ha servido para asegurar precios bajos en los EE.UU. y Canadá de las hamburguesas. Esto ha sido uno, entre otros, de los factores principales de la deforestación en Centroamérica. Cada año desde 1975, las cuotas para exportar a Norteamérica han sido regularmente incrementadas para cada país de la región en un promedio del 15% al 20%.

The demand for cheap beef by the "fast food" trade in the United States has converted enormous areas of forest to pasture, and the number of cattle in Central America has doubled.

Básicamente lo mismo ha pasado y está sucediendo con la explotación de los recursos marinos vivos en ambas costas de la región. La gran mayoría de los camarones, las langostas, el pescado, y las tortugas se exportan fuera de la región.

La combinación de estos tres factores interrelacionados (demandas de productos y materia prima que crecen rápidamente, generadas principalmente fuera de la región, una población que aumenta rápidamente, una tenencia y utilización crónicamente inadecuadas de la tierra, que empeoran rápidamente) crea presiones de explotación, siempre más severas en los recursos naturales. Los resultados son enormes y, en términos prácticos, casi todos irreversibles. Algunos de esos son.

La deforestación en la región es masiva. Si las tasas

actuales estimadas siguen, varios países de la región quedarán efectivamente sin bosques primarios entre 1990 y 2005, dependiendo del caso en cada país. La extinción de especies de plantas y animales sería enorme.

La población rural pobre actual es de aproximadamente 60% del total de la región y está aumentando desproporcionadamente. Esta población es forzada a moverse a terrenos siempre más y más marginales. Este ciclo vicioso de ocupación de terrenos marginales cada año trae más problemas sociales y económicos graves para toda la población, pero especialmente para la rural y pobre.

La productividad agropecuaria de los países, en general, sigue bajando en una escala relativa, a consecuencia de la deforestación, creciente erosión y pérdida de fertilidad de suelos. La dependencia en comida importada ha incrementado notablemente en la última década a pesar de un incremento de casi un 9% de terrenos bajo cultivo (UNEP/CEPAL 1979). Además, en todos los países excepto Honduras, la contribución de la agricultura al Producto Bruto Doméstico disminuyó en ese período (United Nations 1977b).

Una buena proporción de la población rural pobre, al estar forzada a moverse de sus terrenos marginales, finalmente emigra a las ciudades. Esa tendencia contribuye fuertemente al incremento del desempleo y al desorden de las ciudades, la mayoría de las cuales ya tienen serios problemas de contaminación, saneamiento ambiental, etc.

The combination of external demand for raw materials, growing population, and inadequate use of lands has led to large and irreversible results; massive deforestation, disporportionate growth of the rural population, a lowering of agricultural productivity and greater dependence on imported foods, and lessening of the agricultural contribution to Gross National Product.

Avances en Conservación en Centroamérica

Las tendencias anteriormente citadas son obstáculos grandes a la conservación, y se requerirán reformas y acciones bien concebidas y cuidadosamente aplicadas, si se desean cambios significativos en tan corto tiempo como 10 años. No pretendemos reconocer todas las acciones y reformas, ni es nuestro campo de especialización tan amplio y experimentado. Sin embargo, las experiencias obtenidas en diseñar y probar metodologías de planificación y manejo de los recursos naturales renovables, especialmente los silvestres, en los últimos 10 a 15 años en la región centroamericana, en Suramérica y en El Caribe, ofrecen lecciones bien claras sobre cuáles han sido las limitaciones y cuales los éxitos y fracasos y las causas de cada caso. En realidad, se ha logrado mucho y la experiencia ganada es de una magnitud muy notable. Estamos convencidos que esas experiencias han mostrado como organizar y establecer prioridades en las actividades futuras.

In the past ten years the steps taken for conservation have been enormous.

El establecimiento de áreas silvestres o parques nacionales y de otras áreas protegidas en el sentido más amplio de la palabra y, aún más importante, su manejo efectivo es una de las medidas más claras y efectivas para el avance en la conservación. El cambio en los últimos diez años ha sido realmente extraordinario. Desde un punto de vista absoluto, tales cambios no han sido iguales y uniformes en todos los países de la región. Algunos, especialmente Costa Rica, han mostrado ser particularmente fértiles para tales avances y, en otros, especialmente Guatemala, los avances han sido menos notables. Sin embargo, el asunto importante es que, medidos dentro de su propio contexto nacional, los cambios en cada uno de los países han sido realmente notables. No hay todavía definiciones uniformes y cuantitativas para calificar qué es "manejo efectivo", pero juzgado en los contextos nacionales propios es indudable que en todos los países éste se ha dado en mayor o menor grado.

Otro avance muy notable es que en cada país de la región, con excepción de Belice, por los menos un plan de manejo para un área silvestre, (y en algunas de las naciones hasta varias) ha sido preparado, usualmente un parque nacional, por un equipo multidisciplinario e inter-institucional nacional, y el plan se ha puesto en marcha en el terreno. En la mayoría de los países de la región ya existen por lo menos algunos especialistas capaces de preparar tales planes, ponerlos en marcha, criticarlos y revisarlos de acuerdo con los cambios y experiencias ganadas a través del tiempo.

In each country in the area there is at least a small cadre of conservation specialists.

El incremento gradual, especialmente en los últimos años, de áreas silvestres manejadas bajo una variedad de categorías de manejo con las cuales se está experimentando en la región es otra indicación muy positiva. Por razones históricas y tradicionales (a nivel mundial), durante los últimos 15 años el énfasis en el establecimiento y manejo de las áreas protegidas en el terreno fue enfocado casi exclusivamente bajo la categoría de parques nacionales. Durante algunos años la teoría para sistemas, en la práctica no tuvo casi ninguna experiencia con otras categorías de manejo excepto parques nacionales y otras muy similares (monumentos naturales, monumentos arqueológicos). En los últimos 3 a 4 años se ha comenzado a experimentar a nivel nacional con varias otras categorías de manejo para tratar de formar verdaderos sistemas de áreas silvestres. En este sentido, Centroamérica es claramente la sub-región más avanzada en Latinoamérica y El Caribe y, a pesar del trabajo de experimentación y demostración que se ha desarrollado, aún queda mucho más por realizarse. En este "momento" en Centroamérica se están desarrollando, aplicando y probando metodologías de planificación y manejo en el terreno de otras categorías de manejo tales como áreas de uso múltiple, refugios de vida silvestre, reservas forestales, áreas recreativas, reservas de la biósfera y reservas de recursos. Como en el caso de los parques nacionales, tales esfuerzos son realizados por equipos nacionales con la colaboración de especialistas internacionales.

A positive indication is the increase in managed forest lands and the experimentation with forest systems other than national parks.

Mientras se ha avanzado tanto en el establecimiento, planificación y manejo en el terreno de áreas silvestres individuales, hay una tendencia reciente hacia el enfoque de planificación estratégica y programática de sistemas regionales y nacionales de áreas silvestres o recursos silvestres. En gran parte esto ha sido posible, por lo menos en Centroamérica, debido a la experiencia y madurez desarrollada tanto en el personal nacional e internacional en la región, quienes han estado asociados con la planificación y manejo de las áreas silvestres individuales. El enfoque original ha creado una visión y permitido el desarrollo de este nuevo enfoque complementario. Hasta hace pocos años, la colaboración internacional para la conservación en la región latinoamericana y del Caribe ha sido muy esporádica, casi nunca coordinado a nivel regional. Sin embargo, ésto está ya en proceso de cambio mediante el desarrollo de estrategias regionales y nacionales de conservación de sistemas de áreas silvestres, como por ejemplo las estrategias para el medio ambiente marino del Caribe (IUCN 1979) y para las Antillas Menores (Putney 1979), y el desarrollo de un plan de acción regional para el manejo del medio ambiente (UNEP/CEPAL 1979), basado en la experiencia exitosa de un plan de acción similar para la región del Mediterráneo.

Until recently international coordination has been sporadic and local, but there are now regional strategies for forest conservation, successful because of the developmental maturity of the region and its personnel.

Se está comenzando la preparación de estrategias de conservación y planes de sistemas de áreas silvestres a nivel regional en Centroamérica y en dos casos (Nicaragua y Costa Rica) a nivel nacional, utilizando en parte las experiencias obtenidas en esos esfuerzos a nivel global del Caribe ("wider Caribbean") y como experimentos en el desarrollo y aplicación de nuevas metodologías. En Costa Rica, por primera vez en América Latina y El Caribe, el Servicio de Parques Nacionales está desarrollando una planificación programática y estratégica por dos años de todas las actividades, a nivel de oficinas centrales y unidades de apoyo (departamentos) y de cada unidad de conservación en el campo. Se espera que la metodología desarrollada y la experiencia ganada sirva de guía para otros países. Estos esfuerzos son llevados a cabo por equipos inter-institucionales y multi-disciplinarios nacionales con especialistas internacionales, principalmente de ECNAMP (véase abajo) en las Antillas Menores y CATIE en Centroamérica. Cada uno de estos esfuerzos trata de entender a nivel regional o nacional los problemas globales para que la atención y actividades puedan ser enfocadas de una manera coordinada identificando las más críticas y categorizándolos por prioridades.

Planning is beginning at a regional level. Costa Rica is developing a two year general plan making use of methodology from ECNAMP (Lesser Antilles) and CATIE (Costa Rica), which may be a guide for other countries.

Se han logrado avances significativos en la capacitación sobre el manejo de áreas y recursos silvestres en Centroamérica, dado el número de personas capacitadas en la identificación (mediante experimentación y evaluación) de los métodos y tipos de capacitación más eficaces y eficientes y el desarrollo y mejoramiento de tales métodos y tipos de capacitación. Aproximadamente 130 profesionales y técnicos a nivel superior y medio en las instituciones nacionales de manejo de recursos naturales han recibido capacitación en los últimos 15 años, 100 de ellos solamente en los últimos tres años, como parte de las actividades colaborativas de las instituciones nacionales y del Proyecto de Areas Silvestres y Cuencas (PASC) del CATIE. La gran parte de esta capacitación ha sido del tipo de "en servicio. Esto ha sido fortalecido y ampliado secundariamente mediante la realización de capacitación más formales a corto plazo tales como seminarios móviles, talleres de planificación y manejo, cursos y reuniones técnicas. Finalmente, estudiantes a nivel posgrado han recibido o están en proceso de recibir su maestrías en manejo de áreas silvestres en CATIE y unos 50 más han recibido cursos y experiencia práctica en el mismo campo como parte de su carrera de posgrado en CATIE. Aproximadamente 60% de ellos son centroamericanos. Es incalculable el gran beneficio que en términos de su influencia notoria y eficacia ha tenido la capacitación "en servicio", quizás mejor descrita como "aprender mediante el hacer y el experimentar." La capacitación ha sido, por lo menos, igual para los "expertos" internacionales pertinentes. De las actividades de capacitación a corto plazo, es claro que los talleres y seminarios móviles, especialmente a nivel regional, pero también nacional, con fuerte énfasis en trabajos prácticos (v.g. elaborar un plan de manejo de un área silvestre como grupo integral en un taller de 2 a 3 semanas), han tenido una gran influencia notoria y efecto multiplicador. A la vez, es quizás el tipo de capacitación más eficiente y efectiva en términos de costo-beneficio. En general los talleres, cursos cortos, etc., con mayor énfasis en clases con lecciones puras y teoría han sido también efectivos. Finalmente, un acercamiento global al problema de capacitación debe incluír los niveles de grado y posgrado. La experiencia de los últimos años indica que una cantidad limitada de esfuerzo a ese nivel es necesaria para fortalecer a las instituciones nacionales.

In the past 15 years about 130 professionals and upper and middle level technicians have received certification through CATIE, 100 of these in the past three years. About 60% are Central American.

Es importante anotar que, en general, las oportunidades de capacitación se han incrementado en la región en los últimos años. Por primera vez hay un programa de estudio a nivel de grado universitario en la región, enfocado en las especializaciones de manejo medioambiental, incluyendo áreas silvestres y cuencas (Universidad Nacional, Costa Rica). Además, la capacitación en servicio en la región, utilizando los recursos y experiencias de un país para el beneficio de los otros ha comenzando recientemente y está creciendo. Más de 20 personas a nivel medio (superintendente de áreas) y de guardianes han recibido capacitación en servicio por períodos de 15 a 90 días en Costa Rica en los últimos 18 meses mediante el Servicio de Parques Nacionales. Estas personas representaron a todos los otros países de la región (excepto Belice) más la República Dominicana.

Opportunities for training have recently increased.

En general, la cooperación y coordinación regional han mejorado muy significativamente en los últimos tres años. El PASC del CATIE ha funcionado como secretario de una organización regional de agencias nacionales de manejo, manteniendo la comunicación mediante un boletín regional, ("Areas Silvestres en Mesoamérica", 4 ediciones por año desde 1977), reuniones técnicas periódicas y la publicación, conjuntamente con las agencias nacionales, de muchos documentos técnicos basados en los trabajos en los países. Esa documentación, ha sido distribuída ampliamente a las agencias en Centroamérica, junto con publicaciones técnicas claves de otras sub-regiones de América Latina y del Caribe, y han jugado un papel de incalculable importancia en relación a compartir ideas y experiencias, proveer modelos y en general ganar de la experiencia de otras áreas.

PASC (Forest and Watershed Areas Project) of CATIE functions as secretariat for a regional management organization which has issued a regional bulletin (Forest Areas of Mesoamerica) since 1977.

Adicionalmente, aparte de la nueva actividad creciente de compartir recursos para la capacitación ya mencionada, otra señal muy positiva es de que en los últimos dos años han comenzado los primeros casos en que profesionales de un país de la región realizan consultorías técnicas en otro país.

También en los últimos dos años, por primera vez, ha sido parcialmente posible coordinar a nivel regional la ayuda internacional y bilateral de las agencias nacionales de manejo, mediante el PASC del CATIE (aún hay que mejorarlo). Esto se ha logrado con las organizaciones de ayuda como las internacionales (UICN, WWF, RARE, NRDC, Sierra Club, etc.) y muchas gubernamentales bilaterales como USNPS, USFWS, US Forest Service y DDA (equivalente a AID) de Suiza. En menor grado con las agencias de las Naciones Unidas (UNESCO, FAO y UNEP). El reto más grande es lograr una coordinación similar con las instituciones grandes de ayuda financiera como los bancos de desarrollo (BID, Banco Mundial, etc.) y las agencias bilaterales fuertes (AID, CIDA, etc.). En el último caso las líneas de comunicación ya están establecidas y abiertas. Lo importante de esta experiencia es que demuestra que es factible formular una estrategia regional para cada país y luego acercarse a estas instituciones en una forma coordinada y táctica, evitando contradicciones y traslapes entre proyectos, coordinando y concentrando mejor los recursos financieros relativamente escasos y de asistencia técnica enfocando las prioridades. Las nuevas políticas ambientales de AID, el Banco Mundial y BID son causa de optimismo, para lograr una coordinación y forzar cambios de política de los otros bancos grandes y agencias similares.

Recently international and bilateral aid to the area (from BID, World Bank, AID, CIDA, etc.) has been partly coordinated through PASC, and to a lesser degree the agencies of the United Nations.

Se han dado los primeros pasos para fortalecer sustancialmente las organizacones conservacionistas no gubernamentales de manera que jueguen un papel más relevante e importante en cada país. En la primera reunión de tales entidades en diciembre de 1978 fue formada la Federación Mesoamericana de Asociaciones Conservacionistas No Gubernamentales (FEMAC). Ahora se busca financiamiento, con la colaboración de la UICN, para apoyar un proyecto regional de FEMAC, para fortalecerla al igual que a las organizaciones miembros individuales; en otras palabras, que FEMAC juegue un papel para las organizaciones, similar al de PASC con las organizaciones gubernamentales de manejo.

The Mesoamerican Federation of Non Governmental Conservation Association (FEMAC) has been strengthened and is seeking financial support for a regional project.

En los últimos años se ha notado por primera vez una tendencia creciente de incluir en la región a nivel de la planificación nacional y de regiones dentro de los países, la "variable" de conservación de los recursos naturales e impacto ambiental como parte integral de la política y planificación del desarrollo. Todavía estos conceptos están en su infancia, pero por lo menos se están considerando más y más. Hay dos casos concretos de la elaboración de tales políticas. Uno es el de Costa Rica, en donde el nuevo Presidente en 1978 escribió y publicó una política sobre la conservación de los recursos naturales (AID la tradujo a varios idiomas y distribuyó a todas sus misiones en el mundo). El otro es el de Honduras en donde apareció una fuerte sección medioambiental en su Plan Nacional de Desarrollo, 1979–1983.

There is a growing tendency to include conservation in national and regional planning.

Esos avances son el resultado de una combinación particularmente eficaz de esfuerzos y acciones nacionales con colaboraciones catalizadoras internacionales. Una serie de proyectos directamente relacionados, cada uno basado en las experiencias del anterior fueron los responsables de este progreso. Los auspiciadores principales de esos proyectos han sido el Fondo de los Hermanos Rockefeller (RBF), FAO y CATIE. Esos proyectos han sido críticos en catalizar los esfuerzos nacionales y posibilitar tantos avances en tan corto tiempo en Centroamérica. Quizás uno de los aspectos más interesantes es que esos proyectos han funcionado con recursos financieros relativamente pequeños, pero han generado acciones extraordinarias y considerables. La clave de tal éxito ha sido la flexibilidad, pragmatismo y orientación estratégica de su acercamiento a la problemática combinando con la habilidad de responder rápidamente a las situaciones y momentos críticos.

Projects of the Rockefeller Brothers Found (RBF), FAO, and CATIE have catalyzed short term advances.

Finalmente, se ha logrado identificar cuales son los puntos claves y críticos o fallas mayores ("bottlenecks") en las instituciones nacionales en cada país, los cuales limitan el progreso en el manejo de los recursos naturales silvestres. Hay diferencias en cada país, pero las principales limitaciones son comunes:

Bottlenecks or key points limiting progress in the national institutions in each country have been identified.

En cada país hay un número elevado (entre 4 y 8) de instituciones encargadas de la planificación y el manejo de los recursos naturales y existe muy poca coordinación y comunicación entre ellos a nivel nacional y local; en general casi nunca trabajan conjuntamente. Aún dentro de instituciones el trabajar en equipo es generalmente raro.

Each country has several institutions charged with planning and natural resources, but they seldom work jointly.

En cada país, en estas instituciones nacionales hay muy buenos "líderes" potenciales y actuales a nivel alto y medio, pero en general son muy pocos (3 a 4 por país) y les falta suficiente experiencia, capacitación y apoyo para desarrollar mejor su potencial.

In each country these institutions have "leaders" with excellent potential, but they are few and they lack experience, training and support to develop their full potential.

Las instituciones son débiles en infraestructura, especialmente recursos humanos y experiencia para programar y planificar en forma estratégica sus limitados recursos humanos y financieros enfocando las prioridades.

The institutions are weak in infrastructure.

Prioridades y Guías para El Futuro

La experiencia en los últimos cinco años ha proporcionado oportunidades para diseñar y probar muchas ideas, metodologías y técnicas y obtener de ellas guías para actividades prioritarias futuras. La siguiente es una lista de esas prioridades y guías basadas en las experiencias conjuntas de la instituciones y personas nacionales e internacionales las cuales consideramos más críticas.

Las metodología de planificación estratégica y programática a nivel regional y nacional de conservación de sistemas de áreas silvestres y para las instituciones nacionales de manejo necesitan ser mejor desarrolladas, diseñadas, aplicadas y probadas en Centroamérica. Tales estrategias y planes deben ser eventualmente elaborados para todos los países, por equipos nacionales inter-institucionales y multi-disciplinarios con la colaboración de especialistas, pero por pragmatismo se debe comenzar a nivel regional y simultáneamente con uno o dos países como prueba y demostración. A la vez en los otros países se podría comenzar con los primeros pasos (recopilación de información existente e inventarios en el campo) pero a un ritmo más lento. Este acercamiento es absolutamente necesario para que la conservación sea integrada como parte del proceso de desarrollo y para que los esfuerzos se acerquen hacia la conservación de los sistemas biofísicos, ecosistemas y recursos genéticos, los cuales mantienen el desarrollo.

Para lograr que la conservación sea más y más parte del proceso de desarrollo, habrá que darle menor énfasis (pero no excluír totalmente!) a proyectos y programas dirigidos a especies, hábitats o áreas singulares, espectaculares o atrayentes y enfatizar los que orienten, guíen y hagan funcionable la integración del proceso de desarrollo en armonía con el medio ambiente cultural y natural, reconociendo más claramente la dependecia de los sistemas naturales y humanos.

A la vez que se enfatiza la planificación estratégica a nivel regional y nacional y su eventual (corto plazo, 3 a 4 años máximo) ejecución, hay que seguir con la planificación y consecución de áreas silvestres individuales para que sirvan a la vez como experimentos para el desarrollo de herramientas teóricas y métodos prácticos de manejo adaptados a las condiciones de la región, como áreas demostrativas de las posibilidades de manejo, y como áreas para opciones futuras. Es especialmente crítico hacer mayor énfasis en la planificación y realización de un ámbito más amplio de categorías de áreas silvestres, especialmente las del tipo orientado a producir productos y servicios más tangibles (p.e. Areas de Uso Múltiple, Reservas de Producción Hídrica, Reservas de la Biósfera, Reservas de Producción Forestal, Areas de Manejo Integral de Cuencas o Corporaciones de Valle, Centros de Caza o Ranchos de Fauna, etc.) como parte de sistemas de áreas silvestres balanceados que incluyen también las categorías más "clásicas" y proteccionistas como parques nacionales, monumentos naturales, etc. Esto requeriría programas fuertes de protección temporal mientras que se experimenta con sistemas potenciales de producción sostenida en muchos de los casos, p.e. Reservas Forestales de Producción o Areas de Uso Múltiple en zonas de bosques de hoja ancha.

Strategies for conservation and forest management must eventually be developed for each country, and it is essential that conservation be integrated in the national development process. While emphasizing regional and national planning and implementation, local schemes must proceed as trials of theory and practical methods. Emphasis on a greater range of forest categories is critical.

En términos relativos, la conservación de los recursos naturales marinos y costeros de la región han sido casi totalmente ignorados (fuera de IUCN, 1979) comparado con el esfuerzo con los recursos terrestres. Las actividades mencionadas previamente en esta sección, deben ser aplicadas igualmente a estos recursos como una sola actividad integral. Es urgente como un primer paso desarrollar nuevos y mejores conceptos, criterios, metodologías y métodos para la planificación y manejo de los recursos marinos y costeros.

Conservation of marine and coastal resources has been ignored.

Las estrategias indicadas en esta sección, deben incluir como parte integral guías y prioridades para aprovechar y coordinar mejor las "ofertas" financieras y de asistencia técnica de las instituciones internacionales y bilaterales, especialmente las medianas a grandes.

La mayoría de la "tecnología" y los métodos ya desarrollados para el manejo de recursos naturales silvestres, especialmente en los países industrializados y zonas templadas, no son directamente aplicables al medio ambiente natural, cultural e institucional de Centroamérica. Por lo tanto es imprescindible que se trate siempre de buscar soluciones locales como primera prioridad y que la asistencia técnica y tecnologías provinientes de fuera de la región sean adaptadas a las necesidades y circunstancias particulares locales, por aquellas personas que conocen bien a la región.

Most technology developed for resource management, especially for developed countries, is not directly applicable to Central America.

Los programas de asistencia financiera internacional o bilateral deben ser relativamente balanceados a una escala adecuada y no dominar o ahogar las agencias recipientes. A la vez el acercamiento debe ser global para que todos los elementos de manejo (legislación, política, planificación, capacitación, investigación, educación pública, etc.) sean tratados como un solo paquete y no uno por uno.

El enfoque mediante equipos al proceso de manejo debe ser considerado como imprescindible para promover el intercambio e integración de disciplinas, técnicas y puntos de vista y para que una atmósfera de cooperación, coordinación y mediación sea la regla, y no causa conflicto, confrontación y competencia.

La capacitación de personal nacional es quizás el factor clave para que el proceso de manejo avance. Dada las enormes necesidades en la región se debe dar prioridad a la "capacitación" en servicio como parte natural de las actividades. Secundariamente se debe apoyar tipos de capacitación a corto plaza con un elemento fuerte práctico, de experiencia directa. Se debe también proveer una capacitación a nivel superior (grado y posgrado) para profesionales seleccionados. El énfasis en todo debe ser el de proveer capacitación a niveles superiores y medios en forma eficaz (niveles regionales y nacional en general), y dejar que los capacitados resuelvan el problema de capacitar a los niveles más bajos (v.g. guardianes).

Training of national personnel is key to an advance in management, and short term practical training must be emphasized.

La capacitación debe estar basada en la experiencia ganada en proyectos experimentales y demostrativos en la región. Los programas de capacitación tienen que estar relacionados estrechamente con proyectos de acción en el terreno.

Training must be based on experience gained in regional projects.

En el proceso de planificación y ejecución del manejo se debe tomar en cuenta las necesidades y opiniones de los ciudadanos locales directamente afectados por las decisiones de manejo y desarrollo. Esto implica la participación pública activa en una o otra forma.

Public participation in management and development decision making must be active.

Agencias Nacionales Forestales en Centroamérica. 1978–79. Comunicaciones personales. 1. Almanaque Mundial. 1980. Panamá, Panamá, Editorial América S.A. 2. Belize Tourism Board, 1979.

FAO (Food and Agriculture Organization). 1976. Production Yearbook, 1975, FAO. Rome.

Holdridge, L.R. 1967. Life Zone Ecology. Rev. ed. Tropical Science Center, San Jose, Costa Rica.

IUCN (International Union for the Conservation of Nature and Natural Resources). 1979. A Strategy for the Conservation of Living Marine Resources and Processes in the Caribbean Region. IUCN, Morges.

———. 1980. Estrategia Mundial de Conservación. IUCN/UNEP/WWF, Geneva.

Myers, N. 1979. The Sinking Ark. Pergamon Press, New York.

Putney, A. 1979. A preliminary survey of conservation priorities in the Lesser Antilles. Paper presented to 14th Annual General Meeting, Caribbean Conservation Association, Guadeloupe.

Stein, R.E. & G. *Johnson*. 1979. Banking on the Biosphere? Environmental Procedures and Practices of Nine Multilateral Development Agencies. Lexington Books, Lexington.

UNEP/CEPAL. 1980. Development and environment in the wider Caribbean region; a synthesis. Meeting of government-nominated experts to review the draft action plan for the wider Caribbean region. E/CEPAL/PROY. 3/L.INF.3. Caracas.

United Nations. 1977a. World demographic yearbook. 1976. United Nations, New York.

———. 1977b. Statistical yearbook 1976. United Nations, New York.

Research, Tropical Ecology, and the Future of Panama

Peter H. Raven

Panamá es un país privilegiado en cuanto a posición geográfica, riqueza y diversidad de especies, por lo tanto puede contribuir significativamente a la biología y a la agricultura tropical. Si la deforestación continua a la presente tasa, los bosques naturales que aún existen serán totalmente destruidos y con ellos desaparecerán innumerables especies de plantas y animales. En este país las medidas de conservación y de uso racional de los recursos naturales renovables son dictadas por la Dirección Nacional de RENARE. Panamá deberá aprender tanto como sea posible sobre sus recursos naturales para aplicar los conocimientos obtenidos de manera que los recursos naturales sean utilizados racionalmente.

Me es muy grato estar presente esta noche para celebrar la feliz conclusión de la *Flora de Panamá*, una obra que ha ocupado la atención de científicos de varias instituciones por casi cuatro décadas. Iniciada por el Jardín Botánico de Missouri, la *Flora de Panamá* surgió de una larga y fructífera colaboración entre los científicos de la República de Panamá, de los Estados Unidos de Norteamérica, así como de otros paises, y esperamos sinceramente que esta obra servirá para estimular un creciente interés y comprensión de las plantas de este hermoso país. La privilegiada posición geográfica de Panamá, uniendo América del Sur con América del Norte, y bañada tanto por el Atlántico como por el Pacífico, ha producido tal profusión y variedad de plantas, que Panamá reúne por sí sola tantas especies de plantas como todo el continente de Europa, un área doscientos veces más grande.

Las ponencias que hemos escuchado en este interesante simposio, tan hábilmente organizado por la Profesora Mireya Correa de la Universidad de Panamá y por Doctor William D'Arcy del Jardín Botánico de Missouri, dan fe de la gran riqueza biológica de Panamá. Estas ponencias servirán de base para un nuevo libro sobre la historia natural del país, que se editará para celebrar la finalización de la *Flora*, y con miras a contribuir al aprecio y al uso racional de los recursos naturales del país.

Panamá está en condiciones especiales para hacer grandes aportes a la biología y agricultura tropical. Tiene una ubicación geográfica única e instituciones bien conocidas tanto nacional como internacionalmente. Por un lado, el futuro bienestar de sus ciudadanos dependerá en gran parte del uso inteligente de las informaciones disponibles acerca de los recursos naturales del país. Por otra parte, estas mismas instituciones, distinguidas tanto en ciencias básicas como aplicadas, pueden jugar un papel clave en el desarrollo de conocimientos para el beneficio de los seres humanos de toda la faja tropical del mundo.

Special conditions in Panama will allow outstanding contributions to biology and tropical agriculture.

Unas cuantas estadísticas ayudarán a demostrar cuán importantes serán estos conocimientos. Hace un siglo, los bosques tropicales húmedos ocupaban una superficie mucho mayor que Estados Unidos, México, y América Central juntos. Desde entonces, más de la mitad de estos bosques han sido talados o alterados, y con el actual ritmo de conversión, los bosques estarán totalmente destruidos dentro de treintiún años. Por supuesto que el ritmo de destrucción es irregular, pero se calcula que en la mayor parte de la zona tropical, no quedarán áreas significativas de bosques húmedos tropicales dentro de veinte años. Para entonces, las dos terceras partes de la población mundial estará viviendo en zonas tropicales. Esta destrucción de bosques conducirá a una extinción masiva de animales y plantas, muchos de los cuales no pueden reproducirse fuera de los bosques primarios.

If deforestation continues at the present rate, most tropical forest will be totally destroyed within 31 years.

Estas transformaciones de bosques, equivalentes a la desaparición, cada *dos* años, de un área de bosque natural del tamaño de toda la América Central, son la consecuencia del desarrollo de los recursos naturales para el aprovechamiento humano. Si fuese verdad que de estas transformaciones resultase un beneficio humano continuo, éstas serían bienvenidas; sin embargo, para una gran parte del trópico, sobre todo para las zonas actualmente ocupadas por los bosques húmedos tropicales, éste *no* es el caso. Nuestros conocimientos actuales simplemente no alcanzan sino para utilizar una porción mínima de estos suelos para la producción agrícola a largo plazo, y de esta manera, hay que considerar a los bosques y suelos como recursos naturales *no* renovables. Así, nuestro gran desafío como científicos e investigadores en el trópico será el de *aplicar* nuestros conocimientos en el campo práctico para diseñar sistemas agrícolas y agroforestales a largo plazo, mientras luchamos por *concentrar* los sistemas más productivos en las tierras más idóneas. Si logramos acelerar el inventario de la flora y fauna tropical a la vez que adelantamos estudios intensivos de ecosistemas críticos, podremos unir la información necesaria para obtener soluciones prácticas, y ofrecer alternativas a las presiones y necesidades urgentes e inaplazables de una población creciente.

Every two years natural forests the size of all Central America are being destroyed as a result of the development of natural resources for human needs.

La tecnología de las zonas templadas, basada en milenios de experiencia de agricultura contínua, *no se puede* aplicar a la mayoría del trópico; *poco* se hace actualmente para proveer la información tan indispensable para las zonas tropicales. Mientras tanto, la transformación de tierras para usos a corto plazo las hace inútiles para el futuro. De ésto comentaré más adelante. Sin embargo, es precisamente en este área que Panamá puede aportar una contribución insustituible al bienestar humano y a su propia prosperidad.

The technology from temperate zones based upon several thousand years of agricultural experience is not applicable in most of the tropics.

Resumiendo, tenemos por delante una tarea difícil: la de armonizar el desarrollo económico y social de una población en aumento, con la preservación, aunque sea parcial, de un mundo biológico que todavía nos oculta sus mayores riquezas.

It is necessary to correlate economic and social development with conservation of the biological world, at least in part.

Panama, where the former island continent of South America was joined to North America about 3.1 million years ago (Keigwin, 1978; Marshall el al. 1979), is biologically one of the most critical and fascinating areas of the world's tropics. The plants and animals of South America, isolated for more than 100 million years, had an ultimate common origin with those of Africa. As South America converged on North America through the 65 million years of the Tertiary Period, its tropical biota, dispersed across the ever-narrower sea, invaded North America in greater and greater numbers. Panama, which has been occupied primarily by open savanna-type vegetation for much of the past three million years, provided the keystone linkage. For this reason, and because of its geographical situation, its 50,000 square kilometers are occupied by the richest and most diverse assemblage of plants and animals of any area of comparable size.

Cuando América del Sur convergió con América del Norte durante el Terciario, su biota tropical se dispersó e invadió América del Norte. Panamá jugó un papel importante en este proceso de dispersión y por tal razón a pesar de su área pequeña, su flora y fauna es una de las más ricas y diversas.

I have already mentioned that Panama has nearly as many kinds of flowering plants as all of Europe; it may also be pointed out that it has, in just over 52,000 km^2, about a third as many species of flowering plants as all of tropical Africa (30,000 species), an area of 20 million square kilometers (Brenan, 1978). Comparable figures could be provided for most groups of animals that are well enough known to allow such calculations. They illustrate clearly that even by tropical standards Panama is exceptionally rich biologically. It is in fact this rich diversity of the plants and animals of Panama that is so properly a source of national pride, and the reason we have all gathered here for this most enjoyable symposium.

Panamá fue la región en donde la última conección directa entre la biota marina del Atlántico y del Pacifico existió. Esta connección se interrumpió hace aproximadamente tres millones de años.

Adding to its special interest, Panama was the area where the last direct connection between the marine biota of the Atlantic and Pacific existed, and the area where this connection was finally severed three million years ago. As a result, Atlantic and Pacific marine plants and animals not only exist in close proximity, but they exist in such a way that three million years of divergence between the derivatives of some formerly continuous population can be timed accurately—a natural evolutionary experiment without parallel.

Estas y muchas otras razones han sido incentivos para que diversas instituciones al igual que muchos biólogos hayan encontrado en Panamá un lugar excitante e interesante para efectuar sus estudios o investigaciones.

For these and many other reasons, biologists historically have found Panama an exciting and interesting place in which to carry out their studies. Institutions such as the Universidad de Panamá, the Gorgas Memorial Laboratory, and the Smithsonian Tropical Research Institute (STRI) have made significant contributions to our knowledge of the biota of the Republic and their significance for man. A continued dedication to the importance of broadening this knowledge and utilizing it wisely in development has been signaled in recent years by the establishment of RENARE, by the biological and ecological studies carried out under the auspices of IRHE, and by the special position accorded to STRI under the terms of the Canal Treaty.

El Gobierno de Panamá ha reconocido a RENARE como la Institución que debe dictar directrices basadas en los principios de conservación y manejo racional de las tierras.

The Government of Panama has explicitly recognized the principles of conservation and wise management of land as the basis for sustainable development by the formation of RENARE, in which conservation and development are linked for the benefit of all the people of the Republic. Among the more notable recent accomplishments have been the active efforts to protect the watershed of the Canal, including both the projected reforestation schemers and the establishment by President Royo of the adjacent 24,000–hectare Parque Nacional la Libertad; the establishment of Barro Colorado Island as a nature monument and its scheduled enlargement in 1999 at the conclusion of the treaty period; the plans for national parks in Darién along the Colombian border, along the Costa Rican border, and surrounding Volcán Barú, the highest mountain in the Republic, in the western highlands; the consistent efforts by IRHE and other agencies to conduct meaningful biological and ecological inventories before carrying out large-scale public works; and the educational components included in many of these efforts.

Notwithstanding the great biological interest in the natural resources of Panama, the government's clear realization that the prosperity of the people rests upon the wise utilization of these resources, and the current implementation of many outstanding efforts to realize the consequences of this relationship, very serious problems face the Republic. These problems are, for the most part, shared with the other countries of the tropics, where, as already pointed out, some two-thirds of the world's population will be living by the end of the 20th century.

Although the populations of most tropical countries are growing very rapidly—many at a rate at which they will have doubled by early in the next century—population is by no means the only factor involved in the worldwide ecological crisis facing not only Panama but every other tropical country in the world. Panama itself provides a graphic illustration of this relationship. With just under two million people in some 50,000 square kilometers, Panama may be compared not only with the Netherlands, with some 14 million people inhabiting 41,000 square kilometers, but also with Java, which has some 90 million people living in an area only about two-and-a-half times the size of Panama (129,500 km^2). The annual increase in the population of Java is nearly as great as the entire population of Panama! Granted that there are more extensive areas of relatively fertile soil in Java than in Panama, a comparison between the two regions, both located near the equator, illustrates graphically the way in which different patterns of land-use can affect population density. It also must be recognized that the relatively small population of Panama has already destroyed more than half of the original forest resources, and may, unless alternatives fostered by the Government prove effective, make short work of the rest.

A pesar de que la población de muchos de los países tropicales crece rápidamente, ésta no es

de manera alguna el único factor involucrado en la crisis ecológica mundial. La población de Panamá, la cual es relativamente pequeña, ha destruido aproximadamente más de la mitad de los recursos forestales originales.

In Java, the heavy utilization of naturally moist or irrigated rice-lands located on relatively fertile soils is an important element in the nutrition of the people. In addition, the homestead gardens of the rural people in Java provide a rich source of diverse foods, many of them derived from the native forests. Protein is largely provided by fish, some cultured, chickens, goats, and, in non-Moslem areas, by swine. In contrast, land use in Panama, particularly in the long-occupied and once relatively fertile deciduous-forest lands of the Azuero Peninsula, as Stanley Heckadon has emphasized earlier in this symposium, is based on "roza" grain agriculture followed by cattle pasture. These uses are highly destructive of the fertility of the soil when the density of people practicing them becomes too great, and large numbers of *santeños* (from Los Santos Province, on the Azuero Peninsula) are now emigrating to other areas, with potentially serious consequences to which we shall turn our attention later.

Panama is in the fortunate situation of being able to derive much of its income from world trade, but, despite this, the wise use of natural resources is still of basic importance for the country. There is much to be said for regional and local self-sufficiency in a world in which the current energy crisis is just the first symptom of worldwide shortage of raw materials of all kinds that fuels inflation everywhere. No one can guarantee the continued operation of the world economic system under growing pressures of this magnitude, and if Panama is to feed, clothe, and shelter twice as many people early in the next century at least as well as is possible now, or actually to better their lot, the country will need to make extremely wise and forward-looking decisions about the proper utilization of its resources now.

Panamá presenta una posición geográfica privilegiada la cual le permite derivar gran parte de sus entradas económicas a partir del comercio mundial. A pesar de ésto, el uso racional de sus recursos naturales es aún de importancia básica para suplir las necesidades más elementales de una población que crece rápidamente.

These decisions ideally should be made against a background of continued attention to slowing the rate of population growth as much as possible. This will be extremely difficult at first, with about 43 percent of the population under 15 years of age, but the individual prosperity of fewer people utilizing an identical set of resources in the future will obviously be greater than otherwise. For this reason, the continuing efforts of the Government to decelerate the rate of growth of Panama's population appear to be well justified, despite the apparent paradox of the relatively low population level as compared with those of other areas.

The principal elements necessary for improving the lot of the Panamanian people are the same as those necessary throughout the tropics of the world. They are all related to the search for finding ways to implement agricultural, forestry, or other kinds of systems that will sustain human life on areas now occupied with different kinds of natural vegetation, on a diverse array of tropical soils. The most direct reason that the population density (40 people per km^2) of Panama cannot, with the application of present technology, equal that of the Netherlands (340 per km^2), where the per capita gross national product is nearly six times as high as it is in Panama, is that we simply do not know how to utilize most soils of Panama, or most other tropical soils, in as productive a way as is possible in many temporate regions.

Los elementos principales necesarios para mejorar las condiciones de vida de la mayoría de los panameños son los mismos que aquellos que se aplican a otras regiones tropicales. Todos estos elementos están relacionados con la búsqueda de métodos que aumenten y mejoren la agricultura, las ciencias forestales y otros sistemas similares para mantener la vida humana en áreas ocupadas, en el presente, con distintos tipos de vegetación y de suelos.

National schemes have not governed the patterns of land use in Latin America in the past, any more than they have governed the patterns of land use in other regions of the world. What is a problem, however, is that when forest is cleared in most temperate regions, or prairies are plowed, the resulting fields are productive and, with proper management, remain so indefinitely. The kinds of agriculture that function well in temperate regions, however, cannot be transferred wholesale to the tropics, and it is our collective tragedy that we are just starting to act upon the consequences of this fact. The current estimate of Brazilian officials is that no more than 0.3 percent of the Amazon forest can be put into sustainable agriculture, and although the soils of Panama are generally better, the sad truth is that no one really knows how to manage most of those now occupied by moist forest for sustained yield once they are cleared. There is virtually no tropical deciduous forest left in Panama today, and the agricultural practices which have ruined the fertility of this potentially more productive region are now being transferred to the humid lowland forests of the Darién and of Bocas del Toro on the western part of the Caribbean slope, where they will produce food for only a year or two, instead of a decade or two. There is urgency in attempting to restore the fertility and productivity of the ravaged Azuero Peninsula, but technology known at present will not allow the sustained cultivation of the Darién or the Caribbean forests. Where, then, are the hungry santeños to turn?

Virtualmente, en el presente, no existen bosques tropicales deciduos en Panamá. Los bosques húmedos que aún persisten en el noroeste y el este de Panamá están habitados por la más rica y diversa flora y fauna de toda la región centroamericana.

The moist forests that still persist in the northwestern and eastern portions of the Republic are inhabited by the richest and most diverse array of plants and animals of the entire Central American region. Some of them might be useful directly as sources of human food, shelter, energy, medicine, or for other purposes, but the only hope of discovering the extent to which this may be true will be to survey these forests biologically as rapidly as possible and to concentrate in the course of these studies on groups of organisms of known or suspected economic importance. In addition, as much as possible should be learned about the functioning of the undisturbed forest that exists in these regions, because only such knowledge offers hope utilizing them on a sustainable basis or modifying them into some other form from which useful products could be derived. Meanwhile, because these forests cannot, by and large, be put into sustainable agricultural use with currently available technology, the forests should be conserved to the greatest extent possible while we are gaining knowledge, and the modifications that are allowed meanwhile should be as limited as possible in order to help secure the future prosperity of the country. These deceptively rich and productive-looking forests are simply not a suitable locale for most kinds of development, and such development can, in most cases, lead only to their total depletion, often with severe economic loss to those responsible.

Como quiera que estos bosques no pueden utilizarse para actividades agrícolas prolongadas con la tecnología corrientemente disponible, deben conservarse y mientras se obtengan los conocimientos pertinentes, las modificaciones permitidas deben ser mínimas.

For reasons just outlined, both the citizens of Panama and those who live elsewhere have genuine reason to applaud the establishment of the Darién Frontier National Park and the strategic plans that are being developed for the proper management of this area under the guidance and supervision of RENARE. The cooperation between Panama and Colombia that has made possible the progress to date has also been evident in the development of the 200,000 hectare Parque de la Amistad on the Cordillera de Talamanca between Panama and Costa Rica. Of equal importance are the efforts to reforest large denuded sections of the Panama Canal, and for the management of the Volcán Barú park, including the integral management of the Rio Caldera watershed.

El establecimiento del Parque Nacional del Darién en la frontera con Colombia y del Parque La Amistad sobre la cordillera de Talamanca entre Costa Rica y Panamá, al igual que los esfuerzos de reforestación de grandes áreas actualmente desnudas y el manejo del Parque del Volcán Barú son medidas dignas de aplauso.

Commendable as these schemes for preservation and restoration of natural resources are, however, the real problem facing Panama is how to guarantee people an adequate diet without wrecking the natural resources of the country completely—in other words, how to utilize the land on a sustainable basis, and to make the connotation of the word "renovable" in RENARE meaningful. These are global themes, and they have been stated with exceptional clarity and force in the World Conservation Strategy, promulgated by the International Union for the Conservation of Nature and Natural Resources (IUCN) in early March of this year. Throughout the tropics of the world, both multinational corporations, which presumably have a choice, and the peasant poor, who for the most part do not, are rapidly destroying the productive ecological systems on which human livelihood depends, without the creation of viable alternatives. The solution, both for Panama and throughout the tropics, depends upon a reversal of the forces that are destroying these ecological systems. Some of the elements that might be involved include the following:

De igual importancia son los esquemas de reforestación de RENARE y del IRHE para proteger las cuencas hidrográficas del Canal y de la Represa del Bayano respectivamente.

Afforestation schemes for the protection of watersheds, such as those that are currently being conducted by RENARE in the watershed of the Canal and elsewhere, and by IRHE around the Bayano Dam, are of special importance. The establishment of forest plantations and of areas devoted to agroforestry, coupled with soil conservation measures and the improvement of pastures on areas that can sustain cattle grazing, all have immediate desirable consequences, among them the provision of employment for rural populations. National policy can help to prevent the institutional squandering of the national patrimony of forest resources for short-term gain, but only the provision of employment and the ability to feed their children properly can divert the santeños from the moist forests of the northwest and east, where they can cut off the trees and grow a crop or two before the soils are depleted.

Panamá deberá aprender tanto como sea posible sobre sus recursos naturales y aplicar la información obtenida a la utilización racional de estos recursos.

Panama will be well served by learning as much as possible about its existing natural resources and by applying the information gained as directly as possible to their sustainable utilization. In this process, the traditional national institutions such as the universities have a vital role to play, and they should be funded in such a way that they can make the contribution of which they are capable, at the same time that new institutions are established. Meanwhile, the land should be developed as wisely as possible, with highly productive modes of agriculture concentrated in areas, such as perhaps portions of Chiriquí Province, where they can be carried out on a sustainable basis. Large-scale irrigation schemes should be developed elsewhere in appropriate areas to further the concentration of productive agriculture. Much more attention needs to be paid in Panama and everywhere else in the tropics to the improvement of the lot of rural agriculturists, and information needs to be brought as efficiently from other areas of the world with similar climates to be used locally. For example, *Brosimum alicastrum* is a tree crop well known in southern Mexico as a highly productive source of flour that can be grown in many forest soils. Might it not be just as suitable for cultivation in portions of Panama?

Existe una crisis mundial que amenaza todas las áreas de los trópicos. Durante la próxima generación la mayoría de los bosques tropicales serán utilizados por el hombre y miles de especies de plantas y animales desaparecerán.

There is a worldwide crisis that threatens all portions of the tropics. During the course of the next generation, a majority of all undisturbed tropical forest will be modified for human use, and hundreds of thousands of species of plants and animals will become extinct. Only the prompt accumulation of knowledge and its direct application to human welfare will help to ameliorate the situation; the traditional *laissez-faire* outlook on the operation of multinational concerns and neglect of the welfare of rural farmers that characterize all too many tropical countries will only serve to make the situation worse.

Si Panamá continua expandiendo y acelerando sus esfuerzos de autodesarrollo inteligente, este país tiene la potencialidad de contribuir de manera significativa al destino de otros países a través de los trópicos del mundo.

In this situation, it seems possible that Panama, a country blessed with abundant and diverse kinds of plants and animals, relatively fertile soils, a varied topography, the presence of several significant international institutions, the worldwide interest of ecologists and other biologists, a well conceived governmental program that links conservation and the wise use of natural resources with development according to ecologically sound principles, and an impressive array of resident scientists and of national institutions, will be able to make a significant contribution. If Panama can continue to expand and to accelerate its efforts at intelligent self-development, this small country, located at the crossroads of North and South America, of the tropical Atlantic and Pacific, and forming one of the most intense zones of contact between Hispanic and Anglo-American culture, has the potentiality of contributing in a significant way to the destiny of other countries throughout the tropics of the world. We who have come here to celebrate the completion of the *Flora of Panama* pledge ourselves to help as much as we can in striving to attain that transcendently important goal.

Brenan, J.P.M. 1978. Some aspects of the phytogeography of tropical Africa. Ann. Missouri Bot. Garden 65: 437–478.

Keigwin, L.D., Jr. 1978. Pliocene closing of the Isthmus of Panama, based on biostratigraphic evidence from nearby Pacific Ocean and Caribbean Sea cores. Geology 6: 630–634.

Marshall, L.G., R.F. *Drake*, G.H. *Curtis* & R.H. *Tedford*. 1979. Calibration of the great American interchange. Science 204: 272–279.

La Conservación de los Recursos Naturales Renovables en el Desarrollo de Panamá

Irving R. Díaz H.

Following some three decades of development, conservation consciousness has found expression in the decade of 1970–80. In government policy conservation is important but not urgent, and there still is no explicit dedication to conservation of natural resources. However, a number of national projects include conservation elements. The National Development Plan (1976–1980) recognizes that rural poverty has been a major factor in deforestation and urban migration. Several laws permit regulation of natural resources, and some basic inventories have been made. The Bayano dam project has demonstrated many negative effects of engineering projects on conservation and on human welfare. A number of parks have been projected and reforestation of some watersheds has begun. Indiscriminate deforestation causes serious problems for human welfare, land productivity, and success of development projects.

En este trabajo se pretende describir suscintamente la participación de la comunidad panameña en la conservación de los Recursos Naturales Renovables a través de la planificación y de las acciones de desarrollo, del establecimiento de la política y de la legislación y del nivel y magnitud de las realizaciones en las últimas tres décadas.

Puede asegurarse que la corriente conservacionista comienza a expresarse, en la práctica, en Panamá en la década de 1970–1980, por el hecho de que este período se caracteriza por una manifiesta toma de conciencia y participación efectiva de la comunidad en la gestión de promover el uso y desarrollo científico de nuestros recursos.

Si bien estas manifestaciones conscientes son todavía del dominio de una minoría de la población, sobretodo intelectuales y técnicos, a ello han contribuido la integración nacional por modernas carreteras y medios de comunicación social y la descentralización de una gran cantidad de servicios públicos de educación, salud y otros, incluyendo a la Universidad de Panamá, que están permitiendo la popularización de información básica necesaria.

A minority of the Panamanian population, mainly intellectuals and technicians, form the present conservation movement, which began to find practical expression in the 1970's.

Todo ello es el resultado de un proceso histórico y de la voluntad política de inclinar la balanza hacia el campo, para ajustar paulatinamente el desmesurado crecimiento del sector servicios y enfrentar más adecuadamente el déficit de alimentos y el resultado alarmante de la crisis energética, en un país como el nuestro sin petróleo.

Historically public opinion has favored using the natural landscape to reduce food and energy deficits.

El alcance de este trabajo es prácticamente un esquema de realizaciones más importantes que simplemente indican que el camino por recorrer en materia conservacionista apenas se inicia a la luz del devenir histórico de nuestro pueblo.

La Política y Los Planes Nacionales en Relación con la Conservación

Aunque existe el deseo gubernamental de conservar nuestros recursos naturales, esta gestión es considerada por algunos como importante, pero no tan urgente. Y aunque entre los objetivos globales del desarrollo se encuentra el de mejorar la calidad de vida de todos los panameños, no existe una política explícitamente dedicada a la conservación, entendida como "la ordenación de los recursos naturales, el aire, el agua, el suelo y los minerales, las diferentes especies de plantas y animales, inclusive el hombre como sujeto de desarrollo, a fin de lograr la máxima calidad de vida para nuestro país y para la humanidad" y por lo tanto, se requiere una declaración de política que armonice el desarrollo con la conservación y acelere la concientización del pueblo sobre la importancia de su patrimonio natural, y el fortalecimiento institucional consecuente.

A declaration of policy is needed to rationalize development with conservation.

Sin embargo, se nota en los planes nacionales un deseo de lograr la integración nacional, el fortalecimiento de la nacionalidad y la independencia económica y política, un aumento y mejor distribución de la riqueza, sobre todo mediante la participación activa de todos los panameños en el proceso político y en cada área representativa de la geografía nacional.

A pesar de que la tarea de los panameños en la última década, especialmente, se centra en el logro del rescate de su soberanía y de la dignidad nacional, los tratados Torrijos-Carter tienen entre otras cosas, una profunda repercusión sobre el movimiento conservacionista en Panamá, al integrarse el proyecto de Manejo de la Cuenca Hidrográfica del Canal de Panamá y de otras importantes cuencas, de manera de poder administrar adecuadamente y garantizar con visos de renta perpetua el Canal de Panamá, como principal recurso aliado a nuestra posición geográfica. La preparación para este proyecto se inició tres años antes del 1° de octubre de 1979, fecha de iniciación de dichos tratados y tiene asegurado financiamiento para los cuatro años siguientes.

Fuera de esto, el tratado en mención contempla la creación de una Comisión Mixta sobre el Medio Ambiente que debe examinar la situación actual y las repercusiones de los nuevos proyectos.

Además, la altísima prioridad que se expresa en el Plan Nacional de Desarrollo (1976–1980) para dotar a todos los panameños de "iguales oportunidades y niveles de bienestar material más cónsonos con la dignidad del hombre y la justicia social", apunta hacia la solución de las manifestaciones de pobreza "a través de acciones encaminadas específicamente a resolver las causas que motivan estos problemas de pobreza" (Minist. Planific. 1976). El círculo vicioso de la pobreza incluye la destrucción de recursos por la aplicación de tecnología primitiva en ecosistemas frágiles, hacia los cuales la gente menos favorecida ha sido históricamente marginada y ecológicamente desplazada y obligada a subsistir.

Apareciendo entonces, la situación en 1971, en la que el 30% de las familias de más bajos ingresos sólo tenían acceso al 3.8% del total, mientras que el 30% de las familias con más altos ingresos recibían el 74% del total de los ingresos (Minst. Planific. 1976).

Es obvio, también, que la pobreza rural ha sido el principal factor de expulsión de la gente del campo hacia la ciudad, fuera de la influencia de los medios de comunicación modernos como el radio-transistor, condicionándose un serio desequilibrio campo-ciudad que se dramatiza en el hecho de que para esta fecha, la población urbana sobrepasa el 50% de la población total. En un país no industrializado como el nuestro las implicaciones de un proceso de urbanización sin desarrollo son normalmente el desempleo, la delincuencia, los problemas de salud en el área suburbana y en los suburbios y el alto costo social que significa llevar servicios básicos a los migrantes.

Rural poverty has been the principal factor in driving country people towards the city, and in such a country without urban industrial development the results are normally unemployment, delinquency, health problems and high social costs.

De los 3.7 millones de hectáreas desforestadas en todo el territorio nacional a través de su historia, alrededor de 1.2 millones de hectáreas de sabanas antropógenas se encuentran altamente deterioradas por el fuego y el sobrepastoreo. En esta extensión territorial normalmente se ubica la gente más desnutrida y la que más rápida y masivamente se moviliza hacia los centros urbanos. En estos casos, de extrema pobreza y medio deteriorado es difícil tener éxito en la aplicación del crédito y la asistencia técnica, cuando lo que urgentemente clama esa población es salud, educación y mejor alimentación. Mientras estas condiciones no se den, es prácticamente imposible lograr un cambio tecnológico o llevar a cabo proyectos de rehabilitación de tierras mediante la reforestación. En estos casos, el crédito es incluso un incentivo inadecuado. Lo que es necesario hacer es rehabilitar social y económicamente al ser humano y prepararlo para que sea sujeto posterior del crédito y la asistencia técnica normales, en un lapso prudente que en el caso que nos ocupa podría variar entre dos y cuatro años.

Much of the manmade savanna is seriously degraded by fire and overgrazing, and the extreme poverty of such regions makes reclama-

tion difficult without both social and economic rehabilitation of the human stock followed by a lapse of 2–4 years before credit and technical assistance can be applied.

Es así, que con estas premisas básicas se inició en enero de 1980 el Proyecto de "Rehabilitación Social y Económica de Indígenas y Campesinos Marginados", en las estribaciones de la Cordillera Central de Veraguas y Coclé (Buenos Aires, Los Valles, Alto Guarumo, La Yeguada, Huacas de Quije y Olá). Este proyecto beneficiará a 15.000 personas o sea alrededor de 2.660 familias que reforestarán 8.000 has con pinos, fuera de las 4.500 has ya existentes; se organizarán para la producción y el mercadeo de sus productos y producirán alimentos sustitutos de los que provee el Programa Mundial de Alimentos, mediante la filosofía de *"alimentos por trabajo"*.

A pesar de que los planes globales y sectoriales no tratan con la debida profundidad la política de recursos naturales, ella se encuentra expresada en una serie de documentos legales, entre los cuales se encuentra la Constitución Nacional vigente (1972) que en su Artículo 111, llama la atención acerca de la necesidad de conservar intactos muestras importantes de nuestros ecosistemas (manglares, bosques mixtos, cativales, oreyzales, corales, el ecosistema marino de nuestros ríos, lagunas y lagos etc.).

Natural resources policy is contained in Article 111 of the current national Constitution, and also in six other fundamental documents. There is already regulation by the administrations of national parks, equivalent reserves, wildlife and environment, and regulations on the conservation and use of soils.

Aparte de este mandato fundamental, tenemos los siguientes decretos principales:

1. Recursos Forestales—Decreto Ley 39 del 29 de sept. de 1966.
2. Uso de las Aguas —Decreto Ley 35 del 22 de septiembre de 1966.
3. Quemas y Rozas—Decreto 44 del 16 de febrero de 1967.
4. Protección de la Fauna Silvestre Panameña—Decreto 23 del 30 de enero de 1967.
5. Ley 12 del 25 de enero de 1973.
6. Decreto Ejectutivo 70 del 27 de julio de 1973.

Ya existe un anteproyecto de ley para la administración de Parques Nacionales y Reservas Equivalentes, Fauna y Ambiente y se está trabajando en un anteproyecto de ley sobre uso y conservación de suelos.

Levantamientos Multidisciplinarios de los Recursos Naturales e Investigación Básica

El primer levantamiento multidisciplinario que se realizó en la República de Panamá fué el Catastro Rural de Tierras y Aguas, con la participación de un Consorcio de Empresas Norteamericanas bajo la denominación de CATAPAN y un grupo de profesionales panameños por parte del Programa de Catastro Rural de la Comisión de Reforma Agraria. Es cierto, aún a pesar de que para los años 1955 y 1956 hubo coincidencia entre el primer levantamiento detallado de suelos denominado, "Los Suelos y la Agricultura de los Llanos de Coclé" (Mathews & Guzmán 1858) sobre 99.000 hectáreas y el primer levantamiento del mapa ecológico de Panamá por Holdridge y G. Budowski, usando para ello el método de Holdridge, que se ha perfeccionado durante los últimos 25 años. Estos estudios no fueron debidamente integrados, y se realizaron separadamente a base de mapas militares y de reconocimientos de campo, la mayoría por tierra y algunos aéreos.

The first multidisciplinary undertaking effected in Panama was the Rural Cadastral Survey of Lands and Waters (CATAPAN) which by 1955–56 coincided with the first detailed soil study entitled "Soils and Agriculture of the Cocle plains", and the first issuance of the ecological map by Holdridge and Budowski using the Holdridge method.

El Catastro Rural de Panamá fué un levantamiento de recursos naturales y de infraestructura y uso de la tierra, sobre 35.000 km^2, o sea, un 46.4% del territorio nacional, que duró alrededor de 4 años, después de firmado el contrato en diciembre de 1964 (Com. Ref. Agraria 1964). Abarcó la vertiente del Pacífico, hasta los alrededores de Chepo y Chimán en la Provincia de Panamá. Cubrió también parte de Colón y el área del Canal de Panamá. Es decir, se hizo en donde habitaba el 94% de la población nacional.

El Catastro Rural consistió en el estudio del uso actual de la tierra y de los recursos naturales (excluyendo minerales metálicos, bosques y fauna silvestre) en la clasificación de la capacidad de uso de tierra; en la definición y señalamiento en mapas de los linderos de las fincas y propiedad de la tierra, así como el desarrollo de fórmulas del avalúo de las tierras, y la tasación sobre los valores de las tierras rurales.

The cadastral survey consisted of a study of the actual use of the land and natural resources, of a classification of land use capability, the assignment of property boundaries, the development of land valuation formulas, and appraisal of rural land values.

El Catastro Rural de Tierras y Aguas de Panamá fué una entidad Ad-Hoc que desaparece prácticamente después que termina el proyecto. Sin embargo, la integración de la información técnica se expresa en el trabajo sobre avalúo de la tierra y tiene las siguientes repercusiones institucionales:

1. Se crea el Catastro Fiscal, que inicia el manejo de la información por computadora y se ubica en el Ministerio de Hacienda y Tesoro.
2. Se fortalece el Registro Público con el establecimiento del sistema de microfilmes, con lo cual se agiliza el servicio.
3. Se crea la Dirección de Recursos Naturales Renovables en el MIDA.
4. Se crea el Departamento de Catastro Rural en la Dirección de Reforma Agraria.
5. Se producen algunos trabajos aplicados e integrativos de la información entre los que cabe mencionar, "Guía para el estudio de las estructuras Agrarias", "Evaluación de los Recursos Agrofísicos de Panamá" y la "Zonificación Ecológica de los cultivos de consumo básico y tradicionales de exportación de la República de Panamá en condiciones de secano".

Este Catastro fué el segundo realizado en América Latina después del que siguió al terremoto de Chile a principios de la década 1960–1970. Luego siguió el de Nicaragua.

The Panamanian cadastral survey was the first in Latin America since the earthquake in Chile in the early 1960's.

Posteriormente se realizó el inventario forestal del país (1969–1972) a través de la Dirección de Recursos Naturales Renovables (RENARE) y la cooperación técnica de FAO en el proyecto de Inventariación y Demostraciones Forestales FO:SF/PAN 6, con la ayuda del cual se generaron 19 informes técnicos, relativos al sector forestal y se realizó el levantamiento del segundo mapa ecológico del país a base del sistema de L. Holdridge (Minist. Agric. 1971). Tales trabajos, constituyen base de primordial importancia para la planificación del sector forestal del país, aunque no desembocan en un plan de desarrollo. Sin embargo, uno de los trabajos más importantes fué el Plan Nacional de Parques Nacionales que hoy día se ve superado por la inclusión del Parque Nacional La Libertad en la Cuenca del Canal y el Parque Nacional de La Amistad en la frontera con Costa Rica y otros proyectos en estudio, como la Serranía Filo del Tallo en Darién y parte de la Serranía de Majé en Bayano y Chimán.

Later, between 1969 and 1972, a forest inventory was prepared under RENARE which generated 19 technical papers and led to the issuance of the second Holdridge ecological map.

En el período 1975–1978 se realiza un estudio integral del Darién (17.000 Km^2) bajo la coordinación del Ministerio de Planificación y Política Económica y con el apoyo del Programa de Desarrollo Regional de la Secretaría General de la Organización de los Estados Americanos (OEA). Es así como por primera vez, se estudian los recursos naturales y humanos de una región, de manera integral para producir un plan de desarrollo regional y en el cual se considera el impacto ambiental de los proyectos propuestos, incluyendo el proyecto de carretera interamericana en el tramo hasta la población de Yaviza (Minist. Planific. 1978).

Basic investigations by Gorgas Memorial Institute, the Smithsonian Tropical Research Institute and the Missouri Botanical Garden on the ecology, flora, fauna, and epidemiology of Panama along with other studies by notable Panamanian scientists constitute a valuable information resource for Panama.

Se han llevado a cabo también importantes estudios y contribuciones al conocimiento ecológico de Panamá con relación a la construcción de presas para generación de hidroelectricidad, tales como Bayano y Fortuna particularmente, sin que ello haya representado la elaboración de planes para la administración y manejo de sus respectivas cuencas hidrográficas. (Adames 1977).

No hay duda alguna de que los aportes en la investigación básica realizada por el Instituto Conmemorativo Gorgas, el Instituto Smithsoniano, así como el Missouri Botanical Garden, sobre la ecología, flora y fauna de Panamá y epidemiología, aparte de otras investigaciones realizadas por connotados científicos panameños, constituyen un valioso acervo de conocimientos para Panamá. Sin embargo, todo parece indicar que todavía tenemos un serio vacío institucional en materia de fortalecimiento de la investigación, como una tarea rutinaria e imperativa de nuestro país para incorporar eventualmente nuevos y mejorados recursos a la economía nacional. Es quizás en este campo en donde los países desarrollados pueden aportar productivamente su impresionante avance tecnológico a los países que lo necesitan, fortaleciendo y ayudando a la creación de las instituciones de investigación. Esta gestión es la que le permite a los países desarrollados ajustarse más fácilmente a la crisis y a fenómenos como inflación-recesión, que normalmente implican un aumento de la productividad mediante un cambio tecnológico o la creación de nuevas demandas para la población o la solución de algún problema de salud.

El Impacto de las Obras de Ingeniería en la Conservación de los Recursos Naturales Renovables y el Ambiente Humano

Puede asegurarse que los proyectos de ingeniería que mayores discusiones y repercusiones ecológicas han tenido en la última década son, la hidroeléctrica del Bayano y la carretera Panamericana hacia el Darién.

La hidroeléctrica del Bayano inundó 35.000 ha de tierras y bosques, incluyendo 15.000 ha de la mejor clase de suelos para generar 15 megavatios de energía hidroeléctrica.

The largest engineering projects with an impact on the Panamanian ecology were the Bayano hydroelectric dam and the Panamerican highway towards the Darien. The dam flooded 35,000 ha including 15,000 ha of best class soils in order to generate 15 megawatts of hydroelectric energy.

La proliferación de hierbas acuáticas y de mosquitos trasmisores de importante enfermedades tropicales como el paludismo y la fiebre amarilla, así como la movilización de gran cantidad de arbovirus, algunos dañinos al ser humano, son las más importantes repercusiones en el estado potencial de salud del área del Bayano.

No obstante, de todo el proceso de cambio, posiblemente lo más doloroso es la repercusión en las poblaciones tanto indígenas como de migrantes. Es un hecho que desaparecieron 35 comunidades que fueron inundadas y cuyas poblaciones vivían adecuadas al régimen del río Bayano. El régimen del nuevo lago ha producido cambios en la conducta de los indígenas Cunas, al punto de que esta situación merece un estudio que avalúe y determine los cambios.

Por otra parte, los incentivos aplicados a los afectados por parte del Gobierno Nacional han producido adicionales problemas. El pago de las mejoras con el compromiso de que los colonos se retirarían de la cuenca del Bayano produjo efectos desastrosos en la unidad familiar, difíciles de recuperar, adicionándose el hecho de que estos jefes de familia incentivados han vuelto a la cuenca del Bayano en un 80%, después de haberse gastado improductivamente el efectivo que recibieron. Así, se inicia nuevamente el proceso de devastación de recursos por talas, rozas y quemas para la subsistencia en la cuenca del Bayano, con el agravamiento de que también es influído por la carretera interamericana recién abierta hasta Canglón.

Existe un Plan Maestro y un Plan Operativo para la cuenca hidrográfica del Bayano que están esperando financiamiento para llevarlo a cabo.

There is a master plan and an operational plan for the Bayano watershed awaiting funding.

Ya que el embalse del Bayano ha causado serios problemas actuales y potenciales, lo que menos debe permitirse es el desarrollo de todo el potencial que él mismo o parte de él puede desarrollar con las pesquerías, los deportes, la navegación y el turismo.

El lógico afán de lograr la integración nacional y de llevar a todos los rincones del país, la atención que los panameños requieren, ha promovido la construcción de un extraordinario kilometraje de caminos y carreteras, especialmente en los últimos diez años, y entre los proyectos más importantes de las últimas dos décadas se encuentra la carretera interamericana, especialmente en su tramos de Bayano hasta Darién.

To further national integration and to provide for Panamanians in all corners of the country, an extraordinary length of roads and highways was built, especially in the last ten years.

Este proyecto ha dado lugar a nuevas poblaciones; entre las más importantes se cuenta Cañita, Ipetí, Tortí y Cañazas en Bayano y a Santa Fé, Metetí, Iglesias y Canglón en Darién, cuya población total de alrededor de 5.000 nuevos pobladores se encuentra hoy día equitativamente distribuida entre Bayano y Darién. Actualmente esta población crece a un ritmo de diez (10) nuevos pobladores por día, especialmente durante el verano.

Este fenómeno se ha dado en la mayoría de los casos, sin una orientación adecuada de parte del Gobierno que, recientemente, ha propuesto soluciones integrales tanto en Darién como en Bayano e incluyen la participación de la población en forma ordenada en el desarrollo agropecuario y forestal e, incluso, la participación muy específica de la empresa privada en el desarrollo industrial del Darién, aparte de la creación del Parque Nacional del Darién de una extensión de 475.000 ha, como área de prevención en la entrada de la fiebre aftosa y otras zoonosis, y garante de la integración indígena y darienita al desarrollo social, cultural y económico del este del país y para el manejo del complejo hidrográfico Tuira-Chucunaque-Sambú.

Por otra parte, el Decreto Ley No. 71 del 20 de sept. de 1973 que inadjudica 8 km. de cada lado de la carretera interamericana en el Darién, fue diseñado con la sabia proyección de adjudicar la tierra de ese territorio a discreción del organismo rector de la Reforma Agraria que es el MIDA, siempre y cuando las circunstancias lo justifiquen. (Véase Anexo 3).

Actualmente alrededor de 35.000 ha se han deforestado para dar paso especialmente a la ganadería de tipo tradicional, sin que ello haya significado la incorporación de los recursos forestales a la economía de los migrantes, lo cual ha sido una de las grandes fallas encontradas en este proceso, con el agravamiento de que la mayoría de los suelos son superficiales y pobres y que se sigue hoy día talando áreas por encima de 45% de pendiente, para establecer cultivos de subsistencia y, luego, pastos no mejorados. Peor aún, es la desproporción entre lo desforestado y lo realmente necesario para la capacidad del trabajo familiar, de modo que se destruya más de lo que se usa.

Con la apertura oficial de la carretera hacia el Darién este verano de 1980, las perspectivas de una desforestación incontrolada en Darién es aún mayor, a menos que el Gobierno Nacional y Provincial retomen en la práctica la autoridad y el papel orientador que les corresponde en las intervenciones de los colonos, y los apoye con soluciones imaginativas y con una firme y amplia campaña educativa a nivel de las comunidades y comunique las mejores alternativas para su desarrollo, las mejores formas de organización y un sistema de incentivos y restricciones que se complementen para salir del atraso y aumentar la productividad y evitar el despilfarro de los recursos naturales renovables (bosques, suelos, aguas, fauna silvestre).

With the official opening of the Darien highway in the summer of 1980, the prospect of uncontrolled deforestation was even greater unless the national and provincial governments put into practice authority to orient colonists and support them with imaginative solutions and with a strong education campaign at the community level.

Se han dado casos de venta de "mejoras" y de integración de fincas de tamaño superior a las 200 hectáreas. Ambas acciones están prohibidas por el Código Agrario e indican que no todos los colonos califican como agricultores y ganaderos estables y que desde luego, hay especulación con la tierra.

Para acometer detalladamente la tarea de desarrollo ordenado de áreas nuevas, en el Darién trabaja un grupo de técnicos panameños y de FAO en el Proyecto de Aprovechamiento Forestal del Darién PCT/PAN/0001 (I) 1980 que proveerá al Gobierno Nacional y provincial, de la política, las acciones y los proyectos que permitirán dar una respuesta adecuada y dinámica a los problemas encontrados. Este proyecto se inició en enero de 1980 y terminará en julio de este mismo año. La puesta en marcha de sus recomendaciones requerirá, a no dudarlo, un fortalecimiento de las instituciones involucradas y un acrecentamiento de la infraestructura productiva, sin que ello signifique necesariamente, la asignación de mayores y mejores recursos y oportunidades que aquellos que deberían darse en las áreas de origen de los colonos.

The implementation of recommendations for meeting the Darien problems requires a strengthening of institutions and broadening of the productive infrastructure.

Este mismo panorama se ha presentado en otras áreas abiertas o anunciadas por rutas, como Escobal-Cuipo-Tambo- y Pilón-Portobelo-Nombre de Dios-Palenque, para solo citar algunas. Lo más evidente y que es casi axiomático es que, donde hay carreteras no hay bosques y en donde hay bosques, normalmente, no hay carreteras.

La colonización del área de bosques del proyecto del Parque Nacional de Portobelo en la nueva población de Nuevo Tonosí, decayó, por condiciones muy particulares de clima muy lluvioso, topografía accidentada con fuertes pendientes en el minifundismo que ha provocado una mayor inestabilidad de los colonos, siempre ávidos de nuevas tierras para la ganadería en la nueva trocha que se extiende más adelante de Palenque, costa arriba de Colón.

Sin embargo, allí, después de cuatro años de trabajo con la comunidad se logró el entendimiento de trazar el límite de 22.5 km de una nueva reserva forestal que la comunidad hoy respeta, y que lleva el nombre del guardabosque que liderizó esta acción: Línea Milcíades Cedeño.

Apertura de la Frontera Agrícola

El hecho de que la población rural se haya doblado a la fecha, de alrededor de medio millón de personas desde 1950, y que persista un alto porcentaje de analfabetismo en el campo, han sido factores decisivos en los movimientos migratorios de la población rural y en la expansión de la frontera agrícola en los últimos treinta años. Sobretodo, en un país como el nuestro con vocación canalera, en donde mucha gente vive de la agricultura, pero en donde existen relativamente pocos empresarios agrícolas y en donde la mejor tierra agrícola mecanizable podríamos dividirla en 8.000 fincas de 100 ha cada una y en donde por esas razones mucha gente se ha visto en la necesidad de producir alimentos en tierras forestales que constituyen el 70% de la superficie nacional, clasificadas así debido a condiciones muy especiales de clima, suelo y topografía y normalmente de baja fertilidad.

The doubling of population and persistent high illiteracy level have been decisive in rural migration and expansion of the agricultural frontier.

Si para 1950 la superficie en bosques correspondía al 60–70% del total y hoy día contamos con sólo el 45% de nuestra geografía terrestre con cobertura boscosa, significa que se han destruído bosques en los últimos 30 años en incrementos que varían entre 40.000 y 50.000 ha por año. Y un área relativamente insignificante corresponde a suelos en donde puede desarrollarse una agricultura y una ganadería permanentes con los sistemas de tala y quema y ganadería extensiva.

Este fenómeno de destrucción de bosques ha sido más marcado aún en los últimos 15 años con la apertura de nuevos caminos, sin un control y una planificación previa de las áreas influenciadas y con el llamado político a la conquista del Atlántico y de otras áreas difíciles.

Estadísticamente, puede demostrarse que, en la mayoría de los casos, el aumento de la producción nacional de rubros tradicionales, se ha logrado a costa de la expansión de la frontera agrícola en vez de una elevación de la productividad y que ello ha agudizado la destrucción de bosques, el fenómeno de la erosión, el desequilibrio hidrológico, el azolve y contaminación de ríos y lagos, han puesto en peligro nuestra fauna silvestre y ha mantenido los bajos niveles de ingreso de nuestra gente del campo.

Debido al sobrepastoreo en pendientes mayores del 45% y a las quemas de los pastizales, la invasión de malezas y hierbas indeseables como la hierba cabezona (*Paspalum virgatum*), algunos *Andropogon spp*. y escobillas (*Sida spp*.) dominan extensas áreas en donde se practica una ganadería extensiva como es el caso de más de 80.000 ha en uso ganadero en la cuenca del Lago Gatún, en donde el 60% de estas tierras tienen pendientes mayores del 45%. Allí se pierden por erosión cada año no menos de 9 millones de m^3 de suelo por inadecuado manejo de los pastos y del ganado y constituye el área de mayor tasa de erosión en toda la cuenca. Cuando la vegetación de pastos en matojos como la faragua y indiana se degenera, le es muy difícil al ganadero cubrir el costo de reposición de pastos en áreas infestadas por "cabezona". Esto promueve el desaliento de los campesinos que buscan en las áreas con bosques sus esperanza futura. Estos elementos de expulsión son también causas de la movilidad y de las migraciones de agricultores y ganaderos tradicionales en ecosistemas frágiles. No sólo los bosques mixtos sino también importantes áreas de manglares en el Pacífico, han sido y siguen siendo destruídas para favorecer la expansión ganadera tradicional.

Due to overgrazing and firing of pastures weeds now dominate extensive areas such as the Gatun Lake watershed where erosion results in an annual loss of over 9,000,000 cubic meters of soil.

Del mismo modo, los agricultores de "roza" también han destruído nuestros bosques y dejado en su lugar grandes áreas con rastrojos improductivos cuya alternativa en la mayoría de los casos es la reforestación mediante la aplicación de sistemas agrosilvícolas, utilizando para ello recursos internos y transformando, mediante una organización adecuada del trabajo en grupo, la economía de subsistencia del campesino en una economía de escala como ya hemos practicado en 1979, en la Cuenca del Canal de Panamá, con cerca de 600 campesinos en 826 ha en sólo 2.5 meses sin mecanización y utilizando métodos biológicos para el control de malezas.

En esta cuenca se han iniciado acciones tendientes a la introducción de pastos estoloníferos, la estabilización de cárcavas, la reforestación en áreas con pendientes superiores al 45%, a la promoción de una ganadería forestal y al establecimiento de un sistema de baños garrapaticidas para evitar "quemar para controlar las garrapatas" y "para que el ganado apetezca a la faragua cuando empiezan las lluvias".

A beginning has been made in the Canal watershed by introducing stoloniferous grasses, establishing hedges, reforesting slopes, and plac-

ing cattle dips to avoid the need to burn for controlling ticks.

Una característica de nuestros campesinos, en general, es la proyección de su tecnología y de sus costumbres a cualquier espacio de la geografía nacional, así como la proyección del pasado hacia un futuro con pocas esperanzas. Nuevos sistemas y nuevos recursos deben ponerse en manos de nuestra gente del campo.

Otra causa de destrucción de recursos es la falta de conocimiento y convicción sobre la función social del bosque, confundiendo, algunos, la presencia de selvas naturales o bosques en las fincas con la falta de cumplimiento de la función social de la tierra.

Cómo podríamos desatender la función social del bosque en una crisis energética? No sólo desde el punto de vista hidrológico forestal y de conservación de la fauna, sino también desde la perspectiva social de leña y carbón, y económica como metanol, caucho, papel, rayón, celofán, plástico y otros productos alimenticios, medicinales, ornamentales e industriales?

La introducción de la tecnología moderna en la agricultura se inició con mayor agresividad, desde la década 1950–1960 por parte de empresarios panameños, con la mecanización del cultivo del arroz. En la actualidad esta avanzada tecnológica se practica en las áreas agrícolas más importantes del país. Sin embargo, todavía quedan por resolver los problemas relativos a la falta de control de la erosión y al uso indiscriminado de los agroquímicos.

Modern agricultural technology increased in the 1950's with the mechanization of rice culture which occurred in the most important farming areas.

En 1979, por ejemplo, se determinó que la pérdida de suelo por erosión en cuenca del río Caldera fué la de un equivalente a una superficie de 50 ha con 30 cms de profundidad, evidenciándose que la mayor parte provino de la pequeña área relativa, dedicada a la horticultura. De esta manera se justifica objetiva y cuantificadamente un programa de conservación de suelos, reforestación de áreas con pendientes prohibitivas que ya han sido abiertas, el establecimiento de cortinas rompevientos, la promoción de cultivos como el café bajo sombra y el mantenimiento de las áreas actualmente en bosques. El proyecto de administración y manejo del río Caldera tiene asegurado financiamiento por cerca de un millón de balboas para los próximos cuatro años.

El uso indiscriminado de agroquímicos y los desechos de la actividad industrial y los desechos humanos han sido las principales fuentes de contaminación de las aguas. Esta contaminación ha afectado la flora y fauna de los ríos y estuarios incluyendo los manglares como fuente de proteína animal.

Indiscriminate use of fertilizers and both human and industrial waste have affected the flora and fauna of rivers and estuaries, including the mangroves.

Varios miles de hectáreas de suelos se han contaminado con cobre por la aplicación de caldo bordelés en décadas anteriores. Los efectos de los herbicidas a plantaciones de frutales y la excesiva aplicación de insecticidas en las áreas hortícolas poco antes del momento de la cosecha son frecuentes. Así mismo, se requiere más atención a las aplicaciones aéreas de pesticidas y fertilizantes ya que en un número plural de casos el equipo de aspersión se encuentra en malas condiciones, se aplican dosis excesivas de sustancias persistentes y no se toma muy en cuenta la velocidad y la dirección de los vientos en relación a daños a terceras personas y a los recursos naturales renovables. Esta contaminación es también peligrosa en la alimentación animal ya que la acumulación creciente de su toxicidad afecta directamente al hombre. La fauna silvestre se ha visto muy afectada en Panamá y principalmente los camarones y las aves en las áreas de la costa del Pacífico.

Las principales agroindustrias cuentan hoy día con sistemas de control de efluxión y de desechos, aunque todavía es muy ardua la tarea que hay que realizar en los distintos ingenios azucareros, destilerías, beneficios de café, plantas de lechería, fábrica de conservas, curtiembres, aserraderos, etc. . . , de modo que cuenten con los diseños y sistemas sencillos, pero efectivos para controlar la contaminación de las aguas, especialmente en la época seca, para cuando los caudales de agua bajan ostensiblemente y son más notables los efectos de la polución, ya que simultáneamente algunas agroindustrias importantes alcanzan su máxima cúspide de producción para esta época.

The joint activities of various state institutions such as RENARE, SALUD, and local authorities have given excellent results in practice.

La acción coordinada de las diversas instituciones del Estado involucradas (RENARE, SALUD y Autoridades Locales) ha dado excelentes resultados en la práctica de supervisión y control conjuntos.

Las Cuencas Hidrográficas como Enfoque de Desarrollo y Conservación de Recursos Naturales

Situación del Aprovechamiento Integral de las Aguas

En la República de Panamá, a la fecha, no existe un Plan Nacional para aprovechamiento integral del recurso agua; sólo se vienen realizando algunos proyectos de aprovechamiento integral en sectores específicos, que dan solución parcial a problemas locales.

In Panama at present there is no national plan for protection of water resources, but there are several projects for specific sectors that provide partial solution to local problems.

Existen algunas disposiciones legales que dan directrices para normar el aprovechamiento integral de las aguas, así mismo, existe la Institución encargada de hacer cumplir estas disposiciones. Sin embargo, aún no se ha formulado ningún plan en este sentido, porque no se ha tomado la respectiva decisión.

El Gobierno Nacional, en su afán de lograr el aprovechamiento racional e integral de las aguas está realizando, a través del Ministerio de Desarrollo Agropecuario (MIDA), una serie de acciones que permitan la coordinación entre los distintos sectores usuarios del agua para lograr este propósito; con este fin ha formulado la propuesta de un Plan Nacional para el Aprovechamiento Integral de las Aguas, que se considera como de prioritaria necesidad para asegurar su adecuación armónica con el Plan Nacional de Desarrollo del país.

Problemas Principales

Los problemas relacionados con el uso inadecuado de las tierras y las aguas en nuestro país, se caracterizan por el dualismo existente entre lo tradicional y lo moderno, reflejado en la aplicación indiscriminada de la tecnología primitiva practicada por no menos de 100.000 familias campesinas en el medio rural y suburbano y el uso muchas veces inadecuado de equipo y tecnología más avanzada.

Una sociedad como la nuestra orientada especialmente hacia el consumo de productos importados, sufre del efecto de demostración que se da desde afuera, agudizando aún más las tensiones y desequilibrios sicológicos, sociales, demográficos y económicos que actualmente vive el país.

A society like ours, oriented to the consumption of imports, suffers from the effects of examples from abroad which increase psychological, social, demographic and economic tensions present in the country.

El resultado se palpa a través del conocimiento de los factores de expulsión del campo a la región metropolitana y de aquellos factores de atracción que el medio urbano ejerce, provocándose fuertes migraciones internas.

Los factores de expulsión del campo hacia el medio urbano se producen en forma desencadenante así:

1. Tala y quema de los montes.
2. Esta desvastación trae como consecuencia la erosión de los suelos.
3. Igualmente, trae el desequilibrio hidrológico, condicionando inundaciones en época lluviosa y sequía en la época seca.
4. La erosión por la desvastación de los bosques trae como consecuencia la sedimentación de ríos y lagos.
5. Consecuentemente pérdida de la productividad de la tierra.
6. Todo ésto provoca la inestabilidad social, familiar y espacial del medio rural y hacia el medio urbano.
7. Mantenimiento y profundización de la pobreza y agudización de la misma en el área suburbana, que se refleja en el aumento de la criminalidad, ya que el medio urbano no produce el suficiente empleo para satisfacer la demanda creciente de ocupación.

The sequence causing expulsion of humanity from the country to the city begins with cutting and burning the hillsides, followed by erosion of soils, then disruption of runoff causing floods and exacerbating drought, then sedimentation of river courses, then loss of land productivity, then increasing poverty, and then increasing crime because the urban area has insufficient employment.

Todo ello trae como herencia de arrastre que podemos cuantificar como sigue:

1. 1.200.000 ha de suelos ya están degradados por el uso indebido de la tierra.
2. Desvastación de 40.000 ha de bosques por año en los últimos cinco (5) años por valor de 30 millones de balboas anuales, debido a la acción de 60.000 familias campesinas.
3. Pérdidas por inundaciones y sequía por valor de 4 millones de balboas como promedio anual.

The cost of the sequence is 1,200,000 ha of soils

degraded through inappropriate use, devastation of 40,000 ha of forests worth $30 million annually by the action of 60,000 peasant families, and annual average losses of $4 million due to flooding and drought.

Estrategias

Las estrategias a seguir pueden definirse así:

1. A través del Ministerio de Desarrollo Agropecuario (MIDA), propiciar el funcionamiento inmediato del Consejo Consultivo de Recursos Hidráulicos, para cumplir y hacer cumplir las disposiciones legales vigentes relacionadas con el aprovechamiento integral de las aguas.
2. Fortalecer el organismo rector en materia de aguas, para que pueda coordinar entre los distintos sectores usuarios, el aprovechamiento integral de las aguas.
3. Definir proyectos específicos sectoriales con proyecciones a corto, mediano y largo plazo, relacionado con el aprovechamiento de las aguas.
4. Ejecución del plan maestro de aprovechamiento de las aguas del país.

The strategies to follow are to put into place the legality to protect the waters, to enable the state water entity to coordinate the various user sectors, to define specific projects for short, medium and long term use for water protection, and implementation of the master plan for protection of national waters.

Podemos señalar como acciones estratégicas las siguientes:

a. Mantenimiento de parques y reservas forestales.
b. Revegetación, reforestación y aplicación de sistemas agroforestales.
c. Ordenamiento general del uso de la tierra.
d. Normalización de las actividades humanas de manera de aplicar una tecnología apropiada en las urbanizaciones, construcciones de caminos, industrias, agricultura y ganadería.
e. Controlar la erosión, sedimentación y la calidad del agua.

Priorities include maintenence of parks and forest reserves; revegetation, reforestation and application of agroforestry systems; organization of land use; application of appropriate technology to human activities in towns, road construction, industries, agriculture and cattle raising; and control of erosion and water quality.

Criterios Utilizados para Establecer Prioridades en Relación a las Cuencas del País

Para establecer la prioridad en el manejo de las cuencas hidrográficas del país, se tomaron los siguientes factores:

A number of factors should be considered in setting management priorities.

1. *Población*: Este factor incluye tanto la población rural como urbana de cada cuenca.
2. *Superficie total*: Incluye el área de cada cuenca.
3. *Hectáreas de suelos agrícolas*: Incluye las hectáreas de suelos clasificados dentro de las clases II, III y IV, de acuerdo con el sistema de conservación de suelos.
4. *Desequilibrio Hidrológico*: Aquí se incluyeron la frecuencia de inundaciones y sequías y factores relacionados con la degradación del medio, pobre manejo de la tierra, peligro de las poblaciones a los desastres.
5. *Degradación*: Incluye el grado de deforestación producido principalmente por la aplicación de la agricultura primitiva.
6. *Nivel de vida*: Este factor se midió por el nivel de ingreso general de cada cuenca y otros factores determinantes de la calidad de vida.
7. *Inversiones existentes*: Se tomó el valor relativo de los servicios públicos e infraestructura.
8. *Inversiones proyectadas*: Inversiones públicas proyectadas para el futuro (5 años adelante).

Categorización de las Cuencas Hidrográficas por Prioridades

Primera Prioridad:

a. Cuenca del Canal de Panamá
b. Cuenca del Río Santa María
c. Cuenca del Río La Villa
d. Cuenca del Río Grande

Segunda Prioridad:

a. Cuenca del Río Bayano
b. Cuenca del Río Caldera
c. Cuenca del Río San Pablo
d. Cuenca del Río Teribe-Changuinola
e. Cuenca del Río Chiriquí Viejo
f. Cuenca del Río Tonosí

Tercera Prioridad:

a. Complejo Tuira-Chucunaque
b. Río Chico (Chiriquí)

Cuarta Prioridad:
Otras Cuencas del país

Proyectos para Implementarse a Partir de 1980

Generalidades

La Dirección ha preparado proyectos para la conservación de los Recursos Naturales Renovables y Cuencas Hidrográficas por valor de cuarenta millones ($40.000.000), para los próximos cinco a siete años siguientes, a través de los cuales se beneficiarán directamente a siete mil quinientas familias en el medio rural, a cinco mil cuatrocientos balboas ($5.400) por familia para restaurar el equilibrio ecológico e hidrológico perdido. Se espera aumentar la productividad, erradicando la pobreza extrema.

RENARE has prepared projects worth $40 million for the next 5–7 years which will directly benefit 7,000 families to the value of $5,400 each to restore the ecological and hydroelectrical equilibrium. About $18 million has been negotiated, $10 million from a long term loan and the remaining $8 million from the central government.

De todos estos proyectos, se han concluido las negociaciones por aproximadamente dieciocho millones de balboas ($18.000.000), que contempla la Cuenca del Canal de Panamá, Río Santa María, Río La Villa, Río Tonosí y Río Caldera. De este total, 10 millones viene por la vía de un préstamo blando y de amortización de largo alcance, 1.0 millón son una donación del Programa Mundial de Alimentos y los 8 millones restantes constituyen aportes del Gobierno Central.

$16 million is destined for administration and for conservation in the Canal, the Caldera River, and the La Villa River watersheds, and the balance for watersheds of the Santa Maria River and the Chico River of Cocle.

Del total, 16.8 millones de balboas se destinarán a mejorar la capacidad del gobierno para administrar y conservar los recursos naturales renovables del país y para administrar las cuencas hidrográficas del Canal de Panamá, del Río Caldera y del Río La Villa. El resto está dirigido hacia la restauración de las cuencas hidrográficas de los Ríos Santa María y Río Chico de Coclé.

Proyecto para la Administración y Conservación de la Cuenca del Canal de Panamá

El Gobierno Nacional consciente de los compromisos adquiridos por los nuevos tratados del Canal de Panamá, Torrijos-Carter, reconoce que el Canal de Panamá requiere:

1. De la Administración y Mantenimiento del Canal mismo dentro de los límites de la actual Zona del Canal.
2. De la Administración y Conservación de la Cuenca Hidrográfica que alimenta el Canal para garantizar su navegación durante todo el año y para ampliar el desarrollo urbano e industrial de la región metropolitana.

The Torrijos-Carter Canal Treaty requires administration and maintenence of the Canal throughout the Canal Area and administration and conservation of the watershed supplying the Canal to guarantee year round navigation.

Para la primera gran tarea señalada, se creó la Autoridad del Canal de Panamá y, posteriormente, la Dirección Ejecutiva para los Asuntos del Tratado, que, en estrecha colaboración y coordinación con otras entidades del Estado, velará por el buen funcionamiento de esta vía interoceánica y encauzará todo el desarrollo que surgirá en el futuro.

En esta tarea, la Dirección Nacional de Recursos Naturales Renovables del Ministerio de Desarrollo Agropecuario, conjuntamente y en coordinación y colaboración con otras entidades del estado, llevará a cabo el proyecto de Administración y Conservación de la Cuenca Hidrográfica de los lagos que alimentan el Canal de Panamá y a nuestras principales ciudades.

1. Para tal fin se hará una inversión de doce millones de balboas ($12.000.000) para el próximo quinquenio, para financiar los siguientes sub-proyectos.

a. Reforestación de 10.500 ha ($450/ha) por 5.7 millones de balboas.

b. Control de erosión de los suelos por 1.6 millones de balboas.

c. Establecimiento y mantenimiento de Reservas y Parques Nacionales mediante adecuada Administración y vigilancia por $3.7 millones.

d. Mejoramiento de los pastizales por $0.5 millones.

RENARE, with other state agencies, will make an investment of $12,000,000 over five years to: 1) reforest 10,500 ha ($450/ha) for $5.7 million. 2) control erosion with $1.6 million, and 3) es-

tablish national parks and reserves with $3.7 million.

2. Fuera de estos sub-proyectos se coordinarán con todas las entidades gubernamentales, cívicas y privadas aplicación de normas que armonicen con los objetivos que se persiguen.

3. Para realizar toda esta tarea se dará empleo a 1.200 jefes de familia y se fortalecerá institucionalmente el programa de la Dirección Nacional de Recursos Naturales Renovables, con personal, entrenamiento, equipo necesario y edificios, y aumento general de su capacidad técnica y administrativa por $4.3 millones.

Si este proyecto no se lleva a cabo las consecuencias serán las siguientes:

1. La pérdida de todos los bosques naturales y muchas especies de nuestra flora y fauna restante que en su mayor parte cubren la cuenca de los ríos Chagres, Indio, San Juan de Pequení y Boquerón, que desembocan en el Lago Alajuela, con la consecuente disminución de la capacidad de navegación en el Canal de Panamá en la época seca.

2. Un aumento sustancial de la ganadería extensiva hasta el 37% de la cuenca. Este uso provoca el mayor incremento de erosión y sedimentación consecuente.

3. Un incremento de la sedimentación de 9.5 millones de metros cúbicos actualmente a 16 millones de m^3 para el año 2.000. Esto aumentaría el costo del mantenimiento del Canal (1 m^3 de sedimiento removido cuesta en la actualidad $1.30).

4. Reducción de la capacidad del Lago Alajuela en un 80% para el año 2.040, paralizando mucho antes la operación del Canal actual.

5. Degradación de la calidad del agua por sedimentos, nutrientes, agentes patógenos y plantas acuáticas y aumento sustancial del costo de potabilización de las aguas.

6. Un aumento significativo de la pobreza rural. Aunque el plan quinquenal para manejar la cuenca hidrográfica del Canal de Panamá es sólo el inicio del Manejo y Administración adecuados de los recursos de un área, se esperan de él los siguientes beneficios:

Failure to complete this work will result in 1) loss of all forest and wildlife in the watersheds of the Chagres, Indio, San Juan of Pequeni, and the Boqueron Rivers with resulting diminution of Canal navigation capacity in the dry season; 2) increase in grazing lands to 37% of the watershed with resulting increase in erosion and sedimentation; 3) an increase in Canal maintenance cost; 4) reduction of the capacity of Lake Alajuela by 80% by the year 2040 with the Canal operation paralyzed much sooner; 5) substantial increase in rural poverty.

1. La protección de 93.616 hectáreas todavía en bosques y tierras con una gran cantidad de recursos de nuestra flora y fauna nativa y el consecuente mantenimiento de su capacidad reguladora de las aguas, especialmente en la cuenca del Lago Alajuela.

2. La reducción del nivel de sedimentación a 7.9 millones de m^3 para el año de 1983.

3. El mantenimiento del 72% del almacenamiento activo del Lago Alajuela para el año 2.040.

4. La disminución de la contaminación de las aguas.

5. El aumento del ingreso familiar al doble, para varios miles de beneficiarios directos e indirectos que hoy día tienen un ingreso no mayor de $160/año.

Administración de las Cuencas del Río Santamaría y Río Chico de Coclé

Estas cuencas se restaurarán mediante un proyecto de Rehabilitación Social y Económica de dos mil seiscientos sesenta (2.660) familias indígenas y campesinas marginadas en varios puntos de las estribaciones de la cordillera central hacia el Pacífico. Allí se establecerán 8.000 ha de plantaciones forestales destinadas a llenar nuestra demanda de materia prima para producir en Panamá el papel que hoy importamos. Así mismo se creará un conjunto de proyectos agropectuarios para que dichas poblaciones se autoabastezcan y se supere el nivel de desnutrición y de injusticia social prevaleciente.

Este proyecto se financiará con un aporte del PMA de siete millones de balboas en alimentos a precios en Panamá y con 1.6 millones de balboas del Gobierno Nacional.

Se trabajará con la colaboración de los distintos programas del MIDA, de los Ministerios de Salud y de Educación, de las autoridades principales y locales y de otras entidades como la UNICEF.

Todos estos proyectos obedecen a necesidades especiales de manera que para incrementar todas estas acciones dentro de la política de desarrollo y conservación de nuestros recursos básicos, la Dirección Nacional de Recursos Naturales Renovables del Ministerio de Desarrollo Agropecuario, como autoridad responsable y de acuerdo con las disposiciones legales existentes ha confeccionado una propuesta de un Plan Nacional para el Aprovechamiento de las Aguas de la República de Panamá, de manera que se establezca con claridad la política, las prioridades y las acciones conducentes a garantizar, en el plazo más largo posible, una adecuada proporción de la relación demanda-disponibilidad del recurso agua para la navegación en el Canal de Panamá, para el consumo humano y para el desarrollo industrial y agropecuario del país.

Los Parques Nacionales y La Fauna Silvestre de Panamá

Los Parques Nacionales y Reservas Equivalentes son áreas reservadas de los diversos ecosistemas naturales que existen en Panamá como un banco genético de la diversidad de nuestra flora y fauna, que permita incorporar recursos mediante la investigación sistemática y la debida protección a la economía nacional. Además, son los sitios ideales para promover la educación ambiental, la recreación sana y el turismo.

The national parks exist as a genetic bank of the diversity of our flora and fauna. They are ideal places for promoting environmental education, recreation and tourism.

Panamá cuenta hoy día con tres (3) Parques Nacionales declarados oficialmente por Decreto:

1. El Parque Nacional de Campana, cuyas características sobresalientes son su Flora y Fauna Silvestre, sus características geológicas y su clima fresco. Fué declarado mediante Decreto 153 del 28 de junio de 1966—G.O. 15655 y modificado por el Decreto 35 del 28 de abril de 1977—G.O. 18645 del 21 de agosto de 1978, por el cual se crea el Parque Nacional y Reserva Biológica Altos de Campana. El Plan Maestro de este Parque fué redactado en 1976.
2. El Parque Nacional de Portobelo que se crea mediante la Ley 21 del 22 de diciembre de 1976. Las condiciones naturales de este Parque han sufrido serio deterioro por la acción desforestadora de los colonos de Nuevo Tonosí, estimándose la necesidad de discutir si el concepto de Parque Nacional puede mantenerse o cambiarse por el Monumento Histórico de las Ruinas de Portobelo. El Plan Maestro de este Parque se confeccionó en 1977.
3. El Parque Nacional del Volcán Barú se crea mediante el Decreto 40 del 24 de junio de 1976 G.O.—18619 del 13 de julio de 1978. Su Plan Maestro está a punto de concluírse.

Panama's three officially decreed parks are Campana National Park, Portobelo National Park, and Volcan Baru National Park.

Los otros proyectos de Parques son: La Libertad con 22.000 ha de superficie ubicado entre Paraíso, Gamboa y la carretera transístmica. El Plan Maestro de este Parque fué terminado en 1979, así como su primer Plan Operativo anual para 1980 y tiene financiamiento asegurado para los próximos tres (3) años.

El Parque Nacional de Alto Darién de 475.000 ha es un Parque de importancia biológica e histórico-cultural y para la protección de la ganadería nacional contra la fiebre aftosa. El Plan Maestro de este Parque se realizó en 1978 con la cooperación técnica del CATIE.

Existen otros proyectos tales como las islas Zapatilla en Bocas del Toro, las Islas de las Perlas en el Golfo de Panamá y el Parque Nacional La Amistad de 200.000 ha de extensión en la frontera con Costa Rica. Además, se está considerando la conveniencia de incluir la Serranía de Majé en Bayano y la Serranía Filo de Tallo con la Laguna de Matusagratí en la Provincia del Darién.

Other proposed parks are La Libertad and the Upper Darien National Park, and there are other proposals such as Zapatilla Island in Bocas del Toro, the Pearl Islands in the Gulf of Panama, and the La Amistad National Park on the Costa Rican frontier.

Todo ello establece que Panamá está reservando poco más del 10% de su territorio a Parques Nacionales sin incluir la superficie que deben representar las reservas biológicas de La Tronosa, El Montuoso, las estribaciones de la cordillera central hacia el Atlántico y otras reservas que están en proyecto. Esto significa una importante contribución de Panamá al mantenimiento de las mejores condiciones ambientales de la biósfera.

A esto convendría añadir que la Dirección de Recursos Naturales Renovables suspendió la práctica, en febrero de 1980, del ilegal comercio de especies amenazadas de extinción y podemos asegurar que estamos cumpliendo con la Convención sobre el Comercio Internacional sobre Especies Amenazadas de Fauna y Flora Silvestre, firmada en Washington el 3 de marzo de 1973 y fué convertida en la Ley 14 del 28 de octubre de 1977 por la Asamblea Nacional de Representantes de Corregimientos.

La Campaña Nacional de Reforestación

En Panamá se han realizado investigaciones y observaciones en el campo de la reforestación durante los últimos quince años a través de 4.500 ha ya plantadas de *Pinus caribaea* y en 20 parcelas de ensayos y 50 sitios de observación en las más diversas condiciones ecológicas del país y con distintas especies forestales tanto nativas como foráneas.

Las especies más prometedoras hasta ahora encontradas son:

Pinus caribaea	pino
Tectona grandis	teca
Gmelina arborea	melina
Eucalyptus camaldulensis	eucalypto
Eucalyptus deglupta	eucalypto
Terminalia ivorensis	—
Bombacopsis quinatum	cedro espino
Albizia falcata	albizia
Cordia alliodora	laurel

Studies in Panama over the past 15 years suggest several species as most promising for reforestation.

Próximamente se publicarán los resultados de estas experiencias.

Correspondiendo al potencial forestal del país, establecidos en los Inventarios, uno de los pocos con que realmente cuenta Panamá, se hace necesario preparar las acciones tendientes a garantizar un futuro próspero que podría realizarse en gran medida con recursos internos. En el campo forestal estas acciones consistirían esencialmente de lo siguiente:

1. Preparar los planes de administración y manejo, teniendo en cuenta el principio del rendimiento sostenido de los principales bosques puros para someterlos a una sola administración.

TIPO FORESTAL	
a. Cativo (*Prioria copaifera*)	50.000 ha
b. Orey (*Campnosperma panamensis*)	50.000 ha
c. Manglar (varias especies)	600.000 ha

2. Diseñar una fuerte campaña de reforestación nacional con participación mayoritaria de la empresa privada para lo cual deben establecerse un conjunto de incentivos fiscales con sus respectivos controles por parte del gobierno nacional.

3. Fomentar el establecimiento de industrias forestales en las áreas de colonización nueva para lograr incorporar los recursos que hoy día se despilfarran.

In the field of forestry, the following actions are essential: 1) preparation of management plans to be placed under a single administration for sustained yield from the main pure forests; 2) design of a strong national reforestation campaign; 3) the establishment of forestry industries in areas of new colonization.

Para llevar a cabo la tarea de reforestar el país se cuenta con un área estatal de 4.500 ha con pinos (*Pinus caribaea*) en Veraguas y con planes y financiamiento para establecer 40.000 ha de plantaciones adicionales en los próximos cinco años.

La participación de la empresa privada en los planes de reforestación podría garantizar un nivel de generación de empleos e inversiones comparables y aún mayores que los conocidos en la industria de la construcción, con las ventajas adicionales de descentralizar inversiones hacia el campo en área subutilizadas que el gobierno señalaría previamente.

Esta campaña de reforestación es indispensable y urgente para aliviar y aún superar nuestro déficit energético, y permitiría sustituir buena parte del papel que importamos (30 millones de dólares anuales) y exportar productos manufacturados de madera, para lo cual existe un mercado creciente con demanda insatisfecha.

La medida esencial que determinaría el nivel de la inversión privada es el monto total de impuestos que el gobierno nacional estaría en condiciones de devolver a la empresa privada responsable, para que realice esta tarea a razón de $400.00/ha: uno de los costos más bajos de Latinoamérica.

Conclusiones

1. Es indispensable que el Gobierno Central reconozca públicamente una política nacional en el campo de los recursos naturales renovables y del ambiente y la utilice en el diseño de los planes de desarrollo para lo cual podría utilizarse como base la ponencia panameña en la XIII Conferencia Regional de FAO para América Latina que se realizó en Panamá entre el 12 y 23 de agosto de 1974, y en los Lineamientos de Política Forestal .

It is indispensible that the national government publicly recognize a national policy in the field of national resources and the environment and use it in the drawing up of development plans.

2. Es evidente la necesidad de que Panamá diseñe y establezca un sistema completo de investigación básica y aplicada para el desarrollo, en donde participen las distintas instituciones gubernamentales, empresas privada y corporaciones sin ánimo de lucro, bajo la orientación política de la Universidad de Panamá de manera que se recojan, se ordenen y clasifiquen en una sola institución todas las necesidades y los resultados de las investigaciones que en los di-

ferentes campos del conocimiento humano se realicen en nuestro país.

3. Los problemas más grandes de destrucción y deterioro de los recursos naturales renovables de Panamá tienen su causa profunda en las características sociales y económicas de la población especialmente rural, en el crecimiento demográfico y en la fragilidad de los ecosistemas naturales. Sus soluciones deben ser encaradas con la puesta en marcha de proyectos integrados de desarrollo y conservación de los recursos naturales renovables.

The greatest problems of destruction and deterioration of natural resources in Panama stem from social and economic characteristics of the growing population of rural peoples , and from the fragility of natural ecosystems.

4. A diferencia de los países desarrollados los problemas más serios que caracterizan a nuestro país en el campo de los recursos naturales renovables son: la devastación de bosques, la erosión de los suelos y la contaminación de las aguas. Todo ello, esencialmente, debido a la aplicación indiscriminada, tanto de técnicas primitivas, por parte de un importante sector de la población rural, así como por la aplicación de técnicas modernas en los diferentes campos del desarrollo, agudizándose la situación por el enfrentamiento entre lo tradicional y lo moderno.

Adames, Abdiel J. 1977. Evaluación Ambiental y Efectos del Proyecto Hidroeléctrico de Fortuna. Laboratorio Conmemorativo Gorgas. Informe Final. Revista Lotería 254–255–256. Abril-Mayo-Junio 1977.

Comisión de Reforma Agraria. 1964. Contrato 525–L-007–2B Panamá, República de Panamá.

Mathews, E. y *Guzmán*, L. 1958. Los Suelos y la Agricultura de Los Llanos de Coclé. SICAP. Ministerio de Agricultura, Comercio e Industrias, Panamá.

Ministerio de Agricultura. 1971. Informe del Proyecto de Inventariación y Demostraciones Forestales. FO:SF/PAN6, Informe Técnico 1. Panamá.

Ministerio de Planificación y Política Económica. 1976. Plan Nacional de Desarrollo 1976–1980. Objetivos, Políticas y Metas Globales y Regionales. Versión Preliminar. Vol. I.

———. 1978. Proyecto de Desarrollo Integrado de la Región Oriental de Panamá (Darién). (Con la cooperación técnica de la OEA.)

República de Panamá

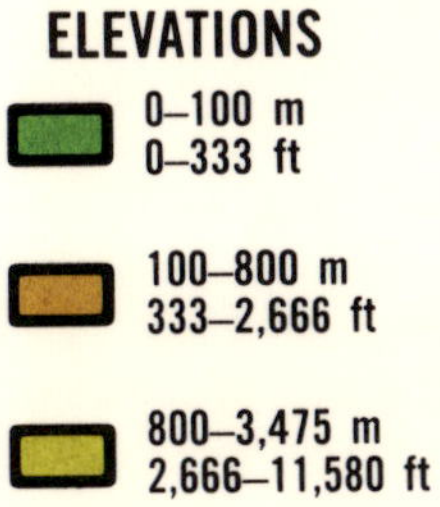

Republic of Panama

GOOD JAYNE! BAD JAYNE!

A Whimsical, bipolar trip to bedlam

Jayne Paynter Moomaw

RED APPLE PUBLISHING

FIRST EDITION

RED APPLE PUBLISHING
15010 113th St. KPN
Gig Harbor, WA 98329-5014
253-884-1450

Printed by Gorham Printing
Rochester, WA 98579

ISBN 1-880222-45-0

Library of Congress Control Number 2001118455

Cover design by Kathryn E. Campbell

Disclaimer:

Names, cities, hospitals, clinics, and descriptions of people have been changed in order to protect their privacy or to safeguard a person's character. The author and publisher assume no responsibility for errors, inaccuracies, omissions, or any inconsistencies. Events are colored by the author's viewpoint and how they affected her.

DEDICATION

To Todd, Chad, and Lisa,
my three children,
who survived my bedlam days
and thrived in spite of it all

PREFACE

It is my hope that this book will be helpful to bipolar and other mentally ill people and my prayer that it will enable mental-health workers to understand what goes on in the minds of those who are struggling. I feel I have given words to the feelings that race through the manic-depressed, bipolar person who is also troubled with personality disorders. The field of mental health has come a long way, but it still has a long way to go. Doctors and staff members with a real understanding of their patients are at a premium, and we need so many more talented people working in this difficult medical arena.

CONTENTS

CHAPTER ONE
CAN'T COPE

The fuzzy fog is creeping in and taking over my troubled, depressed mind. It actually feels good! After contemplating suicide for several years, I finally took the step to attempt it. I vowed I would never *try* self-murder—rather, I would *complete* the task. Being half-conscious in the emergency room was a sure sign I did not accomplish it.

GOOD JAYNE! BAD JAYNE!
Good Jayne, Bad Jayne,
How do you grow?
More depressed and numb
Headfirst in the snow.

Good Jayne, Bad Jayne,
What shall we do?
Send in the hearse?
And the body-bag crew?

"When did you take these pills, and how many did you take?"

"They are my pills, and I don't have to tell you," was my drunken reply to Doctor Jones.

I had meticulously hoarded the pills in a bean pot, sitting out in plain sight on the kitchen counter. I knew Dale, my frustrated husband, wouldn't dream of looking there for my cache.

Doctor Wilson, our family doctor, had given me so many different medications that I had quite a variety to choose from—my very own bean-pot pharmacy from a well-meaning doctor. He had one goal in mind and that was to make me feel better. But, the pills were not the answer—Valium, Elavil, Tofranil, Stellazine, Dalmane, Lithium Carbonate, Fiorinal #3, Midrin, Cafergot. You name it; I had it. I chose Valium and Fiorinal #3 for their fantastic ability to float me to nothingness. They were my trusty friends that helped to relieve the wretched migraines I had had, and I had gotten extra refills of them.

On one occasion Doctor Wilson asked me what medications I had at home. I told him I had my own pharmacy. He thought that was really cute since I usually joked with him when I was in the mood. Little did he know this time I wasn't joking.

BEAN-POT PHARMACY

Pretty pink
Blissful blue
Playful yellow
Wistful white
Groggy gray
They will all take me away.

Blissful beans, hop into my hungering mouth
And float me to oblivion
You lazy beans aren't doing any good just
Sitting there
I'm trying to keep in shape.

Guide me to the big pro training table
Help me keep in training for the big game
The event in which you help me go all the way
The grand finale
The end.

That selfish ol' bean pot has been hoarding
Those beautiful benign beans too long
Let's put them into action with our
Well-trained bodies being the catalyst
We will be pros
We will have a bean blast and determine the champion
Who will win the death cup and make it to
Never-Never Land?

I took those precious painkillers early in the afternoon and kept taking all 100 of them through the evening until Dale, my loving mate, realized what I was up to. That is when he dragged, carried, pushed, and probably even rolled, my petite 160 pounds to the car and on to General Hospital.

"Take her to ICU," barked Doctor Jones, Doctor Wilson's partner.

Going under in the blissful, fuzzy, fog-escape-release-freedom, I pleaded, "Just let me go quietly and softly. Let me be in control of my life and death. But, I really don't want to die; I just want to escape the devastating depression I have been in. Oh, heavenly peace is seeping in."

BLACK KNIGHT! BLACK KNIGHT!
Bad Jayne has taken charge
Oh no, how very sad.
Is there some way we can crush her?
Sorry, that would make Black Knight mad.

Black Knight, Black Knight,
Who is he?
He is the warrior of death
Always in readiness for you and me.

"Don't take my pants off. I haven't shaved my legs."

"That's a new one," responded the nurse. "They usually are worried about their underwear."

When attempting hara-kiri, remove the hair from your hairy legs if you are female. How could I have forgotten that? Guess I have had bigger, more crippling thoughts on my arthritic mind than leg-shaving, make-up, or any other type of self-improvement. Didn't think I would be alive to face the embarrassment of it. But, I did have my usually messy house in order. That was part of my game plan.

As they began to hook up the heart monitor, I stirred enough to make a feeble attempt to get out of there. "I want to go home!" I'm sure I sounded like a drunk, because I felt like one. "I didn't take that many pills."

"Get her in restraints. And, Jayne, lie down and be quiet. There are very sick people in here," exclaimed Doctor Jones. Doctors don't enjoy caring for a dingbat who is drunk from suicidal activities. That is when the precious fuzzy fog took over completely.

SOMEBODY GET HER A GUN

Good Jayne, Bad Jayne,
What about those you love?
Find someone to love them with
A warm caressing glove.

Good Jayne, Bad Jayne,
Where will you go?
To see the home of Black Knight
Because Bad Jayne is running the show.

Does that weird Bad Jayne
Really want to go?
Oh yes, definitely
She's such a hateful thing you know.

Good Jayne, Bad Jayne,
We want to wish you well
Here is a good idea
Send Bad Jayne to hell.

Good idea!
But we are both one
Good Jayne, Bad Jayne,
Somebody get her a gun.

The next evening I awakened from my drugged sleep, had the IV removed and the heart monitor disconnected, and was moved to a hospital bed near the nurses' station. I felt total dejection and embarrassment. I really didn't want to die. I just wanted to escape from my deep, dumpy depression that had been getting to the core of me. The past three years had been especially bad since they labeled me with a manic-depressive personality and put me on Lithium Carbonate and Tofranil. Like the little girl with a curl—when she is good she is very, very good, and when she is depressed she is horrid.

Dale is so angry with me. I'm in big trouble. Wonder what my three kids think? Are they aware of what I did? I had decided that Todd (10), Chad (7), Lisa (6), and Dale would be so much better off without me and the state I had been in the past three years. So, here I was, stuck in General Hospital again.

Hey! Isn't it great? I'm here at the people dump. It is such a delightful place for the sick to come to lie around. Whatever am I doing here? I should be at home being a wife and mother. But, for now, that is too difficult, for it seems so futile and uncertain. I think a more preferable place would be six feet under a watermelon vine. But, people are nice to be around, usually. Now I don't want to be near anybody. I need to go underground. Life goes on; the sun shines; the baby cries; the toilet flushes. That is because life must go on in its fast-paced circle. These days it is too fast for me.

Doctor Wilson, our good-looking, all-business, family doctor, came in and announced he was going to send me somewhere, but he didn't know where yet. I respect him a lot, and he has tried to shape me up in the past. Tonight though he didn't have much to say. Nor did I. But, I thought of him as our body mender.

BODY MENDER, PART I

Hey, mechanic of the liver, ear, throat and anus
What goes through your fastidious mind as you examine and probe
Those ailing deprived bodies laid out before you day after day?
At times these pleading groaning dollies must seem as though
They are passing by on an endless conveyor belt
That confluence of humanity begs for your skill and knowledge
And demands that you repair, restore, rejuvenate.

Think of your numerous tools and helpers, Body Mender,
The stethoscope searching for the rhythm of the body
The many lights you use to check each orifice
Of the human physique
Oh and those abominable white gloves
There are so many instruments and machines waiting to serve you
With your incessant demands.

Your many human helpers, the nurses, technicians, and secretaries
Are all so attentive so eager to respond to your
Slightest gesture or request
They are so like the well-trained dog wanting desperately to
Please his master
And you are so like the master—demanding, clicking your fingers,
Requiring immediate precise obedience.

Body Mender, I would like to share a diagnosis I have made in my
Probe of your active and complex personality
You have some deadly enemies that may ruin your effectiveness
And rapport with those broken dollies that endlessly
Lie, complaining, before you

These enemies could be even more deadly than
The dreaded cancer, drug overdose, or cardiac arrest
You frantically try to battle.

Well, surprise, surprise, it's nothing physical, it is
All mental and could exist in your own medical mind and the
Psyche of your colleagues
Arrogance, la de dah
The feeling of omnipotence, la de dah.
They could be your worst enemies, Body Mechanic.

Humbleness is so much more becoming to a man with your
Medical savvy
You are so vulnerable when prancing around like God himself
Beware, Body Mender, get your feet back on the ground
You are a mere man blessed with intelligence, a competent
Staff and the skill of repairing the broken pleading people
Who seek your ideas for a cure or relief
The ability to heal is not yours, Medicine Man
Nor do you pull the strings of life or death
On your ailing puppets.

God is in charge, my presumptuous fellow
When arrogance is in charge you are so vulnerable to the
Humiliating slap in the face
The great put-down
At times I have seen you so crushed and torn by the loss of a
patient
The boy entering the door of manhood that tried to blow his
head off
The boy just exiting puberty that lost his battle for life
Against a raging rebellious appendix.
Might your pseudo-power get you in trouble?
You do seem to be ubiquitous and you have earned your fancy

Clothes car and home
You are admired and respected by those who trudge before
You day after day
But you are still just a man.

My profound diagnosis is that you are a
Manic-depressive in disguise
There are times you pop into the cubicle
Where we await you in a definite manic state
Your mind is obviously clicking off cures behind you and
Ahead of you in the never ending cubicles of your life
It makes me want to scream "Stop and Listen."

You are a medical and real estate conglomerate magnate
Always scratching reaching and racing
Where are you racing to, Body Mender?
I have heard you say
I am a man of action
I want to be the best doctor possible
When do you slow down?
Who puts you back together and winds you up for your next flurry?
It must be your wife, the person who knows the real you,
The one who has seen you unwind from your powerful manic states.

Yes, Body Mender, you are just a man
Don't let arrogance rear its ugly head
And overwhelm you or even crush you
You have too much on your side
Be humble and grateful
For everything that comes your way,
Vagabond of the
Hospital and clinics.

Remember, physician of the physical,
Let God handle the big tasks
Healing sanity life and death
For omnipotence is not yours
Continue stepping to the steady
Medical rhythm that keeps you on top
Let humility competence and understanding
Be the governors to keep you
On the pathway of success and satisfaction
And quit throwing your pills at people.

Take time
Listen to what those sick dollies are trying to tell you
Listen to those on the phone
Don't let them sit forever on hold
Be perfect with a smile and a thank-you
Don't let your patient patients sit too long
In the sterile cubicle
Be humble, fierce physician,
Be a man.

As the cold January night approached, I came up with a fantastic plan. Since my clothes were next to me in a little plastic bag, I decided to leave the hospital. I very groggily put on everything but my shoes. Just then the young nurse on duty came in to check on me. I tried to tell her I just wanted to go out for awhile. But, she very professionally ordered me back to bed and took my clothes.

I didn't realize she had called Dale, but after a half hour passed, he came bursting in, very angry, wanting to know what was happening. The nurse had told him there was trouble.

"I've had it, Jayne. Either you 'shape up' or I'm leaving you. Doctors Wilson and Paul suggested I divorce you because of all the headaches you have caused me these past three years. Stay in this bed and don't cause any more trouble," ordered Dale.

All I could come up with was a pitiful, "I'm sorry."

"Sorry isn't enough. You've come to the end of your rope, and it's time to shape up!" His blue eyes were shooting fire, and his big fists were clenched in frustration and anger. But, I had seen this before. It was the Moomaw temper coming through.

"No more trouble, Jayne."

"Okay," I mumbled as I fidgeted with my hands, and my body shook (which was an after effect of the pills I had swallowed). I them recoiled into total resignation. Dale left the hospital with the same fury he had entered.

That's fine. Divorce me. Then, I would have some logical reason to create successful suicide. In the past I haven't known why I felt so depressed and suicidal. Loss of husband and three children would be a very valid reason to shorten my life.

Night had settled in and the hospital sounds filtered down the hall—bedpans clinking, people moaning, nurses chatting. But, sleep was easy to come by since I still had some of the effects of the pills in my system.

The next morning I was taken to get a shower and then to a regular hospital room with a roommate and television. I still felt very fuzzy and content to lie around.

My roommate's name was Hattie. She was a retired English teacher, in for platelet trouble. She seemed real nice, but I didn't tell her why I was there. I was too ashamed to announce my desire for immediate terminal life.

Since I'm a retired physical education teacher, at 37, we had something in common, so we chatted with one another when not watching TV.

Just as I was starting to get to know her, she developed a headache and died. Just like that. It only took about ten minutes. The nurses were running around trying to keep her going. Then they asked me to leave the room and go into another room while they waited for the doctor to get there. We had known each other just two nights and two days. Really didn't know that much about her—just that she was an older lady with platelet trouble. It certainly didn't seem bad enough for her to die. But, as God would have it, she died quickly, without too much suffering. But, death was certainly better than leaving her forever sickly and helpless. The stillness and sureness of death just reached down and plucked her away, such a mystery and reality that is with us.

Hattie left the rest of us behind. Death can't be all that bad. It is something definite and sure and is a state we will all enter sometime. So, I feel, Hattie, you fought your fight tonight. Some may think you lost, but I feel you won your fight. Hang in there, Hattie, now that you entered death's door. I pray your soul will go on forever and ever.

Her doctor was my same Dr. Wilson. He came to the other room and asked how I was doing, and I replied, "Fine." Just as it always does, life went on, even without her. After they wheeled Hattie out, they escorted a young girl with abdominal pain into my room. We didn't talk a lot.

During the next few days, Dale would usually come up in the evening to visit, watch TV, and subtly check to see if I was "shaping up." Surprise, surprise, I wasn't.

So, through the day I would sit in my solitude. But, the more I mediated, with a great deal of introspection, the gloomier everything became. I guess I really needed to release some of my unhealthy thoughts, even if in the process my feeble mind blows away.

But, I know I need to go through the motion of thinking and sorting out things, especially with doctors and when family tries to pull me through. But, realistically, solitude is a pain, for I often feel there is nothing to gain.

Dale again made it very clear he didn't want me back home until I "shaped up." That wasn't very encouraging, because I wasn't sure what "shaped up" was, since I had been out of shape for so long. A week passed as I prepared to rot away in General Hospital.

Doctor Wilson showed up in his pert, well-dressed doctor role, to let me know he had reservations for me at the famous Monroe Clinic. He thought they could help me get over this wretched depression. I wanted to go home and pack, since my stay was likely to be a long one. But, Doctor Wilson wouldn't trust me to do even that. My previous two psych ward, or mental health unit, experiences were for three-week periods. Somehow, I sensed this time would be longer. Doctor Wilson dismissed himself with an "I'll see you tomorrow."

BODY MENDER, PART II

So you feel my initial diagnosis was wrong
Well, Body Mechanic, at least I came close
But I guess I will leave the diagnosing to you.

How fantastic it must be to be a perpetual hypomanic
Endless drive and energy has got to be the ultimate
Of gifts to those who possess it
On the other hand the wife or co-workers must be
Terrifically understanding to put up with such
A dynamic whirlwind.

I committed one of the classic human errors
My diagnosis in Part I was based on my own experience
An egocentric judgment was formed
And I tried to fit you into my own mold
My assumption was that no one can be high without falling
Flat on their face after a certain peak is reached
So some personalities don't have to swing as the pendulum swings
Just can't imagine how that would be
But I covet such an immunity to depression.

Oh how I love my highs
Freedom
Accomplishment
Good times
Excess love and good feelings
It's like living in a world made just for me and those I love
I'm in control
No frustrations
No defeat
No death wishes clouding my mind.

Why does God let some people be forever high and yet
Demand that others take a dive into living Hell on occasion?
These dives do make a person humble
I would no doubt have unbearable pride
If I never had those wretched attacks of depression hit me.

Anyway back to you, physician of the physical,
Along with the manic label placed upon you
I warned you about arrogance and omnipotence
Your reply was: I'm humble, aren't I?
Yes I do feel you are humble.
Except the display of a humble psyche is not part of your being
The humility scene isn't characteristic of the doctor image
And definitely not indicative of you
Sorry
If you want humility to be yours
You will have to change.

Just feeling humble is only the beginning
The display of humility does not seem to be one of your gifts
Stop and take a took at yourself from a different view
That of the patient
So many of us have you on such a medical pedestal
Most of your patients and co-workers would like for you
To be proud and yet act humble.

We want you strong demanding indestructible assertive
The majority of us will not allow you to be
Plain common modest unassuming or meek
You must admit, Dr. Competence, none of the common
Labels are yours
Assigned humility cannot belong to masters of medicine
The script is not written that way.

Even so
I'm convinced you are humble caring and proficient
I owe my life to you
And I'm sure many others do
Although abundant life is the answer for me now
I realize this could change
And the no-suicide contract with you will be needed
I'm grateful I am bound to that contract with a
Person who is so positive and self-assured
You are such a dynamic individual
It is impossible to argue with you
For in your eyes you are never wrong.

Maintain that crisp medical beat to which you are stepping
It is very becoming and necessary
You are a well-advanced expert in your field
There are many doctors who could never begin
To step to your cadence, Body Mender
I have been to them
And they are so wrapped up in the typical
American medical play of arrogance self-indulgence golf
Country Club booze and pseudo-competence
That they can't relate
And half the time they don't give a damn
Stay the way you are (except quit pushing pills so hard)
And try to show some of that hidden humbleness you have
We are grateful you are the health commissioner
For our family.

The night before I was to leave, Dale brought Todd, Chad, and Lisa up to say good-bye. Todd wouldn't speak to me and seemed to be very angry. He remarked that he was missing one of his favorite TV programs. What must be going through his pre-adolescent mind? Probably resentment of a mother who was always going to the hospital, and frustration due to having a mother who was terribly moody and mostly down lately.

Chad and Lisa seemed to be unknowing and glad to see me. It was hard to talk to them and Dale, when I knew, only too well, my stupid actions were the cause of our expectant separation. But, we visited for a little while and kissed and bid each other good-bye. What a guilt trip I was taking for being responsible for being in this situation.

Come to think about it, there is nothing good to be found in guilt. It certainly doesn't help self-esteem emerge. Once you feel the guilt, it is so hard to release it, if from your childhood, guilt is part of your emotional inventory. Out of the feelings of guilt the seedlings of doubt, blame, and depression grow. I pray that I don't instill these in our children. I plan to sow seedlings of love, peace, and harmony so, hopefully, our children grow up to be guilt free. I guess if I committed suicide the guilt weeds would grow in abundance in my children.

Dale picked me up at six o'clock the following morning, and we migrated to the Monroe-style, sanity hotel. The ride was a long, quiet one. It should have been a time of problem solving and catharsis for both of us. But, things had gotten out of hand. At this point, communication was impossible. I spent my time doing important things, like inspecting the wires and trees heavy laden with ice and snow. It was a cloudy, then snowy, January morning full of dismay and apprehension.

Dale kept eyes on the road and seemingly his mind on his driving. My mind wandered—thinking again about the pitfalls of depression. There is such a totally empty feeling deep down inside, which is probably typical of those who are depressed. Why must this empty feeling be so constant and acute? It makes the sufferer want to go toward the door of death that Black Knight chaperones.

The empty exists. Even though physical organs remain intact, the empty shell feels stripped of any living or loving. It feels as if a large scoop has penetrated the inner depth of my soul and has carefully scraped all living tissue away.

Maybe I have been reduced to that same emptiness of a cone after a child drops the ice cream.

The doctors and others seem to look for answers, but you get the feeling no one really knows. I just hope I can reach down into my dry, empty self before my mind goes into oblivion.

It just doesn't make sense. I have a good husband whom I love and who loves me, or he would have left me by now. I also have three beautiful children, a nice home, no real big problems, and everything to live for, but I still fall into this black pit of depression. I never was concerned about my high hypomanic times. They had always felt good. Endless energy, incessant talk, increased sexual activity, creativity, devout spirituality, no need for food or sleep. I had had these high and low times since high school days, with normal periods in between. Dale liked the heightened sexual activity during the highs and the fact that the house was always extra clean then, but he didn't particularly appreciate the rest. But, now the lows were getting lower and longer, and the highs were getting a little more wild and action packed.

Three years ago, when Doctor Wilson diagnosed my disorder as manic-depressive illness and started me on Tofranil and Lithium Carbonate, they were supposed to make me feel better but I kept getting worse and more depressed. So, he put me in General Hospital for two weeks for depression, saying I needed a change to lift me out of it. Two months later, in the spring of 1976, he sent me to a psychiatrist in Mitchellville. He took down my history and stuck me in his psych unit for three weeks to observe and then to prescribe the same medication Doctor Wilson had prescribed. Occupational therapy involved painting plaques every day. What a drag it was.

Therapy sessions with Doctor Tarinadski were quite an experience though. He would begin each morning selecting choice patients for therapy. He usually called on me every other morning. Sometimes I looked forward to it and other times I dreaded it. Doctor Tarinadski was put in a concentration camp during his late teens, and he would sometimes bring it up to make a point. Each session would begin with, "Tell me, Jayne, what is on your mind today?" with his thick accent. You had to be on your toes to understand him. Then, he would get out his pipe and puff away as I would respond with some event or feeling I had at the time. So often I would bring up some childhood experience. One day he asked me who was the favorite child in our family. Well, being the youngest of five children, I told him I was. Everybody thought little Jayne Ann was cute.

My father died when I was four years old, and I couldn't remember too much about him. He had been sick and hospitalized several years before he died. I remembered the funeral very vividly. I recall sitting next to my mother saying over and over that I was sorry Daddy had died but I wasn't going to cry. Also, I remembered vividly getting spanked with a willow switch by Mother, and saying, "That hurt, but I'm not going to cry." So, at the age of four I was working hard not to show feeling. Whenever I did actually cry, I hid in a closet trying to hold it all in. I guess I have done that all my life, except now, instead of hiding in a closet, I would just withdraw from everyone. You could call it emotional retardation, using the same technique I had used as a child but just changing it a bit. Feelings were getting so bad that I just couldn't hide them anymore.

One of the many things about the depression that bothered me and especially irritated Dale was my decreasing desire for making love. Mom had always stressed how wonderful sex was in marriage. So, Dale and I waited until we were married before we had intercourse. It drove him crazy on many different occasions but we waited. And, it was worth the wait. We had a very compatible and rewarding love life when I wasn't depressed. We felt that being able to express love through sex was truly a gift of God, especially love sanctioned by marriage, with Christ at the center.

I went to Doctor Tarinadski for a year and then decided it was too expensive and too far to drive for therapy. Sometimes we didn't get along too well.

MIND HEALER, PART I

There you are sitting on your turquoise spinning tower
While your mentally crippled clientele ponder the question
Tell me, what is on your mind today?
Which brings a regurgitation of hurts
Feelings disappointments plans
As you, the Mind Healer, suck platitudes from your pacifying
pipe.
Then you so expertly exhale the stench of repetition in our face.

Yes I know of your horrible experience years ago
I'm sorry it happened
But it has nothing to do with me
Except it seems to be your secret weapon
In the battle of the minds
Yes you did suffer, Mind Healer,
Was it really the ultimate of hurts as you infer?
No one can experience the degree of another person's hurt or joy.

In the past I wanted to please you
Now the rebel in me is taking over
I want to release these innermost thoughts and feelings.

Let me lay them out in front of you
Don' t stifle me with more platitudes
Don't look back
Look forward
Let the mind rule and not the emotions
Stop!
Permit me to be myself and be open
I respect you, doctor of the psyche,
But you say you can t help
So you prefer to ignore these very real problems of mine
That's all right
Your egotistical manner has angered me
Your sarcastic condescending aroma is permeating the room.

Just release me
Hopefully we can compromise
And separate as friends
So march, Mind Healer,
Don't be disgusted with me
Because I'm disgusted with myself

Happy healing to you till we meet again
Hope you enjoyed this much-needed catharsis of mine.

So, I went back to Doctor Wilson for the medication and had psychotherapy at the Ross County Mental Health Center with Doctor Davis, who had a Ph.D. in psychology. Meetings weren't as colorful or beneficial as with Doctor Tarinadski, but it was a form of therapy which I needed. However, I started getting deeply depressed and suicidal. I told Doctor Wilson I didn't went to go back to Mitchellville because I didn't want to paint any more wretched plaques, although therapy with Doctor Tarinadski had been very good for me.

The death wish kept coming back to me. On several occasions when it would get too great, I would call or go see Doctor Wilson, and he would throw me in General Hospital for a few days. The fall of 1977 became really bad, so he sent me to the University Hospital psych unit for three weeks for diagnosis and treatment. They came up with the same manic-depressive illness label, kept me on Lithium and Tofranil, but didn't have any therapy. Dale and I thought maybe the Lithium was keeping me in a depression, but they replied that it wasn't. I had a reaction to the Lithium and was very sick for a few days. It was a bad experience in that loony bin, and my resident doctor didn't give me much support. He even asked me to please not commit suicide until after he finished his term in the psych unit! He advised me to return to Doctor Wilson and Dr. Davis for therapy.

Looking back, I could see the times in my life I had been high and the low times. And, I could see that Mother had high and low periods in her life that were extraordinary. I had heard that manic-depressive illness could be inherited. Or, could it be learned behavior?

High school was a high time. Incessant activity and the drive to be the best in everything—band, softball, student council. I earned All-Conference in basketball and Division I ratings in trumpet and voice solos and groups. You name it, I did it, trying to win Mother's praise and approval and acceptance. But, the spring of my senior year I became slightly depressed. I withdrew from my friends and felt very blue. That was the first low I was aware of. Several more lows followed me through college, and they were always after a high active time, with normal periods sprinkled in.

Honeymoon summer was followed by a depressing fall during our first year of marriage in 1964. Since I started missing work to stay home and cry, I decided to go to a doctor and see about it. He told me I needed to have a baby. Well, we didn't take his advice, and I continued teaching girls' physical education.

Then, after three years of teaching and marriage, we had a baby boy, Todd. I taught a couple more years and had Chad and was normal for that period of time. After Lisa was born, just fourteen months after Chad, we moved to a different town and lived on an acreage. It was great for the kids, but I experienced a pronounced suicidal depression. I wasn't getting out much and missed teaching, and I felt overwhelmed with three pre-schoolers. It lasted about a year and was miserable for all of us. I didn't know anyone in the town of about 400, and I felt all-alone at first.

MUDVILLE, A NICE PLACE TO LIVE

God, you led us to Mudville
To your beautiful hills above the river
But the coolness of some of these people
Makes my lonely heart shiver
So often have I reached out
Only to be cut off
Why can't these people mingle
Rather than separate and scoff?

Thank you, God, that I grew up in a town
That seemed to draw no line
How I long for that acceptance
Rather than friendships so benign
I also thank you, God, for this loneliness
For it brought me back to you
But I pray you will keep me searching for heart friends anew.

The sign at the edge of town says

"Mudville, a Nice Place to Live"
But it seems to me that a town should have so much more to give
I need your help, God, to ease this bitterness and hurt
Please prevent me from withdrawing and treating others so curt.

Bring us together
We who need each other so much
Help us to love mingle reach out and touch
I pray that you would help us
To make Mudville a better place to live
Because we all need love and have much love to give.

The summer after Mom died, in April of 1975, I went to the peak of the roller coaster—super active as the hyped-up manager of a women's softball team, outgoing, boisterous, very spiritual, definitely in high gear. It seemed strange at the time that I wasn't depressed about Mom's death, but that came later.

Mom and I had always been very close, since my brothers and sisters were all quite a bit older. I was the last one to leave the nest, so she clung very close to me and didn't want me to marry Dale. We enjoyed each other and could talk about anything.

Being a Christian, I had accepted Mom's death intellectually, reasoning that with heart trouble and impending hip surgery she was spared these physical traumas through death. I believe there are lots of things worse than death, but I missed her and felt sorry for myself about her absence. Mom was ready to die, for she had accepted Jesus Christ as her Savior and really never did talk as if she feared death. She emphasized that death, for Christians, was not a sad event. I grew up believing this—that when Christians die they are not doomed. Rather, they meet Jesus and account for the way they have lived, and their souls live on forever in the perfect environment of Heaven. One of my first poems was written in 1973 and was read at my mother's funeral, and I want it read at mine.

DON'T CRY FOR ME

Please don't cry for me when I die
For I pray that I will be with God on high.
And it won't be that I was so well behaved
But by Christ's blood that I was saved.

Thank God for making me whole
For giving his only son Jesus to save my soul
This salvation is for you and me
And to think it is given without fee.

But we will lose so much
If we accept Christ merely for salvation
We must bear fruit and give His word to the Nation
Go out and witness and show God's love
Let His light shine through
Share His peace from above.

Find comfort
Knowing God's unselfish love is real
Have faith, give love
Is His Fatherly appeal
When God says
"Come, your time is nigh"
Just as I am should be our reply.

So please, rejoice!
Don't cry for me.

In November, after Mom's death the previous April, I hit rock bottom. That is when I went to Doctor Wilson for help, and he actually gave it a name and started treating me for the depression.

Doctor Tarinadski revealed that when you get high you are actually running away from, and ignoring, an existing depression, until you just finally wear yourself down, and that is exactly what I did.

All of the psychotherapy, Lithium, Tofranil, and trips to the different hospitals had gotten me nowhere. It's no wonder Dale and the kids had had it! So, that is where I am now, heading for another shrink and a different hospital, after this three-year siege with depression.

As we pulled into the parking area of the huge St. Luke Hospital, we couldn't get over the size of the place. It was overwhelming. It was a famous place, and I was starting to see why. Wonder where they keep the twinkies (my new name for the mentally ill)? Dale admitted me. I signed myself in and didn't have to be committed. Then, they sent us to the West locked ward on fourth floor. My bed and room were waiting for me.

CHAPTER TWO
HELLO, BEDLAM

Here I am, folks! My yo-yo personality has escorted me to the granddaddy of them all, Monroe Clinic. Oh, God, I know there must be some purpose for being the lead dingbat in this psychodrama. Is this really my ultimate opportunity to climb, scratch, and claw my way out of this wretched pit called "Depression"? *Good Jayne! Bad Jayne!*—now playing Lead Loony at the famous Fourth Floor Monroe Theater, also known as St. Luke's Hospital.

Gina, one of the many "usherettes," had just skillfully lead Dale and me to my "dressing room." Little did I realize she would prove to be the most compassionate and understanding employee in this crazy retreat for washed-out castaways. The star's "dressing room," with no star on the door, had only a little peephole in the center. What an ugly, morbid cubicle to call home for four to twelve weeks. Dale was releasing feeble remarks about what a nice room it was—with a bed, chair and desk, and better yet, a bathroom and closet all my own.

My mental reply to his attempts to comfort me were, "Yes, dear, it will be just wonderful in this rotten psych ward reeking with the stench of insanity." This "delightful" room was going to be my third closed-psych-ward experience, and I lost track of the number of times my depressed, suicidal moods sent me to the General Hospital at home. I had such a clear view of what fun little activities were awaiting me in this super psych situation.

This drab room, number 4-402, had no curtains but a heavy mesh screen on the *inside* of the window. The switch for the light was out in the hallway of this fourth-floor loony bin—a sterile, protective cubicle designed to preserve the most insane for futurity.

My biggest desire at this moment was to get Dale out of here so I could be alone and start attempting to comply to his desperate command: "SHAPE UP!"

Okay, my foot! Just get out of here so I can concentrate on methods and materials for a successful suicide, taking the soft, quiet route. Which do I really want? Shape-up or suicide? I didn't know for sure. As he left, we gave each other a quick little kiss, resembling that of a junior high couple on their first date. Dale then informed me, "Jayne, I want to talk to your doctor before I leave." Yes, dear, talk to my doctor for me since I am just a discarded rotten vegetable in this garbage heap of spoiled minds.

"I'll call tomorrow night and see how things are going."

"Okay," I responded as he walked out the door and down the long hallway that I knew would become all too familiar. The reality of being imprisoned in this strange place with strange people, as the one I love walks out that locked door, was overwhelming. The door separates the mature sane and the immature insane. Sometimes it is difficult to tell one from the other, although the locked door definitely confirms which team you're on.

It would be so great if this place could provide some solutions or helpful hints that would cure this hopeless depression and stop the lingering obsession I have for self-destruction.

SANE OR INSANE?

Sane or Insane
Which is best?
One is better
The other needs a rest.

Sane or Insane
What is new?
Tofranil Lithium Electroshock too.

Where is Insane?
Stuck in the jug?
That is where she belongs
But she really needs a big hug.

What happened to Sane?
She just ran away
Why doesn't she stay and help?
She'll wait for a better day.

Sane or Insane
Take a long ride
That sounds good
But I would rather go and hide.

Sane or Insane
Good luck.

My first meal was miserable. Who is the craziest in the room? The forks and spoons danced and clicked on the plates like screeching fingers on a chalkboard. No expression on anyone's face . . . all just there. Such a tiny, ugly room to pile the crazies in to partake the hospital goodies. Where have all the sane ones gone? The staff even seems crazy from associating with such spooky clientele. Aimless, dull eyes, so typical of depression, keep wandering across the room with everyone's stare going nowhere. The motion of the forks so methodically moving from plate to mouth. Guess loonies have to eat too. But, maybe I can lose some weight while I'm here! I dawdled with my food for a while and then went directly back to the star's dressing room, 4-402. I have to start shaping up, so I will sit here and plan my next act.

Why do we want to do it? Why do suicide prospects not seem to care? Is it because we are uncaring, selfish, hateful, nasty people? No, it is because we do care, too much, and are thinking that those we really love shouldn't be hurt or ache over and over; one final separation or break might be a solution to that.

People get so angry when they hear about suicide plans. There seems to be contempt because of their love and care; they only fail to understand.

Contemplations . . . what might some of the thoughts be? Everything seems so very hopeless and stagnate forever. Total misery has invaded my whole being. Suicide is a permanent end to all tribulation. Death may possibly be a friend, and it is definitely a reality. What is it all about? Why?

I really feel God knows and understands where each of us is. And, although He isn't pleased with suicide, He does decide to honor that person's desire to die or insist that he remain on earth to work out problems. So, it is our will to take our own lives but God's will whether we live or die. Does God make allowances for mental infirmities as He does for physical ailments? Or, are mentally ill people who can't cope persecuted forever in Hell because of their insanity?

Suicide, another one of the paradoxes in my life. It makes death seem like such a sweet release and such a friend; a definite promise that earthly troubles will end. Yet, it is such a stranger and will take me from those I love. Contemplations of suicide are really not so rare, and they are mostly shared by people who are sensitive and really do care.

Just as I was getting comfortable on my bed at St. Luke's, Doctor Nimrod came in to interview me and get some of my history. He introduced himself as my new Monroe Clinic psychiatrist. I wanted to just lie on my bed and ignore him. But, if I am going to SHAPE UP, I suppose I'd better pay attention and get started. He looked like Mister Casper Milquetoast with his wire-rimmed glasses and red hair, short and trim. Hopefully, he will be my ticket to sanity.

Same old psychiatric questions: "What day is it? Repeat these numbers after me. A rolling stone gathers no moss. What does that mean? Name the last five presidents, starting with the present one." I didn't know what day it was, couldn't concentrate long enough to figure out the saying, but I could repeat all the numbers and name all the presidents. Maybe I'm sane after all. Just bug off and let me get back to bed and insanity. Come and see me when I'm not so foggy, for the fuzzy fog was rolling in.

He talked awhile longer and gave me my physical to see if my body was okay or if it is as screwed up as my mind. The body is a mighty bit fat but otherwise fine. He finally dismissed himself with an "I'll see you tomorrow" promise.

Alone at last. "Hello, bed, my new predictable friend." Always waiting to comfort me. Always silent acceptance. I'll just lie here and soak in the insanity of it all.

After about an hour, Gina came in to check all the items in my suitcase and get all harmful objects segregated: compact, tweezers, all glass objects, razor, etc. "These items are to be kept in a locker for your own protection and the safety of others. Whenever you want anything from the locker, contact one of the staff and they will unlock it and give it to you. When you are finished with it, the staff person will lock it up again. You also need to hand your money to the ward clerk for safekeeping." Just a polite way of saying we don't trust you with any of this stuff, so we're locking it up. Gina sees all right though. I think we will get along.

After getting things put away, I decided it was time to check out the place. Gina took me around and introduced me to the surroundings and some of the people.

"This is May," smiled Gina.

"Hello, my name is Jayne."

Oh boy, do I remember May from lunch. She is the cute, older lady who stuck a big piece of greasy fried chicken in her purse.

"Hello, honey. Can you get me out of here? I have money for you if you can take me home."

"Now, May, we want you to stay here with us," warned Gina with a smile.

I could tell immediately that May would be a fun, interesting individual. Age was playing tricks on her mind and making her quite a colorful person.

"Well, I have to pack my things if we're going," stated May as she wandered on down the hall.

Gina showed me the TV room and the shower and bathroom, which was locked to prevent midday drownings in the tub, although Gina didn't say why the room was locked—that was my idea.

Next was the nurses' station, which you couldn't miss. It was out in the middle of everything. They had quite a vantage point to spy from. The recreation room was right across the hall. Everybody was seemingly busy doing some little project: making candles, working on leather crafts, painting, or working with beads. These were just a few of the many activities available to those who had the stamina left to handle them. Guess I've come to the country club of psych wards. I wasn't in the mood for arts and crafts, but I smiled, thanked Gina for the tour, and then sat down to listen to the stereo in the corner.

At that moment a retarded, black-haired girl sat on my lap and called me a frying pan as she pinched me on the right boob. That is when I knew loony time was here.

Gina grabbed her by the arm. "Come on, Erin, let's sit over here, not on Jayne."

Gina later revealed that Erin was eighteen and put in the hospital for a complete hysterectomy. They couldn't handle her on the surgery floor so they sent her to 4-West. (I didn't think they could sterilize a mentally retarded individual anymore.)

A balding young fellow in glasses and with a deformed hand sat next to me and introduced himself as Jake Collins. He, at least, seemed normal. My first impulse was to ask him what he was in for, but I decided to refrain and just smiled and introduced myself. He told me that his room was next to mine.

I was ready to retreat to my room and become more acquainted with my bed.

"Nice meeting you, Jake. I think I'll look around some more."

But, Jake came with me as I left the rec room. He then introduced me to Paige. She was in her early twenties and had big expressive brown eyes. She was playing a guitar and singing in the lunchroom, which appeared to be a gathering place for informal get-togethers.

I wasn't in the mood for mixing so I went back to my room. Paige and Jake seemed like nice people with whom I could enjoy being witty, but not now. I had to get away and absorb what I had seen and experienced. My bed looked very inviting, but just as I stretched out and closed my eyes, Gina came with a test booklet and answer sheet. "This is the Minnesota Multi-phasic Inventory, Jayne. Work on it whenever you feel like it, and turn it in when you are finished."

"Thank you, Gina. I'll just work on it now."

It was a time-consuming test that was supposed to be an indication of what kind of a person you are. It took several days to complete it since I just worked on it for short periods of time. My concentration wasn't too great, but it was something to do.

The first night in the loony bin was a long one. I listened to the radio all night. Doctor Tarinadski would never let us have a radio in our room at Mitchellville, or even lie around in our room, except at bedtime, so this was a new treat. I stayed in my room the entire evening and night, for I didn't feel like getting involved with anybody in the loony bin just yet.

Time passed without too much excitement. I was slowly reaching out to others. I stayed away from Erin since I didn't enjoy being called a frying pan. Jake and Paige were people I could mingle with, even though Jake was terribly paranoid. He thought the Monroe staff was planning a conspiracy against him and the medication he was taking would harm him in some way.

I always enjoyed May's company, for she was always preparing to go somewhere with somebody. It was fun to fantasize with her.

The psychologist came in on the third morning and wanted to know if I was ready for some testing. Doctor Nimrod had told me she would be coming. She had me sit on the bed and look at various designs, some of color and some black and white. "Tell me everything on your mind as you look at these." I saw plenty, and told about the two dead, bloody chickens hanging upside down, and the monster of ridicule, butterflies of death. They were all there, and more. The results confirmed I was very emotional and suffering deep depression. I could have told them that without the Rorschach. The psychologist thanked me and excused herself.

Doctor Nimrod had been encouraging me to get off my bed and out of the room to take part in activities in the rec room. So, I decided to "motate" and see what was waiting for me. It happened to be a pair of unassembled moccasins. Ned, one of the recreational therapists, showed me how to lace the moccasins together. So, I took the leather pieces and a string over by the stereo and laid the mess in my lap and sat there trying to act sane and busy. It was a tough act to carry out, so I gave up after lacing three stitches. I gave everything back to Ned, and he put it in a special spot with my name on it for a time when I would be a little more handy, creative, and sane. My big desire was to get away from people.

Every time I saw Erin she called me and everyone else a frying pan. I tried to be patient with her, but it was difficult when she would follow me around and say such charming things as *"you fucker hole, shit hole, shit hole."* I suppose some ornery boys in her past had taught her the spicy vocabulary she had.

What is reality?

We would be better off if we could withdraw to our individual thoughts, and if we feel the need, to kill these realities.

Putting mentally ill people together in the same setting is a risky procedure. These people are struggling to get in touch with reality and their true, sane feelings. This is difficult when you are bombarded with insane thoughts and babblings. After so long, it becomes easiest to give in to insanity and to stop struggling for reality.

Right now the loud music in this music room does a fabulous job of drowning out the insanity. The creamy white of the walls brings a pureness and serenity to thoughts. Reality does not have to prevail in this room; it is sufficient to merely exist. That is comforting, and there is hope in merely existing without struggle—just listening to music being the only *criteria* or *must* for existence.

There is something very depressing about the repetitious, unrelenting groanings and complaints of people in a mental ward. They go on without ceasing, a constant barrage of meaningless words expressing boredom and discontent. Erin has intruded and brought her insanity to invade the peace and sanity of this room. "You are a frying pan" has become a reality to me. I am a frying pan. This has passed from insanity to reality. Does something become real if it is repeated often enough? I know it shouldn't be, but in a mental ward it can. I resent her being here in my place of retreat. I would probably resent anyone being here. Such a disappointment. Oh, relief! She has left and once again the music room is becoming sane, where I can be alone with my thoughts and they can be reality or totally insane—it makes no difference here.

I'll go to my room now and wait for Nimrod to make his rounds. I didn't want to disappoint him and not be in bed. Sure enough, in he comes to sit on the chair with his legs crossed, poised for picking my brain. Damn, I wish he wouldn't sit there and yawn. Makes me feel I am terribly boring. He did have good news though. They have another room available, one with curtains and a door without a peephole. I'm really moving up in the world!

He also told me my staffing would be tomorrow. It would be a time when I was asked questions by his superior, a real full-fledged shrink, and all the other residents and nurses would be there. Then, I would move into room 4-438 after the staffing.

Well, tomorrow came, and I wasn't in a mood to sit and have a bunch of people pick my brain. Doctor Preston, the head headshrink, behaved and talked like an arrogant snob. He asked me how it was in the mental health unit, and I declared, "Just wonderful," with sarcasm dripping from it.

"And, what does that mean?" he asked.

"Are you too dense to know what *wonderful* means? I have been here for five days without too much happening and everything is just wonderful."

He asked a few more questions, to which I responded in an unbecoming manner, and then he let the rest of the staff have their turn to ask the dingbat questions. I was happy to see Nimrod's familiar face there and Gina in her professional nurse's role. But, the rest I didn't care about, and I had no reservations about letting them know it. I really acted like a brat, but I didn't care. I just wasn't in a mood to be examined and tested by any type of authority figures. Least of all did I want to act "shaped up." Rebellion was boiling inside me, but they did give me a new label for my yo-yo personality. They called it Bipolar Affective Disorder. I told them I would have preferred *Libby's, Libby's Libby's*, but they didn't think the medical world would understand. I rather like it myself. I felt like they were patronizing me in their condescending way.

Doctor Preston told my mousy social worker that as moody as I was they had better involve my husband in counseling also. I guess she is my connection with my family while here. But, why would he say that in front of me and everybody? I suppose he figures that as insufferable as I was that nothing would bother me. I really felt like the "low person on the totem pole" in that room. Doctor Preston's presence didn't help at all.

After they were through with the inquisition, they dismissed me, and then I suppose they talked about techniques and procedures that would be used to "shape me up." Or, maybe they just talked about how terrible I was.

Then came moving time. Room 4-438, here I come!

CHAPTER THREE
DON'T EMBARRASS THE CLINIC

I was getting used to the hospital routine. It went something like this. First thing in the morning you could hear the nurses and aides wheeling the blood pressure machines down the hall to take blood pressures of all patients taking medication (like Tofranil) that caused the blood pressure to drop. Then, if it was the right day, the "blood man" would come in and take a sample of blood from the arm for a lithane level, to make sure you didn't have too much lithium in your system.

Next, it was time to wash up or shower and go down to breakfast. After breakfast, we went to the recreation room for crafts. Then, if you were in group therapy, you would go to the morning session or wait and go after lunch. Lunch, more group therapy, recreation therapy again, and for the lucky ones who could leave the place, swimming or some sports activity away from the unit. Just think, they would actually let us out for awhile for activity. I hadn't earned that privilege yet. Supper was next, and then the long evening of watching TV or singing with Paige and her guitar in the lunchroom or just talking to other patients. There were visiting hours in the evening and afternoon also, but since I lived so far away, I didn't have many visitors. A snack time and bedtime rounded out a day in the loony bin.

Now I had a room with curtains, which did seem more homey. And, I had a roommate. She was a black lady who was super paranoid. Her name was Wanda, and I've never seen anyone so terrified as she was. She thought they (whoever that was) were going to get her. She kept saying to me, "You won't let them get me, will you?" I assured her they wouldn't.

It's amazing how sane she made me feel. Gina disclosed that Wanda ends up in the psych unit about twice a year when she doesn't take her medication as she should. I now knew what true paranoia was. She didn't want me to go to sleep that night, so she kept wiggling my hand to make sure I stayed awake. Needless to say, it was a long night.

Nimrod came in the next morning to let me know it was time to join a group for group therapy. He thought that would be good for me. The first day in group I didn't say much. I was very content to let the rest of them spill their guts out. And, that is what a couple of them did with Nimrod and Bonnie, a therapist in the craft room and co-leader of our group. She and Nimrod did an excellent job of tearing down facades and getting to the root of loonies' problems.

Mary told about her lousy relationship with her mother. Mary explained that she was reading the book *My Mother, Myself*, which was giving her a new outlook. The others told of their bouts with depression. There were seven of us, and every one of us was depressed. What a lively, fun group we were.

During my next therapy session with Nimrod, I asked when they were going to let me leave the unit and get some physical activity. He responded with "We don't want you to embarrass the clinic."

Attempting suicide or some other dumb stunt, I suppose. I couldn't help it but I just laughed, for it struck me as very funny. He didn't see the humor in it, but I thought his remark was hilarious. "Embarrass the clinic?" Nimrod also informed me he was getting tired of walking into my room and finding me lying on the bed instead of mixing with the other patients and taking part in activities. I decided to take his advice and get out a little more.

Paige and Jake had become interesting friends to have. I enjoyed singing with Paige on occasion and talking with her. She was from Philadelphia and had followed her psychiatrist from there to Monroe Clinic. She had been seeing him three days a week and was very dependent on him. She had gotten suicidal, so he put her in the hospital. She was fun to be with. I had played Ping-Pong with her and giggled at everything she said and did. She was such a vibrant individual I couldn't believe she was depressed. Of course, she probably didn't think I was depressed either. We brought the best out of each other.

And, poor Jake kept talking about conspiracies plotted against him. He thought the clinic was destroying his brain and his future with the medications they were giving him. Who knows; maybe they were.

Then, there was May—sweet and unknowingly ornery. She always had a bundle of clothes under her arm ready to move out with anyone who would help her escape. She just wanted to get home. Yesterday she fired one of the nurses for not keeping the fireplace clean. She did it with such great authority too. Everyone enjoyed her and just took her as she was.

Erin was still there with her repugnant language and constant motion and jabbering. I truly still felt like a frying pan whenever I was around her. I couldn't help but have some compassion for her though. A couple of times I sat with her and helped her with her ABCs and count to ten like a little four-year-old, and then she would take off down the hall following someone, calling them a frying pan. One day when I had my clothes in the Maytag she took them out, soapy water and all, and laid them all over the laundry room. Talk about frustration!

Gertrude, a new patient, had just come in. Her memory was totally gone. The nurses put up signs with arrows saying "Gertrude's room," and the signs led the way for her to find her room. I guess it worked, but loony time really came when Erin would tear them all down and then Gertrude would get lost. Then, the frustrated nurses would verbally fight with Erin, trying to get her to leave the signs alone. Oh, the unbelievable reality of it all.

And, I mustn't forget Herman, the old fellow who was totally disoriented. They usually strapped him in a chair so he wouldn't hurt himself. Well, he called the chair his old Ford, and he would scoot it around and remember when he used to haul coal. Once, I was sitting at the dining room table and Herman scooted up beside me. "This is a nice restaurant. What are you going to order?" I tried to explain to him that this was a hospital, but he didn't believe me.

So goes the life in a psych unit. Never a dull moment—just dull hours, days, and weeks.

Dale had called a couple of times to check in and let the kids say, "Hi, and I love you." I would get terribly homesick and all the more depressed after our phone conversations. They had one phone we could use in a little booth. I'm sure many tears had been shed in that little cubbyhole. I really missed Dale and the kids. I especially wondered how Todd was doing.

TODD

Say, young explorer,
What is filling your mind?
Always scheming probing searching
New discoveries every day
You amaze me.

I'm so glad the establishment hasn't
Sucked away your insatiable curiosity
Your search for reality.

Your eager knowledgeable mind seems to betray
A mere eleven years of life.
But your youthful body confirms that pre-adolescent struggle
Reaching, growing, stretching for manhood
The day you can strike out and
Conquer the world and those in it.

How can I describe you?
Todd, a child of God, an answer to prayer, son of Dale and Jayne,
First-born, leader, thinker, history buff, hunter, bike rider,
Drummer, rocket builder, collector of stamps,
Fire trucks, airplanes, and trivia
Which one of these will you pursue?
What talents will a guy with the nickname of Mooey develop?

Keep up the battle
Hard work and diligence will enable you to be a champion and a winner
So go ahead and prepare for the conflict
But the victory comes from God
For he who loves the wisdom of God loves his own best interest
And will be a success.

We hope while raising you we have provided an anchor for you
Something to hold onto in times of stress, a belief in God
Also we, your authors of life, know you have to sprout wings
So you will be able to fly to new horizons while maintaining an anchor
Whether they will be wings like those of a powerful jet
Or those of a glider in free flight depends on your desire
Either will be fine.

So step on, Todd,
March to the sharp cadence of your own dedicated drumbeat
Which can be synchronized with the beat of God's love
Tune in to life
Soar to the heights of your own choosing
We are with you
But we will also let you go.

The lunchroom had become the place to gather for sitting and talking about almost everything. There was one patient who reminded me so much of one of my nephews, so I told him to call me Aunt Jayne. He was six feet, six inches tall, had blond hair, and wore glasses. He was twenty years old and was a fellow manic-depressive. He and Paige and Jake and I had spent several nights sitting at the dining room table talking about troubles and plans. Gina often joined us. Most of the other female nurses and aides wouldn't, except for some of the young men there as orderlies. Pat, my newly adopted nephew, was very interesting to talk to. He was a college student, into drugs and booze, and had come to Monroe to get straightened out. His father was a gynecologist, and his mother a family doctor. Pat was very creative and clever, always tacking names on people who were ridiculously appropriate.

There were the Syrup Sisters, as he labeled them. All three of them were model patients—never doing or saying anything wrong. Sickeningly sweet. Two were depressed. One, a suicidal college girl named Kim, and Jeanie, a depressed mother of two. The third was Carla, a young mother suffering from anorexia. They thought Jake, Paige, Pat, and I were terribly obnoxious, and they were right. We were. We had such a distorted sense of humor. Everything was always funny. We laughed many a night away. I guess laughing was better than crying. I couldn't cry. We rarely went to bed at ten o'clock, so the night shift didn't appreciate us too much. Although she was getting better, Wanda would not go to bed until I would. I had become her protector. The nurses and aides were always saying, "Now don't upset Wanda." Some nights we would stay up until 1:00 or 2:00 and then want to spend our days in bed and come out when evening came again.

I was becoming a full-fledged mental patient since I actually finished one pair of moccasins for Lisa. It took me a week and a half and should have been done in a couple of days. All we had to do was lace them together. I knew Lisa would like the little moccasins, and I decided I had better make something for the boys. I had made up my mind to refinish some little wooden boxes to put trinkets in. Some days I actually looked forward to going to the craft room and working. At least I was becoming goal oriented.

My emotions were running like a roller coaster. One minute I would be goofing off and giggling with Paige, and then I would go to my room and drop to the bottom. Proverbs 14:13 states, "Laughter cannot mask a heavy heart. When the laughter ends, the grief remains." How true. My room was the place I could go to and not put on any facade. I could just lie down and be depressed.

What a miserable affair not to be able to control the way you are feeling. My mind is like a runaway train destined for a crash. If only I could somehow get things in hand, I know life could be warm and pleasant again. It is really hard to relate to this hopelessness since the union of body and mind are not quite possible. I've heard them say that time will heal. Oh, I hope this saying is for real.

It was Friday morning and Paige and I were engaged in a boisterous game of Ping-Pong. The nurse on duty told us to quiet down. We thought that was ridiculous to have to subdue ourselves while involved in a game. I suppose she thought we would scare Wanda or upset one of the older patients. We continued giggling and missing the ball. It felt good, but I know the staff was upset with us. The whole day we continued to be uncontrolled and funny. By evening we were both definite basket cases. Paige had the loathsome habit of saying "Fuck you" to nurses and aides, whereas here in the Midwest we used "You bet." While I didn't care for her little saying, I did feel she should have the liberty to release her frustrations in this way.

Boobs was on duty one evening. She had a definite dislike for Paige. In fact, I don't think she liked any of the patients. Well, we were sitting in the dining room and Boobs came in. Paige had heard that Boobs had complained to another patient about Paige's actions. So, she stood up and told Boobs if she had any complaints to give them to her personally and not be blabbing to other patients. Then, Paige ended with an untimely "Fuck you." That did it. Boobs declared, "You have had it!

Now, Paige was worried. She whispered, "She is going to put me into seclusion."

"No way. They can't put you in seclusion for that," I assured Paige.

Sure enough, Boobs came back with three other staff members to take Paige to a seclusion room. Paige yelled, "No, I won't go!"

Then, I stood up to protect her from them, and Boobs told me to shut up and sit down. Well, being a honest-to-goodness chicken, I sat down. There was never any warmth or understanding with Boobs. She always had a snarl and abrupt manner, and tonight it was worse than ever.

So, they did take Paige, screaming, down the hall to one of the seclusion rooms. In fact it was 4-402, my old room. She was shouting, "Jayne, Jayne, Jayne. Remember this, Jayne, and tell them what happened in the morning." I was helpless. I felt so frustrated at not being able to help or comfort her.

So, that's what my first room was—a place for seclusion. Just take everything out but the mattress and leave the loony in there to rot or freeze or whatever, with the nice little peephole to check on the offender from time to time.

I thought seclusion would be for patients who were dangerous to themselves or others, not for mouthing off. She was put in there with no heat and only wearing a hospital gown.

Paige kept yelling, "Jayne, help!" through the night, but I could not help her. There was nothing I could do. To say the least, I didn't sleep much that night. The late-night shift came on, which included one nurse, one aide, and one big orderly. His name was Bob, and he definitely was in control. I stayed up, with Wanda tagging along behind me. We watched Johnny Carson and then the late movie.

I had the biggest urge to get my chart and see what they were writing about me. And, I decided I had to keep occupied with something—why not my chart? So, I schemed that the next time Bob was down checking on Paige, and the aide and nurse were doing bed checks, I would slip in and grab my chart.

It was ridiculously easy to do. I quickly slipped into the nurses' station and grabbed the Jayne Moomaw chart, slipped it under my robe and went directly to my bathroom. There was really quite a bit in there—nurses' notes, doctors' notes, etc. Most of it just involved listing my actions of the day, medication taken, moods of the day, and activities I took part in.

It was an exhilarating feeling to be beating the system—my own little silent revolt. Actually, looking at my chart was my legal right. Besides, I was curious. Ah, there is the letter they sent with me from General Hospital. Very interesting. "Husband supportive but stern." How true.

Just about halfway through the chart the night nurse threw open my door and grabbed the chart away from me and told me not to ever do that again. It was worth it, however. At least it kept my mind off of Paige and her dilemma for awhile. There really isn't much they can do to me for chart stealing but get mad.

Morning did finally come. Wanda and I got up, ready for a long Saturday in the unit. Paige was still in seclusion. I decided I would get my laundry done before everybody else came up with the idea. Gina came on that morning, and I told her about our hectic night, which I'm sure she read about in the charts and heard during the morning report—that sacred time when nurses and aides are together at the beginning of the shift, giving the lowdown on all patients. No one can ask them a question, or die, or do anything during report. Gina took all of my story in. It was nice to have a staff person like Gina who seemed to care and took the time to listen.

Paige's shrink finally came in, talked to her, and then dismissed her that very morning. He was a little perturbed with the night staff. I couldn't believe he would discharge her though. I was going to miss her zany personality. She left with promises of calling us, since she did live in town. Here she was, rattling around in her apartment with no job, living off her (mental) disability check. But, I had a feeling her mom and dad from Philadelphia helped her out. And, she is a big girl and can manage on her own.

Dale called that night. I spent most of our talk time actually crying, telling him of Friday night's activities and how upset I felt. He put each of the kids on the phone to say hello and that they loved me. Lisa was in kindergarten, Chad in first, and Todd in fourth. They each told about school activities and how Grandma was doing as substitute mother. Dale's mom always came and took over in my absence and did a very good job with Dale and the kids. I really appreciated her efforts in keeping things going.

Sunday was a long, boring, tiring day. During breakfast I had decided to start saving Tofranil so if things turned too intolerable I would have a painless way out. I just didn't feel secure without a medication cache to plan my "escape" from life. I had hoarded medicine at home so long before finally taking it that I felt downright uncomfortable without an abundant supply. The nurse wasn't at all aware of my trick of holding the cute, little, red pill under my tongue. I decided this was going to be easy. However, that night I tried again and got caught in the act. The nurse's name was Sally. She was short and plump and on the ball. She asked, "Why didn't you take your medication?" I replied that I was going to take it later. I knew I was in trouble.

She then asked me if I had any more stashed away. I told her I had only one. She responded with "Give it to me." I said, "No." So, she started searching around the room and couldn't find it. I had it tucked away in the toe of one of my socks. "You either give it to me or I will take you to seclusion and take off all your clothes and keep you there till you tell us where it is." I hesitated about half a second and showed her exactly where it was and promised there was no more. No way was I going to let myself be subjected to isolation.

The late-night shift came on, which included Bob, the big orderly. He always said "Hi" but that was usually about it. I really respected him (mostly his size). So, I decided to drag Wanda with me to our room and go to bed. Wanda was getting so much better since her medication had taken hold. In fact, she was going to be going home in a day or two. She was kind of fun to kid with. She thought I was something else with my chart stealing and pill saving. We both went to sleep without too much trouble. We had music on the radio, and everybody else in the unit had gone to bed.

Monday came and so did Nimrod in his navy, pinstriped suit. That poor man wore only two suits—the navy one and a brown one. I don't know how much they paid psych residents but it must not be much. He revealed he had been a doctor in the Navy for three years and his dad was a psychiatrist. He still would sit on a chair and yawn while we had our sessions. He would come to my room and talk with me about every other day. He thought group therapy was going really well for me since I was participating more. And, he decided it was time for me to go to the open unit in 4-East. No locked doors, and we could come and go as we pleased. They even let you choose your own menus and handle your own money. 4-East sounded like a good place.

TRIBUTE TO MONROE

After a flirtation with suicide and an affair with depression
Coming to Monroe and yielding has brought on a hint of progression
Here exists a sense of dedication and aroma of pride
Urging all patients toward an optimistic get-well ride.

The attitude seems to be "hang in there; don't give up hope"
It is good to know they are pulling for you, in hopes you can cope
After trying other hospitals and doctors, Monroe's dedication is inspiring
It urges me to continue on the path to see what might be transpiring.

CHAPTER FOUR

BATHROOM ENCOUNTER

Moving day came for me. Doctor Nimrod, in his same brown suit, told me I was ready to go to the open ward—no locks, with a certain degree of freedom. I hated to leave the new acquaintances I had made, but I was ready for the new challenge. So, I packed everything on a cart, said good-bye to Gina, Jake, nephew Pat, May, Erin, and the others, and I went to 4-East ward and freedom. My new room was a single and very pleasant. It didn't take long to get settled—physically, that is.

Well, I had to get used to a whole new staff and different patients. I didn't feel a part of things but felt strange and new. The rec room was nice, and I started working on some leather belts for Dale and the kids. I had made moccasins for not just Lisa but for Todd and Chad and myself, and I had made the wooden boxes for the boys. My mind was constantly on Dale and the kids. I was really a full-fledged mental patient with all of my handicraft to show for it!

One real nice feature about 4-East was the fact they let you pick your own menu from many possibilities. Guess on 4-West that we were too crazy to select our own food!

The first day and evening I spent most of my time in front of the TV. I felt so very insecure and alone, although several of the patients in group therapy were in the open ward. Everyone else seemed polite but involved in their own circle of friends. Along with all my new freedom, I was allowed to leave the unit for sponsored sports activities. Good ol' Nimrod finally had decided I wouldn't "embarrass the clinic" anymore.

The schedule on East ward was very similar to that on West ward. Medication first thing in the a.m., except instead of bringing it to us we went to the nurses' station and took it there. Also, they didn't have a glassed-in nurses' station in the middle of everything. The staff didn't wear uniforms as they did on West, so it seemed much more informal.

After breakfast, it was time for crafts and then lunch and group therapy, which I had actually come to enjoy. I looked forward to going, especially on days when they concentrated on other patients. It was uncomfortable when they would center the attention on me. It was difficult to snow them. You had to pretty much level with them and tell it like it was.

One day the attention was focused on Helen and her depression. She had a diamond on her finger that was huge. Helen had come all the way from Kansas and had money coming out her ears. She let me know that the big reason she was depressed was because she was worried that she wouldn't know what to do with all the property and businesses her husband had if he died. Danny, a younger fellow, a businessman from Indiana, suggested she hire a lawyer to handle all of it. Well, that was too simple for Helen. She didn't think that would be right. She stated that her husband would keep buying her gifts and taking her to different hospitals until she started feeling better.

We decided Helen was enjoying her depression and was getting a lot of mileage out of it. She seemed to enjoy the attention we were giving her. I didn't want to dig too deep, for I knew my day was coming when they would pounce on me and my hang-ups. What a horrible thought. I did such a good job of analyzing other people's behavior and problems! Nimrod even complimented me on my insight. But, I wasn't too great in dealing with my own problems. Consequently, when the chance came to talk about somebody else, we took it and tried to keep *them* the topic of conversation rather than having the attention bounce back to us.

Pat had been kicked out of his group in 4-West so joined us in group in 4-East. He was still too silly and obnoxious to help others or to benefit himself. He didn't have much respect for anything or anybody. However, I couldn't help but like that childish, free spirit he had—typical mixed up manic-depressive.

It is fair to say that the patients in the East ward were not nearly as colorful as those on 4-West. They were all just too sane on 4-East. There were even two doctors as patients—an opthamologist and an internist. Both of them were depressed. There was also a lawyer there who stayed in his room and refused to mix with anyone.

Well, it finally happened! A new patient entered the group, and we were all to tell him our names and what we wanted to accomplish at Monroe Clinic. My simple answer: "To get over my depression and get rid of my suicidal thoughts and actions." It just turned out there were two ladies in the group who were mothers of daughters who had attempted suicide. My statement really set them off. I was now to get the surviving loved ones' point of view. They really let me have it: "Don't you realize how selfish suicide is—showing no regard for your loved ones? It's such a hateful thing to do. You're just thinking of yourself," were some of their comments.

I didn't have a reply for them, for I had to agree with them. I just sat there and soaked it all in. Doctor Nimrod finally started talking about something else. I didn't cry while I was in the group but just sat there feeling numb. Here is another example of my emotional retardation. Behaving as I did when I was four years old, I feel badly but I'm not going to cry.

Bad Jayne has taken charge. They are right. I'm a selfish, no-good person. I don't have a right to stay alive. I went directly to my room after group and lay across the bed and cried in my solitude. It was the first time I was able to cry in ages. Bad Jayne was coming into full bloom. Bad Jayne doesn't love anybody. She's such a hateful thing, you know.

Just as I was getting into my Bad Jayne role, there was a knock at my door. "Are you still going swimming today?" asked a recreation therapy staff member. I guess I did tell them I was going, for it was to be my first trip to the pool. I had better go and try to get out of this Bad Jayne mood. I'll just use that childhood trick of hiding my feelings and put on my "everything is fine" mask.

Swimming took place at a motel on the other side of town. There were six of us going, so they took us in two taxis. I jumped in the same one Pat and Jake were in. It was good to see them again. Jake was just as paranoid as ever and still in 4-West but could be a part of the activities in 4-East. We chattered all the way. In fact, the therapist riding with us became disgusted and told us we should not talk about happenings in the hospital in front of others (the taxi driver in this case).

The pool was in a Holiday Inn motel. The sun was shining but it was still a cold January day, so it felt good to be inside in a heated pool while the wind blew outside.

Keeping busy swimming and talking with the others was good medicine. I had successfully put on my "everything is fine-wonderful" mask. No one had any idea how truly miserable I was inside. I had accomplished this trick at home so very many times.

In my hometown and where we lived now, no one could believe I had trouble with recurring bouts of depression. So many people thought of me as "cheerful and bubbly." I had become a pro at concealing my true feelings. In fact, I had become so good at it that I even fooled myself sometimes.

It felt great to swim back and forth in the pool, for swimming was one of my favorite sports. There were two staff members with us so we had a pretty good group in the pool. But, the time came for us to return to good ol' St. Luke's and the East ward.

My room was waiting for me, and that is where I went when I returned. No supper. Just lay in my room and concentrated on how rotten I was. But, my better half told me to get out of that room. After a couple hours' isolation, I finally came out to get away from myself for awhile.

Nielson, one of the psych residents who was in charge that night, was playing Ping-Pong with one of the nurses. I had decided that keeping busy was my best therapy, so I told him I would play him a game. He was really quite good and had beaten everybody he had played. Since I had become a "professional psych patient," I had become fairly good as a Ping-Pong player, for there was a table in every psych ward I had been in. But, I was no match for Doctor Nielson. He beat me quite handily.

After playing the game, I went to the TV lounge but couldn't get interested in TV or conversation, so I went back to my room, turned out the light, took off my clothes, and crawled into bed and started crying again. I couldn't get the group session out of my mind. Those two women were right. I was a hateful, selfish, uncaring person. Bad Jayne should be disposed of. She has never done any good or cared for anyone. She should be zapped. If I had the right pills here, I could get rid of her.

I didn't go to the nurses' station to get my medication, so Stacey, my nurse for the night, brought my Tofranil and Lithium, the pretty yellow and gray and shiny red friends. She noticed I was crying and wanted to know why. I just told her I felt rotten and if I had the right materials I would put an end to Bad Jayne. She could tell I was dead serious, so she called Dr. Nielson over to 4-East to talk to me.

I told him if I had the right pills I would take them and end it all. He remarked that if I really wanted to die that I could find something to kill myself with without the pills. He thought that maybe I should go back to 4-West if I was really serious about a suicide attempt. Well, I told him I didn't have a way so I would just settle down and forget the whole thing.

He and Stacey left, and there I was alone in the dark with my trusty radio pumping out rock music and God putting up with my moaning and groaning. I knew, at least, that He could understand me. I wasn't sure He approved of my plan, but I did know He understood and loved me in spite of my destructive plan.

ROUND AND SQUARE

God, you made a round world
Then people
You made them round
They even enjoy round positions
The round people like their round positions
The round people are understood—
Love?
They can feel it—
Acceptance?
Always acceptance
They feel warmth and compatibility
The round people like being round
You made them such a pretty color
Oh how I envy the round people.

God, you created some of us square and gray

We know you love us
The round yellow people don't understand
Our gray corners scream for acceptance
No one hears
We plead for understanding
It doesn't come
We send out our gray love
Only to have it stifled by yellow
The round people belittle us
We are square and gray.

The round people know we are here
They pretend not to see us
We don't fit into their round positions
We try but it doesn't work
Forgive us, God, if we don't fit
We gray misfits try to escape from the round world
It makes the round people angry
Please forgive us for quitting
What is heaven like, God?
Is it round or square?

Doctor Nielson is right. If I really wanted to kill myself, I could surely find a way in this room. What do I have? I wanted a method that didn't hurt and was quick and easy and not messy. What could I use? Not the razor blade to the wrist. It would be too painful and messy. Not jumping out the window. It isn't far enough down. I have just the thing—the plastic that lines the waste-basket. "Keep away from small children." This is my answer—not messy or painful. And, it shouldn't take too long. Those two women were right. I wasn't thinking of Dale and the kids, although I had convinced myself they would be better off without me. I was definitely being selfish.

The best place to do this would be in the bathroom. So, I took my radio, pillow, coat to lie on, and plastic bag and locked myself in the bathroom.

Stacey had left, and the night nurse was an Oriental gal named Pearl. She seemed quite nice, and I had no idea I would be hearing more from her.

I lay down on the floor with my feet against the door to bar it with my whole body. It felt good to know I was isolated and no one could see me or know what I was doing. So, I took the bag and put it over my head and held it tight around my neck and started breathing. I started to run out of breath, so after awhile I took the bag off my head and thought about it some more. I was sure I had thought of a successful method for immediate death.

Pearl came to the door and asked me if I was all right. I told her I was fine and to go away.

"I'm coming back," she announced, "and I want you to unlock the door."

Well, I didn't unlock the door. She brought a key and unlocked it, but I wedged myself against it, making it impossible for her to open it. Pearl shouted out if I didn't come out in fifteen minutes she would bring the doctor and take me back to the West ward.

Fifteen minutes—that should be time enough to suffocate. So, I put the bag over my head and started breathing heavily. I felt as if I were climbing a mountain and was out of breath, but I didn't pass out. I just kept gasping for air. Surely a baby would be dead by now. I'm sure the death route is the one I should be taking, if for nothing else but to just get rid of Bad Jayne forevermore.

Well, it wasn't working, and I heard Pearl at the door again, and she had rounded up reinforcements. Somebody started pushing on the door, so I quickly took the bag off my head and put it under my pillow and breathed a big gasp of air.

There was a forceful push on the door and then another one. I could tell it was someone with a lot more strength than Pearl had had. Then, the forceful person came into view. Much to my dismay it was Bob, that big orderly from West ward. I felt very much like a naughty little girl as he stood towering over me. He had pushed right through with his big muscled arm and shoulder and flipped on the light.

I meekly remarked, "Oh, no, not you."

"Yes, it's me." His words were loud and clear.

So much for suicidal activities. It didn't work—but at least no one knew what I was attempting. So, the end came abruptly for bed three. Here I go back to the closed West ward. I walked there with Doctor Nielson and Bob, the friendly neighborhood orderly.

Doctor Nielson seriously asked me if my bathroom encounter and activities were because he beat me in Ping-Pong. I assured him that that wasn't the cause. Maybe I should have told him yes and then maybe he would know for sure that I was crazy. At the present, they probably think I'm just semi-crazy.

CHAPTER FIVE

HELLO AGAIN BEDLAM

As we entered the West ward, there was a young boy with a huge scar on his neck in a spastic manner. He clearly stated, "Hi, crazy people. Everybody here is crazy." I decided right away that they brought me back to the right place! Doctor Nielson and Bob escorted me to room 406. It was another room destined for seclusion.

Some nurses brought all of my clothes and belongings right behind me and locked them in my closet. Here I am in the early morning still alive and doing well. I decided to sit in the window and look at the lights of the city and try to decide if I was glad I didn't get myself killed. I was so mentally ready for death that I was willing to gamble and lose the bet. But, as God willed it, I did not die. My method was terribly primitive and juvenile, but I really thought it would work. Now I was too ashamed to want to admit to anybody what I did. It seemed dumb as I looked back on it. But, it seemed so right at the time. I suppose though I will tell Nimrod if he asks.

Sleep wouldn't come to me, so I decided to take another try at getting my chart since it was the night shift and the staff was spread thin. I walked down to the nurses' station, and no one was there. So, instead of my chart, I grabbed my full history that was sent by Doctor Wilson and the psych units at Mitchellville and University Hospitals. It was sitting out in plain sight above the charts. I slipped it under my robe and walked very casually back to my room. On my way back, Bob saw me and ordered me to get back to my room and stay there. I was very fortunate to have gone as far as the nurses' station without being caught.

I was wide-awake and glad I had so much reading material. Bob had been coming and checking on me about every fifteen minutes. On one of his checks he saw me lying on the bed and asked what I was doing. I replied that I was catching up on my reading. Was I ever! There were carbon copies of letters sent by Doctor Nimrod to Doctor Wilson and Doctor Tarinadski and General and University Hospitals. They all read the same . . . with a form saying, "Thank-you for sending information on this very challenging patient. We hope it will aid us in a diagnosis of her problem."

So, that is what I am to Doctor Nimrod, a challenge—well, how interesting. I had become a challenge to myself too. I didn't know if I was a Good Jayne or an In-between Jayne. The really Bad Jayne had somewhat disappeared. I was definitely in a state of change.

Doctor Wilson had copied every poem I had written and had recorded all the medication prescribed, and each trip to General Hospital was also there. Even a report from Doctor Scott from the University Hospital stating I had a personality disorder, undefined, was part of my medical history. Wonder what "undefined" means? After scanning all the papers in the packet, I thought it was very informative, and I was convinced it was my legal right to see what was written about me.

Now, I had to figure out how to get it all back to the nurses' station. Morning was approaching, and I hadn't slept all night. I decided I would just boldly walk down and lay it on the counter—which I did.

The first person I had to face on the morning shift was Sally, the clever detective who caught me saving pills. She greeted me in a very condescending way, "How are we this morning?"

I told her, "We are just fine."

I hated the way she talked down to me, as if I were mentally retarded—not mentally ill. Sally gave me my trusty Lithium and Tofranil. I had a feeling Nimrod would be disgusted with me for landing in the West ward again. He felt I was doing so well before this. Sure enough, he came in to talk to me. I was, as usual, lying on my bed. His opening statement was, "Tell me, Jayne, what did you find so amusing about the bathroom?"

I explained to him that I wanted to be alone and that "it was my Walden Pond but no one would leave me alone." Then, I told him about my unsuccessful try at suffocating myself and "I could have died of old age before anything would have happened with that method. Why didn't it work with me when it is so fatal to babies?"

Nimrod explained that when babies inhale, the plastic is sucked into their nostrils and mouth and it quickly causes asphyxiation.

I went on to explain I was really ready to die and Bad Jayne had to be killed.

"Do you feel better this morning that you didn't succeed?" inquired Doctor Nimrod.

"No, I had myself ready to go, and I'm still disgusted with myself. I just don't like Bad Jayne, and I want to get rid of her. I'm not going to group today. I just don't want to face any of them."

"Oh, Jayne, I think you had better show up. It will do you a lot of good."

"No way!" I retorted. "Say, I'm getting tired of Sally talking down to me like a child."

"Tell her," urged Nimrod.

"Okay, I will."

Nimrod dismissed himself with an "I'll see you this afternoon in group therapy."

"Don't count on it," I replied.

The next thing on my list of things to do for the day was to have a heart-to-heart talk with Sally. So, I set out looking for her. On my way to the nurses' station, I saw good ol' May walking down the hall with very baggy slacks on (not hers) and a bundle of clothes in her arms. Do you suppose her persistence will pay off? Might she someday escape from here? I admired her spirit.

Ah, there is Sally. I'll get this little discussion over with before breakfast.

"Sally, I have a problem, and it's you."

"Oh, why is that?"

"I feel you talk down to me and the other patients, and I don't like it. You may think I act like a child and should be talked down to, but I'm not a child."

"I didn't realize I was. I'll try to avoid it hereafter."

"Thank you, I would appreciate it." There, that wasn't so difficult.

I strolled to my room, and on the way I asked Gina, the compassionate nurse, to unlock my closet so I could get some clothes and get dressed. With all my things locked up, I was in what they called semi-seclusion. I thanked Gina for helping me out and then dressed and headed down the hall to breakfast.

I joined Pat and Jake at a table and pretended I had never left . . . just like old times. Jake and Pat asked me what I was doing back on the West ward. I didn't want to tell them why, so I just remarked that I missed them and had to come back.

It was late January now and still quite cold. Jake and Pat suggested we go out on the locked screened-in porch after breakfast and get some fresh air. That sounded "refreshing" to me. Sally entered the dining room so I asked her if we could go out on the porch. She replied, "Yes, but we will have to put our coats on."

That reply just about sent me through the ceiling. Number one, I realized it was winter and we needed coats. Number two, I wasn't planning on having her with us.

"Remember our little talk this morning?"

She replied with "Well!" and stomped out of the dining room.

We did get our coats on after breakfast, and she did let us out on the porch. It was a crisp, cold morning, and the fresh air smelled and felt good. We could see children with sleds on the lawn by the hospital, and it made me think about Chad.

CHAD

Who is Chad Moomaw?
You are a guy who God dedicated to Dale and Jayne
You are the playful teasing brother of Todd and Lisa
Diligent worker, bike rider, ball player,
Collector of model horses and bulls, gardener, singer,
Helper, and busy guy
The teacher declares you a leader of the class
And you declare Corey as your buddy.

Such an eager hardworking fella
Is bound to accomplish what he wants

We are just so happy that Chad is in our family
You appear as a strong miniature athlete
Training for the big game
Dark brown, sensitive, affectionate eyes
Tell of your innocence and desire to please
Who can help but enjoy you
And your refreshing sense of humor?
You are special
You are a promise.

At the age of seven
You are well into that fun time called childhood
Enjoy it!
You have been given
Some very impressive gifts and talents
Natural ability for physical activities and sports
Which could make you great with practice.

Don't let people or society stifle
Your desire to meet life head-on
Enjoy life and keep racing to your goal
And you will win the prize
For those who believe
God will renew their strength
They shall mount up with wings like eagles
They shall run and not be weary
They shall walk and not faint. (Isaiah)

At times you seem like a stranger just passing through
Stopping momentarily in our life for a short visit
The manner in which we nurture you while with us
Will help to determine your destiny
You have shown us that food and drink are not enough
Like a young colt you demand freedom, affection, and assurance

You also need faith in God.

We are so proud of you
But more important is your pride in yourself
We are so pleased that God sent you to us
And we love you because you are
Not because you do
The mere fact that you exist and are a part of us
Is enough for our unconditional love
You will know this feeling when you become the
Author, the provider.

So, step on, Chad
Ask God for strength and guidance
Seek and you will find
Knock and the door will be opened.
Faith and success will be your prize
Don't depend on the crowd
Tune in to life and soar to the heights of your own choosing
We are with you but we will also let you go.

While out on the porch, we took the opportunity to release some tension. So, we had a contest to see who could yell the loudest. I think I won, and sure enough, we were acting childish. Just as Nimrod had informed me, I was very impulsive (a polite way of saying I was childish) and I was very ambivalent. Both of these qualities caused me trouble. Sally let us back into bedlam after we spent about fifteen minutes out there. She made no comments; nor did we. Jake and Pat headed for the craft and activity room, but I decided I would go back to my trusty bed.

Lunchtime came, but I chose to stay in my room. One of the younger orderlies came down after lunch to escort me down to group therapy. I told him I was absolutely not going. He tried two more times later but I refused. I was sure group could get along without me. And, it did.

Later, I met the young man who was slightly spastic who greeted me the night before when I returned to bedlam. We started talking. He disclosed that they put him in St. Luke's because he was suicidal. He had been in a motorcycle accident a couple of years before, and it left him brain damaged with a paralyzed right arm and a spastic limp. He walked with a stagger and had a scar on his neck. He was very near death at the time of the accident and wished he had died rather than be in his present condition. He was eighteen and had a lot of life left.

At least he had a good reason for suicide. I could understand why he wanted to die. I felt like telling him all the reasons he had to live but decided that would be a bit hypocritical, so I didn't give any sermons.

It was Friday and the gray boxes were invading 4-West. Everyone quieted down and put on their best behaviors, as if being out of step might win them a gray-box visit. Electroshock, or as the staff called it, *the treatment*, was usually given on Mondays, Wednesdays, and Fridays. After the treatment, the select patients would lie in their rooms with a nurse or orderly in there to help them, and they would stay there until they had regained a portion of their zapped-out sanity.

The Syrup Sisters were all still in 4-West along with May, Jake, and Pat. Wanda and Erin had gone home, and several new people had come. There were always plenty of crazy people to keep the place busy, for life marched on. Martha, a lady who thought she stunk all the time, had tried to drown herself in the bathtub. Kim, one of the Syrup Sisters, saved her by pulling the plug. I doubt if I would have thought of that. I probably would have tried to lift her out. Anyway, Kim was the heroine of the day.

I spent only four days in West ward while Nimrod decided I had passed through my suicidal tantrum. I did manage to get my chart again, but this time, instead of going to my bathroom, I went to Sarah's bathroom. Sarah was a girl I had just met. I read through the complete chart. Again, it didn't hold any surprises for me, but it was interesting—kind of like a scrapbook. When I was finished with it, I walked back to the nurses' station. The nurse was sitting just around the corner so I was hesitant to put it back, for I knew she would hear it. So, I walked up to her with the chart in my hand and was flipping the pages while I asked her if I could have a sleeping pill and that I thought the order was in there somewhere.

"Oh, honey, hand that to me. You shouldn't have that. I'll look it up."

It worked. She thought I was a poor dumb soul trying to find something in my chart rather than an obnoxious twinkie who had just read her chart for the third time.

"Yes, you can have another Dalmane. Wait just a minute and I'll get you one." She gave me the Dalmane and I wandered off to bed. While I lay there in bed, I thought about my chronic blues.

CHRONIC BLUES

Hey, God, I've found these chronic blues troubling me
Are you sending any help?
Where can it be?
Even though I know thinking down can lead to big trouble
I continue to make light troubles deep and double.

Why do I keep contemplating and cheering suicide?
Are there really that many feelings I'm trying to hide?
Even though I would hate to embarrass Monroe Clinic
Frustration has grown and I've become a cynic.

God, I know you could help me reach out and grasp
That person I would like to be
But you know the chronic blues have gotten to the core of me
I'm trying to do the right thing
It will be interesting to see what the future will bring.

Doctor Nimrod came in the next day to scold me for not showing up for group therapy and informed me a Dixieland band in which he played the drums was going to entertain the patients tonight. He and five other doctors from the clinic had formed a group for their amusement and to entertain others. I told Nimrod I was sure group could manage fine without me and I would like to hear his Dixieland band.

Lily sat by me to listen to the band. She always seemed to be in a stupor, sitting with her face in her lap, or while eating, her face was in her plate. About halfway through the concert, I asked Lily how she liked it. She raised her head and declared the drums were too loud. Then, she dropped her head back to her lap again. (I decided Nimrod played a pretty good set of drums, even if Lily did think he was too loud.)

It always amazed me that she would respond at all, for she always seemed to be out in space. Never could figure out what her problem was, but she was interesting somehow. One morning she was sitting at the breakfast table by herself, with her face in her plate, and I said, "Good morning, Lily." She picked up her head quickly and she responded with a very lively "Good morning" and then her face went directly back to her plate. It really struck me as funny she could be so spry in her replies.

Sunday came and I attended church services in the rec room. It consisted of a couple of songs, a prayer, and a short sermon for crazy people. I found it hard to get too caught up in it, although I did enjoy singing hymns.

In the afternoon I was surprised by a visit from Jeanie, a childhood friend, and her husband. We grew up and graduated from high school together. I tried to put on my sane mask for Jeanie and tell her some of the activities and craziness that went on. She had also visited me when I was at University Hospital, so she was getting quite an education about loony bins. We had a good time chatting and bringing up old memories and good times. They stayed for about an hour, which is long enough for any sane person to spend in 4-West.

Paige called that evening and tried to think of some way we could rendezvous the next day. Former patients were not allowed to visit in the unit. I told her I could meet her in St. Luke's basement cafeteria after my group session. They had been trusting me to return on my own, so I knew I could slip away and meet Paige. It gave me something to look forward to.

As another night settled in at 4-West, I watched TV. I couldn't believe the need most people on this ward had to smoke. The staff doesn't let anyone have matches, so they all used the electric lighters placed in the dining room and day room where the TV was. Most would nearly turn blue while trying to light up. Paige had always jokingly revealed she thought it was a conspiracy to kill all smoking twinkies. The cigarettes really hit me. My smoking days were long over. I smoked in college and when we were first married, but I quit when I became pregnant with Todd. I didn't miss it at all, even though I did gain a lot of weight.

Monday came and so did group therapy. I was very preoccupied during the whole session, planning my getaway without attracting too much attention. After group, I left quickly while everyone was still talking and standing around. It was good to see Paige again. We talked and sipped on Cokes and laughed a lot. Paige was her usual comical self, and I enjoyed it thoroughly. After about an hour, Paige left, and I returned to good ol' bedlam.

Nimrod came to see me about 5:00 o'clock to let me know he was going to try to put me on the East open ward again. I told him I was ready. So, the next morning I packed all of my stuff on a cart and headed for the open, more sane, East ward.

CHAPTER SIX

TRYING TO GET IT ALL TOGETHER

I was put in a single room, since I didn't want to be in with a smoker. It was a nice room at the end of the hall with a south window. It didn't take long to get organized and ready to settle in.

Nimrod had even told me Dale could come for a weekend and I could check out and go with him. Sounded pretty neat to me. If he does come up though, he will be checking me out to see if I have "shaped up." I'm afraid I haven't gotten to that point yet. I had started saving pills again, the pretty red and white Dalmane and my trusty coral-colored Tofranil. It was really easy to do. They never did check to see if I had taken the pill. So, after I held it under my tongue, I went to my room to deposit it. No one knew what I was up to. I just felt I needed to have the pills as a crutch to ensure an escape if I needed it. I still had the feeling of not being in control.

It was nice to be able to leave the unit when I felt like it. I really did like that free feeling it gave me. The only stipulation was for us to sign out whenever we left the unit, which was something I didn't care to do. I felt I didn't have complete freedom if I had to tell them where I was going.

Being on the East ward was also nice since patients could check out and go to a movie in town. Five of us decided to call a taxi and go see *Good-bye Girl*. We had a great time. After we saw the movie we went out for pizza. It was great—just like being out on our own. Susan, Sam, Danny, George, and I made the trip together and returned to St. Luke's on time.

I had met Susan on closed ward where she had an IV she walked around with. She had overdosed and had ingested vomit into her lungs. This caused pneumonia, and the IV contained antibiotics. She was interested in Sam, who was about 45. Susan was 22 and a nurse at St. Luke's. She was wondering if they would take her back after her overdose fiasco since she acquired the medication while she was at work. Sam, a truck driver from Chicago, suffered from severe headaches and depression. He and Susan had gotten quite chummy. Danny and George and Susan and Sam were all members of our therapy group. We had all gotten to know each other pretty well. It was fun to spend an evening together away from our zany surroundings.

In the afternoon we went to a gym and played basketball and watched the hockey team practice. We still went swimming at the Holiday Inn and had recreational therapy in the rec room. Monroe was definitely the "Country Club of Psych Units."

I had Marcia and Robert assigned to me as my primary counselors on open ward. Marcia was heavyset and a nurse. She was easy to talk to, and everything seemed more informal since they didn't wear uniforms on the East ward. Robert was a psych teacher and very understanding, sensible, and easy to talk to. He and I had several long, late-night talks. Since I wasn't taking my meds, I had a hard time sleeping. Every night I would ask for Dalmane at bedtime and later would ask for another. But, since I was tucking them under my tongue, I stayed awake half the night. Robert was fairly young and was contemplating getting more schooling to become a counselor. He was very understanding and usually came up with the right comments during our marathon late-night talks.

Doctor Nimrod informed me Dale was coming for the weekend and I could be in his care. It was kind of scary since I knew with my pill saving and rotten attitude that I hadn't "shaped up" yet. I wondered if Dale could figure this out. We were still having group therapy, taking turns spilling our guts, and I was able to tell the others in group that I was very nervous about seeing Dale.

However, when I did see him Friday afternoon, it was good to come face to face with him. He had gotten a motel room across the street from St. Luke's. We ate out and then returned to the motel for some serious talk and long-overdue lovemaking. We got along really well. However, I couldn't be totally truthful or open with him. When Sunday came and he had to leave, we handled it okay. We had gone to see *Good-bye Girl*, since I thought he would like it. Then, we met with my cousin, Bryan, who lived only twenty miles away, and went out for pizza with him and his wife. Bryan enjoyed my psych unit tales and liked to share them with his junior college students in biology.

Dale phoned he would be coming up one more weekend and Nimrod had suggested he come up for an entire week and sit in on group therapy. I had a feeling it could prove to be very interesting as it would be close to the time I would be discharged.

Saturday night arrived, and it was a very quiet, boring evening. I suggested we have elevator races. It was very juvenile but much fun. I won three out of five races.

The following Monday Pat came down to East ward to visit me. He had gone to my room but I wasn't there at the time. As I came down the hall, I could hear him sputtering and saying, "God dammit!" Evidently, he decided to brush his teeth and accidentally used my Head and Shoulders shampoo that was in a tube. I asked him why he was using my toothbrush. He didn't say—just that he wanted to freshen his breath. We went to the other end of the hall to the rec room and listened to music. As he was sitting next to me, he scooted a little closer. I let him know I wasn't romantically involved with him and if he came down to see me to ignite some kind of relationship to forget it. We sat and talked small talk and listened to music. Then, he had to return to West bedlam for the night. I really giggled about his brushing his teeth with shampoo and then trying to put the make on me. I was old enough to be his mother, and I didn't think of myself as sexy with anyone but Dale.

I continued to save pills and stay up talking to Robert after everyone else had gone to bed. One night he confronted me on the fact I would talk and laugh a nervous laugh. Then, I would get a real sad look on my face. I told him I didn't feel like I had everything under control and I still hadn't "shaped up" for Dale yet.

The next evening Marcia was working. She would pop into my room and visit and then check on her other patients. I decided I was due to get my chart again and see what was going on. It was so easy to do since no one was at the nurses' station, and the charts were sitting there so handily in the chart cart, just waiting for me. I grabbed it and headed straight for my room. It didn't take Marcia long to discover I had it with me. She stormed in. "This is it. We are going to lock the chart up and not permit you to get ahold of it again." And, sure enough, that is what she did.

That same evening I received a call from a couple from Mudville who went to the same little Friends church that we did. They shared how they thought a lot about me and just called to let me know they cared. After I hung up I felt guilty and decided to tell Marcia I had pills saved.

I asked her to come to my room and that I wanted to talk with her. I decided what I was doing was definitely wrong and I needed to bring it out in the open. I disclosed I had some pills I had saved. Then, I stopped and could not tell her any more. That is all the further my guilt would take me. I had them in a pocket of my jeans and decided that was where they would stay. She stated she would have to tell Doctor Nimrod, and I okayed it but that I couldn't tell her any more. The on-call doctor came to see if I had the pills that I had mentioned to Marcia. I responded with "No, I got scared and flushed them down the toilet." He actually believed me.

FRIEND OR FOE

I have these pills that I have saved and stashed away
And I'm wondering if they will be friend or foe on good-bye
day
They look so innocent and benign lying there
I know they could deliver me to the land without care.

Isn't it something that such little things can join together
And bring grief or elation?
It is amazing that they alternate from friend to foe
During my contemplation.

I have a yearning to get rid of these pills and
Tell suicide good-bye
But there is still that desperate need to hang on to them
Until death or peace is nigh.

The pills are friends if they bring Heaven and serenity and
peace
But they could also bring destruction and hell with no release
Ask the ones I love which is most suitable for me
Friend or foe, which one should it be?

The next day in recreational therapy I noticed the chaplain, Father Bryska, was talking to Robert. I was confused and needed some answers to be given. So, I walked up to him and asked if I could talk to him. He replied, "Sure," and we sat in a corner of the rec room and had a talk about God, and me, and suicide. He finally declared it sounded like "my" God approved of suicide. I responded by saying, "If Jesus cares for the physically impaired, he surely has the same compassion for the mentally impaired or someone with emotional problems."

Father Bryska suggested I make a no-suicide contract with someone and abide by it. I chose Robert and Marcia as my contract companions and stated I would contact them before I tried to harm myself.

Little did I know Nimrod had advised them to search my room for my stash of pills. Evidently, he didn't believe what I had told the on-call doctor. The pills were still in my jeans pocket, and when they came marching in to announce their search, I sat back and told them they wouldn't find them in my room. They looked absolutely everywhere—the curtain hems, the light fixture, under and between the mattress, all of my drawers, inside socks, and clothes.

I knew I was had when Robert left and Marcia told me to take off my clothes and empty my pockets. My first thought was to take off in a dead run. Then, I went to the bathroom and dug deep in my pocket to pull out my pretty friends. I thought about throwing part of them in the stool, but I knew I was caught. So, I just handed them over. My pocket was full of all of the Dalmane and Tofranil I had been saving. She couldn't believe it as I kept pulling out more and more pills. Astonished, she could only say, "Jayne, you've done the right thing." I told her I didn't think so. After she left, I had a desperate feeling of not having a ticket to death.

Robert was angry with me that I had assured him the pills weren't in my room when I had them in my pocket in my room. I really didn't care if he was angry—I had just given up my death insurance and I felt somewhat panicky. And, this coming weekend Dale was coming up to spend the week with me and to sit in on group therapy with me the following week.

Friday came and Dale had reserved a room and kitchenette about a block from the hospital. It was good to see him, but I crossed my fingers in hopes he wouldn't find out about my pill saving. He brought notes, pictures, and valentines from the kids. Lisa wrote a note as only a first grader could:

> *Dear Mom*
>
> *I really miss you and I wish I could see you but I can't. To switch the subject in church to day we are having a poster contest and in The King James there is something you said I shood put on it can you find it for me please?*
>
> *I'll try to call I miss you very very much I hope you come home soon! Oh grandma gave me some chocolates I'm going to save you one even if it takes 3 weeks I'll save it. It will be a strawberry one.*

VALENTINE
Happy
all your wishes come true
pink
pretty as a rose
you and me

Valentine
a happy Valentines Day
lisa and jayne together
everybody
never live my side
todd and jayne
i love you
nice is like you

everybody together with you
super mom

Dad and mom
always nice to me
you

I could tell that my little daisy, Lisa, was missing me very much. I missed her and the boys too.

LISA

Daisy, oh daisy delight,
How and why do you grow?
Always reaching, stretching sunward.
Our damsel dancing to and fro.

You are a dream come true
Born with ease.

We were friends from the beginning
Those first days in the hospital were empty
When your tiny body wasn't next to mine
Having flourished from a minute seed
Given in love by your father
You emerged as a promise, a possibility, a seedling
Your tiny roots grew down deep into the soil
Of God's marvelous love
Giving you a firm foundation.

Who is Lisa Moomaw?
Child of God, daughter of Dale and Jayne,
Sister of Todd and Chad.
Baker, singer, piano player, helper, lover of cats and dogs,
Book reader, TV fan, girl with lots of love and enthusiasm

We're glad you are our girl with the golden hair and sunny smile
The joy and warmth that smile emits is something precious
So, let your light shine, daisy delight.
Damsel Lisa, so feminine, fun and fine.

My prayer for you is that you will overflow more and more
With love for others
And at the same time keep on growing
In spiritual knowledge and insight
Serve God with gladness.

Be yourself, and most important, like yourself
Let contentment and acceptance come from within
Not depending upon your achievements
Or attention and worship from others
The only one you have to please is God.

We are proud of you
But more important is your pride in yourself
We are so pleased that God sent you to us
And we love you because you <u>are</u>, not because you <u>do</u>
The mere fact that you exist and are part of us
Is enough for our unconditional love.

So, dance on, Lisa,
Don't depend on the crowd
Tune in to life
Soar to the heights of your own choosing
With God as your guide.

Keep the right goals and work hard
And success will be your prize
We are with you but we will also let you go.

The weekend with Dale was great, but I felt uncomfortable not having an escape from life available. We ate out and made love numerous times.

While we were shopping in a large drugstore, I noticed some different medications I could use to overdose on. There was a $9.00 bottle of 100 cold tablets that looked like just what I needed, but I didn't have the money for them and didn't want to ask Dale for it. I then started scheming how I could get those pretty little pink pills. What a rotten person I was to be planning like this.

Monday came and so did group therapy with Dale. It was a disaster. Dale sat there and assured the group he wasn't depressed and didn't have any problems. The air seemed tense and you knew they all wanted to hit him over the head with a baseball bat as he said all of the group members had problems but that he was just fine. This type of attitude prevailed all week long. Dale certainly didn't win any prizes for congeniality from the group.

I was able to spend each night with Dale, which was really nice. We had time to talk about feelings and problems, and he had a great opportunity to check out whether or not I had "shaped up."

About the middle of the week, I slipped off to the drugstore and actually shoplifted the cold pills I had noticed earlier. It was so easy, but I felt so horrible for stooping to do such a thing. But, my guilt feelings couldn't outweigh the desperate need I had for my death insurance.

Dale's last night with me was fantastic. We made love as if we would never get to again, for I truly thought it would be our last. I had ambivalent feelings about returning home. I wanted to be with Dale and the kids, and yet it scared me too. It was a lot of responsibility to keep family and home together, and I didn't know if I was ready for it.

When Dale left Sunday afternoon, I thought it might be the last time I would see him. I just didn't know for sure. Nimrod thought I would probably be going home in a week or so, but he had told me that a couple of times before. It was the first part of March already, and I had spent about six weeks at Monroe.

At the Monday group, most of them complained about Dale's attitude. The group was videotaped about twice a week. Then, we would have to return in the evening and watch ourselves hash out problems. They picked up on my nervous mannerisms, such as twiddling and tapping my fingers and fidgeting around. It was terribly obvious I was uncomfortable. However, I did feel more satisfied knowing I had an escape planned. I kept my little pink gems in their bottle in the far reaches of my bed-stand drawer.

Wednesday morning Marcia told me I had a slight temperature. I told her I had diarrhea too and informed her that going home was scaring the shit out of me! Then, I laughed as if it was a joke, but I was really serious.

At about 3:00 p.m., after group, I decided it was time to ingest my little friends. I had a big bottle of Pepsi and started taking just a mouthful at a time along with the Pepsi.

In order to keep my contract with Robert and Marcia, I told them I was contemplating problems and maybe we could talk later when they had time. The contract stated that if I broke the contract I would be returned to the West closed ward. Well, I kept popping pills with Pepsi until I had downed all one hundred of the little dears. It took about two hours to get all of them taken.

Paige called me around 6:00, and she wanted to know what was wrong with my voice. I told her I was taking some pills for a cold. However, I didn't tell her how many, but I started getting sick. Marcia came in soon after. She asked me what was wrong, and I told her I didn't feel well. She took one close look at me and figured out what I had done. She called Robert and told him to get the on-call doctor.

To my surprise, Nimrod was the on-call person. He had told Robert to give me some Ipecac to induce vomiting, and as Nimrod arrived, I was barfing up my toenails. He admonished me and then brought in a wheelchair. Robert wheeled me to the Intensive Care Unit on first floor.

Robert, Marcia, and Nimrod were all very angry with me, and they told me so!

CHAPTER SEVEN
INTENSIVE CARE

Two doctors, or interns, and one nurse started working me over. They started inserting a tube down my throat, but it gagged me so that I pulled it out as fast as they put it in. They inserted it again, but I pulled it out again. Then, one of the interns put it down my nose, which was uncomfortable but worked better. As they were pumping my stomach, I started coughing and then vomited all over everybody. I heard one intern compliment the other on his expertise in stomach pumping, and the reply was that he had done it a lot in Chicago.

After they finally finished that unpleasantry, they had me drink two glasses of charcoal water. When I downed that, they hooked me up to all of the beep-beep machines and left. There was an older lady in the bed next to me who was also hooked up to a monitor.

Well, here I was again with a botched up overdose. I didn't even feel groggy anymore—just lying there with my usual mixed feelings, as if I wish I were dead and then glad I wasn't. I felt as if I were "between a rock and a hard place." I didn't want to go home, and yet I didn't want to stay at Monroe. So, I guess my decision was not to do either but to just "fade off into the sunset." I laid awake the whole night trying to sort out my priorities and mistakes. A nurse came in and removed the IV early in the morning.

Dr. Nimrod would be coming down to check on me and demand I go to group in the afternoon. After not sleeping all night and going through my whole ordeal, I felt a little frazzled and I wondered what the day ahead would hold. Sure enough, Nimrod came to see me bright and early and advised me

I had no choice—I was to attend group.

Father Bryska also came to visit in the morning, but I didn't feel like talking to him. So, as he sat there in silence, except for a few questions, I just let him know I wasn't in the mood to talk.

Around noon one of the nurses brought me an outfit to wear to group. I knew I was doomed to attend our friendly little group therapy meeting. One o'clock arrived and so did I arrive at group. They all started firing questions at me—like, "Why did you do it?" The only answer I could think of was that I didn't want to go home and I didn't want to stay here . . . so, I chose suicide. They all thought that was really dumb.

Danny wanted to know why I started taking the pills in the afternoon instead of taking them all at the same time. I told him I took them over a two-hour period so the meds couldn't be pumped out of my stomach and some of it would have a chance to go through my system.

I truly felt as if I were on trial and was happy to get back to my little corner in intensive care. Doctor Nimrod told me as soon as there was an empty bed in closed ward he would have me transferred up there. So, bedlam, here I come again . . . when you're ready for me!

CHAPTER EIGHT

GOOD-BYE, FLIGHT CONTROL

Finally about 6:00 p.m. they came and escorted me to West ward and crazy time. I was assigned to my old room. There were some new patients and some ol' standbys like Pat and Jake. That evening I found Bertie (a middle-aged female) in my closet. She had peed her pants in there and was hiding so no one could get her. Bertie and I were true bedlamites. It was Thursday night, and I was to go home Saturday morning. There would be one more group to sit through.

That evening Pat, Jake, Brother David (a manic-depressive monk) and I sat in the dining room to chat. Pat had thought of a new term for the nurses' station. He called it "Flight Control," because of their observing us and checking on our whereabouts through their glassed-in vantage point. I really liked the Flight Control idea Pat had created. Brother David informed us he came from a monastery in California that made wine, was the choir leader, and had a beautiful singing voice. He appeared to be slightly manic with his rapid speech and wanted to sing all of his favorite songs to us. When I returned to my room to go to bed, Bertie was in my closet again, so I escorted her to one of the nurses.

The next day would be my last full day at Monroe, and, thankfully, it would be my last group session. Group was not my favorite activity at Monroe. The morning arrived, and I was actually starting to like the idea of going home. It would be great to see the kids and Dale, and I knew I needed to get back into my mother role for the kids' sake. It had been such a long two months since I had been home that I was nervous about getting there, and I still hadn't straightened up or "shaped up" or whatever!!

That afternoon, at my last group, was the last day for both Danny and me. I sat in silence the whole session observing everyone's talk and body language. I could feel the staff staring at me through their observation window. As soon as group was dismissed and everyone started saying their good-byes, I slipped out the door and walked quickly to freedom and the cafeteria in the basement. I had a Coke and some chips and then headed for the hospital library looking for something to occupy my time on my last night in bedlam. It took a long time to pick just the right selections, and I still wasn't ready for the West ward. So, I spent about an hour in the waiting room of the Intensive Care Unit. Finally, after about three and a half hours of freedom, I decided to return to bedlam. As I was taking the elevator up to West ward, a nurse from East ward noticed me on the elevator. She sounded angry and jumped in to let me know my freedom was over, and she escorted me to closed West ward. She informed me they were very concerned about my disappearance.

Nimrod was scheduled to go on vacation the same day I left. I think he had had enough of my antics. As the night settled in, I spent one more evening with Jake's paranoia, Pat's jokes and analogies, and Brother David's songs. It was about time to say good-bye to Flight Control.

Saturday morning arrived and so did Dale and our friend Ron, who had ridden up with him for the long, five-hour trip. As they were in my room gathering all of my paraphernalia, Brother David came in to sing a few hymns for them, and he promised to send me a bottle of wine when he returned to the monastery. I had my doubts and decided not to plan on any bottles coming to our house.

Another doctor came to bid me farewell and answer any questions we might have. He gave me my prescriptions for the good ol' Lithium and Tofranil, and then we were off and running toward home and sanity.

CHAPTER NINE

HOMECOMING WITH ITS UPS AND DOWNS

It was so good to get home and see the kids. I hugged and kissed each one of them and dragged out all of the craft articles I had made for them. Belts, moccasins, jewelry, and wooden trinket boxes were all given with my love, but I don't think they could make up for my long absence.

The house looked nice since Grandma had been here with her efficient organization and cleaning skills. She had been so good with the kids too.

It was the middle of March, the spring of 1978. I didn't have the heart to tell Dale I hadn't "shaped up" yet. In fact, I felt even more insane since they gave me three labels to identify my craziness. I now had an excuse to be totally insane! The labels were Bipolar Affective Disorder (depressed), Histrionic (highly emotional) Personality Disorder, and Marriage Dysfunction. Now anybody with all of those disorders has a license to be crazy. Right? The whole year of '78 was spent in setting out to let everybody know I could live up to my labels.

I found out Nimrod had called Doctor Wilson and told him he should not deal with my psychiatric problems but with physical ailments only and Doctor Tarinadski should be my psych doctor. In looking back, since my original rescuer had died (my mom), Doctor Wilson had become my rescuer, and I didn't want to lose him. When he told me he was instructed to leave the psych to Doctor Tarinadski, I panicked.

After I was home for only two weeks, I did the dirty deed again. It was so easy! I went to Les, our friendly pharmacist, and told him I wanted a refill of the Fiorinal #3. He called the doctor's office, and Doctor Wilson's nurse okayed the refill. I was really working myself into the Bad Jayne role again. I was so into myself that I didn't think about my role or responsibility as a wife or mother. I just knew I wasn't being very capable with either role. I had come to the conclusion Bad Jayne needed to be stamped out.

I didn't realize at the time that I was so desperate to have a rescuer in my life. The big rejection is what I felt and that no one understood the emotional pain I was going through. I couldn't explain it to anyone and couldn't begin to even understand my emotions, but I did know the oblivion and peace that overdosing provided. It was kind of like getting stone drunk and losing all of my cares and worries.

Dale went to the Lions Club; Chad and Lisa had gone to bed; so there was just Todd and me. It was so easy to do. I just started popping those pretty yellow and navy-blue capsules with some Coke. I was truly being Bad Jayne again—thinking only of myself and ways to get rid of my frustration, even though I didn't know what was causing the frustration. I just know I felt depressed and didn't like feeling that way.

TRY TO UNDERSTAND

Please try to understand
I know of your beautiful sunset
Also of your fragrant spring blossoms
But try to understand
The gift of peace
Eternal
Forever
The suicide does not ask why
Why is not important
The method is of most concern
Pills, multiple pills
Bloody wrists, limp and red
Lethal fast cars
Lethal still cars

A gun to the head
Now for the place and time
Don't follow me
Let me do it my way
Alone
Forever
Don't ask why
Try to understand
Grief will pass away
Love is forever
I don't want to hurt
I just want out
Eternity
Escape
Peace
Why is not important
We all must one day die
Pretend it was God's idea
I love you all!!!
Try to understand

When Dale came home, he knew right away what I had been up to. So, away we went to good ol' General Hospital. I don't remember the ride, or getting there. I just remember waking up and wanting to go home.

Dale was in a frenzy trying to track down who was responsible for giving me the medication. First he went to Les and released his anger on him. When Les said he was not in the wrong since Doctor Wilson's office had approved it, Dale unleashed his wrath on Doctor Wilson. Dale and Doctor Wilson decided to send my picture, with a warning, to all pharmacists in town, that I was not to receive any meds from them. They really fixed me good. Dale took my drug card from me also. I started to panic again after I was released from General Hospital, and on one of my trips to the grocery store I bought four bottles of Sominex. While checking them out, the clerk asked, "You aren't going to take these all at the same time are you?"

I replied, "No, I just heard they work really well so I wanted to get stocked up." I had that obsession to have a ticket to release, and the Sominex pills were it.

The remainder of the spring I found every book on suicide I could. I really started having a death wish. The books I couldn't find in our library I ordered from other libraries. I was in hopes I would find some answers.

July came and we went on vacation to the Black Hills in South Dakota. Before I left I called my friend Maida to tell her we were going to be leaving, and I jokingly told her to not be surprised if she saw me hanging from George's nose on Mount Rushmore on the evening news. That would be my escape route if I felt suicidal while on vacation, jumping off George's head, so I didn't need to take the pills along.

The vacation went well. We had a fun time and rented a cabin in the vicinity of Mount Rushmore. The kids enjoyed it, and I behaved and didn't do anything dumb to ruin our trip. I returned home safely.

I had signed a contract with Doctor Wilson that if I felt suicidal I promised I would call him before I did anything to harm myself. And, the day came when I felt I needed to take all of the friendly little Sominex pills. It was a nice sunny day in late July, and at lunch I yelled at the kids for fighting. My Bad Jayne thoughts went crazy, and I decided I was not a fit mother. I sent all of the kids to do some activities at a friend's house, and after they all left, I called Doctor Wilson's office and told the secretary I was going to have to break the contract.

SUNSHINE

Hello, sunshine, why do you seem to evade me so?
My crippled psyche just can't seem to get in the show
Struggle
Confusion
Ugliness
Death wishes clouding my mind.

Hey, sunshine!
Let me feel your warmth

Fill me with warm love
Get this empty mood that has enveloped me
Get it out
Fill me with something.

All four bottles of the Sominex pills went down easy. I decided I would go out in our pasture to lie down to die. But, there was one problem. The kids' pony, Cinnamon, walked over to where I was lying and started nibbling on me. After about fifteen minutes of that, I staggered to the front porch and laid myself down on the couch. I blacked out completely, and the next thing I knew I was vomiting all over myself.

Dale came home from work to find vomit all over me and the porch. He called Maida to take care of the kids, and then he escorted me to General Hospital again. He had changed my clothes and had put me in a pair of jeans that were too small. I don't remember the trip at all, but I do remember trying desperately to zip up the jeans and I couldn't. I barfed some more and then just slept it off. Later, Dale told me how he and the attending nurse laughed at me every time I tried to zip the jeans up. It was another episode where I was drunk from my suicide-attempt activities.

CHAPTER TEN

THE DAY-CARE CENTER

August came and I applied to work in a day-care center. I thought maybe if I dug up a job I would get my mind off myself and do something more constructive. I had quit taking my Lithium, and what was to be the first (and last) week I worked, I started getting high. I had told my employer I was a manic-depressive but I had it under control.

As the week progressed, I was appalled at the emotional abuse the head teacher was dishing out. She and another teacher would talk to each other about a child, in front of the child, as if he weren't there. An example was Billy, who had forgotten his glasses. Wilma, the head teacher, would say to Jim, another teacher, "Jim, do you know Billy forgot his glasses today? What do you suppose we should do with him?"

Jim responded, "I know. Let's not let Billy watch cartoons with the other kids."

Wilma continued, "That is a good idea. He will miss cartoons for not bringing his glasses."

I took Billy aside and read him a book. Later, as I put Billy down for a nap, he said, "They hate me, don't they?" I assured Billy it was not hate but they were just kind of dumb. By the way, Billy was just three years old.

THE CHILDREN, DEAR GOD, THE CHILDREN

Is there some way we can help the children?
These precious miniature boys and girls are so vulnerable
I can't stand and witness this emotional rape much longer

It can't be any less harmful than physical rape
Often the individual suffers physical abuse and trauma
But, Dear God, wouldn't emotional rape of a three-year-old
Be devastating to his whole being, if done day after day?

Why don't the parents know?
Why don't the children tell Mom and Dad what is happening?
Ah, they don't want them to know what "rotten kids" they are
Being told incessantly that you are naughty is not the kind of
Thing you run home to tell Mom and Dad.

I can see it, God, and it nauseates me to witness these adults'
Regurgitation of insults, threats, put-downs,
Sprinkled occasionally with praise
It's happening right here in this town
Where parents are bringing up their children in good faith
For loving care a child needs love and praise to grow into a healthy individual.

I am seeing emotional invalids being formed in the bud stage
Please, God, what can any of us do to prevent the blossom of the ruined ego
An emotional cripple
Those children who have been at day-care for a year or more
Have a "day-care mask" worn only at the day-care center
This mask says "I don't want to listen or watch anymore."

The zest for life while in our care has been sucked away
From some with ridicule
Where is the unleashed joy and laughter so indicative of happy children?
Who can tune these dense adults, the "image destroyers,"
Into what they are doing
To these precious vulnerable human beings?

Do they know?
No one would purposefully "clone" preschool children into
Zombie-like individuals
Would they?

Thank you, God, that the new younger children haven't been
pushed
And shoved and ridiculed long enough to put on their tuned
out
"Day-care mask" with the lackluster gaze
These children are a promise, a possibility.

The children, dear God, the children,
What can we do?

The poem was written when I was on the manic side. As the week progressed, I seemed to be more and more in a manic phase and told one teacher, who was a college student, that I was wondering if he was as dumb as he looked. He had red hair pulled back in a long ponytail and had an arrogant demeanor. Arrogance is not one of the favorite attributes that I see in others.

I had seven three-year-olds to take care of and teach and was able to mold them into polite little people during snacks and lessons throughout each morning.

On Friday of that week I received a paycheck and had told Dale I would be going to the Newton library to pick up some books. However, I didn't tell him about the appointment I had made with Doctor Monte, who had delivered Lisa. I was up to my old tricks of scheming and conniving.

I knew Doctor Monte quite well. I informed him I was angry with Doctor Wilson for some things he had done, so I was going to switch back to him. It was so easy to convince Doctor Monte that I was in need of my trusty Fiorinal #3 and Valium for the wretched migraines. He prescribed both with no questions asked and even gave six refills.

Dale and Doctor Wilson had not given my picture to pharmacies in Newton, so I had my pick. I chose a pharmacy on the square and walked right in and bought my "trusty friends." I had taken our drug card out of Dale's billfold. It was all so easy. Again, no questions asked.

On Labor Day the director of the day-care center called. She informed me I was fired—just like that—no warnings or discussion. I informed her I was going to return anyway. She just let me know I would be paid for thirty days but she did not want me back there. By this time I was taking my Lithium, and my manic phase was letting up only a little bit. So much for my career as a day-care-center teacher. I told them I wanted a hearing.

CHAPTER ELEVEN

BAD JAYNE TAKES CHARGE AGAIN

As the month of September passed, I made a couple more trips to Newton for refills of the dreaded prescriptions.

A hearing was set up with all my fellow former staff and three panel members who would vote on whether I should have been fired or whether I was treated fairly. I didn't want my job back; I just wanted vindication. The hearing was a total disaster. Dale went with me to keep me acting properly. I felt as if they had all ganged up on me. It felt more like a crucifixion than a hearing. I did not say the right things and let all of my former staff members complain about the way I behaved. I felt so overwhelmed that I let Bad Jayne take charge. The next day, which was a Friday, I felt the need to overdose again.

GALLANT BLACK KNIGHT

Why do you elude me, gallant Black Knight?
My flirtations have been to no avail
Have I not been trying hard enough?
I trust you are not resentful of my affair with depression
Although that affair has been lasting
My real lust has been for you, Black Knight,
Only you.

Most members of the Kingdom try to avoid you
They get so indignant and angry with me for seeking you
But I have aimlessly struggled with life

It seems that no matter how much I give it isn't enough
Life has such an insatiable demand on my whole being
My fumbling attempts to divorce life have been fruitless.

You are real, Black Knight,
Inevitable
Why do you keep prodding your black steed beyond my grasp?
The power of you and your steed is tremendous
You seem to have many of the prizes I am competing for
Eternity
Escape
Release
Peace
Satisfaction (of a job well done)
Rest forever
Destination complete
You have them all, Black Knight.

Will you continue to deny me these trophies of yours?
Yes I realize most members of the Kingdome meet you reluctantly
God's will is no doubt the powerful determinate
Suicide attempts have failed to reach you
What a clever elusive fellow you are, Black Knight.
Do you view me as a court jester, a fool?
Am I not worthy of having your name?
Let me know when you change your mind
I feel so cheap and foolish after getting on my knees
And begging for your company
You know how rejection frustrates me so.
Why do you continuously reject such a willing candidate, Black Knight?
What a clever gallant fellow you are, Black Knight.
When can we all go the way?

I wakened two days later at home. Then, I felt total frustration. I was both sad and glad I woke up. I felt desperate. Dale had asked my sister Mary to come and help take care of me since he had had it. I went slightly insane and had finally decided I should have succeeded. I stumbled outside and said and did a lot of weird things. Dale called Doctor Tarinadski to get me admitted to the Mitchellville psych unit. He had told Dale I liked being in the hospital, so he had not admitted me the times I had overdosed earlier in '78. This time Dale convinced him I needed help.

Mary watched the kids, and Dale took me to Mitchellville in my drunken state. This was the fifth time I had overdosed in '78, and the last for that year. Doctor Tarinadski tried to do an intake with me, but I was so out of it he just called the nurse to put me to bed.

After the effects of the pills wore off in a couple of days, I started getting in a hypomanic phase. Doctor Tarinadski had taken me off Lithium and put me on Thorazine and continued with Tofranil.

There were two women in the psych unit who left some definite impressions on me. One of them was a fellow manic-depressive in a manic phase. She and I had quite a team. Our rooms were next to each other, and what she didn't think of, I did. Her name was Gloria, and I think we about drove the nurses insane.

Another was a woman in her early 30s who was suicidal. She had been like me in having a death wish, or at least a desire to hurt herself with pills. Her daughter and son came to visit her, and I could see the love and hurt in her daughter's eyes as she hugged her mom. Judy's doctor was Doctor Benson, and he was going to send her to another psych unit for more intensive therapy. After seeing her two kids, I felt guilty about what my own kids would be feeling and what they had gone through.

Lisa was the one who found me after this last time with the dreaded overdose, and I started thinking what a mean, insensitive thing it was to do that to her and the boys.

I could see myself mirrored in both of these women. It was scary. I had not stopped to really look at myself during the past year. I was too busy being crazy. But, it was there in both of them.

Since Halloween was coming, Judy and Gloria and I were busy decorating the psych unit. I was into drawing all kinds of pictures. They even took some of the drawings down to the reception lobby and more to the pediatric unit. It kept me busy and out of trouble. And, I decorated the whole hospital in my manic state.

I had started saving my Tofranil, and one of the aides caught me and they searched my room and found the pills. Doctor Tarinadski very adeptly ordered shots for me. The shots could only be given in 25 mg. doses, so it took four of them at night for my 100 mg. and four of them during the day. He also gave the Thorazine in a liquid form that was putrid. He really took care of that. I decided I was dealing with a real pro, not like it was at Monroe.

MIND HEALER, PART II

Mind Healer, do you ever tire of the incessant
Moaning of your clientele
If you do
You conceal it well
You are such a clever healer of the troubled psyche
I'm truly a benefactor of your survival of the Third Reich.

You provided aid to me in my battle against Black Knight
Even though most times I wanted to surrender
You claimed you couldn't help my suicidal brain
Without more effort from me
Now I realize it was a bluff in your battle of wits
Well you are definitely a clever mender of arthritic brain cells
But I still wonder how long my soul can tolerate this
Manic-depressive personality with these mood swings
And evade Black Knight himself.

But never mind for now my lust is for life
I realize how insolent I was
Hopefully you can see that my impertinence displayed
In Mind Healer Part I was a result of misplaced anger
Yes I bestowed most of my fury stored for someone else

On you, my medical scapegoat.

That dreaded pool of defiance has turned into a
Fountain of gratitude
I really appreciate your adept handling of my
Rebellious crippled mind
You have reflected a refreshing pleasant image to me
It has enabled me to crawl out of the depth of depression
To once again become a promise
A possibility
Thank you, Mind Healer.

Although I complained of your self-appeasing pipe
I must add that along with cliches you share a
Vapor of wisdom and insight
The timbre of your shrewd calculating mind transcends
That of your psychiatric colleagues
As you know I have encountered many.

March on as a leader in your field, Mind Healer,
I revere the sharp ingenious cadence to which you are stepping
It is all your own
Enjoy your upcoming respite from your mind-twisted customers
You have earned it
Snare some more of your proverbs for me on your holiday
And once more happy healing till we meet again.

However, I did get my chart stolen once to check things out. Doctor Tarinadski also gave me an "as needed" shot of Thorazine if I acted a little too high or wild. I actually asked for it a couple of times.

Frantically drawing pictures and pacing the halls is what kept me busy. Often Gloria would touch up my drawings with obnoxious, insane marks. It would make me angry, and then we two would go round and round. It kept the unit lively.

Dale and I had a chance to go to our friends in Nebraska for a weekend visit if I could get a pass. I told Pam, one of the nurses, that if I didn't get to go I would lie in front of the door and scream and kick till they let me go. She took me seriously and scolded, "Jayne, you shouldn't make threats like that." We did go and had a normal, fun, sane time.

CHAPTER TWELVE
BACK HOME AGAIN

Doctor Tarinadski released me the following Monday, and I was angry with him.

MIND HEALER, PART III

You are too egotistical, Mind Healer.
I have been angered by your platitudes
Such repetition
I am so pleased to be home and away from your mental hotel
"You are acting like a child," has been uttered by you
A senile old man too many times
I know of many children who are a pleasure to be around
Not so with senile old men.

You acted like a trainer clicking your fingers when you said
"Jayne, come here," as I walked down your crazy hallowed hall
But I didn't respond as a well-trained obedient dog
No!
Not obedience but rebellion boils in me
I am not a member of your well-trained staff
That jumps at your slightest gesture
No I am one of your rebellious cripples
That came aboard your ship of crazies.
Stop your platitudes

I am so incensed by platitudes spewing from your repetitious
mouth
Your cliches sometimes flow as if you
Truly do have diarrhea of the mouth
Your platitude plague is too much
Get some new ideas.

Thank you for trying to help me
And please be aware that my pill saving and chart reading
While aboard your tightly run ship Scream Control
Was not a game
It was a need and my legal right.

Now my lust is again for life
And I can make my plans for the future
Even though I do still have my ticket to death
As a precaution to insure terminal life.

Trudge on you arrogant diplomat of the mind
I hope to cool down some day
And come to see you on your turquoise spinning tower.

The new medication proved to be right for me. I was able to go from 1978 to 1982 without being in the hospital. The Thorazine caused me to have blackouts when I first took it. When I would get up too fast, I would shake and then black out. After about six months, my body was able to adjust, and Doctor Tarinadski put me on a maintenance dosage of 150 mg. of Tofranil and 150 mg. of Thorazine a day. I would get depressed on occasions but not to the point that I had to be admitted to the hospital. The new medication definitely relaxed me and kept me sane, but very fat.

In 1980, I applied for a job as a youth worker at the Ross County Youth Shelter. The job and the meds gave me an incentive to stay normal, even though I had gained a lot of weight. It felt good to feel halfway normal for four years.

In the spring of '82, I hit the pits again. I had promised Lisa that I would not overdose again. When things grew worse, I conjured up the most insane plan of using Dale's rifle and killing all of us. That way there would be no one left to feel guilty or lonesome or depressed, no survivors to suffer the pain and agony of my craziness.

When I could handle it no more, I let Doctor Tarinadski know I was in a mess. I wanted to be rescued from my morbid thoughts, but I did not tell him about the plan with the gun. He admitted me to the psych unit. This time though I would not talk to him, and we had come to know each other pretty well through monthly meetings.

He advised me to seek another psychiatrist if I would not confide in him. That scared me into talking, since I did not want to go to another shrink. When I told him of the rifle plan, he said he would have to let Dale know about this. I about died right there and was afraid Dale would want to divorce me if I was thinking of doing all of us in.

The day came for Dale to come to the hospital to hear the news. I told Doctor Tarinadski I would prefer he tell Dale of the plan. But, Dale did not panic or get angry; nor did he want to get rid of me. I decided it must be true love if he would tolerate that. He had put up with a lot of my crap.

Since I started taking the Thorazine, I was better, even with the extra 60 pounds. I looked like a butterball. This frustrated Dale, and he would often comment about my now-ugly figure. He also thought I slept too much since taking the Thorazine. I was also sun-sensitive, so Doctor Tarinadski put me on Mellaril.

I continued working at the shelter on the late-night shift. We had a ten-bed facility, and I would work alone on the 10:30 p.m. to 7:00 a.m. shift.

About two weeks after leaving the Mitchellville psych unit, I had a very weird experience with a frying pan. It wasn't Erin calling me a frying pan. It was getting beaten over the head with the dreaded frying pan. A large, fourteen-year-old girl just came out and beat me as I was sitting in a chair. All I could do was scream, "What are you doing?" After five whacks, she quit and went to her room. The first hit stunned me so that I didn't even defend myself. I asked for the on-call and told her I was bleeding and had been pounded on the head. She and another worker hurried out to the shelter, but Yvonne, the whacker, just stayed in her room. I drove to the hospital and had to have eighteen stitches. They couldn't believe I had been beaten with a frying pan. Maybe it was a secret message from Erin to remind me I was a frying pan.

CHAPTER THIRTEEN
ELECTROSHOCK

In the spring of '83, I found myself down in the pits again, wanting to visit Black Knight. This time they decided I should have electroshock treatments, and I had nine of them. Lisa sent a letter to me during my hospitalization.

Lisa was eleven at this time. I had stressed to her and the boys that they were not the cause of my illness and that they shouldn't blame themselves for my craziness. She was very worried about the shock treatments I would receive, in spite of her silly letter.

Due to the treatments, I couldn't remember a thing about that trip to Mitchellville. Later, they told me I had overdosed when I arrived at the hospital with pills I had smuggled in inside my bra. I remembered planning but couldn't remember doing it. Nor could I remember getting shock treatments.

On one occasion Dale and the kids came up to take me out to eat. They told me my hair was so scraggly that they were embarrassed to be seen with me. I don't even remember going out with them. They thought my hair hadn't been washed in a week. They still talk about it sometimes, like "Remember the time we took Mom out to eat with her greasy hair?" The treatments stripped my mind into oblivion.

From '83 to the spring of '86, I did quite well again. I even acted halfway normal again for those three years.

Letter from Lisa

(1)

go to page 2
follow
the
arrows→ → → → → → → → → → → → → → → so long! → →

(4) But really I love
you and I will miss
you very much!

I
hope
this
says
good bye!

(2) But I'll bet you will have a shocking experience.

(3) I L♡ve
you!

But just remember
if you die I get
your stuff.

Lisa

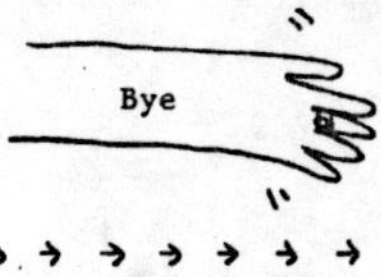

→ →

NEW BEGINNINGS

Thank you, God, for new beginnings that you provide
After this bout with depression I'm beginning my uphill ride
It is a ride that will boost my self-esteem
And bring me closer to you
I'm so glad to have passed through that hellish dark avenue.

New beginnings in life are such wonderful things
I'm so thankful that I'm still here to enjoy the contentment life brings
You know how I wanted to stop the journey of life
But now I realize that once again I have managed
To conquer that horrible strife.

It helps to have family, friends, and doctors
Whose love and understanding are rare
Even though I'm sure they knew at times I just didn't care
So thank you, God, for this new beginning you've given to me.
Help me to remember new beginnings
If ever again I go on a wretched depression spree.

Visits with Doctor Tarinadski had been very beneficial since we started working on my personality and didn't have to hash over problems with my depression. I saw him once a month while I was feeling good, and we uncovered a lot of my problems. He helped me see what needed to be changed.

CHAPTER FOURTEEN

HISTRIONIC PERSONALITY DISORDER

I really didn't think I had a full-blown Histrionic Personality Disorder. But, Doctor Tarinadski explained to me that I became histrionic when I was under a lot of stress and couldn't handle it. In the DSM III (Diagnostic Statistics Manual), the disorder is explained in this way:

A. Behavior that is overly dramatic, reactive, and intensely expressed as indicated by at least three of the following:
   1. Self-dramatization, e.g., exaggerated expression of emotions
   2. Incessant drawing attention to oneself
   3. Craving for activity and excitement
   4. Over reaction to minor events
   5. Irrational, angry outbursts or tantrums

B. Character disturbances in interpersonal relationships as indicated by at least two of the following:
   1. Perceived by others as shallow and lacking genuineness even if superficially warm and charming
   2. Egocentric, self-indulgent, and inconsiderate of others
   3. Vain and demanding
   4. Dependent, helpless, constantly seeking assurances
   5. Prone to manipulative suicidal threats, gestures, or attempts.

The underlining shows which behaviors I would exhibit when depressed or when under a lot of stress. A lot of people couldn't believe I would get depressed, for I was able to conceal it well. Nor could they believe I was suicidal, since Doctor Tarinadski was the only one I would bare my soul to.

Following is an article from the Ann Landers column that I cut out:

> *Dear Ann:*
>
> *"I am a sixteen-year-old girl who had a happy, carefree life until a very good friend of mine committed suicide. He was good looking, smart, came from a well-to-do family, and had a bright future.*
>
> *"Everybody was shocked. No one could figure out why he didn't want to live any longer. His death reminded me of a poem I once read in your column. It was called* "Richard Cory." *Will you please run it again?*
>
> *"People need to know that some teenagers who seem to have everything going for them can have terrible problems that no one knows about."*
>
> —*A Canadian Reader*

Ann Landers reply: "The poem by Edwin Arlington Robinson (1869–1935) is an excellent reminder that even keen observers can be fooled. The masks people wear—the facades they construct! How little we know about what goes on in the hearts and minds of those around us.

RICHARD CORY

Whenever Richard Cory went downtown,
We people on the pavement looked at him;
He was a gentleman from sole to crown,
Clean favored, and imperially slim.

And he was always quietly arrayed,
And he was always human when he talked;
But still he fluttered pulses when he said,

'Good-morning' and he glittered when he walked.

And he was rich—yes, richer than a king—
And admirably schooled in every grace:
In fine, we thought that he was everything
To make us wish that we were in his place.

So on we worked, and waited for the light,
And went without the meat, and cursed the bread;
And Richard Cory, one calm summer night,
Went home and put a bullet through his head.

I continued working the late-night shift at the youth shelter, and my three kids had grown to be teenagers, in spite of their crazy mom. In '86 things started piling up on me again.

MIND HEALER, PART IV

Thank you for giving me a positive reflection
Even though at times I'm full of self-rejection
Your wisdom and advice is usually good for me
But at times it doesn't put a stop to my misery.

There you are in your comfy chair
You listen absorb then reflect
With comments like, "You have a good brain
You need to use it."
That sounds like a good idea
But my histrionic mind runs wild within me.

You're right—I don't want to die—I just want oblivion
Deep dark everlasting nothingness
Bad Jayne thinks she wants to go to the home of Black Knight
She lacks insight and the desire to push on.

Thank you for reflecting to me
A Jayne who is capable and intelligent
But for now I ache inside
And I don't handle pain too well.

You're right, Mind Healer,
I'm the only one who can help me
But I've reached the "I don't give a damn" stage
You did some brain dusting for me
As the competent housekeeper of my mind
But you have barely touched the surface
Since my brain is enduring a dust storm that shows no sign of letting up
I'm overwhelmed and don't know where to begin.

CHAPTER FIFTEEN
COPE-GROPE

We had a big turnover in our administrative staff at work, which was trying and frustrating.

Dale decided we should split up since he had had a hard time dealing with my fatness and wanting to sleep all the time. However, he had supported me up till then.

Doctor Tarinadski was on vacation in April, so I called Doctor Benson, his partner, and made an appointment with him. I let him know I was having a real problem with suicidal thoughts again. This time my plan was to ram my car into a bridge on the way home from work. That way Dale could claim the insurance and they wouldn't have to feel guilty if I had a car wreck and it didn't appear like suicide.

Doctor Benson had me admitted to Mitchellville on the 11th anniversary of Mom's death. Work and Dale were all getting to me.

I had decided, after reading the DSM III about my problems that Dale had a few problems of his own. He too had symptoms of a couple of personality disorders.

Doctor Benson threw me in the "people dump." I went straight to the hospital from his office, and I took a couple of my trusty Fiorinal #3 for the headache I had. Then, I decided to smuggle in the rest of the pills for a way out if I need it. I still had that need for a ticket to death.

Well, they faked me out this time in the hospital. Jenny, the psych tech, had me remove all of my clothes, and there were the Fiorinal in my bra. After she took those, I really panicked. Doctor Benson called my pill-saving "playing games." I called it a necessity.

When Doctor Tarinadski returned from vacation, Dale wanted to have a counseling session with the three of us. During the session, Doctor Tarinadski did not pick sides, but I promised I would try to lose weight and try not to sleep as much. A week later I was released from the hospital.

The "Cope-Grope" continues, but I have come a long way in my battle with mental illness. A faith in Jesus has helped.

THE ULTIMATE FRIEND

What a wonderful Lord you are
Thank you for your love
I feel your warmth when others are cold
How greatly I appreciate your acceptance
Especially when many seem to think of me as an untouchable
Your understanding is so complete
Even when no one else can
Or even wants
To understand my weary psyche.

Lord, you are such a good companion during my loneliness
You never forsake me
Even though I sometimes lose track of you
You are still there with your perfect love
How lovely is the feeling of communion with you
Help those I love to realize your love is real
I plead that they would be open to your healing love.

When my spirit needs escape I can turn to you, Lord, and be at peace
You are always there
In fact you were there all the time
Even during my darkest hours when life seemed so futile
You were still there to listen to my grumbling and moaning
The ultimate of friends
What a wonderful Lord you are!

There is always hope, and I have a lot of it. I feel good keeping busy with work and the house and keeping up with Chad and Lisa's activities. Todd is a junior in college. Chad is a junior in high school. And, Lisa is a sophomore. All three of them appear to be normal, in spite of my craziness the last eleven years.

Dale is still his cantankerous self, but I have learned to release my anger toward him instead of saving it up. When he gets cranky, I just speak up and let him know how I am feeling. This detours a lot of depression. We have been married twenty-three years, and we're both learning to cope with one another and to cope with life itself. Dale has stuck by me and has never given up.

For you fellow manic-depressives, there is hope. I have decided I'm very fortunate I survived all of my suicide thoughts, plans, and gestures. I'm thankful I am here to go to school activities for the kids and can work and maintain a home and a job and can be a loving, faithful wife. I continue taking my Tofranil 150 mg. and Mellaril 150 mg. per day and have continued seeing Doctor Tarinadski once a month. He said I will probably take the medication the rest of my life. I can handle it. The seven trips to psych units since 1976 have all been beneficial and the Cope-Grope is succeeding. I now can cope!!

The main reason I never did complete the suicide act was that deep down I didn't want to step out on my kids. That was the main deterrent that kept me halfway in line. I always had several reasons for wanting to escape, but I felt my sense of motherhood is what usually saved me.

Sometimes, in my crazy mind, I would resent them and felt they were holding me prisoner here on earth, but now I'm so glad I didn't complete. Also, I came to realize I was always idolizing a permanent solution for temporary problems.

Sometimes Doctor Tarinadski would say some of my suicidal plans could be incomplete and leave me in a vegetable state. I would always say I was willing to take that chance. Oh how stupid I was. Doctor Tarinadski would continually remind me I needed to let my brain take charge rather than let my feelings be in control. I think I finally figured that out. My recollections of "Flight Control" are in the past.

IT'S OVER

A gentle breeze has whisked me away from that
Wretched stagnation of the mind.
Was it really me trying to kill me?
Will that depth of depression and passion for Black Knight return?
Maybe it was all just a nightmare.
That helpless begging decaying soul couldn't have been Good Jayne.

I don't like that weird person who wrote all those crazy poems
There is nothing good about her
She is a helpless nonproductive slob
It doesn't seem right that such a capable pleasant individual
So full of love and life can rot away like a stinking putrid spoiled tomato
Nor does it seem possible that the same rotten fruit
Can be transformed into a fresh flowering possibility
But it has happened—rejuvenation
Hope
Live
Or die
But don't spoil everything.

There is real potential bursting from Good Jayne
She has really returned
How did we stifle Bad Jayne?
Is there a way she can be stamped out forever?
Oh forget it
Let's make up for lost time
Get into the life of contentment and accomplishment
Doesn't it feel great?

Maybe life can go on like this forever
Everything has a precious new quality
A refreshing scent has taken over
Will that stench of death and depression return?

Please, somebody tell me it won't be so
I hate Bad Jayne and all she represents
For now the choice is life:
Precious, abounding, giving, loving, thriving
I'm almost afraid to enjoy it too much
However for now I will live
The wretched depression trip is concluded
It's over!
Good Jayne is alive and well.

EPILOGUE

Hey, you think it's over, Jayne? What a joke. With bipolar type II (a depressed type manic-depressive), it is never over.

Enjoy your good times because after the fun you may find yourself back in the land of black-cherry-gelatin land. Can't move. Can't see the sun. Can't laugh. Ridicule and intimidation crush your very being.

Hang in there, Good Jayne. Love, and love hard and fast as you can. Love those competent, dependable, and sweet offspring. They are your legacy, your miracles, your blessings. Hold them tight. Let them know it is not their fault that you dip into crazy land off and on. And, pray that they do not have the dreaded gene wrapped up in their being.

Good Jayne, you can do it. Shine while you can shine. Overflow while your cup can overflow. But, just "stay the course." You are like the crops of the field. They spring from seed in the spring, nurtured by the sun and rain, and they prosper in the summer—just as you do. But, when fall comes, they provide the crop and then fade and die. By winter they are gone. Bad Jayne, you have had so many winters.

Well, thank you, my inner psyche, for letting me know my destiny. I will hang in there like the sloth on a tree. But, when Black Knight shows his ugly face, it just crushes me. He has that come-hither way about him that grabs ahold of my soul.

I will let my light shine and strive to become strong like silk, physically and emotionally. I'm still struggling with Bad Jayne, but I pray that Good Jayne will prevail.